MARE Publication Series

Volume 13

The MARE Publication Series is an initiative of the Centre for Maritime Research (MARE). MARE is an interdisciplinary social-science network devoted to studying the use and management of marine resources. It is based jointly at the University of Amsterdam and Wageningen University (www.marecentre.nl).

The MARE Publication Series addresses topics of contemporary relevance in the wide field of 'people and the sea'. It has a global scope and includes contributions from a wide range of social science disciplines as well as from applied sciences. Topics range from fisheries, to integrated management, coastal tourism, and environmental conservation. The series was previously hosted by Amsterdam University Press and joined Springer in 2011.

The MARE Publication Series is complemented by the Journal of Maritime Studies (MAST) and the biennial People and the Sea Conferences in Amsterdam.

More information about this series at http://www.springer.com/series/10413

Svein Jentoft • Ratana Chuenpagdee
Editors

Interactive Governance for Small-Scale Fisheries

Global Reflections

Springer

Editors
Svein Jentoft
Norwegian College of Fishery Science
UIT – The Arctic University of Norway
Tromsø, Norway

Ratana Chuenpagdee
Department of Geography
Memorial University
St. John's, Newfoundland and Labrador
Canada

Too BIG To IGNORE
Global Partnership for Small-Scale Fisheries Research

Too Big To Ignore (TBTI; toobigtoignore.net) is a global research network and knowledge mobilization partnership, funded by the Social Sciences and Humanities Research Council of Canada, and supported by 15 partner organizations and over 100 members from around the world. The network aims at elevating the profile of small-scale fisheries, arguing against their marginalization in national and international policies, and developing research and governance capacity to address global fisheries challenges.

ISSN 2212-6260 ISSN 2212-6279 (electronic)
MARE Publication Series
ISBN 978-3-319-17033-6 ISBN 978-3-319-17034-3 (eBook)
DOI 10.1007/978-3-319-17034-3

Library of Congress Control Number: 2015942138

Springer Cham Heidelberg New York Dordrecht London

Printed on acid-free paper

Springer International Publishing AG Switzerland is part of Springer Science+Business Media (www.springer.com)

Cover image: Small-scale fishers from the Manayaaweli village in the Trincomalee district on the East coast of Sri Lanka, maneuvering their vessel through the surf. By permission of Gayathri Lokuge, winner of the photo contest at the 2nd World Small-Scale Fisheries Congress, Merida, Mexico, Sept 21-26, 2014

For Jan Kooiman, whose work inspired this book.

Foreword

The examination of issues of governance has become pervasive in several fields of scholarship including political science, economics, and sociology due to the growing plurality and power of actors beyond the state engaged in decision-making processes (e.g. multinational corporations, private foundations, civil society organizations, formal and spontaneous social and political networks) and the growing complexity and global reach of societal dilemmas.

Fisheries are beset by complex societal dilemmas that render governance difficult and usually time and cost intensive. The reasons include the multiplicity and diversity of fishery resources and habitats, the vastness of their spatial distribution in the inland water and oceanic environment, and their common pool characteristics; the high vulnerability of fishing communities to the impacts of climate change and disasters; growing competition with other economic sectors; and the political, economic and social marginalization of many fishing communities and indigenous peoples.

While economically marginal in all but a few countries, the importance of small-scale fisheries can hardly be overestimated for food security, nutrition, livelihoods, rural development, and poverty reduction.

It has taken several decades of advocacy, education, and mobilization for small-scale fisheries to be appropriately recognized at the international level and within countries. The culmination of this recognition has been the adoption by the members of the United Nations Food and Agriculture Organization (FAO) of the Voluntary Guidelines for Securing Sustainable Small-Scale Fisheries in the Context of Food Security and Poverty Eradication (SSF Guidelines) in June of 2014. Citing from the Preface, the Guidelines seek "..... to support the visibility, recognition and enhancement of the already important role of small-scale fisheries and to contribute to global and national efforts towards the eradication of hunger and poverty."

Governance issues are extensively addressed in the Guidelines whose guiding principles are based on international human rights standards, responsible fisheries practices and standards, and sustainable development and other relevant instruments. The principles include human rights and dignity; respect of cultures; non-discrimination; gender equality and equity; equity and equality; consultation and participation; rule of law; transparency; accountability; economic, social and environmental sustainability; holistic and integrated approaches; social responsibility; feasibility; and social and economic viability. Systems and practices of good governance are informed by a similar set of principles.

The promotion of human rights and good governance has become prominent within the UN system. The Rio +20 outcome document "The Future We Want" recognizes the need to involve in decision-making processes all major groups including women, children and youth, indigenous peoples, non-governmental organizations, local authorities, workers and trade unions, and others. FAO's former technical committee on food security has been reformed into a multi-stakeholder and multi-agency body to advise on and oversee global governance on food security. In its deliberations, international civil society organizations and private sector associations engage nearly on par with government representatives even though governments remain the ultimate decision-makers. It is this new form of "*interactive public reasoning*" (Amartya Sen) that has informed and made possible the SSF Guidelines and other recently adopted progressive international soft laws such as the Right to Food Guidelines and the Voluntary Guidelines for the Responsible Governance of Tenure of Land, Fisheries and Forests in the Context of National Food Security.

The development, negotiation, and adoption of the SSF Guidelines would have been unthinkable without the marked changes in the systems of governance at local, national, and international levels during the last four decades. In the 1970s, the gap between the policy and grass-roots levels was still very pronounced. In many countries, small-scale fishers – as other small-scale producers – were not organized, and the formal and informal structures were few to exert influence on policy-making. I became acutely aware of this when FAO convened in 1984 a World Conference on Fisheries Management and Development, a conference held in the context of the adoption of the new international Law of the Sea in 1982. As a young staff member who had just joined FAO Headquarters from a FAO/UNDP project on the promotion of small-scale fisheries in South Asia, I was placed in the team to develop background materials on small-scale fisheries that led to what I believe continues to be until today a valuable document. However, my concern at that time had been that the primary actors and interest groups, i.e. small-scale fishers and their supporters in civil society, were unlikely to be represented at this conference. Sharing my concern with John Kurien – my first and foremost teacher on small-scale fisheries – he concluded that this required an appropriate response. In a short period of less than 1 year, he managed with the help of his wide network of contacts among concerned activists, scientists, and sympathetic government staff to organize the International Conference of Fishworkers and their Supporters held from 4 to 8 July 1984 in parallel with FAO's Conference. About 100 participants from 34 countries attended this conference. Half of them were fishworkers (men and women engaged as crew members, small-scale fishers, processing workers, and sellers) from all continents. Most of the fishworkers were small-scale fishers operating in coastal and inland waters. The supporters were individuals and representatives of organizations who identified themselves with the cause of fishworkers.

Salient recommendations from this conference – all of which touch upon governance issues – continue to resonate today:

- "*governments to be responsive to the demands of the local fishworkers organizations and respect and guarantee the fundamental rights of fishworkers to free association.*
- *attitudes and values towards women be changed in order to get their full participation in decision making at all levels.*
- *the positive contribution of non-governmental organizations in the development of technology and forms of participatory management should be recognized to ensure the future of small-scale fisheries.*

  The Conference requested that all scientists who recognize the importance of conserving and enhancing the person-nature relationship take a strong stand on behalf of the small-scale fishermen. They were urged to work in collaboration with local fishermen's organizations to complement their knowledge of the sea and enable them to regain their rights over the sea."

Subsequent to this conference, we have seen some significant developments that have impacted on fisheries governance nationally and internationally. These included the formation of the Collectif National des Pêcheurs Artisanaux du Sénégal (CNPS), the Confederacion National de Pescadores Artesanales de Chile (CONAPACH), federations such as Bikis Lakas in the Philippines, and, perhaps most importantly, the formation in 1986 of the International Collective in Support of Fishworkers (ICSF). ICSF has become one of the foremost promoters and defenders of the causes of small-scale fisheries at the international level. Assisted by ICSF, in 1996–1997 national fishworkers organizations federated internationally to form the World Forum of Fish Harvesters and Fish Workers (WFF) and the World Forum of Fisher Peoples (WFFP). All three organizations have played a pivotal role in the consultation and negotiation processes of the SSF Guidelines.

The 2008 Global Conference on Securing Sustainable Small-Scale Fisheries, co-organized by FAO and the Royal Government of Thailand, laid the groundwork for the adoption of a human rights-based approach in the SSF Guidelines. In her plenary presentation to the Conference, Chandrika Sharma, Executive Secretary of ICSF, who tragically was aboard flight MH 370 that disappeared on 8 March 2014, stated that adopting a human rights approach for improving the life and livelihoods of fishing communities – and indeed all marginalized groups – was not really a matter of choice but an obligation. The SSF Guidelines have been dedicated to Chandrika Sharma who deeply cared for people, worked tirelessly for the betterment of the lives of fishworkers all over the world, and contributed invaluably to the formulation of the Guidelines.

The statement of the preparatory civil society workshop to the Global Conference called upon FAO, Regional Fisheries Management Organizations, and national governments to secure access, post-harvest, and human rights of small-scale and indigenous fishing communities. The 106 participants from 36 countries called on FAO's Committee on Fisheries (COFI) to include a specific chapter in the Code of Conduct for Responsible Fisheries (CCRF) on small-scale fisheries, recognizing the obligations of states towards them.

In a parallel development, concerned fisheries academics convened in 2010 the first World Congress on Small-Scale Fisheries (WSFC). An impetus for the Congress had been the realization that small-scale fisheries were too important to continue to be treated on the margins of the World Fisheries Congresses held every 4 years. An important outcome of the WSFC has been the formation of an international small-scale fisheries research network which has come to be known as Too Big To Ignore (TBTI). TBTI has become the foremost global governing structure and network of small-scale fisheries research. TBTI also provides an important conduit for the implementation of the SSF Guidelines. In collaboration with FAO, it assembled a large number of ideas and suggestions to promote implementation of the SSF Guidelines from more than 400 participants, all of whom enthusiastically welcomed the adoption of the SSF Guidelines, at its 2nd WSFC held in Merida, Mexico, in September 2014.

Nearly all the contributors to this fascinating book on small-scale fisheries governance were participants of the Merida Conference and are members of TBTI. Several of them including the editors of this book have provided critical inputs into the development and negotiation processes of the SSF Guidelines and helped in avoiding the derailing of a politically sensitive matter.

The book and its authors truly do justice to the diversity and cultural richness of small-scale fisheries. Contributions are from 36 countries and from all regions of the globe. As the book amply demonstrates, the minds of the 69 well-known academics are set on transdisciplinary and innovative research and their hearts are with the daily struggles of small-scale fishing communities for a better future. The case studies cover small-scale fisheries in a great number of bio-physical, social, economic, cultural, political, and governance settings. An idea of the variety of situations analyzed by the case study authors can be gleaned from their geographical spread: going roughly westwards around the globe, case studies are from Hawaii, Solomon Islands, Japan, South Korea, Philippines, Indonesia, Cambodia, Thailand, India, Sri Lanka, Cyprus, Zanzibar/Tanzania, South Africa, Malawi, Sierra Leone, Senegal, Netherlands, Denmark, Norway, France, Portugal, Canada, USA, Brazil, Uruguay, Nicaragua, Barbados and St. Lucia, Jamaica, Belize, Mexico, Colombia, Ecuador, and Chile. Preceded by an introductory chapter of the editors on the concepts of governance and governability, the case studies are grouped around seven themes focusing on governability challenges, aligning modes in governing system complexity, rights and justice concerns, securing space, cross-boundary governance, governance in transition, and meta-governance. The case studies cover small-scale fisheries governance through self-governance, pure and hybrid forms of traditional community-based and modern co-governance, hierarchical top-down governance modes, and many forms in-between. They provide strong evidence of the complexity of the systems-to-be-governed and the wickedness of the problems of small-scale fisheries, especially in the widely found situations of declining resources, high livelihood dependence on fisheries, growing pressure from other economic sectors, and social and political marginalization. The case studies confirm findings from the body of common-property resources research that solutions are highly context specific and, thus, there are no one-size-fits-all solutions. But as highlighted by the book's editors in their concluding chapter in which they urge positive changes in governance, there are some commonalities in the changes needed in governance processes to improve the governability of small-scale fisheries. These include interventions that improve the information flow and quality of interactions both within and between the governance system and the system-to-be-governed, the transition from hierarchical modes of governance to co-governance and self-governance, and enhancement of the capacity and capability of the governors and subjects of governance to participate in interactive public reasoning and decision-making on an equal footing and in a responsible, transparent, and accountable manner.

My hope for the future is the widespread implementation of the SSF Guidelines. This requires in most situations transformative changes in governance. The book's case studies provide important leads on what it takes in practice to bring about such changes.

For anyone associated with and interested in empowering small-scale fisheries, improving their governance, and realizing their full potential to contribute to food security, nutrition, and poverty eradication – be it in government, civil society, or academia – this book is a must!

Kressbronn, Germany Rolf Willmann

Preface

This compendium on small-scale fisheries and governance is the apex of a series of scholarly efforts. My version of the story commences with the establishment of the social-science Centre for Maritime Research (MARE) in the year 2000 in Amsterdam, the Netherlands. Although small-scale fisheries was not an explicit theme of research and action, MARE's initiators all had a background in the field, with in-depth experience from Asia, Africa, and Europe (this has now expanded to include the Americas). Besides an academic interest in small-scale fisheries, the group also shared an anxiety about their future. John Kurien – an eloquent champion of the cause of small-scale fishers and co-founder of the International Collective in Support of Fishworkers (ICSF) – spoke forcefully at MARE's first People and the Sea Conference (2001) about 'a tropical majority world perspective' in which small-scale fisher peoples play a core role.[1] Triggered by his speech, the MARE team decided subsequently to establish a 'tropical maritimes lecture' at each of its biennial international conferences.

Although MARE is obviously more than small-scale fisheries, its interest in this field has continued over the years. This is expressed in its engagement with the Too Big To Ignore (TBTI) project, to which I return below. It is also reflected in the MARE Publication Series, which has devoted great attention to fisheries issues in general and small-scale fisheries in particular. The *Fish for Life* (2005) and *Governability of Fisheries and Aquaculture* (2013) volumes have highlighted four concerns that are of importance to fisheries governance, namely ecosystem health, social justice, livelihoods, and food security. The authors of the present volume note that small-scale fisheries make a meaningful contribution to each of them.

The earlier two books and the current one connect with yet another initiative – the Fisheries Governance Network (FGN) – that was inspired by Jan Kooiman of Erasmus University. Kooiman had conceived of a highly original interactive governance approach in the 1990s and was keen to develop an application. Chance events led to an association with a group of fisheries specialists and ultimately to

[1] The keynote address was subsequently published in *Maritime Studies* (2001, 1, 9–26) and can be accessed via http://www.marecentre.nl/mast/documents/JohnKurien.pdf

MARE. The European Union (5th Framework Program) provided means for a group of 15 social and natural scientists from all over the world to gather and apply Kooiman's holistic and thought-provoking framework to the fisheries field under the MARE banner. This resulted first in the *Fish for Life* book. Having decided not to close shop, the network then moved to elaborate the concept of governability, exploring the overall quality of fisheries governance. The *Governability of Fisheries and Aquaculture* book was then published in 2013. And now another important step has been taken by applying the governability framework to the world's small-scale fisheries and taking a governance modes perspective. This showpiece could only be realized, however, in close collaboration with the TBTI network and its many initiatives.

TBTI is another branch on the tree. Inspired and directed by Ratana Chuenpagdee of Memorial University, Canada, who is also a member of FGN and a co-author/editor of previous volumes, the TBTI network has made impressive strides in putting small-scale fisheries on the map. Joining up with FAO's effort to establish an international Voluntary Guidelines for Securing Sustainable Small-Scale Fisheries – an effort that finally paid off in 2014, when the Guidelines were adopted – TBTI has launched an information system for small-scale fisheries and initiated a range of collaborative research efforts on core issues. MARE conferences and policy days have provided a useful platform for TBTI deliberations, and we are pleased that many of its academic outputs are also being released in the MARE Publication Series and the journal of *Maritime Studies* (MAST).

This book is a major achievement and an important contribution to the precious field of small-scale fisheries. It will hopefully receive wide attention.

Amsterdam, The Netherlands
January 2015

Maarten Bavinck

Acknowledgements

This book would not have been possible without the effort of the 68 authors from around the world. Thank you so much! Several other people have helped us with the production of the book. Ajit Menon provided excellent language editing. Vesna Kerezi was instrumental in communicating with the authors and in the formatting of chapters. We thank Gayathri Lokuge for allowing us to use her photo for the book cover and Eric Rogers for producing the map. Judith Terpos, Naomi Portnoy, A. Pavitra and Fritz Schmul took care of the publishing process. We are enormously grateful for the effort of all the 64 reviewers, whose names are listed in the Appendix. Special thanks to Andrew Song who, besides contributing two chapters, also reviewed several chapters. We are indebted to Maarten Bavinck and Rolf Willmann for their support and encouragement. Finally, this work would not have happened without the inspiration and scholarship of Professor Jan Kooiman to whom we dedicate this book.

The book is a product of the Too Big To Ignore – Global Partnership for Small-Scale Fisheries Research (TBTI), funded by the Social Sciences and Humanities Research Council of Canada, headquartered at Memorial University, St. John's, Newfoundland, and directed by Ratana Chuenpagdee. Most of the contributors to the book are members of TBTI. Svein Jentoft wishes to acknowledge the Center of Sami Studies, UIT – The Arctic University of Norway for the logistical support while working on this book.

Tromsø and St. John's	Svein Jentoft and Ratana Chuenpagdee (editors)
January 2015

Contents

Contributors

Adam L. Ayers Department of Urban and Regional Planning, University of Hawai'i at Mānoa, Honolulu, HI, USA

Mauricio Castrejón Interdisciplinary PhD Program, Dalhousie University, Halifax, Canada

Inés Chinea-Mederos Institute of Social and Political Sciences, Universidad de La Laguna, Tenerife, Spain

Ratana Chuenpagdee Department of Geography, Memorial University, St. John's, Canada

Philippa Cohen WorldFish, Honiara, Solomon Islands

Australian Research Council Centre of Excellence for Coral Reef Studies, James Cook University, Townsville, Australia

Shelly-Ann Cox CERMES, University of the West Indies, Cave Hill, Barbados

Larry B. Crowder Hopkins Marine Station, Stanford University, Pacific Grove, CA, USA

Center for Ocean Solutions, Stanford University, Pacific Grove, CA, USA

Fabio de Castro Centre for Latin American Research and Documentation, University of Amsterdam, Amsterdam, The Netherlands

Raquel De la Cruz-Modino Institute of Social and Political Sciences, Universidad de La Laguna, Tenerife, Spain

Maricela de la Torre-Castro Department of Physical Geography, Stockholm University, Stockholm, Sweden

Birgit De Vos LEI Wageningen UR, Wageningen, The Netherlands

Omar Defeo Facultad de Ciencias, UNDECIMAR, Montevideo, Uruguay

DINARA, Montevideo, Uruguay

Alyne Elizabeth Delaney Innovative Fisheries Management, Aalborg University, Aalborg, Denmark

Steve Donda Department of Fisheries, Lilongwe, Malawi

Jorge Euan CINVESTAV, Merida, Mexico

Louisa Evans Geography, College of Life and Environmental Sciences, University of Exeter, Exeter, UK

Alice Joan G. Ferrer Division of Social Sciences, University of the Philippines Visayas, Miagao, Iloilo, Philippines

Elena M. Finkbeiner Hopkins Marine Station, Stanford University, Pacific Grove, CA, USA

Julia Fraga CINVESTAV, Merida, Mexico

Katia Frangoudes UMR M101, AMURE, OSU-IUEM, Université de Brest, Brest, France

Clément Garineaud Département Hommes, Nature et Société, Eco-anthropologie et ethnologie, Muséum d'Histoire Naturelle, Paris, France

Sunny George Kerala Institute of Local Administration (KILA), Thrissur, India

Leopoldo Cavaleri Gerhardinger Environmental Studies and Research Centre, University of Campinas, Campinas, Brazil

Hugh Govan Locally Managed Marine Area Network, Suva, Fiji

Tim Gray Department of Politics, Newcastle University, Newcastle upon Tyne, UK

Maria Hadjimichael Innovative Fisheries Management, Aalborg University, Aalborg, Denmark

Mafaniso Hara Institute for Poverty, Land and Agrarian Studies, University of the Western Cape, Cape Town, South Africa

Jeppe Høst Department of Ethnology, SAXO-Institute, University of Copenhagen, Copenhagen, Denmark

Michael Hurley LiveWater International, Portland, OR, USA

Moenieba Isaacs Institute for Poverty, Land and Agrarian Studies (PLAAS), University of Western Cape, Cape Town, South Africa

Svein Jentoft Norwegian College of Fishery Science, UIT – The Arctic University of Norway, Tromsø, Norway

Jahn Petter Johnsen Norwegian College of Fisheries, UIT – The Arctic University of Norway, Tromsø, Norway

Derek Johnson Department of Anthropology, University of Manitoba, Winnipeg, Canada

Estelle Jones School of Marine Science and Technology, Newcastle University, Newcastle upon Tyne, UK

Ronald W. Jones Learning Institute, Phnom Penh, Cambodia

Ahmed Khan UNEP-IEMP, Beijing, China

School of Business & School of Environment, Saint Mary's University, Halifax, Canada

John N. Kittinger Center for Ocean Solutions, Stanford University, Pacific Grove, CA, USA

Conservation International, Betty and Gordon Moore Center for Science and Oceans, Honolulu, HI, USA

Marloes Kraan IMARES Wageningen UR, Wageningen, The Netherlands

Lars Lindström Department of Political Science, Stockholm University, Stockholm, Sweden

Silvia Lourenço Divisão de Modelação e Gestão de Recursos da Pesca, Instituto Português do Mar e da Atmosfera (IPMA), Portugal, Lisbon, Portugal

MARE – Marine and Environmental Sciences Centre, Faculdade de Ciências, Universidade de Lisboa, Lisboa, Portugal

Camille Manel Département de Peche Maritime, Dakar, Senegal

Patrick McConney CERMES, University of the West Indies, Cave Hill, Barbados

Iris Monnereau CERMES, University of the West Indies, Cave Hill, Barbados

Friday Njaya Department of Fisheries, Lilongwe, Malawi

Paul O. Onyango Department of Aquatic Sciences and Fisheries, University of Dar es Salaam, Dar es Salaam, Tanzania

María José Barragán Paladines Department of Geography, Memorial University of Newfoundland, St. John's, Canada

Sölmundur Karl Pálsson Department of Anthropology, University of Manitoba, Winnipeg, Canada

José J. Pascual-Fernández Institute of Social and Political Sciences, Universidad de La Laguna, Tenerife, Spain

João Pereira Divisão de Modelação e Gestão de Recursos da Pesca, Instituto Português do Mar e da Atmosfera (IPMA), Portugal, Lisbon, Portugal

Graham J. Pierce Centre for Environmental and Marine Studies (CESAM) and Departamento de Biologia, Universidade de Aveiro, Aveiro, Portugal

Oceanlab, University of Aberdeen, Newburgh, Aberdeenshire, UK

Cristina Pita Centre for Environmental and Marine Studies (CESAM) and Departamento de Biologia, Universidade de Aveiro, Aveiro, Portugal

James Prescott Australian Fisheries Management Authority, Darwin, NT, Australia

Olivier Randin Department of Geography, Memorial University, St. John's, Canada

James Riwu Office of Marine Affairs and Fisheries, Kompleks Perkantoran Bumi Ti'i Langga Permai, Nusa Tenggara Timur, Indonesia

Silvia Salas CINVESTAV, Merida, Mexico

Jyothis Sathyapalan Centre for Economic and Social Studies, Hyderabad, India

Joeri Scholtens Department of Geography, Planning and International Development Studies, University of Amsterdam, Amsterdam, The Netherlands

Heidi Schuttenberg U.S. Agency for International Development, Washington, D.C., USA

University of Aberdeen, Aberdeen, Scotland, UK

Sheku Sei Statistics and Research Unit, Ministry of Fisheries and Marine Resources, Freetown, Sierra Leone

Cristiana Simão Seixas Environmental Studies and Research Centre, University of Campinas, Campinas, Brazil

Say Sok Royal University of Phnom Penh, Phnom Penh, Cambodia

Adam Soliman Director, The Fisheries Law Centre, Vancouver, Canada

Adjunct Professor, Faculty of Law, The University of British Columbia, Vancouver, Canada

Carlos Sonderblohm Centre of Marine Sciences (CCMAR), University of the Algarve, Faro, Portugal

Andrew M. Song Department of Geography, Memorial University, St. John's, Canada

Merle Sowman Department of Environmental and Geographical Science, University of Cape Town, Cape Town, South Africa

Natasha Stacey Research Institute for the Environment and Livelihoods, Charles Darwin University, Darwin, NT, Australia

Selina Stead School of Marine Science and Technology, Newcastle University, Newcastle upon Tyne, UK

Dirk J. Steenbergen Research Institute for the Environment and Livelihoods, Charles Darwin University, Darwin, NT, Australia

Part I
Governance and Governability – Introducing the Concepts

Chapter 1
Exploring Challenges in Small-Scale Fisheries Governance

Ratana Chuenpagdee and Svein Jentoft

Abstract This chapter sets the stage for the book about small-scale fisheries governance and governability, which draws lessons and reflections from 34 case studies about small-scale fisheries in 34 countries around the world, including north and south, east and west. The diversity, complexity, dynamics, and scale of small-scale fisheries and their governance are described. Characteristics of small-scale fisheries and their wicked problems call for a shift from management to governance, under whatever governing mode that aligns best with the particularity of the system that is being governed, and in corresponding with the governing system capacity (instrumental) and quality (normative). The need for and contribution of research on governance, particularly from social and transdisciplinary sciences are underscored, given the human presence and interactions that dominate small-scale fisheries. Interactive governance theory and the governability concept and assessment framework underpinning this investigation are perspectives that emphasize the systemic nature of small-scale fisheries, calling thus for a holistic and integrative analysis that transcends sectoral approaches often employed in management discourse and practice.

Keywords Interactive governance • Governability • Governing modes • Small-scale fisheries • Transdisciplinary research

---

R. Chuenpagdee (✉)
Department of Geography, Memorial University, St. John's, Canada
e-mail: ratanac@mun.ca

S. Jentoft
Norwegian College of Fishery Science, UIT – The Arctic University of Norway, 9037 Tromsø, Norway
e-mail: Svein.Jentoft@uit.no

S. Jentoft, R. Chuenpagdee (eds.), *Interactive Governance for Small-Scale Fisheries*, MARE Publication Series 13, DOI 10.1007/978-3-319-17034-3_1

Introduction

The problems of fisheries governance are well recognized. They constitute what is sometimes referred to as wicked problems (Rittel and Webber 1973; Jentoft and Chuenpagdee 2009), meaning that there is no easy solution to the problems faced in sustaining the resources and providing viable livelihoods to people involved. In fact, there is no general consensus as to what the challenges are, why they occur, and how to address them. For the most part, fisheries policies and governance are directed at what the interactive governance perspective recognizes as the 'first order,' i.e. the daily tasks that technical and routine solutions may be able to handle (Kooiman 2003; Kooiman et al. 2005). This has resulted in a limited fisheries management vision in which there have been few attempts to question fundamental issues and perspectives that effective and legitimate governance requires. The consequences have been felt in industrialized fisheries around the world, with numerous examples of fisheries collapse and several cases of fish stocks being over-exploited (Bavinck 2011), along with destruction of marine habitats and a high level of incidental catches and other collateral damage (Chuenpagdee et al. 2003). At the receiving end of these calamities are fishing people who experience threats to their livelihoods, food security, and wellbeing. There are no other places where these effects are felt more significantly than in small-scale fisheries, which account for over 90 % of the fishing people in the world, many of whom are impoverished, vulnerable and marginalized.

In this book, we posit that a broader lens to fisheries governance is needed, one that also takes into account the second order and the meta-order (Kooiman 2003; Kooiman et al. 2005). The second order of governance is related to the design of institutions, how they limit and enable social interactions, including problem solving and decision-making. This goes beyond analyzing what makes institutions work in a particular context, and what they actually do, to asking questions about what images give rise to how they are constituted, and in turn how institutions help to frame those very problems and challenges that governors are confronting within small-scale fisheries. Discussion about images takes us to the meta-order of governance, which, along with principles, norms and values, offers avenues and mechanisms to think beyond existing institutional framing, in order to be innovative and transformative in our approach. The meta-order concept invites us to step back and revisit issues that have been typically taken for granted in governance discourse and which governors do not normally tread into. For instance, the tragedy of the commons is often associated with too many small-scale fishing people for whom fisheries are bound to be the occupation of last resort given their poverty. This framing, we argue, is limiting our way of thinking about small-scale fisheries and leads to policy developments that have further negative consequences on the ability of small-scale fishing people to innovate and escape the poverty trap (Béné 2003; Jentoft and Eide 2011). For the most part, this reflects a path dependency determining decisions and actions at the first order. Rather than thinking anew when facing day-to-day problems, governance just does more of the same of what has brought it there in the first place. The interactive governance perspective, which is central to this book, provides us the possibility of an exploration in a broader realm of what governance involves and how it can be improved, to address the concerns and challenges confronting small-scale fisheries, be they food security, wellbeing, livelihood viability, social justice or environmental sustainability (Chuenpagdee et al. 2005; Bavinck et al. 2013). It is in this context that we introduce governability as a concept in a comprehensive assessment of the governance system, which includes the natural and social system-to-be-governed, the governing system and their interactions.

Governability is a composite concept related to the overall quality of governance, both in terms of its instrumental and normative dimensions. First, it is about the inherent and constructed characteristics of the system-to-be-governed that may contribute to making the system more or less governable. It is also about the capacity of the governing system to address societal concerns, given its own structure and function. How the governing system performs its function, whether it corresponds and responds to the system-to-be-governed, and how the two systems interact are therefore key aspects of governability. Further, the governability analysis helps understand whether governing instruments, be they marine protected areas (MPAs), individual transferrable quotas or catch shares, prohibit or facilitate sustainability of small-scale fisheries. Finally, governability involves asking whether governance processes and outcomes are aligned with universal meta-order values and standards, such as those stated in the Code of Conduct for Responsible Fisheries (FAO 1995) and the Voluntary Guidelines for Securing Sustainable Small-Scale Fisheries (or SSF Guidelines) recently adopted (June 2014) by member states of the Food and Agriculture Organization of the United Nations.[1]

These multiple facets of the governance system taken together are what a governability assessment would focus on in order to identify where governance interventions are required and where opportunities for improving governability exist. Such an assessment would also reveal the appropriateness of the particular mode of governance, given the context of small-scale fisheries that are the focus of governance. As such, governability places a strong emphasis on the importance of examining in detail how governing institutions are designed, and how they work, to facilitate interactions between the governing system and the natural and the social systems that it aims to govern. Interactive governance theory recognizes three ideal types of governing modes, i.e. hierarchical governance, co-governance, and self-governance (Kooiman et al. 2005). Notably, these analytical constructs are not a true reflection of empirical reality as illustrated in this volume of case studies of small-scale fisheries governance around the world. In reality, governance occurs along a continuum of mode, from the "top-heavy" rigid governing mode with governments at the apex on one end of the spectrum to the "bottom-hard" functioning associated with the strong self-governing mode on the other end, often in the form of customary local institutions. It can also be a mixture of modes, a certain form of hybrid, or in transition from one mode to another, as a way to respond to governability issues in particular situations. As reiterated in interactive governance theory, the diversity, complexity and dynamics are prominent features of the governing

[1] http://www.fao.org/fishery/ssf/guidelines/en

system, as much as they are to the system-to-be-governed. Therefore, governability assessment must account for all these characteristics, even as the emphasis is on the governing mode(s), as it is in this volume.

In what follows, we first present a brief overview of small-scale fisheries and governance, exploring their wicked problems and challenges. Second, the chapter identifies the need for and contribution of research to enhancing the governability of small-scale fisheries globally. Finally, it introduces the content and the structure of the book and how these issues have been dealt with in a series of in-depth case studies from around the world.

Small-Scale Fisheries and Their Governance

Around the world, small-scale fisheries are diverse in terms of pre-harvest, harvest and post-harvest conditions, activities and technology. They also occur in a wide range of aquatic environments, from river, lake and lagoon to estuarine, inshore and open sea. Small-scale fisheries are closely connected to communities, with kinship and other unique relationship networks, providing them with the sources of resilience and safety nets (Johnson 2006). It is therefore difficult and inappropriate to consider small-scale fisheries as a distinct sector in governance efforts (Sunde 2014). In fact, small-scale fisheries are not isolated from other activities, some of which may pose a threat to them, while some are complementary and synergistic. Also, fisheries governance, whether traditional or modern, formal or informal, does not exist in a vacuum but is often part of a larger governing system that includes other societal sectors and resources. This connectivity is important to recognize, even if it adds challenges to fisheries governance, particularly because of the embeddedness of small-scale fisheries in communities and sectors where solutions and opportunities to address crises and concerns may be found (McCay and Jentoft 1998).

We argue that any discussion about small-scale fisheries should begin with the recognition that they are different from their large-scale counterpart, not only in terms of size and scale of their technology, but also in the social organization of fishing units, economic motivation, and market linkages, to name a few. Small-scale fisheries differ from region to region, north to south, and often from community to community. Even poverty in small-scale fisheries, which is observed in many places, does not look the same around the world (Jentoft and Eide 2011). Thus, small-scale fisheries governance needs to take this diversity into account, in the formulation of the governance system that corresponds with the specificities of these fisheries. For this reason, small-scale fisheries governance cannot happen from afar, but must be situated in proximity to where small-scale fishing takes place, be sensitive to unique features and situations, and be clear about its effect on people's livelihoods, life prospects and human rights (Allison et al. 2012). The subsidiarity and dexterity principles are called in to play here in improving governability in small-scale fisheries (Jentoft et al. 2011). Subsidiarity refers to scale, the location of decision-making, and the first-right to fisheries resources (Bavinck and Jentoft 2011), while dexterity is about being perceptive to contextual details, which subsidiarity is meant to facilitate.

Unlike large-scale, industrial fisheries where harvest and post-harvest activities may either be part of vertical integration within large corporations or conducted by separate entities, harvest and post-harvest activities within small-scale fisheries are people-intensive, involving men, women and children, often the entire family and communities. This means that small-scale fisheries are the backbone of coastal livelihoods, contributing to income, jobs, food security, and sustainability for people in fishing places and beyond. The involvement of family members and local people in small-scale fisheries means that income generated from the sector remains in the communities and contributes to wellbeing, growth and local economic development. Catches from small-scale fisheries bring nutritious food to the table, sometimes the most important source of protein (HLPE 2014). Competition from large-scale fisheries, especially for small pelagic used in fishmeal production, often means less fish to meet the protein requirement of small-scale fisheries households. In addition, small-scale fishing people in many instances experience encroachment of their fishing grounds by large-scale operators. Moreover, access to landing places and work and living space is restricted by newcomers such as tourism operators, other industries and conservation organizations promoting spatially-based protection efforts like MPAs. Small-scale fisheries are thus suffering from what some term land and ocean grabbing.[2]

While small-scale fishers may exert pressure on fisheries resources by their sheer number and by their fishing practices, compared to industrial fisheries, their environmental impact, in terms of bycatch, discards, and overall effect on the local ecosystem, is far less (Kolding et al. 2014). This is especially the case when considering the limit of their range, the scale of their operation, and the subsistence nature of their harvest. Importantly, small-scale fishing people are often very active in stewardship initiatives and conservation efforts to sustain their immediate surroundings (Chuenpagdee and Juntarashote 2011). As shown by many examples from around the world, when small-scale fishing people are involved in the design and operation of MPAs, these initiatives have a higher chance of succeeding because governability obstacles can be dealt with in a way that small-scale fisheries stakeholders can be satisfied with (Chuenpagdee et al. 2013; Caveen et al. 2015). Unfortunately, many of these conservation efforts, as well as other zoning initiatives, do not appreciate the presence and contribution of small-scale fishing people, thus undermining the sustainability of both the ecological and the social system-to-be-governed.

Globally, fisheries are commonly governed by the hierarchical governing mode, with nation states as the dominant actors. This has particularly been the case after the declaration of the United Nations Convention on the Law of the Sea (UNCLOS) which extended the state's jurisdiction and responsibility for fisheries to 200 nautical miles. The fisheries crisis around the world (McGoodwin 1990) has also led to a more ambitious and authoritative role for the state in fisheries governance, often at

[2] See http://worldfishers.org/wp-content/uploads/2014/08/The_Global_Ocean_Grab-EN.pdf

the expense of the local governance system and its jurisdiction. The marginalization of small-scale fisheries is not so much the cause but the consequence of state interference, resulting in the disembedding of local and customary norms and rules, thus undermining the self-governance of fishing communities. In other words, traditional stewardship responsibility has been taken away from the community and given to the state bureaucracy. There has been a growing appreciation in recent years of the need to re-embed the responsibility and function of fisheries governance to local and regional organizations as a tool for enhancing governability. The SSF Guidelines express support for this transition. Moving from hierarchical to co-governance, and in some instances, self-governance, may therefore be something to consider from a governability enhancement perspective.

In this book, the institutional arrangement capturing this move is termed 'co-governance.' By employing this term, rather than 'co-management,' we underscore the fact that there are no technical solutions to governability problems, regardless of how inclusive the sectors and the processes are. This perspective also emphasizes the move from first-order governance to higher orders of governance, meaning that the day-to-day governing activities must be underpinned by considerations about institutions and their founding normative principles. This is particularly important in small-scale fisheries where the human dimension is salient and where issues of social justice and human rights are prevalent. Those things cannot be left to the bureaucracy and scientists but require active participation of small-scale fishing people, not only in co-management but also in co-governance of fisheries resources. It should be noted, however, that in some contexts, for example in developed countries, small-scale fisheries stakeholders already have representation in fisheries governance. The process of broadening the participation of stakeholders means that new and often more powerful groups are included in decision-making processes at the expense of small-scale fisheries actors, who increasingly feel overwhelmed and threatened (Jentoft and Knol 2014; Jentoft and Chuenpagdee 2015).

With 20 % of the world's population relying on fish as the main source of protein, and at least 135 million people depending on fishing for their livelihoods (FAO 2009), it is imperative to get governance right. The SSF Guidelines include key governance principles for their sustainability, rooted strongly in human rights standards and tenure rights. They cover principles related to human dignity, respect for cultures, non-discrimination practices, equity and equality, also related to gender, meaningful participation, rule of law, transparency, and accountability. In the context of small-scale fisheries governance, protecting and encouraging people's right to express their opinion, to organize, and be involved in the governance processes is essential (Sen 2009).

Small-Scale Fisheries Research

Fisheries governance under the command of the state relies mainly on knowledge and information produced by scientific communities, particularly those in the natural science and economic disciplines. Given the strong human presence in small-scale fisheries and the related multifaceted concerns of society at large, research into this sector and its governance systems should have a strong social science emphasis with interdisciplinary and transdisciplinary orientation. Capturing the richness of small-scale fisheries is difficult without thick description and in-depth study. Such a study should also be done in ways that consider small-scale fisheries as a system, not just an isolated sector (Jentoft 2014).

The contribution of social science, and indeed science in general, extends the scope of the investigation of small-scale fisheries and their wicked problems and opportunities. Enhancing governability does of course require rigorous research and comprehensive data, but the theoretical perspectives and concepts that science provides are no less important. In order to address the governability challenges of small-scale fisheries, we also need to be able to find a way to look at them and talk about them. The perspectives and concepts that governors and stakeholders employ in their discourses with much ease do in fact have their origin in academic discourse. In other words, scientists provide governors and stakeholders with a language that they could not be without, if they want to enhance governability. This contribution of science is not as valued as it should be and is often dismissed as too abstract. There is already a huge literature on concepts in academic discourse long before they became part of mainstream discussion. For instance, concepts like ecosystem-based management, stakeholder participation, empowerment, local ecological knowledge and governance have long histories, contained in research articles and books for many decades already, before they reached public political discourse. With better communication and information sharing and exchange mechanisms, this time lag could be reduced. It should, however, not be forgotten, that scientists, particularly social scientists, draw from observations, interactions and communication with local people in order to derive their concepts and theories. Knowledge production is therefore attributed to both parties. For these reasons, integrating scientific and local knowledge is a barrier that must be overcome, especially if one is to embrace a holistic and comprehensive form of interactive governance.

Corresponding to the perspective presented in interactive governance, transdisciplinary research draws on the broadest possible knowledge, including social and natural sciences and local and indigenous knowledge. In other words, both from the instrumental and normative perspectives of governability, interactive governance requires interaction among scientists of various backgrounds and specialities, as well as with practitioners, fishers, and community members. The experience of those who draw fish out of the water and who process and sell it must be valued and incorporated in the governance process. A level playing field is necessary for a good process to take place. The transdisciplinary perspective underlines what global research networks like Too Big To Ignore (toobigtoignore.net) aims to do and of which this book is an output. Issues and problems in small-scale fisheries governance are one of the main research priorities of that initiative. This book is thus a culmination of a 2-year effort to bring out experiences from around the world about how small-scale fisheries are governed and with what outcomes. The aim of this exercise is to shed light on governability challenges and illustrate what can be done to overcome limitations and create opportunities for small-scale fisheries systems to become vibrant and sustainable. Transformation and innovation in small-scale fisheries governance is of particular interest to Too Big To Ignore, as it recognizes the sophistication and advanced knowledge of small-scale fishers and their communities, and their potential contribution to address larger societal problems in the world. Research must not only focus on internal problems of small-scale fisheries, which are undoubtedly serious, urgent and wicked, but extend its scope to broader issues of governance at all levels. Through this lens, it is possible to examine both the internal and external origin of favorable and unfavorable governance outcomes in small-scale fisheries.

Research on governing modes begins with an understanding of the continuum and the non-steady or discrete nature that the systems to be studied pose. As highlighted in many chapters, several governing modes may be at play at the same time and in the same area, and they may also be transitional in nature. This phenomenon needs to be understood in its particular context, and in response to the demands and concerns of the systems that exist throughout the world. Special attention is paid in this book to the specificity of small-scale fisheries in individual locations and the extent to which the choice of a governing mode adheres to the meta-order principles promoted in interactive governance. Clearly, the dynamics of the governing mode adds to governability problems. These actually demand that the governing modes are flexible, adaptive and responsive, and even transformative and innovative. Thus, research must capture the plurality and mutuality of governing modes relative to governability. The comprehensive lens offered by governability assessment means not only that the details are not lost but that they are explicitly considered in thinking about what governing mode(s) may work best, according to the systems that are being governed.

About the Book

The book contains 37 chapters, 34 of which are case studies about small-scale fisheries in 34 countries around the world (see Map 1.1). The chapter authors are encouraged to employ the interactive governance and governability concept to explore any aspect(s) of fisheries governance that are most pertinent to their case studies, keeping in mind that the main emphasis of the book is the collective experiences about governing modes in operation. They are free to make references to small-scale fisheries in the way that works best for their case studies and there is no restriction in how small-scale fisheries are defined. This is to recognize the diverse and complex characteristics of small-scale fisheries, as well as the specific contexts of their existence. We also refrain from imposing any standardization about the terms used to refer to people involved in small-scale fisheries. Thus, some chapters use fishers, while others prefer fishing people, fisher people, fisherfolk or fish harvesters. In all instances, it is well acknowledged that small-scale fisheries governance is about men, women and children actively fishing, as much as about those involved in different post-harvest activities.

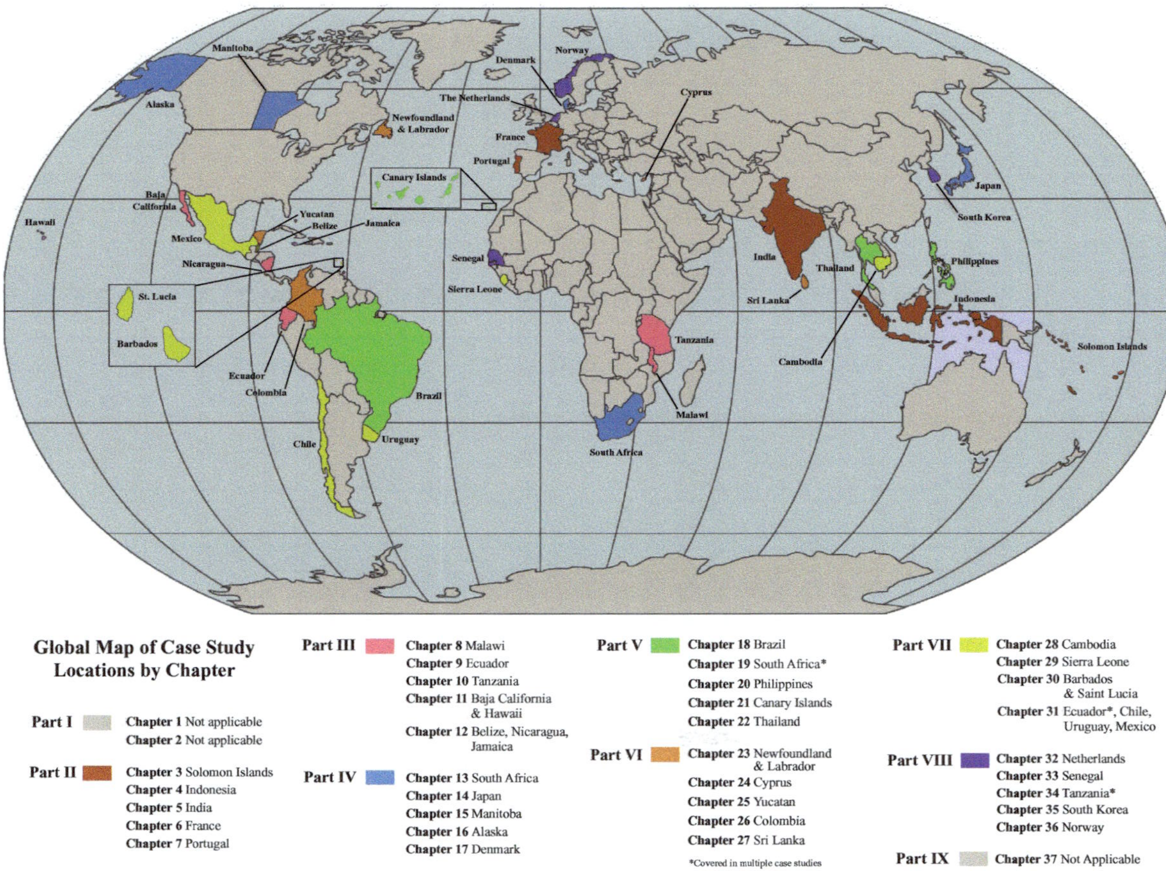

Map 1.1 Global map of case study locations by chapter

The book is organized into nine parts. Together with this introductory chapter, *Part I: Governance and governability – introducing the concepts* includes a chapter that introduces the concept of governability and the elements and procedures of governability assessment. It offers a coherent lens that helps to understand how the following chapters in this book make use of concepts in interactive governance theory and of the governability assessment framework, and how both together provide empirical evidence pertaining to governability in small-scale fisheries. It also serves as a guide to exploring possible options and opportunities in addressing these challenges.

Small-scale fisheries are diverse, complex, and dynamic. This is the case even within a single community, not to speak of between communities across countries and regions. These traits pose certain demands on governance, but they are also where opportunities and innovation to address local concerns and demands can be found. In other words, small-scale fisheries can mean many things and so it should remain in order for them to be viable and adaptive to changes. This implies that governance efforts should not aim to unify small-scale fisheries, but rather strengthen and empower them to embrace the social, cultural, economic and ecological contexts in which they are situated. This is a recurring theme in the five chapters included in *Part II, Small-scale fisheries diversity – identifying governability challenges*. Geographically, small-scale fisheries in these chapters include the Solomon Islands in the Southwest Pacific, Rote Island in Eastern Indonesia, the backwater of Kerala, India, the Iroise Sea of France, and coastal Portugal. In all cases, the social and cultural richness of fishing and coastal activities shines through, along with the vulnerability of small-scale fishing people who are exposed to pressure and threats within their surroundings, ranging from population growth, urbanization, market development, and climate change. Traditional small-scale fishing communities, in particular, are at risk of losing their identity with all these pressures, exacerbated by governance shortcomings. Understanding the small-scale fisheries and their governability in face of diversity and complexity is therefore necessary, as well as recognizing the dynamics triggered by internal and external stressors and drivers.

Similarly, small-scale fisheries are often governed by systems that are as diverse and complex as the fisheries themselves, if not more. In many instances, changes in the governing systems have brought about multiple challenges to small-scale fisheries. How to structure the governing system such that it fits with the property of the small-scale fisheries is a key governability issue that needs to be addressed. We see examples of this in *Part III: Governing system complexity – aligning modes*, with Lake Victoria in Tanzania, the lake fisheries of Malawi, the wider Caribbean lobster fisheries, Ecuador and the Galapagos, and Baja California, Mexico and the Hawaiian Islands. The self-governing capacity of small-scale fishing communities, which is often the mechanism for dealing with issues at the local scale, is undermined when they are drawn into the larger system of governance, including markets. As far as small-scale fisheries are concerned, the fish chain can stretch past community and country boundaries, thus aggravating governability problems beyond the control of the state agency. Multi-national governance is sometimes required, as seen in the wider Caribbean and Lake Victoria where different modes of governance are in operation. The dynamics between self-, co- and hierarchical governance pose governability problems and new opportunities for small-scale fisheries.

Part IV: Rights and justice concerns – securing access deals directly with one of the most contested issues in fisheries governance. Access refers not only to the allocation of resources but the basic right to food, livelihoods and human dignity, which makes fishing rights in actual practice human rights. How issues of rights and access impact governance and governability can be seen in the case of snoek and rock lobster fisheries in South Africa, in the context of fishing cooperatives in Japan, in the Lake Winnipeg fishery in Canada, in Halibut and Sablefish fisheries in Alaska, and in the case of small-scale fisheries in Denmark. These rights-based systems are often perceived as synonymous with private property rights, such as individual transferrable quotas or even community quotas, which from the perspective of small-scale fishing communities at large may have detrimental effects. Risks involved in privatization as far as small-scale fisheries are concerned are well illustrated in Part IV. Providing open access to all is essential for securing livelihoods and food security, and for sustainable communities, as long as an appropriate and responsible governing system is in place to ensure ecological sustainability, economic viability and social justice. What this governing system may look like is what interactive governance theory explores.

While privatization is a technical 'global' fix of an economic nature, MPAs have been introduced to fix ecological problems related to fisheries. For the most part, they are created in areas where small-scale fisheries are in operation, thus resulting in considerable obstacles to their success. Involving communities in the governance of MPAs is essential not only to their governability but also to ensuring that they do not have a similar negative effect to privatization, in terms of restricting small-scale fisheries access. Although MPAs may have positive long-term effects on small-scale fisheries livelihoods in so far as they result in improving ecosystem health, concerns about the short-term effects are equally important in the analysis of governability. The case for careful consideration of short-term and long-term effects on small-scale fisheries in the discussion of MPAs is taken up in *Part V: Marine protected areas – securing space* of the book, with illustrations from Southern Brazil, South Africa, the Philippines, Canary Islands, and Thailand.

Next, low governability is often the result of problems related to boundaries and scale at national and international levels. In some instances, small-scale fisheries are located in areas where major conflicts, including armed conflicts, are present, which results in restrictions on the fisheries. Thus, issues of boundary and scale are pertinent in small-scale fisheries governance where governability challenges extend the fisheries sector into areas where small-scale fisheries governing systems have no reach. The chapters in *Part VI: Cross-boundary governance – fostering interactions* bring us from Palk Bay between India and Sri Lanka, to Colombia and Nicaragua, with Cyprus, Newfoundland and Labrador of Canada, and Mexico, in between. Together they illustrate that small-scale fisheries often find themselves in a very complex political environment where issues are politicized beyond the local level and beyond fisheries, making the governability of small-scale fisheries even more challenging. In some cases, it is hard to see where solutions lie, as they may be found outside of the small-scale fisheries system.

Governance arrangements for some of these problems tend to evolve around reform and transformation. Crises often lead to the questioning of the quality and the capacity of the existing governing order and modes. In other words, it is the governability of the prevalent system itself that is under scrutiny. While these crises tend to make reform urgent and legitimate, governance reform may in itself trigger further governability challenges, which illustrate the inherent "wickedness" of such reform. These challenges are illustrated in *Part VII: Governance in transition – reforming institutions* with case studies from Cambodia, Sierra Leone, the Eastern Caribbean, and a number of countries in Latin America. The issue of involving multiple stakeholders in partnership arrangements, such as co-governance, the second governing order, is explored in all cases.

In Zanzibar, Tanzania, Senegal, the Netherlands, Norway and South Korea, the discussion about small-scale fisheries governance centers around fundamental issues with regard to values, principles and worldviews. Critical examination of this meta-order governance is sometimes necessary for building trust in order to improve governability and governance outcomes. Basic assumptions and ideas behind governance are often implicit and assumed to be valid. These create tunnel visions and hamper the innovation potential for governance reform because they limit the range of alternatives and opportunities that governing actors can envisage. The chapters in *Part VIII: Meta-governance – realizing the possibility* illustrate how understanding the difference in the fundamental mindset of governors and those who are governed can make in enhancing governability. Interactions, coordination, and matches between institutional aims and people's perceptions of what the problems are have been identified as key features for better governance.

In the final part, *Part IX Governability challenges – urging change*, a single chapter highlights the lessons learned and the reflections made by chapter contributors about small-scale fisheries governance and governability. As such, the chapter can be read as a collective statement from all the authors about what we can do to improve governance and governability, thus securing the sustainable future of small-scale fisheries around the world. The heart of small-scale fisheries concerns is governance. Within governance, the governing modes and the meta-order principles that underlie their decisions are of great significance. They unfortunately rarely receive sufficient consideration since it is normally minor adjustments to the first-order governance of a technical nature that seems to get the attention of governors and the financial and scientific resources of the state. We argue that governance reform is necessary in all cases, and that for it to be practical and functional it must emphasize the relationship between meta-order and the second-order governance. In other words, it must address basic values, norms, principles and worldviews, and the institutional implications that are drawn from them. Interactive governance offers a systematic and holistic perspective on the need to investigate these issues whereas the governability framework helps examine where problems and opportunities lie. Governability can also be seen as a "playful" lens that is required to deal with the diversity, complexity, dynamics, and scale issues surrounding small-scale fisheries. This means that rather than offering a cookbook solution on how to fix small-scale fisheries governance problems, a governability analysis invites examination and exploration of explanations and possible solutions that can be further investigated. This may frustrate people who prefer short answers. But with the future of small-scale fisheries at stake, certain commitment is required. At the end of the day, people who make decisions about small-scale fisheries do not necessary live the consequences. The basic principle for responsible decision-making, i.e. accountability, is largely lacking in the world of small-scale fisheries today. The contributions in this book are illustrations of how change is necessary and indeed possible.

References

Allison, E. A., Ratner, B. D., Asgard, B., Willmann, R., Pomeroy, R., & Kurien, J. (2012). Rights-based fisheries governance: From fishing rights to human rights. *Fish and Fisheries, 13*(1), 14–29. doi:10.1111/j.1467-2979.2011.00405.x.

Bavinck, M. (2011). The mega-engineering of ocean fisheries: A century of expansion and rapidly closing frontiers. In S. D. Brunn (Ed.), *Engineering earth: The impacts of mega-engineering projects*. Dordrecht: Kluwer.

Bavinck, M., & Jentoft, S. (2011). Subsidiarity as a guiding principle for small-scale fisheries. In R. Chuenpagdee (Ed.), *Contemporary visions for world small-scale fisheries*. Delft: Eburon.

Bavinck, M., Chuenpagdee, R., Jentoft, S., & Kooiman, J. (Eds.). (2013). *Governability in fisheries and aquaculture: Theory and applications*. Dordrecht: Springer.

Béné, C. (2003). When fishery rhymes with poverty: A first step beyond the old paradigm on poverty in small-scale fisheries. *World Development, 31*, 949–975.

Caveen, A., Polunin, N., Gray, T., & Stead, S. M. (2015). *The controversy over marine protected areas. When science meets politics*. Dordrecht: Springer.

Chuenpagdee, R., & Juntarashote, K. (2011). Learning from the experts: Attaining sufficiency in small-scale fishing communities in Thailand. In S. Jentoft & A. Eide (Eds.), *Poverty mosaics: Realities and prospects in small-scale fisheries*. Dordrecht: Springer.

Chuenpagdee, R., Morgan, L. E., Maxwell, S. M., Norse, E. A., & Pauly, D. (2003). Shifting gears: Assessing collateral impacts of fishing methods in the U.S. waters. *Frontiers in Ecology and the Environment, 1*(10), 517–524.

Chuenpagdee, R., Degnbol, P., Bavinck, M., Jentoft, S., Johnson, D., Pullin, R., & Williams, S. (2005). Challenges and concerns in fisheries and aquaculture. In J. Kooiman, M. Bavinck, S. Jentoft, & R. Pullin (Eds.), *Fish for life: Interactive governance for fisheries*. Amsterdam: University of Amsterdam Press.

Chuenpagdee, R., Pascual-Fernandez, J. J., Szelianszky, E., Alegret, J. L., Fraga, J., & Jentoft, S. (2013). Marine protected areas: Re-thinking their inception. *Marine Policy, 39*, 234–240.

FAO. (1995). *Code of conduct for responsible fisheries*. Rome: Food and Agriculture Organization of the United Nations.

FAO. (2009). *The state of world fisheries and aquaculture 2008*. Rome: Food and Agriculture Organization of the United Nations.

HLPE. (2014). *Sustainable fisheries and aquaculture for food security and nutrition*. Rome: A Report by the High Level Panel of Experts on Food Security and Nutrition of the Committee on World Food Security.

Jentoft, S. (2014). Walking the talk: Implementing the international voluntary guidelines for securing sustainable small-scale fisheries. *Maritime Studies, 13*, 16.

Jentoft, S., & Chuenpagdee, R. (2009). Fisheries and coastal governance as a wicked problem. *Marine Policy, 33*, 553–560.

Jentoft, S., & Chuenpagdee, R. (2015). The 'new' marine governance. Reality or aspiration. In M. Gilek & K. Kern (Eds.), *Governing Europe's marine environment: Europeanization of regional seas or regionalization of EU policies?* London: Ashgate Publishing.

Jentoft, S., & Eide, A. (Eds.). (2011). *Poverty mosaics: Realities and prospects in small-scale fisheries*. Dordrecht: Springer.

Jentoft, S., & Knol, M. (2014). Marine spatial planning: A view from the North Sea. *Maritime Studies*. Retrieved from http://maritimestudiesjournal.com.content/12/1/13

Jentoft, S., Eide, A., Bavinck, M., Chuenpagdee, R., & Raakjær, J. (2011). A better future: Prospects for small-scale fishing peoples. In S. Jentoft & A. Eide (Eds.), *Poverty mosaics: Realities and prospects in small-scale fisheries*. Dordrecht: Springer.

Johnson, D. S. (2006). Category, narrative, and value in the governance of small-scale fisheries. *Marine Policy, 30*(6), 747–756.

Kolding, J., Béné, C., & Bavinck, M. (2014). Small-scale fisheries: Importance, vulnerability and deficient knowledge. In S. M. Garcia, J. Rice, & A. Charles (Eds.), *Governance of marine fisheries and biodiversity conservation: Interaction and coevolution*. Hoboken: Wiley Blackwell.

Kooiman, J. (2003). *Governing as governance*. London: Sage.

Kooiman, J., Bavinck, M., Jentoft, S., & Pullin, R. (Eds.). (2005). *Fish for life. Interactive governance for fisheries*. Amsterdam: Amsterdam University Press.

McCay, B. J., & Jentoft, S. (1998). Market or community failure? Critical perspectives on common property research. *Human Organization, 57*(1), 21–29.

McGoodwin, J. R. (1990). *Crisis in the world's fisheries: Peoples, problems, and policies*. Stanford: Stanford University Press.

Rittel, H. W. J., & Webber, M. M. (1973). Dilemmas in a general theory of planning. *Policy Sciences, 4*, 155–169.

Sen, A. (2009). *The idea of justice*. London: Penguin.

Sunde, J. (2014). *Customary governance and expressions of living customary law at Dwesa-Cwebe: Contributions to small-scale fisheries governance in South Africa*. Doctoral thesis, University of Cape Town, Cape Town.

Chapter 2
Assessing Governability of Small-Scale Fisheries

Svein Jentoft and Ratana Chuenpagdee

In principle, all societal systems can be looked upon from the point of view of their governability. (Jan Kooiman 2008, 187)

Abstract This chapter presents interactive governance as an analytic perspective in small-scale fisheries research. It serves as an introduction to the conceptual framework that chapter authors apply and describes the overall research questions raised in this book. A key concept that will be elaborated is governability, which was also central to other books on interactive fisheries governance preceding this one (Kooiman et al. 2005; Bavinck et al. 2013). Given the diversity, complexity and dynamics that characterize small-scale fisheries at various scales, and the multitude of concerns that are associated with them, the governance and governability of small-sale fisheries is a challenge worth exploring. We argue that there is no standard answer to the many problems facing small-scale fisheries globally but governance solutions must always be attuned to the particularities of the sector and circumstances within which they exist and operate. The governability assessment, as outlined in this chapter, is developed to assist the researcher to locate the problems and opportunities and thus address the complex issues and demands in small-scale fisheries governance.

Keywords Governability assessment • Interactive governance • Small-scale fisheries • Governing modes • Co-governance

S. Jentoft (✉)
Norwegian College of Fishery Science, UIT – The Arctic University of Norway, 9037 Tromsø, Norway
e-mail: Svein.Jentoft@uit.no

R. Chuenpagdee
Department of Geography, Memorial University, St. John's, Canada
e-mail: ratanac@mun.ca

S. Jentoft, R. Chuenpagdee (eds.), *Interactive Governance for Small-Scale Fisheries*, MARE Publication Series 13, DOI 10.1007/978-3-319-17034-3_2

Introduction

Small-scale fisheries, which involve millions of people around the world, contribute significantly to food security, livelihoods and employment. They also represent cultural heritage and identity, and help sustain coastal communities whose existence is dependent on them. Still, small-scale fishers often find themselves in a situation of marginalization, and in many instances, poverty. Rarely do they have a voice in forums where their fate is decided. It is for this reason that the FAO member states in June 2014 endorsed the "Voluntary Guidelines for Securing Sustainable Small-Scale Fisheries in the Context of Food Security and Poverty Eradication", hereafter "SSF Guidelines".[1] The guidelines speak primarily to national governments, who are urged to take greater responsibility for the livelihoods and sustainable development of small-scale fisheries through targeted policy initiatives, followed by legal and organizational reform and economic support.

However, small-scale fisheries do not always lend themselves easily to government intervention. They are simply too diverse, complex and dynamic to allow one-dimensional governance solutions. Small-scale fishers value their freedoms and are often suspicious of ambitions articulated on their behalf. In some instances, the relationship between small-scale fishing communities (e.g. the people who fish, process and bring the produce to the market) and government is even characterized by hostility. Small-scale fishing communities do not always welcome government interference, especially when government is not perceived to be on their side. Development of small-scale fishing activities often takes place independent from government. In many instances, government has actually made its entrance into the life of small-scale fishing people fairly recently. It is also for this reason that policy makers ignore them. The implementation of the above-mentioned guidelines is therefore likely to be less than straightforward (Jentoft 2014).

The absence of government rule and the resistance to external control speak to two general issues, which are the key focus of this book: governance and governability. Both concepts allude to the observation that there is no single governance recipe that would work in all settings (Degnbol et al. 2006; Ostrom et al. 2007) and that many of the problems governors, whoever they are, face can truly be termed as "wicked" (Rittel and Webber 1973; Jentoft and Chuenpagdee 2009). The interactive governance perspective emphasizes the need to look for governability problems within the system-to-be-governed, the governing system as well as in the ways the two interact, which is what the governability assessment approach suggests, as shown below.

Although small-scale fisheries share many characteristics globally, problems are not the same everywhere. There is always something unique about the problems and characteristics that small-scale fisheries face in a given locality. Therefore, contextual factors must be taken into account when governing. This is not only the

[1] http://www.fao.org/fishery/ssf/guidelines/en

case for small-scale fisheries activities but also for their institutions. The basic assumption underpinning the effort behind this volume is that there are lessons to learn from various case studies, from what they have, or do not have, in common. These lessons can help improve small-scale fisheries governance, which is a global concern.

The "new governance" concept, as originally coined by Rhodes (1996), acknowledges that governance is not only for government but also for civil society and markets, to the extent that one may imagine governance beyond the role of the government. The composite concept of governance is also one identified by "interactive governance" as defined and discussed by Kooiman and Bavinck (2005), emphasizing the different roles and capacities of different governing actors and how they together may contribute to enhancing governability in a way that each of them separately would be incapable of.

It is still relevant to discuss the role of government in socio-political governance, but more in terms of a proper division of labor between state, market and civil society. This division of responsibilities cannot be decided without recognizing the actual capacities of each of these actors in concrete situations. It is well known that the capacity to buttress and enforce policies varies a lot from country to country, including within small-scale fisheries governance. In some instances it is relevant to talk about "soft" or "failed" states (Thorpe et al. 2009), where poor or ineffective governance is widespread and does not only affect small-scale fisheries but society as a whole. In other instances, the state apparatus not only works well, but also enjoys considerable legitimacy among small-scale fishers.

In any case, there are limits to the effectiveness of small-scale fisheries governance, for instance due to knowledge constraints, poor legislation, missing or inadequate data. There are also ways to assess the success of any governing system, which would also apply in the context of small-scale fisheries, for example in terms of "good governance" or human rights. The governance of small-scale fisheries is therefore subject to both specific and general demands, which are not always easily harmonized, and which add to the governability challenge.

Governability problems also exist within civil society and local communities, where often poor organization hinders policy up-take and collective action. Improving governability of small-scale fisheries requires that people are empowered, something that better organization can provide. Still, this is not a universal bottleneck. In some instances, governing systems work well in the absence of government because there is someone other than government there already to shoulder the governing functions. In such circumstances, one does not start from scratch; rather there is something to take advantage of, build on and support. Introducing new organizations in situations where organizations already exist and are operative may actually increase rather than alleviate the governability problem.

Small-scale fisheries are often upheld by customary organizations that predate current nation-state formations and which continue to be operative, often side by side with government regulations. Customary institutions can be highly formalized, but in many instances, they are informal and tacit. They operate in a situation of legal pluralism, where several normative orders are active at the same time, which may create governance confusion but can also be a source of institutional innovation (Jentoft and Bavinck 2014). One may expect to find situations where customary institutions are filling a void in the chain of governance, thus increasing governability (Sunde 2014). One may also find that legal pluralism restricts governability, such as when formal and informal rules and regulations are incompatible. Governance decisions must therefore be informed by whatever pluralism that exists in particular situations (Jentoft 2014). If not, governance risks misfiring, failing or damaging small-scale fisheries and communities. Evaluation must be part of a broader governability assessment that should occur prior to, or in concurrence, with small-scale fisheries policy implementation.

This chapter aims to clarify the governability concept and what a governability assessment may entail in the context of small-scale fisheries. As shall be seen, it is a composite concept, containing multiple elements. We argue that governability is not only about the extent to which a governing system addresses current concerns, but also about the way it actually goes about it. In assessing governability, the aim is about understanding under what conditions a particular governing system addresses certain basic concerns, such as environmental health, food security, sustainable livelihoods and social justice. It is also about examining the process through which governance solutions are arrived at, i.e. the degree to which the process has been inclusive, transparent and socially just.

The multiple dimensions of governability make the concept as complex as the governance effort itself. Nonetheless, the governability assessment framework (Chuenpagdee and Jentoft 2013) is used as a common thread in all chapters in order to assist researchers in tracing the complex governance challenges of small-scale fisheries and thus locating the problems and opportunities that exist in their particular contexts. We start by describing what governability is, and what it implies, for small-scale fisheries governance research. After that we present the key governing modes as they appear within this sector and suggest how best to phrase relevant governability research questions around them. Finally, we summarize the key elements of the governability assessment and the steps it involves, noting the need for flexibility in its application.

The Governability Concept

In a previous publication we argued that assessing governability is part of a reality check that governors must perform in order to improve governance effectiveness and legitimacy (Chuenpagdee and Jentoft 2009). It is thus integral to governance itself. The improvement of governance goes through the enhancement of governability – and vice versa. Their linkage is, in other words, mutual and reinforcing. This leaves governing actors with a number of difficult questions, such as what governability is and what governability assessment should focus on.

For Kooiman (2003, 193), governability is an "overarching concept, as the quality of a social-political entity as a whole." This would imply that governability is not only about the capacity of the governing system to address certain basic concerns and deliver on its goals. Rather, governability depends also on the characteristics of the governing system, the natural and social systems-to-be-governed and the governing interactions, seen together. More specifically, the focus of the analysis would be on: (a) diversity, e.g., the number of species, actors, interests, images and values that make up the natural and social systems that are being governed and the governing system; (b) complexity, or how the mentioned components are related; (c) dynamics, i.e., how they interact, move and change relative to each other; and (d) scale or boundary issues, whether or not they are part of larger systems that establish conditions and channels for interaction at the margin of systems.

Governability is not a static phenomenon, but will change with the interactions that occur within and between the governing system and the system-to-be-governed. These dynamics are largely, but not completely, shaped and facilitated by the way these systems are structured, as there will always be room for intentional action and choice by those involved that will make systems more or less governable.

Analytically, governability constitutes two complementary but necessary dimensions: (1) the capacity to govern, which depends in part on the structure and function of the governing system, but also on the inherent and constructed characteristics of the system-to-be-governed that may either lend itself to governance or inhibit its functioning; and (2) the quality of governance processes and outcomes and the values that they express, whether or not they are in accord with a set of agreed-upon principles, following certain norms and values. In the first instance, the emphasis is on what the governing system can possibly do, given the characteristics of the natural and the social systems that it aims to govern, and based on its own effectiveness and efficiency. Governability recognizes that certain aspects of the system-to-be-governed may make governance challenging, no matter how capable the governing system is. In other words, capacity must be seen relative to what is actually being governed. Small-scale fisheries may be inherently difficult to govern; they may be too diverse, complex and dynamic for any governing system to meet expectations. Established goals may therefore not be attainable in practice, also because they are unstable, and difficult to realize.

In the second instance, governability depends on how the governance system actually performs and delivers on its goals, given financial, political and legal constraints. But governability is not just instrumental; it is also normative, meaning that it can be subject to philosophical reasoning and political argument. Thus, governability in terms of the quality of governance must be scrutinized with regard to what matters are being pursued, why and how. The normative aspect is about what goals are good, which goes beyond the mere technicalities and instrumentality of governance, and includes ethical choices. The instrumental aspect is about process, which stakeholders are listened to, and how they are involved. One can think of governability as a two dimensional graph, with capacity on one axis and quality on the other. "Good governance" requires high scores on both axes. Some authors in this book put more emphasis on instrumental capacity, some on normative quality, and some on both. But together they contribute to a fuller understanding of what governability entails. Below, we discuss what may occur in the analysis of governability along both dimensions.

Capacity What is attainable or not is often a function of what means are available, and whether they include the stick or the carrot. Governors can do more for small-scale fisheries if they have legal means to enforce policies and when they can hold those who are being governed accountable. Second, their degree of professionalization is important, for instance their experience and education level. Do they really know the system-to-be-governed, its diversity, complexity and dynamics? They can also be more effective if they have resources to offer, such as economic support, that are hard to resist. This is largely an issue of legitimacy and power. But the effectiveness of power is dependent on the counter powers it confronts. Thus governments may have plans for small-scale fisheries, but they are not able to implement them effectively because they are up against powerful interests within or outside small-scale fisheries that resist them. For example, small-scale fishers might be well organized and resist policy changes they deem adverse to their interests. Or other stakeholder groups, such as industrial operators, may also not agree with what the government is trying to do for the small-scale sector.

Empowerment in the context of governability should however not necessarily be seen as a zero sum game. Governability may be enhanced through a constructive partnership, which would largely hinge upon strong and capable organizations on both sides, as when trade unions stand in the way of "wildcat" strikes. It is not likely that governability benefits from a fragmented, disorganized, and disenfranchised small-scale fisheries system-to-be-governed, even if it may provide the agencies of the governing system a freer hand to do as they please.

Thus, capacity is partly a function of the enabling properties of the system-to-be-governed, in this case of the small-scale fisheries chain as it is situated in communities, markets and civil society, and those of the governing system, i.e. the institutions, steering mechanisms (rules) and incentives (rewards, penalties) available. Notably, the governing system may well be integral to the system-to-be-governed, as in self-governance, whereby small-scale fishing peoples rule themselves without interference by an external authority. In another publication we have argued that marine protected areas (MPAs) can fruitfully be analyzed as a system-to-be-governed and a governing system all in one, but not necessarily at the same time, as the questions confronting each are different. As a system-to-be-governed the issues would be related to the actors and activities that occur within the boundaries of the MPA and the eco-system within it. As a governing system one would focus on the decision-making processes, the representation of stakeholders therein, and the rules and regulations that apply to access and use (Jentoft et al. 2007).

One should not forget, however, that MPAs are rarely isolated units, but typically part of a larger environmental or fisheries governing system. Despite that, they tend to have their own autonomous governing system within which various user-groups including small-scale fishers have representation. The degree and practice of this self-governance practice is an issue to be considered in any governability assessment as it suggests the extent to which the two systems are integrated or overlapped. Whether governability is enhanced or inhibited by such integration is a research question. This is also where the issue of legal pluralism, mentioned earlier, comes in.

The limits and opportunities of governability must however also be sought in the governing system, which can be similarly diverse, complex and dynamic. Fisheries governance is typically multi-organizational and multi-scalar and the chain of command can run from the top down as well as from the bottom up. Governability can suffer because of too many fragmented institutions with overlapping and contradictory mandates at inappropriate scales, as when small-scale fisheries are being governed by an authority distant from where the problems are being felt, leading to insufficient or missing contextual knowledge. Small-scale fisheries governability can also suffer from lack of political priority and attention because the governing system is geared to serve industrial fisheries and export. The governing system may be inadequate, for instance understaffed, financially weak, legally unauthorized, data-poor, and ill-informed about what small-scale fisheries are and what conditions they are working under (Mahon and McConney 2004).

The lack of muscle of the governing system and the size, shape and texture of the system-to-be-governed can hamper governability. But so too can the interaction between the two systems. A man's ability to lift a stone is certainly dependent on his own strength as well as the weight and shape of the stone, but his ability can be augmented by the tools that he has and can use. The tool, a winch for instance, will enhance his capacity to move the stone. Thus the winch is the means through which he interacts with the stone. The point being made is that governability also depends on the interaction that takes place between the person (the governing system) and the stone (the system-to-be-governed), and that the means of communication between the two systems are part of the governability equation. If the stone was alive, as the small-scale fisheries system-to-be-governed is, the relationship between the two systems and hence governability would be again different; it would make the interaction more or less difficult depending on how cooperative the interaction is. What would also matter is the governors' belief in their ability to govern, their willingness to involve and share governance functions and responsibilities with those who are being governed, in this case small-scale fishing people. Governability would also depend then on the self-confidence of the latter, their preparedness to participate, and their readiness to assume responsibility.

Interactive governance (Kooiman 2003) operates with essentially three modes; hierarchical, co-, and self-governance, each with distinct features pertaining to relationships and forms of communication. These modes are to be perceived as "ideal types", i.e. theoretical constructs meant for empirical comparison. In reality, interactive governance is generally exercised through mixed modes, depending on function and scale. But such mixtures are hardly stable; rather they are in flux over time, for instance as a consequence of adaptive learning.

The idea of adding the prefix "interactive" to governance is not to suggest that interaction necessarily enhances governability and therefore is inherently good, as interaction can also be in the form of domination and misuse of power. Interaction should rather be seen as a regulative idea for research, i.e. researchers should investigate whatever interactions take place and how interactions have implications for governance outcomes, such as governability.

Quality This governability attribute and its assessment is about to what extent the governing system design, culture and practice are in accordance with established and generally accepted global and local normative standards,– be they accountability, inclusiveness, coherence, social justice, and those that are frequently associated with the concept of "good governance," as understood by the World Bank[2] or as enshrined in international codes of conduct, including those pertaining to human rights. A low score on such indicators would not only suggest low quality, but also reduce the capacity and effectiveness of governing institutions (see Fukuyama 2013).

A governing system that does not live up to such standards is vulnerable, at least in the long run, as it may not receive the legitimacy and support it needs to be sustainable. To be sustainable it must be institutionally robust, but such robustness is not only controlled by those who are formally in charge of it, but also to a variable extent by those who are subject to it. In fisheries, rules and regulations need acceptance among users, as the possibilities to evade them are many, as in the case of discards and use of illegal fishing gears (Jagers et al. 2012). Inspectors cannot be everywhere all the time. Revolts happen because subjects no longer accept and respect what institutions are and do, and therefore decide to exit them, through boycott. To put it in interactive governance terms, this happens because of a growing dissonance between the system-to-be-governed and the governing system. Revolts are a way for the system-to-be-governed to communicate that dissonance. It is important to remember in this regard that legitimacy of institutions is never guaranteed permanently. Institutions survive as long as they continue to meet the expectations that users have of them. Once they fail to do so, they will not be able to enhance the governability they were set up for, unless they employ means to force compliance.

It is possible for a governing system to have a high governability score for capacity and a low score for quality, as would be the case for a repressive, undemocratic system which is not living up to the "good governance" criteria. One can envisage that such a system can still be effective in "getting the job done" if rule enforcement is sufficiently harsh. For instance, it could be argued that Iraq was more governable under the regime of Saddam Hussein than it has been after his overthrow, but the way that regime operated was still highly questionable. Thus, governability can never be a good thing if it has a low score for the qualitative dimension. Governance is also about social justice and human rights, which are goals in themselves, not least in labor intensive small-scale fisheries. Governing small-scale fisheries is often about people who tend to be poor, marginalized and hence vulnerable (Jentoft and Eide 2011). This is also why the human rights approach plays such a prominent role in the SSF Guidelines and in academic publications (Allison et al. 2012; Ratner et al. 2014). Such standards and principles when implemented must always be

[2] http://info.worldbank.org/governance/wgi/

operationalized for concrete situations, but without relativizing them to the extreme. These norms and principles are intended to improve the performance of small-scale fisheries, which may in some instances represent a break with established custom, for instance those that discriminate against women.

Thus governability cannot be a goal in itself unless it is qualified. Conditions under which governability is obtained must be specified, and the costs of it taken into account. Governability that is obtained through voluntary consent has a higher score, certainly in terms of the qualitative dimension but most likely also in terms of capacity, than one that is secured through force. Co-governance has a higher potential for producing voluntary consent than hierarchical governance, as the process of involvement and participation is more conducive than one of command and control, partly because of the interactive learning it facilitates. That is, however, something to be investigated, as both hierarchical and co-governance can mean different things as the actual institutional design relative to the task at hand and the context within which it is supposed to work always matters (see for example Wilson et al. 2004).

Small-scale fisheries as a system-to-be-governed would probably be more governable if all the diversity and complexity that we associate with them are removed, the sector is down-scaled and standardized, and is finally run by a single company. One cannot, however, "save" small-scale fishing by killing it, which some of the modern governing mechanisms such as individual transferable quotas (ITQs) have a tendency to do, even if they reduce effort, relieve pressure on the fish stocks and increase the resource rent that can be generated as a consequence. It may be good, in other words, for other things but not for small-scale fisheries as an activity one aims to support and sustain. Governability must therefore first be related to the basic concerns of fisheries governance, such as those listed by Chuenpagdee et al. (2015) and Bavinck et al. (2013), i.e., ecosystem health, social justice, livelihood and employment, food security and food safety. These are concerns that governing systems convert into meta-principles, like those defined and operationalized in section 6 of the Code of Conduct for Responsible Fisheries[3] or section 3 of the SSF Guidelines, and subsequently into operational goals when this code and these guidelines are being implemented in concrete contexts.

These concerns, principles and goals and their attainability must be part of the governability assessment baseline. One can obviously enhance governability if one concentrates on only one of these concerns and ignores the rest of them. But that would largely be to make the governance task easier for oneself. One should acknowledge a fisheries governing system that succeeds in rebuilding or sustaining fish stocks, but what if it happens in a way that undermines the well-being of small-scale fisheries communities? The quality of small-scale fisheries must obviously be evaluated with their degree of holism in mind, as stressed in the SSF Guidelines. The holistic perspective will also establish the criteria for assessing capacity.

[3] http://www.fao.org/docrep/005/v9878e/v9878e00.htm

Governability Assessment

The diversity, complexity and dynamism of small-scale fisheries systems require systematic, iterative and contextualized assessments of governability. Thus the governability assessment of small-scale fisheries is concerned with how the various structural features and processes of the governing system, the system-to-be-governed and their interactions affect governability. The normative aspect of governability requires an analysis of meta-order governance, which is about values, images and principles, and how these translate into institutional design and practices at the second-order. Several tools and approaches to examine the various governability characteristics of the fisheries systems are available (see Chuenpagdee and Mahon 2013). Given the often informal nature of small-scale fisheries and their governance, it is important to have an eye for subtleties, things that are tacit, also to the participants involved. Governability assessment cannot just be *etic*; its perspective must also be *emic*. The latter stresses the fact that people have their own images and conceptualization of the world they live in which guide their actions and interactions (Headland et al. 1990).

Interactive governance theory as a regulative idea suggests that the researcher must strive to understand how people (those that are being governed and those who are governing) themselves regard and act upon demands pertaining to the diversity, complexity, dynamics and scale that characterize the system. Governance is after all not a natural but a social process. Given that governability assessments should be integral to the governance process itself, they should be inclusive, participatory, transparent and interactive. The contribution of stakeholders also has functional merits, since they possess knowledge and means of control that are relevant for the enhancement of governability as capacity. Governability assessment helps facilitate a holistic analysis, one that does not leave substantial gaps in understanding what small-scale fisheries are, what they contribute to, and where their governability challenges exist. It also helps phrasing key research questions in the context of governability and how they fit into the overall governance research agenda.

Assessment Framework

We have in a different publication (Chuenpagdee and Jentoft 2013) described a stage-wise governability assessment approach, which is only summarized here. The various stages are outlined in Table 2.1. Case study contributors were encouraged to use the framework to guide their analysis, but not necessary to feel restricted by it.

Stage 1 is devoted to problem definition; how stakeholders within both the system-to-be-governed and the governing system perceive the essence of the governability challenges that they face; what they are, what they are caused by, and how they see them being addressed. For instance, do they perceive the problem as a structural or a process one, something that requires fundamental or marginal

Table 2.1 Governability assessment framework

Assessment stage	Targets (Where to look)	Features (What to look for)	Measures (What to look at)
Stage 1	Fisheries governance problem	Degree of wickedness of the fisheries problem	Stakeholders' images of the problem
			Existence of stopping rules
			The embedded nature of the problem
			Cost and reversibility of prescribed solutions
Stage 2	Natural & social system-to-be-governed; Governing system; Governing interactions	Prevalence of system properties (i.e., diversity, complexity, dynamics and scale)	Components
			Relationships
			Interactions
			Boundaries
Stage 3	Governing system	Goodness of fits of elements (i.e. images, instruments and actions)	Behaviour, decisions, mental models, institutional arrangements, implementation
		Responsiveness of modes (i.e. self-, co-, and hierarchical)	Awareness, learning, sensitivity, conflicts
		Performance of orders (i.e. first, second and meta)	Consistency, effectiveness, transparency, justice
Stage 4	Governing interactions	Presence and quality of interactions	Information sharing, co-learning, adaptiveness
		Enabling and restrictive role of power relations	Inclusiveness, representativeness, participation

Source: Chuenpagdee and Jentoft (2013)

change? Where in the overall governance system exactly do they think the problem sits? It is to be expected that stakeholders would always have an eye for context (embeddedness), i.e. that governability is part of bigger problems that may or may not be specific to fisheries and that they would feel that they are at the receiving end of forces beyond their control. This is basically about the low level of empowerment of small-scale fisheries that is a central issue in the SFF Guidelines.

Next, in Stage 2, the governability assessment involves a description of the natural system (ecosystem, species, resources, and habitats), the social system (stakeholders, communities, organizations and institutions, social practice, culture, and others) and the governing system. This description focuses on the inherent and constructed systems properties, i.e., diversity (components, actors), complexity (relationships, horizontal and vertical linkages and networks), dynamics (interactions – process), and scale (spatial and administrative boundaries –boundary traffic and control). For the natural system, these properties are influenced by human interference but are not constructed by humans. In the case of the social system, these properties may be inherited but are still part of the social construct, which implies that normative qualities are key. Similarly, the characteristics of the governing system contain both instrumental and normative aspects, both of which give rise to governability challenges.

The book has a particular focus on how different governing modes (i.e. hierarchical, co-, and self-governance) respond to the problems and opportunities that arise from the system properties. For stage 3, under each mode, the assessment involves answering questions related to goodness of fit, responsiveness of modes and performance of orders. The first refers to the degree to which the governing system match the traits of the system that it aims to govern. For instance, a highly diverse system-to-be-governed would require a decentralized governing system in order to achieve governability (Jentoft 2007). Responsiveness of modes is related to how the governing system deals with conflicting and sensitive issues and its ability to enhance awareness. Finally, the performance of order is about the effectiveness and legitimacy of the governing system as it executes and implements its principles and functions. Here the analysis is both about the capacity and the quality of governance.

Stage 4 is about the assessment of governing interactions, i.e. those that occur between the systems that are being governed and the governing system. Here it is essential to describe the nature, type and quality of governing interactions and what difference the various modes make relative to the characteristics of the natural and social systems to be governed and their interactions. At the forefront are issues pertaining to power relations and discourses, representation and participation, information sharing and interactive learning. How are science and local knowledge informing interactions?

Governing Modes

Governability can be contextualized in various ways according to the interactive governance framework (see Kooiman and Chuenpagdee 2005; Kooiman et al. 2008). For the purpose of this book, it is done with respect to governing "modes." The modes refer to the locus of governance, where governance actually takes place. Interactive governance theory distinguishes between three modes, which are to be thought of as ideal types in the sense of Weber (1978); self-governance co-governance, and hierarchical governance. All fisheries governing systems demonstrate, and require, mixes of these three modes.

Self-governance Under certain conditions, particularly at lower scales, small-scale fisheries systems have the ability to govern themselves without (much) external interference or support. This may occur in ways that also live up to some or all quality standards associated with the governability concept. It is essential to understand what these conditions are, which of them are sufficient and which are needed. Ostrom's (1990, 90) list of eight institutional design principles could be a vantage point, but may not be exhaustive. There is a vast literature documenting how small-scale fisheries governance occurs at the level of the household and the community, often through informal instruments but not always so. For instance, Acheson's analysis of the functioning of the lobster fishery of Maine, US, is now a classical example of a largely informal decentralized governance system (Acheson 2004).

Whether formalizing a governing system that is essentially and traditionally informal is a good thing from a governability perspective or not is a research question. While recognizing the fact that small-scale fisheries are often governed through institutions and by mechanisms that are informal is essential, the answer to their governability problems is not necessarily formalization, although that may also help in certain situations. For instance, bringing informal use-practices under legal control may be helpful, like when tenure rights are secured by law. Lack of secure communal property rights might bring about encroachment from the outside, which may lead not only to detrimental outcomes for small-scale fisheries but also to a tragedy of the commons. This would be a clear sign of a governance deficiency, either as cause or effect. The latter is a scenario where government imposes regulatory regimes that change customary law and situates control outside small-scale fisheries, which would be an example of the state "colonizing the life-world" of local communities (Habermas 1984). Government interference is sometimes also described as a dis-embedding process (Granovetter 1985; Hanna and Jentoft 1986), as when governance is undertaken by the state bureaucracy at the expense of community management.

Informal governance in small-scale fisheries may be a problem or an opportunity. It is therefore essential to explore the capacity and quality of the self-governing mode in particular situations, as contextual factors are likely to influence governance outcomes. In interactive governance terms, the issue is how diversity, complexity, dynamics or scale creates adequate conditions for self-governance. A particularly important research question is how government interference in self-governing systems influences governability, and how conflicting norms and principles (legal pluralism) are understood and addressed.

In *co-governance*, the government is not seen as a threat and a troublemaker, but a potentially constructive partner in interactive governance. Governability is enhanced by drawing on the capacities of both small-scale fishers and the government, while compensating for the inherent disabilities of both. In fisheries, co-governance is synonymous with co-management, where management is perceived broadly to also involve institutional matters. Whereas co-governance is meant to enhance the governability of issues pertaining to scale and complexity, it is also seen to be a qualitative governability measure, one that enables stakeholder participation, power-sharing and democracy. But co-governance may in itself contribute to complexity in fisheries by increasing the number of possible relationships, interactions and transaction costs. The broader the participation, the more cumbersome is the process. Enhanced governability is at best a possibility and not a given.

Finally, *hierarchical governance* is the third governing mode in interactive governance theory. This is perhaps the most common form of fisheries governance. The actual and potential roles of the state in small-scale fisheries governance, what its limits are, and under what circumstances government can work in support of small-scale fisheries are important research issues. Here, it is important to note that hierarchical governance is not tantamount to state governance by the national government, but a mode that can be found also within local government, even within communities, corporations and cooperatives. It is in other words about the way governance occurs rather than about who is exercising it.

Moreover, hierarchical governance is not necessarily illegitimate, as when it is exercised on the basis of a mandate arrived at through a democratic process. An important governability research issue is analyzing when the use of state power is legitimate in fisheries governance and what power relationships are conducive to governability. Democracy and participatory processes are not cost-free, as it may be cumbersome, time-consuming and ineffective (Mikalsen and Jentoft 2003). The SSF Guidelines have national governments as the most important addressee, and thus clearly recognize the responsibilities that state governments have vis-à-vis small-scale fisheries. The guidelines do not differentiate between state agencies and levels. It is clear that this would be needed when the guidelines are implemented, given the holistic agenda extending beyond the responsibility of a typical fisheries ministry or department. However, the guidelines do not envisage an omnipotent state that governs fisheries only through the hierarchical approach. Similarly, state institutions may in themselves deviate from the ideal Weber (1978) bureaucratic model, assuming more of an organic feature in order to be responsive to shifting circumstances and demands, thus aiming to enhance governability.

Kooiman (2003) argues that the governing system must be isomorph to the system-to-be-governed; i.e. if small-scale fisheries as a system-to-be-governed are diverse, complex, dynamic and multi-scalar, so must the governing system. Still, one may imagine that the latter cannot be a goal in itself, that there are limits to how complex and dynamic a governing system can be, if this leads to reduced transparency and predictability. Institutional stability and robustness usually go together. In practice, fisheries governance often involves a mixture of elements of the three modes, forming hybrid institutions and sharing of various governance functions. Small-scale fisheries research should therefore examine from a governability perspective the relationships and dynamics that occur between different governing modes in different contexts and how they perform and develop over time.

Assessment Questions

Table 2.2 provides a set of key research questions that emerge from the interactive governance framework and are aimed at assessing modes. The rows depict the three modes of governance outlined above, whereas the columns contain the three systems, the system-to-be-governed, the governing systems and governing interactions, along with questions related to the three features in Table 2.1. The important thing to stress is that the conditions for governability would reside within all three systems and in all modes. Thus, the governability assessment should look into all nine cells. How is the match between the governance problems and opportunities within small-scale fisheries and the selected modes of governance in the particular context under investigation?

The governability assessment framework allows for comparative research, assuming that small-scale fisheries despite their diversity, share many of the same characteristics wherever they are, but that governing modes tend to differ. Thus, this volume attempts to draw from empirical experiences to generate general insights into what works or not from a governability perspective in different natural, social and political settings. Interactive governance theory argues that we must then take into account the diversity, complexity, dynamics and scale issues associated with each of the systems as they characterize small-scale fisheries globally. Their internal and external linkages are multiple and intricate, thus forming relationships and interdependencies that must be discerned and comprehended.

Furthermore, small-scale fisheries usually display a capacity for adaptation and change that is essential for their continued existence, as well as a certain degree of robustness that makes it possible for them to withstand pressure both from within and outside, be they natural or social (political, institutional, economic or cultural). This capacity is largely determined by the interactions that occur between the system-to-be-governed and the governing interactions where the governing modes can make a big difference depending on how they are designed. In other words, "third-order" – or meta-governing – principles are essential but so too are their concrete operationalizations into institutional formation and day-to-day decision-making, which interactive governance labels "second"- and "first-order" governance (see for example Kooiman 2008). Thus, in order to respond to the inherent dynamics of small-scale fisheries, the governing system must also be flexible, adaptive and innovative in the way they relate to the three modes.

Governors here have a choice to make for which governability assessment would be helpful. Indeed, choosing between the modes and making them work better in particular settings constitute a "wicked problem," where no easy solution exists, as there will often be several preconceived ideas at play about what mode functions better. There will rarely be a consensus about which mode works either in general or in concrete situations, which is as much a political as a technical or scientific issue. Small-scale fisheries do not exist in a social, economic or political vacuum. Rather, they form an open system with permeable boundaries. This is also why scale is an important variable in a governability assessment of this sector.

Table 2.2 does not pretend to be exhaustive, only illustrative, regarding governability research questions that the interactive governance framework inspires. The case studies in this volume address the issues and questions emphasized in one or more cells, but research questions are not necessarily the same across chapters. In general, assessing governability involves asking questions to help identify key instrumental and normative issues pertaining to the three modes that interactive governance theory emphasizes. From the perspective of the system-to-be-governed,

Table 2.2 Governing modes and governability

Governability research questions	System-to-be-governed (goodness of fit questions)	Governing system (responsiveness of mode questions)	Governing interactions (performance of order questions)
Self-governing mode	Are communities at odds within and among themselves, and how are these issues handled? Do small-scale fishers think that self-governance institutions are up to task?	What local governing institutions exist and what roles they play? Are they in harmony or conflict with those of governments?	What conditions and drivers encourage compliance and free-riding behavior? What is the level of adherence to locally-made rules and regulations?
Co-governing mode	What tradition and culture for cooperation and collective action exists? What actions have been taken to build capacity and create participatory environment?	Are government and fisher institutions open to cooperation and sharing of power and responsibility? Has there been any co-production of knowledge, and awareness to inform decision-making?	How has experience informed interactions and relationships? Has collaboration resulted in trustful interactions, mutual understanding and high compliance?
Hierarchical governing mode	What are the characteristics of small-scale fisheries that call for top-down intervention? How do the governing images correspond with those of the communities?	How sufficient are resources allocated to implement policies supporting small-scale fisheries? How effective are policies at addressing small-scale fisheries concerns?	What are the various forms, frequencies and instruments used to interact with small-scale fisheries? What is the legitimacy associated with the top-down orders?

the governability of each mode is mostly about the goodness of fit, whereas when examining the governing system along the modes, the most relevant measures are those related to how responsive these modes are. Finally, the perspective of governing interactions, the analysis of the modes is about performance of order.

To further elaborate on how best to assess governability with regard to the modes, other questions may be added. For instance, with respect to *self-governance*, an additional question may be what governing institutions exist at the level of community? How has the small-scale fisheries community built its autonomous capacity for governance, such as organizations, rules and enforcement mechanisms, in absence of the government? How do they live up to the qualitative or normative criteria of good governance, such as transparency, inclusiveness and equity, and to what extent do they succeed in addressing major governance concerns such as food security, poverty alleviation and sustainability? The governability issue is also related to legal pluralism, i.e., the prevalent local normative orders existing on their own or alongside those of government, often in conflict, but sometimes in harmony. The hypothesis is that the greater the conflict between legal orders, the lower the governability may be.

With regard to *co-governance*, the investigation should focus on the degree to which the system-to-be-governed and the governing institutions are open to cooperation and sharing of power and responsibility. Is the existing legal system conducive to such an arrangement? What lessons have been learned, and how have they informed interactions and relationships between the governing system and small-scale fisheries in a way that has enhanced governability?

When looking at *hierarchical governance* from the governability perspective, we would, for instance, be interested in how the governing system interacts, or communicates in a broad sense, with small-scale fisheries that are being governed, in what forms, through which channels, by which instruments and at which frequency? What is the level of compliance at the local level to the normative orders constituted and enforced by the governing system? To what extent are these rules considered to be legitimate, appropriate and socially just among small-scale fisheries stakeholders? Here the hypothesis is that the better the quality of these interactions, the higher the governability is.

Endnote

Interactive governance theory and the governability concept form the integrating perspective of this volume. They are employed to provide focus and coherence in order to address the overall common research questions with regard to the choice and design of governing modes and governability outcomes. We draw mostly on the research experience that authors already have from places where they have done extensive fieldwork. In some cases authors were already familiar with and have themselves used interactive governance theory in their research. In other cases, authors have applied it post facto for their contributions to this volume, as an analytical reflection and not as a research design instrument. Readers should therefore not expect, and neither have we as editors, that authors follow the interactive governance theory and the governability assessment framework slavishly but only as they see fit. Interactive governance theory is still a work in progress, and readers should appreciate and learn from its eclectic usage, as we have done.

The aim of this volume of small-scale fisheries case studies from around the world is to demonstrate their diversity and what can be learned from this diversity; that they are different from north to south, from country to country and from situation to situation, and that these differences have governability implications. It is also for this reason that it is difficult to agree on a universal definition, one that will fit the entire world of small-scale fisheries. This can also be seen from the SSF Guidelines, which leave it to the countries themselves to develop a workable definition. Still, small-scale fisheries display many similar characteristics, and play a role that is not that different from country to country, such as making contributions to food security, livelihoods and community well-being. They are commonly important for their cultural heritage, and they struggle with many of the same pressures caused by globalization, overfishing, marginalization and decline, and in many instances poverty and deprivation. They are therefore in need of policy and better governance structures and practices. It is for this reason that we draw on interactive governance theory and propose a procedure for governability assessment, which we believe would be relevant regardless of geography. The quintessence of this approach is the idea that small-scale fisheries are a system rather than a sector (see Chap. 1 in this volume; Jentoft 2014), linked up internally and externally in ways that are not always clear but which need to be thoroughly investigated before one can advance their governability. The advantage of this analytical framework, as we see it, is that it does not only allow the researcher to penetrate the issue of governance in their particular cases but that it also enables generalization from comparison. Such comparison would make possible broad and systematic reflections about the state of small-scale fisheries and how to enhance their governability in a way that is ecologically sound, socially just, politically acceptable and economically viable.

References

Acheson, J. M. (2004). *Capturing the commons: Devising institutions to manage the Maine lobster industry*. Lebanon: The University Press of New England.

Allison, E. H., Ratner, B. D., Åsgård, B., & Willmann, R. (2012). Rights-based fisheries governance: From fishing rights to human rights. *Fish and Fisheries, 13*, 14–29.

Bavinck, M., Chuenpadee, R., Jentoft, S., & Kooiman, J. (Eds.). (2013). *Governability of fisheries and aquaculture: Theory and applications*. Dordrecht: Springer.

Chuenpagdee, R., & Jentoft, S. (2009). Governability assessment for fisheries and coastal systems: A reality check. *Human Ecology, 37*, 109–120.

Chuenpagdee, R., & Jentoft, S. (2013). Assessing governability: What's next? In M. Bavinck, R. Chuenpagdee, S. Jentoft, & J. Kooiman (Eds.), *Governability of fisheries and aquaculture: Theory and applications* (pp. 335–350). Dordrecht: Springer.

Chuenpagdee, R., & Mahon, R. (2013). Approaches and tools in examining governability. In M. Bavinck, R. Chuenpadee, S. Jentoft, & J. Kooiman (Eds.), *Governability of fisheries and aquaculture: Theory and applications*. Dordrecht: Springer.

Chuenpagdee, R., Degnbol, P., Bavinck, M., Jentoft, S., Johnson, D., Pullin, R., et al. (2015). Challenges and concerns in fisheries and aquaculture. In J. Kooiman (Ed.), *Fish for life: Interactive governance for fisheries*. Amsterdam: University of Amsterdam Press.

Degnbol, P., Gislasson, H., Hanna, S. S., Jentoft, S., Raakjær Nielsen, J., Sverdrup-Jensen, S., & Wilson, D. C. (2006). Painting the floor with a hammer: Technical fixes in fisheries management. *Marine Policy, 30*, 534–543.

Fukuyama, F. (2013). What is governance? *Governance: An International Journal of Policy, Administration, and Institutions, 26*(3), 347–368.

Granovetter, M. (1985). Economic action and social structure: The problem of embeddedness. *American Journal of Sociology, 91*(3), 481–510.

Habermas, J. (1984). *The theory of communicative action*. Boston: Beacon.

Hanna, S., & Jentoft, S. (1996). The human use of the environment. An overview of social and economic dimensions. In S. Hanna, C. Folke, & K.-G. Mähler (Eds.), *Rights to nature*. Washington, DC: Island Press.

Headland, T., Pike, K., & Harris, M. (Eds.). (1990). *Emics and etics: The insider/outsider debate*. London: Sage.

Jagers, S., Berlin, D., & Jentoft, S. (2012). Why comply? Attitudes towards harvesting regulations among Swedish fishers. *Marine Policy, 36*, 969–976.

Jentoft, S. (2007). Limits of governability: Institutional implications for fisheries and coastal governance. *Marine Policy, 31*, 360–370.

Jentoft, S. (2014). Walking the talk. Implementing the International Guidelines for securing sustainable small-scale fisheries. *Maritime Studies (MAST)*. Retrieved from http://www.maritimestudiesjournal.com/content/13/1/16

Jentoft, S., & Bavinck, M. (2014). Interactive governance for sustainable fisheries: Dealing with legal pluralism. *Current Opinion in Environmental Sustainability, 11*, 71–77.

Jentoft, S., & Chuenpagdee, R. (2009). Fisheries and coastal governance as a wicked problem. *Marine Policy, 33*, 553–560.

Jentoft, S., & Eide, A. (Eds.). (2011). *Poverty mosaics: Realities and prospects in small-scale fisheries*. Dordrecht: Springer.

Jentoft, S., van Son, T., & Bjørkan, M. (2007). Marine protected areas: A governance system analysis. *Human Ecology, 35*, 611–622.

Kooiman, J. (2003). *Governing and governance*. London: Sage.

Kooiman, J. (2008). Exploring the concept of governability. *Journal of Comparative Policy Analysis, 10*(2), 171–190.

Kooiman, J., & Bavinck, M. (2005). The governance perspective. In J. Kooiman, M. Bavinck, S. Jentoft, & R. Pullin (Eds.), *Fish for life. Interactive governance for fisheries* (pp. 11–24). Amsterdam: Amsterdam University Press.

Kooiman, J., & Chuenpagdee, R. (2005). Governance and governability. In J. Kooiman, M. Bavinck, S. Jentoft, & R. Pullin (Eds.), *Fish for life. Interactive governance for fisheries* (pp. 325–350). Amsterdam: Amsterdam University Press.

Kooiman, J., Bavinck, M., Jentoft, S., & Pullin, R. (Eds.). (2005). *Fish for life. Interactive governance for fisheries*. Amsterdam: Amsterdam University Press.

Kooiman, J., Bavinck, M., Chuenpagdee, R., Mahon, R., & Pullin, R. (2008). Interactive governance and governability: An introduction. *The Journal of Transdisciplinary Environmental Studies, 7*(1), 2–11.

Mahon, R., & McConney, P. (2004). Managing the managers: Improving the structure and operation of small fisheries departments, especially in SIDS. *Ocean & Coastal Management, 47*(9–10), 529–535.

Mikalsen, K. H., & Jentoft, S. (2003). Limits to participation? On the history, structure and reform of fisheries management in Norway. *Marine Policy, 27*, 397–407.

Ostrom, E. (1990). *Governing the commons: The evolution of institutions for collective action*. Cambridge: Cambridge University Press.

Ostrom, E., Janssen, M. A., & Anderies, J. M. (2007). Going beyond panaceas. *Proceedings of National Academy of Sciences, 104*(39), 15176–15178. Retrieved from www.pnas.org#cgi#doi#10.1073#pnas.0701886104

Ratner, B. D., Åsgård, B., & Allison, E. H. (2014). Fishing for justice. Human rights, development, and fisheries sector reform. *Global Environmental Change, 27*(1), 120–130.

Rhodes, R. A. (1996). The new governance: Governing without government. *Political Studies, XLIV*, 652–667.

Rittel, H. W. J., & Webber, M. M. (1973). Dilemmas in a general theory of planning. *Policy Sciences, 4*, 155–169.

Sunde, J. (2014). *Customary governance and expressions of living customary law at Dwesa-Cwebe: Contribution to small-scale fisheries governance in South Africa*. Doctoral thesis, University of Cape Town, Cape Town.

Thorpe, A., Whitmarsh, D., Ndomahina, E., Baio, A., & Kemokai, M. (2009). Fisheries and failed states: The case of Sierra Leone. *Marine Policy, 33*, 393–400.

Weber, M. (1978). Economy and society: An outline of interpretive sociology. Edited by G. Roth & C. Wittich. Oakland: University of California Press.

Wilson, D. C., Raakjær Nielsen, J., & Degnbol, P. (Eds.). (2004). *The fisheries co-management experience: Accomplishments, challenges and prospects*. Dordrecht: Kluwer Academic Publishers.

Part II
Small-Scale Fisheries Diversity – Identifying Governability Challenges

Introduction

Globally small-scale fisheries display enormous diversity as a result of differences in natural, social, cultural and political factors. This means that small-scale fisheries must always be considered in their particular context and that governability challenges be dealt with at lower scales when appropriate, including at the community-level. Small-scale fisheries are also a vibrant sector, comprised of linkages to land-based activities and markets, which make them part of a complex system of social relations and institutions. As a human activity and a way of life that has been around for millenniums, small-scale fisheries are not stable but subject to drivers and stressors, now increasingly caused by globalization, including climate change. In terms of governance, therefore, small-scale fisheries sustainability is a moving target. Governance interventions cannot stop changes that are taking place but can attempt to steer them in directions that benefit small-scale fishing people and communities.

This part of the book draws on case studies from different parts of the world. *Chapter 3* by Philippa Cohen, Louisa Evans and Hugh Govan, is situated in the Solomon Islands in the Western Pacific. They argue that decentralizing governance to the community level permits responsiveness to local dynamics and customary practices and rights. They hold that while community-based, co-management helps to increase the governability of Pacific small-scale fisheries, this mode still needs innovation and adaptation given the exceptionally high diversity and complexity of small-scale fisheries in the region. However, finding the 'best mix' of governance modes and management responses is a challenge.

Chapter 4 by James Prescott, James Riwu, Dirk J. Steenbergen and Natasha Stacey keeps us in the same region, more precisely in the island of Rote in southern Indonesia. The small-scale purse seine fishery there is under threat from large scale operators and dwindling resources. Due to poor information impeding serious conversations about alternative governance modes, low compliance and ineffective hierarchical governance remains a problem that needs to be addressed for governability to improve.

Chapter 5, by Jyothis Sathyapalan and Sunny George, takes us to the Estuarine System of Cochin, the so-called "backwater", in Kerala state of India, where small-scale fisheries are threatened by urbanization and industrialization, and the problems that accompany such processes, water pollution being one. The two authors argue that, as a governability problem, the situation calls for a new platform that allows all stakeholders to participate and deliberate on equal terms. The latter suggests that it is important to prohibit the most powerful stakeholders, who in this case are not small-scale fishers in the Cochin backwater, to set the terms for others.

The following two chapters are case studies from Europe, France and then Portugal. In *Chap. 6*, Katia Frangoudes and Clément Garineaud assess the governability of small-scale kelp harvesting in the Iroise Sea in north-west Brittany where people, many of them women, have been harvesting kelp for centuries, and where co-governance regulations have prevented competition and conflict between local communities and avoided over-exploitation of resources. A new marine protected area, however, has been created that threatens local livelihoods. The authors hold that it is important to build on the cooperative governance capacity of fishers and the processing industry, developed over a long period of time. Collective action and a balance of power should also be promoted.

Chapter 7, authored by Cristina Pita, João Pereira, Silvia Lourenço, Carlos Sonderblohm and Graham J. Pierce, discusses governability of the traditional small-scale octopus fishery in Portugal. This fishery is excluded from quota regulations under the Common Fisheries Policy and instead is managed nationally. As of late, fishers have become part of decision-making but poor organization, lack of trust, and little cooperation between small-scale fishers yield little influence. Recent initiatives aimed at improving the governance framework give hope for a more sustainable future. The authors argue that the high social and economic dependence of fishers on this resource and lack of monitoring and assessment make it imperative to involve fishers in management decision-making through co-governance arrangements. For this to be effective, communication channels between the authorities, industry, and fishers must be created.

Chapter 3
Community-Based, Co-management for Governing Small-Scale Fisheries of the Pacific: A Solomon Islands' Case Study

Philippa Cohen, Louisa Evans, and Hugh Govan

Abstract The discourse on solutions to address small-scale fisheries concerns in the Pacific tends to focus heavily on community-based forms of co-management. Decentralizing governance to the community level permits responsiveness and specificity to local dynamics, not possible through hierarchical governance. It also allows for proper recognition of the (often legally backed) customary rights of local resource owners, common throughout the Pacific. Partnerships between communities and governments, NGOs or research organizations draw together knowledge, expertise and institutions to develop and implement co-management arrangements. In exploring Solomon Islands as a case study we find that interactions between community-based, co-management (a form of co-governance), and self-governance (particularly customary institutions) are fundamental for contextualizing and 'fitting' management to the community level – and that this helps to account for the exceptionally high social and ecological diversity and complexity of Solomon Islands. Community-based, co-management represents a hybrid of traditional and contemporary, local and higher level images, instruments and actions. Interactions between community-based, co-management and hierarchical governance can bolster and inform local management and governance solutions. This is particularly true (and necessary) for pressures (e.g., population growth and commercial, export-orientated exploitation) that extend beyond the local scale or have not before been encountered by customary institutions. While these relations can increase governability, they can also be contradictory and undermining, particularly when objectives are dynamic

P. Cohen (✉)
WorldFish, Honiara, Solomon Islands

Australian Research Council Centre of Excellence for Coral Reef Studies,
James Cook University, Townsville, Australia
e-mail: p.cohen@cgiar.org

L. Evans
Geography, College of Life and Environmental Sciences, University of Exeter, Exeter, UK
e-mail: Louisa.Evans@exeter.ac.uk

H. Govan
Locally Managed Marine Area Network, Suva, Fiji
e-mail: hgovan@gmail.com

S. Jentoft, R. Chuenpagdee (eds.), *Interactive Governance for Small-Scale Fisheries*, MARE Publication Series 13, DOI 10.1007/978-3-319-17034-3_3

and differ across scales. Finding the 'best mix' of governance modes and responses is a moving and elusive target. Nonetheless, we conclude that while community-based, co-management is an appropriate and fitting mode for increasing the governability of Pacific small-scale fisheries in some contexts, in its current form it alone is not up to the task of realizing fisheries sustainability objectives. We recommend that small-scale fisheries policy more explicitly seeks, and tests, new forms of governance interactions amidst the diversity and complexity of Pacific small-scale fisheries.

Keywords Co-governance • Decentralised • Coral reef • Food security • Customary • Network • Hybrid

Introduction

Small-scale fisheries provide food, income, and a way of life for a high proportion of the largely rural and coastal dwelling populations of Pacific Island Countries and Territories. Within the Pacific region small-scale fisheries are deemed important for their role in maintaining self-sufficiency and for their potential to fuel development in rural areas. Pacific populations commonly demonstrate high levels of participation in small-scale fisheries and very high consumption rates of fresh fish as the major source of protein (Bell et al. 2009) where Pacific people consume on average around 34–37 kg of fish per year compared to 16.5 kg for people worldwide (Gillett 2009). Coastal marine resources provide the Pacific Islands with US$262 million in annual revenue, and are a major contributor to many national economies (World Bank 2000). Although cash-based economies are expanding, in many Pacific Island countries where human development is low, the subsistence economy, including small-scale fisheries, plays an important role in human well-being (Adams 2012). However, there are concerns for the sustainability of small-scale fisheries in the light of rapid population growth, increased connectedness to global markets, intensifying interactions with commercial enterprise and projected effects of climate change (Gillett and Cartwright 2010).

The discourse on solutions to address small-scale fisheries concerns in the Pacific tends to focus on co-management, and in particular on community-based, co-management. This reflects a more global trend that is based on the growing realization that resource status and exploitation are driven by social and economic factors, and therefore that governability will be increased when resource users are actively involved in designing and implementing management solutions (Pomeroy 1995; Berkes 2009). As a result, co-management strategies are now globally a mainstream approach to managing many natural resources, including those utilized by small-scale fisheries (Evans et al. 2011). Resource-user involvement supports social justice, equity and empowerment, and legitimacy, which can lead to improved 'fit', better acceptance and enhanced compliance with management (Pomeroy 1995; Berkes 2009). Fisheries co-management is defined by relationships between a resource-user group (e.g., local fishers) and another entity (e.g., a government agency or non-government organization) in which management responsibilities and authority are shared (Pomeroy and Berkes 1997; Evans et al. 2011). In practice, co-management arrangements vary according to the degree of authority and influence the resource users have over management, relative to partners (Sen and Nielsen 1996). In this chapter, we focus on the 'collaborative, community-based' end of the co-management spectrum (Pomeroy 1995), that many initiatives within the Pacific region aspire towards (Govan et al. 2009).

In many Pacific Island Countries and Territories national governments are poorly resourced (most resources that are tagged to fisheries are directed towards industrialized fisheries such as tuna) relative to the scope of small-scale fisheries. Small-scale fisheries are typically diverse, but particularly in the Pacific due to the region's exceptionally high cultural diversity and marine biodiversity (Veron et al. 2009). Small-scale fisheries support subsistence of the predominantly rural and coastal dwelling populations, and so most fishing activities are hundreds to thousands of kilometers from urban centres where governments are based. Further, in many Pacific countries, there are legally recognized systems of self-governance involving customary tenure and traditional management. Resultantly, from the perspectives of resourcing and logistics, well-being, and legal rights community-based, co-management is a mainstream and popular strategy for addressing small-scale fisheries concerns (Jupiter et al. 2014). However, concerns about small-scale fisheries sustainability, economic performance, and governance more broadly, persist widely across the Pacific.

Research Questions and Methods

We present Solomon Islands as a Pacific Island case study of coastal small-scale fisheries. We dig deeper into a series of examples of community-based, co-management within Solomon Islands by drawing on published and unpublished work, particularly (Cohen et al. 2012, 2013; Cohen and Alexander 2013; Cohen and Steenbergen 2015). The methods include semi-structured interviews, fish catch surveys, and key informant interviews. Methods are described in detail in the articles from which we cite results.

Our overarching research question is; *In Pacific Island developing country contexts, can community-based, co-management (a) deliver culturally appropriate and locally governable measures adequate to deal with contemporary pressures on resources, (b) be locally governed in an equitable and participatory manner, and (c) influence and be influenced by higher scales of governance and learning?* To put Solomon Islands small-scale fisheries under the microscope we use the Interactive Governance Framework to: (i) analytically examine features (diversity, complexity, dynamics across scales) of the *system-to-be-governed* (section "System-to-be-governed"), and the *governance system* (section "Governing system"); (ii) concentrate explicitly on interactions between the hierarchical governance, co-governance and self-governance modes, and interactions with the *system-to-be-governed* (section "Governance interactions and outcomes"), and (iii) examine the goodness of fit, responsiveness and performance (Chuenpagdee and Jentoft 2013) of the *governing system* in addressing challenges faced by small-scale fisheries (section "Governance interactions and outcomes"). In section "Governance interactions and outcomes", we are broadly asking; what characteristics and interactions between the features of the Solomon Islands' small-scale fisheries *system-to-be-governed* and the *governing system* render it more or less governable? What does this mean for the future of small-scale fisheries governance in Solomon Islands and the Pacific Islands region?

In answering our research question we also align our responses to the three conditions that are the main focus of the Framework (Chuenpagdee and Jentoft 2013). First, by examining local governability and cultural appropriateness we explore 'goodness-of-fit' between the *governing system,* including governance institutions, and the *system-to-be-governed.* We then look at how the creation of institutions is influenced from across scales and how 'responsiveness' of institutions differs between modes of governance. Third, in examining the adequacy of management for dealing with contemporary pressures on resources, and whether governance is equitable and participatory, we start to understand the 'performance' of the governing system. Chuenpagdee and Jentoft (2013) highlight that decentralized governance faces its own set of challenges; in this chapter we unpack some of the particular challenges associated with our case and critically analyze community-based, co-management as a principle vehicle with which to govern small-scale fisheries of the Pacific.

System-to-Be-Governed

The Solomon Islands form a double-chained archipelago (Fig. 3.1a) spanning 1,500 km, and is comprised of large and small islands and atolls surrounded by reef, mangrove and seagrass habitats and extending rapidly into deep ocean (Fig. 3.1b). Solomon Islanders predominantly (80 %) reside in rural areas, and 94 % of the rural population live within 5 km of the coast (Fig. 3.1d; Foale et al. 2011). While population densities are relatively low on a global scale, the population growth rate is amongst the highest i.e., 2.3 %. The population is incredibly *diverse* in terms of history, language (>70) and culture, ranking 6th in the world in terms of biocultural diversity (Table 3.1; Harmon and Loh 2004). Land and marine tenure rights and inheritance are dictated by complex social relationships; customary matrilineal or patrilineal descent systems (depending on region), kinship or clan affiliations, and other social relationships and transactions. Cultural relationships and social norms are embodied in *kastom.*[1] Many aspects of *kastom* have proven resilient to external pressure from colonization, Christianity, ethnic conflict and international

[1] Kastom is the pidgin English word for 'custom'; referring generally to cultural norms and institutions.

Fig. 3.1 Panel figure depicting Solomon Islands, showing (from *left* to *right*, *top* to *bottom*); (**a**) the double-chained archipelago, (**b**) low lying islands, (**c**) fish critical in diets, (**d**) coastal housing, (**e**) multi-species fin fisheries, (**f**) small-scale agricultural and fisheries marketing, (**g**) participation in fisheries, (**h**) transportation of fish to markets, and (**i**) mangrove invertebrates for household subsistence

development agendas even where these have acted to undermine traditional arrangements (Ruddle 1994, 1998; Hviding 1998). *Kastom* and associated customary tenure can both foster and hinder resource management and rural development (Hviding 1998), and remain critical dimensions of the governability of small-scale fisheries in the Pacific and Solomon Islands. Additionally, population growth, urbanization and market integration have been influential to differing degrees in different locales and underpin highly *dynamic* social landscapes (Table 3.1).

People's livelihoods in Solomon Islands are diverse and dynamic: the norm is that rural dwelling people are non-specialist fishers and farmers (Fig. 3.1c, f), and opportunistically participate in other forms of livelihood activities for food and income (Govan et al. 2013b). While participation in small-scale fisheries is high (i.e., estimated to be over 80 % of households), it is also variable in space and time, depending on conditions within the fishery and external opportunities and constraints. Fisheries are viewed as a mainstay of subsistence lifestyles and a foundation of food security (Bell et al. 2009), but also as a potential engine to drive local and national development. Men, women and youth participate in small-scale fisheries (Fig. 3.1g) – harvesting, processing and marketing. Men's fishing activities often

Table 3.1 Summary of the diversity, complexity and dynamics of the social and natural subsystems of small-scale fisheries in the Pacific across national, provincial and local levels

	Social	Natural
Diversity	Ethnicity	High coral diversity (486 species)
	Language (70+) i.e., a proxy for cultural diversity	High fish diversity (1,019 species)
	Non-specialized fishers	Extensive seagrass and mangrove habitats
	Multi-species fishery	Relatively low value, subsistence fisheries
	Gear diversity	
	High participation – men, women, children	High value, commercial small-scale fisheries
Complexity	*Kastom* (cultural norms and institutions)	Geography (Volcanic archipelago, low-lying islands)
	Intermarriage	
	Conflict (Ethnic tensions 1999)	Connectivity between land and sea (Impacts from logging and mining)
	Colonial history (Independence 1979)	
	Gender dynamics	
Dynamics	Very high population growth	Increasing rates of environmental decline
	High rates of urbanization	Climate change
	Market integration	
	Changing aspirations	
	Changing place of *kastom* and decreasing respect for customary authority	
	Fisheries (highly dispersive or migratory) cross tenure boundaries (small governance units)	
	Importance of fisheries at individual, household and community level are dynamic, relative to other opportunities or issues	

involve the use of small boats, fishing close to home exploiting coastal habitats to deeper waters further afield. Women more frequently harvest from shore or in shallow reef and mangrove areas (Fig. 3.1i) proximate to their home village (Kronen and Vunisea 2009). The benefits derived from fisheries are mediated not just by direct participation in harvesting, but also by factors ranging from intra-household gender relations to urbanization to international markets, as articulated by Foale and colleagues (2013) in the context of future food security. The cross-scale complexities of the social *system-to-be-governed* in Solomon Islands, and the Pacific more widely, provide both opportunities and challenges to the governability of small-scale fisheries, as expanded below.

Solomon Islands hosts coral reef systems that are globally recognized for their exceptionally high biodiversity (Veron et al. 2009). Small-scale fisheries operate in coral reef, mangrove, lagoon and near-shore pelagic environments, but concern for small-scale fisheries, resource and habitat decline and impetus for management tend to focus heavily on coral reefs (e.g., Bell et al. 2009; Solomon Islands Government 2009). In fact, Solomon Islands' reefs and reef-associated resources are in relatively good shape if considered globally (Green et al. 2006). Nonetheless, fishers, managers and researchers are concerned about the resource declines they observe in all coastal zones. For example, based on estimates of coral reef fisheries production relative to future demand for reef fish, Solomon Islands is projected to be one of 11 Pacific Island countries for which there will be a food security shortfall by 2030 (Bell et al. 2009).

Rich biological diversity translates into a high diversity of small-scale fisheries target taxa (Fig. 3.1e). Small-scale coastal fisheries focus on 20 reef-associated families of fish for food (Pinca et al. 2009) or around 180 species (Skewes 1990); however, invertebrates, near-shore pelagic and mangrove-associated species also contribute to small-scale fisheries. Invertebrates play an important role in diets, and trochus (*Trochus niloticus*) and sea cucumbers dominate small-scale fisheries exports. However, sea cucumber fisheries of the Pacific commonly operate on "boom and bust" cycles, iterating between national bans implemented in response to concerns about over-exploitation, and relatively intense harvesting to capture benefits from the lucrative trade (Friedman et al. 2011). Common small-scale fishing methods include spear fishing, handlining, netting and gleaning. The composition of catches and the prevalence of fishing methods vary substantially between villages, regions and fishers. Vessels are most commonly dug-out paddle canoes (Fig. 3.1g), and to a lesser extent boats with outboard motors. Finfish and invertebrates harvested from inshore areas are commonly consumed locally, though any opportunities for sale locally or at provincial centres may be seized depending on proximity and accessibility to transportation (Fig. 3.1h).

Governing System

Governance of small-scale fisheries is influenced by formal national and provincial level governing bodies and institutions, as well as informal cultural and local institutions that operate at the community or clan level. The relative influence of state versus local institutions varies depending on social group, geographic location, resource of concern, exploited habitat and fishing method, but is also dynamic depending on local or state responses to resource decline, harvesting opportunities or conditions external to the fishery. In Solomon Islands, coastal ecosystems and fisheries are formally governed by the state through environment and fisheries legislation administered by their respective government ministries (Lane 2006). Additionally, nine Provincial governments are recognized, in theory (Lane 2006) and in policy (e.g., Solomon Islands Government 2009), as key units for decentralization of resource management and development. In reality, the financial, technical, and human resources required for delivering services or governing in rural areas far exceeds those made available to Provincial Governments (Lane 2006; Govan et al. 2013b). The national government concentrates on managing commodity invertebrates (e.g., trochus and sea cucumber) at points of export. Management instruments include size restrictions, export licensing and (in the case of sea cucumber) indefinite moratoria; instruments that are implemented to optimize economic efficiency, profitability, resource rent and/or sustainability. Rural communities are legally required to adhere to these regulations, but awareness and enforcement in rural areas is minimal. In practice, national and provincial governments have had low levels of success in affecting management on non-exported, small-scale fisheries (Ruddle 1998; Govan et al. 2013b).

This governance and management 'gap' has, in effect, been filled by numerous non-governmental organizations (NGOs) and research agencies working in Solomon Islands to support conservation of coastal ecosystems and management of small-scale fisheries. While these organizations hold no formally legitimized governing role, they commonly act as co-management partners to coastal communities and have been recognized as government 'partners' since 2007. In Solomon Islands, these partnerships have led to the formation of at least 137 community-based, co-managed areas (Cohen et al. 2012). In most situations the national and provincial governments have relatively little direct involvement in these management efforts, in part because their capacity has been prohibitively low. Yet, in the last 5 years several national level, government-led policies have sought to capitalize on this emergent model by (1) explicitly recognizing and promoting community-based, co-management as a principle, national approach for resource management and rural development, and (2) creating and investing in mechanisms (e.g., governance networks) to coordinate the 'partner' agencies involved, and to improve alignment with national policies and strengthen relationships with government agencies to build and supplement their capacity (discussed further in section "Can community based management of small-scale fisheries influence, and be influenced by, higher scales of governance and learning?").

At the village or local level, customary governance systems remain intact and influential to varying extents. In any one village or community there may be several clans, each with its own leaders and leadership structure, as well as elected village chiefs (White 2004). Since the introduction of Christianity into Solomon Islands in the early 1900s, the Church has also emerged as important in village governance (White 2004). The church is influential in deciding, declaring and enforcing rules, including those associated with community-based, co-management (Cohen and Steenbergen 2015).

Throughout much of the Pacific customary land and marine tenure systems persist; in Solomon Islands for example 87 % of land falls under customary tenure (AusAID 2008), which also frequently extends to coastal marine areas (Hviding 1998). Customary land and marine tenure align to different clans who have the rights to decide when and how resources are accessed, used and managed, and by whom (Hviding 1988). As a result, customary marine tenure is highly influential, and in fact foundational (Polunin 1984; Govan et al. 2009), in crafting and implementing contemporary small-scale fisheries management and development strategies in Solomon Islands, and many other Pacific Island countries.

In association with customary tenure, coastal societies throughout the Pacific have developed other norms and institutions that influence the way marine resources are used and governed (e.g., Johannes 1982). Scholars draw analogies between these customary instruments (e.g., bans on consuming or harvesting certain species; temporary reef closures; restrictions on fishing methods) and contemporary resource management instruments (Colding and Folke 2001). And in fact, customary instruments are commonly adapted, and integrated into contemporary community-based management efforts in Solomon Islands (discussed further in section "Can community-based, co-management of small-scale fisheries deliver culturally appropriate and locally governable instruments adequate to deal with contemporary pressures on resources?"), and throughout the Pacific (Johannes 2002; Govan et al. 2009; Cohen and Steenbergen 2015).

Governance Interactions and Outcomes

In this section we use our description of the *system-to-be-governed* and the *governance system* to analyze governability. It is argued co-governance is generally the better suited mode for governing highly diverse *systems-to-be-governed*, and will be more responsive to localized or relatively rapid change (Chuenpagdee and Jentoft 2013). Our case illustrates elements that fit with this generalization, but in this section we also highlight particular characteristics leading to reduced governability. We go beyond the typical focus on the implementation of management tools or institutions – taken to indicate effective community-based, co-management – to consider social and ecological outcomes of governability. In our discussion we provide examples and unpack ideas around flexibility, adaptation, sustainability, innovation and hybridization – which tend to be treated fairly favorably and uncritically in the literature. Here we explore what they mean for governability.

Can Community-Based, Co-management of Small-Scale Fisheries Deliver Culturally Appropriate and Locally Governable Instruments Adequate to Deal with Contemporary Pressures on Resources?

'Self-governance' of land and marine resources is prolific throughout the Pacific through local and customary governance structures (e.g., chiefly systems) and institutions (e.g., customary tenure, and area, method and consumption taboos). Yet, scholars argue that customary institutions evolved for social reasons and were not intended or necessary (due to low resource-use pressure on marine ecosystems) to promote ecological sustainability (Polunin 1984; Foale et al. 2011). This differentiation is important because it suggests fundamentally different *images* – whereby customary resource rules are *social* institutions (underpinned by complex, often opaque and dynamic social relations and with the purpose of building and maintaining social capital), with the *potential* to be adapted and applied in a way that can address contemporary sustainability concerns. Extending this argument would suggest that, without some adaptation, self-governance is a poor *fit* in today's contexts and illaligned with *images* of longer-term sustainability or economic efficiency. In other words, in their current form customary institutions or self-governance are an important and fitting foundation and starting point, but not the *end point* for increasing governability of small-scale fisheries today. Integration of self-governance authorities and institutions with other governance arrangements needs to acknowledge their social and cultural foundations so as not to romanticize their promise for conservation purposes (Foale et al. 2011; Fabinyi et al. 2014).

To improve sustainability outcomes in contemporary contexts where pressures on resources are more diverse and intense, scholars argue that the application, design and intent of customary institutions will need to evolve (with various caveats and degrees of caution) – incorporating scientific information, modern management principles, and cross-institutional bolstering (Cinner and Aswani 2007; Foale et al. 2011). Interactions between self-governance and co-governance are common place in that local and customary institutions are explicitly and formally recognized as the foundations of contemporary community-based, co-management of coastal resources and fisheries (e.g., Apia Policy 2008). Co-governance inherently encourages the combination of knowledge sources, but also provides a platform for deliberation, identification and hybridization of values, norms and institutions (Jentoft and Chuenpagdee 2009). In practice partners might provide management advice, facilitate processes to integrate forms of knowledge and practice, and pursue mechanisms to bolster local management efforts and governance arrangements. In the Pacific a hybrid between local (customary) and science-based management and conservation practice is often sought (Govan et al. 2009; Aswani and Ruddle 2013). Yet, the resultant governance and management arrangements are rarely well-described or critically appraised. Here we ask, is the "resultant 'hybrid' form of management socially appropriate, locally governable and able to deliver benefits to fisheries and fishing communities?"

We focus our analysis on a specific *instrument* (and associated *governance system*); periodically-harvested marine closures or area "taboos", which are commonly employed in community-based, co-management in the Pacific (Govan et al. 2009; Cohen and Foale 2013). Taboos are cultural institutions that were historically implemented for relatively short periods to control use and access to resources for social objectives, e.g., to mark the death of a prominent community member, protect sacred sites, or "save-up" stocks prior to harvests for feasts or trading (Hviding 1998; Foale et al. 2011). We found that in contemporary community-based, co-management, taboos are closed for prolonged periods (whereas historically taboos were instated only on special occasions for finite periods) reflecting attempts to alleviate fishing pressure and enhance ecological sustainability. In practice, area openings were flexible, in response to local social and economic needs (e.g., school fees, contributions to social events), and were opened more frequently than originally planned (Cohen et al. 2013). The flexibility to change management practices in response to altered conditions or new knowledge is an important element of adaptive co-management (Armitage et al. 2008). Social and economic triggers to harvest make this measure locally acceptable and culturally fitting – yet too much flexibility may not align with longer term objectives of sustainability. While Kooiman and Chuenpagdee (2005, 327) suggest that "governing diversity takes a broad and long-term view on fisheries and incorporates fine-tuning and feedback", our case highlights that such fine-tuning may represent adjustments between longer and shorter term goals, rather than tuning towards better achievement of the longer term goals. It is arguable that "local fishers…often have a longer temporal perspective than government" (Kooiman and Chuenpagdee 2005, 338), but in developing country contexts in particular, necessity and desire for (relatively modest) improvements to livelihoods may take precedence over a longer-term view. This is an example of a hard choice between two desirable goals that may, in some cases, be contradictory (Kooiman et al. 2005).

Taboo areas are commonly employed to manage the diversity of multi-species and multi-method fisheries – something for which spatial management is often promoted. However, any particular harvesting cycle – whether planned or unplanned, annual or more frequent, intense or light, for hours or weeks – may fit with social objectives but may not allow for sufficient replenishment for some taxa (Cohen and Foale 2013). Taboos in the Pacific have, in some cases, been seen as a panacea, and managers may be asking too infrequently "What [particular] problem is this instrument supposed to solve? Why was this particular instrument chosen and not another one?" (Kooiman and Chuenpagdee 2005, 331–332). A more comprehensive suite of instruments might better fit the diversity of small-scale fisheries, yet other measures tend to be less readily accepted or implemented than taboos in Pacific community-based, co-management (e.g., Cohen et al. 2013; Léopold et al. 2013).

Implicit in community-based, co-management is that communities will be able, to some extent, to deal with enforcement and sanctioning locally. Partners may therefore prioritize their efforts towards communities where local governance is relatively robust and functional, and where customary institutions are intact. However, given there is declining respect for local authority and *kastom* in many regions of the Pacific (Macintyre and Foale 2007) governance may need local strengthening and external bolstering. Where enforcement and sanctioning are localized a range of traditional, religious and formal legal institutions and associated sanctions may be invoked; providing institutional diversity and redundancy that reflects Jentoft and colleagues' (2009) notion of legal pluralism. For example in our community-based, co-management cases, we found that sanctions might include the customary payment of food or shell money (customary governance), 'bad luck' from forces beyond the human realm (customary and Christian belief systems), or a warning, and a monetary fine (state or NGO-supported governance) (Cohen and Steenbergen 2015). Yet, even in situations where local governance might be considered intact, respected and strong, it is almost inevitable that communities will appeal for enforcement support from the government, particularly for repeated infringements, or infringements by "outsiders" (Govan et al. 2009); illustrating community perceptions and experiences of the limits to local governability. In situations where tenure is unclear or disputed, self-governance is perceived as "weak" and resource use is intense, particularly when driven by lucrative international markets (i.e., situations that are becoming more common across the Pacific), community-based, co-management may be unviable. In fact, many community-based, co-management efforts fail, but tend to go unreported. In these situations hierarchical governance or greater involvement of the government as a co-management partner may be required to address small-scale fisheries concerns.

Taboos are employed within most Pacific community-based, co-managed systems (Govan et al. 2009), and emerge as a socially acceptable and locally implementable measure with potential to enhance sustainability – something that has largely been untested to date. We used an interdisciplinary approach to examine four periodically-harvested taboos; firstly looking at harvesting dynamics (i.e., fishing effort, gear and method use, periodicity of harvesting) that would affect sustainability, and secondly looking at indicators of fisheries performance (i.e., catch rates, yield, fish length and displacement of fishing effort) for multi-species, multi-method fisheries (Cohen and Alexander 2013; Cohen et al. 2013). We made comparisons between four taboos and 55 nearby continuously-fished reefs. We found total annual effort and catch in taboos was low to moderate compared to reefs continuously open to fishing. When taboos were opened, effort in the area was very intense, but because taboos were only opened for a small proportion of the year total yield did not exceed annual benchmarks of sustainability but nor was it appreciably different from reefs continuously open to fishing (Cohen et al. 2013). Catch rates of invertebrates during openings were significantly improved suggesting that the periods of closure or overall relief from fishing pressure were sufficient to allow some recovery, however we did not find evidence that the strategy had substantially benefited multi-species finfisheries (Cohen and Alexander 2013). While taboos may alleviate fishing pressure in a small area of fishing grounds, it is unlikely they provide substantial benefits to broader fishing grounds. Further, openings of long duration, high frequency and intense exploitation, may lead to unsustainable harvesting within the area. While the instrument *fits* well with the local *governance system*, in many forms its *performance* will be inadequate to deal with all the diversity and complexity of the *system-to-be-governed*.

Can Community Based Management of Small-Scale Fisheries Be Locally Governed in an Equitable and Participatory Manner?

Co-management helps to ensure that benefits from small-scale fisheries remain at the local level, rather than being accumulated by few in more centralized, 'wealth-based' models of management (Béné et al. 2010). While many community-based, co-management initiatives aim to improve community-wide wellbeing, the reality is that inequitable distribution of benefits or involvement in decision making is not uncommon when initiatives work within customary governance structures (Béné et al. 2009; Cinner et al. 2012). In short, chiefly systems and customary tenure are not built on western ideas of participatory decision-making and equitability. A chief's or clan's propensity for participatory process and equitability are influenced by leaders' and communities' worldviews and characteristics. While customary tenure can be used to *exclude*, it may also be used to build social capital with non-tenure holders by *permitting* them access and use rights (Carrier 1987). Many land and marine areas might effectively operate as open-access, at least for some resources, despite there in fact being a mosaic of different tenure claims. Exclusivity via tenure is more frequently enacted when competition or interest in resources intensifies (e.g., for commercial exploitation or other uses) (e.g., Carrier 1987; Macintyre and Foale 2007). In efforts to build equitable processes and to ensure management instruments (such as those that create exclusive access) are not to the (substantial) detriment of particular people within a community, many community-based, co-management partners promote democratic and participatory processes and structures e.g., resource management committees with representatives from different sectors of society. Nonetheless, tenure is often bolstered and used as a mechanism for access exclusivity (Jupiter et al. 2014). The ability to restrict access is a foundation of effective resource governance (Ostrom 1990). However, this illustrates the wickedness of small-scale fisheries governance (Jentoft and Chuenpagdee 2009); where resources are limited, but people's reliance on resources is high "solutions for one group of stakeholders may cause problems for other stakeholders" (Chuenpagdee and Jentoft 2013, 346).

A key assumption of co-governance is that "no single actor is in control" and that between actors there is some degree of equality (Kooiman and Chuenpagdee 2005, 336). However, in practice power asymmetries may cause governability challenges (Jentoft 2007), and as discussed in the previous paragraph equitable benefit and cost sharing, or involvement in decision making can be problematic when working with customary governance structures (Béné et al. 2009). We found that in some cases community-based, co-management arrangements supported the use of the benefits from harvesting for communal, village-wide purposes, but in other cases decision making and distribution of benefits were based on genealogy or social standing. For example, particular clans benefited from the potentially more profitable early stages of harvesting newly opened taboos, and in the most extreme case benefits of harvests were appropriated largely by particular elites (Cohen and Steenbergen 2015). As competition for resources intensifies, scenarios of "elite capture" or inequitable distribution of benefits may become more common and have greater implications for non-elite or marginal groups. Despite explicit efforts to enhance inclusion of women and other marginalized groups in decision making, males who held traditional leadership roles took the lead in decisions in the cases we examined. In summary, the distribution of benefits and costs, and representation in resource related decision making are highly variable and dynamic depending on local context and the processes employed to establish management. Community-based, co-management partners face a significant and ongoing challenge to understand and align with local governance and social structures without compromising equitability objectives.

Can Community Based Management of Small-Scale Fisheries Influence, and Be Influenced by, Higher Scales of Governance and Learning?

To differing extents, community-based, co-management interacts with hierarchical governance via (a) government policies and actions that validate, support or undermine community-based, co-management, (b) national fisheries and environment legislation that simultaneously regulate the use of coastal habitats and resources, and (c) mechanisms and structures such as networks that seek to foster learning, institutional support and spread of improved management. Community-based, co-management in Solomon Islands and the Pacific involves interactions between national organizations, and regional and international initiatives and networks. While supporters of co-management place much emphasis on improving management practices at local levels (e.g., Jupiter et al. 2014), there has been less attention on how the institutional and cross-scale governance environments enhance localized outcomes. In this section we consider those cross-scale and cross-institutional interactions that create an environment that enables, constrains or otherwise interacts with community-based, co-management.

The constitutional recognition of customary tenure and governance provide the foundations for community-based, co-management in Solomon Islands. Current environmental (Solomon Islands Government 2009) and fisheries (MFMR 2008) policies recognize and promote a "*people-centred and integrated resource management approach that relies on a core of community based management as a national strategy to improve food security, adaptive capacity and conservation*" (Solomon Islands Government 2009, 9). Even regional scale objectives such as those articulated in the Coral Triangle Initiative[2] are delivered through community-based co-management (Solomon Islands Government 2009). In line with this there have been amendments to recent environmental and proposed fisheries legislation to provide further legal backing. In practice, however, operational budgets of government agencies remain low and do not trickle down to local level engagement or support (Govan et al. 2013a, b). To what extent strengthening legal backing will actually bolster the implementation and enforcement support that communities seek remains uncertain. Where there has been some success is in communities restating some national regulations in their local management plans (Cohen et al. 2013), as a form of 're-regulation' (sensu Chuenpagdee et al. 2013). However, there are also examples where interactions between hierarchical governance and

[2] The Coral Triangle Initiative on Coral Reefs, Fisheries and Food Security focuses on six countries (Solomon Islands, Indonesia, Malaysia, Philippines, Papua New Guinea and Timor Leste) based on their exceptionally high marine biodiversity.

community-based, co-management were beyond the control or influence of local governors, and experienced by communities as acute shocks. For example, sea cucumber fisheries had been an important source of cash in an otherwise largely subsistence economy of the community of Kia, and were the focus of Kia's early management efforts. However, in 2006, in response to concerns about overexploitation and resource decline, the government imposed, with very little warning, a national export ban and the 'fishery' ceased to exist. This created substantial hardship in the community until they were able to adapt their livelihoods and their management efforts.

Networks are a prime example of co-governance, providing a platform for deliberation, communication, cooperation and coordination (Kooiman and Chuenpagdee 2005). The Solomon Islands Locally Managed Marine Area Network (SILMMA; comprised of NGO, research agency, government ministry, and local community managers) was explicitly designed to facilitate cross-scale, cross-institutional learning, coordination and local-level representation associated with community-based, co-management (Cohen et al. 2012) – mirroring an option highlighted for improving governability (Kooiman and Chuenpagdee 2005). Solomon Islands' small-scale fisheries is an example of a highly diverse and complex *system-to-be-governed*; for which, it is argued, co-governance is the best suited mode for high diversity (Siry 2006), whereas highly complex systems are better suited to more coordinated, and even centralized approaches (Chuenpagdee and Jentoft 2013). The SILMMA network allowed community-based, co-management initiatives to account for diversity and dynamics at local scales, but simultaneously promoted multi-actor, cross-site learning on fit, responsiveness and performance. The network drew the government into a lead coordinating role, something that is reportedly rare (Kooiman and Chuenpagdee 2005). Yet, in practice, coordination was minimal, and the prime interactions were directly between co-management partners, and communities (Cohen et al. 2012). Durable and meaningful interactions with outcomes of learning and coordination are challenged by both the costs of interaction (relatively rarely recognized in theory, but commonly faced in practice) and by the intangible, but often stated, objectives of 'learning', 'sharing information' and 'coordination'. In contexts defined by high complexity, diversity and dynamism across scales these ideas can be hard for governors to pin down and progress. Further, while the network provided a critical pathway for higher-level representation of local issues in national and international policy arenas, there were still substantial challenges in improving representation and downward accountability (Ratner et al. 2013). Autonomy of stakeholders is an important characteristic of co-governance (Kooiman and Chuenpagdee 2005), but because community representation was financially and logistically reliant on their partners, autonomy to represent community interests may be jeopardized (Ratner et al. 2013).

In addition to national and sub-national interactions, networks can also link community-based, co-management to regional and international partners. The other network that mediates local to national to international linkages is the National Co-ordinating Committee formed in 2009 under the Coral Triangle Initiative. Committee representatives include the environment and fisheries ministries, other government stakeholders (e.g., Ministries of Development Planning and Aid Co-ordination, Finance, and Provincial Government) and those NGO and research organizations operating nationally for small-scale fisheries and marine conservation. This committee has participated in formulation of regional policy associated with the Coral Triangle Initiative, but has also buffered the impact of this policy by translating it into a national context by developing its own targets and definitions contrary to those outlined in regional policy. For example, the network prioritizes community-based, co-management rather than the pervasive model of permanently closed Marine Protected Areas (Solomon Islands Government 2009). While designed to co-ordinate the Coral Triangle Initiative the National Co-ordinating Committee is increasingly becoming the *de facto* structure through which conservation and natural resource management initiatives are co-ordinated, and as such plays an important role for enhancing the fit of external-supported initiatives to the Solomon Islands context.

Discussion

In this chapter we have used the Interactive Governance Framework to examine governability of Pacific Island small-scale fisheries, focusing in particular on Solomon Islands community-based, co-management. In our discussion we draw out key learnings from our engagement with the framework, and highlight noteworthy aspects of Pacific small-scale fisheries that have emerged from our analysis.

The framework facilitates a novel analysis of Pacific small-scale fisheries in two main ways. First, in examining the *system-to-be-governed* the approach requires assessment of the diversity, complexity and dynamics across scales of both the social and ecological subsystems of the fishery. With these descriptions at hand we were able to analyze the goodness-of-fit, responsiveness and performance of community-based, co-management (as the *governance system*) relative to the *system-to-be-governed*. Our analysis highlighted some relatively unique aspects of governability of Pacific small-scale fisheries. The marine biological and cultural diversity within the Coral Triangle region (in which Solomon Islands is situated) is unmatched. Further, few places retain such extensive customary tenure systems as those in the Pacific that form the foundations of contemporary efforts to establish community-based, co-management. In the past, many places with tropical small-scale fisheries had customary management systems, but many have experienced their erosion or dissolution. For example, in the Philippines customary management is no longer considered a viable foundation of contemporary governance (Aswani et al. 2012). In other cases, a legacy of customary management remains and efforts to resurrect some of these traditional institutions continue e.g., *Sasi Laut* in Eastern Indonesia (Cohen and Steenbergen 2015). Nonetheless, we have shown that even in the Pacific these foundations can offer challenges to equitability in decision making and in the distribution of costs and benefits. The Pacific is rapidly changing through population growth (at a rate among the highest globally), urbanization and increased market integration. In the Pacific this rapid change is often operating beyond the local scale, but nonetheless presents challenges to local governability of small-scale fisheries.

The second source of novelty in the Interactive Fisheries Governance framework is its explicit focus on three modes of the *governance system* – self-governance, co-governance and hierarchical governance – as interacting rather than as mutually exclusive pathways. In Pacific small-scale fisheries this analytical feature captures the different emphasis on these modes of governance for different parts of the small-scale fishery (e.g., finfish versus higher value invertebrate export fisheries) at different times thus better explaining the evolution of the governance approach, and its effects on increasing or decreasing governability. In Pacific small-scale fisheries community-based, co-management systems are founded on self-governance enabling them to remain culturally relevant and locally governed, however we found that some aspects of sustainability may be compromised. Hierarchical governance plays an important role in protecting high value invertebrate fisheries (which is a gap in self-governing modes), but plays a relatively small part in subsistence fisheries management. So while the community-based, co-management model appears to represent more idealized forms of co-management as outlined in Sen and Nielsen's (1996) spectrum of possible arrangements, in reality the government does not act as a service provider that responds to communities' governance support needs (Govan et al. 2013b). We highlight that in certain circumstances (i.e., where self-governance is weak, resource use is intense etc.) greater involvement of government may be required to effectively increase governability. Further, while mechanisms for promoting cross-scale coordination and learning exist, these are faced with the practicalities of working in diverse, dynamic and complex contexts – particularly in developing countries where there is commonly a deficit of technical resources, finances and capacity.

Conclusion

Within the Pacific region capacity limitations of national governments have meant that the hierarchical mode of governance has been challenged to address small-scale fisheries concerns. Hierarchical governance, throughout the Pacific, has also faced difficulties in reconciling top-down authority with the constitutionally protected rights of local resource owners to govern their own marine resources. Yet, in the face of intensifying pressures on fisheries resources, there have been increasing concerns from local resource users, the state and civil society alike, that the self-governance institutions, even where intact, are not up the tasks of ensuring fisheries sustainability, or realizing contemporary development objectives such as equality and broad participation. The Solomon Islands model of co-governance has similarities with co-governance practices throughout Melanesia and the broader Pacific region, and has emerged from global theory and local context, as the most appropriate model to compensate for the inherent short comings of hierarchical and self-governing modes. The Interactive Governance Framework allowed us in this chapter to examine the hybridization of traditional and contemporary, local and state models of management and governance. We have highlighted that interactions between co-governance and self-governance modes are fundamental for improving the 'goodness-of-fit' to the community level, being highly diverse and complex on national and regional scales. Localizing governance permits responsiveness to local dynamics, not possible through hierarchical governance. Further, interactions between co-governance and hierarchical governance both bolster and inform local management and governance solutions, and are ultimately anticipated to improve governance performance. Yet, while community-based, co-management (or the co-governance mode) is recognized as an appropriate and necessary mode for governing Pacific small-scale fisheries, it is certainly not without a suite of challenges, and in certain situations will not be up to the task of increasing the governability of small-scale fisheries. To better address these challenges and shortcomings, we recommend that Pacific small-scale fisheries policy and practice more explicitly seeks, and tests, new forms of governance interactions, as they are starting to do with arrangements such as cross-scale governance networks.

References

Adams, T. (2012, May/August). The characteristics of Pacific Island small-scale fisheries. *SPC Fisheries Newsletter, 138*, 37–43.

Apia Policy. (2008). *Pacific islands regional coastal fisheries management policy and strategic actions (Apia policy) (2008–2013)*. Noumea: Secretariat of the Pacific Community.

Armitage, D., Marschke, M., & Plummer, R. (2008). Adaptive co-management and the paradox of learning. *Global Environmental Change-Human and Policy Dimensions, 18*(1), 86–98. doi:10.1016/j.gloenvcha.2007.07.002.

Aswani, S., & Ruddle, K. (2013). Design of realistic hybrid marine resource management programs in Oceania. *Pacific Science, 67*(3), 461–476. doi:10.2984/67.3.11.

Aswani, S., Christie, P., Muthiga, N. A., Mahon, R., Primavera, J. H., Cramer, L. A., Barbier, E. B., Granek, E. F., Kennedy, C. J., Wolanski, E., & Hacker, S. (2012). The way forward with ecosystem-based management in tropical contexts: Reconciling with existing management systems. *Marine Policy, 36*(1), 1–10. doi:10.1016/j.marpol.2011.02.014.

AusAID. (2008). *Making land work: Reconciling customary land and development in the Pacific*. Canberra: AusAID Pacific Land Program.

Bell, J., Kronen, M., Vunisea, A., Nash, W. J., Keeble, G., Demmke, D., & Andréfouët, S. (2009). Planning the use of fish for food security in the Pacific. *Marine Policy, 33*, 64–76.

Béné, C., Belal, E., Baba, M. O., Ovie, S., Raji, A., Malasha, I., Njaya, F., Na Andi, M., Russell, A., & Neiland, A. (2009). Power struggle, dispute and alliance over local resources: Analyzing 'democratic' decentralization of natural resources through the lenses of Africa inland fisheries. *World Development, 37*(12), 1935–1950. doi:10.1016/j.worlddev.2009.05.003.

Béné, C., Hersoug, B., & Allison, E. H. (2010). Not by rent alone: Analyzing the pro-poor functions of small-scale fisheries in developing countries. *Development Policy Review, 28*(3), 325–358.

Berkes, F. (2009). Evolution of co-management: Role of knowledge generation, bridging organizations and social learning. *Journal of Environmental Management, 90*(5), 1692–1702. doi:10.1016/j.jenvman.2008.12.001.

Carrier, J. G. (1987). Marine tenure and conservation in Papua New Guinea: Problems in interpretation. In B. J. McCay & M. Acheson (Eds.), *The question of the commons: The culture and ecology of communal resources* (pp. 142–167). Tucson: The University of Arizona Press.

Chuenpagdee, R., & Jentoft, S. (2013). Assessing governability – What's next. In M. Bavinck, R. Chuenpagdee, S. Jentoft, & J. Kooiman (Eds.), *Governability of fisheries and aquaculture: Theory and applications* (MARE publication series, Vol. 7, pp. 335–349). Amsterdam: Amsterdam University Press.

Chuenpagdee, R., Jentoft, S., Bavinck, M., & Kooiman, J. (2013). Governability – New directions in fisheries governance. In M. Bavinck, R. Chuenpagdee, S. Jentoft, & J. Kooiman (Eds.), *Governability of fisheries and aquaculture: Theory and applications* (MARE publication series, Vol. 7, pp. 335–349). Amsterdam: Amsterdam University Press.

Cinner, J. E., & Aswani, S. (2007). Integrating customary management into marine conservation. *Biological Conservation, 140*, 201–216.

Cinner, J. E., McClanahan, T. R., MacNeil, M. A., Graham, N. A. J., Daw, T. M., Mukminin, A., Feary, D. A., Rabearisoa, A. L., Wamukota, A., Jiddawi, N., Campbell, S. J., Baird, A. H., Januchowski-Hartley, F. A., Hamed, S., Lahari, R., Morove, T., & Kuange, J. (2012). Co-management of coral reef social-ecological systems. *Proceedings of the National Academy of Sciences of the United States of America, 109*(14), 5219–5222. doi:10.1073/pnas.1121215109.

Cohen, P. J., & Alexander, T. J. (2013). Catch rates, composition and fish size from reefs managed with periodically-harvested closures. *PLoS ONE, 8*(9), e73383. doi:10.1371/journal.pone.0073383.

Cohen, P. J., & Foale, S. J. (2013). Sustaining small-scale fisheries with periodically harvested marine reserves. *Marine Policy, 37*, 278–287. doi:10.1016/j.marpol.2012.05.010.

Cohen, P., & Steenbergen, D. (2015). *Social dimensions of local fisheries co-management in the Coral Triangle*. doi:10.1017/S0376892914000423.

Cohen, P., Evans, L., & Mills, M. (2012). Social networks supporting governance of coastal ecosystems in Solomon Islands. *Conservation Letters, 5*, 376–386. doi:10.1111/j.1755-263X.2012.00255.x.

Cohen, P., Cinner, J., & Foale, S. (2013). Fishing dynamics associated with periodically-harvested marine closures. *Global Environmental Change, 23*(6), 1702–1713. doi:http://dx.doi.org/10.1016/j.gloenvcha.2013.08.010

Colding, J., & Folke, C. (2001). Social taboos: "Invisible" systems of local resource management and biological conservation. *Ecological Applications, 11*(2), 584–600.

Evans, L., Cherrett, N., & Pemsl, D. (2011). Assessing the impact of fisheries co-management interventions in developing countries: A meta-analysis. *Journal of Environmental Management, 92*(8), 1938–1949. doi:10.1016/j.jenvman.2011.03.010.

Fabinyi, M., Evans, L., & Foale, S. J. (2014). Social-ecological systems, social diversity, and power: Insights from anthropology and political ecology. *Ecology and Society, 19*(4), 28.

Foale, S., Cohen, P., Januchowski-Hartley, S., Wenger, A., & Macintyre, M. (2011). Tenure and taboos: Origins and implications for fisheries in the Pacific. *Fish and Fisheries, 12*, 357–369. doi:10.1111/j.1467-2979.2010.00395.x.

Foale, S., Adhuri, D., Aliño, P., Allison, E., Andrew, N., Cohen, P., Evans, L., Fabinyi, M., Fidelman, P., Gregory, C., Stacey, N., Tanzer, J., & Weeratunge, N. (2013). Food security and the Coral Triangle Initiative. *Marine Policy, 38*, 174–183. doi:http://dx.doi.org/10.1016/j.marpol.2012.05.033

Friedman, K., Eriksson, H., Tardy, E., & Pakoa, K. (2011). Management of sea cucumber stocks: Patterns of vulnerability and recovery of sea cucumber stocks impacted by fishing. *Fish and Fisheries, 12*(1), 75–93. doi:10.1111/j.1467-2979.2010.00384.x.

Gillett, R. (2009). *Fisheries in the economies of Pacific Island countries and territories Pacific studies series*. Mandaluyong City: Asian Development Bank.

Gillett, R., & Cartwright, I. (2010). *The future of Pacific Island fisheries*. New Caledonia: Secretariat of the Pacific Community.

Govan, H., Tawake, A., Tabunakawai, K., Jenkins, A., Lasgorceix, A., Schwarz, A.-M., et al. (2009). *Status and potential of locally-managed marine areas in the South Pacific: Meeting nature conservation and sustainable livelihood targets through wide-spread implementation of LMMAs*. Suva: Secretariat of the Pacific Regional Environment Programme/Worldwide Fund for Nature/WorldFish-Reefbase/Coral Reef Initiative of the South Pacific.

Govan, H., Kinch, J., & Brjosniovschi, A. (2013a). *Strategic review of inshore fisheries policies and strategies in Melanesia – Fiji, New Caledonia, Papua New Guinea, Solomon Islands and Vanuatu – Part II: Country reports. Report to the Secretariat of the Pacific Community for the Melanesian Spearhead Group* (33pp.). Noumea: Secretariat of the Pacific Community.

Govan, H., Schwarz, A.-M., Harohau, D., Oeta, J., Orirana, G., & Ratner, B. D. (2013b). *Solomon Islands: Essential aspects of governance for aquatic agricultural systems in Malaita Hub*. Vol. Project Report AAA-2013-19. CGIAR.

Green, A., Lokani, P., Atu, W., Ramohia, P., Thomas, P., & Almany, J. (2006). *Solomon Islands marine assessment: Technical report of survey conducted May 13 to June 17, 2004*. Brisbane: The Nature Conservancy.

Harmon, D., & Loh, J. (2004). *A global index of biocultural diversity*. Discussion paper for the international congress on ethnobiology. University of Kent.

Hviding, E. (1988). *Marine tenure and resource development in Marovo Lagoon, Solomon Islands*. Bergen: Centre for Development Studies, University of Bergen.

Hviding, E. (1998). Contextual flexibility: Present status and future of customary marine tenure in Solomon Islands. *Ocean and Coastal Management, 40*(2–3), 253–269.

Jentoft, S. (2007). In the power of power: The understated aspect of fisheries and coastal management. *Human Organization, 66*(4), 426–437.

Jentoft, S., & Chuenpagdee, R. (2009). Fisheries and coastal governance as a wicked problem. *Marine Policy, 33*, 553–560.

Jentoft, S., Bavinck, M., Johnson, D. S., & Thomson, K. T. (2009). Fisheries co-management and legal pluralism: How an analytical problem becomes an institutional one. *Human Organization, 68*(1), 27–38.

Johannes, R. E. (1982). Traditional conservation methods and protected marine areas in Oceania. *Ambio, 11*, 258–261.

Johannes, R. E. (2002). The renaissance of community-based marine resource management in Oceania. *Annual Review of Ecology and Systematics, 33*, 317–340.

Jupiter, S. D., Cohen, P. J., Weeks, R., Tawake, A., & Govan, H. (2014). Locally-managed marine areas: Multiple objectives and diverse strategies. *Pacific Conservation Biology, 20*(2), 165–179.

Kooiman, J., & Chuenpagdee, R. (2005). Governance and governability. In J. Kooiman, M. Bavinck, S. Jentoft, & R. Pullin (Eds.), *Fish for life; interactive governance for fisheries* (pp. 325–349). Amsterdam: Amsterdam University Press.

Kooiman, J., Bavinck, M., Jentoft, S., & Pullin, R. (Eds.). (2005). *Fish for life: Interactive governance for fisheries*. Amsterdam: University of Amsterdam Press.

Kronen, M., & Vunisea, A. (2009). Fishing impact and food security – Gender differences in finfisheries across Pacific Island countries and cultural groups. *SPC Women in Fisheries Information Bulletin, 19*, 3–10.

Lane, M. B. (2006). Towards integrated coastal management in Solomon Islands: Identifying strategic issues for governance reform. *Ocean and Coastal Management, 49*(7–8), 421–441. doi:10.1016/j.ocecoaman.2006.03.011.

Léopold, M., Beckensteiner, J., Kaltavara, J., Raubani, J., & Caillon, S. (2013). Community-based management of near-shore fisheries in Vanuatu: What works? *Marine Policy, 42*, 167–176.

Macintyre, M. A., & Foale, S. J. (2007). Land and marine tenure, ownership and new forms of entitlement on Lihir: Changing notions of property in the context of a goldmining project. *Human Organization, 66*(1), 49–59.

MFMR. (2008). *Solomon Islands national strategy for the management of inshore fisheries and marine resources*. Honiara: Ministry of Fisheries and Marine Resources.

Ostrom, E. (1990). *Governing the commons: The evolution of institutions for collective action*. Cambridge: Cambridge University Press.

Pinca, S., Vunisea, A., Lasi, F., Friedman, K., Kronen, M., Awira, R., Boblin, P., Tardy, E., Chapman, L., & Magron, F. (2009). *Solomon Islands country report: Profiles and results from survey work at Ngella, Marau, Rarumana, Chubikopi*. Noumea: Pacific Regional Oceanic and Coastal Fisheries Development Programme (PROCFish/C/CoFish).

Polunin, N. V. C. (1984). Do traditional marine "reserves" conserve? A view of Indonesian and New Guinean evidence. In K. Ruddle & T. Akimichi (Eds.), *Maritime institutions in the Western Pacific* (pp. 267–283). Osaka: National Museum of Ethnology.

Pomeroy, R. S. (1995). Community-based and co-management institutions for sustainable coastal fisheries management in Southeast Asia. *Ocean and Coastal Management, 27*(3), 143–162. doi:10.1016/0964-5691(95)00042-9.

Pomeroy, R. S., & Berkes, F. (1997). Two to tango: The role of government in fisheries co-management. *Marine Policy, 21*(5), 465–480. doi:10.1016/s0308-597x(97)00017-1.

Ratner, B. D., Cohen, P., Barman, B., Mam, K., Nagoli, J., & Allison, E. H. (2013). Governance of aquatic agricultural systems; analyzing representation, power and accountability. *Ecology and Society, 18*(4), 59.

Ruddle, K. (1994). External forces and change in traditional community-based fishery management systems in the Asia-Pacific Region. *Maritime Anthropological Studies, 6*, 1–37.

Ruddle, K. (1998). The context of policy design for existing community-based fisheries management systems in the Pacific Islands. *Ocean and Coastal Management, 40*(2–3), 105–126.

Sen, S., & Nielsen, J. R. (1996). Fisheries co-management: A comparative analysis. *Marine Policy, 20*(5), 405–418.

Siry, H. S. (2006). Decentralized coastal zone in Malaysia and Indonesia: A comparative perspective. *Coastal Management, 34*(3), 267–286.

Skewes, T. (1990). *Marine resource profiles: Solomon Islands*. Honiara: Forum Fisheries Agency.

Solomon Islands Government. (2009). *Solomon Islands national plan of action; Coral Triangle Initiative on coral reefs, fisheries and food security*. Honiara: Ministry of Environment Conservation and Meteorology.

Veron, J., Devantier, L. M., Turak, E., Green, A. L., Kininmonth, S., Stafford-Smith, M., & Petersen, N. A. (2009). Delineating the coral triangle. *Galaxea, Journal of Coral Reef Studies, 11*(2), 91–100.

White, G. (2004). *Indigenizing local governance: Chiefs, church, and state in a Solomon Islands society*. Honolulu: East-west Center.

World Bank. (2000). *Voices from the village: A comparative study of coastal resource management in the Pacific Islands* (Pacific Islands Discussion Paper Series No 9 East Asia and the Pacific Region 22247, p. 85). Washington, DC: World Bank.

Chapter 4
Governance and Governability: The Small-Scale Purse Seine Fishery in Pulau Rote, Eastern Indonesia

James Prescott, James Riwu, Dirk J. Steenbergen, and Natasha Stacey

Abstract Rote is Indonesia's southern-most island with a population of approximately 128,000 people. Largely unregulated small-scale fisheries are integral to local livelihood strategies. Local catches are highly diverse, which reflects regional biodiversity and mixed fishing strategies. Rote's four mile coastal marine zone open to local small-scale fisheries is porous, resulting in competition against fishers from outside the district. Beyond these four miles local fishers compete against large-scale fishing operations for declining resources. To maintain fisheries sustainability and improve fishing-dependant livelihoods, improved governance is needed. Aligning with the interactive governance framework, this chapter examines a small-scale purse seine fishery operating around Rote waters, looking in particular at the implications of governance change through a coherent, carefully prioritized, reform scheme of investment and management. We argue that the major challenges to effective governance frameworks for small-scale fisheries in Rote include: (i) poor information flow that impedes new discourses on the comparative advantages of alternative arrangements leaving governing bodies consistently confronted by wicked problems; (ii) local attitudes towards compliance with fisheries laws and a limited capacity for enforcement; and (iii) a hierarchical governance system characterized by insecure tenure and competing governance priorities. We also present and argue for some likely pathways to improved governance.

Keywords Indonesia • Rote • Interactive governance • Governability • Small-scale fisheries • Purse seine

J. Prescott (✉)
Australian Fisheries Management Authority, Darwin, NT, Australia
e-mail: jim.prescott@afma.gov.au

J. Riwu
Office of Marine Affairs and Fisheries, Kompleks Perkantoran Bumi Ti'i Langga Permai, Rote Ndao District, Nusa Tenggara Timur, Indonesia
e-mail: j_riwu@yahoo.com

D.J. Steenbergen • N. Stacey
Research Institute for the Environment and Livelihoods, Charles Darwin University, Darwin, NT, Australia
e-mail: dirk.steenbergen@cdu.edu.au; natasha.stacey@cdu.edu.au

S. Jentoft, R. Chuenpagdee (eds.), *Interactive Governance for Small-Scale Fisheries*, MARE Publication Series 13, DOI 10.1007/978-3-319-17034-3_4

Introduction

Globally, small-scale capture fisheries are important components of food production, international economies and the social, economic, and cultural fabric of many societies (McGoodwin 2001; FAO 2008; Mills et al. 2011; Harper et al. 2013; Rockefeller Foundation 2013). Southeast Asia, hosting rich bio-diverse seas and an estimated ten million people directly involved in fisheries (Williams and Staples 2010), is a region where the importance of fisheries is comparatively high (White and Green 2014). Fisheries resources are intensively exploited throughout the region and there is a general consensus that fish stocks have been and continue to be exploited beyond the limits of sustainability, leaving many depleted[1] (Butcher 2004; Williams 2007; Williams and Staples 2010) and some vulnerable species extirpated (Anon 2014). In response to the threats to marine biodiversity, and to economically valuable and nutritionally important fish stocks in the region, six countries with a direct interest and the wider international community have allocated considerable financial and human resources to address these issues. The establishment of the regional Coral Triangle Initiative on Coral Reefs, Fisheries and Food Security (CTI-CFF) in 2009 has provided funding to the six participating countries[2] to achieve a sustainable future through an Ecosystem Approach to Fisheries Management (EAFM) of the Asian-Pacific shallow seas (CTI 2009; Foale et al. 2013; White and Green 2014). However, despite the significant investments that have been made towards developing effective fisheries governance, evidence of more sustainable fisheries across the relevant seascapes remains scarce. Much of the area under the CTI-CFF is still exposed to high numbers of (licensed and unlicensed) fishing vessels, uncontrolled use of fishing gear, growing numbers of fish aggregating devices deployed by industrial fisheries and opportunistically utilized by small-scale fishers, and regular reports of destructive fishing practices (Stobutzki et al. 2006; Wagey et al. 2009; Green et al. 2014).

Although fisheries sustainability is a high-priority agenda in the Asia-Pacific (White and Green 2014), fisheries governance, particularly *local* governance, is characteristically under-represented in the literature, including eastern Indonesia. In contrast, fisheries management and increasingly EAFM is routinely part of the fisheries and marine conservation dialogue (Foale et al. 2013). EAFM clearly has a place as best practice fisheries management, however the disproportionate preoccupation we encountered in the region on improving management (EAFM or other-

[1] We define ‘sustainable fish stocks’ as stocks that exist at levels of abundance which safeguards their ongoing existence, maintains high surplus production, and supports high catch rates and favorable economic return. By ‘depleted stocks’ we mean stocks that have, through over-fishing, lost these characteristics. Consequently depleted stocks may involve loss of profitable catches, amplified risks to food security and livelihoods, and potential weakening of ecological resilience to pressures such as climate change.

[2] The Coral Triangle is comprised of Indonesian, Malaysian, Philippine, Timor Leste, Papua New Guinea and Solomon Island marine waters.

wise) over strengthening governance, has prompted this study on the dynamics of local (district-level) governance around Rote Island in eastern Indonesia.

Located in the largest and most densely populated archipelagic country, where average household dependency on marine resources is high, with an estimated national per capita fish consumption of 12.8 kg in 2011 (FAO 2014).[3] Rote's case provides a useful platform to examine the challenges and opportunities in district-level fisheries governance and moreover offers insights into better practices that may be applicable across others sites in the region. Particular focus is put on the purse seine fishery that is carried out around Rote, as this fishery is typical of what is regarded as small-scale fisheries in the archipelagic region of Indonesia, and shares many of the same management and governance features and complexities existing in dozens of other Indonesian provinces. We seek to analyze Rote's purse seine fishery through revealing challenges, tensions and lessons learnt from past interventions. As well, we offer insights into potential pathways towards improved governance, as a contribution to wider initiatives that are aiming to develop improved local governance and fisheries outcomes across the Southeast Asian region.

After outlining the methods applied in this study, the chapter will proceed by presenting Rote's purse seine fishery, as the analytical "system-to-be-governed". In doing so both human (social) and natural dimensions of the system will be addressed. From here the second analytical domain will be examined, namely the 'governing system'. This will involve consideration of the (formal and informal) governing structures that are in place that attempt to control or manage Rote's purse seine fishery. Finally an examination of the governance interactions between these systems will highlight particular tensions and ultimately provide important insights into potential pathways towards improved governance in Rote and the wider regional context.

Methods

The study draws from several years of research with fishers and stakeholder groups involved in small-scale fisheries, both in Rote (Stacey 2007; Stacey et al. 2012) and beyond, across Australian and eastern Indonesian fishing zones (Fitriana and Stacey 2012; Prescott et al. 2013; Steenbergen 2013a, b). Information on the purse seine fishery was obtained directly from the Rote district fisheries office, where one of the authors (Riwu) is a senior officer, in combination with further observations and field-work that was carried out during visits to Rote in 2013 and 2014. Semi-structured interviews with 40 fishing crews from Rote Island were also carried out between 2009 and 2012 to understand the various perceptions among Rotinese fishers of how (formal and informal) small-scale fisheries governance is functioning in local and

[3] This is an average taken from very heterogeneous patterns of fish consumption across the archipelago, and which included a large inland population where fish consumption is much lower than coastal areas.

trans-boundary contexts, and what determines local practice (i.e. compliance versus non-compliance).

Rote's Small-Scale Pelagic Purse Seine Fishery

The small-scale purse seine fishery has developed in Rote since 1988 when the first purse seine fishing vessel started operating in the area. The growth of the fishery has been an organic process whereby coastal resource harvesting practices have been partially shaped as a result of the political shift in Indonesia towards decentralization after the 1999 reforms, giving the fishery its 'district identity and character'. The fishery is seasonal with fishing taking place mostly during the southeast monsoon (April to November) when the western side of Rote is sheltered and fishing conditions are favorable. However, even during this period strong south-easterly winds may cause significant interruptions to fishing activities.

In 2014, 43 operational purse seine vessels of about 15 m in length and less than 10 gross tonnes were reported to be operating in the fishery around Rote.[4] Although in other regions such boat dimensions may be typically associated with larger scale fisheries, these have a limited range, and more importantly all net handling is done manually and normally involves a crew of about 10–12 men. Its relatively low technological input and high labor input in comparison to what are regarded as large scale fisheries in the Indonesian context is what makes this fishery a small-scale fishery. The fishery is loosely divided into two overlapping categories of fishing; daytime fishing involving smaller scale operations (*lampara siang*) and a night time fishery which tends toward marginally larger scale vessels and nets (*lampara malam*). Both of these categories use purse seine nets with a stretched mesh size of 1 inch. The nets range in length from 100 to 300 m and 20 to 60 m deep with the *lampara malam* at the upper end of these ranges. Purse seining generally has little impact on the physical environment as nets are either set where they do not touch the bottom or they are set over sandy substrates so as not to damage habitat or the net. The vessels operating from Rote have all been registered and licensed by the district, however there are no exclusive rights associated with a license, i.e. it is effectively an open access fishery.

The vessels operate out of four subdistricts (*Kecamatan*) including the district's capital town and port of Ba'a. Vessels, costing around 20,000 USD are generally owned by individuals who have purchased them through informal financial means as is common in Indonesia. Banks rarely provide loans to fishers to purchase vessels and nets as it is considered a risky investment and boat owners are either unwilling or unable to provide financial or physical assets (such as land or houses) to secure loans.

[4] This figure is based on data from the Office of Marine and Fisheries, Rote. However the BPS 2014 reports that in 2014 there were 59 registered purse seiners http://rotendaokab.bps.go.id/?hal=publikasi_detil&id=11 (accessed 13 October 2014.)

The arrangements on a vessel around distribution of financial benefits are often organized on the basis of hierarchical links between the boat owner, captain and crew. Typically the captain keeps a written running journal of expenses, type and sale price of catch, and gross earnings per month. At the end of each month the profits are divided amongst the boat owner, captain and crew. In most cases the owner gets half of the earnings while the other half is split across all crew members. A captain typically claims 2 shares plus 10 % of the owner's half. Estimating monthly income per boat is difficult however on average boat owners reportedly earned about 2.5 million IDR, a captain about two million IDR and crew members could expect about one million IDR (about USD 250, 200 and 100, respectively).

All registered Rote-based purse seine operators receive gear subsidies from the district government and as many as eight purse seine units (i.e. vessel, engine, and net) have been provided at no cost to fishers in the last 10 years. These were provided based on proposals by fishers, which need to be endorsed by local village leaders, and assessed by the fisheries department. Fuel subsidies are not provided to Rote fishers as there is no qualifying fuel distributor; however local fuel prices are already low because of the national fuel subsidies.[5]

The contribution of this fishery to food security on the island is unknown however, it is suggested that the estimated landings (up to 3,600 t per year) and the comparative affordability (fish sold to consumers at points of landing or via traders at local markets cost between one and two USD kg^{-1}) makes the catch important to many households. It is estimated that the fishery collectively produces between 70 and 90 % of the fresh fish landed on Rote (Riwu, unpublished data). Although there have been no stock assessments on small pelagic species specifically within Rote's jurisdictional area, within the surrounding fishery management area they are assessed as fully-exploited but not over-exploited (Anon 2010). Economic data that have been collected indicate, besides a relatively low unit value for the catch, that the fishery is locally important in economic terms with annual net revenues possibly as high as three million USD.

With the significant expansion of marine protected areas (MPA) across the region, most of the Rote Ndao district sits in the large Savu Sea MPA, gazetted and formalized in 2014 under the National Ministry of Marine Affairs and Fisheries Decree No. 5/2014 on the establishment of the National Marine Protected Area in the Savu Sea and Nusa Tenggara Timor Province (*Keputusan Menteri Kelautandan Perikanan No. 5/2014 tentang Kawasan Konservasi Perairan Nasional Laut Sawudan Sekitarnya di Provinsi Nusa Tenggara Timur*). There is legislation indicating several no-take areas in Rote's coastal waters which would theoretically exclude purse seine fishery access however these are not yet fully implemented protected areas so boats commonly fish the 4 nautical mile zone without access restrictions.

Being Indonesia's southern-most island, Rote also sits on the maritime border between Indonesia and Australia and has a long history of trans-boundary activities,

[5] Since June 2013, when petrol prices increased by 44 % in effort to cut subsidy bills, the Indonesian government has continued to budget downwards on subsidy spending, however we are reporting on periods prior to the subsidy cuts.

including both legal and illegal fishing (Fox 1977a; Campbell and Wilson 1999; Stacey 2007). During most years hundreds of fishers from Rote make the trip to an area in the Australian Exclusive Economic Zone where, according to a Memorandum of Understanding between Indonesia and Australia, they are allowed to fish using traditional methods and non-motorized vessels (Stacey 2007). However, others have engaged in illegal fishing in waters or over seabeds that came under Australian jurisdiction since Australia's maritime borders expanded following the United Nations Convention on the Law of the Sea (Fox 2009; Fox et al. 2009). The jurisdictive boundaries and the associated legislative responses to illegal activities on either side of the border have had profound impacts on livelihoods and fishing behavior of local fishers in Rote.

Natural System

Rote Ndao district includes the main Island of Rote in Nusa Tenggara Timor Province, located off the southwest tip of Timor and about eight smaller, mostly uninhabited islands (Fig. 4.1). Its location means that Rote's northwest and southeast facing coasts are exposed to two major outflows of the Indonesian Throughflow, which accounts for 80 % of the total water volume transported by the Throughflow from the Pacific into the Indian Ocean (Gordon 2005).

This area makes up part of the southern margin of the highly bio-diverse Coral Triangle region (Wilson et al. 2011). Nevertheless, most fish taken by the fishery belong to a comparatively small number of species found within several families. The principle species in the catch are Indian mackerel (*Rastrelliger spp.*), herring (*Herklotsichthys spp.*), sardines (*Sardinella spp.*), and small tunas of the family Scombridae (e.g. *Auxis thazard*). In some years there are also significant catches of squid (*Uroteuthis bartschi*). In addition, fishers also catch a number of other less common species, including flying fishes (Exocetidae), halfbeaks (Hemiramphidae), needlefish (Belonidae), scads (Carangidae), fusiliers (Caesionidae), and ponyfish (Leiognathidae). Occasionally protected marine turtles are also taken (for consumption or sales) if caught as bycatch.

The life-history characteristics of the primary fish species taken in the fishery, including high population growth rates (Coll et al. 2013; Froese and Pauly 2014), accounts for their resilience to exploitation and acts to reduce complexity. This is expressed through a weaker connection between the fishery's governance and the productivity which favors governability (Chuenpagdee et al. 2013). However, the same characteristics that lend themselves to greater resilience are also likely to contribute other complexity in terms of spatial scale (boundary issues) and the dynamic behavior of the stocks. Inter and intra-seasonal variation in the abundance and sizes of the fish belonging to these relatively short-lived species is not well documented, however they are likely to vary in response to the environment (i.e., through environmentally driven recruitment determined by prey availability and other biotic and abiotic factors). Further complexity arises when considering the connectivity of the fishery in

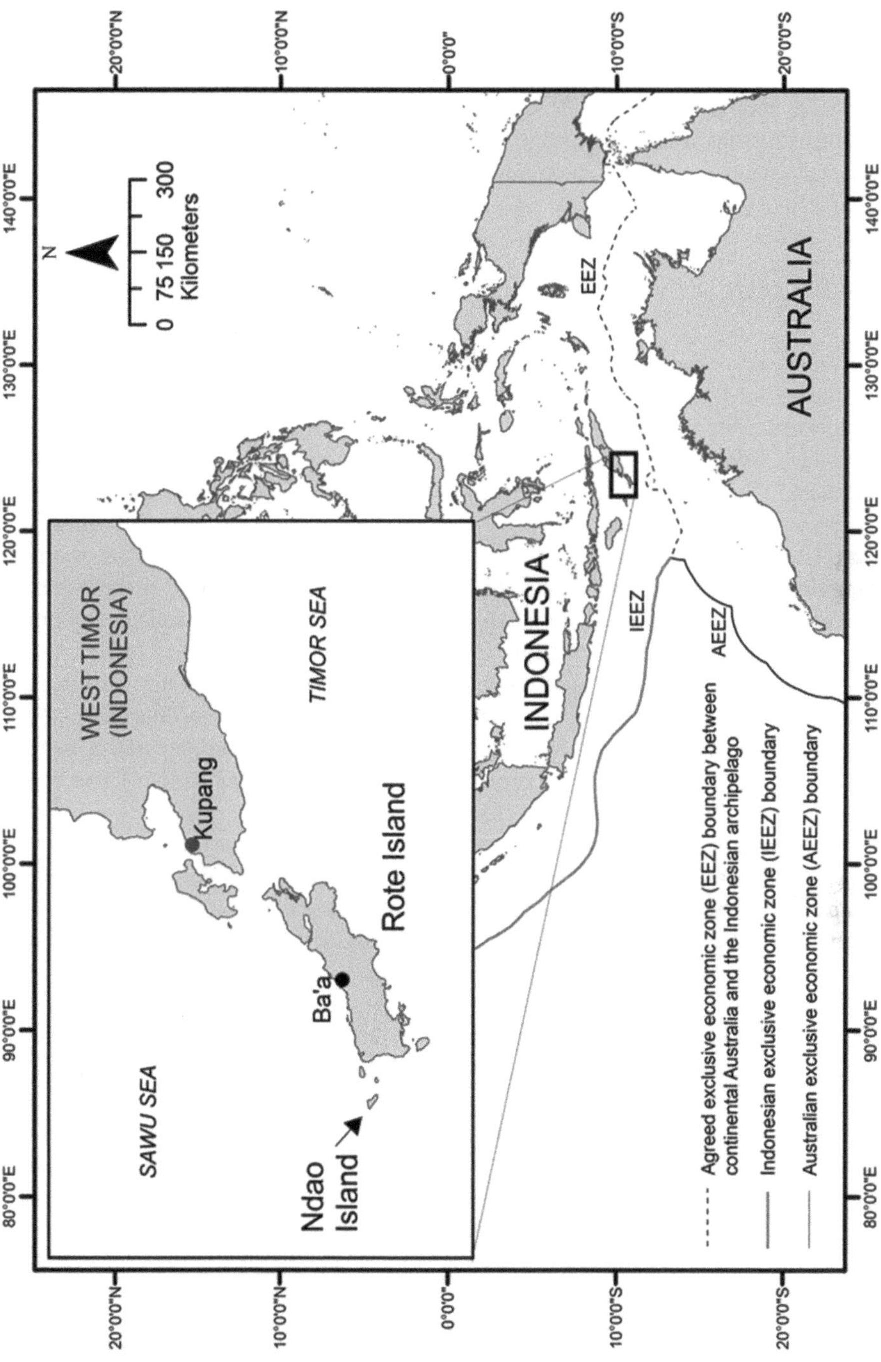

Fig. 4.1 Map of the district of Rote Ndao and its wider geographic setting

Rote with those around it. Given that the coastal-pelagic species targeted by the fishery are not connected to coastal features (e.g. coral reefs), they belong to ecosystems and interact with fisheries across wider seascapes that are an important feature of these stocks. Any declines in fish stocks surrounding Rote are thus likely to be linked to both local and external trends. This complexity and dynamics in the natural system, as a result of its open and connected nature, in combination with its interplay with human systems, contribute to uncertainty and unpredictability of the fishery.

Human System

Rote's population of approximately 128,000 people is spread across 82 administrative villages (*desa*), of which 58 are located within its coastal zone (BPS 2014a, 37). The majority of the population are ethnic Christian Rotinese (Paul et al. 2014), but also includes recently settled maritime-orientated Muslim migrants from other parts of Indonesia (Stacey 2007; Carnegie 2008, 2013). Like many other coastal communities, residents exploit both terrestrial and marine resources to derive their livelihoods. Households on Rote engage in various farming activities; including arable farming that primarily focuses on rice production next to a wide range of secondary crops and livestock production. Both forms of farming however are highly constrained by the climate on Rote making the annual farming calendar highly dichotic, with a monsoon season of concentrated high rainfall followed by a dry season of little or no rainfall (Fox 1977b). Although figures for Rote are not available, the Nusa Tenggara Timor province, of which Rote is part, is considered one of the poorest in Indonesia in which 22 % of rural households are reported to live under the poverty line (Fitriana and Stacey 2012; BPS 2014b).

Fewer than 500 fishers are estimated to be directly involved in purse seining. However a far larger population is involved or dependent through either consumption or as part of the value chain. Stakeholder groups have not been formally defined or analyzed but they might be broadly grouped as fishers, buyers, consumers, and processors. Less clear is who might consider themselves to be the resource owners, either informally or under customary laws. Although the sector (beyond the fishers) forms a significant wider contextual group, given that the scope of our analysis is on governance dynamics around purse seine fisheries, our study focuses on groups directly involved in the fishery.

The ethnic diversity of the resident fishers is low with two groups of vessel owners being either Rotinese (including from the island of Ndao) or intermarried and living in Rotinese communities or from the district capital of Kupang. A third 'outsider' group of fishers operates opportunistically in Rote's fishing zone and travels from as far as east Java and Bali. The latter group of purse seine fishers uses larger vessels, nets and may use fish aggregating devices to attract fish. While this group can cause tensions, their activities have not been the cause of any inter-ethnic fisheries conflicts as experienced in some other parts of Indonesia (Butcher 2004; Atmaja and Nugroho 2011). The fishery's porous border clearly increases scale and complexity, which in turn impacts its governability (Gunawan and Visser 2012; Chuenpagdee et al. 2013).

Governing System

High value fisheries in Indonesia (e.g. pearls and mother-of-pearl; trepang; and tortoise shell) have a very long history of private control in Indonesia dating back to colonial rule in the seventeenth century (Boomgaard 2005). In his historical account of fisheries in Indonesia Boomgaard (2005) draws attention to early foreign involvement in fisheries and to the export of marine products, indicating expansive trade networks in the fisheries sector at the time. He notes however that subsistence fisheries occupied the vast majority of fishers from the many small island communities across the Archipelago. Controls placed on resource access were almost exclusively imposed for either the collection of taxes or to secure exclusive right-of-use of resources.

With support from international donors recent decades have seen significant development in fisheries policy in Indonesia, driven by central government visions to boost exports and domestic consumption in what was announced as a "Blue Revolution" (Bailey 1988). The rapid modernization of fisheries that ensued left small-scale fishers highly marginalized in the wake of new and more efficient trawling vessels. In response to this, the central government banned trawling in all waters but the Arafura Sea in passing the Presidential Decree No. 39. Bailey (1988) notes how the very effective enforcement of the ban led to re-investments in other unregulated fisheries, in particular the Java Sea purse seine fishery. Despite indications of severe over-fishing in many fisheries, central government fishery policy remained (and still remains) largely focused on increasing fisheries production.

Decentralization of the many government functions in Indonesia was legislated in 1999 and has been implemented progressively since 2001 to the present day. Prior to this the fisheries sector in Rote received little support from the government by way of subsidies or infrastructure. This is reflected in allocated funding for fisheries growing by more than two orders of magnitude since decentralization. The decentralization process transferred responsibility of many previous national government functions, including fisheries, to provincial and district level governments (Gunawan and Visser 2012). This was a means to partially ease tensions over the natural resources being exploited without a satisfactory share of the benefits accruing to local economies (Resosudarmo 2005). It was also designed to make government more accountable by exposing it to greater public scrutiny at the local level which was a particular concern of then President Habibie in 1999 (Buehler 2010).

Under this new decentralized system, responsibility for fisheries management over marine areas within four nautical miles from coastlines fell to district-level authorities (a district is known locally as regency or *Kabupaten*). Additionally, "municipalities" (*Kota*), such as nearby Kupang, were given the same level of responsibility for governing fisheries within their four nautical mile zone. It was the responsibility of the districts and municipalities to pass regulations necessary to control fishing in their jurisdictions, including all inshore waters which were effectively reserved for the use by small-scale fishers through controls on vessel size and gear. Although during this study, district and municipal fisheries administrations maintained mandate over fisheries governance within the four miles zone, in October 2014 amendments to Law 32/2004 on decentralization reassigned this responsibility to provincial administration. However, this has yet to be formally endorsed and until such time as it is, the District fisheries department reports that they will continue to assert responsibility for the area. The intention of the amendment was to provide greater control over the issuance of licenses by removing the potential economic incentive that districts may have to allocate excessive numbers of licenses to generate revenue (Ria Fitriana, personal communication, 2014). The districts and municipalities under the recent amendment would retain a share of the revenue based on fishing activities in their previous zones from the coast to four nautical miles which maintains one of the initial purposes for which the zones were established (Resosudarmo 2005).

Provincial fisheries already had responsibility for in fisheries the waters between 4 and 12 miles from the coast however this recent amendment means that they also have primary mandate over the four mile zone. The national government remains responsible for fisheries from the outer provincial limit to the outer limit of the Republic of Indonesia's exclusive economic zone and for all vessels exceeding 30 gross tonnes. In developing more localized and relevant governance systems, the process of decentralization also created a highly complex array of jurisdictions with intersecting maritime boundaries.

Compounding governance complexity are zones established by national Ministerial Regulation, including most recently *Peraturan 02/MEN/2011* (Fishing Land and Placement of Fishing Tools and Auxiliary Fishing Tools in the Fishery Management Area of the Republic of Indonesia: Regulation of the Minister of Marine and Fishery 31 January 2011),[6] which defines two additional "fishing lanes" within the four mile inshore marine zones. Zone 1 forms the low tide mark to the two mile limit, while Zone 1b extends a further two miles to sea up until the four mile limit. The purpose of these zones is to regulate the gear and vessels used in inshore waters based on the 'selectivity and capacity of the gear' or the size of the vessel. These national regulations effectively created a duopoly of responsibility for the jurisdiction (and governance) over inshore small-scale fisheries.

A further issue with potentially profound implications for governance is Article 33(3) of the Indonesian constitution, that states that 'the land, the waters and the natural resources within shall be under the powers of the State and shall be used to the greatest benefit of the people' (ILO 2012). This is often interpreted as meaning not only that the state controls the use of terrestrial and marine resources, but also

[6] The regulation can be found at http://faolex.fao.org/docs/pdf/ins107228.pdf accessed 15/10/2014.

that access rights cannot be assigned exclusively to any one person or group of people (Sarah Waddell, personal communication, 2012; Imran Amin, personal communication, 2013). So, although rights-based management, in one form or another, has a world-wide reach (Scott 2000), and is usually considered to be an important element (if used appropriately) of fisheries governance, allocating fishing rights in Indonesia does not seem possible under the prevailing interpretation of the constitution. We also note that fishing rights, as pointed out by Allison et al. (2012) in contexts such as the one in Rote, are unlikely to perform their intended purposes unless other basic needs of fishers, such as food security or access to healthcare, are also ameliorated.

As the elected head of the district government of Rote Ndao, the Regent (also known as the *Bupati*) tops a highly hierarchical governance system. The regent is ultimately responsible for all the functions of the district government including the management of the fisheries resources within the district's fishing zone. In Rote this position carried both responsibility and influence, and continues to do so in respect of many other important government functions. Below the regent are the elected representatives who comprise the district's legislative assembly, however, at least in Rote, fisheries are not specifically allocated as a portfolio to an individual elected member. Instead, members vote on fisheries laws put before the assembly but otherwise have little involvement in the system of governance. Between 2003 and 2013 the legislative assembly passed only two laws pertaining to fisheries, despite fisheries being an important sector of the local economy. The most important district laws as they existed until October 2014 included: Rote Ndao District Regulation No. 33 and 34 of 2004 permitting the collection of, respectively, business licensing fees and levies on the catch and fishing license fees; Rote Ndao District Regulation No. 6 of 2010 regarding operations of fisheries, tariff and fisheries landings management; and Rote Ndao District Regulation No. 4 of 2012 concerning licensing levies. While the district regulations were focused on the collection of fees of one kind or another, other national legislation also applied to fishing activities which would be administered by the district, such as prohibited (destructive) fishing methods. However, operationalizing these laws at sea also is limited by the ambiguity that persists regarding enforcement roles, as there is little detail in the Fisheries Law about how enforcement is to be implemented (Waddell 2009).

Public servants working in the district Marine Affairs and Fisheries office provide technical advice and carry out their assigned functions which include monitoring control and surveillance, licensing, data collection and fisheries development activities including distribution of subsidized fishing equipment and aquaculture development (Fig. 4.2). Communities are only peripherally engaged in any formal fisheries governance however there is little doubt that internal community dynamics do influence the operation of the fishery. As noted above, community leaders can exert influence through the process of applying for subsidies for purse seine equipment. Communities are also important in other normative ways such as influencing fishing activity during or for family, community or religious events.

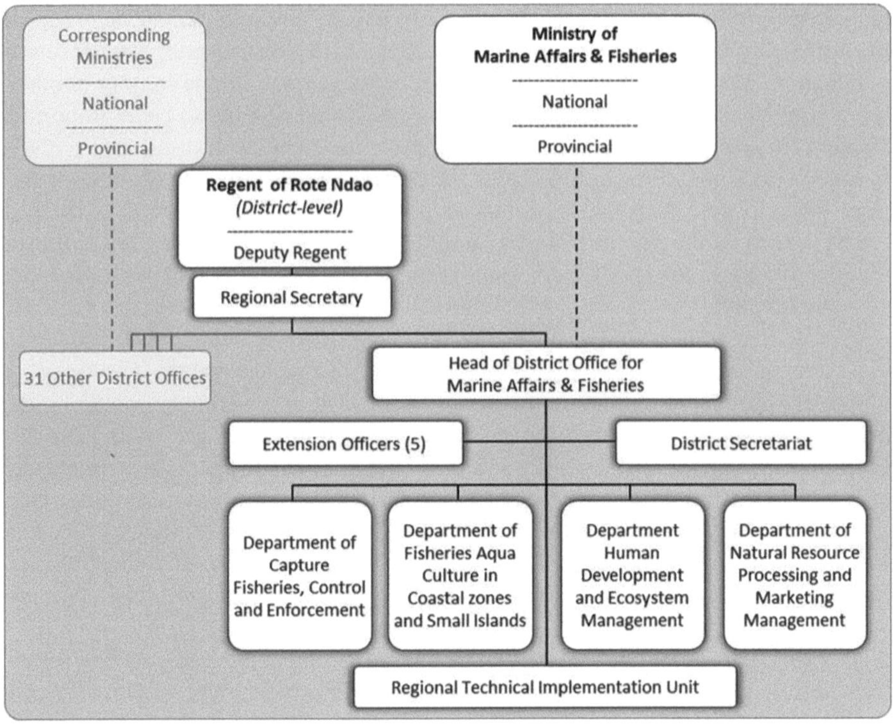

Fig. 4.2 The formal governing structure of the Rote Ndao's Marine Affairs and Fisheries Department

Similar to other parts of the tropics, in Rote many of the livelihood opportunities people have involve activities that occur outside formal management arrangements but instead may be part of traditional systems maintained by communities. A particularly relevant part of traditional systems of governance in Indonesia is local law (*hukum adat*) whereby communities in many parts of Indonesia used to (and in some parts still) control access to and extractions from their local waters and landscapes. Until colonial times, the traditional governance system on Rote island comprised 18 states (*nusak*) presided over by a ruler, each with varying forms of customary laws (Fox 1977a, 82). A local customary law system was formed from these early governance arrangements, and is known as *haholok* or *papadak* (ATSEA 2014). Under this system *deak* (east Rotinese language) and *lutu* (west Rotinese language) are practices associated with regulating access to coastal ecosystems and resources (ATSEA 2014). These are still recognized to a certain extent however, these practices are not well documented in published literature, and it was out of the scope of this research to detail subtle interplays with other more contemporary practices. However, in 2014 the district Marine Affairs and Fisheries office established a pilot program in four villages in Rote to establish local laws for fishery management. From this, local regulations (perda) were expected to be established under the regency.

Governance Interactions

Governance interactions in this fishery could be described by local managers and stakeholders with the narrative "it's a long way to the top". While this describes quite well the hierarchical nature of the governing system and its distant link to the natural and human system-to-be-governed, it also alludes to a failure on the part of the governing institutions to capture a complete consideration of what constitutes sustainable fisheries management. The governors' experience of fisheries governance is largely confined to what might be best termed "fisheries development". The governors' primary interests are to see economic activity increase and poverty decrease; and, to see food production increase and food insecurity decrease. To this end, the governors obtain budgets which are used to foster a growing fishing industry through mechanisms such as fuel, gear or vessel subsidies. There is little evidence to show that governors have concerns about the long-term sustainability of the fishery or indeed that present or increased rate of exploitation are perceived as problematic.

Many fisheries officers are well trained and are fully aware of the pressing issues of sustainability however high ranking governors, who have the ultimate decision-making powers, generally have no specialist training. Governors must therefore rely upon, among other things, good policy advice from their officers to help them make decisions. This creates a critical link in the governance system where the transmission of advice must have an open path for the system to operate on a technically informed basis.

Increasingly, effective governance of fisheries requires governors to have a more involved role in overseeing the implementation of management interventions that reduce fishing capacity and catches to ensure sustainability and the Rote purse seine fishery is arguably approaching the point where such interventions will be needed. These measures are usually associated with consequential losses of employment opportunity, lower revenue (including for the system of governance where landings may be levied), less food and considerably more dissatisfaction by those affected. Although it is usually argued that such measures will provide long-term benefits, transmission of this difficult message through the system is easily frustrated. Informing an extremely hierarchical governance system in a democratic political system that unpalatable decisions need to be made for future sustainability, is a risky endeavor for any government official at the base of the pyramid. Western notions of 'frank and fearless' advice in such highly culturally defined arenas are unlikely to be an effective means to communicate issues and measures pertaining to fisheries sustainability.

The elected regent in Rote does meet his constituent stakeholders on an ad hoc basis. The close proximity of the communities from which the purse seine fishery operates to the offices of the regent allows fishers to access the 'governor'. Fishers usually draw upon their experience in the fishery in their interactions with the governor, which is likely to reinforce the view that the fishery needs further development and growth rather than control. With no secure rights to the fishery and with a perception of an uncertain fishery future many purse seine fishers' incentive is (consciously or not) to gain the greatest immediate benefit (Teh et al. 2014), which typically leads to conditions of Hardin's (1968) "tragedy of the commons". However, it is also possible that other fishers may not perceive any future risks to the sustainability of their fishery – a collapsed purse seine fishery is not part of their life experience, yet. Unfortunately, either perception is likely to foster demand for more development and send a message to the governor that is inconsistent with the fishery's future sustainability. Given these conditions the interactions within the governing system and between it and the system-to-be-governed provide a poor conduit for information dissemination that may support more effective governance under other conditions.

Effective governance should be understood as an evolutionary process that responds to its environment, whereby the signals from systems are received and visions are formed accordingly (Mahon et al. 2005). Experience, training, partnerships are just some elements of clear importance. However, governance examples also often provide important points of reference (good or bad). The purse seine fishery, as the system-to-be-governed, and its governing system are bordered by systems much like we documented in Rote, and consequently offer little opportunity to learn by example and influence actors to pursue alternative policy routes that might help to steer governance in a better direction. Rather, these surrounding systems normalize Rote's own systems.

Discussion

Is There a Need for Better Governance?

The Java Sea purse seine fishery is one of several examples of fisheries in Indonesia that have all but collapsed and where over-exploitation and environmental degradation has pushed small-scale fishers who use smaller and less efficient vessels into states of poverty (Purwanto 2003). Like the fisheries in Rote, the Java Sea fishery was "de facto" open access. The Rotinese purse seine fishery, unlike the Java Sea purse seine fishery, does not appear to be under any immediate threat of collapse however catch rates, an imperfect but the singular index of abundance, have suffered serious declines since the fishery began, possibly falling by as much as 90 % (Riwu, unpublished interview data; Yan Rohi, personal communication 2014). However this has not caused economic collapse of the fishery and it continues to operate.[7] Enhancing the governability of the fishery through processes that promote interactive communication and learning (Jentoft and Chuenpagdee 2009) should promote improvements in sustainability, production and profitability. We argue in moving towards such mechanisms there is a strong potential for more effective governance.

[7] In the absence of formal management objectives and fishery reference points or a quantitative assessment, current catch rates can only be qualitatively evaluated against historic ones. The fact that the vessels are still active suggests the stocks still support an economically viable fishery.

Contrasting Views

From an external fisheries perspective justifications to shift the governance system away from the status quo appear straight forward; since a more consistent supply of fish for people, greater food security and improved resources and ecosystem integrity are to be gained. Arguments are also made for potentially higher revenues for the fishers, more jobs and inroads to be made in reducing poverty and maintaining livelihood strategies.

However, the status of this fishery, and indeed many of the world's fisheries, suggests that this is either less obvious or far more difficult to achieve than often appreciated by outside observers. In some cases promoting more effective governance may be subordinate or contrary to the other interests, legitimate or otherwise, of those with greatest responsibility and influence over the governing system. In many instances there may be a lack of awareness of the principles and values of interactive governance (Mahon et al. 2005) and the mechanics by which governance systems run. How governance systems develop or fail to develop for many small-scale fisheries is rarely reported from the 'inside', i.e. from the perspective of those who govern or those who are governed. Most often governance is studied and reported on from 'outside' the system, by individuals with different cultural frames of reference and different conceptual understandings of governance and, quite probably, entirely different experiences to draw upon.

Any governing institution aspiring to improve capacity for governance adaptation by creating opportunities to learn-by-doing or by instigating strategic change in one or both the governing system and system-to-be-governed assumes some risk that the changes made may actually or perceptually not produce the expected outcomes and could even appear to make things worse. Stock recovery, on which good fisheries outcomes will often rely, is an uncertain and "multifaceted" process that can be influenced by a suite of controllable and natural variables (Neubauer et al. 2013). Moreover, regardless of whether there is a positive response within the fish population to the 'experiment', there is a real risk that the intervention and a positive outcome will not be associated with one another because of the time lag that is necessarily involved (Hutchings 2000; Neubauer et al. 2013). And, such time lags may simply be too long for political systems to cope with, given the election cycles of governments (Mahon et al. 2005).

The Ecosystem Approach to Fisheries Management and Its Governance

Contemporary best practice governance often incorporates governance systems based around EAFM. Adequately underpinning EAFM is likely to require legislative tools that are more sensitive to local complexities than previous management systems have been. Fisheries in Rote and the surrounding region invariably involve multiple species and complex, dynamic marine ecosystems. Moreover, many types of fisheries operate in the waters around Rote, often separated by very porous boundaries. With such different fisheries operating within the same highly biodiverse waters, fisheries governance must accommodate, control and secure complex and wide-ranging interests and biophysical interactions.

Decision-making in management under EAFM must be an inclusive, 'genuine', and participatory process with all relevant stakeholders. However the common use of such terminology alluding to desirable levels of participation warrants more critical examination. While there is little gain to be made from trying to reach consensus on 'how much participation is enough', there are certain important key elements of consultation that are widely cited as necessary. One of these is a legislated right to be consulted (e.g. Routel and Holth 2013; FAO 2014). The Indonesian Law No. 31 of 2004 Concerning Fisheries which states that 'Fishery Management for capture fisheries and fish-culture must take into account any existing customary laws and local wisdom, including community participation' (Article 6(2)), attempts to fulfill this requirement, at least through mention in higher level legislation. However, in the purse seine fishery identifying with whom the consultation must occur and recognizing that interests may vary across individual actors, customary groups, and potentially across numerous other stakeholders is yet to be done. Moreover, free expression of interests can be muffled by political or social marginalization of individuals, genders or groups, exacerbated by extremely hierarchical systems as exists in this fishery. Processes of inclusion may therefore benefit from mechanisms that secure confidentiality but simultaneously capture and disseminate views of less powerful or marginalized actors.

Challenges to Governance

Although a more thoroughly integrated governance process appears essential, many of the local customary systems that might provide a framework for a more inclusive process are being eroded through external globalizing influences from government, markets and religion (Barlow and Gondowarsito 2009). Internal tensions are emerging through competing perspectives on modernization. With most of eastern Indonesia characterized by high rates of poverty, and exhibiting some of the country's lowest economic development rates (Resosudarmo and Jotzo 2009), these conditions have focused attention on actively developing the economy through mainstream activities or passively allowing a wide variety of activities (e.g., small-scale mining, extractive forest enterprises) to flourish that are viewed as developmental. Under these conditions it is no surprise that development has often come at the expense of the marine or terrestrial environment (Henley and Osseweijer 2005; Dutton et al. 2009) and has distorted traditional systems of governance along the way.

We have also noted that although the complexity of the system-to-be-governed is not high in some respects, it is complex in others. The population dynamics that contribute to species' resilience also expose them to both anthropogenic and natural effects that increase both the scale and dynamics and unpredictability of the fishery which can reasonably expected to exacerbate the governability problem.

It would be speculative to suggest how livelihood diversification would interact with the fishery governance and more importantly how it would impact governability. Attractive alternatives could potentially improve governability by reducing dependence on the fishery resource. This has occurred in other eastern Indonesian cases where the emergence of seaweed cultivation alleviated pressure on local coral reef fisheries (Steenbergen 2013b). However such alternatives could potentially also free highly diversified stakeholders from the economic (and other) constraints that might otherwise force change at the level of the individual or collective in the form of greater understanding and awareness for fisheries management for the future. In such cases, developing alternative livelihoods would ironically promote solutions that circumvent the problem at hand rather than address it directly.

Pathways Towards Better Governance

Describing the state of governance around a purse seine fishery in Rote is complex enough however offering pathways that may materially outperform the status quo is far more challenging. This has perhaps become even more challenging in the short term with the most recent shift of fisheries governance responsibility from the district of Rote to the province of Nusa Tenggara Timur.

In keeping with Indonesia's strengthening democracy and the continued drive towards local empowerment (aside from the recent amendment), co-governance probably offers an important opportunity. Sharing governance across levels and actor groups would not only improve decision making but equally improve compliance and governability through "culturally matched" (Cornell and Kalt 1992; Scholtens and Bavinck 2013) social control and self-regulation mechanisms that could be implemented at community level, at least initially. However, given the connectivity of the fishery with those that surround it and developmental pressures from further afield in Indonesia, it is also imperative to have a strong governing support, in form of legitimate legislation that recognizes the need to safeguard fisheries boundaries and maintain a governable scale which the district appeared to do.

Experimentation or 'sensible foolishness' (March 1976, in Jentoft 2007) might be one way in which new directions to pursue for the fishery could be identified. Yet we have noted that any experimentation involves inherent risk. This approach is thus only likely to be practiced by the most confident and bravest governors and, those who acknowledge that other alternatives exist and are worth pursuing despite the risk. For many governors then, including those responsible for the governance of Rote's purse seine fishery, improving governance presents itself as a wicked problem whereby change implies risk given that much of the information they have is incomplete and that the human and natural systems involved are inherently complex and dynamic (Jentoft and Chuenpagdee 2009). Enabling confidence to act for change under these circumstances appears critical. Empowering governors and stakeholders through facilitation and partnerships (Mahon et al. 2005) is one way of reducing the wickedness of the problem but the provision of more and better information should not be neglected.

Making the connection between information (scientific or local ecological knowledge) and policy remains an obstacle and is one that is well documented (for example see Daw and Gray 2005 and McConney and Haynes 2011). Although this divide is not emphasized as an important one in small-scale fisheries where information is often poor, in some larger fisheries the use of management strategy evaluations that incorporate intuitive computer interfaces is growing as one solution to this problem. One promising example for such an application in small-scale fisheries is a recent study where the approach, modified appropriately to accommodate social issues, was used to support informed discussions about management options for a small-scale tropical rock lobster fishery (Plagányi et al. 2013). Cast in relevant contexts, for example livelihoods,[8] we argue that such tools could radically improve the communication of knowledge about complex systems that might be critically important to improving governance and opening substantial new pathways. Such tools could creatively provide stakeholders with opportunities to learn by 'experimenting' with alternative systems in a risk free virtual environment. This approach may be one way of forming new "images" that may otherwise never surface because of shifting baselines among stakeholders or within institutions. Of course such methods will not provide 'the solution' but they should be able to help stakeholders find better paths through confusing a maze of possibilities.

How successful governance in the fishery is seen, as we have noted above, is a matter of perception, for example from inside either the governing system or system-to-be-governed, or as perceived from the outside. Perceptions of performance are also formed by judging this against goals or targets. Chosen carefully these can both move the fishery in the right direction and reduce the likelihood of governance failure by setting useful but attainable targets. The common concepts of maximum sustainable or economic yield are clearly not appropriate targets for this fishery now or possibly ever. Hilborn (2009) suggested a target he called "pretty good yield" and while this was still technically based it demonstrated that much more easily reached targets still assist fisheries to perform 'pretty well' and for some fisheries at least might lead one to conclude that the governance was successful when more conventional targets would have led to a different conclusion. Thus we suggest meaningful but modest targets and steps to achieving them to minimize the risk of failure. We also suggest a shift in governance away from models focused on fisheries expansion and modernization that were often enabled through international development agencies in past decades (Bailey 1988), to one that recognizes

[8] An example of a relevant strategy evaluation tool for a small-scale fishery might be a Livelihood Strategy Evaluation where alternative outcomes are reported in terms of livelihoods (number, quantum or other relevant metric) rather that more conventional metrics like biomass or catch per unit effort. Such tools could be developed (with experts) and used interactively in the fish chain thereby contributing to the interactive governance process.

limits to sustainable production and places greater emphasis on securing long term local food security.

Under ideal conditions and with the best information at hand, metaphorically speaking, effective governance is a product of a ripe seed (a robust concept) planted in fertile ground (receptive stakeholders at all levels), followed by years of care (consistent reflection and response, monitoring and adaptation). In the wider, regional, context we therefore contend that programs to improve fisheries governance must be highly strategic and carefully prioritized to ensure best use of limited funds. This requires careful analysis of context, the political circumstances in each jurisdiction, identification and testing of champions for change, and knowledge of the system complexity permitting those with lower complexity to be targeted. Of critical importance in initiating co-governance is finding places where the need for better governance outcomes have been locally identified, i.e. where those who are governing small-scale fisheries recognize independently the need for change rather than being told so. In addition, reliable funding and human resources remain imperative. Where removing excess fishing capacity that, if left in the system, would undermine a process of change, significant funding must also be found and applied. In their book 'Saving Global Fisheries: reducing fishing capacity to promote sustainability' Barkin and DeSombre (2013) highlight excess capacity around fisheries systems as a critical problem for sustaining or rebuilding fisheries. We are unaware of any programs in the Southeast Asian region that are expending funds to remove capacity (vessels). On the contrary subsidized vessel building programs continue to increase fishing capacity in Indonesia. The 'Inka Mina' program for example is set to produce up to a 1,000, 30 gross tonnes vessels (http://simantap.djpt.kkp.go.id/inka-mina accessed on 07 October 2014) with the intention to shift fishing operations away from the intensively exploited coastal waters, while smaller programs, like the one in Rote, have provided a number of smaller purse seine as well as other types of vessels. It is likely that such programs will continue to impair governability in the purse seine and other fisheries unless other acceptable alternatives are identified.

If we accept that fishery subsidies are a politically important and an entrenched part of the governing system a sensible way forward may be to reallocate "bad" subsidies (Barkin and DeSombre 2013) to more productive purposes, e.g. improving governance interactions. Supporting stakeholders to create the frameworks that empower them to play a more inclusive role the governing system is one option, however a careful stakeholder analysis would be necessary to optimize the utility of any re-allocated subsidies. There are potentially many ways that subsidies could be both more productive and still satisfy some of the other socio-political purposes they may serve now. Taking this approach would be working with the existing governance system rather than challenging it and in doing so reducing the risks that may be perceived by the governors. Governability could be served by both improved pathways for sustainability and continued support for the governors by the stakeholders.

It would be easy to fall into trap where stakeholders other than the fishers, and other actor groups in the governing system, were marginalized in a process of moving to more effective governance. As Daw et al. (2011) point out trade-offs in prioritizing certain ecosystem services or human benefits over others are an inevitability of such a process. Therefore, these must be understood and catered for, perhaps using subsidies that were previously applied elsewhere to ensure that "losers" through the process are minimized.

Early and sustained success in delivering more effective governance and demonstrable benefits to the stakeholders is almost certainly necessary for a system's ongoing performance. These outcomes are more likely under conditions that are opposite to the crisis situations in which many small-scale fisheries find themselves today, and potentially where the purse seine fishery in Rote could eventually be headed. As fish stocks diminish and human populations in this region grow (Sale et al. 2014), and the social, economic and biological 'pressures' increase effective fisheries governance is only likely to be challenged further. In noting this, addressing governance short falls by promoting inherent change, rather than pursuing mitigating strategies, is imperative in ensuring long term use of fishery resources despite risks involved in working towards such change. The critical question to ask then is whether there is sufficient critical mass to support a pursuit of change towards more governable systems, or whether contemporary political, economic and social accountabilities stand in the way of such a pursuit.

Acknowledgements We would like to thank the staff of the Rote Ndao district fisheries office for the assistance in data collection about the fishery from purse seine fishers and Rusman B.H. Sayidin from the National Ministry of Marine Affairs and Fisheries for his explanations of Ministerial regulations and the interplay between national and district level legal instruments.

References

Allison, E. H., Ratner, B. D., Asgard, B., Willmann, R., Pomeroy, R., & Kurien, J. (2012). Rights-based fisheries governance: From fishing rights to human rights. *Fish and Fisheries, 13*, 14–29. doi:10.1111/j.1467-2979.2011.00405.x.

Anon. (2010). *Laporan: Komisi Nasional Pengkajian Stok Sumberdaya IkanTahu 2010*. Jakarta: Kementerian Kelautandan Perikananbadan Litbang Kelautandan Perikanan, Pusat Penelitian Pengelolaan Perikanandan Konservasi SumberdayaIkan.

Anon. (2014). *State of the Coral Triangle: Indonesia*. Unpublished report (66p). Manila: Asian Development Bank.

Atmaja, B. A., & Nugroho, D. (2011). Impact of the increasing catchability coefficient of the large purse seiner to the depletion of the small pelagic fish biomass in the Java Sea. *Fisheries Research Journal, 17*(1), 13–20.

ATSEA. (2014, April 21–27). *Report on exchange visit of Timor-Leste Indigenous communities to Rote Ndao District (Eastern Nusa Tenggara – Indonesia)*. Unpublished report. ATSEA Program.

Bailey, C. (1988). The political economy of marine fisheries development in Indonesia. *Indonesia, 46*, 25–38.

Barkin, J. S., & DeSombre, E. R. (2013). *Saving global fisheries*. Cambridge: MIT Press.

Barlow, C., & Gondowarsito, R. (2009). Socio-economic conditions and poverty alleviation in East Nusa Tenggara. In B. P. Resosudarmo & F. Jotzo (Eds.), *Working with nature against poverty* (pp. 94–121). Singapore: Institute of Southeast Asian Studies.

Boomgaard, P. (2005). Resources and people of the sea in and around the Indonesian Archipelago. In P. Boomgaard, D. Henley, & M. Osseweijer (Eds.), *Muddied waters: Historical and contemporary perspectives on management of forests and fisheries in island Southeast Asia* (pp. 97–119). Leiden: Koninklijk Instituut Voor Taal-, Land- En Volkenkunde Press.

BPS. (2014a). *Rote Ndao Dalam Angka 2014, Badan Pusat Statistik, Kabupaten Rote Ndao*. Retrieved October 10, 2014, from http://rotendaokab.bps.go.id/?hal=publikasi_detil&id=11

BPS. (2014b, March). *Badan Pusat Statistik 2014. Number and percentage of poor people, poverty line, poverty gap index, poverty severity index by Province*. Retrieved November 27, 2014, from http://www.bps.go.id/eng/tab_sub/view.php?tabel=1&id_subyek=23

Buehler, M. (2010). Decentralisation and local democracy in Indonesia: The marginalisation of the public sphere. In E. Aspinall & M. Mietzner (Eds.), *Problems of democratisation in Indonesia elections, institutions and society* (pp. 267–285). Singapore: Institute of Southeast Asian Studies.

Butcher, J. G. (2004). *The closing of the frontier: A history of the marine fisheries of Southeast Asia c. 1850–2000*. Singapore: Institute of Southeast Asian Studies.

Campbell, B., & Wilson, V. E. (1999). *The politics of exclusion: Indonesian fishing in the Australian Fishing Zone*. Perth: Ocean Centre for Peace Studies and the Australian Centre for International Agricultural Research.

Carnegie, M. (2008). Development prospects in Eastern Indonesia: Learning from Oelua's diverse economy. *Asia Pacific Viewpoint, 49*(3), 354–369.

Carnegie, M. (2013). Sailing-trading livelihoods in Southeastern Indonesia: Adapting to change. *Asian Journal of Social Science, 41*, 543–579.

Chuenpagdee, R., Jentoft, S., Bavinck, M. J., & Kooiman, J. (2013). *Governability – New directions in fisheries governance. Governability of fisheries and aquaculture: Theory and applications*. New York: Springer.

Coll, M., Navarro, J., Olson, R. J., & Christensen, V. (2013). Assessing the trophic position and ecological role of squids in marine ecosystems by means of food-web models. *Deep Sea Research Part II: Topical Studies in Oceanography, The Role of Squids in Pelagic Ecosystems, 95*, 21–36.

Coral Triangle Initiative. Regional plan of action. (2009). *Coral Triangle Initiative on Coral Reefs, Fisheries and Food Security (CTI-CFF)*. Manado: Coral Triangle Initiative.

Cornell, S., & Kalt, J. P. (1992). Reloading the dice: Improving the chances for economic development on American Indian reservations. In S. Cornell & J. P. Kalt (Eds.), *What can tribes do? Strategies and institutions in American Indian economic development*. Los Angeles: American Indian Studies Center, University of California.

Daw, T., & Gray, T. (2005). Fisheries science and sustainability in international policy: A study of failure in the European Union's Common Fisheries Policy. *Marine Policy, 29*, 189–197.

Daw, T., Brown, K., Rosendo, S., & Pomeroy, R. (2011). Applying the ecosystem services concept to poverty alleviation: The need to disaggregate human well-being. *Environmental Conservation, 38*(4), 370–379.

Dutton, I. M., Djohani, R., Sastrapradja, S. F., & Dutton, K. D. (2009). Balancing biodiversity conservation and development in eastern Indonesia. In B. P. Resosudarmo & F. Jotzo (Eds.), *Working with nature against poverty: Development, resources and the environment in eastern Indonesia* (pp. 125–146). Singapore: Institute of Southeast Asian Studies.

FAO. (2014). *Voluntary guidelines for securing sustainable small-scale fisheries in the context of food security and poverty eradication*. Food and Agriculture Organization of the United Nations. COFI/2014/Inf.10.

FAO & World Fish Center. (2008). *Small-scale capture fisheries: A global overview with emphasis on developing countries*. Food and Agriculture Organization of the United Nations (2008) The WorldFish Center. Penang, Malaysia. 63 p.

Fitriana, R., & Stacey, N. (2012). The role of women in the fishery sector of Pantar Island, Indonesia. Gender in aquaculture and fisheries: Moving the agenda forward. *Asian Fisheries Science, S25*, 159–175.

Foale, S., Adhuri, D., Aliño, P., Allison, E. H., Andrew, N., Cohen, P., Evans, L., Fabinyi, M., Fidelman, P., Gregory, C., Stacey, N., Tanzer, J., & Weeratunge, N. (2013). Food security and

the Coral Triangle Initiative. *Marine Policy, 38*, 174–183. doi:http://dx.doi.org/10.1016/j.marpol.2012.05.033
Fox, J. J. (1977a). *Harvest of the palm: Ecological change in eastern Indonesia*. Cambridge: Harvard University Press.
Fox, J. J. (1977b). Notes on the southern voyages and settlements of the Sama-Bajau. *Bijdragen Tot de Taal-, Land-en Volkenkunde, 133*(4), 459–465.
Fox, J. J. (2009). Legal and illegal Indonesian fishing in Australian waters. In R. Cribb & M. Ford (Eds.), *Indonesia beyond the waters edge: Managing an archipelagic state* (pp. 195–220). Singapore: Institute of Southeast Asian Studies.
Fox, J. J., Adhuri, D. S., Therik, T., & Carnegie, M. (2009). Searching for a livelihood: The dilemma of small-boat fishermen in eastern Indonesia. In B. P. Resosudarmo & F. Jotzo (Eds.), *Working with nature against poverty: Development, resources and the environment in eastern Indonesia* (pp. 201–225). Singapore: Institute of Southeast Asian Studies.
Froese, R., & Pauly, D. (Eds.). (2014, August). *FishBase*. World Wide Web electronic publication. www.fishbase.org
Gordon, A. L. (2005). Oceanography of the Indonesian seas and their through flow. *Oceanography, 18*, 14–27.
Green, A. L., Fernandes, L., Almany, G., Abesamis, R., McLeod, E., Aliño, P. M., White, A. T., Salm, R., Tanzer, J., & Pressey, R. L. (2014). Designing marine reserves for fisheries management, biodiversity conservation, and climate change adaptation. *Coastal Management, 42*(2), 143–159. doi:10.1080/08920753.2014.877763.
Gunawan, B. I., & Visser, L. E. (2012). Permeable boundaries: Outsiders and access to fishing grounds in the Berau Marine Protected Area. *Anthropological Forum, 22*(2), 187–207.
Hardin, G. (1968). The tragedy of the commons. *Science, 162*(3859), 1243–1248.
Harper, S., Zeller, D., Hauzer, M., Pauly, D., & Sumaila, U. R. (2013). Women and fisheries: Contribution to food security and local economies. *Marine Policy, 39*, 56–63.
Henley, D., & Osseweijer, M. (2005). Introduction. Forests and fisheries in island Southeast Asia. Histories of natural resource management and mis-management. In P. Boomgaard, D. Henley, & M. Osseweijer (Eds.), *Muddied waters: Historical and con-temporary perspectives on the management of forests and fisheries in island Southeast Asia* (pp. 1–42). Leiden: KITLV Press.
Hilborn, R. (2009). Pretty good yield in exploited fisheries. *Marine Policy, 31*, 193–196.
Hutchings, J. A. (2000). Collapse and recovery of marine fish. *Nature, 406*, 882–885.
ILO. (2012). *The 1945 Constitution of the Republic of Indonesia: As amended by the First Amendment of 1999, the Second Amendment of 2001, the Third Amendment of 2001 and the Fourth Amendment of 2002*. Unofficial translation by the International Labour Organisation. Retrieved November 22, 2014, from http://www.ilo.org/wcmsp5/groups/public/---ed_protect/---protrav/---ilo_aids/documents/legaldocument/wcms_174556.pdf
Jentoft, S. (2007). Limits of governability: Institutional implications for fisheries and coastal governance. *Marine Policy, 31*, 360–370.
Jentoft, S., & Chuenpagdee, R. (2009). Fisheries and coastal governance as a wicked problem. *Marine Policy, 33*, 553–560.
Mahon, R., Bavink, M., & Roy, R. N. (2005). Fisheries governance in action. In J. Kooiman, M. Bavinck, S. Jentoft, & R. Pullin (Eds.), *Fish for life: Interactive governance for fisheries* (pp. 353–378). Amsterdam: University of Amsterdam Press.
March, J. G. (1976). The technology of foolishness. In J. G. March & J. P. Olsen (Eds.), *Ambiguity and choice in organizations*. Bergen: Universitetsforlaget.
McConney, P., & Haynes, C. (2011). Communication. In R. S. Pomeroy & N. L. Andrew (Eds.), *Small-scale fisheries management: Frameworks and approaches for the developing world* (pp. 178–195). Wallingford: CABI.
McGoodwin, J. R. (2001). *Understanding the cultures of fishing communities: A key to fisheries management and food security* (FAO fisheries technical paper). Rome: FAO.
Mills, D. J., Westlund, L., Graaf, G., Kura, Y., Willman, R., & Kelleher, K. (2011). Under-reported and undervalued: Small-scale fisheries in the developing world. In R. S. Pomeroy & N. L.

Andrew (Eds.), *Small-scale fisheries management: Frameworks and approaches for the developing world* (pp. 1–15). Wallingford: CABI.

Neubauer, P., Jensen, O. P., Hutchings, J. A., & Baum, J. K. (2013). Resilience and recovery of overexploited marine populations. *Science, 340*, 347–349.

Paul, L. M., Simons, G. F., & Fennig, C. D. (Eds.). (2014). *Ethnologue: Languages of the world, Seventeenth edition*. Dallas: SIL International. Retrieved October 10, 2014, from http://www.ethnologue.com

Plagányi, E. E., van Putten, I., Hutton, T., Deng, R. A., Dennis, D., Pascoe, S., Skewes, T., & Campbell, R. A. (2013). *Integrating indigenous livelihood and lifestyle objectives in managing a natural resource*. Retrieved from www.pnas.org/cgi/doi/10.1073/pnas.1217822110

Prescott, J., Prescott, J., Vogel, C., Pollock, K., Hyson, S., Oktaviani, D., & Panggabean, A. S. (2013). Estimating sea cucumber abundance and exploitation rates using removal methods. *Marine and Freshwater Research, 64*, 599–608.

Purwanto. (2003). Status and management of the Java Sea fisheries. In G. Silvestre, L. Garces, & I. Stobutzki (Eds.), *Assessment, management and future directions for coastal fisheries in Asian countries* (pp. 793–832). WorldFish Center conference proceeding 67, 1:120.

Resosudarmo, B. P. (2005). Introduction. In B. P. Resosudarmo (Ed.), *The politics and economics of Indonesia's natural resources* (pp. 1–9). Singapore: Institute of Southeast Asian Studies.

Resosudarmo, B. P., & Jotzo, F. (2009). Development, resources and environment in Eastern Indonesia. In B. P. Resosudarmo & F. Jotzo (Eds.), *Working with nature against poverty: Development, resources and the environment in eastern Indonesia* (pp. 1–18). Singapore: Institute of Southeast Asian Studies.

Rockefeller Foundation. (2013). *Securing the livelihoods and nutritional needs of fish-dependent communities*. New York: Rockefeller Foundation. Retrieved from http://www.rockefellerfoundation.org/blog/securing-livelihoods-nutritional-needs

Routel, C., & Holth, J. (2013). Toward genuine tribal consultation in the 21st century, University of Michigan. *Journal of Law Reform, 46*, 417–475.

Sale, P. F., Agardy, T., Ainsworth, C. H., Feist, B. E., Bell, J. D., Christie, P., Hoegh-Gulberg, O., Mumby, P. J., Feary, D. A., Saunders, M. I., Daw, T. M., Foale, S. J., Levin, P. S., Lindeman, K. C., Lorenzen, K., Pomeroy, R. S., Allison, E. H., Bradbury, R. H., Corrin, J., Edwards, A. J., Obura, D. O., Sadovy de Mitcheson, Y. J., Samoilys, M. A., & Sheppard, C. R. C. (2014). Transforming management of tropical coastal seas to cope with challenges of the 21st century. *Marine Pollution Bulletin, 85*, 8–23.

Scholtens, J., & Bavinck, M. (2013). South Indian trawl fisheries – Assessing their governability. In M. J. Bavinck, R. Chuenpagdee, S. Jentoft, & J. Kooiman (Eds.), *Governability of fisheries and aquaculture: Theory and applications* (pp. 177–199). New York: Springer.

Scott, A. (2000, November 11–19). Moving through the narrows: From open access to ITQs and self-government. In R. Shotton (Ed.), *Use of property rights in fisheries management, Proceedings of the FishRights99 conference, Fremantle, Western Australia* (pp. 105–117). Mini-course lectures and core conference presentations. Rome: FAO Fisheries Technical Paper, FAO, 404(1), 342.

Stacey, N. (2007). *Boats to burn: Bajo fishing activity in the Australian Fishing Zone* (Asia-Pacific Canberra: Environment monograph series). ANU E Press. Retrieved from http://epress.anu.edu.au/boats_citation.html

Stacey, N., Karam, J., Meekan, M., Pickering, S., & Ninef, J. (2012). Prospects for whale shark conservation in eastern Indonesia through Bajo traditional ecological knowledge and community-based monitoring. *Conservation and Society, 10*(1), 63–75.

Steenbergen, D. J. (2013a). The role of tourism in addressing illegal fishing: The case of a dive operator in Indonesia. *Contemporary Southeast Asia, 35*(2), 188–214.

Steenbergen, D. J. (2013b). *Negotiating the future of local 'backwaters': Participatory marine conservation on small Islands in Eastern Indonesia*. Doctoral thesis, Murdoch University, Perth.

Stobutzki, I. C., Silvestre, G. T., & Garces, L. R. (2006). Key issues in coastal fisheries in South and Southeast Asia, outcomes of a regional initiative. *Fisheries Research, 78*, 109–118.

Teh, L. S. L., Teh, L. C. L., & Sumaila, U. R. (2014). Time preference of small-scale fishers in open access and traditionally managed reef fisheries. *Marine Policy, 44*, 222–231.

Waddell, S. (2009). Rising to the challenge of providing legal protection for the Indonesian coastal and marine environment. In R. Cribb & M. Ford (Eds.), *Indonesia beyond the waters edge: Managing an archipelagic state* (pp. 147–169). Singapore: Institute of Southeast Asian Studies.

Wagey, G. A., Nurhakim, S., Nikijuluw, V. P. H., Badrudin, & Pitcher, T. J. (2009). *A study of illegal, unreported and unregulated fishing in the Arafura Sea, Indonesia*. Jakarta: Research Centre for Capture Fisheries, Agency for Marine and Fisheries Research, Ministry of Marine Affairs and Fisheries.

White, A. T., & Green, A. L. (2014). Introduction. *Coastal Management, 42*(2), 81–86. doi:10.1080/08920753.2014.877758.

Williams, M. J. (2007). *Enmeshed: Australia and Southeast Asia's fisheries* (Lowy Institute paper 20). NSW: Lowy Institute for International Policy.

Williams, M., & Staples, D. (2010). Southeast Asian fisheries. In R. Q. Grafton, R. Hilborn, D. Squires, M. Tait, & M. Williams (Eds.), *Handbook of marine fisheries conservation and management* (pp. 243–257). Oxford: Oxford University Press.

Wilson, J., Darmawan, A., Subijanto, J., Green, A., & Sheppard, S. (2011). *Scientific design of a resilient network of marine protected areas*. Lesser Sunda Ecoregion, Coral Triangle, Asia Pacific Marine Program, Report 2/11.

Chapter 5
Governability Challenges in Sustaining Small-Scale Fisheries in an Urban Context: A Study of Cochin Backwaters, India

Jyothis Sathyapalan and Sunny George

Abstract The water body that stretches from Azhikode in the north to Alappuzha in the south in Kerala, the southernmost state of India, is generally known as the Cochin Estuarine System or the Cochin backwaters. This backwater system provides a nursery ground for many marine species, making it a productive fishing area for many small-scale fishing communities living in the vicinity. Evidence shows, however, that this water body has been subjected to both horizontal and vertical shrinkages due to various reasons associated with urbanization and industrialization. This chapter examines the challenges that modern development is posing for small-scale fisheries and how small-scale fishers cope with it. We argue that governance interventions to address the threat of urban and industrial development on backwater fisheries requires a deliberative process involving all stakeholders, particularly local organizations (panchayats) in which small-scale fishers play an important role. In order to improve governability of the Cochin backwaters, a platform needs to be created for small-scale fishers so that they can negotiate on equal terms with their urban counterparts, who hold economic and political power, and whose interests centre around industrialization, tourism, and urban infrastructure development.

Keywords Urbanization • Backwater • Fisheries • Deliberative • Process • Governability

J. Sathyapalan (✉)
Centre for Economic and Social Studies, Hyderabad, India
e-mail: sjyothis@cess.ac.in

S. George
Kerala Institute of Local Administration (KILA), Thrissur, India
e-mail: georgesgv@gmail.com

S. Jentoft, R. Chuenpagdee (eds.), *Interactive Governance for Small-Scale Fisheries*, MARE Publication Series 13, DOI 10.1007/978-3-319-17034-3_5

Introduction

The Cochin backwaters, a shallow semi-enclosed body of brackish water running parallel to the coastline of Kerala state, the southernmost state of India, is believed to have attained its present configuration in the fourth century A.D. It was primarily a marine environment bounded by an alluvial bar parallel to the coastline and interrupted by the Arabian Sea at intervals. It is said that as a result of a catastrophic deluge which took place in 1341 A.D, a number of islands arose, thus separating a distinct water body from the sea with connecting channels at different locations (Menon 1913, cited in Gopalan et al. 1983; Gopalan 1991). Historically, these backwaters extended from north to south, passing well beyond the boundary of the erstwhile princely state of Cochin ruled by the *Cherama* Kingdom in the fifteenth century. The history of Cochin shows that it was one of the first spots visited by Europeans in India for trade. The princely state succumbed to the Portuguese in 1599, to the Dutch in 1662 and to the British in 1799 (Hunter 1885). The shallow backwaters made the city one of the main trade centres of India in those days as it had openings to the sea suitable for navigation. The imperial gazetteer of India observes that the area surrounding the backwaters was rich in rice, coconuts and fishes and that a considerable amount of fish trade was carried out with Sri Lanka. In Cochin, sufficient fish was available locally at nominal prices which had a positive impact overall on the well-being of local people (Hunter 1885). Fishers of the Cochin backwaters used interesting and diversified fishing methods that were not generally known in other parts of India, but which had parallels in the Malay Archipelago and other Asian countries (Gopinath 1953). Historically, small-scale fishing in Cochin backwaters was very advanced.

Today Cochin is an expanding coastal city with a population of around 601,000. The human use of the backwaters has increased significantly for purposes such as tourism, and industrial and infrastructure development. The traditional uses of the backwaters have become more vulnerable in the urban context due to various types of negative externalities arising out of modern uses of the backwaters. Conflicts have arisen due to competing uses of the backwaters and fishers have been adversely affected. Development has created negative externalities for the backwaters fishery at the pre-capture stage of the life cycle. The life cycle of many migratory species has been disturbed, ultimately affecting the livelihoods of fishers. Fishers see themselves as victims of industrialization and predict a bleak future. Consequently, out of desperation, fishers themselves engage in ecologically unsustainable fishing practices that are often illegal as well.

As suggested by Jentoft and Chuenpagdee (2009), fisheries governance is challenging when competing interests are at play, with multiple stakeholders in conflicting positions, and even more so when these groups have different levels of economic and political influence on policy and decision-making. We argue from the interactive governance perspective that the nature and quality of interactions between stakeholders play an important role in the governability of the natural resource system, in this case the Cochin backwaters. These interactions need to be fostered and improved through appropriate governing interventions, based on the knowledge of where each stakeholder group stands in terms of power and influence in the decision-making process. We illustrate through this case study that an even platform needs to be created to facilitate involvement of local small-scale fisheries organizations in the process of governance. This may be achieved by designing co-governance interactions where societal parties join hands with the common purpose of saving the backwaters from further ecological deterioration and improving adaptive capacity of the small-scale fishers. Such efforts would then lead to improving small-scale fisheries livelihoods, promoting coastal sustainability and hence enhancing governability.

Using the governability assessment framework, we first describe the natural system of the Cochin backwaters. This is followed by a detailed description of the social system and the recent development-related changes that have taken place in the area. We discuss these changes in the context of threats and pressures on small-scale fishing and how small-scale fishers cope with them. Next we identify options and opportunities that small-scale fishing people have in order to be better able to interact with urban developers and large-scale industries, including tourism operators, in the governance of the area. We conclude with a discussion about the role of the local self-governance institutions, *panchayats*, in enhancing governance capacity of small-scale fishing people, and in improving the overall governability of the Cochin backwaters.

This chapter is based mostly on secondary information, gathered and synthesized from various publications, including unpublished government documents. We undertook primary data collection through a rapid fisheries assessment method and focus group discussions in March; 2014.

Cochin Backwaters Natural System Under Multiple Stressors

The Cochin backwaters in Kerala state of India stretch from *Alappuzha* (*Alleppey*) in the south to *Azhikode* in the north (Fig. 5.1). It is a shallow, semi-enclosed body of brackish water, about 231 km^2 in size, bounded by an alluvial bar parallel to the coastline and interrupted by the Arabian Sea at intervals (Gopalan et al. 1983). Several rivers (*Periyar, Pamba, Meenachil* and *Muvattupuzha*) and a network of canals empty themselves into the Cochin backwaters, discharging large quantities of freshwater, particularly during the monsoon months (June to August). The constant mixing of fresh water with seawater through tidal exchange gives the backwaters the characteristics of a typical estuary in a few locations. The productivity of the biotic subsystem and the richness of the ecosystem are closely linked to non-living or abiotic components. For instance, the availability of water, amount of dissolved oxygen and nitrogen in the water, and the acidity of the water or soil are crucial for the vitality of living organisms. Zooplankton biomass and its distribution fluctuate seasonally and with the salinity of the water. After heavy rainfalls, and because of a large influx of freshwater into the estuarine system, many marine organisms migrate

Fig. 5.1 Location of Cochin backwaters in Kerala state of India (Source: Srinivasan 2002)

from the estuarine environment (Silas and Pillai 1975). The balance between biotic and abiotic elements of this water ecosystem needs to be maintained since alterations can lead to depletion of fish stock and even extinction of many species (Gopalan et al. 1983; Gopalan 1991).

The Cochin backwaters are a highly complex wetland ecosystem and are part of the larger system of *Vembanad-Kole* Wetlands, a Ramsar site. Mangrove forests lining the area contribute to the richness in biodiversity; there are close to a 100 species in the backwaters (Thomson 2001), including economically valuable fish and invertebrates such as penaeid prawns, pearl-spots, mullets, crabs and clams.

Given that these species contribute substantially to small-scale fisheries, concerns have been raised about development activities in the area that negatively impact the ecosystem. Firstly, the water body is shrinking horizontally (in terms of surface area), due mainly to land reclamation from the backwaters, something that started as early as 1834. Land reclamation has taken place for agriculture, industry, housing, husk retting, harbors and urban development (Gopalan et al. 1983). Vertically, the area has also been reduced in size due to siltation, occurring as a result of river discharge and tidal inflows, and accelerated by direct human alterations in upland Kerala such as deforestation, construction of dams and reservoirs. The magnitude of siltation in the backwaters is reflected in the removal of 2.5 million cubic yards of silt every year in order to maintain the shipping channel at Cochin Harbor (Kurup 1971). In the course of 50 years the average depth of Cochin backwaters has been reduced from 6.7 to 4.4 m (Gopalan et al. 1983). The reduction in the area and volume of the backwaters has obvious negative impacts on fish production and the livelihoods of fishers. But so far, small-scale fishers have not received any compensation.

Second, reclamation of land for infrastructure projects has adversely affected the rich mangroves of Cochin backwaters. A report on the state wetlands inventory pointed out that in certain locations of Cochin mangroves are fast depleting (Centre for Earth Science Report cited in Krishnakumar 2013). The study estimated that nearly 129 acres of the mangroves were destroyed recently for road building, resorts and villas and other constructions, despite the prohibition. Fishers in the region had also alleged that large tracts of mangroves near the Cochin Liquid Natural Gas terminal were destroyed after the dumping of waste and construction materials. As mangroves provide ecological services by supporting the life cycle of juvenile fish and crustaceans, their destruction has direct effects on small-scale fishing livelihoods.

Finally, in addition to domestic wastes, organic fertilizers and residues from agricultural lands, and industrial pollutants discharged from a large number of heavy industries (such as plywood, paper and textile) located on the banks of the river Periyar that drains directly into the Cochin backwaters adversely affect the backwaters and its marine organisms. As shown by Joseph (2004), three species of edible fishes in the Periyar River have significant levels of radioactive contamination, whereas 16 fish species have completely disappeared from the Periyar River in the last 50 years. Despite this evidence, government authorities do not seem concerned about the cumulative impact of pollution on the ecosystem and its effect on fish species depletion and livelihood loss. The governance of this water system is consequently very challenging.

The Small-Scale Fishers of Cochin Backwaters

Kerala has achieved high quality of life indicators as compared to many other states in India. With a population of about 35 million, Kerala is one of the leading maritime states. As per the population census of 2011, the fisher folk population in Kerala was one million, namely 771,000 in the coastal area and 231,000 thousand inland. It was also estimated that about 74,100 people are engaged in allied fishery activities (Government of Kerala 2013). The state has made a transition from a low quality of life to a higher one without high economic growth and with low levels of energy and other natural resource use, resulting in a development model widely discussed in academic and administrative circles (Parayil 1996). Indian Planning Commission estimates put Kerala's poverty level at only 7 % as compared to almost

23 % for the whole country as a whole. Nevertheless, small-scale fishers of Kerala are one of the most backward communities in India even though fish is an important part of the Kerala diet (almost 34 kg/person/year). Indebtedness, poor housing, and low quality of life characterize fishers' life in contrast to the high quality of life achieved by the state as whole. Fishers remain at the margin of society (Kurien and Achari 1988; Kurien 1995), although Kurien (2004) observed that fishing communities are slowly catching up with the rest of the population.

The inland fishers of Alappuzha and Ernakulum districts depend on the Cochin backwaters for livelihoods. The total inland fisher population in Alappuzha and Ernakulum districts is 60,500 and 62,400 respectively or a total of 123,000. Around 73,000 depend directly on the backwaters (Government of Kerala 2013). A rapid assessment of small-scale fisheries conducted in a few landing centres and fishing hamlets of Cochin backwater revealed that most of the active fishers are male between 40 and 50 years of age. Female participation in the fish retail trade is an important source of income for families. Many of these women are part of community self-help groups (*Kudumbashree*). The average literacy rate of fisher folk in Kerala is 72.5 % as compared to the all- India literacy level of 52 % for fishers (Government of India 2010). Fisher literacy in the Cochin backwaters stands at 87 % as compared to the state average of 95 %. The state stands first amongst Indian states in terms of literacy and quality of life. Notwithstanding this, many fisher families are daily wage earners in the fishing sector. They earn and spend on a daily basis and are hardly able to save money, hence remaining poor.

Geographically, Muslim fishing communities are concentrated in the northern region of Kerala while Hindus in the central part and Christians in the southern part. Cochin backwaters fishers include people from all three religions. Fishers are deeply religious and also organized along caste lines. Historically, *Valan* and *Arayan* were the two dominant castes (Iyer and Krishna 1909). Later they, along with a couple of other castes, came together under the name of *Dheevara.* This was done mainly to protect fishers' rights and to negotiate with the state for caste-based reservations in government jobs. Irrespective of their religion, fishers are treated as one social group, namely 'other backward classes (OBCs)', for all government benefits, particularly for reservation in government jobs.

Kerala is known for its long history of political radicalism and trade-unionism. Fishers were, for the most part, not involved in such political activism except for in some places like Alappuzha. It was only much later that Catholic priests brought them into politics in struggles against the first communist ministry of Kerala state (Halfdanardottir 1993). These struggles led to the formation of fishers unions and grass root fishers cooperatives Unions were united under an umbrella body called the *Kerala Swathanthra Malsya Thozhilali Federation* in 1980. This was an independent and secular body fighting for fishers' rights. Today, party-based trade unions have increasingly taken root and the *Kerala Swathanthra Malsya Thozhilali* Federation has lost its clout as an independent union.

Another important characteristic of the Cochin backwaters fishing communities is their local knowledge pertaining to fishing techniques and their diversity of gears employed. According to Thomson (2001), there are at least 25 types of gill nets, 10 types of seines, and more than 10 other gears such as stake nets, dip nets, cast nets, hook and lines, and six to seven gears that are considered as indigenous methods of fishing. The Chinese nets (*Cheena Vala*)[1] and stake nets (*Oonni or Kutti Vala*) are fixed gears commonly used in backwaters fishing. Based on information provided by government officials, there were about 700 Chinese nets and 1,450 stake nets operating in the area in 2014. A sizable proportion of these nets (230 of the former and 300 of the latter) are being used within the Cochin Corporation limit. These diverse fishing gears target both pelagic and demersal species such as shrimp, mullets, pearl spots, crabs, oysters, clams, milk fishes, scampi, and cat fishes. The diversity of techniques highlights both the importance of fishing to these communities and fishing as an important source of local livelihoods (Thomson 2009).

The clam fishery, in particular, and the associated lime shell collection support livelihoods of about 4,163 fishers in the backwaters (Suja and Mohamed 2011). Clams are picked with a collection basket known as *unda vala* or *kilungam*. A more effective method locally known as *varandi* or *kolli* is employed for combing the bottom, where shells are recovered. The flesh of the clam is consumed by local people and also used as raw material in aqua-feed and fishmeal production. Shells are used for the manufacture of hydrated lime, soda ash, bleaching powder and precipitated calcium carbonate. It is also used in the paddy fields to neutralize the acidity of soil.

One issue within the clam fishery that offers a governance challenge is that of illegal dredging of lime shell. The Government of India listed lime shell as a minor mineral under the Mineral Concession Rules, 1949. Industries, using this rule, Illegal dredge lime shell using large scale suction pumps. The issue of unauthorized mining has resulted in tension among fishermen, cooperative societies and local communities. The acquisition, sale, supply and distribution of lime shell in the state are at present controlled by the Kerala Lime Shells Control Act, 1958. According to this act, licenses for collection of lime shells should be issued only in favor of co-operative societies formed exclusively of lime shell fishers. In practice, as suggested above, co-operative societies are often undermined.

Governing System

Historically, the backwaters fishery was governed collectively through traditional systems, for example, the *padu* system (Lobe and Berkes 2004). The padu system is a traditional system of granting entitlements to eligible members of a particular community to undertake specific fishing activities in certain designated fishing grounds during specified seasons. Padu rights are transferred across generations. This system of resource allocation facilitates equitable distribution of resources, and

[1] Chinese fishing nets (Cheena vala) are fishing nets that area fixed land installations for fishing. While commonly known as "Chinese fishing nets" in India, the more formal name for such nets is "shore operated lift nets".

helps in minimizing resource use conflicts as well as promoting sustainable management of fishery resources. In the recent past, however, it has eroded due to changes in backwaters boundaries, technological interventions etc.

State intervention in the backwater fishery started with the introduction of the Travancore Cochin Fisheries Act of 1952. Under this Act, the backwater fishery was to be regulated by the Department of Fisheries, something that continued after Kerala was formed in 1956. This Act empowered the government to make rules to protect fish and prohibit illegal fishing in specified waters. Rules were formed that regulated both access to fishing and banned particular destructive fishing practices such as use of fishing nets with small mesh size and fishing during high tide, especially near the estuarine bar mouth. In 2010, the Government of Kerala enacted the Kerala Inland Fisheries and Aquaculture Act (Government of Kerala 2010), which was aimed at protecting the traditional rights of fishers and to ensure the availability of nutritious fish to ensure food security to the people. The term 'fishery' was more widely defined in the new Act, namely as "any activity or occupation related with the conservation, development, propagation, protection, exploitation, and utilization of fish and fish products" (Government of Kerala 2010). The recognition of fish and aquatic life as government property went further than in previous act which were confined to stipulating access as opposed to ownership. The Fisheries Department, moreover, became the second layer of governance. Another striking feature of the Act was the recognition of the role of local self-governing organizations in fisheries governance. In India, under the 73rd and 74th amendments to the Constitution in 1992, state governments are obliged to decentralize powers to local level bodies, i.e., panchayats for rural areas and municipalities for urban areas, and endow them with "such powers and authority as may be necessary to enable them to function as institutions of self-government". The 11th Schedule and 12th Schedule of the Constitution of India lists the subjects that fall under the jurisdiction of self-governments (Article 243G of the Constitution of India) (Government of India 1992). The Constitution further stipulates that members governing these institutions are to be elected every 5 years. One-third of the seats are to be reserved for women, and proportional reservation of seats is to be given for socially and economically backward castes.

Interestingly, fisheries are one of the 29 subjects that come under panchayat control in Kerala. In principle, panchayats play an important role in protecting rights of small-scale fishers. As per the Act, development and management of the inland fishery rests with the state government but these are subject to the rights vested with local self-governing institutions. The Act also highlights the significance of involving peoples' representatives in decision-making and promoting a two-tiered advisory committee for the management of fisheries. Thus, at the grassroots level, local self-governments appear to be important on paper.

Nevertheless, evidence shows that panchayats have no official mandate either to interfere in resource management decisions at the state and district levels or to arbitrate conflicts related to resource uses. These tasks still lie with the state fisheries department, which is hierarchical in nature. There are, moreover, many concerns with hierarchical governance of the system pertaining to the backwaters. Studies have pointed out, for instance, that the 'command and control' method of resource management, through the registration and licensing system, has not worked well in the case of backwater fisheries, due among other things to informal and local rules exercised by different traditional stakeholders operating in the area. For example, small-scale fishers historically exercised their rights to brackish water territories and agrarian communities to wetlands along low-lying fields (Thomson 2001). While the exercising of traditional rights is not in itself a bad thing, inter community interaction was extremely poor among traditional stakeholders. Moreover, Srinivasan (2005) has found that fishers do not confine themselves to the physical boundary of their respective princely states for fishing and therefore monitoring and enforcement was difficult. This has left a legacy that remains today. Today's governance system suffers, therefore, from a lack of interaction between the competing governance structures, namely (a) the structures formed by the customary organizations (b) the structures incorporated in the various inland fishery Acts and (c) the local self-government (Panchayat) structures. Thus co-ordination between these structures is an important governance challenge in backwater fishery.

In this context, it is also important to ask how the present governing system addresses the adverse impacts of development such as the destruction of mangroves and pollution. Historically, the Fisheries Department has been silent about these issues. However, recently it has started a project on a pilot basis to restore mangroves on the Kerala coast as it has realized that the removal of mangroves leads to decline in fish recruitment and adversely affects biodiversity of fish resources. The program for afforestation of mangroves is being implemented through the Fisheries Resource Management Society (FIRMA). This initiative suggests that the department is open to go beyond conventional projects that usually focus on direct fisher welfare activities. It also needs to, however, involve other government agencies, and local self-governing institutions. The extent of their involvement will depend on their capacity.

There have been some success stories such as the *Malsya Keralam* (Fish Kerala), a project sanctioned by the state government. The department is particularly proud of the functioning of this project as it resulted in improved management practices, information exchange, timely supplies of quality inputs, crop insurance coverage, and product sales (Government of Kerala 2009). Importantly, the program has benefited many fish farmers in the state. The success of the *Malsya Kerala* Project shows the potential for local self-governing institutions to address concerns through co-operation with local people. Unfortunately, there is no clear mandate given to panchayats to deal with large commons such as the Cochin backwaters.

Addressing Governance Challenges

Shrinking backwaters, depleting mangroves, high industrial pollution, depleting fish stocks, all combined with the low social status of fishers and a lack of basic social infrastructure contribute to making fishers life highly vulnerable. Fishers become helpless when under a hierarchical governance system the state supports industrial marine fisheries, through mechanization of fleets, trawling, foreign fishing vessels in Indian waters, and fish farming. In this process, the traditional small-scale fishers depending on estuaries and backwaters suffer. Their suffering is accentuated because they are considered socially backward and landless and hence without political influence. Their contribution to the local economy is also not accounted for. How do they then cope with the fast changing society and economy? Does this situation also create governance and governability challenges? We explore these questions in the last section of this chapter.

Fishers have their own rationale in dealing with these changes. In the words of one fisher, "addressing today's hunger is more important than addressing tomorrow's". This is the case because they are daily income earners trying to maximize today's income rather than relying on tomorrow. They are well aware of the fact that the capacity of the system to provide adequate amount of fish to all fishers has been reduced and they are in a vulnerable position in this fast changing local economy. Traditional fishing communities, therefore, have started responding to the crisis in many ways. Firstly, they have intensified fishing by using smaller mesh sizes and resorted to indiscriminate fishing of whatever resources they can catch. Non-compliance and use of illegal fishing practices has become the norm. In fact, fishers' non-compliance with fisheries regulations in the backwaters goes back several decades. In 1989 there were more than 3,000 unlicensed nets operating as compared to 1,692 licensed nets. Of the 1,727 stake nets operating in the Cochin backwaters, only 794 were licensed (Srinivasan 2002; Thomson and Berkes 2006). In our interviews, fishers said it was difficult to control illegal nets and illegal operations because many stake nets are operating on a customary basis and hence not officially registered. Srinivasan (2005), moreover, explains that the license for operating a fixed engine at a particular location in the estuary already existed and the officials had no power to bypass the rule to deal with situations where, in actual practice, no fixed engines were operated. Transferability of licenses, especially in the case of fixed engines, has been another obstacle in monitoring and enforcement by state authorities. Fishers have often not cancelled or revoked licenses when giving up fishing and instead transferred their license to another person without official consent. As a result, the government has had difficulties both in monitoring the issuances of new licenses for fresh applicants. Additionally, since fishers prefer to operate stake nets during high tide, it leads to conflicts between those fishing upstream and downstream. Generally speaking, however, "we close our eyes and compromise," stated one fisher. This does not mean that fishers are unaware of the problems they generate, but rather are unable to do anything about it given the situation of helplessness that exists among fishers.

Another way that fishers are dealing with hardships is by reducing the number of workers they employ in fishing operations. Fishers operate Chinese nets with the help of five to six people. After paying their wages and giving them a share of the catch, nothing is left for them (the owner of the gear). One fisher said that "if the government is ready to give free electricity to fishers, we can fix a high power electric motor to operate the net and save on labor costs. But we know this will not

Table 5.1 Different stakeholder of the Cochin backwaters

Stakeholders	Interest	Backwaters use
Fishers	Livelihood	Catch fish
Clam fishers	Livelihood	Clam collection
Rice farmers (Aquaculture)	Income	Prawn filtering
Coconut husk retting	Livelihoods	Soaking coconut husk
Industries	Profit	Coir processing
Port	Profit	Space for ships
Tourism	Profit	Space for resorts and house boats
Inland navigation	Transportation	Space for boats
Government departments	Development	Infrastructure development
Fishery departments	Resource management	Licensing fishing

happen because for the last three years we have been waiting after arranging electric poles and other materials". Reducing labor input privileges individual rationality over the good of society as a whole. The same fisher also said "if we get a chance we invite tourists to operate the net; we teach them how to do it. When they realize the difficulty in operating these nets and the poor returns in terms of fish caught, they extend a helping hand." The fisher also showed us a card which said: "To our friendly tourists: Help our fishing industry survive; nets are taken up every five minutes. Please help us pull up the net and take your holiday photo. Any other financial help would be much appreciated. If every tourist gives a tip, we will survive! Thank you". These stories explain the challenges small-scale fishers face in adapting to a fast growing local economy.

Reasons for the ungovernability of the backwater system are not only attributable to the non-compliance to rules by small-scale fisheries. Many other people also use the backwaters, including the local elite, politically influential and economically better off (Table 5.1). An example of the latter group is that of paddy farmers who undertake aquaculture in the low-lying areas they own, after the harvest of 'Pokkali,' an indigenous variety of rice grown between June to October. Aquaculture is extensive in nature. Prawns enter the pond during the high tide when no feeding is required.

The issue of property rights and ownership has also been discussed extensively in the past (Thomson 2001, 2009; Srinivasan 2002, 2005; Lobe and Berkes 2004). Property rights should be linked to discussion of customary rights? When discussing customary rights, the first concern is that of 'who are real fishers'? In a caste based society, fisher identity is attributed to their caste. The Dheevara community, for example, claim they are the real fishers of the Cochin backwaters but others do not agree. The Dheevara Maha *Sabha* is a politically powerful and empowered institution that can negotiate with the state regarding fishing rights and even reservation in government jobs. Their fishing *sangas* (societies) allocate fishing sites through the *padu* system (Lobe and Berkes 2004).

The Church is an important player in the socio-political and economic life of fishers. Church based organizations support small-scale marine fishers' struggle for dignity and their rights to fish, especially in a context where mechanized fishers are assuming a larger role. While supporting small-scale fishers through welfare programs and job reservations, the church has played less or a role in resolving conflicts in the backwaters.

Small-scale fisher voices have been re-energized through the influence of trade unions. Recently, traditional fishers marched to the Cochin Port Trust demanding compensation for the loss of Chinese and stake nets due to the construction of the Vallarpadam container terminal in Cochin. The march was organized by the Fisheries Coordination Committee of all the fishing trade unions. Other examples of fisher activism include protests against mechanized dredging for shells in the backwaters and a sequence of conflict between fish workers of Perumbalam village and Travancore Cement Limited (Thomson 2009). The latter struggle for livelihoods in Perumbalam has been seen as one of the major environmental movements in modern Kerala's history. Fisher society, therefore, appears to be politically alert and conscious about its rights.

Improving Governability Through Self-Governance

Kerala is well known for its participatory planning and budgeting at the local level. Panchayats (in rural areas) and urban Local Bodies (in urban areas) are involved in the process of planning for economic development and social justice (George and Martina 2013). As mentioned earlier, the 11th Schedule and 12th Schedule of the Constitution of India lists the subjects that fall under the jurisdiction of local governments (Article 243G of the Constitution of India). Fisheries were one of the 29 subjects enlisted for Panchayats.

The Government of Kerala has earmarked one-third of its development funds to local governments. This amount plus resources raised by the local government are used for implementing local plans. Based on the principle of subsidiarity, institutions such as the Agricultural Department, Fisheries Department, Health Centres, Anganwadis (women and child care centres), etc. have been placed under the jurisdiction of local governments. As a result of this, the number of activities undertaken by local governments and the efficiency of them has increased substantially. In the case of fisheries, local governments having coastal areas and backwaters have initiated many projects such as fish farming and provided equipment for fishing. Women have formed groups, known as Kudumbashree[2] (self-help groups), which are engaged in shell collection. This provides livelihood support for many families.

The decentralization process in Kerala is aimed at establishing community governance including over common property resources. At the grassroots level, there are neighborhood groups whereas at ward level (each local body is geographically divided into wards) there are constitutionally mandated gram sabhas/ward sabhas.

[2] Kudumbashree means prosperity through development of family. It is the name of the poverty reduction programme of government of Kerala.

Suggestions from the gram sabhas/ward sabhas are taken up at the local body level where annual plans are formulated annually and a development report once in 5 years. A five year plan is also prepared. Notwithstanding these steps towards decentralized governance, the impact of local bodies on planning vis-à-vis productive sectors has been limited. As has been illustrated earlier, when it comes to priorities such as industrial development, they seem to take precedence even if there are concerns from local bodies or local constituents including fishers. For example, in the case of shell dredging in Perumbalam Island for industrial purposes, the local panchayat, which could have intervened through policy decisions, remained silent. Where grama sabhas have been more vocal is in demand articulation and beneficiary selection for developmental projects. Local governments have also been active in supporting disadvantaged sections (women and poor) in grama sabhas (Government of India 2006).

Strengthening of institutional and organizational changes is needed at the panchayat level. Governance needs to move from a strictly a hierarchical system to a more collaborative system that involves local governance institutions such as panchayats. Despite powers including that of fisheries being decentralized to panchayats, panchayats are not part of fisheries management in large water bodies such as the Cochin backwaters. Panchayat involvement in backwater management can be achieved only through radical institutional and organizational changes that focus on linking governance structures, namely those of the customary system, Fisheries Department and local self-government. This will require significant political will not only from the government but also from the grassroots.

Conclusion

The Cochin backwaters are part of an internationally recognized wetland system known as the Vembanad-Kole wetlands. Historically, the backwater played a crucial role in sustaining the local economy and trade with neighboring countries and Europe. Fishers employed interesting and novel techniques for catching fish.

Small-scale fishers continue to depend on the ecosystem for their livelihoods. Today, however, they face hardships because of multiple other users and competing uses including tourism, industry and infrastructure development. Negative externalities created by these competing activities affect fish recruitment which in turn adversely affects the livelihoods of thousands of small-scale fishers. Lack of education and social and economic backwardness in terms of caste, income and assets makes them vulnerable to urban development. Small-scale fishers find it difficult to adapt to the fast growth of other sectors, be it tourism, infrastructure, or industries. The adaptive capacity of fishers is weak since they are poor and lack skills for employment in other sectors. There is, moreover, no common platform for fishers and other stakeholders – such as medium and large industrialists, port authorities, tourist operators, and inland navigation -- to negotiate issues pertaining to backwaters use and conservation.

The present hierarchical governance system has limitations in addressing these issues effectively. There are no links between customary organizations, the Fisheries Department and local self-governments, thus co-ordination is an important governance challenge. The complexity of governance of the inland water resource system also arises from the fact that the different stakeholders exercise varying influence in the three different governance structures. However, because of the overlap of the three governance structures, it is difficult to have any focused and meaningful governance of the backwater fishery resource system unless linkages between these structures are established. Equally important is translating the powers given to panchayats under the Constitutional amendment into practice. The challenge at the local level is to create a platform for small-scale fishers to negotiate with other stakeholders (industry, tourist operators, and infrastructure developers). Such a platform could then lead to the improvement of small-scale fisheries livelihoods, consequently promoting coastal sustainability and enhancing governability.

References

George, S., & Martina, N. (2013). Democratic decentralisation and participatory budgeting: The Kerala experience. In Y. Sintomer, R. Traub-Merz, & J. Zhang (Eds.), *Participatory budgeting in Asia and Europe: Key challenges and of participation*. London: Palgrave Macmillan.

Gopalan, U. K. (1991). *Kayal Nammude Sampath (In Malayalam)*. Kozhikode: Kerala Shasthra Sahitya Parishad.

Gopalan, U. K., Vengayil, D. T., Udhayavarma, P., & Krishnankutty, M. (1983). The shrinking backwaters of Kerala. *Journal of Marine Biological Assessment, 25*(1&2), 131–141.

Gopinath, K. (1953). Some interesting methods of fishing in the backwaters of Travancore. *Journal of Bombay Natural History Society, 51*, 466–471.

Government of India. (1992). *Extracts from Constitution of India of Articles 243G, 243H, 243I, 280*. Retrieved November 25, 2014, from http://www.panchayat.gov.in

Government of India. (2006). *Evaluation report on decentralised experience of Kerala*. New Delhi: Planning Commission, Government of India New Delhi.

Government of India. (2010). *Marine fisheries census 2010 India*. New Delhi: Department of Animal Husbandry Dairying and Fisheries New Delhi/Central Marine Fisheries Research Institute Cochin.

Government of Kerala. (2009). *Success story of Matsya Keralam: First phase report*. Thiruvananthapuram: Department of Fisheries.

Government of Kerala. (2010). *The Kerala Inland Fisheries and Aquaculture Act. 2010*. Thiruvananthapuram: Law Department.

Government of Kerala. (2013). *Economic review 2013*. Thiruvananthapuram: State Planning Board Thiruvananthapuram.

Halfdanardottir, J. (1993). Social mobilization in Kerala: Fishers, priests, unions, and political parties. *MAST, 6*(1/2), 136–156.

Hunter, W. (1885). *The imperial gazetteer of India: Cochin to Ganguria* (2nd ed.). London: Trubner & Co.

Iyer, A. K., & Krishna, L. (1909). *The Cochin tribes and castes*. Madras: Higginbotham & Co.

Jentoft, S., & Chuenpagdee, R. (2009). Fisheries and coastal governance as a wicked problem. *Marine Policy, 33*, 553–560.

Joseph, M. L. (2004). *Status report on Periyar River: The declining trend of biodiversity and fish production in consequence of pollution in the Lower Reaches of Periyar River, report of the*

Kerala Research Programme for Local Level Development. Thiruvananthapuram: Centre for Development Studies.
Krishnakumar, G. (2013, February 3). Puthuvype's mangroves losing ground to rampant urbanisation. *The Hindu*, Kochi.
Kurien, J. (1995). The Kerala model: Its central tendency and outlier. *Social Scientist, 23*(1/3), 70–90.
Kurien, J. (2004). *Changing profile of poverty among Kerala's marine fishing communities: Are they still outliers?* Retrieved from https://www.academia.edu
Kurien, J., & Achari, T. T. (1988). Fisheries development policies and fishermen struggle in Kerala. *Social Action, 38*(1), 15–36.
Kurup, P. G. (1971). Silting in Cochin harbour. *Sea Food Export Journal, 3*(1), 11–13.
Lobe, K., & Berkes, F. (2004). The padu system of community-based fisheries management: Change and local institutional innovation in south India. *Marine Policy, 28*, 271–281.
Menon, C. A. (1913). *Cochin state manual*. Trissur: Mangaloddayam Company.
Parayil, G. (1996). The Kerala model of development: Development and sustainability in the third world. *Third World Quarterly, 17*(5), 941–957.
Silas, E., & Pillai, P. P. (1975). Dynamics of zooplankton in a tropical estuary (Cochin Backwater), with a review on the plankton fauna of the environment. *Bulletin of Department of Marine Sciences, 7*(2), 329–355.
Srinivasan, J. T. (2002). *Economic and institutional factors in the use and management of wetlands: A case study of the Cochin backwaters*. Doctoral thesis, Institute for Social and Economic Change, Bangalore.
Srinivasan, J. T. (2005). State regulation versus co-management: Evidence from the Cochin Estuarine Fisheries in India. *Environment and Development Economics, 10*, 97–117.
Suja, N., & Mohamed, K. S. (2011). Role of co-operative societies in black clam fishery and trade in Vembanad Lake. Marine Fisheries Information Service 2011 (Issue 207), 6–8.
Thomson, K. (2001). *Economic and social issues of biodiversity loss in Cochin Backwaters, report of the Kerala Research Programme for Local Level Development*. Thiruvananthapuram: Centre for Development Studies.
Thomson, K. (2009). Development policies, state interventions and struggles for livelihood rights in coastal communities in Kerala, India: A case study of the Cochin clam fishery. *Ocean and Coastal Management, 52*, 586–592.
Thomson, K., & Berkes, F. (2006). *The role of public-private cooperation in the management of estuaries in South India*. New Delhi: SHARP Technical Report, Canada, Shastri Indo-Canadian Institute.

Chapter 6
Governability of Kelp Forest Small-Scale Harvesting in Iroise Sea, France

Katia Frangoudes and Clément Garineaud

Abstract The Iroise Sea, at the northwest end of Brittany in France, hosts one of the largest kelp forests in Europe. Kelp is large brown marine seaweed that has been used for centuries by small-scale coastal communities for various agricultural and industrial purposes. For the past few decades, the kelp fisheries have been managed through a co-governance system by kelp harvesters in collaboration with processing industries, fisheries administration and scientists, to provide a sustainable supply to local alginate factories. The recent creation of a marine protected area in the Iroise Sea has resulted in the emergence of new actors and new concerns have arisen, modifying thus the system-to-be-governed and its governance. This case study is an example of how socio-ecological system and their governance can rapidly evolve according to changing circumstances. From the governability perspective, we examine how such change affects the viability of small-scale kelp harvesters.

Keywords Kelp harvesting • Evolution • Institutional dynamics • Governability • France

Introduction

This chapter examines the governability of small-scale kelp harvesting in the Iroise Sea in France. Kelp forests are vast and critical species for the ecosystems in which they are found. Consequently, to fully comprehend the governability of small-scale kelp harvesting, one must analyze the interactions, not only among the institutions and social actors, but also between the social and the natural systems.

K. Frangoudes (✉)
UMR M101, AMURE, OSU-IUEM, Université de Brest, Brest, France
e-mail: Katia.Frangoudes@univ-brest.fr

C. Garineaud
Département Hommes, Nature et Société, Eco-anthropologie et ethnologie, Muséum d'Histoire Naturelle, Paris, France
e-mail: cgarineaud@mnhn.fr

S. Jentoft, R. Chuenpagdee (eds.), *Interactive Governance for Small-Scale Fisheries*, MARE Publication Series 13, DOI 10.1007/978-3-319-17034-3_6

Kelp harvesting is a traditional activity and harvesters have been part of the management of this ecosystem for centuries. Today, together with scientists, the fisheries administration, and seaweed processing industries, kelp harvesters manage kelp harvesting so as to meet their economic needs as well as ensure the ecological sustainability. Thirty-five small-scale boats currently exploit kelp seasonally; the remainder of the year they target scallops.

In this chapter, we will examine and analyze the synergies among actors using interactive governance theory and the governability concept. The use of governance theory and the governability concept highlight the interactions between actors as well as the principles guiding them. Here, interactive governance is defined as "the whole of public as well as private interaction taken to solve societal problems and create societal opportunities. It includes the formulation and application of principles guiding those interactions and care for institutions that enable them" (Kooiman and Bavinck 2005, 17). Governability is seen as "the overall capacity of governance of any societal entity or system" (Kooiman 2008, 173). In this chapter, we examine the governance of the kelp system. By "kelp system" we refer to the social and ecological processes and interactions that combine extraction and conservation concerns. The natural and social system of kelp management will be detailed and an analysis of the interactions between the two systems undertaken. These interactions are the key element of the evolution of management rules and the governability of kelp. Stakeholders cannot manage this resource sustainably on an individual basis. Collaboration and negotiation are key to achieving sustainable management of kelp. The socio-economic system is composed of the following stakeholders' groups: organizations of kelp harvesters and owners of the local seaweed processing industry, public agencies representing the state, and scientists. Relations and exchanges among different stakeholders are solid as they have come together and overcome different crises in the past. Applied science researchers joined the governing system in the mid-1960s, bringing their knowledge about resource conservation, the mechanization of vessels and harvesting to the decision making table. They also brought with them innovative uses of seaweed. The transfer of decision making to fishers' organizations illustrates the fact that societies are polycentric with multiple centers (Rhodes 1996, 657).

In 2008, a new institution joined the governing arena, namely the Marine Natural Park of Iroise Sea (PNMI). This new body has the responsibility of governing the ecosystem and also human activities within the park's territory with the objective of maintaining and increasing local biodiversity. Given that most of the kelp forests are located in the park's territory, the arrival of PNMI meant a significant change in the governance structure and governability of the kelp system.

The interactions between stakeholders within the governing system are based on power relations, which will be described in further detail below. Decisions taken by the kelp commission, part of the fisheries organization, are the outcome of negotiations with regard to regulation of catches, quotas and later individual quotas, and number of kelp permits including boat characteristics (size, capacity power, etc.). Decisions are validated by the public authorities as long as they comply with European and national laws. The presence of scientific knowledge and a local fisheries administration in the social system is taken by national authorities to mean that decisions are scientifically made.

Within the kelp governing system, the negotiation process requires time for an agreement to be reached because different stakeholders try to get the best deal for themselves. For example, agreement on the implementation of a total allowable catch (TAC), as advised by scientists, took several years of negotiations because all stakeholders wanted to ensure that their interests and power was ensured in the new system. An analysis of decisions taken by the kelp commission will contribute to better understand the evolution of the governing system and what makes different stakeholders accept changes.

This chapter will first describe the natural system and its rich biodiversity. That will be followed by a description of the social system and the main stakeholders. Next the governing system and governing interactions will be analyzed. Traditional stakeholders share one overarching vision: the need for adequate seaweed as their livelihoods depend upon it. The sustainable harvesting of seaweed is, therefore, critical to them. Kelp harvesters have always been organized and had rules to manage kelp. At first, local knowledge was key to their management system, but over time scientific knowledge has supplemented it and improved resource management. The chapter will also focus on how national decisions pertaining to kelp management and power dynamics between the processing industry and help harvesters have shaped the system. Finally, the concern of kelp harvesters and the processing industry, namely that PNMI will be an obstacle to seaweed harvesting will be discussed.

The System-to-Be-Governed

The system-to-be-governed has two components, the natural and the social system. The importance of the kelp ecosystem to the biodiversity of the area will be discussed first.

Natural System

The Iroise Sea is characterized by a great variety of landscapes and species. Many species with high ecological importance are found in the area: grey seals, basking sharks, common bottlenose dolphins, and little terns amongst others. Some of these species are legally protected by the EU habitats and birds directives (Natura 2000 network). The presence of kelp forests, rocky drops off the coast and shallow bays provide spawning grounds and nurseries for many marine species. The natural and architectural heritage of the area were the main reasons for the establishment of the marine protected area (PNMI).

The main kelp species found in the area are *L. digitata* and *L. hyperborea*. Kelp is the common name given to the large, brown seaweeds found in the shallow coastal waters of temperate parts of the world. Kelp forests have a higher productivity than most productive land systems and provide food and habitat for many animals (e.g. invertebrates and fishes) (Steneck et al. 2002; Vasquez 2008). Within kelp forests there is rich interaction between species. Many natural scientists are undertaking research in kelp forest on a number of issues, including the genetics and biodiversity of kelp forest, the relationship of kelp to other species, and food networks. Geneticists have shown that kelp, from the Iroise Sea, is the genitor for *L. digitata* stocks for the whole of Brittany (Valero et al. 2011). The impact of climate change on kelp is also getting attention as kelp is very sensitive to increases in temperature. An increase of two degrees Celsius in the water temperature of the Iroise Sea could destroy this rich natural system.

Socio-economic Context: Historical Overview

Traditionally, the local coastal population harvested kelp for domestic purposes, such as for fuel, manure for agriculture, and food for animals (Arzel 1987; Frangoudes 2011). Until the nineteenth century seaweed was used for glass production primarily. In the first half of the eighteenth century, burning kelp was allowed along the French coastline and algae ashes replaced wood in the production of glass. The use of kelp ash for making glass ended around 1789 because they were better markets for kelp in iodine manufacturing. In 1829, a local chemical engineer developed an industrial process to produce iodine from kelp. About 30 iodine factories were set up in northern Finistère and they played an important role within the local economy. They employed a great number of kelp harvesters and skilled workers. In 1944, the number of kelp harvesters registered in the national social security system was estimated at between 3,000 and 4,000. In practice, however, probably 15,000 people were involved in kelp harvesting and seaweed foot gathering (Muller 1944). From the early 1950s, iodine was increasingly produced from chemicals (nitrates from Chile), resulting in the closure of all but three local factories and hardships to kelp harvesters. Local fishers' organizations helped kelp harvesters and their families cope with their economic difficulties by distributing subsidies.

The use of kelp for industrial purposes had forced local and national authorities to formulate rules to regulate harvesting and gathering of kelp. This was necessitated because earlier harvesting techniques with regard to seaweeds, namely gathering seaweed that was cast ashore, or cutting seaweed with rakes, were replaced in the eighteenth by sickles fitted on a four meters shaft called the "guillotine". Kelp was loaded on flat-bottomed boats called "pigouillers" and brought to shore (Arzel 1987).

In 1952, the three remaining local processing industries created an association to promote research on new uses of seaweed. What emerged from this was the possibility of producing alginate acid from kelp as opposed to iodine. Doing so would help avoid closure of local processing industries and maintain seaweed harvesting activity in the area. In 1960, the kelp economy was further boosted by the knowledge that extraction of new molecules such as alginate acid, carrageenan and agar-agar from seaweed was possible. These new materials were used for textiles, food processing, and cosmetics.

Until the end of the 1970s seaweed was dried on the dunes. Women and old men were the main labor used to dry seaweed. It was a hard job because if it rained the seaweed needed to be collected and stored. Motorization of boats and mechanization of harvesting occurred between 1962 and 1971. These important changes resulted in increased boat size and production of seaweed. At the end of the 1970s new processes to dry seaweeds were introduced.

Socio-economic Context: Harvesting Today

Each fishing boat is manned by the owner of the boat and some owners employ a crew member who is often a family member. The kelp fleet is composed of 35 boats between 8 and 12 m in length, all having significant storage capacity to stock seaweed. Boats are equipped with electronics (GPS, VMS), deck gear to handle fishing gear and equipment for processing and packaging the catch (IFREMER 2007).

In 2012, the local production of *L. digitata* reached 57,000 tonnes (2012) whereas that of *L. hyperborea* was 14,000 tonnes. All 35 boats have permits issued by the Regional Fisheries Committee of Brittany. Two main gears are used: the "*scoubidou*" (a curved iron hook attached to the end of a hydraulically driven mechanical arm) for *L. digitata* harvesting and a type of comb (gear originated from Norway) for *L. hyperborea* (Figs. 6.1 and 6.2).

Harvesting of *L. hyperborea* has intensified in the last few years due to high demand from the two main local processing industries (30,000 tonnes in 2014). Only 11 of the 35 boats harvest *L. hyperborea* in the Iroise Sea. During winter the smaller boats shift to harvesting shellfish in the Rade of Brest, an area well known for scallops and warty venus (*Venus verrucosa*), or fishing for other fish species (Alban et al. 2001, 2004, 2011). Seaweed harvesting is the main source of income for boat owners and a reduction in harvesting would impact them significantly. Kelp harvesters cannot shift easily to fishing for other species full-time as other small-scale boats already occupy the limited space at sea.

Most kelp harvesters live in towns near the Iroise Sea. When referring to their area of operation, they speak about "their" fields. Boats leave early in the morning and return to the harbors around 4 o'clock in the afternoon to land the seaweed. Trucks wait for the boats and as soon as the seaweed is weighed and loaded the trucks depart for the two main processing industries both located near the harvesting area. In today's context women are less involved in harvesting and more involved in other jobs or in charge of administrative work related to the boat. It is quite common to find women selling scallops or fish during the fishing season. Sons often follow their fathers into kelp harvesting.

Fig. 6.1 Gear for *L. digitata* scoubidou (Source: Clément Garineaud)

Fig. 6.2 Gear for *L. hyperborea* (Source: Clément Garineaud)

Socio-economic Context: Seaweed Processing

Two main industries buy 95 % of the total harvested *L. digitata* and *L. hyperborea* (Garineaud 2012). Both of these industries are foreign companies, namely Cargyll and Dupont de Nemours. Each of them employ around a 100 persons. There are other smaller enterprises located in the area that require less quantity of the two kelp species. These processing industries also procure other harvested species (*Chodrus Crispus, Porhyra sp., Ascophylllum nodosum, Palmaria Palmata,* etc.) from foot gatherers who procure them from the shore.

Evolution of Governing System

During Charles the Great, King of Frank's rule (768–814) the seashore became royal domain. The exception to the rule was the seashore of Leon in North Finistère where local people belonging to coastal parishes had the right to use cast ashore kelp as well as other algae growing on the shore (*Ordinance d'usance de la principauté de Léon* (Arzel 1987).

The first national legal ordinance regulating kelp harvesting, was decreed in 1681. It defined the rules governing harvesting (15/1–15/4) and stipulated the number of harvesting days (30). Harvesting of kelp at night as well as outside of the authorized season and territory of the community were illegal. The gathering of cast ashore kelp, on the other hand, was practiced freely by the local population. In 1692, Louis XIV (King of France) gave property rights to kelp for 20 years to "Saint Gobin", the main factory producing glass. Only ashes obtained from seaweed, however, were authorized to be transported to factories (Arzel 1987).

After the French Revolution (1789), kelp was declared a public good and accessible to all citizens, and fishing rights were restored to the coastal population. During the season, farmers living around the area gathered kelp to use in their fields. Farmers added significantly to the number of gatherers. Limiting the number of harvesters was the objective of the authorities. The national bylaw of 1852, revised in 1990, is the main law regulating small-scale fisheries activities and seaweed harvesting. Since 1853, the harvesting of kelp could be practiced only by professional fishers (fishers contributing to the social security system for fishers). In French law algae are considered as "marine vegetables" and hence fall within the purview of fisheries regulation (Mesnildrey et al. 2012).

The 1990 revised law distinguished between three main categories of algae and determined the conditions for harvesting. The first, shore seaweed, includes all seaweed reachable by foot at low tide and growing on the shore. The second category includes seaweed growing at sea and harvested by boat. The last category includes detached biomass found on the shore or beaches (wrecked seaweed). The harvesting of the first two requires a permit issued by the Regional Fisheries Committee and legally endorsed by the regional state representative. Detached biomass can be gathered by anyone.

Institutional Dynamics

In 1945, the Local Fisheries Committees and the National Fisheries Committee were established. The committees were created in all areas where fishing was important. Members were not elected by nominated by fishers, crew or national trade unions. More often than not, selected leaders represented the industrial fleet.

The aim of the law was to transfer "resource management" to these Committees and place them under the control of the fisheries administration. Changes in the structure and objectives of the Committees were introduced by the Fishers Organization Law 1991-411 and Law 2010-874 for the modernization of agriculture and fisheries. Fisheries Committees were allowed to formulate rules for the management of the resource. One of the first rules these Committees introduced was the need for a fishing permit to catch species such as langoustines, sole, algae, and scallops. Each boat would be issued a permit for harvesting specific species after the payment of fees. The cost of a permit is variable; those for algae are 100 Euros per season and per species.

Organization of the Seaweed Processing Industry

In 1952, the three local processing industries created an association aimed at furthering research on new uses of seaweed. In 1961, a common committee called CIAM (Inter-professional Committee of Marine Seaweed), consisting of kelp harvester representatives and members of the processing industries was established to promote the modernization of the processing industry and implementation of resource management rules. The board of the committee was elected for 3 years and was chaired alternately by a fisher and a representative of the processing industry (Arzel 1998). The Committee's role ended in 1992 when decision making power with regard to resource management was given to the Regional Fisheries Committee. While processing industry representatives could attend meetings held by the Kelp Commission, established by the Regional Fisheries Committee, they had no voting rights.

The processing industry set up their own organization with the main objective of promoting the industry's interests within the Kelp Commission and at the local level. The name of the new organization was "*Chambre Syndicale des algues et des végétaux*" (the Union chamber of seaweed and marines vegetables processors). Nineteen processing industries, mainly those that used algae (alginate production, cosmetics, products for human consumption) and seaweed farmers are members of the Union.

Natural Marine Park of Iroise Sea

The other institution part of the governing system is the PNMI established in 2007. Its aim is the conservation of marine biodiversity and the management of human activities within its territory. By law, this protected area authority does not have any legal mandate to manage fisheries activities but its management council has become the forum in which all users meet and discuss issues related to the marine area and its future. Fishers and kelp harvesters, processing industry representatives, recreational fishers, diver club members, tourism operators, territorial authorities, local environmental NGOs, and scientific institutes are all members of the management council. The fisheries administration and other state agencies also participate in meetings of the council but do not have the right to vote. The main issues for discussion pertain to fisheries and kelp harvesting.

Interactions of Governing System: From National to Local Decisions

The increase in the production of kelp, due to the mechanization of kelp harvesting, forced CIAM to discuss the formulation of rules to regulate kelp harvesting and manage the resource. Issues for discussion included matters related to delimiting the harvest season, regulating the exploitation of the resource, and implementing a global annual quota and permit system. CIAM also fixed the price of kelp.

The following rules were adopted by CIAM. In 1985, a permit system for boats harvesting *L. digitata* was established, and in 1986 maintaining a logbook was made mandatory. Boats without logbooks could lose their permit. In 1987, it was decided to allow only one landing per day with the objective of limiting extraction. Discussions about boat-based individual quotas started in 1987 but took 5 years to reach decision.

All decisions taken by CIAM were the result of scientific advice and had the approval of the fisheries authorities. Fisheries scientists participated in the discussions related to technical innovations and resource management played. Collaboration between kelp harvesters and other stakeholders was impacted by the implementation of the 1991 national law on fishers' organizations. The law strengthened the role of Fisheries Committees in decision making and transferred responsibilities from district level committees to regional level committees. CIAM was dissolved and a commission was established within the framework of region-based management.

The processing industry lost its key role in the new legal framework. While, as mentioned above, they were able to participate in discussions pertaining to fisheries management, they had no right to vote. It is for this reason that they established their own organization. The processing industry, by establishing its own organization, demonstrated their wish to be part of the decision-making process and to have a leadership role within it.

The PNMI is the last institution in the kelp governing system. Although it may not have any direct power in fisheries management, PNMI conducts scientific studies aimed at increasing knowledge about the marine ecosystem and about economic activities within its jurisdiction. The law gives power to PNMI to exercise its veto if economic development is likely to negatively impact the ecosystem. Since its establishment,

PNMI has exercised its veto twice: (1) to oppose a Regional Fisheries Committee decision to increase the number of purse seines permits for catching sardines and (2) to oppose the extension of a pig farm located near the marine park.

Interactions and Evolution of Governing System

Since 1979, the CIAM has discussed the benefits of applying individual quotas to fishing boats as a means to avoid over-exploitation. Quotas had been tried before, in the form of global seasonal quotas, weekly quotas or individual quotas. However, these quotas were not fixed permanently, but rather set yearly before the beginning of the season.

The individual quota system was introduced on a permanent basis in 2008 as a solution to difficulties faced by the two main buyers. The processing plants, due to an EU regulation, could not treat more than 600 tonnes of raw materials weekly as their sewage purification plants did not have the capacity to process greater quantities. The processing industry, therefore, imposed individual quotas, calculated from historical catches and the technical characteristics of each boat. While fishers were not happy with this quota system, they accepted them as no other solution could be found. Boats having a small carrying capacity were affected the most as they faced economic difficulties and had to shift to other métiers. Larger boat owners whose boats had a bigger carrying capacity were not satisfied either because the allocated quantities were too low.

During this period, another external event modified the interactions between traditional stakeholders. An inspection undertaken by the General office of Concurrence, Consumption and Fraud, found that the setting of a seaweed price did not respect the principle of free competition. Prices were agreed upon by kelp harvesters and the processing industry at the beginning of each harvesting season and were common to all harvesters. This was like forming a cartel. A new system has now been implemented whereby each enterprise signs a contract with individual harvesters for a given quantity at a given price. The new system has considerably altered the role of the Kelp Commission of the Regional Committee of Fisheries. This was the case decision making powers related to resource management were shifted from the Kelp Commission to the processing industry. Kelp harvesters felt that the commission was no longer required as they could solve their problems directly with the industry.

Tensions between fishers appeared rapidly because kelp harvesters unable to fulfill their contracts moved to other areas "belonging" to other harvesters. The Commission had to intervene to find collective solutions which would satisfy all kelp harvesters and avoid conflicts. In April 2014, new management rules were introduced to deal with this new situation. Kelp harvesters, participating at the Kelp Commission, voted for the following two rules: (1) harvesting areas would be based historical harvesting rights, and (2) moving to another harvesting zone required authorization from the Kelp Commission. These new management rules were approved by scientists and have been brought into law by the regional fisheries administration.

Fishers demanded the intensification of *L. hyperborea* harvesting on an experimental basis. It would guarantee better incomes as demand for it was greater given new extraction and application possibilities for the processing industry. Harvesting of *L. hyperborea* would also reduce the pressure on *L. digitata* as this would guarantee better incomes especially to bigger boats.

While not against promoting the harvesting of *L. hyperborea*, PNMI ordered scientific studies to evaluate the impact of increasing *L hyperborea* on the species and second on associated species. The PNMI also tried to obtain data on the impact of fishing gear on other species such as dolphin and sharks. One study showed that the frequency of noise produced by the fishing gear used for *L. hyperborea* harvesting disturbed dolphins and sharks. Despite concerns by other stakeholders, traditional kelp gatherers tried to convince the park authority to accept intensified harvesting. The park authority tried its best to raise environmental concerns. However, the concerned scientist working on population dynamics and management measures for seaweed supported the intensification of *L. hyperborea* harvesting on a rotation basis. He was inspired by the Norwegian management system that advocated closure of harvesting areas for 5 years (based on the species life cycle). The processing industry also pushed for the opening of *L. hyperborea* harvesting and increased their demand for *L. digitata* so as to maintain current employment levels. The local administration and political class had only objective in mind, namely "*the creation of new jobs in the area*" and hence supported efforts at increased production of kelp.

Intensification of *L. hyperborea* harvesting was authorized by PNMI. In return, kelp harvesters accepted the closure of dolphins' refuge areas and *L. digitata* forests to kelp harvesting. The Regional Fisheries Committee viewed the acceptance of these closures as "a sign of a good will".[1]

Analysis of Kelp Harvesting Governability

Qualitative data collected through semi-structured interviews conducted with key stakeholders form the basis of the governability analysis. The interviews dealt with the following issues: kelp harvesters' views about existing management rules, the sustainability of current harvesting from kelp forests, and the role of PNMI in the management of kelp and fisheries harvesting.

[1] Oral communication by the secretary of the District Fisheries Committee.

1987 Decision About "Only One Landing Per Day"

The official reason given to limit landings to one per day was to ensure fairness in the share of catch between boats. The following quotations highlight kelp harvesters' opinions about the decision. "*It allows for a better monitoring of the resource and regulates harvest quantities for the processing plants*" and "*we cannot harvest as much seaweed as we want because plants cannot treat these quantities*". In other words, kelp harvesters felt that the decision was made with the interests of the processing industry in mind not theirs. Not surprisingly, therefore, the processing industry felt the decision "facilitated their work" because it limited the supply of seaweed. The scientist involved felt the decision benefited conservation because it limited global harvesting effort.

The decision adversely impacted fisher cohesion as it divided fishers into two groups: boats with bigger carrying capacity and those with less capacity. For kelp harvesters belonging to the second group the rule "*penalized*" them and they wanted to have two landings per day. Kelp harvesters belonging to the first group were satisfied and were against the proposal for "*two landings per day*" because "*we will never have the time to return back to sea to harvest more*". Bigger boats were allocated 70 tonnes per day while smaller boats could only store 25 tonnes.

Decision About a Harvesting Permit and Individual Quotas (IQ)

The harvesting permit system is perceived as a "*good system*" because "*it limits the total capacity of the fleet.*" At present, 35 permits are allocated. However, fishers feel that the main aim of reducing kelp boats was not met. A smaller boat owner said: "*We are too many and everybody cannot earn enough money from seaweed harvesting*". As the permit system and one landing per day were not adequate instruments to regulate catches, the processing industry in 2008 decided to introduce the individual quota system by way of the individual contracts system. The global weekly quota was fixed at the tonnage that the two industries were able to treat. The initial allocation of IQ was based on a boat's historical rights, technical characteristics and number of persons working on board. At the beginning this system did not satisfy the owners of small boats because they were able to harvest larger quantities per day than their quota. A small boat owner said: "*I got 24 tonnes per week but I can harvest 47*" *(kelp harvester/smaller boat)*. The IQ system, moreover, is viewed by many small-scale fishers "*as difficult to manage because it is impossible to accurately limit harvesting quantities*".

Another small boat owner/kelp harvester said: "S*ome weeks I have 3 tonnes more than the quota which means that one week at sea actually brings no benefits*". Perceptions of the IQ system changed after a few years. One harvester said that "*from the beginning of the season, we know how much we can harvest and also the price given by the factory. We are able to plan our harvesting which is much better for us.*" Kelp harvesters are also able to know how much money they will earn and plan accordingly.

Fishers' View of the PNMI

Local fishers participated actively in the implementation of the PNMI so as to prevent fishers from other areas entering their territory (Frangoudes and Alban 2004). However, fishers feel the park constrains their activities. One harvester said: "*I am convinced that the park will impact negatively on our livelihoods. There is always the possibility that all types of fishing will be restricted. I am convinced that in the future they will ban seaweed harvesting.*" Many fishers have similar views because they feel that the PNMI is primarily aimed at conservation (Frangoudes et al. 2011). The recent decision to close some areas to harvesting confirms fishers' worst fears.

Not all kelp harvesters have this negative view. Some harvesters feel the PNMI can have a positive impact on their profession: "*In the past, we faced a lot of opposition from environmental NGOs, recreational fishers, and other users of the sea. Even research institutes did not support us. Today, with the PNMI we don't have the same problems. The park authorities realize our activities are not anti-conservationist. We are better placed today than in the past*".

Governability of the kelp system has improved after the establishment of PNMI. Fishers and kelp harvesters' organizations are members of the park council and the PNMI participates in the Kelp Commission. Both the PNMI and the Kelp Commission are sensitive to the economic needs of the harvesters and the conservation of the resource. Kelp harvesters are informed about the findings of studies aimed at increasing knowledge about kelp forests and are aware of management rules discussed by the Kelp Commission. Overall, research is used to improve the governability of the whole system. For example, the results of recent research undertaken by the PNMI on the location of kelp forests and the abundance of biomass in each area contributed to formulation of recent management rules for *L. digitata*. The allocation to each boat of a specific harvesting zone was based on the fact that each boat is equipped with a Vessel Monitoring System (VMS/satellite recording). The mapping of *L. hyperborea* has also been done by the PNMI. The acceptance by kelp harvesters of the closure of some areas is probably based on the fact that they have been informed by park authorities as to the reasons for it. The park authorities also support kelp harvesters better manage the resources. The increased use of scientific knowledge does however mean that local ecological knowledge is playing a lesser role now.

Discussion

Seaweed harvesting in the Iroise Sea has been going on for centuries. From the beginning, the local population and later kelp harvesters regulated harvesting so as to avoid conflicts among themselves and with other users. Under national law, fisheries authorities regulated fisheries activity at first, but this responsibility was passed on to fisheries committees. Fishers' organizations take decisions pertaining to resource management which are endorsed by the state. The management and development of the seaweed industry has been a result of interactions between the two main actors, fishers and processing plants. For more than 50 years, kelp harvesters and the processing industry have collaborated with the objective of improving landings and sustaining the resource. By and large, their collaboration can be viewed as a success. After 50 years of harvesting, the resource is still abundant and not over-exploited.

The processing industry has been the dominant partner. Harvesters must follow the rules set forth by the processing industry if they wish to provide algae to them. The introduction of individual contracts and quotas recently has been the most significant change in the management system. This has contributed to reinforce a trend towards the segmentation of the harvesting sector between bigger boats, allowed to harvest both species of kelp with larger quotas and the smaller boats exclusively harvesting *L. digitata* with smaller quotas. This creates tensions based on the feeling of inequity.

Conservation of the marine ecosystem has also become a priority, leading to the creation of a natural park in the area where kelp is mostly found. The Park Authorities have an important role to play in resource management and in ensuring that the economic needs of harvesters are met. In doing so, it needs to recognize the capacity of fishers and the processing industry, built over decades, if not centuries, of cooperation, in resource governance.

The case of kelp exploitation in Brittany well illustrates how the natural system, the social system and the governance system are changing over time. The natural system-to-be-governed is clearly part of a social construct. Initially concerned with sustaining the exploitation of one species of kelp, the natural system-to-be-governed now includes two key kelp species but also a wide range of biodiversity elements and interactions. Change in the end product of kelp harvest and technical innovation in harvesting and processing have modified, sometimes with a rapid pace, the social system-to-be-governed. The governance system, both formal institutions and interactions, has also changed over time to respond to new needs. Despite doubts and criticism, there is a view among the leaders that the present governance, based on close collaboration between the fishers, the park authority and the industry, has a potential to improve the overall governability of the system. This relates to the facts that there is more management capacity at the local level but there are also more resources to produce knowledge about the socio-ecological system, its components, interactions and dynamics.

Acknowledgement This work benefited from the support of the French government through the National Research Agency with regard to an investment expenditure program, IDEALG (ANR-10-BTBR-04).

References

Alban, F., Kervarec, F., Le Lec, G., Le Floc'h, P., & Boncoeur, J. (2001). *Contraintes socio-économiques de l'activité des navires goémoniers de la région Bretagne, Rapport Final*. Etude réalisée pour le CRPM de Bretagne, Brest, CEDEM, UBO/UEM, 105 p.

Alban, F., Le Floc'h, P., & Boncoeur, J. (2004). The impact of economic and regulatory factors on the relative profitability of fishing boats. A case study of the seaweed harvesting fleet of Northwest Brittany (France). *Aquatic Living Resources, 17*, 185–193.

Alban, F., Frangoudes, K., & Frésard, M. (2011). Kelp harvesting fleet dynamics and the fleet's dependence on the Laminaria forests in the Iroise Sea (North Finistère, France). *Cahiers de Biologie Marine, 52*, 507–516.

Arzel, P. (1987). *Les goémoniers, Le chasse-marée, éditions de l'estran.*

Arzel, P. (1998). Les laminaires sur les côtes bretonnes, éditions IFREMER, Plouzané.

Chuenpagdee, R., & Jentoft, S. (2013). Assessing governability – What's next? In M. Bavinck, R. Chuenpagdee, S. Jentoft, & J. Kooiman (Eds.), *Governability of fisheries and aquaculture: Theory and applications*. Dordrecht/New York: Springer.

Frangoudes, K. (2011). Seaweed fisheries management in France, Japan, Chile and Norway. *Cahiers de Biologie Marine, 52*, 517–525.

Frangoudes, K., & Alban, F. (2004). Fishermen and the creation of Marine Parks: Northern Sporades (Greece), northern cap of Creus (Cataluna), Iroise Sea (France). In J. Bossevain & T. Selwyn (Eds.), *Contesting the foreshore: Tourism, society, and policy on the coast* (MARE publication series, Vol. II). Amsterdam: Amsterdam University Press.

Frangoudes, K., Alban, F., Daurès, F., Roussel, M., Raux, P., Roncin, N., Le Grand, C., & Van Iseghem, S. (2011). Etat de lieux initial sur la perception et l'acceptabilité sociale du PNMI par les pêcheurs professionnels, Université de Brest, IFREMER, UMR Amure, Brest, 11 p.

Garineaud, C. (2012). Des hommes et des algues, techniques et savoirs naturalistes des goémoniers de Nord Finistère, Mémoire de Master 2, Spécialité Environnement, Développemen, Territoires Sociétés, 2011–2012, MNHN –UMR 7206 Eco-anthropologie et Ethno-biologie, INRA UR 1326 Sciences en Société.

IFREMER (coord). (2007). *Small scale coastal fisheries in Europe*, final report of the contract n° FISH/2005/10, 447 pages.

Jentoft, S. (2007). Limits of governability: Institutional implications for fisheries and coastal governance. *Marine Policy, 31*, 360–370.

Kooiman, J. (2008). Governability: A conceptual exploration. *Journal of Comparative Policy Analysis, 10*(2), 171–190.

Kooiman, J., & Bavinck, M. (2013). Theorizing governability–The interactive governance perspective. In M. Bavinck, R. Chuenpagdee, S. Jentoft, & J. Kooiman (Eds.), *Governability of fisheries and aquaculture: Theory and applications*. Dordrecht/New York: Springer.

Kooiman, J., & Chuenpagdee, R. (2005). Governance and governability. In J. Kooiman, M. Bavinck, S. Jentoft, & R. Pullin (Eds.), *Fish for life: Interactive governance for fisheries* (pp. 325–349). Amsterdam: Amsterdam University Press.

Mesnildrey, L., Jacob, C., Frangoudes, K., Reunavot, M., & Lesueur, M. (2012). Seaweed industry in France, Report Interreg Program NETALGAE. Les publications du Pole halieutiques n°9, Agrocampus Ouest, Rennes, 32 p.

Muller, C. R. (1944). *Pêches et Pêcheurs de la Bretagne Atlantique*. Paris: Armand Colin.

Rhodes, R. A. (1996). The new governance: Governing without government. *Political Studies, XLIV*, 652–667.

Steneck, R. S., Graham, M. H., Bourque, B. J., Corbett, D., Erlandson, J. M., Estes, J. A., & Tegneret, M. J. (2002). Kelp forest ecosystems: Biodiversity, stability, resilience and future. *Environmental Conservation, 29*, 436–459.

Valero, M., Destombe, C., Mauger, S., Ribout, C., Engel, C. R., Daguin-Thiébault, C., & Tellier, F. (2011). Using genetic tools for sustainable management of kelps: A literature review and the example of Laminaria digitata. *Cahier de Biologie Marine, 52*, 467–483.

Vasquez, J. (2008). Production, use and fate of Chilean brown seaweed: Resources for sustainable fisheries. *Journal of Applied Phycology, 20*, 457–467.

Chapter 7
The Traditional Small-Scale Octopus Fishery in Portugal: Framing Its Governability

Cristina Pita, João Pereira, Silvia Lourenço, Carlos Sonderblohm, and Graham J. Pierce

Abstract The common octopus (*Octopus vulgaris*) fishery is of substantial importance in southern Europe. This is the case in Portugal where the octopus fishery has considerable social and economic value, with small-scale fishing being increasingly economically dependent on this resource. The octopus fishery in the European Union is excluded from quota regulations under the Common Fisheries Policy, and hence Portugal is responsible for managing its own fishery. This chapter describes the current status of the small-scale octopus fishery in Portugal, its governing system, challenges faced by the fishery and implications of these challenges for the governability of the fishery. The Portuguese octopus fishery faces several challenges including those inherent to the biological features of the species and its

C. Pita (✉)
Centre for Environmental and Marine Studies (CESAM) and Departamento de Biologia, Universidade de Aveiro, 3810-193 Aveiro, Portugal
e-mail: c.pita@abdn.ac.uk; c.pita@ua.pt

J. Pereira
Divisão de Modelação e Gestão de Recursos da Pesca, Instituto Português do Mar e da Atmosfera (IPMA), Portugal, Lisbon, Portugal
e-mail: jpereira@ipma.pt

S. Lourenço
Divisão de Modelação e Gestão de Recursos da Pesca, Instituto Português do Mar e da Atmosfera (IPMA), Portugal, Lisbon, Portugal

MARE – Marine and Environmental Sciences Centre, Faculdade de Ciências, Universidade de Lisboa, Campo Grande, 1749-016 Lisboa, Portugal
e-mail: slourenco@ipma.pt

C. Sonderblohm
Centre of Marine Sciences (CCMAR), University of the Algarve, Campus de Gambelas, Faro, Portugal
e-mail: c.sonderblohm@gmail.com

G.J. Pierce
Centre for Environmental and Marine Studies (CESAM) and Departamento de Biologia, Universidade de Aveiro, 3810-193 Aveiro, Portugal

Oceanlab, University of Aberdeen, Main Street, Newburgh, Aberdeenshire AB41 6AA, UK
e-mail: g.j.pierce@abdn.ac.uk

S. Jentoft, R. Chuenpagdee (eds.), *Interactive Governance for Small-Scale Fisheries*, MARE Publication Series 13, DOI 10.1007/978-3-319-17034-3_7

sensitivity to environmental influences. Limited understanding about the resource has led to inappropriate monitoring and assessment, and a lack of intervention by authorities in the management of octopus. Although formal participation of fishers in the decision-making process for octopus management is a recent phenomenon, the management measures implemented over the years were the result of pressure from fishers worried about decreasing economic returns. However, poor organization, lack of trust and little cooperation between fishers results in them having little influence in the decision-making process. Several recent developments aimed at improving the governance framework and increasing the profitability of the octopus fishery, however, have the potential to improve the current management system and increase sustainability.

Keywords Cephalopods • Europe • Governability • Governance • Octopus • *Octopus vulgaris* • Portugal

Introduction

The common octopus (*O. vulgaris*) is the most important commercially harvested octopus species worldwide. In Europe, the common octopus is fished in both the northeast Atlantic and the Mediterranean Sea, mainly by small-scale fishing fleets using hand-jigs, pots, traps, fyke-nets, trammel nets and bottom trawls (Lefkaditou et al. 2002; Tsangridis et al. 2002; Pierce et al. 2010; Sobrino et al. 2011).

In Portugal, the common octopus has long been an important target species for small-scale fishers. Nowadays, it is an increasingly important fishery resource in terms of quantities landed and particularly in terms of commercial value. The small-scale fishing fleets targeting octopus in Portugal, as well as in other southern European countries, are of considerable socio-economic importance, as the octopus fishery plays a major role in providing employment and income to coastal fishing communities.

The common octopus fishery, like other European Union (EU) cephalopod fisheries, is excluded from quota regulations under the Common Fisheries Policy (CFP) and hence management of this fishery has evolved under the tutelage of national and/or local governments. In Portugal, fisheries management is carried out through a top-down system based on a series of input and output control measures, with little participation from the fishing industry in the decision-making process.

In this chapter we examine the governability of the traditional small-scale common octopus fishery in Portugal using the governability assessment framework, as originally put forward by Kooiman et al. (2005, 2008) and further developed by Bavinck et al. (2013). As such, we describe the natural and socio-economic systems in which the fishery operates, and follow that with a description of the governing system for the common octopus fishery and governing interactions. We then highlight the key challenges these pose to the system-to-be-governed. Finally, we discuss the implications of these challenges to the governability of the octopus fishery and provide some concluding remarks.

Methodology

The data presented in the chapter is based on information elicited through semi-structured questionnaires administered to 18 representatives of fisher associations, Producer Organizations and traders of fishery products, as well as from participant observation in official governmental and non-governmental meetings and workshops about the octopus fishery. Some authors, as members of task-forces organized for the octopus fishery, were also privy to discussions in these committee meetings. Questionnaires were administered between May 2013 and September 2014, in mainland Portugal. They solicited information about interviewees' opinions pertaining to the main challenges and opportunities for the small-scale fishing sector in Portugal, the most appropriate management measures for small-scale fisheries, the best measures to increase the value of the catch, and how best to increase participation in the decision-making process. Data was also collected from official national statistics on landings, fleets and fishers. Figure 7.1 shows the location of the study site.

The System-to-Be-Governed

The Natural System-to-Be-Governed

The Portuguese fishing takes place in a transition area between subtropical and temperate environments in the northern part of the Canary Current upwelling system, one of the four major eastern boundary upwelling systems in the Atlantic (Aristegui et al. 2009). This system is highly dynamic and productive with marked seasonality. Moreover, the system supports a complex food web (Bode et al. 2004) and a particularly rich marine biodiversity (Sousa et al. 2005).

The decrease in landings of finfish in Portugal, since the 1970s, has directed fishers to search for alternative resources (Pereira 1999; Erzini 2005). In the same period, therefore, landings of cephalopods, and mostly common octopus, increased (Fig. 7.2). Since the 1970s, the octopus fishery has been one of the most important fisheries in the country. The yield of common octopus more than doubled, from an average annual catch of 4,000 tonnes for the period 1970–1986 to 8,800 tonnes for the period from 1987 to 2013.

It appears that common octopus catches in Portugal remain at sustainable levels (Baeta et al. 2005) as landings have continued to rise in response to increasing effort. Whether this will continue to be the case in the long-term is unclear. The

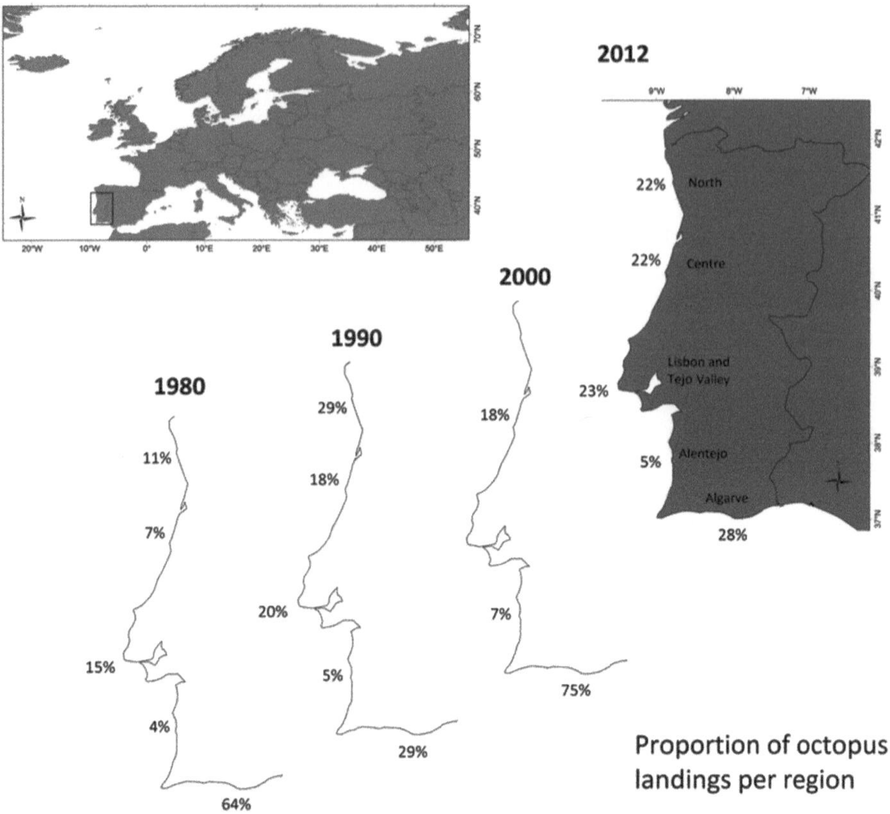

Fig. 7.1 Map of mainland Portugal and proportion of landings (in quantity) of octopus per region over time (Source: Instituto Nacional de Estatística (INE; Portuguese official statistics bureau), and predecessor official statistics bureaus)

upward trend in landings is possibly a consequence of the fact that the octopus "benefits" from the current high level of fishing pressure and from discards of other species. As in the case of other cephalopods, there are huge annual fluctuations in octopus populations (fluctuations varying as much as 40 % a year), implying that abundance varies from year to year. Figure 7.1 shows the landings per region over time, reflecting these fluctuations over time.

Octopuses are terminal breeders, with non-overlapping generations, a rapid non-asymptotic growth (i.e. high individual variability in increases in length or weight), a short lifespan, and a high sensitivity to environmental influences (Alford and Jackson 1993; Sobrino et al. 2002; Otero et al. 2008; Pierce et al. 2010; Sonderblohm et al. 2014). Age estimates for the eastern Atlantic demonstrate that octopuses live for 1–2 years (Domain et al. 2000; Perales-Raya et al. 2014) and annual abundance of the resource depends on the level of survival of the pelagic paralarvae, which in turn is strongly related to environmental conditions, such as

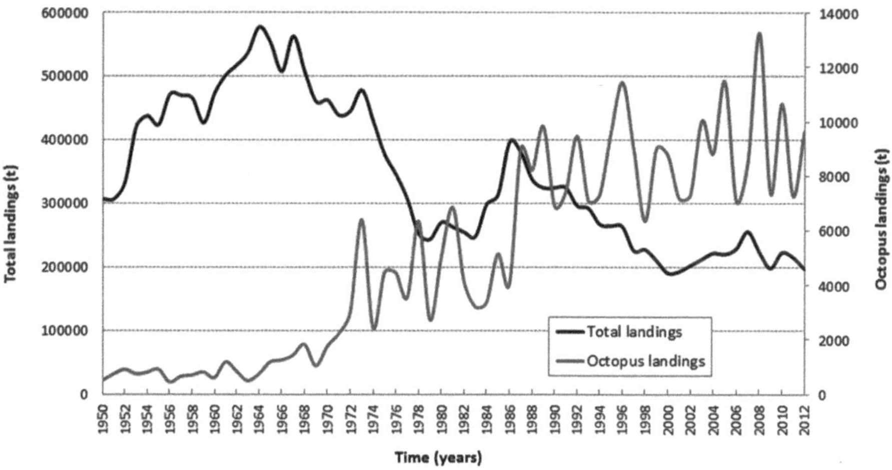

Fig. 7.2 Time series of total landings and octopus landings in Portugal, in quantity (Source: Instituto Nacional de Estatística (INE; Portuguese official statistics bureau), and predecessor official statistics bureaus)

upwelling intensity, temperature and the input of fresh water (Otero et al. 2008; Sonderblohm et al. 2014). Aside from the high fluctuations in abundance due to environmental factors, the fact that octopuses are terminal breeders and have non-overlapping generations also has implications for fisheries. Breeding females subsequently die and are thus no longer available to the fishery, just as every female removed by fisheries will not breed. Thus, if fisheries remove a whole generation of recruits there could be a shortage of adults to repopulate the stock. However, this is mitigated by a long breeding season, and variable growth which ensures that not all adults are present at the same time in the fishery, and by paralarval immigration and repopulation from other areas.

Octopuses are widespread and dwell upon a variety of bottom types (Pereira et al. 1995). Common octopuses are quite sedentary benthic organisms as adults, with their only well-known migration occurring when juveniles leave the pelagic realm and adopt the benthic near-bottom life style (Villanueva and Norman 2008). It is at this final stage of the life cycle that they become vulnerable to fishing. Considering this, disturbances on the sea bed and human activities (including trawling fisheries) can potentially affect the success of octopus recruitment. Moreno et al. (2014) identified eight distinct recruitment grounds for common octopus in Portugal. In some cases these grounds are under intense fishing pressure, both by small-scale fisheries using static gear and by bottom trawling (Pilar-Fonseca et al. 2014), potentially damaging the seabed habitat of the octopus, as well as directly removing the animals.

Current fishery management under the EU-CFP is very much focused on relatively long-lived fish and shellfish. Multispecies assessment and management is already difficult, and this is made worst by the lack of knowledge and experience of dealing with short-lived species with highly variable abundance or indeed with small-scale fishing that operates on a different spatial scale to trawling. The biological features of this species and its sensitivity to environmental influences present particular problems that make management a real challenge.

The Socio-economic System-to-Be-Governed

Portugal is a traditional fishing nation and fishing has long been an economically important activity for many coastal communities. Portuguese fisheries are generally characterized as multi-gear and multi-species small-scale fisheries. Sardine, horse mackerel, chub mackerel and common octopus usually form the largest volume of landings (INE 2014). The small-scale sector is a major component of Portuguese fisheries, accounting for 90 % of all vessels registered (12 % of GT, 40 % of kW) and 68 % of all fishers (INE 2014) (Table 7.1).

Nowadays, the common octopus is one of the most important fishery resources in mainland Portugal in terms of value. In 2013, octopus landings generated 15 % (€37.6 million) of the total official first sale revenue generated by fisheries, second only to sardines (16 %; €39.7 million) (INE 2014). However, octopus catches are frequently unreported (anecdotal information suggests a minimum of 20 % of unreported landings), easily making octopus the most valuable species caught in Portugal. Around 90 % of officially landed common octopus is captured with pots and traps (Moreno et al. 2014) by the "local fleet" (comprised of small-size boats, less than 9 m in length) and the "coastal fleet" (comprised of vessels generally ranging from 9 to 15 m in length). These two fleets are of the utmost importance for a large proportion of fishing communities.

Table 7.1 Socio-economic indicators for the Portuguese fishery (data for 2013)

Socio-economic indicators	Total fisheries	Small-scale fisheries (percent of total)
Landings, quantity (thousand ton)	144.7	60.5 (42 %)
Landings, value (million €)	253.2	167.1 (66 %)
Fleet, number of boats	8,232	7,409 (90 %)[a]
Fleet, tonnage (GT)	99,917	12,241 (12 %)[a]
Fleet, power (kW)	366,279	147,443 (40 %)[a]
Number fishers	16,797	11,481 (68 %)[b]
National per capita fish consumption (kg/person/year)[c]	56.7	–

Data Source: INE (2014)
Notes:
[a]Refers to vessels employing static gear <12 m in total length
[b]Fishers' employed in the local and coastal multi-gear (polyvalent) fleet
[c]Data for 2010 (Data source: EC 2014)

The fishing of octopus dates quite far back. It is a traditional fishery, primarily in the south (Algarve region), where octopus has been caught (and even exported) since, at least, the fifteenth century (Godinho 1963). It is still the main species landed in the region (INE 2014), and the local small-scale fishing industry is highly dependent on octopus. The commercial fishery of octopus was also originally mainly based in the Algarve region and used clay pots of various types and shapes, possibly derived from the ancient amphora. These were later replaced by dedicated pots which were hand-hauled, necessarily slow to handle and limited in capacity, and which resulted in a low catch rate and low annual landings. Official statistics show that average annual landings for the period 1927–1972 were approximately 1,025 tonnes (+/−50 %). Between 1973 and 1986 the fishery was modified by the introduction of mechanical hauling devices (winches), which resulted in a sharp increase in effort and a fourfold increase in landings. The fishery also gradually expanded northwards and by 1992 octopus was being caught all over the Portuguese coast. To add to this, at the same time, the main gear in use shifted to baited wire mesh traps, resulting in the technological capacity multiplying. Effort is also believed to have increased substantially all along the coast while the proportion of unreported landings skyrocketed. Figure 7.3 shows a timeline analysis for the octopus fishery in Portugal.

Nowadays, the octopus fishery supports small-scale fishing communities all over the Portuguese coast as it is a source of guaranteed income for most small-scale fishers. This increased economic dependence of coastal communities on the octopus fishery is not exclusive to fishers operating creels and pots; the trawling fleet has also increased its economic dependence on octopus (Pilar-Fonseca et al. 2014).

Octopus catches, as most other catches, are required to be landed for first sale at DOCAPESCA S.A., a state-owned company (under the Ministry of Agriculture and

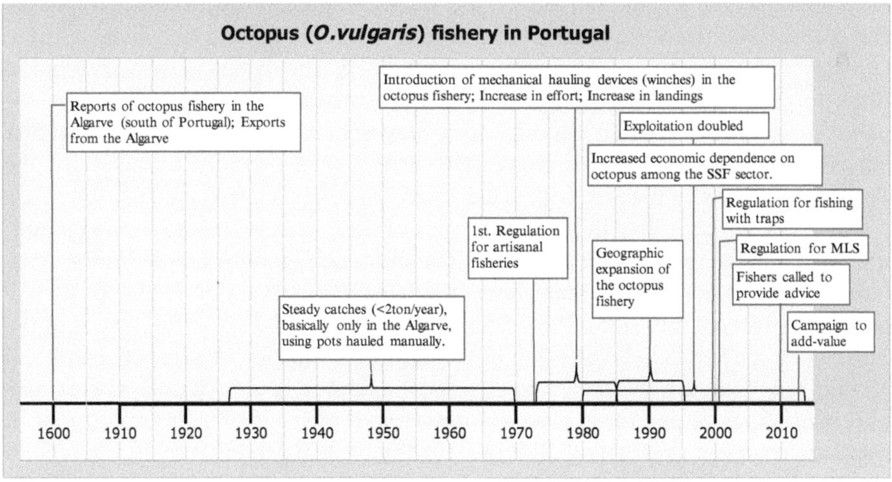

Fig. 7.3 Timeline analysis of the common octopus fishery in Portugal

the Sea) responsible for organizing the first sale of fish and managing fish ports. Fishery products are sold using a descending-bid type of auction, also known as the Dutch auction system. Octopus is sorted by size class – a system of four sizes, from T1 (for "Tamanho", i.e. size) to T4 – in order to set the initial price at auction. Weight and initial price decrease from T1 to T4 (bounded by the legal minimum landing weight of 750 g). The weight range within each category has been loosely defined by DOCAPESCA according to local fishers' perceptions of "large" and "small" at different ports, thus eluding standardization for a number of years. Most interviewees were unsatisfied with the current auction model, accusing a few retailers of manipulating and significantly lowering auction prices. Fluvia et al. (2012) point to the fact that the Dutch auction system may not be the most appropriate mechanism for fishers to extract a high revenue from their catch. Only around half of the octopus caught in Portugal is ultimately sold in the Portuguese market. The largest buyer of octopus in Portugal estimates that it purchases around 80 % of all octopus sold in auction, and exports half of its product wholesale to Spain and Italy, where octopus is then processed. Many other buyers at first auction also export their product, for example to Spain; several buyers in the Algarve region are Spanish.

The socio-economic systems in which the Portuguese common octopus small-scale fishery operates give rise to challenges for the governability of this fishery, in particular related to illegal effort deployment, landings of undersized octopus, and potential problems of economic viability of the fishery (at least temporarily).

The Governing System

Under the EU-CFP there is still no regulation for cephalopod fisheries, meaning that octopus fishery management in Portugal derives exclusively from government specific and general legislation, under the direct responsibility of the Directorate General for Natural Resources and Maritime Services and Safety, which, in turn, belongs to the Ministry of Agriculture and the Sea. The technical measures put in place to manage octopus generally derive from research advice provided by the national fisheries research institute, local governmental counterparts and higher education research institutions. Since 2010 fishers have also been called on, by the Secretary of State for the Sea, to provide advice on octopus management (Table 7.2).

Management measures in place consist essentially of regulations defining a minimum landing weight and the gear used. The minimum landing weight for the common octopus is 750 g. The legislation stipulates a maximum of 3,000 non-baited pots per vessel (of any size). Baited trap limits vary according to the length of the boat: 750 traps per vessel under 9 m in length, 1,000 traps for vessels between 9 and 12 m in length, and 1,250 traps for vessels over 12 m in length. The legislation also puts restrictions on the mesh size for traps, trap dimensions for different types of traps, and the mesh size for trawl fisheries. Finally, the legislation sets spatial-temporal constraints on the fishery, by setting a minimum distance from shore at which the gear can be deployed (e.g. 0.5 nm for vessels below 9 m in length using pots and

Table 7.2 Governance of the common octopus fishery in Portugal

Octopus (*O. vulgaris*) fishery governance	
Governance mode	Hierarchical
Key management institutions and organizations	Ministry of Agriculture and the Sea; Directorate General for Marine Resources and Maritime Services and Safety
Main stakeholders involved in the decision-making process	Management bodies (above), fisheries research institute (Portuguese Institute of the Sea and Atmosphere), higher education research institutions
Other stakeholders	Higher education research institutions, local governmental, fishermen associations
Main management measures	Minimum landing size of 750 g for *O. vulgaris (*main measure); Gear restrictions
Main legislation	Minimum landing size for octopus (Portaria n° 27/2001). Regulation for fisheries using traps (Decree Law 43/87 and Ordinance 1102-D/2000, emended by Ordinance 447/2009, altered by Ordinances 774/2009, 193/2010, 1054/2010, 132/2011, 97-A/2012 and 230/2012); Interdiction of using live bait in traps (Ordinance 230/2012)
Enforcement and control	Maritime police and harbour authorities, under the responsibility of the Directorate of Maritime Authority and the National Republican Guard; Armed forces (navy and air force)

1 nm for vessels over 9 m in length using pots and/or traps, although this varies according to season). Regional rules have also been put in place, for example the prohibition of using live bait (Common green crab, *Carcinus maenas*) in the south coast fishing grounds (Algarve region).

Control and law enforcement are carried out by the maritime police and harbor authorities, while the armed forces (navy and air force) are also involved in monitoring and control at sea. The effectiveness of the monitoring and control system at sea is very limited, due mostly to lack of human resources to patrol large areas. This means that the number of traps deployed is in practice under no control. There is also limited enforcement on land, reducing the efficacy of the minimum landing weight legislation. According to fishers, compliance with rules and regulations is minimal, especially at times of economic difficulties.

The excessive amount of static gear in the water is a problem and leads to, or potentially increases, conflicts among fishers, raising questions of social justice. The accumulated investment in fishing gear is reportedly enormous and tends to be increasingly unbalanced, as some fishers continuously invest in gear (mostly plastic pots due to their low cost) and deploy all this gear in the water to secure fishing areas. Some static gear users accuse others of occupying all of the fishing grounds, and anecdotal reports have emerged of commercial exchange of gear in the water in order to secure access to certain fishing grounds. Simultaneously, encroaching on mobile gear areas results in accusations of foul play.

Monitoring of fisheries in Portugal is carried out through satellite tracking devices (vessel monitoring systems), fishing logbooks, dock-side monitoring and digital record keeping at auctions. There are also research programs (national fish-auction sampling programs and stock assessment research cruises) to provide further support for management advice and guide legislation. The current fisheries monitoring system and stock assessment practices in Portugal have long been identified as inadequate for the management of octopus as the data are collected so as to meet the requirements of finfish assessments and not those of short-lived resources mostly caught by static gear (see Pereira 1999). There is also the issue of lack of data, including socioeconomic data, on the small-scale octopus fishery, and indeed on small-scale fisheries in general.

In Portugal, small-scale fishers participate in management initiatives through a large number of small (and generally geographically arranged) fisher associations. Formal participation by small-scale fishers in the decision-making process is a recent innovation and is still rare. However, fisher participation in management appears to be developing rapidly.

The high dependence on octopus by the Portuguese small-scale fleet and the adverse socio-economic impacts in years when catches are low have been the main reasons that fishers have pressurized government bodies, leading to changes in legislation, and to increased interest by the industry to participate in the decision-making process. This was the case in 1996, when representatives of small-scale fishers approached the fishery management body and the national fisheries research institute with concerns about increasing effort and a potential future decrease in catches. They were of the opinion that the increase in effort over the previous years was putting the stock at risk of overexploitation and requested protective measures for the octopus (Pereira 1999). This resulted in the implementation of new legislation setting a minimum landing weight for octopus (Ordinance 27/2001). While fishers' concerns were taken into account, fishers were not formally involved in the decision-making process. A "bad" fishing year in 2010 was again the reason for fishers pressurizing the government. Low catches in 2009, after a particularly good fishing year (2008), resulted in some fishers in the Algarve region accusing others of employing methods (namely the use of live bait in traps) that they considered to be responsible for the decline in octopus abundance. The affected fishers demanded a change in legislation and one was introduced, forbidding the use of live bait in traps. The legislation was, however, subsequently retracted only to be reintroduced once again in the Algarve (Ordinance 230/2012). In this instance fisher associations were involved in an *ad-hoc* expert-group convened by the Secretary of State for Fisheries. Their advice was considered alongside the scientific advice. However, after two task forces and four pieces of legislation over a period of 4 years (2009–2012), the issue of employing live bait in traps remains controversial (Nov. 2014), and a source of much debate and disagreement between associations in the Algarve region.

Recently, several measures have been put in place that aim at improving the long-term sustainability and profitability of the octopus fishery. One such measure was a campaign by DOCAPESCA S.A. to promote the consumption of Portuguese fresh octopus, develop new recipes and uses for octopus (e.g. ready-to-eat meals) and, as such, increase the visibility and presence of Portuguese octopus in the

Portuguese market so as to increase the economic value of the species. Another initiative of interest was DOCAPESCA's decision to hand over the running and management of several first sale auction sites ("lotas") to fisher associations. Octopus is the main species sold in these "lotas" in the Algarve region. The lotas now managed by fisher associations were previously amongst the least profitable, due mostly to illegal and unreported landings. It is believed that management by fisher associations will lead to increased compliance and added revenue for small-scale fishers.

Further measures are also being explored, such as an initiative to implement an adaptive management system, referred to as "responsive fisheries management system" (developed through the EU-funded EcoFishMan project; www.ecofishman.com), for the octopus fishery in the Algarve. This initiative, developed by the Centre of Marine Sciences (CCMAR) of the University of the Algarve, consists of a bottom-up co-management system based on the optimization of results. Over its first year monthly workshops, with representatives from fishing authorities, fishing associations and researchers, to propose and discuss management policies, were organized. The initiative, still in its early stages, has been well-received by management bodies and fishers alike. The fishery management authority is also exploring the possibility of establishing exclusive area-based concessions for the exploitation of octopus; in other words a form of Territorial Use Rights in Fisheries (TURF), a system already successfully implemented to achieve sustainable management of benthic resources, including for the management of octopus in Chile and Madagascar (see. Martín et al. 2010; Gelcich et al. 2012; Raberinary and Benbow 2012).

Although all the initiatives described above are still at early stages of development and hence not ready for performance evaluation, representatives from fisher associations tended to see these measures as a step in the right direction and as having the potential to increase the profitability and economic viability of small-scale fisheries. Still, major challenges remain for the governability of the octopus small-scale fishery, related to the governing system in place. These include lack of a viable level of monitoring and assessment, poor control and enforcement, lack of respect for rules and regulations by fishers and lack of trust of fishers in management bodies. In addition, the low levels of organization, trust and cooperation between fishers themselves result in them having a limited influence on the decision-making process.

Governing Interactions

The management of the Portuguese octopus small-scale fishery is undertaken through a top-down system and, in general, formal participation by small-scale fishers in the decision-making process is still scarce. Nevertheless, based on the interviews conducted, most new measures put in place so far have emerged from fishers' demands. Moreover, although formal fisher participation in the decision-making process only started in 2010, participation has increased significantly.

Governmental fisheries bodies, research institutions and small-scale fisher associations are attempting greater and more effective involvement and participation of fishers in the management of the octopus fishery. However, a general lack of internal consensus amongst fisher associations means that they lack a single voice and therefore still have little influence. Nevertheless, participation is increasing with more and more fishers attending meetings and workshops, and showing an interest and willingness to participate in management. Increased participation of fisher associations has, sometimes, resulted in increased expectations from fishers, some of which are not very realistic and are unlikely to be achievable (such as the establishment of subsidized closed seasons and areas).

In short, governing interactions are changing in Portugal, with increasing attempts by the small-scale octopus fishing sector to participate in management, as a result of the socio-economic turmoil the sector has been facing and due to a growing socio-economic dependence on the octopus fishery. Authorities are increasingly discussing concerns with fishing communities and appear to be open to fishers' opinions and proposals. There is no doubt that this openness to increased participation presents an opportunity to improve the governability of the fishery, since both monitoring and control are more effective with the active support and participation of fishers. However, the structure and procedures needed to achieve a fully participatory approach remain to be determined. In addition, there remains a concern that a lack of a broad-scale vision, one which considers the octopus fishery in the context of existing socioeconomic conditions, biological sustainability considerations and the evolution of the marine and maritime sectors in Portugal and the EU, may threaten the sustainability of the resource.

Discussion and Conclusions

Any attempt to improve governance inevitably requires an assessment of the system's governability and this, in turn, requires the understanding of the system's basic qualities (Bavinck et al. 2013). The governability assessment framework provided a good basis for defining and assessing a data poor fishery such as the small-scale octopus fishery in Portugal.

The Portuguese small-scale octopus fishery is faced with many challenges which need to be taken into account for successful management and governance. Particularly important challenges for management relate to the biology of the species itself, the lack of a viable level of monitoring and assessment, and a general lack of understanding about the resource (i.e. about octopus as a species) and knowledge about the fishery by managers. These current limitations have resulted in the inappropriate management of the resource.

Under the EU-CFP there is still no assessment program for cephalopod fisheries (ICES 2013) and no routine assessment of *O. vulgaris* stocks in EU waters. In addition, there is a general lack of detailed data collection on cephalopod fisheries, something that Pierce et al. (2010) identified as a major impediment to the routine assessment of southern European cephalopod stocks. Biological data collection for the common octopus in Portugal is done in a way that meets the requirements of finfish assessments but is not suitable for assessing octopus and, as such, existing datasets are inadequate to support the assessment and management of octopus (Pereira 1999). The biological features of the species (e.g. short lifespan, terminal breeders) make it simultaneously difficult to assess abundance and regulate catch levels. To add to this, the sensitivity of octopuses to environmental influences (such as variations in temperature and salinity/rainfall), and the resulting natural inter-annual variability of abundance, presents further problems for management.

Nonetheless, new assessment approaches are currently being refined and trialed for EU cephalopod stocks (see ICES 2014), including the use of production models which incorporate environmental effects, and population biomass models (Gras et al. 2014). Although wide fluctuations in abundance occur (typically averaging around 40 % a year), studies in Galicia (Spain) suggest that abundance of the upcoming cohort is potentially highly predictable based on knowledge of environmental conditions (Otero et al. 2008). These are promising new developments for a better assessment and management of octopus stocks. However, even if future abundance can be to some extent predictable, natural abundance variations require fishers to be adaptable, for example to switch to other target species in years of low abundance. However, the dominant rhetoric within the fishing sector is that octopus will continue to provide increasing volumes of landings at a high price. Not surprisingly, therefore, coastal communities are content to specialize in octopus fisheries.

The nature of the small-scale octopus fishery, including the high social and economic dependence of fishers on octopus, and the aforementioned lack of routine monitoring and assessment, make it imperative to involve fishers in the implementation of effective monitoring solutions on board vessels and in monitoring landings. In addition, the involvement of fishers in decision-making and management is probably the only way to increase compliance with rules and regulations. A vast body of literature exists on the advantages of involving stakeholders, mainly fishers, in the decision-making process (Pita et al. 2012). All this literature suggests that stakeholder involvement facilitates common understanding, contributes to establishing trust, increases stakeholders' responsibility and accountability, enhances the legitimacy and acceptance of management policies and decisions, and increases the likelihood of compliance (e.g. Jentoft and McCay 1995; Coffey 2005; Marshall 2007; Berghofer et al. 2008; Pita et al. 2010) thus improving governance.

The octopus fishery was systematically ignored by management bodies in Portugal for a long time. However, this trend seems to be changing and there appears to be openness on the part of management bodies to support new management initiatives for small-scale octopus fishing. This provides an excellent opportunity for the development of new management frameworks. Several initiatives to implement new marketing strategies to increase the added-value of catches and facilitate co-management are already being explored. These initiatives can be particularly promising for the management of octopus fisheries by small-scale fishing communities. Moreover, the new opportunities created with recent marketing strategies, and the development of mechanisms which put fishers in direct contact with the market, can result in younger generations once again being attracted to fishing. All these new developments have the potential to increase empowerment of the fishing community and their sense of ownership of resources, as well as to enhance stewardship.

The future of the octopus fishery in Portugal depends on more appropriate stock assessment and monitoring as well as on the successful implementation of management measures in cooperation with the fishing industry. Such measures would help reduce fishing effort (in particular the deployment of excessive numbers of pots), improve compliance with rules and regulations and increase the added-value of the catch. It is important to note that all the changes to octopus fishery legislation so far have resulted from pressure from the small-scale fishing sector. Small-scale fishers are increasingly more organized and interested in taking part in the decision-making process. This willingness to participate, together with the increasing openness from management bodies to fishers' participation, constitutes the minimal conditions for governance interactions (Bavinck et al. 2013). The empowerment of small-scale fishers and active participation of the fishing community in the management of the fishery is essential as it leads to an increased sense of ownership and thus compliance with rules and regulations. A shift to co-management, a requirement of the newly reformed CFP, could be the best, and indeed the only effective, way to achieve long-term sustainability for the octopus fishing fleet. However, there is still a way to go to move forward co-governance arrangements in the small-scale Portuguese fisheries. Improving communication channels between authorities, industry and fishers (and indeed within the small-scale fishing sector) is extremely important. A simpler and clearer framework for participation at local/regional levels and technical assistance programs to aid fishing associations could be key for empowering coastal communities to face the upcoming challenges that the recently reformed CFP will bring, such as the landing obligation, the decentralization of governance, the empowerment of the fishing sector, and the implementation of differentiated management arrangements for small-scale fisheries.

Acknowledgements CP was supported by the "Cátedra do Mar" UA-CGD grant "Sustainability of small-scale fisheries: Alternative approaches to fisheries management and governance". SL was supported by a grant under the GesPe project (PROMAR -31-03-01-FEP-0017). GJP thanks Caixa Geral de Depósitos (Portugal) and the University of Aveiro for financial support. The fishery statistics reported here are from the INE and DGPA official statistics. Part of the information used in this book chapter resulted from interviews carried out as part of the "Cátedra do Mar" grant. The authors also wish to thank all the representatives from Fishermen's Associations, Producers' Organizations and companies trading in fishery products who took part in the survey.

References

Alford, R. A., & Jackson, G. D. (1993). Do cephalopods and larvae of other taxa growth asymptotically? *American Naturalist, 141*, 717–728.

Aristegui, J., Barton, E. D., Álvarez-Salgado, X. A., Santos, A. M. P., Figueiras, F. G., Kifani, S., Hernández-Leon, S., Mason, E., Machú, E., & Demarcq, H. (2009). Sub-regional ecosystem variability in the Canary Current upwelling. *Progress in Oceanography, 83*, 33–48.

Baeta, F., Pinheiro, A., Corte-Real, M., Costa, J. L., de Almeida, P. R., Cabral, H., & Costa, M. J. (2005). Are the fisheries in the Tagus estuary sustainable? *Fisheries Research, 76*, 243–251.

Bavinck, M., Chuenpagdee, R., Jentoft, S., & Kooiman, J. (Eds.). (2013). *Governability of fisheries and aquaculture: Theory and applications* (MARE publication series, Vol. 7). Dordrecht: Springer.

Berghofer, A., Wittmer, H., & Rauschmayer, F. (2008). Stakeholder participation in ecosystem-based approaches to fisheries management: A synthesis from European research projects. *Marine Policy, 32*(2), 243–253.

Bode, A., Álvarez-Ossorio, M. T., Carrera, P., & Lorenzo, J. (2004). Reconstruction of trophic pathways between plankton and the north Iberian Sardine (Sardina pilchardus) using stable isotopes. *Scientia Marina, 68*(1), 168–178.

Coffey, C. (2005). What role for public participation in fisheries governance? In T. S. Gray (Ed.), *Participation in fisheries governance. Reviews: Methods and technologies in fish biology and fisheries* (pp. 27–44). Dordrecht: Springer.

Domain, F., Jouffre, D., & Caverivière, A. (2000). Growth of Octopus vulgaris from tagging in Senegalese waters. *Journal of the Marine Biological Association of the United Kingdom, 80*, 699–705.

EC. (2014). *Facts and figures on the common fisheries policy: Basic statistical data* (2014th ed., 44p). Luxembourg: European Commission.

Erzini, K. (2005). Trends in NE Atlantic landings (southern Portugal): Identifying the relative importance of fisheries and environmental variables. *Fisheries Oceanography, 14*, 195–209.

Fluvia, M., Garriga, A., Rigall, R., Rodríguez-Carámbula, E., & Saló, S. (2012). Buyer and seller behavior in fish markets organized as Dutch auctions: Evidence from a wholesale fish market in Southern Europe. *Fisheries Research, 127–128*, 18–25.

Gelcich, S., Fernández, M., Godoy, N., Canepa, A., Prado, L., & Castilla, J. C. (2012). Territorial user rights for fisheries as ancillary instruments for marine coastal conservation in Chile. *Conservation Biology, 26*(6), 1005–1015.

Godinho, V. M. (1963). *Os descobrimentos e a economia mundial* (Vol. 2). Lisboa: Editora Arcádia.

Gras, M., Roel, B. A., Coppin, F., Foucher, E., & Robin, J. P. (2014). A two-stage biomass model to assess the English Channel cuttlefish (Sepia officinalis L.) stock. *ICES Journal of Marine Science*. doi:10.1093/icesjms/fsu081.

ICES. (2013, October 8–9). *Report of the workshop on the necessity for Crangon and Cephalopod management (WKCCM)*, Copenhagen, Denmark. ICES CM 2013/ACOM:82. Copenhagen: International Council for the Exploration of the Sea.

ICES. (2014, June 16–19). *Report of the working group on Cephalopod fisheries and life history (WGCEPH)*, Lisbon, Portugal (in press).

INE. (2014). *Estatistica da Pesca 2013* [Fisheries statistics 2013]. Lisboa: Instituto Nacional de Estatistica.

Jentoft, S., & McCay, B. (1995). User participation in fisheries management: Lessons drawn from international experiences. *Marine Policy, 19*(3), 227–246.

Kooiman, J., Bavinck, M., Jentoft, S., & Pullin, R. (Eds.). (2005). *Fish for life: Interactive governance for fisheries*. Amsterdam: Amsterdam University Press.

Kooiman, J., Bavinck, M., Chuenpagdee, R., Mahon, R., & Pullin, R. (2008). Interactive governance and governability: An introduction. *The Journal of Transdisciplinary Environmental Studies, 7*(1), 1–11.

Lefkaditou, E., Kavadas, S., & Papaconstantinou, C. (2002). *Cephalopod fisheries statistics in Greek waters (NE Mediterranean)*. ICES Living Resources Committee, ICES CM 1999/G: 4, Working group on Cephalopod fisheries and life history, Lisbon, Portugal, 4–6 December 2002. Copenhagen: International Council for the Exploration of the Sea.

Marshall, N. A. (2007). Can policy perception influence social resilience to policy change? *Fisheries Research, 86*(2–3), 216–227.

Martín, G., Parma, A., & Orensanz, J. (2010). The Chilean experience with territorial use rights in fisheries. In R. Q. Grafton, R. Hilborn, D. Squires, M. Tait, & M. Williams (Eds.), *Handbook of marine fisheries conservation and management* (pp. 324–337). Oxford: University Press.

Moreno, A., Lourenço, S., Pereira, J., Gaspar, M., Cabral, H., Pierce, G., & Santos, A. (2014). Essential habitats for pre-recruit Octopus vulgaris along the Portuguese coast. *Fisheries Research, 152*, 74–85.

Otero, J., Álvarez-Salgado, X., González, A., Mirand, A., Groom, S., Cabanas, J., Casas, G., Wheatley, B., & Guerra, A. (2008). Bottom-up control of common octopus (Octopus vulgaris) in the Galician upwelling system, northeast Atlantic Ocean. *Marine Ecology Progress Series, 362*, 181–192.

Perales-Raya, C., Jurado-Ruzafa, A., Bartolomé, A., Duque, V., Carrasco, M. N., & Fraile-Nuez, E. (2014). Age of spent Octopus vulgaris and stress mark analysis using beaks of wild individuals. *Hydrobiologia, 725*, 105–114.

Pereira, J. (1999). Control of the Portuguese artisanal octopus fishery. In C. Nolan (Ed.), *Proceedings of the international conference on integrated fisheries monitoring* (pp. 369–378). Rome: FAO.

Pereira, J., Cunha, M. M., & Moreno, A. (1995). Os cefalópodes e o seu habitat: substratos preferenciais. In J. M. G. Matoso (Ed.), *8° Congresso do Algarve* (pp. 829–837). Vilamoura.

Pierce, G. J., Allcock, L., Bruno, I., Bustamante, P., Gonzále, Á., Guerra, Á., Jereb, P., Lefkaditou, E., Malham, S., Moreno, A., Pereira, J., Piatkowski, U., Rasero, M., Sánchez, P., Santos, M. B., Santurtún, M., Seixas, S., Sobrino, I., & Villanueva, R. (Eds.). (2010). *Cephalopod biology and fisheries in Europe* (ICES cooperative research report no. 303). Copenhagen: International Council for the Exploration of the Sea.

Pilar-Fonseca, T., Campos, A., Pereira, J., Moreno, A., Lourenço, S., & Afonso-Dias, M. (2014). Integration of fishery-dependent data sources in support of octopus spatial management. *Marine Policy, 45*, 69–75.

Pita, C., Pierce, G. J., & Theodossiou, I. (2010). Stakeholders' participation in the fisheries management decision-making process: Fishers' perceptions of participation. *Marine Policy, 34*, 1093–1102.

Pita, C., Chuenpagdee, R., & Pierce, G. J. (2012). Participatory issues in fisheries governance in Europe. *Management of Environmental Quality: An International Journal, 23*(4), 336–347.

Raberinary, D., & Benbow, S. (2012). The reproductive cycle of Octopus cyanea in southwest Madagascar and implications for fisheries management. *Fisheries Research, 125–126*, 190–197.

Sobrino, I., Silva, L., Bellido, J. M., & Ramos, F. (2002). Rainfall, river discharge and sea temperature as factors affecting abundance of two coastal benthic cephalopod species in the Gulf of Cadiz (SW Spain). *Bulletin of Marine Science, 71*(2), 851–865.

Sobrino, I., Juarez, A., Rey, J., Romero, Z., & Baro, J. (2011). Description of the clay pot fishery in the Gulf of Cadiz (SW Spain) for Octopus vulgaris: Selectivity and exploitation pattern. *Fisheries Research, 108*, 283–290.

Sonderblohm, C., Pereira, J., & Erzini, K. (2014). Environmental and fishery-driven dynamics of the common octopus (Octopus vulgaris) based on time-series analyses from leeward Algarve, southern Portugal. *ICES Journal of Marine Science*. doi:10.1093/icesjms/fst189.

Sousa, P., Azevedo, M., & Gomes, M. C. (2005). Demersal assemblages off Portugal: Mapping, seasonal, and temporal patterns. *Fisheries Research, 75*, 120–137.

Tsangridis, A., Sánchez, P., & Ioannidou, D. (2002). Exploitation patterns of Octopus vulgaris in two Mediterranean areas. *Scientia Marina, 66*(1), 59–68.

Villanueva, R., & Norman, M. D. (2008). Biology of the planktonic stages of benthic octopuses. *Oceanography and Marine Biology – An Annual Review, 46*, 105–202.

Part III
Governing System Complexity – Aligning Modes

Introduction

If the small-scale fisheries system-to-be-governed is diverse and complex, it is likely that the governing system will be the same. This is what interactive governance theory refers to as the "goodness of fit." As with part II, this part focuses on experiences from different regions, in this case Africa, the Caribbean and Latin America. *Chapter 8*, by Mafaniso Hara, Steve Donda and Friday Njaya about the governance of Malawi's small-scale fisheries, describes the experience of a number of governing arrangements introduced in several of the countries' lakes since the 1990s. The authors analyze the varying management outcomes under the three existing modes of governance – hierarchical, co-governance and self-governance. In *Chap. 9*, María José Barragán Paladines compares the small-scale fisheries governing system of mainland Ecuador and the Galapagos Islands. She describes a situation where lack of trust, leadership and cohesion among small-scale fisheries entities are among several factors that reduce organizational capacity and hence limit governability. The chapter concludes that hierarchical and co-governance modes for small-scale fisheries have underperformed. She believes nonetheless that both models have potential to strengthen the sector's governability.

Back on the African continent, Paul O. Onyango in *Chap. 10* compares governing modes in the Lake Victoria small-scale fisheries of Tanzania. His chapter aims to broaden understanding of governance by focusing on interactions among governors within the wider parameter of good governance variables. Traditionally this fishery was largely self-governed at the community level, where local chiefs played an important role, until the government introduced a co-governance system to deal with the poor resource situation, partly induced by the introduction of new and alien species to the lake, but also due to heavy fishing pressure. He argues that in order to achieve high governability when introducing formal co-governance institutions, governing systems must allow for the integration of customary practices and local knowledge.

The next two chapters are situated in the Western hemisphere. *Chapter 11* by Elena M. Finkbeiner, Adam L. Ayers, John N. Kittinger and Larry B. Crowder focuses on governability of small-scale fisheries in Baja California Sur, Mexico and the Hawaiian Islands. The authors compare the governing systems of the two regions, both of which have a history of collective action and self-governance. In recent years these traditions have eroded and the system has become more open access, thereby causing governability problems. Their chapter provides lessons about why this has occurred through a comparative analysis of similarities across small-scale fisheries in developed and developing countries. They argue that transition towards co-governance must be thought of as a continuously evolving non-linear process, requiring patience by the parties who must also be prepared for unintended consequences and even failures. In *Chap. 12*, Iris Monnereau and Patrick McConney undertake a similar, cross-country comparison, of lobster fishery governance in different countries of the Wider Caribbean, namely Nicaragua, Belize and Jamaica. In all three cases, the species and the market are the same. The lobster fisheries were also developed roughly at the same period of time. Yet the governing modes display very different outcomes and implications for small-scale fisheries. The choice of modes, they argue, reflects the different political and institutional histories of the three countries and the interactive linkages that have developed between the governing system and the system-to-be governed in the three cases. This, they argue, is also where governance reform must occur in order to improve governability and the wellbeing of small-scale fishers, and by extension that of society at large.

Chapter 8
Lessons from Existing Modes of Governance in Malawi's Small-Scale Fisheries

Mafaniso Hara, Steve Donda, and Friday Njaya

Abstract The small-scale sector lands over 95 % of Malawi's fish catch and provides income and employment to about 65,000 fishers and over 500,000 people engaged in pre and post-harvest activities. The sector contributes 4 % to GDP and fish is the cheapest source of animal protein in the country. Annual production for the commercially important Chambo (*Oreochromis species*) has plummeted from 8,000 to less than 1,000 tonnes annually since the 1980s although recent years have witnessed a 60–70 % increase in catch mainly of low value usipa. Most of the management problems in Malawi can be traced to problems of governance. Fisheries management still remains largely centralized under the Department of Fisheries. A number of co-management arrangements had been introduced in the 1990s such as in Lakes Malombe, Chilwa and parts of Lake Malawi with mixed results. Examples of organic community based management also exist such as on Lake Chiuta, which appear to be yielding very positive results from fishers' perspective. This chapter attempts to analyze the varying management outcomes under the three existing modes of governance (hierarchical, co-governance and self-governance) using the interactive governance framework's three components – governing system, system-to-be-governed and governing interactions. Such a critical analysis will contribute towards finding possible solutions to current management failures in Malawi fisheries and other small-scale fisheries with similar characteristics.

Keywords Malawi • Chambo collapse • Governance modes • Governing interactions • Lessons • Small-scale fisheries

M. Hara (✉)
Institute for Poverty, Land and Agrarian Studies, University of the Western Cape, Cape Town, South Africa
e-mail: mhara@plaas.org.za

S. Donda • F. Njaya
Department of Fisheries, P.O. Box 593, Lilongwe, Malawi
e-mail: stevedonda@gmail.com; fnjaya@gmail.com

S. Jentoft, R. Chuenpagdee (eds.), *Interactive Governance for Small-Scale Fisheries*, MARE Publication Series 13, DOI 10.1007/978-3-319-17034-3_8

Introduction

With 20 % of Malawi's surface area covered by water, fisheries are an important contributor to livelihoods of the rural population and economy of the country. The major fishing areas include Lake Malawi, Lake Chilwa and other smaller water bodies such as Lake Malombe and Chiuta and the Lower Shire River system. Capture fisheries directly employ nearly 65,000 fishers and over 500,000 people indirectly in fish processing, fish marketing, boat building and other ancillary activities (Department of Fisheries 2013a). Except for a small commercial sector made up of pair trawlers and larger stern trawlers, Malawi's capture fishery is largely small-scale in nature. The small-scale sector lands over 95 % of the catch and employs over 95 % of those involved in fishing activities. The industry is estimated to contribute about 4 % to the Gross Domestic Product. Although fish contributed 70 % of the animal protein in the 1970s and 1980s, this has declined to around 40 % as a result of declining catches against population growth. Thus the per capita fish supply also fell from 14 kg in the 1970s to about 5.6 kg per person per year by 2011, a 60 % decline. Because of the importance of small-scale fisheries in Malawi, there are concerns about the implications of the continuing dissipation of fisheries rent for the livelihoods of dependent communities, ancillary industries and post-harvest value chain actors resulting from poor or dysfunctional governance.

This chapter looks at the governability of small-scale fisheries in Malawi. According to Kooiman (2008, 3) and Kooiman and Bavinck (2013, 12) governability *is* 'the overall quality for governance of any societal entity or system'. The quality of governance is a function of: the *System-to-be-Governed* (SG) which in the case of small-scale fisheries refers to both the natural and social systems; the *Governing System* (GS) which refers to the institutions and organizations that have a steering role; and the *Governing Interactions*, that is how the GS and the SG interact with each other (Bavinck et al. 2005, 2013; Kooiman et al. 2005, 2008; Chuenpagdee et al. 2008; Song and Chuenpagdee 2010). The conceptualization of governability by these various authors suggests that governability depends on qualities of the GS and SG and the relationship between the two. Thus what influences governability of a fishery are the inherent traits and constructed capabilities of the two systems and their interactions (Chuenpagdee and Jentoft 2013). Some of the key characteristics of GS and SG that influence governability are their diversity, complexity and dynamism (Song and Chuenpagdee 2010; Chuenpagdee and Jentoft 2013; Kooiman and Bavinck 2013). These traits point to underlying uncertainty and unpredictability of the behavior of (natural and social) systems, which cautions reductionist approaches to governability assessments (Bavinck and Kooiman 2013). Song and Chuenpagdee (2010, 237) posit that "governing interactions are related to the presence of interactive attributes such as participation, communication, collaboration and adaption". In reality, there are numerous types of interactions between the governors and the governed. Interactive governance distinguishes three main modes of governance: hierarchical, co-governance and self-governance (Bavinck and Kooiman 2013). Assessing governability therefore requires taking into consideration system properties and the quality and type of interactions between the systems. Interactive governance also proposes that in order for governors to execute their governing tasks, they use governing elements namely images, instruments and actions. "Images are sets of governing ideas. Instruments give the ideas substance and action puts these instruments to work" (Kooiman and Bavinck 2013, 18). Song et al. (2013) and Kooiman (2003) suggest that governance is value-ridden. It is therefore important to understand 'meta-level' governance elements – comprising of values, images and principles – that resource governing actors (GS) and users (SG) hold about how the world works.

In order to evaluate the governability of small-scale fisheries in Malawi, this chapter firstly reviews the management and regulatory framework (principles, images and values) that have been used and applied for management. Secondly, we select and describe three case studies that closely relate to the three modes of governance – hierarchical, co-governance and self-governance – in order to look at performance of governance under each. We then discuss and draw lessons from governance of small-scale fisheries in Malawi.

Historical Basis of Management and Regulatory Framework

The first scientific basis for deriving fisheries management measures for Malawi fisheries was the research by Lowe from 1945 to 1947 whose report was published in 1952 (Lowe 1952). Lowe's work laid the foundation for the axiomatic (Song et al. 2013) scientific knowledge of the chambo and thus the biological basis for the management measures for the fishery. Most of what she had recommended remains the backbone and substantive basis of the scientific knowledge-base for the regulations on Lake Malawi. The background to the colonial government's commissioning of the surveys by Trewavas (1942)[1] and Lowe (1952) was the worry that chambo (*Oreochromis spp.*),[2] the most valuable species, was being overexploited. Lowe's brief (Terms of Reference by the colonial government) was to come up with recommendations on the status and management of the chambo. Her recommendations led to the enactment of the 1949 Fisheries Ordinance (Government of Nyasaland 1949) by the Governing Council, 3 years before her report was published. Lowe used samples, data and information from fishers and other useful information Trewavas and her team had collected a few years earlier, in order to study the systematics, breeding, growth and life histories of the chambo. She also looked at the effects of fishing on the chambo.

[1] The survey by Trewavas and her team was disrupted by the Second World War.

[2] Chambo is a general local name for three (*squamipinnis*, *lidole* and *karongae*) closely related species of *tilapine cichlids* presently placed in the subgenus *Nyasalapia* of the genus *Oreochromis* (Trewavas 1983).

Lowe's recommendations and rationale had their basis in the theory of Maximum Sustainable Yield, a concept and governance principle that was becoming prominent in the 1930s and 1940s when the emerging and dominant governance vision was the conservation of wild fisheries (Gulland 1977). Lowe reiterated that "in Lake Nyasa (the present Lake Malawi), the long term policy should aim at allowing all the Tilapia the chance to breed and rear one brood of young before they are caught" (Lowe 1952, 61). She noted that the "Tilapia was vital to the whole economy of the lakeshore. Their loss could have constituted a real disaster and that it could not have been too strongly emphasized that drastic steps needed to be taken to guard against such a happening" (Lowe 1952, 61). Her evidence showed that there were three key species of chambo, namely *Tilapia squamipinnis*, *T. saka* and *T. lidole* and that the three species took 3 years to reach breeding maturity, meaning that they needed to be protected from being caught for 3 years. In addition, the *Tilapia* brood its young, meaning that whole broods of young may be destroyed with the parent if these are caught at this stage.

Based on her findings, she recommended the following management measures: minimum takable size; minimum mesh sizes; protection of the chambo in the breeding grounds; and prohibition of fishing during the breeding seasons (Lowe 1952, 61). Lowe also suggested that the amount of fish to be caught and/or intensity of fishing needed to be controlled.

Conversely, Lowe realized that the complete protection of the larger species, in particular the chambo, would impede the utilization of the smaller species, something that would be nonsensical as the smaller species were also of great importance to the fishers. Thus measures for the 'protection' of the *Tilapia* had to be a compromise. An independent Malawi enacted a new Fisheries Act in 1973 through an act of parliament (Government of Malawi 1973) leading to the establishment of the Department of Fisheries (DoF) in 1974.

Management Framework Post Independence

The 1973 Fisheries Act (Government of Malawi 1973) provided the basis and framework for fisheries management in Malawi after independence in 1964. The fisheries sectoral objectives were stated as being; '*to maximize the sustainable yield from fish stocks that can be economically exploited from the national waters; improve the efficiency of exploitation, processing and marketing; promote investment in viable fish farming units, and exploit existing and develop new aquatic resources*'. The Act outlined the general restrictions and prohibited fishing methods and the offences and penalties thereof. Most of the regulations largely returned the fervor of conservation of the chambo using the standard technical regulations for indirect control of effort. To a large extent these were those that had been recommended by Lowe (FAO 1993; Hara 2006b). What is important to note is that the Act and regulatory framework did not provide for restrictions on effort nor output despite Lowe's warning that these needed to be controlled.

As can be seen from the above discussion, most of the regulations are based on the biological conservation of the chambo. The assumption was that a combination of all these technical regulations would achieve biologically sustainable patterns and levels of exploitation of the chambo, even without regulating the amount of fish taken, effort applied and access into the fishery. Most of these techniques and management approaches that had been developed for the chambo on Lake Malawi have been transposed to other lakes and water bodies in Malawi.

Towards Participatory Management

The 1973 Fisheries Act (Government of Malawi 1973) was revised into the Fisheries Conservation and Management Act No. 25 of 1997 (Government of Malawi 1997). The objective of the revised Act was stated as being "*An Act to make provision for the regulation, conservation and management of the fisheries of Malawi and for matters incidental thereto or connected therewith*". The resulting National Fisheries and Aquaculture Policy (Government of Malawi 2001) shifted towards a more people centered (Maximum Social Yield) approach stipulating the sectoral policy as being "*to sustain the contribution of the national fish resources for the upliftment of the quality of life in Malawi by conserving the resources for the benefit of the present and future generations*".

The key aspect of the revised Act was inclusion of participatory management under subsidiary legislation (Conservation and Management (Local Community Participation) Rules 2000; Government of Malawi 2000). Some of the key revisions particularly in relation to participatory management are as follows:

- The Director may appoint suitable persons to be *honorary fisheries officers* who can exercise the powers of fisheries protection officers to assist with effecting of the provisions of the Act (section 4(1) & (2));
- Section 7 provides for *local community participation* in conservation and management of fisheries in Malawi; while
- Section 8 makes provision for the Director of Fisheries to enter into *Fisheries Management Agreements* with a fisheries management authority.

In particular, the subsidiary legislation Conservation and Management (Local Community Participation) Rules, 2000 provides for establishment of Beach Village Committees (BVCs) as management authority bodies (part II) and the formation of Fishermen's Associations (part V) and that such an association be required to have legal personality. With regard to consultation, it is stated that no subsidiary legislation related to the conservation and management of fisheries resources shall be amended without consultation with the relevant association(s) unless it is not necessary or impracticable to do so. In terms of fees and levies, the Subsidiary Legislation (section 59) provides for the issuing of licenses and permits, the levying of license and permit fees and the administration and expenditure of the district fees by local management bodies.

Methodology

Using the governability assessment framework (Chuenpagdee and Jentoft 2013), we examine three governance modes that are (formally or informally) in practice in three water bodies in Malawi (Fig. 8.1), namely hierarchical (Lake Malawi), co-governance (Lake Malombe), and self-governance (Lake Chiuta). For each lake, we analyze the performance of governance by looking at SG and GS, and how these interact with each other (governing interactions). We used the following questions to guide the research: what are the (natural and social) components and characteristics (diversity, complexity, dynamism and scale) of the SG?; Who are the stakeholders that comprise both the GS and SG?; What images, values and principles had been/are being used by the GS (and SG)?; What institutions (formal and informal) are used by the GS and the SG for/and in the interactions?; and how do the SG and the GS interact and what is the quality of these interactions? We used secondary material (both published and unpublished) and supplemented it with our knowledge based on our long experience in working in these systems.

Descriptions of the System-to-Be-Governed, the Governing System and the Governing Interactions in the Three Lakes

This section outlines and describes the three case study lakes in terms of the three governance components, namely GS, the SG and governing interactions. This will lay the ground for analyzing performance of governance in each lake. We hypothesize that Lake Malawi is in practice governed under the hierarchical mode, Lake Malombe formally under the co-governance mode and Lake Chiuta by default under the self-governance mode.

Lake Malawi

System-to-Be-Governed

Lake Malawi, which lies between 9° 30′S and 14° 30′S, is the southernmost of the African rift lakes. It is the third largest lake in Africa and is 360 km in length, has a 50–60 km mean width and an average depth of 292 m (maximum recorded depth 700 m). Its total surface area is 28,000 km^2. The lake is shallower and more productive in the south but gets deeper and less productive northwards. Lake Malawi has over a 1,000 species, most of which are endemic belonging to the Cichlidae family, that have evolved over millions of years. As such it is classified as an 'ancient' lake of high biological significance (Turner et al. 2001). The Lake is reputed to have more endemic fish species than any other lake in the world. Fishing is categorized into commercial and small-scale (Banda et al. 2005).

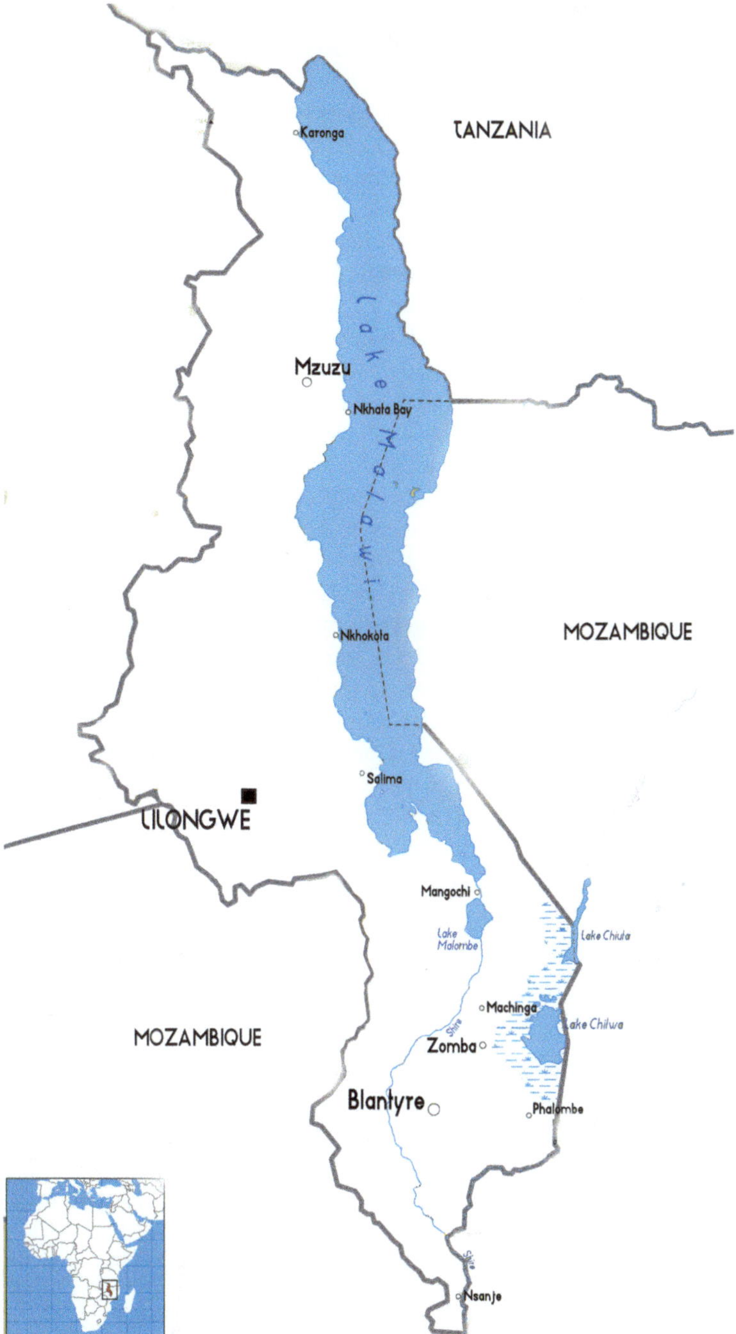

Fig. 8.1 Location of Lake Malawi, Lake Malombe and Lake Chiuta (Source: www.nationsonline.org/oneworld/map/malawi_map.htm)

Lake Malawi's small-scale fishery is based on a multi-species and multi-gear fishery with no designated landing points for fishers. Migratory fishing patterns are also a key characteristic of the fishery. Fishing units are organized around a gear owner employing crew members with sharing of benefits based on the share cropping approach (Hara 2006a). The small-scale fishers use canoes and plank boats with or without engines to catch fish intended primarily for sale but also for own consumption. Gears used in this sector include beach seines, open water seines, gill nets, fish traps, long lines and hand lines. The most important fishing gears are offshore seines (chilimira and nkacha), gillnets and beach seines. The majority of fishers operate throughout the year but switch to different gear types depending on availability or seasonal variation of specific species. It is common therefore for most gear owners to have a number of different gear types. Most gear owners also raise investment capital privately beyond government or banks (Hara and Jul Larsen 2003).

One of the characteristics of the small-scale fisheries on Lake Malawi are gear innovations and technological developments within fishing that are taking place all the time in reaction to decline in productivity of existing gear types (Hara 2006b). For example, in the last 20 years, fishers have developed new methods of fishing such as *nkacha,*[3] *kandwindwi,*[4] *chalira,*[5] *wogo*[6] and *ngongongo*,[7] some of which are not recorded in fishery statistics by the DoF as they do not exist in the legal context. As a result, these are taken to be illegal and most have been banned.

The huge catches of *usipa* in recent years (Department of Fisheries 2013b) resulting in the general overall increase in total catches actually masks the problem that is being experienced with the catchability of other species. In practice, there has been a decline in profitability of most of the larger species, especially the most valuable such as *chambo*. This is largely due to rampant use of illegal gears and uncontrolled increase in fishing effort (Hara 2006b, 2011). Most importantly, other than technical input regulations, the management framework does not include the use of limits on access, effort and output.

Governing System

The DoF is the responsible agency for management of fisheries in Malawi. The principal statute for management is the Fisheries Conservation and Management Act and its subsidiary legislation, which provide for the following specific rules and regulations:

[3] Open water seine net that was developed by an artisanal fisher, Mr. Paudala, based at Lake Malombe in the late 1980s and introduced Lake Malawi in the 1990s.

[4] A gillnet in the form of a trawl with a long rope that is sometimes over 1,000 m operated by several fishers as a beach seine.

[5] A mosquito net operated behind a beach seine to catch any escapees.

[6] A small chambo beach seine usually described as such because of its long warps usually over 1,000 m.

[7] A smaller meshed gillnet of less than the recommended size of 95 mm usually operated during cold season (May to August) to target *Copadichromis* spp. (utaka).

- *Licensing*: Licenses are required for all gears that operate on the lake.
- *Mesh restrictions*: Gillnets with meshes less than 95 mm are prohibited in Southeast Arm and those with less than 89 mm in the area lying south of latitude 12° 15′ of Lake Malawi. There is no mesh restriction for the rest of the lake lying north of latitude 12° 15′. Regulations on length and depth size do not exist.
- *Fish size limits*: It is prohibited to catch species of *chambo* and *mpasa* of less than 15 cm and 30 cm respectively.
- Closed season for chambo seines from 1st November to 31st December.
- Participatory management.

Important management activities include catch data collection, outreach (extension) programmes and enforcement of regulations that are still mainly executed by the DoF. The DoF is severely hampered by inadequate funding leading to poor execution of its management mandates. In particular, poor enforcement of regulations has led to rampant illegal fishing, in effect leading to depletion of key stocks such as the *chambo* (Hara 2006a, b; Song and Chuenpagdee 2013). Under administrative decentralization, it is proposed that fisheries should be managed jointly between the DoF and District Councils, with the latter to have powers to formulate and pass by-laws while the DoF would play the role of technical advisor (Hara 2008).

Governing Interactions

BVCs are supposed to be the main vehicle for formal interaction between the DoF and fishers. Currently, interactions are mainly through *adhoc* meetings called by the DoF, which are rarely conducted anyway. Sometimes public notices are placed in the print media or made on radio to reach out to fishers with specific extension messages. Under administrative decentralization, Village Development Committees will be the main organizational structure for interaction between village communities, government ministries/departments, NGOs and donors. The fate of BVCs under decentralization remains unclear (ibid.).

Lake Malombe

System-to-Be-Governed

Lake Malombe (Fig. 8.1) lies between latitude 14° 21′ to 14° 45′ south and longitude 35° 10′ to 35° 20′ East. The Lake is about 30 km in length, 15 km maximum width and has an average depth of 4 m. The Lake is approximately 390 km^2 in total surface area and is fed by water from Lake Malawi through the 12 km stretch of the Upper Shire River. At its peak in 1988, the lake produced about 15,500 tonnes of fish, which was approximately 17 % of Malawi's total production. The Lake's fishery was dominated by two major fish groups, namely, *Oreochromis* spp. and *kambuzi* (*Haplochromis* spp.) (FAO 1993). The fishery experienced acute decline in catches of *chambo* (from about 8,300 tonnes in 1982 to less than 1,000 tonnes at present), which represents a considerable loss in income levels for the fishers. The main gears in use are nkacha and to a less extent gillnets. The use of beach seines and kambuzi nets has declined dramatically following the depletion of the chambo.

Governing System

Participatory management on Lake Malombe was launched in 1993 as a reaction to the collapse of the *chambo* and decline in other species (Hara et al. 2002). It was the first formal and institutionalized participatory management arrangement in Malawi. The program was facilitated by the DoF through funding contributions from several donors (Hara et al. 2002). After consultative workshops between government and the user communities, the overall objective of the program was agreed as being '*To promote recovery of the fisheries of Lake Malombe and Upper Shire River to catch levels of the mid 1980s through cooperation between the government and the user communities*' (Bell and Donda 1993).

BVCs (31 in total representing 69 villages) are elected community level institutions for representing communities in the participatory management set-up. In 1997, the BVCs elected an umbrella organisation for all the committees called the Lake Malombe/Upper Shire River Fishermen's Association. They are mainly composed of gear owners and crewmembers, although effort was made to include other interest groups like traders/processors and women. Village headpersons are made *ex-officio* members of BVCs in their areas of jurisdiction. BVCs are supposed to be the vehicles for granting specific access rights so as to turn the fishery into limited access. It was hoped that the communities, through their BVCs, would progressively assume greater responsibility for management of the fishery leading to community-based management. While funds for launching the programme were from donor projects, it was envisaged that in the long term funding for the programme would come from gear license fees as stipulated in the Fisheries Conservation and Management Act (Government of Malawi 1997).

Governing Interactions

Following the end of project funding, formal and active interactions between BVCs and the DoF declined dramatically due to inadequate funding. The DoF could not afford continuation of the programme at a high activity level as was the case during the period of project funding. As a result, the partners (BVCs and the DoF) rarely interact. Because of the role they are being asked to play, BVCs are major power brokers. The powers of BVCs in relation to those of traditional leaders are an area of contestation (Hara et al. 2002; Njaya et al. 2012). An aspect of contention between village headpersons and BVCs is that the former insist upon continuing to exercise their customary powers independent of the BVCs, for example around issues such as accepting in-coming migrant fishers who have no transfer letters without the permission or knowledge of BVCs, and/or collection of *mawe* (honorarium) from fishers. In most instances, BVCs are not able to successfully confront and challenge the village headpersons over such issues. For BVCs to play their intended role, they need some semblance of independence from the local leaders without, at the same time, infringing on the traditional and historical powers and privileges of local leaders (Hara et al. 2002; Hara 2008).

Lake Chiuta

System to-Be-Governed

Lake Chiuta (Fig. 8.1) is the smallest of the four lakes in Malawi. The Lake is located at an altitude of 620 m in the southern part of Malawi (Machinga District). It is a shallow lake with a mean depth of 5 m and a total surface area of about 200 km^2, of which 40 km^2 lies in Mozambique (FAO 1993). The southern part of the Lake is mostly permanently covered with emergent vegetation penetrable only by canoes. The Lake's waters are clearer and less saline than those of Lake Chilwa. The fishery is based on small-scale fishers using either dug-out or planked canoes. Since 1985, the estimated annual fish production from the Lake has fluctuated between 700 and 4,000 tonnes.

Following the severe drought of 1992 that dramatically reduced the level of Lake Chilwa a large number of Nkacha fishers from that lake moved to Lake Chiuta. The Nkacha fishers were catching huge quantities of *Oreochromis* juveniles using these small meshed open water seine nets, which devastated the *Oreochromis* gill net fishery that local (resident) fishers solely depend upon. In addition, operation of these open-water seines damaged gill nets set by local fishers. The operation of nkacha nets also required and resulted in removal of macrophytes on which productivity of the lake is largely based. The catches by local fishers began to decline causing tension between the local fishers and the migrant nkacha fishers.

Governing System

The DoF never had any real presence on Lake Chiuta. It was the growing tensions resulting from the migrant nkacha fishers that brought fisheries management on the Lake to prominence. Following expressed concerns by local fishers about the negative fishing methods and other social activities of the migrant nkacha fishers, a meeting for all fishers was called by the Traditional Authorities of the area and the

Malawi Congress Party (MCP) (the ruling party at the time) to resolve the simmering conflict (Njaya 2005). It was decided at this meeting that nkacha nets should be banned from Lake Chiuta forthwith. Instead of leaving, the nkacha fishers bribed the MCP Chairman and the Chiefs of the area and continued fishing. As a result, the chiefs and local politicians lost all credibility in the eyes of local fishers. It dawned on the local fishers that they had to deal with this problem on their own. The local fishers formed their own BVCs (13 in total), having observed that such bodies had been formed on Lake Malombe by the government. Membership of the BVCs was extended to all primary stakeholders – the local fishers. In addition, a Lake Chiuta Fisher's Association was formed by the BVCs. Of significance was that unlike the BVCs in Lake Malombe, those formed by the fishers on Lake Chiuta excluded chiefs (with the exception of two village headmen who were also fishers) due to their history of having received bribes from nkacha fishers (Njaya 2005). In addition, all BVC members pay a membership fee twice a year to their BVC. Not only does this fee provide operational funds for BVCs, it also confers a sense of ownership of the BVCs on members. In turn, the BVCs pay an agreed annual membership fee to their Association.

On 19th May 1995 another meeting of all concerned (Fishers and their BVCs, Traditional Authorities, the Member of Parliament for the area, representatives from the District Assembly, representatives from the Police and the DoF) was convened to address the issue of the nkacha nets still operating in Lake Chiuta despite having been banned. As previously, it was again agreed that the nkacha should be banned from Lake Chiuta. The following day, the BVCs together with their members (local fishers) violently evicted all nkacha fisheries (ibid).

Governing Interactions

BVCs hold regular meetings. An annual general meeting of the Association, which is open to all BVCs members, is held every year. At the annual meeting elections for the Association office bearers for the coming year are held using a secret ballot.

A very strong enforcement programme, which includes punitive sanctions, has been developed by BVCs without the involvement of the DoF. As a result of these strong locally based management operations by the local fishers the *Oreochromis* fishery has recovered and been restored to its status before the invasion of the nkacha fishers.

When Group Village Headperson Njelwa was suspected of having been compromised by the nkacha fishers to support their return to Lake Chiuta in 2002, the Fisheries Association went to his house and beat his wife (he had been warned and had run away) and pulled down his house. A similar incident occurred in 2009 when TA Ngokwe died and his nominated successor publicly announced that he intended to allow back nkacha fishers. The Fishers Association went to his house and beat him up and forced his appointment to be rescinded.

Governance Outcomes on the Three Lakes Under the Three Modes

Lake Malawi

While we consider Lake Malawi as characterized by hierarchical governance, we are mindful of the fact that since the late 1990s, the DoF has been increasingly attempting to shift away from centralized management. This is evidenced by formation of over 300 BVCs on the Lake. Even then, most of the power, authority and responsibility for management still largely reside with government. Although the Fisheries Conservation and Management Act promotes devolution of power to BVCs and their Associations, there is reluctance to relinquish power and responsibility by government. Even if government was willing to devolve power and responsibility, most BVCs lack the capacity and resources to assume and exercise such power. One of the problems is that legislation for ploughing back gear license fees for management costs has never been operationalized. Thus the existing type of co-governance on Lake Malawi could be characterized as instructive/consultative (Sen and Raakjaer Nielsen 1996). A number of historical antecedents of the GS and natural and social characteristics of the SG could also be contributory to governance problems on Lake Malawi.

Malawi's fisheries had been managed from a developmentalist vision since independence. In this context, government viewed fishing as an economic activity that had to be open to all who were able to and could afford to invest and enter the fishery, in effect promoting open access. Secondly, there are no (and have never been) limits on effort or output as part of the management strategy both in commercial and small-scale fisheries. Thus although the Maximum Sustainable Yield principle has been used as a management goal since the 1950s (Lowe 1952; Government of Malawi 1973, 1997; FAO 1993), this has never been applied in practice. In addition, the Maximum Sustainable Yield principle assumes that biomass estimates for the target species can be done every year to set the Maximum Sustainable Yield for each species for a given year and then the requisite effort allocated and controlled. Although Catch Assessment Survey and Malawi Traditional Fisheries (MTF) sampling systems are used to collect catch data in the small-scale fishing sector systematically every year,[8] the data is never used to calculate and set Maximum Sustainable Yields for the key target species. It is merely used to compute total annual catches. The biggest problem though is both technical and practical; that even if the DoF obtained good time series data based on these sampling surveys and was able to compute the biomasses and then Maximum Sustainable Yields for the key species, expecting the fishers to keep within proposed Maximum Sustainable Yield levels through the use of technical regulations without limits on levels of effort and output would be very difficult if not impossible (ibid).

[8] Because of lack of resources and poor supervision of the data collectors, the data from the two systems is very unreliable.

Lake Malawi exhibits variable ecological and biological diversity, complexity and dynamism (Song and Chuenpagdee 2010). In addition, the fishery is based on multiple species and multiple gears. Migratory fishing strategies are used by the majority of fishers. The fishing units are organized around a gear owner and crew-members using variable and sometimes complex benefit sharing systems, who sell their catch to fish processors and traders who in turn take the fish to retail markets and dispose of it to retailers. This variation in diversity, complexity and dynamism among the (natural and social) SG and the way all these influence each other internally and within the value chain means that the (potential) governability of the system even if the GS was adequately functional is likely to be low. Song and Chuenpagdee (2010) suggest the need to adopt a more cautious approach in application of any management strategies (including co-governance) in the face of such incomplete knowledge and information.

If power and responsibility still resides within the DoF then the department needs to be able to exercise this in fulfillment of its mandates. The DoF lacks the capability to enforce the existing (technical) regulations due to inadequate technical capacity and inadequate resources. Monitoring and controlling activities of fishers is made even more difficult in that small-scale fishers are not required to operate from specific points or harbors. In effect, the DoF is unable to enforce the existing regulations leading to rampant illegal fishing activities and widespread use of illegal fishing gears and methods (Song and Chuenpagdee 2013). The BVCs that had been formed also lack both an enforceable mandate and resources to undertake enforcement activities despite the stipulations of the Fisheries Conservation and Management Act and its subsidiary legislation.

Lake Malombe

Co-governance has not achieved the intended objectives of arresting the general decline of the fishery and the specific recovery of the chambo in Lake Malombe despite 20 years of implementation of the participatory fisheries management strategy. The fishery remains depressed and well below its productivity at its peak in the late 1980s (Department of Fisheries 2013b). A number of arguments can be suggested for the below-par performance of co-governance.

The co-governance strategy in Lake Malombe was built around BVCs taking over law enforcement and other management responsibilities from the DoF in their local areas while the DoF retained the final say in what changes could be made to rules and regulations (Hara et al. 2002; Njaya et al. 2012). Although the Fisheries Conservation and Management Act empowers fishing communities to make by-laws that could be applicable in specific districts and possibly in the local areas within districts through District Assemblies/Councils when these become functional, such by-laws would have to be in line with national fisheries legislation, thereby reducing potential for real changes. In addition, the Fisheries Conservation and Management Act still empowers the Director of Fisheries to develop local management plans that s/he can subsequently unilaterally impose. Although in this process, the local communities are supposed to be formally consulted, they cannot influence change of the management plans.

At the beginning of the programme, BVCs had the confidence and the boldness to undertake the various management tasks and responsibilities required of them such as checking gears for the right mesh sizes, ensuring that patrols take place especially during the closed season, collecting license fees from fishers, etc. and applying sanctions on their own such as confiscation of gears during the closed season or confiscation of under mesh sized nets. Increasingly though, BVCs rarely undertake these management tasks nor apply such sanctions locally. The main reason given by BVCs for reluctance to confiscate gears is that they fear the consequences of gears being stolen while in their hands. In general, there is fear of reprisals from owners of confiscated nets and their crew members. Another factor for decline in willingness to undertake operational management activities is the feeling among BVC members that they are doing risky work without pay. The best that most BVCs currently do is to expel wrongdoers from their areas of jurisdiction. Usually, they also report the problem areas to the DoF, which is then expected to come and patrol and arrest any wrong doers. The only task that most BVCs currently perform on their own is the issuing of residence letters to in-migrant fishers and also issuing transfer letters to their en-migrant fishers.

Another problem is that the Fisheries Conservation and Management Act does not fully empower communities to assume full management responsibilities. For example, Sections 5(1) and 4(1) retain powers with the Director of Fisheries to appoint members of the Fisheries Advisory Board and Honorary Fisheries Protection Officers. Issues of control of beaches, funding mechanisms for BVCs and issuing of fishing permits and licenses by BVCs have not been implemented either. All in all, the Director of Fisheries retains most of the management powers s/he had under the former formal centralized hierarchical mode. As the saying goes; 'The more things change, the more they stay the same!' No wonder one of the fishers sarcastically remarked at the end of the donor funded PFMP in 1995 that "now that comanagement is over, we can go back to the way things were. No more harassment by BVCs!".

Lake Chiuta

Given the isolated and rural setting of this small lake, fishing is the main source of livelihood for the Chiuta community. The majority of fishers are thus keenly aware of the value of the lake and the importance of utilizing the fishery sustainably. It is this knowledge and perception that is conducive to a locally derived strong management and stewardship ethos among local fishers, trust, ownership and loyalty for their BVCs. Voluntary participation in both collective and operational management activities is high among both fishers and the local community. Fishing rules and regulations are locally formulated and enforced by the fishers through their

BVCs. The level of adherence to these locally made and instituted regulations is very high. When action needs to be taken, there is cooperation among the local stakeholders even without external (government or local leaders) support. A good example of such locally organized and executed management action is the banning of the nkacha and eviction of the nkacha fishers who had refused to leave voluntarily in 1995 and the successful exclusion of this fishing gear until now. The BVCs are also responsible for enforcing the locally agreed 2¾ inch minimum mesh size for gillnets and the licensing of all gill nets operating on the lake among their members. The commercialization and resultant high value of the fishery within the local economy, in terms of income from fishing and employment, results in high regard for and compliance with the regulations upon which the continued success of the fishery is based. This success is also based on a locally based efficient and sometimes violent system of enforcement of the regulations and sanctions for wrong doers.

Governability of Small-Scale Fisheries in Malawi

If we assume that the quality of governance is related to whether the management objectives are being achieved and also the means and processes by which they are being achieved, then self-governance would appear to be the most proficient in Malawi's context. The Chiuta system is producing outcomes that fishers are happy and satisfied with in terms of level of catches and benefits. The violent means by which regulations and alignment to community agreements are sometimes enforced ought to be seen and understood from the vision and perspective that fishing is very important for this rural community so much so that they cannot afford to have the resource destroyed by the selfish actions or greed of a few, let alone outsiders. One of the underlying characteristics of Lake Chiuta is that this is a simpler natural and social 'closed system' located in a rural and remote area where the user community can define itself and create secure rights. At the same time, the community is highly aware that because of the low natural and social resilience of the SG, they need to protect the natural and social systems from destruction through strong management actions.

Unlike Chiuta, Lake Malawi is variably diverse, complex and dynamic as a natural and social 'open system' especially with most of the fishing taking place offshore. In this system, it would be very difficult to create secure geographic, ecological and social borders for locally enforceable communal rights. How does one divide and apportion such an open system? In this instance, secure rights could probably work best by basing them on a quota controlled fishery that would limit both participants and also the output, at the risk of creating socio-economic inequities. Without 'closing the commons' and creating a system of limited access and limited output, sustainable governance is unlikely to be achieved. These are the hard choices (Song et al. 2013) that will need to be made. The issue of the ability to create a governance system based on limited access and limited output is also at the heart of the failure of the Lake Malombe co-governance initiative to achieve its objective of recovering the fishery. Thus while Part V (Fishing Licenses) sections 15 (No person shall fish in fishing waters falling within the jurisdiction of a BVC unless s/he is authorized to do so by a license or written authority of a BVC) and 16 (Written authority of a BVC shall be in a form set out in the Second Schedule) of the subsidiary legislation provides for the capability for BVCs to limit access, these local management bodies have so far not been able apply this legislation because of lack of support from the DoF in terms of operationalizing the relevant aspects of the legislation. Administrative decentralization is likely to complicate matters by adding another layering of bureaucracy, legislation and new competing organizational structures at the local level. As a result, the intended co-governance strategy has gravitated back towards the hierarchical mode by default. The type (continued use of nkacha nets) and level of fishing effort in Lake Malombe remains too high to permit recovery of the *chambo*, *kambuzi* or the macrophyte flora on which the *chambo* depend. This serves as a warning that without such capabilities and a governance system based on demarcation of the Lake into zones or areas that BVCs can exercise control over, co-governance on Lake Malawi might not achieve the objective of arresting the decline of the major key species and eventually the fishery as a whole. The fear is that this is the direction that Lake Malawi is moving towards – that is, practices based on use of destructive fishing methods and uncontrolled increase in levels of effort that BVCs have no control over. Together this will lead to the decline of the fishery. This can be seen in the depressed productivity of the most valuable (indicator) species, the *chambo,* in most parts of the Lake (Department of Fisheries 2013b). Overall therefore, unless there is ability to institute a GS and SG based on clear fishing tenure and stewardship practices and operationalization of the Fisheries Conservation and Management Act, co-governance on both Lake Malombe and Lake Malawi will not work for functional governance.

The Chiuta and Malombe cases show that local chiefs can be a hindrance to positive governance outcomes if they are corruptible or unwilling to accommodate new governance arrangements that seem to threaten their powers and authority through which they derive benefits from fishing. It is only through the exclusion of corruptible local leaders that the fishers in Lake Chiuta have successfully excluded the nkacha fishers leading to recovery of the fishery. At the same time, chiefs in Lake Malombe have been known to disempower BVCs for personal benefit or due to power struggles to the detriment of the co-governance arrangement.

The government's main incentives for introduction of co-governance in Lake Malombe and Lake Malawi were reduction in "transaction costs", change of user behavior towards sustainable exploitation and improvement in the legitimacy of the management system. For some members of BVCs, the main reason to attend meetings and workshops still appears to be monetary. This divergence in incentives for co-governance largely remains. Since the number of meetings and workshops declined after the initial implementation stages, most fishers shun away from taking up positions as BVC members, arguing that it is *thangata* (work without pay). Part of the reason for the decline in willingness of people to take up voluntary positions of responsibility is the political change in Malawi from dictatorship to democracy. Whereas people could be forced to do self-help work under the one party rule of Dr.

Kamuzu Banda's dictatorship, people now demand that they be paid for any work that they do.

One of the questions that this review raises is whether government (or other external actors) should be conceiving, incubating, giving birth to and then babying local Management Units for co-governance. If so, for how long should these be nurtured before they are weaned? The fact that government formed and technically and materially supported the BVCs in Malombe (compared to those in Chiuta that were formed by the fishers on their own) made them dependent on government to the extent that it has proved difficult to wean them off government support to the detriment of their performance as independent co-governance or self-governance units. This should act as a warning for the DoF not to repeat the same mistake in other areas such as Lake Malawi.

Conclusion

This chapter looked at governability of small-scale fisheries in Malawi. Although legislation promotes co-governance, problems underlay a shift towards this approach. For an open system such as Lake Malawi where most of the fishing takes place offshore, the challenge for governance is creating secure communal rights based on geographic territorial integrity that BVCs can enforce. The solution to this could be an access limitation and individual quota based management approach with powers for distribution of rights devolved to or exercised together with BVCs. This approach could still pose its own problems given that centralized landing points are not used nor do they exist. Another problem would be shifting the fishers' ingrained attitude for open access that has been promoted through the historical developmentalist vision. Increasingly though, it has become clear that such an approach is disastrous both in terms of conservation and socio-economically. Retracting from this path lies at the heart of moving towards sustainable governance of fisheries. In addition Maximum Sustainable Yield, which would be key for a quota based management approach, has never been put to practical use given the technical difficulties of operationalizing this in a multi-species, multi-gear fishery and also the poor quality of data. On Lake Malombe, co-governance has not resulted in a positive change of behavior of fishers nor limitation of access. One of the key reasons for poor co-governance outcomes is that the enabling legislation (Fisheries Conservation and Management Act) has never really been fully operationalized. It does not help that the contestation for power between BVCs and local chiefs undermines the functioning of BVCs. One of the underlying characteristics of Lake Chiuta is that it is a 'closed system' located in a rural and remote area where the user community can define itself and create secure rights. As a result, locally initiated self-governance has functioned well resulting in secure exclusionary rights for the local communities and positive management outcomes.

Government as the GS finds itself in the position of having power legally but not being able to make effective use of it for positive governance outcomes, in effect resulting in low governability of fisheries. As a result, the GS has grown increasingly negligent by default through a lack of capacity and resources to fulfill its legal mandate. This leaves the social SG to its own devices. Where the social system (of the SG) has the capability to take charge of governance through self-governance, positive results can be achieved on the natural system as the Chiuta case demonstrates. But where the social system cannot institute and achieve the requisite self-controls as in Lake Malombe and increasingly on Lake Malawi, the effects on the natural system can be devastating.

The governance challenge in Malawi's small-scale fisheries is thus whether the GS can develop the capacity and humility to deal with and interact with the (natural and social) SG for positive management outcomes given the variable diversity, complexity and dynamism of the SG. In addition, there is need for the social system component of the SG to be increasingly included as part of the GS in co-governance arrangements. This will need more than just enabling legislation on paper. It will require a GS that is willing to empower and strengthen the fishing communities as partners in co-governance rather than continuing to view them only as part of the SG. It is only fishers and their communities who are increasingly empowered that could move co-governance arrangements towards self-governance, which is the best way to optimize sustainable socio-ecological benefits from small-scale fisheries.

References

Banda, M.C., Kanyerere, G. Z., & Rusuwa, B. (2005). The status of the chambo in Malawi: Fisheries and biology. In M. C. Banda, D. Jamu, F. Njaya, M. Makuwila, & A. Maluwa (Eds.), *The chambo restoration strategic plan: Proceedings of the national workshop held 13–16 May 2003, Mangochi* (WorldFish Center conference proceedings 71, pp. 1–7).

Bavinck, M., & Kooiman, J. (2013). Applying the governability concept in fisheries – Explorations from South Asia. In M. Bavinck, R. Chuenpagdee, S. Jentoft, & J. Kooiman (Eds.), *Governability of fisheries and aquaculture: Theory and applications* (pp. 131–153). Amsterdam: Springer.

Bavinck, M., Chuenpagdee, R., Diallo, M., van der Heijden, P., Kooiman, J., Mahon, R., & Williams, S. (2005). *Interactive fisheries governance*. Delft: Eburon Publishers.

Bavinck, M., Chuenpagdee, R., Jentoft, S., & Kooiman, J. (Eds.). (2013). *Governability of fisheries and aquaculture: Theory and applications*. Amsterdam: Springer.

Bell, R., & Donda, S. J. (1993). *Community participation consultancy report* (Vol. I). Lilongwe: Fisheries Department.

Chuenpagdee, R., & Jentoft, S. (2013). Assessing governability – What's next. In M. Bavinck, R. Chuenpagdee, S. Jentoft, & J. Kooiman (Eds.), *Governability of fisheries and aquaculture: Theory and applications* (pp. 335–349). Amsterdam: Springer.

Chuenpagdee, R., Kooiman, J., & Pullin, R. S. V. (2008). Assessing governability of capture fisheries, aquaculture, and coastal zones. *Transdisciplinary Environmental Studies, 7*(1), 20.

Department of Fisheries. (2013a). *National frame survey results (1998–2012)*. Lilongwe: Department of Fisheries.

Department of Fisheries. (2013b). *Annual catch report*. Lilongwe: Department of Fisheries.

FAO. (1993). *Fisheries management south east Lake Malawi, upper Shire River and Lake Malombe*. Prepared by GOM/FAO/UNDP Chambo fisheries research project. CIFA technical paper 21. Rome: FAO.

Government of Malawi. (1973). *Fisheries act (Government of Malawi laws cap. 66:05)*. Zomba: Government of Malawi.
Government of Malawi. (1997). *Revised fisheries act*. Zomba: Government of Malawi.
Government of Malawi. (2000). *Conservation and management (Local Community Participation) rules*. Lilongwe: Government of Malawi.
Government of Malawi. (2001). *National fisheries and aquaculture policy of 2001*. Lilongwe: Government of Malawi.
Government of Nyasaland. (1949). *Fisheries ordinance No. 17: An ordinance to provide for the control and protection of fish and for the regulation of fisheries*. Zomba: Government of Nyasaland.
Gulland, J. A. (1977). *The management of marine fisheries*. Seattle: University of Washington Press.
Hara, M. M. (2006a). Production relations and dynamics among user-groups in the artisanal fisheries of Malawi: Implications for representation in co-management arrangements. *MAST, 4*(2), 53–71.
Hara, M. M. (2006b). Restoring the chambo in southern Malawi: Learning from the past or re-inventing the wheel? *Aquatic Ecosystem Health & Management, 9*(4), 419–432.
Hara, M. (2008). Dilemmas of democratic decentralisation in Mangochi district, Malawi: Interest and mistrust in fisheries management. *Conservation and Society, 6*(1), 74–86.
Hara, M. (2011). Community response: Decline of the chambo in Lake Malawi. In S. Jentoft & A. Eide (Eds.), *Poverty mosaics: Realities and prospects in small-scale fisheries* (pp. 251–273). Amsterdam: Springer.
Hara, M., & Jul-Larsen, E. (2003). The "lords" of Malombe: An analysis of fishery development and changes in fishing effort on Lake Malombe, Malawi. In E. Jul-Larsen, J. Kolding, R. Overa, J. Raakjaer Nielsen, & P. A. M. van Zwieten (Eds.), *Management, co-management or no management? Major dilemmas in southern African freshwater fisheries* (FAO fisheries technical paper 462/2, pp. 179–200). Rome: FAO.
Hara, M. M., Donda, S. J., & Njaya, F. J. (2002). Lessons from Malawi's experience with fisheries co-management initiatives. In K. Geheb & M.-T. Sarch (Eds.), *Africa's inland fisheries: The management challenge* (pp. 31–48). Kampala: Foutain Publishers.
Kooiman, J. (2003). *Governing and governance*. London: Sage.
Kooiman, J. (2008). Exploring the concept of governability. *Journal of Comparative Policy Analysis, 20*(2), 171–190.
Kooiman, J., & Bavinck, M. (2013). Theorizing governability – The interactive governance perspective. In M. Bavinck, R. Chuenpagdee, S. Jentoft, & J. Kooiman (Eds.), *Governability of fisheries and aquaculture: Theory and applications* (pp. 9–30). Amsterdam: Springer.
Kooiman, J., Bavinck, M., Jentoft, S., & Pullin, R. (Eds.). (2005). *Fish for life: Interactive governance for fisheries*. Amsterdam: Amsterdam University Press.
Kooiman, J., Bavinck, M., Chuenpagdee, R., Mahon, R., Pullin, R., & Salagrama, V. (2008). Governability of aquatic resources: Conceptual exploration and applications. *Journal of Transdisciplinary Environmental Studies, 7*(Special Issue), 1.
Lowe, R. H. (1952). *Report on the tilapia and other fish and fisheries of Lake Nyasa 1945–47*. Colonial Office Fishery Publications (Vol. 2). London: Colonial Office.
Njaya, F. (2005). *Challenges of co-management on shared fishery ecosystems: The case of Lake Chiuta*. Commons southern Africa occasional paper series No. 9. Harare/Cape Town: Centre for Applied Social Sciences & Land and Agrarian Studies.
Njaya, F., Donda, S., & Béné, C. (2012). Analysis of power in fisheries co-management: Experiences from Malawi. *Society and Natural Resources, 25*, 652–666.
Sen, S., & Nielsen, J. R. (1996). Fisheries co-management: A comparative analysis. *Marine Policy, 20*(5), 405–418.
Song, A., & Chuenpagdee, R. (2010). Operationalizing governability: A case study of Lake Malawi. *Fish and Fisheries, 11*, 235–249.

Song, A., & Chuenpagdee, R. (2013). The damage schedule approach. In M. Bavinck, R. Chuenpagdee, S. Jentoft, & J. Kooiman (Eds.), *Governability of fisheries and aquaculture: Theory and applications* (pp. 279–299). Amsterdam: Springer.

Song, A., Chuenpagdee, R., & Jentoft, S. (2013). Values, images, and principles: What they represent and how they may improve governance. *Marine Policy, 40*, 167–175.

Trewavas, E. (1942). Nyasa fishes of the genus Tilapia and a new species from Portuguese East Africa. *Annals and Magazine of Natural History, 11*, 294–306.

Trewavas, E. (1983). *Tilapiine fishes of the genera Sarotherodon, Oreochromis and Danakilia.* London: British Museum (Natural History).

Turner, G. F., Seehausen, O., Knight, M. E., & Allender, C. J. (2001). How many species of cichlid fishes are there in African lakes? *Molecular Ecology, 10*(3), 793–806.

Chapter 9
Two Rules for the Same Fish: Small-Scale Fisheries Governance in Mainland Ecuador and Galapagos Islands

María José Barragán Paladines

Abstract The small-scale fisheries sector in the Ecuador mainland and Galapagos Islands face major challenges triggered by large scale human and climatic events, which compromise their sustainability. Lack of trust, leadership and cohesion among small-scale fisheries entities, limited organizational skills, and social problems within fishing communities reduce this sector's governability. Further, lack of willingness of fishers to observe rules, limited governing capacity and lack of political attention to small-scale fisheries often contribute to making the overall system less governable. According to interactive governance theory, different governing modes would be suitable for different systems, depending also on the types and quality of governing interactions. Using the interactive governance framework, this chapter explores the performance of the hierarchical and co-governance mode for small-scale fisheries governance in the Ecuador mainland and Galapagos Islands, respectively. These two case studies, by using empirical evidence and triangulation-based methods, analyze the small-scale fisheries sector and some mechanisms through which the governing system, the system-to-be-governed, and governing interactions are operating. The chapter highlights that both hierarchical and co-governance modes of governance have underperformed and consequently that their governability has been reduced for three main reasons: the mismatch between legal frameworks, the undefined social system's borders, and the use of inappropriate mechanisms for information mobilization. The chapter concludes by suggesting that both models need to contribute to national initiatives, to strength the social system, and to increase the small-scale fisheries sector governability. Only then can fisheries sustainability be achieved.

Keywords Small-scale fisheries • Management • Governability • Ecuador • Galapagos Islands

M.J. Barragán Paladines (✉)
Department of Geography, Memorial University of Newfoundland, St. John's, Canada
e-mail: m.j.barraganpaladines@mun.ca; majobarraganp@yahoo.es

S. Jentoft, R. Chuenpagdee (eds.), *Interactive Governance for Small-Scale Fisheries*, MARE Publication Series 13, DOI 10.1007/978-3-319-17034-3_9

Introduction

The majority of the world's fisheries are fully exploited, over-exploited, depleted, or in need of recovery (Worm et al. 2009; FAO 2014). Well-grounded scientific evidence links the poor state of the world's fisheries to the overexploitation and mismanagement of large- and small-scale fisheries (Andrew et al. 2007), which have additionally been worsened by large scale human and climatic events such as markets globalization, pollution and environmental change (Pauly et al. 2002, 2005; Perry et al. 2011). All these external drivers of change threaten small-scale fisheries governance at different temporal and spatial scales, reducing their governability and weakening the likeliness of achieving sustainability (Defeo and Castilla 2012).

In the Ecuador mainland and Galapagos Islands, small-scale fisheries are confronted with similar problems (Arriaga and Martínez 2008, 10) that have led to the systems' uncontrollability, unpredictability, and reduced governability (Jentoft and Chuenpagdee 2013). In Ecuador (mainland and Galapagos Islands), the challenge exists with regard to all three governance dimensions alluded to by interactive governance theory: the governing system, the system-to-be-governed, and the governing interactions (Kooiman et al. 2005; Bavinck et al. 2013). Among the most important challenges confronting small-scale fisheries management in Ecuador (mainland and Galapagos) are the absence of a specific fisheries-ruling body, the reduced technical capacity of the governing entities, the lack of political will to enforce regulations and take hard decisions, and the absence of reliable, systematic and accurate information regarding the small-scale fishery systems (Campbell et al. 1991; Coello and Fundación Natura 1993; Martínez et al. 1997).

Additionally, the natural system-to-be-governed suffers from decreasing fishing stocks, uncontrolled coastal development and deteriorating marine environments, whereas in the social subsystem, willingness to obey rules, trust, leadership, and cohesion among fishers, and agency skills are limited. Furthermore, the small-scale fishing communities currently are being negatively affected by the social processes occurring as a direct consequence of the inadequate attention given to this (e.g., migration, occupation displacement of younger generations of fishers, addictions, etc.) (Campbell et al. 1991; Domínguez et al. 1991; Gaibor et al. 2002).

In Ecuador, historically, fishing has been important culturally and socially (and more recently economically). There is evidence of pre-Hispanic communities consuming and trading fish products at a mid and low scale, locally and regionally (Norton 1985; McEwan and Silva 1998, as cited in Josse and Iturralde 2002). From the early 1950s, fishing developed as a commercial sector, aided by international bodies (Allsopp 1985; Williams 1998, as cited in Symes 2000). Since then, small-scale fisheries have been identified to be critical for the economic growth of fishing communities in both the Ecuador mainland and Galapagos Islands as well as for other sectors such as construction and tourism.

Despite no official estimates existing about the exact number of people dependent on fishing and its economic contribution regionally (Béné 2006), it is believed that more than 100,000 people are directly depending upon this sector for their livelihoods in Ecuador (SETEMAR 2013). The small-scale fisheries sector in

Ecuador (mainland and Galapagos Islands), moreover, contributes socially to the vitality and identity of fishing communities locally and regionally (CPPS 2000). It is also central to the culture of the majority of Ecuadorian coastal and some of the island's population (Coello and Fundación Natura 1993).

Small-scale fisheries governance has evolved differently in the Ecuador mainland and the Galapagos Islands. Prior to March 1998, when the Galapagos Marine Reserve was created, fishing resources were nationally managed by a hierarchical centralized structure, ruled by one ministry. After the creation of the Reserve, the small-scale fisheries in the mainland remained under a hierarchical governing system but not so the fisheries of the Galapagos Islands. The islands have from that date been ruled under a Special Law for Galapagos (i.e., *Ley Orgánica de Régimen Especial para la Conservación y Uso Sustentable para la Provincia de Galápagos*, or *LOREG*). Unlike the hierarchical governance of the mainland, there is now a co-governance system in place in the Galapagos Islands.

The governance of the small-scale fisheries sector in Ecuador (mainland and Galapagos Islands) has rarely been assessed. This chapter attempts to address this gap. Based on the claim that overall governance quality depends first and foremost on the inherent characteristics of the human and natural systems that are being governed and of the governing system (Chuenpagdee and Jentoft 2009, 2013), this chapter applies an interactive governance framework to assess both modes of governance (Kooiman et al. 2005, 2008; Bavinck et al. 2013).

The research is based on multiple methods (i.e., triangulation, Clifford and Valentine 2003) and consisted largely of detailed semi-structured e-mail surveys and interviews or guided conversations (Walmsley et al. 2005) with a wide range of key informants and actors representing the small-scale fisheries sector, nationally. Field observations and relevant published documents and grey literature were also central to the research. The research was conducted between July 2011 and September 2013 and only included the marine-small-scale coastal fisheries, not aquaculture nor inland small-scale freshwater fisheries. Purposive or theoretical sampling (Mays and Pope 1995) was used to select sample units (Teddlie and Yu 2007). This technique enabled the inclusion of a breadth of relevant perspectives (Kerr and Swaffield 2012) reflecting the diversity within a given population (Kuzel 1992).

The first section describes the system-to-be-governed and the governing system in mainland Ecuador and the Galapagos Islands. The second section illustrates some of the governing interactions that take place in both case study areas. The final section reflects upon the current status of the two different governance modes.

Study Area

Two broad regions were covered in the study: the coastal mainland including five provinces with direct access to the sea, and the Galapagos Islands. Ecuador is located in one of the ten regions of global priority for conservation and has a rich abundance of natural and marine resources (Olson and Dinerstein 1998; Olson et al. 2002). This is due to the presence of upwelling systems in the Pacific caused by southeasterly trade winds, which shift relatively cold and nutrient rich waters to the euphotic zone along the coast (Charles et al. 1994; Hannah et al. 2013).

The Galapagos Islands are one of the twenty-four Ecuadorian provinces and the only island region in the country. The islands are of volcanic origin, located about 1,000 km off the Ecuadorian coast. The islands are located in one of the most complex, diverse and unique marine ecoregions in the world and are considered a natural laboratory to study and understand evolutionary processes (Stone et al. 2006). The Galapagos Marine Reserve (GMR) is high in biodiversity, productivity, endemism, and ecosystem richness (Danulat and Edgar 2002) and contains four representative marine ecosystems: wetlands, littoral, subtidal, and open waters (Dirección Parque Nacional Galápagos 2014). This diversity is derived from geo-biophysical and oceanographic features and from the convergence of three major oceanic current systems in the area (i.e., Peru-, Cromwell-, and Panama Currents) (Bustamante et al. 1999; Bensted-Smith et al. 2002; Edgar et al. 2004; Baine et al. 2007) (Fig. 9.1).

System-to-Be-Governed

Mainland Ecuador

As in other places, the small-scale fisheries sector in this area has been continuously deadlocked and marginalized by national and local governments because of isolated and inappropriate policy measures (Agüero 1992, 2007; Pomeroy and Berkes 1997). The priorities of the sector either have been ignored or other more powerful interest groups have been privileged (Béné et al. 2007). Hence, small-scale fishing communities remain poor, with an estimated average monthly income of about US$ 250 (FAO 2013).

Fishers usually start fishing at an early age. Fishing lies within the family and many fisher children give up school so as to contribute to their family's livelihood and/or because employment elsewhere is hard to come by (Agüero 1992; Beltrán-Turriago 2001; Ramírez 2004). Thus, mainland small-scale fishers remain heavily dependent on fishing as their main (or only) livelihood source. The cultural importance of fishing means that many continue fishing despite the low profitability of it (Pollnac and Bavinck 2008, as cited in Palacios and Schuhbauer 2012; Pollnac and Poggie 2008; Cinner et al. 2009, as cited in Palacios and Schuhbauer 2012).

Until the 1950s, small-scale fisheries in the mainland were entirely of subsistence nature. From the 1950s onwards, international aid agencies (e.g., FAO) promoted the development of local fishing-related organizations through the institutionalization of local management and research bodies (e.g., *Instituto Nacional de Pesca,* INP) (personal communication, October 20, 2013). International initiatives supported the conversion of wood-made hand-carved sail or paddle-aided boats and canoes into fiberglass boats. Additionally, outboard engines and improved

Fig. 9.1 Small-scale fishing grounds in the coastal region of the Ecuadorian mainland and the Galapagos Marine Reserve (Source: Courtesy of the University of Texas Libraries, The University of Texas at Austin)

fishing gears (e.g., surface and bottom gillnets, surface and bottom long-line, and hand-line) were provided so that more species could be targeted including small-, large-pelagic-, demersal-species, and sharks by using improved fishing gears

(Beltrán-Turriago 2001; Gaibor et al. 2002). Other initiatives to boost the sector included training for fishers and programs for fish handling and processing techniques as well as alternative product production (e.g., fish meal), and business skills development (personal communication, October 20, 2013).

In 1995s, there were about 30,300 fishers and 14,355 boats operating along the coast of the Ecuadorian mainland (Arriaga and Egas 1998; Beltrán-Turriago 2001) with 32,000 tonnes of fish landed in 105 landing sites along the coast (Charles et al. 1994). The number of active fishers, living in 138 small-scale coastal fishing villages (FAO 2013), increased to more than 50,000 in the 2000s and to more than 80,000 in 2003 (Arriaga and Martínez 2008). According to the last national population census (INEC 2010), the mainland coastal region hosts a population of *ca.* 6.8 million inhabitants.

Galapagos Islands

Fishing in the Galapagos Islands is said to not be a "traditional" activity since human occupation of the islands is relatively "new" (Ospina 2001; Grenier 2007); fishing has said to have started in the early twentieth Century (Marder and Arcos 1985; Rodríguez 1987 unpublished; Ramírez 2004). After the Galapagos Islands were officially annexed to Ecuador in 1832, migrant workers in Galapagos were temporarily hired by fishing fleet owners to work in fish processing (Anón 1983, as cited in Ramirez 2004). Later in the 1920s, the first "permanent" commercial fishing initiative led to some Norwegian fishers setting up a canning industry that failed a few years later (Reck 1983; Marder and Arcos 1985). Fishing in Galapagos continued during the 1930s, with U.S. and Japanese long-liners and purse seiners (Reck 1983) coming to fish in Galapagos, attracted by economic interests but influenced by geopolitics factors (Finley 2009). That interest was evidenced during the 1940s, when a U.S. Navy Base was established in Baltra Island during the Second World War. The setting up of a U.S. Navy Base led to a considerable reduction in subsistence fisheries in the islands while boosting commercial small-scale fisheries (Stewart 2009) as local fishers preferred to sell their fresh catch to military personnel who paid higher prices (Ramírez 2004; Jobstvogt 2010).

During the 1960s, industrial fleets targeted mainly the Galapagos Grouper (*Mycteroperca olfax*) and the Spiny Lobster (*Panulirus penicillatus* and *P. gracilis*). In the 1970s, a 20-ship industrial tuna fleet fished in Galapagos waters regularly, targeting the Pacific big-eye (*Tunnus obesus*) and the yellow-fin tuna (*T. albacares*). Finally, during the late 1980s and early 1990s, the sea cucumber (*Isostichopus fuscus*) fishery blossomed in Galapagos after the depletion of their populations in the coastal region of mainland Ecuador (Ramírez 2004; Stone et al. 2006; Jobstvogt 2010).

As of late, economically valuable species (e.g., lobster and sea cucumber) have reduced significantly in the islands (Wolff et al. 2012). This, along with quota allocations, fishing restrictions, and limited prices in local and international markets, has reduced fishers' motivation to continue fishing. As a result, fishers have been displaced from fishing and transited to more lucrative businesses such as tourism or more stable jobs within administrative positions (Defeo et al. 2014). Currently, the population of the Galapagos Islands is approximately 30,000 (INEC 2010), out of which about 1,000 are officially registered fishers living in three fishing communities, namely San Cristobal, Santa Cruz, and Isabela islands. However, only 400–470 of these fishers are commercially active small-scale fishers (Palacios and Schuhbauer 2012), suggesting that the number of small-scale fishers actually fishing is decreasing.

The Governing System

Mainland Ecuador

Since the second half of the twentieth Century, there has been a hierarchical form of governance in place for fisheries management in mainland Ecuador. Hierarchical governance implied that both harvest and post-harvest fishing-related institutions and fishing activities (Breton et al. 2006) were governed by the National Fisheries Law (or *Ley de Pesca y Desarrollo Pesquero* in Spanish) since 1974. This law was updated in 1985 and 1992 (Beltrán-Turriago 2001). Additionally, other legal instruments (i.e., National Fiscal Authority and Penal Code) ensure that the harvest, processing, and trade of fish products, nationally and regionally, under this law are complied with. Illegal or abnormal practices within small-scale fisheries, such as the non-regulated exploitation of protected species or destructive fishing activities (Jacquet et al. 2008), are prosecuted based on instruments created to support those actions (e.g., the National Action Plan for sharks).

In 1995, the Fisheries Resources Program was created as an adjunct division of the Presidency of the Republic in order to promote the sustainable use of fishing resources. Since then, the regulations request two licenses to fish. First fishers need to get a *fishing permit* from the fisheries authority (i.e., Undersecretary of Fisheries Resources) to allow them to fish in the continental fishing ground (i.e., up to 12 miles off the coast). Additionally, fishers are required to have a *vessel permit* (issued by Navy authorities) which authorizes them to use their fishing boats (SETEMAR 2013). In, 1996 the newly created Environment Ministry, along with the Foreign Trade-Fisheries-and-Tourism Ministry, was given two tasks: protecting national water bodies and overseeing the fisheries sector. In 2008, the fisheries sector in the Ecuador mainland was put under the jurisdiction of the Ministry of Agriculture, Livestock, Aquaculture, and Fisheries – still no doubt under a hierarchical centralized governing system (Fig. 9.2). In that same year, after intense lobbying by the provincial authorities and the local fishing sector in Manabi (where Ecuador's biggest fishing harbor is), the headquarters of the national authority for fisheries resources (i.e., the Undersecretary of Fisheries Resources), traditionally located in the biggest Ecuadorian harbor city (i.e., Guayaquil), was relocated to Manta (Fig. 9.1).

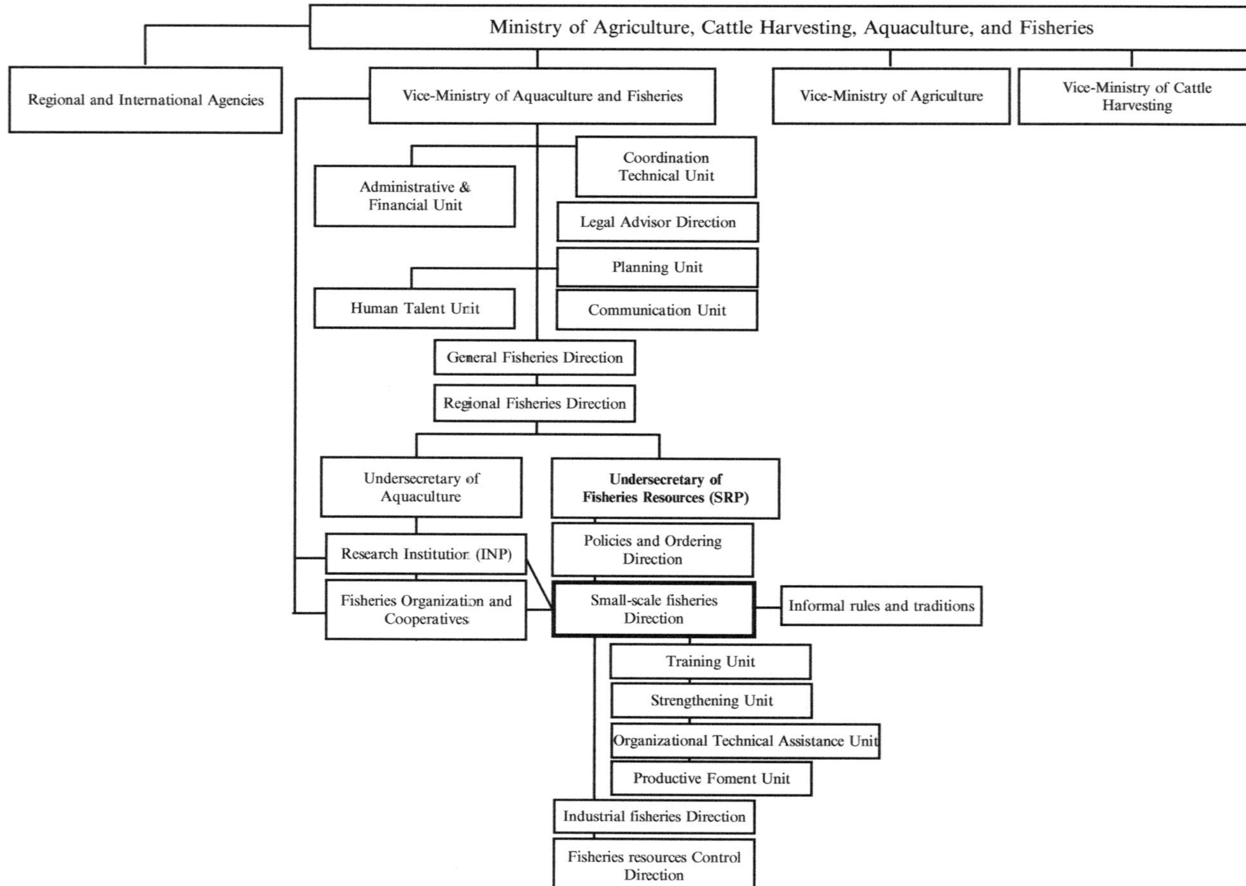

Fig. 9.2 Governing bodies in the fisheries sector of the Ecuador mainland (Source: SETEMAR 2013, SRP 2013; Vera, F. *personal communication,* 16.07.14)

Despite efforts to regulate small-scale fishing activity, the management system in place has been unable to deliver a sustainable fishery mainly because of the reduced technical and logistic capacity of the bodies in charge of law implementation (Arriaga and Martinez 2008). Consequently, the fishery has become a *de facto* open access one.

Cooperatives of small-scale fishers are supported by the National Federation of Small-scale Fisheries Cooperatives of Ecuador operating under the National Cooperatives Law. Workers engaged in fish-related activities other than fishing (e.g., processors, transporters, traders associations) are fostered by other associations. Despite the key role played by these organizations in supporting and giving a voice to the small-scale sector, their actual power to influence management decisions is rather limited (Arriaga and Martínez 2008).

Galapagos Islands

The Galapagos National Park (GNP) created in 1959 encompassed only the terrestrial environment. In 1986, the first marine protected area (MPA) was declared in Galapagos. In 1998, after the passing of the Special law for Galapagos, a modified version of the MPA became the Galapagos Marine Reserve (GMR). With the formation of the GMR, a new participatory co-management system for small-scale fisheries came into effect within an area of 40 miles around the Galapagos Islands. This was the first participatory process vis-à-vis marine resources management in Ecuador's history. This co-governance model provided space for the different interest groups in GMR to legally participate in decision and policy making (Heylings and Bravo 2007). However, it also added complexity and diversity to the already complex and diverse small-scale fisheries sector, locally and nationally. Within this new management initiative, industrial fisheries were banned, exclusive fishing rights were allocated to local small-scale fishers (Parque Nacional Galápagos 1996, 2006; Moreno and Hearn 2007), and fishing areas were restricted through a marine zoning system.

The GMR management model includes two participatory management bodies. The first body is the Participatory Management Board (locally known as "*La Junta*"), which is a local executive forum that gives advice and is open for consultation. It comprises representatives from the local small-scale fisheries, tourism industry, Naturalistic Guides Association, science and education establishment (initially represented by the Charles Darwin Research Station), and the management sector (represented by GNPS, the executive arm of GMR in charge of implementing the management plan) (Parque Nacional Galápagos 2006; Baine et al. 2007). Interest groups can submit proposals regarding issues that require deliberations and consensus. The second body is the Inter-institutional Management Authority which is a ministerial forum for decision making, based in Ecuador's mainland. It is comprised of the Ministries of Environment (acting as President), Agriculture-Cattle-Aquaculture-and-Fisheries, Tourism, and Defence, the Ecuadorian NGOs Network,

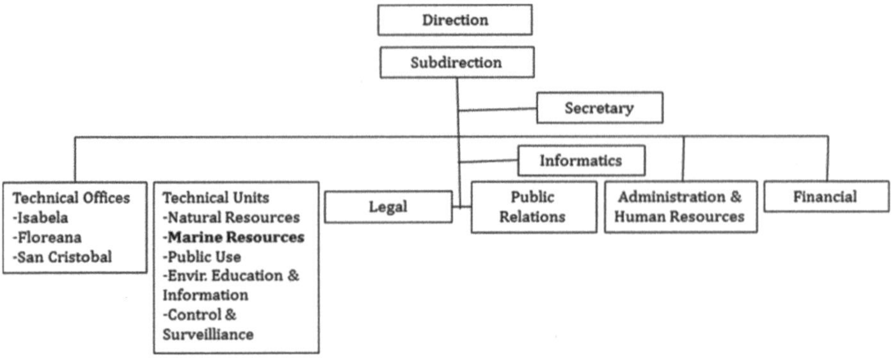

Fig. 9.3 Administrative structure of the Galapagos National Park, including small-scale fisheries within the marine resources technical unit (Parque Nacional Galápagos 1996)

local sectors (i.e., the small-scale fisheries and the Galapagos Chamber of Tourism), the Charles Darwin Research Station (acting as Technical Advisor), and GNPS (acting as Technical Secretariat for the Environment Ministry). In cases where consensus is not achieved at the Participatory Management Board level, the proposal is forwarded to the Inter-institutional Management Authority for resolution. A majority decision is required to pass a resolution. This resolution must be enforced by GNPS and/or its advisor(s). However, when urgent actions are needed, GNPS can take decisions by direct resolutions independently from both boards (Parque Nacional Galápagos 2006; Baine et al. 2007). Institutionally, GNPS has operative departments such as the Marine Resources Technical Unit under which fisheries-related issues are addressed (Fig. 9.3).

Since 1998, the Galapagos fisheries and other marine resources have been ruled by the Special Law. In 2008, the new National Ecuadorian Constitution (2008) recognized Galapagos as a Special Territory which implied that ordering, planning, and development in the islands must be undertaken by the Government Council. This political authority has embraced the idea of "living well" (i.e., *buen vivir* in Spanish; *sumaq kawsay* in Quichua). According to the new constitution, Galapagos would have a new governing system under the purview of the Government Council. This Council would replace the Inter-institutional Management Authority as the main decision making body in Galapagos province, though the functions and relationship between the Government Council, the GNPS, and the Participatory Management Board have not been clarified yet. According to Castrejón and Charles (2013), the future of the Galapagos co-governance system will remain unpredictable until the new governing structure is in place and new processes are in order.

Fishing in Galapagos is restricted to local small-scale fishers holding the PARMA license (Spanish acronym for Small-scale Fisher of the Galapagos Marine Reserve) under three main circumstances. First, the resource is managed as a common property resource (i.e., fisheries) by the local regime (Stone et al. 2006). Second, fishing rights to fish in Galapagos are allocated based on certain conditions (e.g., the legal migratory status of fishers, their inherited rights to fish, and the optional membership to local fisher's cooperatives). Third, the fulfillment of the total allowable catch (for sea cucumber and lobster) combined with other measures to control and limit fishing (e.g., fleet and boats size, the number of fishers, and type of gears used). Additionally, other management instruments are applied (e.g., opening/closure and extension of fishing seasons, and target species) through a fishery management plan called "*Capítulo Pesca*" (Castrejón et al. 2014). The above mentioned measures are established after biological assessments of the stocks are taken on a periodic basis through the Participatory Monitoring and Research Program (i.e., Charles Darwin Research Station staff, the GNPS, and the Galapagos fisher's cooperatives representatives). The final decision as to whether a fishery is to be opened is decided by consensus at the Participatory Management Board level, and then approved by the Inter-institutional Management Authority (Stone et al. 2006).

Small-scale fishers in Galapagos Islands are organized under the cooperative model, supported by the National Federation of Small-scale Fishers Cooperatives functioning under the National Cooperatives Law. Basically, these cooperatives are similar to the fishers' cooperatives of the mainland. The Galapagos cooperatives, unlike the ones in the mainland, are much more involved in decision and policy making.

Governing Interactions

Mainland Ecuador

In the Ecuador mainland, communication channels within the small-scale fisheries sector differ. Fisheries authorities have representatives (e.g., inspectors and officers) and technical staff in the coastal region who regularly monitor and check on the species and size of catches at the landing sites and fish markets (especially during ban periods) (Arriaga and Martínez 2008). This is one type of formal interaction at the post-harvest stage within the fish chain between the governing system representatives, the fishers, and their organizations. Other formal interactions between the governing bodies and fishers take place at the pre-harvest stage instance, namely when permits or licenses are being allocated (Campbell et al. 1991). Informal interactions between the governing system and fishers very seldom occur and are limited to private interactions mostly. Whereas formal interactions between fishers occur more with regard to economic transactions, informal interactions are shaped more by social and familial dynamics (Gaibor et al. 2002). Small-scale fishers have little direct interaction with global markets. High value fish species are mostly sold to processors and exporters (Béné et al. 2007).

Policy is communicated to fishers mostly through formal channels such as regulations, laws, amendments to laws, bans, and legal notifications. Informally, fishers obtain information and receive training from fishing governing bodies through capacity building workshops and courses. Public media is used as a mechanism to spread information (e.g., press) and to relay urgent information quickly (e.g., bad weather at sea, accidents on the sea, call for meetings, etc.).

Galapagos Islands

Interactions between the governing system and the small-scale fisheries sector in Galapagos, like in the mainland, takes place at formal and informal events, at pre- and post-harvest phases of the fish chain, locally and globally. Examples of formal interactions between the governing system and the fishers at the pre-harvest stage include patrolling activities within the GMR jurisdiction and the provision to fishers of fishing permits and boats licenses, activities by the GNPS and the Navy authority respectively. Other forms of formal interaction occur at the post-harvest stage, namely when concerned authorities undertake surveillance of landing sites and fish markets to check about species and size of catches (especially during ban periods). Informal interactions between the governing system and fishers occur individually or at family gatherings since GNPS staff members and fishers are sometimes part of the same family.

Interestingly, interactions between small-scale fishers at the individual or family level can take place locally or nationally and contribute as well to interactions between fishers from the mainland and islands. For years, fishers from Galapagos, who are originally from the mainland, maintained active relationships with their former communities and fishing peers, despite having migrated to the archipelago long ago. Family bonds remain as well as communication links. This is the case when fishers holding the PARMA license are permitted to fish in Galapagos when the sea cucumber season starts, regardless of whether they live permanently in Galapagos or in the mainland. However, illegal hiring of fishers from the mainland (without PARMA license) still occurs because they demand lower wages.

The main communication channel available to small-scale fishers in Galapagos is the management boards. Fisher leaders, who take part in these board meetings, communicate afterwards to their peers about resolutions taken. Additional ways to communicate to fishers include public media, such as street-located boards, local radio and TV programs, and local and national press. Information shared with fishers include that pertaining to fishing season opening/closure, ban periods and new regulations established, or requirements for the PARMA license renewal. As in the case of the mainland, information about global markets and their dynamics are normally communicated to fishers through their representatives or middlemen.

Discussion

We have argued that small-scale fisheries sector in the Ecuador mainland and Galapagos Islands faces a number of governance challenges that affect its governability. While Agüero (1992) and Arriaga and Martínez (2008) have highlighted critical issues affecting small-scale fisheries governance, they have not analyzed the causes for poor governance and how it affects the possibilities of improving governability. Additionally, little has been said and done to address issues such as

Table 9.1 Hierarchical and co-governance models for small-scale fisheries in mainland Ecuador and Galapagos Islands

Governing system	Hierarchical (mainland Ecuador)	Co-governance (Galapagos Islands)
Policy		
Nationally concerted policy	Yes (fisheries law)	No
Provincially concerted policy	No	Yes
Management style		
Management actions	Reactive/passive[a]	Preventive/active[b]
Instruments		
Global (UNCLOS[c])	Yes	Yes
International (CITES[d])	Yes	Yes
National MPAs[e]	Yes	Yes
Specific laws (e.g., Fishing Law; LOREG[f])	National	Local
Fishing ordering instrument	Yes (national)	Yes (local)
Regulations for tourism activities in PAs[g]	Yes (legal gaps)	Yes
Management plan	Yes (for areas/species)	Yes (integral)
Decision making		
Participative	No	Yes

[a]Passive management actions: short-term approach in terms of decisions taken as a response to problems or crisis (Castrejon 2011)
[b]Active management actions: preventive decisions taken as a precaution against crisis
[c]United Nations Convention for the Law of the Seas
[d]Conventin on International Trade in Endangered Species
[e]Marine Protected Areas
[f]Special Regime Organic Law for the Conservation and Sustainable Use of Galapagos Province, by its Spanish acronym
[g]Protected Areas

marginalization and poverty, and societal priorities and values that could help improve policy and governance (Andrew et al. 2007; Béné et al. 2007).

One key finding of this chapter is that the small-scale fisheries sector is significantly different in the mainland and islands in terms of fishing practice and management (Table 9.1). Geographic, political, and administrative attributes shape the respective fisheries.

Interactive governance, the theoretical framework applied in this case study, has helped address key questions asked about governability: (1) why do governance problems persist?, (2) where do they arise from?, and (3) what can be done? (Chuepagdee and Jentoft 2013). Three issues have been identified as critical in answering the above questions. First, governance problems persist because of incongruent jurisdictional and legal frameworks operating in parallel. Second, more attention needs to be given to social and cultural dimensions in fisheries governance as opposed to technical aspects. Third, information needs to be mobilized and shared better with fishers to improve governance.

Conflicting Legal Frameworks

The overriding principle governing fisheries in both areas is that of sustainability and social wellbeing. There is one single national constitution that at the end of the day dictates actions, priorities, and strategies that authorities, in both areas, must follow. In both areas, at least formally, the laws dictate norms, rules, ensure compliance, and punish violators. However, the existence of two governing bodies for fisheries resources, with overlapping jurisdictions and also legal gaps, facilitate a poor legal observance and a limited agency. All these factors make small-scale fisheries a difficult domain to govern. Consequently, governing systems have not been "*contextually sensitive and structurally diverse*" as Chuenpagdee and Jentoft (2009, 113) have argued, is necessary.

Borders of Fishing Communities

The Galapagos Islands and the mainland region are clearly defined by their geography and natural boundaries, though the social dynamics of small scale fisher communities is much less evident. It has been argued by others that the governance shortcomings of the fisheries sector in the Ecuador mainland and Galapagos Islands are due to social attributes of the local society (Jara 1987). According to Jobstvogt (2010), a key reason for failure in the governance of small-scale fisheries in the Galapagos Islands is the heterogeneity of the local fishing communities, an attribute that reduces the likeliness to agree on "common-interest" issues. According to Habermas (1997, as cited in Keulartz et al. 2004), consensus can only exist within single homogeneous communities but not between communities in pluralistic societies, holding competing views of the good life. Keeping that in mind, Ospina (2001) posits that different understandings of what constitutes a "fishing community" has critically influenced their conceptualization within Galapagos society and by doing so, has influenced the core construct of the social system of small-scale fisheries in the islands, which in Camhi's (1995) words has been shaped by the "get-rich-quick" mindset. Additionally, Galapagos inhabitants remain linked to their provinces of origin, despite their self-defined or imposed status of, early settler, resident, or illegal migrant in the Galapagos. Consequently, governing small-scale fisheries in this socially poorly defined system is challenging. In the Ecuador mainland, on the other hand, there are 312 fishing communities (i.e., *caletas*[1] *pesqueras*), clearly identified and spatially distributed in the five coastal provinces (SETEMAR 2013). It seems that these communities exist generally within beach community units that provide these fishing-related spatial areas or *caletas* with their own dynamics (ESPOL et al. 1987). However, their reduced governance seems to be

[1] *Caleta* is any village or town closely related and highly dependent on fishing activities.

influenced by weak social cohesion, competing interests for scarce resources, and weak law enforcement.

Economic boundaries are also not very clearly defined in terms of region and profession. Tour operators as well as the crew from cruise ships from the Galapagos have economic interests (e.g., investments) in both the mainland and islands simultaneously. Similarly, fishers in the islands own boats, fishing-gears, and also tourism licenses to operate in Galapagos, suggesting that they earn income in both the fisheries and tourism sectors. It is difficult, therefore, to estimate how much fisheries actually contributes to the local, national, or global economy as well as to food security, making governance choices more complicated.

Information Production and Mobilization

Irregular and unsystematic sharing of information with fishers also affects the governing system. These problems arise because scientific findings remain largely with the people and institutions who procure them. Information is rarely translated to make it accessible for fishers and other users, limited to public libraries, shared only in training courses, workshops, and meetings, and rarely displayed on public hoardings. Furthermore, there is limited institutional commitment to promote information flow. For example, in the Galapagos Islands, fishery-related research is conducted either by the GNPS or by the Charles Darwin Station (a scientific non-governmental organization) and mainly pertains to the biological and ecological dimensions of fisheries. Research findings are produced, stored, and circulated only as a scientific publication or as "grey literature". Fishers rarely can access this information. Equally importantly, fishers often do not trust this research and have little faith in research institutions, perhaps as a sign of loss of institutional resilience of those generating the information themselves (Castrejón et al. 2014).

Another problem that limits the applicability of science to fisheries governance is the limited role that fishers play in the production of knowledge. In the Ecuador mainland, fishers are not involved in research activities at any stage. As already pointed out, research produced by the National Institute of Fisheries and the National Center for Maritime and Aquaculture Research mostly pertains to the biological and ecological dimensions of fisheries, and that too of high value species. Even worse, the data generated by fishing inspectors (e.g., at monitoring landing sites) remains as raw data because using and publishing it involves such complex and bothersome bureaucratic processes that create disincentives for staff to use it (F.V. personal communication, July 15, 2014). While in Galapagos, fishers' do participate in scientific research, it is limited to data collection. Seldom are fishers consulted in terms of analysis and interpretation of data, i.e. in the production of knowledge. Fisher participation is, therefore, more about ticking the boxes pertaining to public participation than actual participation (Arnstein 1969), making the participatory process largely irrelevant and dysfunctional (Stone et al. 2006; Usseglio et al. 2014).

What Can Be Done?

As seen in the previous sections, governance of small-scale fisheries is vital to the future of small-scale fisheries in the Ecuador mainland and Galapagos Islands. Despite the actions taken by the National Government to partially mitigate social constrains by addressing issues of social justice, empowerment, human rights, and decent working conditions, significant challenges remain. Achieving governability understood as the "overall quality for governance" (Kooiman 2008) does not require fixed recipes nor "to-do-lists". What it requires is specific interventions in the different regions that address the social wellbeing of fishing communities and sustainable use fishing resources.

Gender

Key governance challenges described by the literature (Bavinck et al. 2005; Chuenpagdee et al. 2005) such as poverty, ecosystem health, social inequalities, livelihood enhancement, food safety and security, social justice, and addressing tradeoffs have been partially identified as missing pieces in the Ecuador and Galapagos fisheries discourse (Arriaga and Martínez 2008). However, issues such as gender equity have not. The national training program for small-scale fishers includes varied components: environmental education, business management, fishing processing techniques, leadership and entrepreneurship skills development. They seem aimed, however, exclusively to fisher men not women since different training programs are offered to women (e.g., baking and jewelry work). It is necessary that gender issues are included as a mandatory aspect within small-scale fisheries discourse when addressing fisheries sustainability and social wellbeing.

The "Perfect" Governance Mode

The co-governance mode used in Galapagos is promoted as the ideal model for marine resource management (Honey and Littlejohn 1994) since it foregrounds sustainability (Kosamu 2015). Co-governance certainly adds legitimacy to decision making processes, since varied interest groups are represented albeit unequally. However, participation from multiple interest groups adds complexity to the system and hence increases the potential for conflict and reduces the possibility of cooperation (Suárez de Vivero et al. 2008).

While co-management might be the preferred mode of governance, disagreements about fishing gear prohibition remain in GMR. For example, regarding the use of the artisanal vertical longlines, different opinions between the authorities, conservation groups, and fishers still exist. This highlights the fact that consensus-based decisions are not fully accepted by all involved. Moreover, users often feel that their representatives do not represent them adequately, complicating decision making. Corruption becomes rife in the system and real agreements are not achieved. Consensus based decisions in Galapagos constituted only two per cent of decisions taken by the Participatory Management Board between 2003 and 2007 (Bravo 2003 unpublished). When the numbers are disaggregated, it becomes clear that this two per cent includes not only consensus-based decisions but also compliance to decisions taken, and commitments being fulfilled.

Differently, in the mainland, the hierarchical model is claimed to be inadequate, authoritarian and centralized, and so, is argued to inhibit the user's willingness to take responsibility over the resource governance. However, good examples within this governing mode are found when addressing social justice (e.g., social insurance for fishers), and labor conditions (e.g., fishing harbor facilities building) issues. At the end, as Berkes et al. (2001), Pinkerton (1994, as cited in Hauck 2007) and Jentoft (1989) have suggested, for effective governance to take place participation of users must be genuine and all-encompassing.

Conclusions

Improved governance and increased small-scale fisheries governability require more than just management. Fisheries matters must be seen as community issues where communities are involved in defining what development means. Moreover, a nationally concerted approach to fisheries governance must take into account the particularities of governance modes and differences between regions while addressing common issues such as gender equity. It is important to go beyond thinking that co-governance is good and the hierarchical mode bad. In fact, there are no right/wrong responses in addressing the needs of the social and natural systems within fisheries. Both the hierarchical and co-management modes have positive and negative sides. It is important to note that there are no prescriptive models or recipes to solve the complexities of this sector.

It is also necessary to go beyond the rhetoric and see how actual systems function. Participation is often plagued with power dynamics, complicating sound policy making (Stone et al. 2006). While co-governance in the Galapagos is positively addressing issues of public participation, its success is not complete. Similarly, the hierarchical mode in the mainland in its more recent decentralized form facilitates governance, but perhaps weakens the national authorities' commitment. What all this says is that the sustainability of small-scale fisheries, as Kosamu (2015) posits, "solely depend on how strong the collective social capital of the local communities is at the resource scale". As Kosamu goes on to say "...with weak local social capital, the degree of government involvement is irrelevant; the fisheries result unsustainable in all cases." The point is that for small-scale fisheries governance in Ecuador (mainland and Galapagos) to improve communities must be strengthened.

Acknowledgements Thanks to the Fisheries Under-minister G. Morán, to other SRP staff members (V. Alvarado, R. Caiche, E. Castro, G. Cedeño, J.F. Bittar, I. Juez, G. Loor, J. López, and

F. Vera), to E. Barragán, and to L. Maridueña who kindly contributed with valuable information to the development of this case study. Thanks to the two anonymous reviewers who greatly contributed to the improvement of this manuscript.

References

Agüero, M. (1992). La pesca artesanal en América Latina: Una visión panoramic. In Agüero (Ed.), *Contribuciones para el estudio de la pesca artesanal en America Latina* (pp. 1–27). ICLARM Cod. Proc. 35. International Center for Living Aquatic Resources Management: Manila

Agüero, M. (Ed.). (2007). *Capacidad de pesca y manejo pesquero en América Latina y el Caribe.* FAO Documento Técnico de Pesca 461. Roma: Organización de las Naciones Unidas para la Agricultura y la Alimentación.

Allsopp, W. H. L. (1985). *Fishery development experiences.* Farnham: Fishing New Books Ltd.

Andrew, N. L., Béné, C., Hall, S. J., Allison, E. H., Heck, S., & Ratner, B. D. (2007). Diagnosis and management of small-scale fisheries in developing countries. *Fish and Fisheries, 8*, 27–240.

Anón, H. (1983). *Enciclopedia de las provincias del Ecuador (Galápagos).* Quito: Científica Latina Editores, Cia. Ltda.

Arnstein, S. R. (1969). A ladder of citizen participation. *Journal of the American Planning Association, 35*(4), 216–224.

Arriaga, L., & Egas, E. (1998). *Taller Pesca, Acuicultura y Desarrollo Sustentable. Plan Estratégico de Desarrollo y Seguridad Nacional, Ecuador.* Secretaría General de Planificación-VECEP. Informe Técnico. Guayaquil.

Arriaga, L., & Martínez, J. (2008). *Plan de Ordenamiento de la Pesca y Acuicultura del Ecuador.* Informe de Consultoría. Subsecretaría de Recursos Pesqueros-Ministerio de Comercio Exterior, Industrialización, Pesca, y Competitividad. Convenio de Préstamo BIRF-No. 43-46 EC.

Baine, M., Howard, M., Kerr, S., Edgar, G., & Toral, V. (2007). Coastal and marine resource management in the Galapagos Islands and the Archipelago of San Andres: Issues, problems and opportunities. *Ocean and Coastal Management, 50*, 148–173.

Bavinck, M., Chuenpagdee, R., Diallo, M., van der Heijden, P., Kooiman, J., Mahon, R., & Williams, S. (2005). *Interactive fisheries governance.* Delft: Eburon Publishers.

Bavinck, M., Chuenpagdee, R., Jentoft, S., & Kooiman, J. (2013). *Governability of fisheries and aquaculture. Theory and applications* (MARE publications series 7). Dordrecht/New York: Springer.

Beltrán-Turriago, C. S. (2001). *Promoción de la ordenación de la pesca costera. 2. Aspectos socio-económicos y técnicos de la pesca artesanal en El salvador, Costa Rica, Panamá, Ecuador y Colombia* (FAO Circular de Pesca. No. 957/2). Roma: FAO.

Béné, C. (2006). *Small-scale fisheries: Assessing their contribution to rural livelihoods in developing countries* (Fisheries Circular No. 1008, 46 pp). Rome: FAO.

Béné, C., Macfadyen, G., & Allison, E. H. (2007). *Increasing the contribution of small-scale fisheries to poverty alleviation and food security* (FAO Fisheries Technical Paper. No. 481, 125 p). Rome: FAO.

Bensted-Smith, R., Powell, G., & Dinerstein, E. (2002). Chapter 1: Planificación para la Ecoregión. In Fundación Charles Darwin para las Islas Galapagos (FCD) y WWF (Eds.), *Visión para la Biodiversidad de las Islas Galápagos. Taller Internacional de Biólogos de la Conservación* (pp. 11–16). Puerto Ayora: Galapagos.

Berkes, F., Mahon, R., McConney, P., Pollnac, R., & Pomeroy, R. (2001). *Managing small-scale fisheries: Alternative directions and methods.* Ottawa: International Development Research Center.

Bravo, M. (2003). *Sistema de seguimiento de la Junta de manejo participativo documento de trabajo no. 8* (Proyecto No. 1274 / 0C-EC). Programa de Manejo Ambiental de las Islas Galápagos – PMAIG. Mayo, 2003.

Breton, Y., Brown, D., Davy, B., Haughton, M., & Ovares, L. (Eds.). (2006). *Coastal management in the wider Caribbean: Resilience, adaptation and community diversity*. (ebook) Ian Randle Publishers/IRDC. Retrieved August 6, 2008, from http://www.idrc.ca/openebooks/336-4/

Bustamante, R., Espinoza, E., Nicolaides, F., Murillo, J. C., Chasiluisa, C., Ruttemberg, B., Andrade, R., Torres, S., Toral, M. V., Barreno, J., & Piú, M. (1999). *Fishing in Galapagos: A summary of the main indicators for 1998*. In Fundación Natura & WWF (Eds.), *Galapagos report 1998–1999* (pp. 35–49). Trama Publishers: Quito.

Camhi, M. (1995). Industrial fisheries threaten ecological integrity of the Galapagos Islands. *Conservation Biology, 9*(4), 715–724.

Campbell, R., Fallows, J., Scott, I., Ortiz, J., Rodríguez, T., & Mora, Y. (1991). Una revisión del sector pesquero artesanal en el Ecuador y los factores de consideración para el desarrollo. *Boletín Científico y Técnico, Instituto Nacional de Pesca, 11*(8), 1–4.

Castrejón, M. (2011). *Co-manejo Pesquero en la Reserva Marina de Galápagos: tendencias, retos y perspectivas de cambio*. Mexico: Tinker Foundation/ECCD/Kanankil.

Castrejón, M., & Charles, A. (2013). Improving fisheries co-management through ecosystem-based spatial management: The Galapagos Marine Reserve. *Marine Policy, 38*, 235–245.

Castrejón, M., Defeo, O., Reck, G., & Charles, A. (2014). Fishery science in Galapagos: From a resource-focused to a social-ecological systems approach. In J. Denkinger & L. Vinueza (Eds.), *The Galapagos Marine Reserve, social and ecological interactions in the Galapagos Islands* (pp. 160–185). New York: Springer Science+Business Media.

Charles, A. T., Brainerd, T. R., Bermudez, M. A., Montalvo, H. M., & Pomeroy, S. R. (1994). *Fisheries socioeconomics in the developing world: Regional assessments and an annotated bibliography*. Ottawa: IDRC.

Chuenpagdee, R., & Jentoft, S. (2009). Governance assessment for fisheries and coastal systems: A reality check. *Human Ecology, 37*, 109–120.

Chuenpagdee, R., & Jentoft, S. (2013). Assessing governability – What's next. In M. Bavinck, R. Chuenpagdee, S. Jentoft, & J. Kooiman (Eds.), *Governability of fisheries and aquaculture. Theory and applications* (MARE publications series 7, pp. 335–349). Dordrecht/New York: Springer.

Chuenpagdee, R., Degnbol, P., Bavinck, M., Jentoft, S., Johnson, D., Pullin, R., & Williams, S. (2005). Challenges and concerns in capture fisheries and aquaculture. In J. Kooiman, M. Bavinck, S. Jentoft, & R. Pullin (Eds.), *Fish for life. Interactive governance for fisheries* (pp. 25–37). Amsterdam: Amsterdam University Press.

Cinner, J. E., Daw, T., & McClanahan, T. R. (2009). Socioeconomic factors that affect artisanal fishers' readiness to exit a declining fishery. *Conservation Biology, 23*, 124–130.

Clifford, N. J., & Valentine, G. (Eds.). (2003). *Key methods in geography*. London/Thousand Oaks: Sage.

Coello, S., & Fundación Natura. (1993). *Diagnóstico de la Actividad Pesquera en la Zona de Influencia del Parque Nacional Machalilla*. Estudios en Áreas Protegidas. Volumen 5. Fundación Natura.

Comisión Permanente del Pacífico Sur (CPPS). (2000). *Plan de Acción para la Protección del Medio Marino y Áreas Costeras del Pacífico Sudeste*. Quito: Estado del Medio Ambiente Marino y Costero del Pacífico Sudeste.

Danulat, E., & Edgar, G. (Eds.). (2002). *Reserva Marina de Galápagos. Línea Base de la Biodiversidad*. Santa Cruz: Fundación Charles Darwin/Servicio Parque Nacional Galápagos.

Defeo, O., & Castilla, J. C. (2012). Governance and governability of coastal shellfisheries in Latin America and the Caribbean: Multi-scale emerging models and effects of globalization and climate change. *Current Opinion in Environmental Sustainability, 4*, 344–350.

Defeo, O., Castrejón, M., Pérez-Castañeda, R., Castilla, J. C., Gutiérrez, N., Essington, T. E., & Folke, C. (2014). *Co-management in Latin American small-scale shell fisheries: Assessment from long-term case studies*. Fish & Fisheries Journal.

Dirección Parque Nacional Galápagos (DPNG). (2014). *Plan de Manejo de las Áreas Protegidas de Galápagos para el Buen Vivir*. Galápagos: Puerto Ayora.

Domínguez, B., Casquete, C., & Suárez-Avilés, J. (1991). *Análisis socioeconómico de los pescadores artesanales de la zona sur de Manabí, Caso específico: Puerto Cayo, Machalilla, Puerto López y Salango*. Tesis de Economista, Universidad de Guayaquil.

Ecuadorian National Constitution. (2008). Supra note 5, at art. 71.

Edgar, G. J., Banks, S., Fariña, J. M., Calvopiña, M., & Martínez, C. (2004). Regional biogeography of shallow reef fish and macro-invertebrate communities in the Galapagos archipelago. *Journal of Biogeography, 31*, 1107–1124.

ESPOL, CEPLAES, & ILDIS. (1987). *La pesca artesanal en el Ecuador*. Quito: CEPLAES.

FAO. (2013). *Fisheries in Ecuador*. Retrieved October 3, 2013, from http://www.fao.org/fi/oldsite/FCP/es/ecu/body.htm

FAO. (2014). *The state of world fisheries and aquaculture. Opportunities and challenges*. Rome: FAO Fisheries and Aquaculture Department/Food and Agriculture Organization. Retrieved December 1, 2014, from www.fao.org

Finley, C. (2009). The social construction of fishing, 1949. *Ecology and Society, 14*(1), 6. Retrieved June 22, 2012, from http://www.ecologyandsociety.org/vol14/iss1/art6/

Gaibor, N., Rosero, J., & Altamirano, M. (2002). *El Impacto de la Migración Humana en las Artes Pesqueras Artesanales y Semi-industriales utilizadas en los Parques Nacionales Galápagos (Isla Isabela) y Machalilla*. Informe de Consultoría, The Nature Conservancy/Fundación Natura, La Unión.

Grenier, C. (2007). *Conservación Contra Natura. Las Islas Galápagos*. Quito: AbyaYala.

Habermas, J. (1997). *Between facts and norms*. Cambridge: Polity Press, as cited in Keulartz, J., van der Windt, H., & Swart, J. (2004). Concepts of nature as communicative devices: the case of Dutch nature policy. *Environmental Values, 13*(1), 81–99.

Hannah, L., Ikegami, M., Hole, D. G., Seo, C., Butchart, S. H. M., Townsend Peterson, A., & Roehrdanz, P. R. (2013). Global climate change adaptation priorities for biodiversity and food security. *Plos One, 8*(8), e72590.

Hauck, M. (2007). Non-compliance in small-scale fisheries. A threat to security? In P. Beirne & N. South (Eds.), *Issues in green criminology. Confronting harms against environments, humanity, and other animals* (pp. 270–290). Willan Publishing, Cornwall.

Heylings, P., & Bravo, M. (2007). Evaluating governance: A process for understanding how co-management is functioning, and why, in the Galapagos Marine Reserve. *Ocean and Coastal Management, 50*, 174–208.

Honey, M., & Littlejohn, A. (1994). Paying the price of ecotourism. *The Americas, 46*(6), 40–47.

Instituto Nacional de Estadísticas y Censos (INEC). (2010). *Censo Nacional de Población y Vivienda*. Retrieved December 10, 2013, from http://www.ecuadorencifras.gob.ec/censo-de-poblacion-y-vivienda/

Jacquet, J., Alava, J. J., Ganapathiraju, P., & Henderson, S. (2008). In hotsoup: Sharks captured in Ecuador's waters. *Environmental Sciences, 5*(4), 269–283.

Jara, C. J. (1987). Y si la pesca artesanal desaparece? In ESPOL, CEPLAES, & TLDIS (Eds.), *La pesca artesanal en el Ecuador* (pp. 129–137). Quito.

Jentoft, S. (1989). Fisheries co-management: Delegating government responsibility to fishermen's organizations. *Marine Policy, 13*, 137–154.

Jentoft, S., & Chuenpagdee, R. (2013). Concerns and problems in fisheries and aquaculture exploring governability. In M. Bavinck, R. Chuenpagdee, S. Jentoft, & J. Kooiman (Eds.), *Governability of fisheries and aquaculture. Theory and applications* (MARE publications series 7, pp. 33–44). Dordrecht/New York: Springer.

Jobstvogt, N. (2010). *Fish stock assessment of top-predator wahoo, Acanthocybium solandri, in the Galapagos Islands*. Diploma thesis, Department of Biology, Chemistry & Pharmacy. Faculty of Biology, Freie Universität Berlin.

Josse, C., & Iturralde, M. (Eds.). (2002). *Compendio de Investigaciones en el Parque Nacional Machalilla*. Fundación Natura/Centro de Datos para la Conservación, CDC-Ecuador. Quito: Nuevo Arte.

Kerr, G. N., & Swaffield, S. R. (2012). Identifying cultural service values of a small river in the agricultural landscape of Canterbury, New Zealand, using combined methods. *Society & Natural Resources, 25*(12), 1330–1339.
Keulartz, J., van der Windt, H., & Swart, J. (2004). Concepts of nature as communicative devices: The case of Dutch nature policy. *Environmental Values, 13*(1), 81–99.
Kooiman, J. (2008). Exploring the concept of governability. *Journal of Comparative Policy Analysis: Research and Practice, 10*(2), 171–190.
Kooiman, J., Bavinck, M., Jentoft, S., & Pullin, R. (Eds.). (2005). *Fish for life: Interactive governance for fisheries*. Amsterdam: Amsterdam University Press.
Kooiman, J., Bavinck, M., Chuenpagdee, R., Mahon, R., & Pullin, R. (2008). Interactive governance and governability: An introduction. *The Journal of Transdisciplinary Environmental Studies, 7*(1), 171–190. ISSN 1602–229. Retrieved from http://www.journal-tes.dk/
Kosamu, I. B. M. (2015). Conditions for sustainability of small-scale fisheries in developing countries. *Fisheries Research, 161*, 365–373.
Kuzel, A. J. (1992). Sampling in qualitative inquiry. In B. F. Crabtree & W. L. Miller (Eds.), *Doing qualitative research* (pp. 31–44). London: Sage.
Marder, R., & Arcos, C. (1985). *Normas Societarias, Actitudes, Grupos de Poder y Conflicto en Galápagos*. Documento para Discusión, Cuaderno No. 7. Pontificia Universidad Católica del Ecuador, Facultad de Ciencias Humanas, Departamento de Sociología.
Martínez, J., Cobeña, M., & Domínguez, C. (1997). *Las pesquerías artesanales de Puerto López (provincia de Manabí-Ecuador) durante 1996*. Realizado por el Área de Pesca del Programa de Cooperación Técnica para la Pesca Unión Europea-VECEP ALA92/43. Inf. Téc. PESCART 03/97. Ecuador. In Gaibor et al. (2002).
Mays, N., & Pope, C. (1995). Rigour in qualitative research. *British Medical Journal, 311*, 109–112.
McEwan, C., & Silva, M. I. (1998). Arqueología y comunidad en el Parque Nacional Machalilla. In C. Josse & M. Iturralde (Eds.) (2002). *Compendio de Investigaciones en el Parque Nacional Machalilla*. Quito: Fundación Natura/Centro de Datos para la onservación, CDC-Ecuador. Editorial Nuevo Arte.
Moreno, A., & Hearn, A. (2007). Investigación Pesquera. Investigación para el Manejo. In Fundación Natura and WWF (1999) *Galapagos report 1998–1999* (pp. 19–29). Quito.
Norton, P. (1985). *Boletín de los museos del Banco Central N° 6*. Simposio 45 Congreso Internacional de Americanistas. Universidad de los Andes, Bogotá.
Olson, D. M., & Dinerstein, E. (1998). The global 200: A representation approach to conserving the Earth's most biologically valuable ecoregions. *Conservation Biology, 12*(3), 502–515.
Olson, D. M., Dinerstein, E. D., Wikramanayake, N. D., Burgess, G. V. N., Powell, E. C., Underwood, J. A., D'Amico, I., Itoua, H. E., Strand, J. C., Morrison, C. J., Loucks, T. F., & Allnutt, T. H. (2002). The global 200: Priority ecoregions for global conservation. *Annals of the Missouri Botanical Garden, 89*, 199–224.
Ospina, P. (2001). *Identidades en Galápagos. El sentimiento de una diferencia*. Quito: Trama.
Palacios, P., & Schuhbauer, A. (2012). *Tourism as an economic alternative for Galapagos fishers: Opportunities and lessons learned* (pp. 109–113). Galápagos report 2011–2012.
Parque Nacional Galápagos (PNG). (1996). *Management plan for conservation and sustainable use of the Galapagos marine reserve*. Galapagos National Park Service, Puerto Ayora: Galápagos Islands.
Parque Nacional Galápagos (PNG). (2006). *Plan de Manejo*. Ministerio del Ambiente/Parque Nacional Galápagos. Quito: Arte Digital.
Pauly, D., Christensen, V., Guénette, S., Pitcher, J., Sumaila, U. R., Walters, C. J., Watson, R., & Zeller, D. (2002). Towards sustainability in world fisheries. *Insight Review. Nature, 418*, 689–695.
Pauly, D., Watson, R., & Alder, J. (2005). Global trends in world fisheries: Impacts on marine ecosystems and food security. *Philosophical Transactions of the Royal Society B, 360*, 5–12.
Perry, R. I., Ommer, R. E., Barange, M., Jentoft, S., Neis, B., & Sumaila, U. R. (2011). Marine social-ecological responses to environmental change and the impacts of globalization. *Fish & Fisheries, 12*, 427–450.

Pinkerton, E. (1994). Summary and conclusion. In C. L. Dyer & J. R. McGoodwin (Eds.), *Folk management in the world´s fisheries: Lessons for modern fisheries management* (pp. 317–337). Colorado: University of Colorado Press.
Pollnac, R. B., & Poggie, J. J. (2008). Happiness, well-being and psycho-cultural adaptation to the stresses associated with marine fishing. Special Section on Vulnerability and Resilience in Fisheries. *Human Ecology Review, 15*(2), 194–200.
Pollnac, R. B., & Bavinck, M. (2008). Alternative livelihoods and job satisfaction among fishermen: a cross-national study. In *EOCST: Ecosystems, societies, consilience, precautionary principle: Development of an assessment method of the societal cost for best fishing practices and efficient public policies*. D4SA Ecoregion report on sociological aspects: Results of the Job Satisfaction surveys of all case-studies (85 pp).
Pomeroy, R., & Berkes, F. (1997). Two to tango: The role of government in fisheries co-management. *Marine Policy, 21*, 465–480.
Ramírez, J. (2004). *La Pesca Artesanal en la Reserva Marina de Galápagos. Dinámica Laboral y Conflictos Socio-Ambientales*. Disertación Licenciatura en Antropología Social, Facultad de Ciencias Humanas, Pontificia Universidad Católica del Ecuador, Quito.
Reck, G. K. (1983). *The coastal fisheries in the Galápagos Islands, Ecuador: Description and consequences for management in the context of marine environmental protection and regional development*. Doctoral thesis, Christian-Albrechts-Universität, Kiel.
Rodríguez, W. T. (1987). Pesquerías Artesanales en las Islas Galápagos, Ecuador. en: J. Martínez, A. Ansaldo, & M. Hurtado (1987) *La Pesca Artesanal en el Ecuador*. Guayaquil: Instituto Nacional de Pesca.
SETEMAR. (2013). *La Pesca en el Ecuador. Cap. 1* (48 pp). Technical report.
SRP. (2013). http://www.subpesca.gob.ec/subpesca. Accessed 04 Oct 2013.
Stewart, M. L. F. (2009). *Of fish and men: An economic analysis of the Galapagos Marine Reserve Resources Management Plan*. Doctoral thesis, Agricultural and Resource Economics, University of California Davis.
Stone, S., Viteri, C., Conrad, J., Nøstbakken, L., & Franklin, H. (2006). *Fisheries management in the Galapagos Marine Reserve: A bioeconomic perspective*. Region 3 Barbados, Colombia, Ecuador, Guyana, Jamaica, Peru, Suriname, The Bahamas, Trinidad & Tobago, Venezuela. Economic and Sector Study Series. Washington: Inter-American Development Bank.
Suárez de Vivero, J. L., Rodríguez Mateos, J. C., & Florido del Corral, F. (2008). The paradox of public participation in fisheries governance. The rising number of actors and the devolution process. *Marine Policy, 32*, 319–325.
Symes, D. (Ed.). (2000). *Fisheries dependent regions*. Oxford: Fishing New Books Blackwell Science.
Teddlie, C., & Yu, F. (2007). Mixed methods sampling. A typology with examples. *Journal of Mixed Methods Research, 1*(1), 77–100.
Usseglio, P., Schuhbauer, A., & Friedlander, A. (2014). *Collaborative approach to fisheries management as a way to increase the effectiveness of future regulations in the Galapagos Archipelago*. The Galapagos Marine Reserve. New York: Springer.
Walmsley, S. F., Howard, C. A., & Medley, P. A. (2005). *Participatory fisheries stock assessment (ParFish) guidelines*. London: MRAG.
Williams, M. (1998). Aquatic resources. Education for the development of world needs. In D. Symes (Ed.) (2000). *Fisheries dependent regions* (pp. 164–174). Oxford: Fishing new Books Blackwell Science.
Wolff, M., Schuhbauer, A., & Castrejon, M. (2012). A revised strategy for the monitoring and management of the Galapagos Sea cucumber (Isostychopus fuscus). *Revista de Biologia Tropical, 60*(2), 539–551.
Worm, B., Hilborn, R., Baum, J. K., Branch, T. A., Collie, J. S., Costello, C., Fogarty, F. G., Fulton, E. A., Hutchings, J. A., Jennings, S., et al. (2009). Rebuilding global fisheries. *Science, 325*(5940), 578–585.

Chapter 10
Governability, Self-Governance and Co-governance in the Context of Lake Victoria Fisheries, Tanzania

Paul O. Onyango

Abstract Policy makers are faced with the challenge of improving how small-scale fisheries are governed. Governance is here understood beyond the rule of law, transparent and accountable government and a vibrant civil society which capacitates its citizens to claim their rights. The chapter broadens this understanding to include interactions among governors. By presenting a case study of Lake Victoria small-scale fisheries in Tanzania, the genesis of how the fisheries were governed and how fisher communities around the lake have been self-governing are traced. Using the interactive governance framework, I discuss the quality of local fishers self-governing on the one hand and co-governance on the other hand. The latter is tracked from when the government started to manage the fisheries. The chapter therefore examines the impact of the government's intervention on the self-governance of fisher communities. The argument made is that achieving high governability will require an arrangement that borrows institutional understanding from established local practices in fisher communities and combining these, in a practical manner, with professional understanding of management institutions, which are easily adopted by governments.

Keywords Governability • Interactive governance • Self-governing • Co-governance • Tanzania • Lake Victoria

P.O. Onyango (✉)
Department of Aquatic Sciences and Fisheries, University of Dar es Salaam, Dar es Salaam, Tanzania
e-mail: onyango_paul@yahoo.com

S. Jentoft, R. Chuenpagdee (eds.), *Interactive Governance for Small-Scale Fisheries*, MARE Publication Series 13, DOI 10.1007/978-3-319-17034-3_10

Introduction

Since the early 1990s, small-scale fisheries around the world have not performed as expected and some have become threatened (McGoodwin 1994; Béné 2003; Béné et al. 2004; FAO 2005; Clark 2006). There has been a persistent decline in environmental quality evidenced by deteriorating water quality, increasing poverty, landings that have either remained stagnant or have been declining, and fish stocks going down (Béné 2003; Green et al. 2003; Béné et al. 2004; McClanahan et al. 2008). These changes have occurred despite management efforts being in place. Why is this so? Do the efforts in place lack the needed tenacity? Are these fisheries inherently difficult to govern? Is it a problem with the institutions governing them? Or is it a lack of understanding about how fisheries work or about the nature of how authorities interact with fishers?

This chapter applies the interactive governance approach to examine the questions raised above. By looking at Lake Victoria fisheries governance mechanisms in Tanzania, the chapter explores the capacity of fishers and their institutions to take on governance functions and responsibilities so as to address the problem of declining environmental quality. It notes that fisher communities' self-governing system is able to govern itself and confront problems facing the fishers as well the natural system; i.e. it has a high capacity to govern. However, governability reduced when the colonial government started to manage the fisheries in the late 1890s and continued to remain low after independence. Not even the introduction of a co-governance system through the establishment of Beach Management Units increased governability (BMUs) (URT 1997; Fisheries Division 2005). The chapter therefore examines the causes and limits of this low governability, investigating, for instance, how the decision-making process enables or restricts fishers and fisheries authorities to execute their governance roles. It also examines whether or not the decision-making process is transparent, responsive to fishers, inclusive and based on consensus building.

It is argued in this chapter that achieving high governability for Lake Victoria fisheries will require an arrangement that borrows institutional understanding from established local practices in fisher communities and combines these with a professional understanding of management institutions, which are easily adopted by governments (Booth and Cammack 2013). In other words, enhancing governability will require taking institutional ingredients from local communities and expert institutional designs and combining these in a co-governance arrangement that is acceptable to both groups and is practically workable.

The chapter first presents a theoretical understanding on governability based on the interactive governance perspective. This is followed by a discussion of the different components of interactive governance in the context of Lake Victoria, focusing especially on questions of diversity, complexity, dynamics and scale. The next section analyzes the effectiveness of self-governance and co-governance. The final section is devoted to the interactions that go on within the fishers' self-governance system and the co-governance institution that the government has initiated. It discusses the nature, type and quality of governing interactions and the differences between the two governance modes in terms of how they deal with the natural and social systems-to-be-governed. It also examines the impacts of the two systems on fish stocks.

Governability in the Context of Interactive Governance

What is governability from an interactive governance lens (Kooiman 2003; Kooiman and Chuenpagdee 2005; Kooiman et al. 2005; Bavinck and Salagrama 2008; Chuenpagdee and Jentoft 2009)? Interactive governance is an analytical framework that operates on the assumption that societal systems are made up of three components, namely a system-to-be-governed, a governing system and governing interactions (Bavinck et al. 2005; Chuenpagdee and Jentoft 2009; Jentoft and Chuenpagdee 2009). A societal system, as Kooiman and Bavinck argue, are "the whole of interrelations among a given entities belonging to the natural and social worlds" (2013, 13). It is part of larger events and structures. The system-to-be-governed is made up of natural and social systems and is where activities that meet most basic human needs are located. A governing system, on the other hand, is made up of a set of governors, mechanisms and processes, that are used to manage, steer and control the system-to-be-governed. As discussed below, it is within this system where decision-making takes place. The governing interactions are made up of mutually influencing relationships that go on between two or more actors in a governance environment (Bavinck et al. 2005; Jentoft and Chuenpagdee 2009; Chuenpagdee and Jentoft 2013; Kooiman and Bavinck 2013).

The components of the societal system are characteristically diverse, dynamic and complex. Society also operates at various scales, for example, at local, village, divisional, district, regional, national, and international or global scales. These characteristics are what pose challenges to the system's governability. Governance demands that governors tackle inherent problems arising from the systems-to-be-governed. Furthermore, the ability of the governing system to deliver on the challenges of the system-to-be-governed constitutes a key aspect of governability (Kooiman et al. 2008; Kooiman 2010).

Within interactive governance, governing activities take place at three orders: first-, second- and third or meta-governance. In the first-order governance, governors address the daily problems that society faces or that exists within the three related systems. In so doing, they may create opportunities, as they get involved in a process of decision-making.

At second-order governance, institutions are established to tackle societal problems daily. As suggested by Kooiman and Bavinck (2013) institutions—which are sets of rules and regulations for stimulating, regulating and controlling human behavior—is part of governance. Institutions comprise organizations where decision-making takes place, and rules and regulations that are implemented. Kooiman and Jentoft (2005) argue that there is a lot of interactive learning that goes on as problems are tackled and opportunities created. The learning takes place at the three orders of governance. Interactive learning they state "is a process in which participants learn from each other and from each other's learning" (2005, 289).

As governors make decisions, they are often influenced by values and principles, or what governance theory calls meta-governance (Kooiman 2003; Kooiman and Jentoft 2005, 2009). They guide the governors or actors involved in governing interactions always directing governing from a distance (Kooiman and Chuenpagdee 2005). Governing interaction is about bringing the governing system and systems-to-be-governed together for easy coordination and harmonization. Therefore governability will depend on the qualities of the system-to-be-governed, governing system and governing interactions (Kooiman and Bavinck 2013).

Methods

In order to understand governability in the fisheries of Lake Victoria (Fig. 10.1), eight interviews—two each in Magu, Ukerewe, Bukoba and Rorya districts—with fishers were carried out between December 2013 and June 2014. These interviews focused on how fisher communities govern themselves, solve conflicts among themselves and ensure proper use of the fisheries resources. Data from several interviews conducted prior to 2013 were also used. These were used to understand the history of the fisher communities after Tanzania's independence. Secondary data and literature on how decisions were made in the traditional riparian fishing communities of Lake Victoria were collected and analyzed between December 2013 and June 2014. Data and literature were retrieved from other published and unpublished sources located at libraries in the Lake region as well as at the University of Dar es Salaam. Some additional literature was retrieved from websites. I also draw from my experience and knowledge gained in working in the area for over 20 years and interacting with several ethnic communities including the Sukuma, Kerewe, Luo, Jita, Haya and Wakabwa. These are fishing communities who have lived along the lake for an unknown period of time.

Lake Victoria System

The Natural System

The lake is among the most diverse natural systems in the world. Since approximately 14,000 years ago (Johnson et al. 2000), the lake has witnessed an unusually rapid speciation which has been described as 'spectacular' (Okeyo-Owuor 1999). Okeyo-Owuor argues that "the rate of evolution and speciation of the fishes of the Great Lakes of East Africa warrant the use of the terms such as evolutionary avalanche, explosive evolution and explosive speciation" (1999, 10).

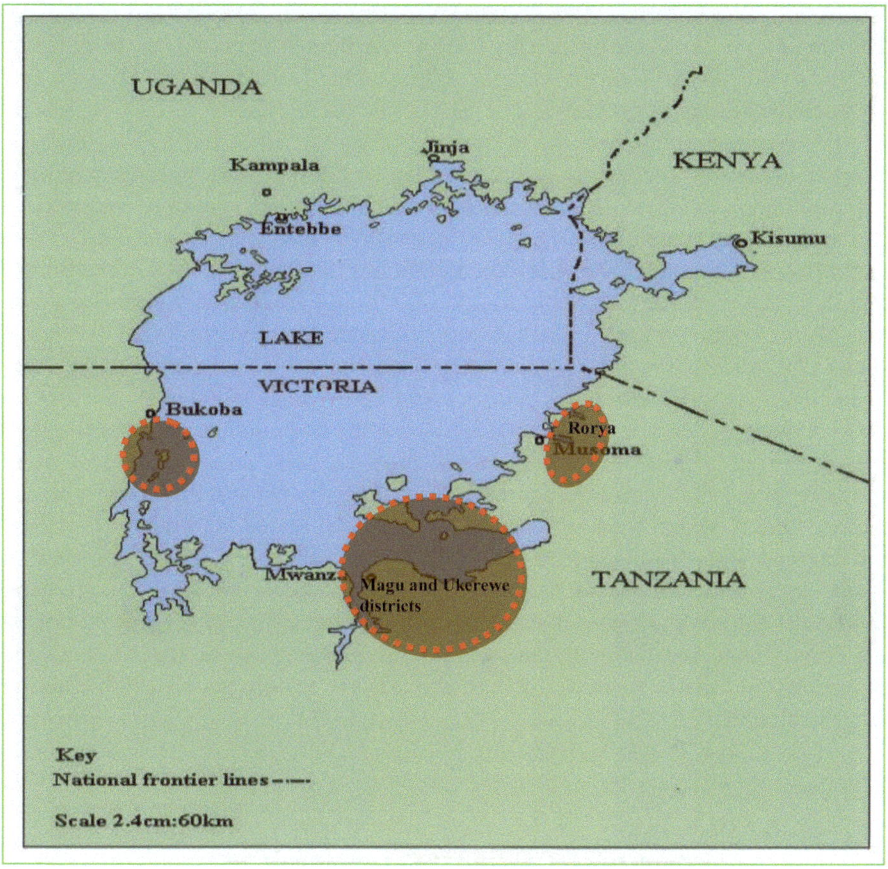

Fig. 10.1 Map of Lake Victoria showing study area (in *red circles*) (Source: LVFO)

Numerous species have been recorded in the lake (Witte et al. 1992, 1995, 2000; Ligtvoet et al. 1995). Johnson et al. (2000) indicate that over 400 endemic species of cichlids occupied the lake after the Pleistocene period. However, species diversity has not been static. Nonetheless, as late as the 1970s, the lake was still a home to a large number of endemic fish species comprising more than 28 genera with more than 350 species (Witte et al. 1992, 1995, 2000). Cichlids—mainly haplochromines and two tilapiine species, most notably *Oreochromis esculentus* and *O. variabilis*—were amongst the most abundant, contributing 300 species. The tilapiine species (Graham 1929) constituted the main fishery between the 1980s and early 1990s when they were overtaken in importance by the Nile perch fishery. Besides for the cichlids 38 other species, belonging to other families, were present in the lake (Graham 1929). In the 1950s, a decrease in the tilapiine species and *Labeo victorianus* were observed. This decrease was attributed to intensive gill-net fishing. Thus in order to boost the decreasing fishery, five new fish species that included *O. niloticus* (Nile Tilapia), *Lates niloticus* (Nile perch), *Tilapia rendalii, T. zillii, and O. leucostictus* were introduced.

The natural system overall has been deteriorating since the 1970s, due to a number of factors including pollution from agricultural and mining activities, increased pressure from urban population growth, and the spread of the invasive weed, the water hyacinth (*Eichhornia crassipes*). Fish catches have also been declining. For example, the Nile perch yield has decreased from a maximum of 338,000 tonnes in 1990 to about to 234,000 tonnes in 2007. Its biomass which was estimated at two million metric tonnes has dropped to below safe biological limits ($B_{lim} = 0.25$; $B_0 = 500{,}000$ t) (Marriot and Halls 2013). This has made management extremely challenging and 'wicked' (Rittel and Webber 1973; Conklin 2006; Chuenpagdee and Jentoft 2013). There has been debate whether to allow the lake to go back to its former state of high diversity and healthy ecosystem status or allow Nile perch to continue to change the ecosystem. Even with the boom of the Nile perch fishery, concerns have been raised with regard to directing management objectives specifically at increasing the targeting of commercial species only or also allowing fishing of indigenous species. In addition, there are proposals that advocate that management should not restrict fishers from targeting certain specified fish sizes and weights, for example those that are between 50 and 80 cm long or those that weigh above 2 kg (Kolding et al. 2008; Kolding and Zweiten 2010).

The Social System

Ever since the potential for fisheries was known, fishing in the lake has been undertaken through various means and by a number of ethnic communities. Historically the riparian fisher communities of the lake included the Luo, Jita, Sukuma and the Wakerewe. However, other riparian communities such as the Haya and Sinza and non-riparian communities from across the country also joined the fishery (Onyango Forthcoming). Fisher communities used gillnet and beach-seines made out of sisal to fish. These were used for a long time before an observed reduction in fish sizes led to mesh size regulations in the early 1900s.

The fishery has also witnessed an influx of fishers as a result of migration from other sectors such as agriculture and mining (URT 2010). The increased number of fishers has also meant increased fishing pressure and number of conflicts (Heck et al. 2004). Fishers use different gears such as gillnets, long lines and driftnets. A fisher can own one to several boats with several gillnets or long lines or a driftnet. Some fishers own up to 3,000 hooks of long lines and up to 10 gillnets. Fishers are organized into groups or camps, which vary by size, depending on the number of boats owned by the group or camp owner. A group or a camp may consist of one boat or up to 50–60 boats.

Fishers have reported conflicts ranging from entanglement of gears in fishing grounds, to theft of fishing gears, to transboundary conflicts. Other conflicts include conflicts as a result of using gears meant for commercial fish species even for indigenous species. In other words, conflicts arise because fishers use gillnets of above five inch mesh size to target species which mature at sizes smaller than five inches and hence are illegally caught. With respect to dagaa and Nile perch fisheries, complex and ever changing fish chains have been established. In the Dagaa fishery, for instance, once fishers land their catches, there are women fish traders who buy the fish, dry it and either sell it directly to consumers or to other traders from distant markets. There are also collectors who collect the fish and sell it to other traders. Others buy the fish specifically to sell to chicken feed making plants. The Nile perch fish chain is not so different except that the chain ends in overseas markets. Networks have been established in the chain; for example, there are several women and marketing groups whose main activities have been to trade fish locally as well as with buyers for fish processing plants.

The Governing System

The governing system involves several players. During the pre-colonial period, local communities managed the lake resources through their clan and kinship system where tribal chiefs and clan leaders and elders ruled (Onyango 2004). Tribal leaders managed the sections of the lake where their communities resided. There was no lake wide coordinated management. Management of the lake resources was integrated in their overall management system. In other words, fishing was practiced as an activity that formed part of a livelihood system. There were fishing, farming, harvesting and resting periods. Clan elders had exclusive rights to make decisions on virtually all matters concerning the lake's resources, livelihood activities as well as other aspects of community life. In most cases the decisions they made related to fisheries pertained to when fishers could go fishing.

Tribal were charged with the responsibility of deciding on all matters affecting the tribe and concerning the sub-tribe in relation to other sub-tribes. These responsibilities included decisions about whether a tribe had to go to war with other tribes, maintenance of peace, and adjudicating on offences against the community and suits between parties. In other words, these communities were self-governing but also hierarchical. Tribal members were compliant to the rules and regulations set and overseen by their leaders. This was because tribes shared common traditions and each member knew these traditions. If one did not follow them it was easy to know. All members of a clan were known to each other and hence if any member broke any rule it was quickly reported to the clan elders who called the culprit and reprimanded him or her immediately. Clan members found it easy to point fingers at those who broke rules. This tribal governance system was however done away with when the communities were colonized during the late 1890s to mid- 1990s. The colonial government abolished tribal government as they sought to expand their rule over several tribes and bigger territories (Alemazung 2010). The colonial government also managed fisheries resources.

After independence in 1960, fisheries management became the responsibility of the Tanganyika state (URT 1998, 2003). Local communities became users rather than governors. In the 1980s, fisheries became lucrative with the Nile perch boom and a ready market overseas. A number of actors entered the picture, reducing the role of those fishers who had local knowledge. Markets exerted great influence over the fisheries. The government enacted various laws and regulations to manage the fisheries. Fisheries officers were employed at various institutional levels related to fisheries, including within research institutes and fisheries colleges. The government also decentralized its administrative system and established district governments, with specific roles and responsibilities and some autonomy given to them. District Fisheries Officers were posted within the district governments.

However, due to the observed decline in fish catches and health of the lake, the government re-introduced community participation in 1997 through a co-governance regime (Hoza and Mahatane 2003). The government felt that they could not manage the fisheries alone due to widespread environmental degradation resulting from the use of destructive fishing gears including beach seines and undersized gillnets. Inadequately trained fisheries staff and budgetary constraints posed further problems.

The government through the Ministry of Livestock and Fisheries Development (MFLD) was charged with policy formulation and guidance. The ministry worked with district governments—district fisheries officers and ward fisheries staff—whose responsibilities were to implement fisheries regulations. At the regional level, the Lake Victoria Fisheries Organization (LVFO), (See Fig. 10.1) was established through a tripartite interstate agreement involving Tanzania, Kenya and Uganda so as to enable joint management (LVFO 2001).

Under the co management regime, the government, whether central or district, has the final authority on matters regarding fisheries (URT 2003). The day-to-day decisions (first-order) and activities are made in ministry and district offices. If a law is to be repealed or reviewed, the ministry has to decide on how and when to do it. According to the law, the ministry basically refers to the Minister or the Permanent Secretary who executes his functions through the Director of Fisheries. Decisions made at this level have minimal involvement of fishers. However, BMUs have become major players and must be consulted on changes to fisheries laws. The only limitation in terms of the involvement of BMUs is that given the size of the lake and the fact that there are over 430 BMUs existing in Tanzania alone, it has not been possible to consult all BMUs but only their national and regional/lake wide leaders. Once an agreement has been reached at that level, the Minister prepares a Bill for Parliament to discuss and then recommends it to the President.

Others involved in the decision making process include researchers, fish processing plant owners, and civil society organizations such as Tweyambe (Onyango and Jentoft 2011), Environmental Management and Economic Development Organization (EMEDO), Tanzania Coalition of Sustainable Development and East

African Communities Organization for the Management of Lake Victoria Resources (ECOVIC).

The laws and regulations which have been used to manage fisheries from the colonial period onwards include the Fish Protection Ordinance of 1950, the Tanzanian Fisheries Policy of 1997, and Act of 2003. These laws all focused a lot on input controls in fisheries, especially mesh size controls, fishing gear types, and numbers of fishers permitted to fish As well as a fisheries license. The licenses are issued by the District Fisheries Officers. The requirement of having a fishing license issued by district authorities, once fulfilled, enables fishers to move from one place to another in search of fish. Moreover, fishers can move from one landing site to another and practice illegal fishing. If they are noticed by the authorities, they move to another landing site. Also fisheries law does not restrict entry into the fisheries. As a consequence of these factors, fisheries governance involves many challenges. Catching fishers who engage in illegal fishing is difficult. Getting fishers to report illegal activities are equally difficult as they do not know what harm will come to them. As a result of these challenges, these laws have undergone several reviews since they were enacted.

Managing the lake's fisheries through these laws has also been difficult because of differences in interpretation. Since the lake is a shared resource among three countries, each country has enacted its laws and implements them according to their own jurisdiction. While LVFO helped to harmonize these laws, there was a period in which aspects of the law, namely those pertaining to mesh size restrictions, was understood differently in each country. For instance, in Tanzania mesh size restrictions were read as referring only to gillnets, but in Kenya fishers understood it to refer to the cod-end of a beach seine. Currently, there is a harmonized mesh size regulation for dagaa nets and gillnets for Nile perch fisheries, and beach seines are banned in the entire lake. Each country is advised to make sure its laws comply with the harmonized regional laws. While in theory countries have done this, in practice minimum mesh size varies from country to country, with Tanzania requiring the largest size mesh. Tanzanian fishers have asked why they should be subjected to stricter regulations than their colleagues in other countries.

Governing Interactions

The above discussion with regard to interactions around governance of the lake raises questions about governability. Is a system involving many governors (Fig. 10.2) more governable than a system with fewer governors such as those in traditional societies? As the Fig. 10.2 diagrammatically shows, interactions in Lake Victoria begin right from the beach level, where BMUs have established interactions between and among themselves. The BMUs have formed divisional, district, national and regional committees where decisions at the beach level are discussed and coordinated. For instance, divisional BMU committees ensure that decisions agreed upon within the district or division, which are based on decisions at the

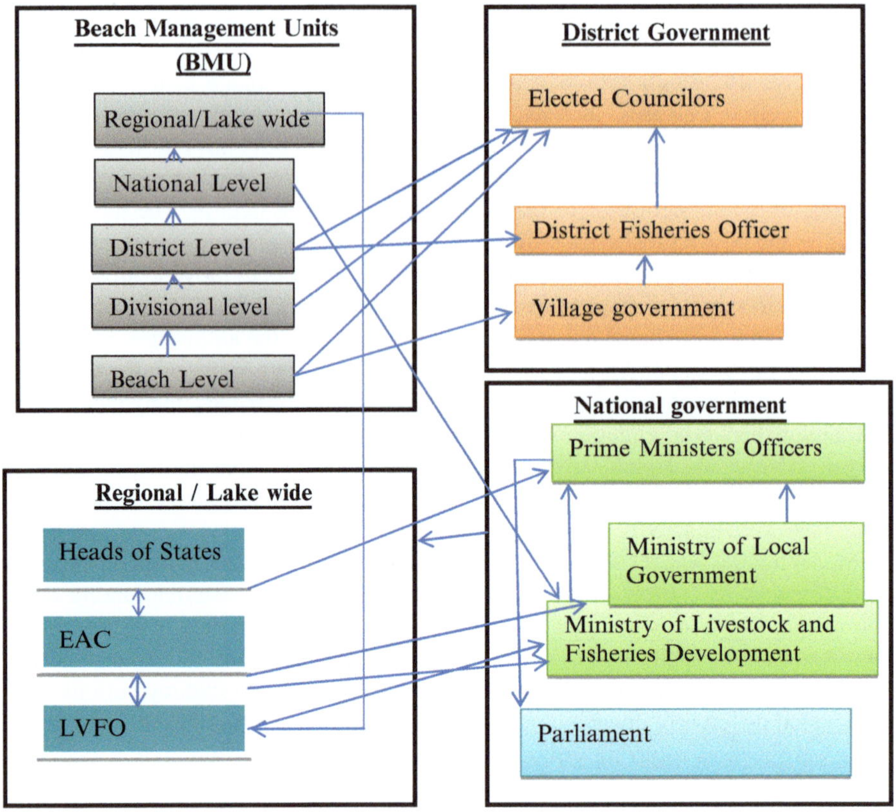

Fig. 10.2 Governing interactions in the Lake Victoria fisheries

beach level, are coordinated and adhered to. So there is a flow of information from the BMUs at the beach up the ladder to the regional committee. This is in complete contrast to what existed in the pre-colonial period where clan elders took charge of everything. There was no authority beyond the clan elders especially with regards to fishing and other livelihood activities. Elders from one clan however networked with neighboring clan elders.

BMU decisions and/or deliberations are supposed to flow to the districts through the village government. The latter is required to act as clearing house where BMU decisions requiring their action are implemented even as they send the decisions to the districts. Once all information from BMUs in the district is received at the district level, the District Fisheries Officer is required to relay the same to the Ministry of Livestock and Fisheries Development where there is an established BMU desk. The ministry can change a decision reached by a BMU and or district government if such a decision is in conflict with national fisheries or other laws. Changes could also be made to take care of different interests groups. In traditional society, once a decision was reached by the clan elders, it was final. Only the same elders could change it.

Ministry of Livestock and Fisheries Development is required to provide answers to questions, address challenges and give guidance and solutions to any identified problems to BMUs. The ministry can also generate and send requisite information BMUs. Such information is sent through the District Fisheries Officers. Lately, the Ministry of Local Government has come into the picture as custodian of decentralization in the country. Although Ministry of Livestock and Fisheries Development has direct communication with the district officials, such communication is based on a gentleman's agreement. However, communication from the Ministry of Local Government is taken to be a directive. All district authorities are answerable to the Local Government Ministry directly, and not to Ministry of Livestock and Fisheries Development.

Information flows from the national government to the regional level. It should be understood that the national government is sovereign and is not obligated to report to regional institutions like LVFO. However, the countries sharing the lake have agreed in principle to share information for the purposes of jointly managing the lake and its resources so as to facilitate, for example, joint patrols and monitoring of stocks. At the regional level, the LVFO has clear working mandates. Researchers are required to provide scientific information on a particular issue to managers for implementation. Once laws have been harmonized at the LVFO level, the secretariat communicates the same to the ministries responsible for fisheries in the partner countries for action. Civil society and the private sector participate in decision-making at the LVFO level through their representatives.

How does this governing interaction affect governability of the lake's small-scale fisheries? Whereas this is a subject of continuous debate, what is observed is that fish stocks have and continue to decrease (Marriot and Halls 2013; Mgaya forthcoming), the lake's littoral zones are polluted (Kolding et al. 2008) and there is a continued use of illegal fishing techniques (URT 2010; LFVO 2011). In other words communication and implementing responsibilities of governors, those charged with ensuring sustainable fisheries exploitation on the one hand and those monitoring pollution on the other hand, in this complex governing interaction have not operated effectively. Subsequently this has negatively affected the systems-to-be-governed leading to low governability.

Modes of Governance

Two different interaction patterns have emerged with regard to the governing of Lake Victoria fisheries. First is self-governance which refers to fishers managing their own fisheries resources. Second is co-governance in which fisher communities and fisheries authorities together make decisions.

Self-Governance

Fisher communities operate as if they are completely dependent on the fisheries authorities but in reality they are not as they operate with some level of independence. They do not want to admit that they are operating as an independent entity because this is likely to jeopardize the kind of relationship they have developed with fisheries officials working in their villages and districts. On the other hand fisheries officials do also at times allow illegal undersized fish to be caught in their areas of jurisdiction.

The current relationship between the fishers and fisheries officials is a symbiotic one where each party protects the interest of the other.

It is required by law that all fishing vessels be licensed and provided with numbers, which must be written on the vessel. The license has to be renewed annually. There is a patrol unit charged with ensuring adherence to fisheries laws and regulations. They make unannounced visits to landing sites. On one occasion, the patrol team arrived in one landing site at a time when fishers had already landed their catch. They found several unlicensed boats and gears, but when they asked for the owners, no one claimed ownership. Every person they talked to claimed that they were visiting, hence the officials were unable to track the owners of the boats and gears. Even children who were found playing in the village claimed that they did not know the owners. Everybody they talked to was a visitor.

After their narration I asked them why they behaved that way. They told me that they did not want to be blamed by their fellow fishers and community members for having been the person to divulge information that would bring down a fisher or a community member whose life depends on fishing. On further examination, it seems that fisher communities have formulated an informal ethical mode of operation, which defines acceptable behavior in their community, a system of ensuring that outside interference does not affect their operations until what they aim at is achieved. They have also instituted a system of conflict resolutions and discipline such as gossip, warning, fines, censure and oath taking. This is an illustration of self-governing qualities within them. These qualities are important to their fishing activities as they control behavior and how they relate to the fisheries resources.

Fishers have immense knowledge about the natural system of the lake, especially when it comes to fish species diversity. They know which fish to catch and when to fish for it. They are knowledgeable about the ecosystem, breeding areas, fishing seasons, and fishing gears required for different species. This level of knowledge has not been adequately tapped by the government. However, it has enabled the self-governance system to be governable. Fisher communities' governing system at the local level involves very few institutions—historically their tribal institutions and nowadays their community informal institutions—whose obligations are clearly spelled out. There are no overlaps as each institution functions according to its expectations. In contrast, the government has neither been able to fully understand the natural nor the social systems. Neither have research biologists who provide advice.

In the fisher community, the relationship with the natural system was mediated through clans. Today it is mediated through groups or fishing camps. Each camp consists in most cases of fishers from the same community. The relationship between groups and camps happens with regard to various fronts ranging from sharing knowledge about fishing spots that have been discovered, collecting baits together, assisting one another while fishing, providing fish for food for those who did not go fishing on a particular day, and assisting fellow fishers in times of need and bereavement. Although the relationship is fairly complex, the system works well for fishers.

Historically people who migrated to the Lake zone learned to keep together as a community as this was the only means of winning wars and gaining access to land and resources. As a consequence these communities developed habits of keeping secrets—silent behavior—always designing strategies to beat their opponents. This silent behavior included taking decisions in places only accessible to community members and reprimanding community members from discussing community matters with strangers. Even intermarrying was encouraged as a strategy of winning over opponents. During discussions, all members were given opportunity to talk, either to present a suggestion or provide information. No one wanted the community to be defeated and made subjects of other communities. Every community member participated in one way or the other to ensure that community objectives were achieved. Through these strategies, the communities managed to continue governing themselves.

Some of these strategies continue to be used by fisher communities. It is often the case that a fishing boat comprises four people that may normally either come from one family or are from one tribe. In a fishing village, it is normal for one tribe to form a welfare group to assist all the members in times of need. During fieldwork in these communities, I observed several instances where community members gathered publicly to solve different forms of conflicts such as those related to theft of gears, fighting over fishing grounds and physical fights. Thus these societies have autonomous systems, which function well and provide for their needs through collective action.

Co-governance

Although co-governance was formally introduced in 1997, government efforts to manage the fisheries can be traced back to 1908 when the first colonial legislation for the lake known as the Fish Protection Ordinance was enacted. This law legalized the gillnet fishery after it was observed from landings by colonial administrators that *O. esculentus* had reduced in size (Geheb 1997). Changes in fish size resulted from both intensive fishing to the demands from far away markets made possible by the newly constructed railway line. In 1947, a Lake Victoria Fisheries Service (LVFS) was established to enforce the gillnet mesh size as stipulated in the Fish Protection Ordinance (Hoza et al. Forthcoming). Local communities were not subjected to this law because it was thought that they would disapprove of it as they would find it illegitimate.

Fisheries continued to decline until the 1950s when Nile perch was introduced. However, LVFS collapsed in 1960 and was transformed into the East Africa Fresh Water Fisheries Research Organization (EAFFRO) that itself came to an end along with the East Africa Community in 1977–1978 (EAFRO 1958; EACSO 1967). The Tanzanian government then formed a Fisheries Division which up to 1999 solely managed the fisheries. The government's presence in the lake was, however, very low until Nile perch attracted an overseas market in the 1980s. This was because there were inadequate staff and resources. Fishers continued to perceive fish resources to be government property meant for exploitation rather than community sustenance.

Due to widespread environmental degradation resulting from the use of destructive fishing gears and methods, inadequately trained fisheries staff and budgetary constraints, co-governance was introduced in 1997 by establishing BMUs (RK et al. 1995; Hoza and Mahatane 1998). BMUs were then rolled out across the lake in the succeeding 7 years. Evaluations of their effectiveness (Luomba and Mhagama 2007; NCO 2007) in discharging their responsibilities have been mixed. There are areas in which BMUs have been somewhat effective, for instance in resolving conflicts within fisher communities, fostering collaboration with other BMUs and fisheries authorities, participating in joint patrols and maintaining cleanliness in landing sites as well as fish weighing buildings.

A major challenge that co-governance has faced is cultural insensitivity. Local community institutions were built on social relations. However, co-governance is based more on meeting certain economic and fishery related goals. The responsibilities they are charged with such as beach cleanliness, use of badges or racks for receiving fish from fishermen during landing, keeping records of fishers' gears and boat licenses amongst other things, have an economic bearing rather than social. Consequently, local fisher communities have had to struggle with ensuring that the BMUs address their objectives, most importantly making sure that they get fish to their tables. This is why members of fishing communities do not reveal any illegalities committed by their own community members. What all this implies is that socio-culturally sensitive institutions are more likely to result in successful fisheries resources governance. Fisheries governance reform, therefore, should emphasize the socio-cultural dimension more.

Discussion

Factors that Make the System Less Governable

Is there a correlation between multiple actors being involved in governing and high governability? The empirical information presented in this chapter suggests otherwise. Whereas, at present there are many governors in the management of the lake's fisheries, the fishery is currently under threat. Experts predict that if the current effort level is not reduced then the Nile perch fishery will collapse by 2025 (Marriot and Halls 2013). While the fishery has in the past received substantial funding (Mgaya Forthcoming), to reverse the decline in fish catches observed since the mid-1990s, amongst other things, it has not resulted in such a reversal (Marriot and Halls 2013). Thus the state of the fisheries and the level of governability discussed suggest that there is need to re-examine what has gone wrong.

As described above, problems associated with governing small-scale fisheries can be traced to the systems-to-be-governed, the governing system and the governing interaction. It is also within these systems that a governability assessment should be undertaken. It has been suggested in this chapter that fisher community systems are more governable when not interfered with by the government. Once the government comes in, governability reduces. Why is this so? What are the limiting factors and what can be learnt from this? One of the limiting factors lies in the governing system and has to do with what fisheries laws aim to achieve. In the fisher communities, their informal laws aimed at strengthening interactions and relationships among community members. This is why fisher communities cannot easily divulge information to outsiders. In the case of government laws, the focus is different. Government laws are focused on fish stocks. The emphasis on mesh size regulations, types of fishing gears, types of fishing practices has created a perception among fisher communities that laws are meant to prevent them from fishing and consequently obtaining food for survival. For instance, when laws prevent fishers from catching certain species, fishers see this as being denied access to resources. This can also explain why fishers protect their fellow fishers who practice illegal fishing.

Second, a complex interactive governing system that involves several players is also a limiting factor. Figure 10.1 shows a diagrammatic simplification of interactions that have been established in governing the fisheries of the lake. However, failure to adhere to the laid down reporting channels has reduced the quality of interactions. For example, meetings and communication between governors is irregular. Even at the LVFO level, meetings of the statutory organs as stipulated in their operations have not always been convened. The communication system has also not been as inclusive and transparent as was the case within self-governing systems of fisher communities. The feedback system is also very weak. There is no clearly defined mechanism that representatives can use to inform those they represent. At most, they end up disseminating the information to those who are interested or those who ask for it.

Factors that Make the Systems More Governable

A number of factors appear to have contributed to making the systems more governable. First, when looking at how fisher communities operate, one cannot help but notice that they are concerned with social justice, equity, and participatory decision-making (Hernes et al. 2005; Jentoft 2013a). These factors are also the foundation upon which co-governance was formulated. The kind of social justice fishers are concerned with in these communities relates to access to fish resources for food and survival. Fisher communities emphasize that everybody who has the ability to fish should not be denied the opportunity, regardless of where one comes from, type of gear one uses or the number of fishers already fishing in the lake (SEDAWOG 2000). This is why fishers do not provide information to patrol teams or to fisheries officials. As Hernes et al. (2005, 104) argue "fisheries management should not be reduced to a technical exercise that should only be seen from a means-end perspective …… but must be subjected to moral scrutiny". In other words, issues such as social justice, legitimacy and access rights should be taken into consideration.

One other factor that appears to have contributed to high governability is simplicity of the decisions-making process. As argued above, local fisher communities were earlier subject to the decisions of their clan leaders. Currently also, their decisions are made primarily by community leaders or opinion leaders. In the co-governance system decisions are made at various levels including the BMUs, village government, district government and central government. In other words, there are more players in the decision making process within the co-governance system than in the self-governance system of fishers. Although the self-governing of Lake Victoria fisher communities shows that the less the number of decision-making players the higher the governability level, is this an arrangement that can be borrowed for the co-governance of the lake's fisheries?

Conclusion

Two main lessons can be drawn from this chapter. The first lesson has to do with recognizing that the low governability of small-scale fisheries is part of a bigger problem. Although observed problems of fisheries governance are to be found in the systems-to-be-governed, these problems need to be seen in a broader perspective that takes into account the entire system, namely the system-to-be-governed, governing system and governing interactions and their collective interactions. To identify the problem requires finding out when the problem begun and then establishing what caused it. Governors must stay conscious of and open minded about emerging challenges, be able to notice them and examine how they affect governability in small-scale fisheries.

Second, governability of fisheries in the Lake Victoria case is affected by two systems operating side by side. On the one hand, there is the self-governing fisher community system and on the other hand he government system. The latter has found its way into the former and has been powerful enough to demand compliance from the former. The consequence of this has been low governability of these fisheries. For this reason it is important to rethink how best to make the system more governable. From the case study what emerges is the need to draw ingredients from different sources—for instance from the self-governing system and the government system—combining them in a practical hybrid form (Bavinck 2005; Booth and Cammack 2013; Jentoft 2013b; Kooiman and Bavinck 2013). In other words, a good arrangement would be that takes on institutional understanding from fisher communities and marrying it with the government system, constituting a new form of practical hybridity. This can be made practical, especially where local fishers are reluctant to divulge information as is the case in Lake Victoria, through among other things building fishers' capacities in intelligence gathering and prevention of illegal fishing.

The idea is to understand governability beyond following a 'best practice' method adopted elsewhere. That is to say, good governance can be measured in terms of how accountable leaders are to the public, transparent government, periodic free and fair elections and a vibrant civil society. The lessons learned from this case study also suggest that governance should be contextualized in order to achieve high governability. Co-governance, as presently practiced in Lake Victoria, can be made more effective if it borrows, some if not all, characteristics that made self-governance more governable.

The systems-to-be-governed present challenges which may be similar in different places but still require home grown solutions. For example in Lake Victoria, fishing indigenous species using gears meant for commercial species has presented a challenge to governors. It could be argued that a simple solution is to allow multi species gear fishing. However, this may create conflicts as evidenced in the lake and even impact upon the stocks, thereby making the fisheries not perform as expected. Hence, the concern as to how to address this without impacting fisher communities and commercial species stocks still remains. It is for this reason that high governability will require an arrangement that borrows institutional understanding from established local practices in fisher communities and combining these with professional understanding of management institutions and practice.

Acknowledgement This chapter has been produced with the help of various people whose contribution cannot go without mention. First, I thank the fishing communities in Rorya, Bukoba and Magu districts for their willingness to provide information used here. I am grateful to the Fisheries Officials at the ministry and, district and village levels who took their time to accommodate me despite their busy schedule. I am particularly indebted to all those who reviewed this chapter. I thank them for putting up with my ideas and the way they have shaped and made them readable. Finally and most importantly I thank the Tanzania Commission for Science and Technology (COSTECH) for supporting field data collection.

References

Alemazung, J. A. (2010). Post-colonial colonialism: An analysis of international factors and actors marring African socio-economic and political development. *The Journal of Pan African Studies, 3*(10), 62–84.

Bavinck, M. (2005). Understanding fisheries conflicts in the South-A legal pluralist perspective. *Society and Natural Resources, 18*(9), 805–820.

Bavinck, M., & Salagrama, V. (2008). Assessing the governability of capture fisheries in the Bay of Bengal – A conceptual enquiry. *The Journal of Transdisciplinary Environmental Studies, 7*(1), 13.
Bavinck, M., Chuenpagdee, R., Diallo, M., van der Heijden, P., Kooiman, J., Mahon, R., & Williams, S. (2005). *Interactive fisheries governance: A guide to better practice*. Delft: Eburon.
Béné, C. (2003). When fishery rhymes with poverty: A first step beyond the old paradigm on poverty in small-scale fisheries. *World Development, 31*(6), 949–975.
Béné, C., Bennet, E., & Neiland, A. E. (2004). The challenge of managing small-scale fisheries with reference to poverty alleviation. In E. Neiland & C. Béné (Eds.), *Poverty and small-scale fisheries in West Africa* (pp. 83–102). Dordrecht: FAO/Kluwer Academic Publishers.
Booth, D., & Cammack, D. (2013). *Governance for developing countries: Solving collective action problems*. London: Zed Books Publishers.
Chuenpagdee, R., & Jentoft, S. (2009). Governability assessment for fisheries and coastal systems: A reality check. *Human Ecology, 37*, 109–120.
Chuenpagdee, R., & Jentof, S. (2013). Assessing governability. What is next? In M. Bavinck, R. Chuenpagdee, S. Jentoft, & J. Kooiman (Eds.), *Governability of fisheries and aquaculture: Theory and applications* (MARE publication series. 7, pp. 335–342). Dordrecht: Springer.
Clark, W. C. (2006). *The worldwide crisis in fisheries: Economic models and human behaviour*. Cambridge: Cambridge University Press.
Conklin, J. (2006). *Dialogue mapping: Building shared understanding of wicked problems*. West Sussex: Wiley.
EACSO. (1967). Annual report 1966. *East African Freshwater Fisheries Research Organisation*. Kampala: Uganda Press Trust Ltd.
EAFRO. (1958). *East African Fisheries Research Organization (EAFRO)*. Annual report 1956/1957 (45 pp). Jinja: East Africa High Commission.
FAO. (2005). *Increasing the contribution of small-scale fisheries to poverty alleviation and food security* (FAO Technical Guidelines for Responsible Fisheries No. 10, 97 pp). Rome: FAO.
Fisheries Division. (2005). *National guidelines for beach management units* (33 pp). Government of Tanzania: Dar es Salaam.
Geheb, K. (1997). *The regulators and the regulated: Fisheries management, options and dynamics in Kenya's Lake Victoria Fishery*. Unpublished doctoral thesis, University of Sussex, Falmer, Brighton, U.K. Reprinted as LVFRP Technical Document No. 10. LVFRP/TECH/00/10. Jinja: The Socio-economic Data Working Group of the Lake Victoria Fisheries Research Project.
Graham, M. (1929). *The Victoria Nyanza and its fisheries. A report on the fishing surveys of Lake Victoria (1927–28)*. London: Crown Agents Colonies.
Green, S. J., White, A. T., Flores, J. O., Carreon, M. E., & Sia, E. A. (2003). *Philippines fisheries in crisis: A framework for management* (77 pp). Cebu City: Coastal Resources Management Project of the Department of Environment and Natural Resources.
Heck, S., Ikwaput, C., Kirema-Mukasa, T., Lwenya, C., Murakwa, D. N., Odongkara, K., Onyango, P. O., Owino, J. P., & Sobo, F. (2004). *Cross-border fishing and fish trade on Lake Victoria*. Fisheries Management Series 1. IUCN/LVFO Socio-economics of the Lake Victoria Fisheries Phase II (97 pp). IUCN Eastern Africa Regional Programme & Lake Victoria Fisheries Organization (LVFO).
Hernes, H. K., Jentoft, S., & Mikalsen, K. H. (2005). Fisheries governance and participatory decision-making. In T. Grey (Ed.), *Participation in fisheries governance* (pp. 103–118). Dordrecht: Springer.
Hoza, R., & Mahatane, A. T. (1998). *Co-management in Mwanza gulf*. A Lake Victoria environmental management project report. Fisheries Management Component. Dar-es-Salaam Ministry of Natural Resources and Tourism July 1998.
Hoza, R. B., & Mahatane, A. T. (2003). Establishment of collaborative fisheries management in the Tanzania Part of Lake Victoria. In S. G. M. Ndaro & M. Kishimba (Eds.), *Proceedings of the LVEMP – Tanzania 2001 scientific conference, 6–10 August 2001* (pp. 15–36). Mwanza: Tanzania Lake Victoria Environmental Management Project Dar es Salaam.
Hoza, R. B., Mgaya, Y. D., & Bwathondi, P. O. J. (Forthcoming). Historical trends in fisheries management. In Y. Mgaya (Ed.), *Lake Victoria fisheries*.

Jentoft, S. (2013a). Social justice in the context of fisheries – A governability challenge. In M. Bavinck, R. Chuenpagdee, S. Jentoft, & J. Kooiman (Eds.), *Governability of fisheries and aquaculture: Theory and principles* (MARE publication series 7, pp. 45–65). Dordrecht: Springer.

Jentoft, S. (2013b). Walking the talk: Implementing the international voluntary guidelines for securing sustainable small-scale fisheries. *Maritime Studies, 13*, 1–16.

Jentoft, S., & Chuenpagdee, R. (2009). Fisheries and coastal governance as a wicked problem. *Marine Policy, 33*, 553–560.

Johnson, C. T., Kelts, K., & Odada, E. (2000). The holocene history of Lake Victoria. AMBIO. *Journal of Human Environment, 29*(1), 2–11.

Kolding, Z., & van Zweiten, P. (2010). *The tragedy of our legacy: How do global management discourses affect small-scale fisheries in the south?* Public lecture given at the University of Tromsø.

Kolding, J., van Zwieten, P., Mkumbo, O., Silsbe, G., & Hecky, R. (2008). Are the Lake Victoria fisheries threatened by exploitation or eutrophication? In G. Bianchi & H. R. Skjoldal (Eds.), *The ecosystem approach to fisheries* (pp. 309–354). Oxfordshire: CAB International.

Kooiman, J. (2003). *Governing as governance*. London: Sage.

Kooiman, J. (2010). Governance and governability. In S. P. Osborne (Ed.), *The new public governance* (pp. 87–104). London: Routledge.

Kooiman, J., & Bavinck, M. (2013). Theorizing governability – The interactive governance perspective. In M. Bavinck, R. Chuenpagdee, S. Jentoft, & J. Kooiman (Eds.), *Governability of fisheries and aquaculture: Theory and principles* (MARE publication series 7, pp. 9–30). Dordrecht: Springer.

Kooiman, J., & Chuenpagdee, R. (2005). Governance and governability. In J. Kooiman, M. Bavinck, S. Jentoft, & R. Pullin (Eds.), *Fish for life: Interactive governance for fisheries* (pp. 325–349). Amsterdam: Amsterdam University Press.

Kooiman, J., & Jentoft, S. (2005). Hard choices and values. In J. Kooiman, M. Bavinck, S. Jentoft, & R. Pullin (Eds.), *Fish for life: Interactive governance for fisheries* (pp. 285–299). Amsterdam: Amsterdam University Press.

Kooiman, J., & Jentoft, S. (2009). Meta-governance: Values, norms and principles, and the making of hard choices. *Public Administration, 87*(4), 818–836.

Kooiman, J., Bavinck, M., Jentoft, S., & Pullin, R. (Eds.). (2005). *Fish for life: Interactive governance for fisheries*. Amsterdam: Amsterdam University Press.

Kooiman, J., Bavinck, M., Chuenpagdee, R., Mahon, R., & Pullin, R. (2008). Interactive governance and governability – An introduction. *The Journal of Transdisciplinary Environmental Studies, 7*(1).

Lake Victoria Fisheries Organization (LVFO). (2001). *The convention for the establishment of the Lake Victoria fisheries organization* (40 pp). Jinja: LVFO Secretariat.

Ligtvoet, W., Mkumbo, O. C., Mous, P. J., & Goudswaard, P. C. (1995). Monitoring fish stocks from survey data. In F. Witte & L. T. van Densen (Eds.), *Fish stocks and fisheries of Lake Victoria: A handbook for field observations* (pp. 119–134). Cardigan: Samara Publishing Limited.

Luomba, J., & Mhagama, F. (2007). *Report on governance in co-management survey in Lake Victoria, Tanzania. Implementation of a fisheries management plan (IFMP)* (42 pp). Jinja

LVFO. (2011). *Technical report*. Stock Assessment Regional Working Group (29 pp). Jinja: LVFO.

Marriot, S. P., & Halls, A. S. (2013). *Nile perch fishery management plan for Lake Victoria project ref. number ea-1.2-b3* (53 pp). Jinja: Lake Victoria Fisheries Organization.

McClanahan, R. T., Castilla, J. C., White, A. T., & Defeo, O. (2008). Healing small-scale fisheries by facilitating complex socio-ecological systems. *Reviews in Fish Biology and Fisheries, 19*, 33–47.

McGoodwin, J. R. (1994). *Crisis in the world's fisheries: People, problems and policies*. Stanford: Stanford University Press.

Mgaya, Y. (Forthcoming). *Lake Victoria fisheries*.

National Coordinating Office (NCO). (2007). *Post training evaluation report on BMU orientation training, 21st–31st May 2007. Implementation of fisheries management plan (IFMP)* (77 pp).

Okeyo-owuor, J. B. (1999). *A review of biodiversity and socio-economics research in relation to fisheries in Lake Victoria. Socio-economics of the Lake Victoria fisheries* (55 pp). IUCN Eastern Africa Programme Report 5.

Onyango, P. O. (2004). *Reforming fisheries management: A case study of co-management in Lake Victoria Tanzania*. Master's thesis, Norwegian College of Fisheries Science, University of Tromsø, Tromsø, 133 pp.

Onyango, P. O. (Forthcoming). The socio-economics of Lake Victoria fisheries. In Mgaya (Ed.), Dar es Salaam: Lake Victoria Fisheries.

Onyango, P. O., & Jentoft, S. (2011). Climbing the hill: Poverty alleviation, gender relationships, and women's social entrepreneurship in Lake Victoria, Tanzania. *MAST, 10*(2), 117–140.

Republic of Kenya (RK), United Republic of Tanzania (URT), & Republic of Uganda (RU). (1995). *Lake Victoria Environmental Management Project (LVEMP)*. Final Project Proposal document submitted to World Bank.

Rittel, H. W. J., & Webber, M. M. (1973). Dilemmas in a general theory of planning. *Policy Science, 4*, 155–169.

SEDAWOG. (2000). *The co-management survey: Co-managerial perspectives for Lake Victoria's fisheries*. LVFRP Technical Document No. 11. LVFRP/TECH/00/11. Jinja: The Socio-economic Data Working Group of the Lake Victoria Fisheries Research Project.

United Republic of Tanzania (URT). (1997). *National fisheries sector policy and strategy statement* (24 pp). Dar es Salaam: Ministry of Natural Resources and Tourism.

United Republic of Tanzania URT. (1998). *The constitution of Tanzania*. Retrieved from http://www.issafrica.org/cdct/mainpages/pdf/Corruption/Legislation/Tanzania/Tanzania%20Constitution%20in%20English.pdf

United Republic of Tanzania URT. (2003). The Fisheries Act. No. 22 of 2003.

URT. (2010). *Report on Lake Victoria fisheries frame survey results 2010 – Tanzania*. National Frame Survey Working Group.

Witte, F., & van Densen, W. L. T. (1995). *Fish stocks and fisheries of Lake Victoria. A handbook for field observations*. Cardigan: Samara Publishing Limited.

Witte, F., Goldschmidt, T., Goundswaard, P. C., Ligtvoet, W., Oijen, M. J. P., & van Wanink, J. H. (1992). Species extinction and concomitant ecological changes in Lake Victoria. *Netherlands Journal of Zoology, 42*, 214–232.

Witte, F., Msuku, B. S., Wanink, J. H., Seehausen, O., Katunzi, E. F. B., Goudswaard, P. C., & Goldschmidt, T. (2000). Recovery of the cichlid species in Lake Victoria: An examination of factors leading to differential extinction. *Reviews in Fish Biology and Fisheries, 10*, 233–241.

Chapter 11
A Comparison of Small-Scale Fisheries Governability: Baja California Sur, Mexico and the Hawaiian Islands

Elena M. Finkbeiner, Adam L. Ayers, John N. Kittinger, and Larry B. Crowder

Abstract Most research on small-scale fisheries governance derives from developing countries, limiting our understanding of small-scale fisheries dynamics within developed nations. This knowledge gap creates disparities between these systems, limiting comparisons of similarities and differences across ranges of socioeconomic development and deriving solutions that can be feasibly applied to small-scale fisheries governance. Our chapter compares small-scale fisheries governance systems in the Hawaiian Islands and Baja California Sur, Mexico, using the interactive governance framework. Historically, collective action and self-governance were important in both regions, yet over time, have eroded due to a variety of factors. The current state of small-scale fisheries in the Hawaiian Islands and Baja California Sur, Mexico is generally characterized by open-access with low governability, but with opportunities to hybridize customary and centralized management systems through co-management. This chapter will provide insight into why this has occurred by drawing on similarities across small-scale fisheries in developed and developing country contexts, in addition to highlighting divergences occurring

E.M. Finkbeiner (✉)
Hopkins Marine Station, Stanford University, Pacific Grove, CA, USA
e-mail: elenamf@stanford.edu

A.L. Ayers
Department of Urban and Regional Planning, University of Hawai'i at Mānoa, Honolulu, HI, USA
e-mail: alayers@hawaii.edu

J.N. Kittinger
Center for Ocean Solutions, Stanford University, Pacific Grove, CA, USA

Conservation International, Betty and Gordon Moore Center for Science and Oceans, Honolulu, HI, USA
e-mail: jkittinger@gmail.com

L.B. Crowder
Hopkins Marine Station, Stanford University, Pacific Grove, CA, USA

Center for Ocean Solutions, Stanford University, Pacific Grove, CA, USA
e-mail: larry.crowder@stanford.edu

S. Jentoft, R. Chuenpagdee (eds.), *Interactive Governance for Small-Scale Fisheries*, MARE Publication Series 13, DOI 10.1007/978-3-319-17034-3_11

along the trajectory from high governability, self-governance of fisheries resources to low governability and open-access tragedies.

Keywords Co-management • Developing • Developed • Small-scale fisheries • Governance

Introduction

Small-scale fisheries around the world can be characterized by diverse ecological resources, different social and ecological attributes, as well as a wide variety of management, policy, and institutional contexts. Small-scale fisheries are globally important in terms of food and livelihood security, but they also face key sustainability challenges. These fisheries are classic collective-action problems with divergences between group and individual-level interests, which complicate efforts to avoid the deterioration of a shared resource. Currently, there are disagreements about the effectiveness of different fisheries governance approaches, including state regulation (e.g., a fishery management agency formulates rules and enforces them, such as in MPAs); privatization and rights-based approaches (e.g., resource users are allocated ownership rights to fisheries resources, such as through catch shares and TURFS); collective action approaches, whereby resource users self-organize to manage the resource (e.g., through fishing collectives or associations that self-organize and self-regulate); and co-management approaches (e.g., decision-making, monitoring, and enforcement rights are distributed between user groups and government). With limited understanding of local communities, some of these tools tend to exclude or reduce highly resource-dependent small-scale fisheries with direct impacts on livelihoods and well-being. But when social and ecological conditions, local institutions, and external drivers are considered, these tools can advance sustainability of small-scale fisheries.

Here we focus on one particular governance approach: co-management arrangements to manage common pool resources such as fisheries (Armitage et al. 2009; Cinner et al. 2012). Co-management can take many forms, but generally involves shared management authority and responsibility between resource users or community groups at local levels and central government authorities (Berkes 2010). In the context of coastal communities and small-scale fisheries, co-management is associated with increased collaboration and learning among partners, integration of traditional ecological knowledge and place-based approaches, higher compliance with regulations, community empowerment, and increased stakeholder buy-in and stewardship (Jentoft et al. 1998; Jentoft 2005; Berkes 2009; Gelcich et al. 2010). But co-management can also lead to increased social conflict, perverse incentives for resource overexploitation, and other undesirable outcomes (Castro and Nielsen 2001; Gelcich et al. 2006).

Research on governance in small-scale fisheries has typically focused on developing countries, so it is less clear whether our evolving understanding of small-scale fisheries dynamics also applies to developed countries. Furthermore, much of the discourse on small-scale fisheries, and other common pool resources, has dichotomized how we perceive these systems: artisanal versus industrial, pre-capitalist versus capitalist, and first world versus third world (Martin 2006). This potentially creates disparities between these systems, limiting understanding of similarities and differences across ranges of socioeconomic development and the solutions that can be feasibly applied to transition both developed and developing country small-scale fisheries to sustainability. Here we compare small-scale fisheries governance systems in the Hawaiian Islands and Baja California Sur, Mexico, using an approach modified from the interactive governance framework (Kooiman and Bavinck 2005). Historically, collective action and self-governance were important in both regions. In Hawai'i, a sophisticated, customary tenure system managed resources within an entire watershed, from 'ridge-to-reef' including inshore fisheries. Similarly, after the Mexican revolution, large land grants were allocated to groups for communal governance and fishing rights were distributed to fishers formally organized into cooperatives. Over time, communal arrangements and self-governance regimes have eroded in both cases due to a variety of factors, including internal conflict, misguided state intervention, insufficient resources allocated to governance, and the effects of an increasingly globalized economy. We assess the current state of small-scale fisheries in both regions as *de facto* open-access with low governability, but with opportunities to hybridize customary and centralized governance systems through co-management. Here we provide insights into why this has occurred by drawing on similarities across small-scale fisheries in developed and developing country contexts, in addition to highlighting divergences occurring along the trajectory from high governability, self-governance of fisheries resources to low governability and open-access tragedies.

Specifically, we begin by reviewing historical trends of small-scale fisheries in the two geographies, leading to present day governance challenges. Second, we provide more context to these case studies through discussion of their social-ecological dimensions. Next, we discuss the emergence of co-management as a governance alternative in these systems, and evaluate the institutional capacity for this transition. We end by evaluating the relative success of both co-management regimes, and by outlining pathways forward.

Historical Trends and Current Governance Challenges in Small-Scale Fisheries

Fisheries governance problems in Hawai'i and Baja Peninsula, Mexico are particularly 'wicked', i.e., merely defining the problem is difficult, a well-defined suite of solutions does not exist, and their issues tend to resurface even after solutions are found (Jentoft and Chuenpagdee 2009). Wickedness in these two geographies is evidenced by the high costs associated with prescribed solutions (e.g. more enforcement officers, inspectors, supply-chain analysts), the dynamics of the resource systems (it is difficult to definitively determine whether solutions are effective), and differing images of the problems that exist among stakeholders. Below we track and describe some historical trends in Hawai'i and Baja Peninsula, Mexico that have compounded the governance problems in both areas.

Hawai'i

Native Hawaiians employed a sophisticated marine tenure system that managed resources sustainably and provided a stable food source for a thriving population in pre-contact times (McClenachan and Kittinger 2013). This tenure system was slowly eroded after western contact and colonization until the last of the traditionally managed fisheries were annulled during the transition to statehood in the late 1950s (Friedlander et al. 2013). The health of Hawai'i's coral reefs and fisheries declined significantly during this time. Under the purview of Hawai'i's Fish and Game agency from the 1950s through the mid-1980s, fisheries were targeted and developed for economic purposes. This strategy has contributed to poor resource outcomes and further depletion of marine resources in Hawai'i. Although the Fish and Game Division was reorganized in the mid-1980s into separate management (The Division of Aquatic Resources) and enforcement (Division of Conservation and Resource Enforcement) entities, in many ways the culture of fisheries overexploitation continues.

However, as a response to perceived failures in centralized or hierarchal management, many communities heavily dependent on marine resources for sustenance across Hawai'i began expressing interest in reviving traditional marine tenure systems or strengthening local management. This interest, coupled with a window of opportunity, paved the legal pathway for co-management in 1994. Since the enabling legislation was passed, however, communities across Hawai'i have received little support from their partner agency at the state level, The Division of Aquatic Resources.

Baja Peninsula, Mexico

Historically, communal governance of land and other resources was an important institutional characteristic of Mexico. After the Mexican Revolution in 1917, large land grants, *ejidos*, were allocated to groups of individuals to be held and managed in common. Additionally, after the Revolution, fishing rights for the most lucrative species were only distributed to fishers formally organized into cooperatives (Vasquez-Leon 1999). The first fishing cooperatives organized in the State of Sinaloa in northwestern Mexico in the 1930s were seen as major successes in securing livelihoods for rural communities, improving living standards, increasing food security, and generating income from exports (McGoodwin 1980).

Over time, communal arrangements and fishing cooperatives have eroded due to a variety of factors, including internal conflict, misguided state intervention, insufficient resources allocated to governance, local and state-level corruption, and neoliberal reform (i.e. free trade, open markets, and deregulation) (Vasquez-Leon 1999; OECD 2006; Cinti et al. 2009; Espinoza-Tenorio et al. 2011). The existing mechanism for effort control and regulation, an opaque and inefficient licensing scheme, has tended to concentrate the legal right to fish in the hands of the wealthy and powerful, and consequently many fishers around northwestern Mexico continue to fish without the legal right (Cinti et al. 2009, 2010; Basurto et al. 2012). As a result, the fisheries of northwestern Mexico are generally characterized by *de facto* open-access (Cint et al. 2009) and it is increasingly evident that these regional small-scale fisheries and associated ecosystems are in decline (Sala et al. 2004; Saenz-Arroyo et al. 2005; OECD 2006; Lluch-Cota et al. 2007; Peckham et al. 2007; Sagarin et al. 2008). However, there are documented examples of fishers who have continued to sustain collective action, through enduring self-governance or co-management regimes, despite being embedded in a larger system characterized by illegality and corruption (Basurto 2005; Ponce-Díaz et al. 2009b; Basurto and Speer 2012; McCay et al. 2014).

Despite significant sociopolitical, cultural, and geographical differences, Hawai'i and Baja Peninsula of Mexico share similar governance trajectories. Self-governance was once important in both geographies until increased state intervention undermined locally-tailored governance institutions. Coupled with demographic shifts, technological change, and globalization, ineffective centralized management (and *de facto* open-access conditions) led to dramatic declines in catch and ecosystem health. In recent years, efforts have been made in Hawai'i and the Baja Peninsula to reintroduce or reinforce locally-tailored institutions in some areas via co-management. Both areas have created legal pathways for co-management and these pathways have attracted significant community interest.

Social-Ecological Dimensions of Small-Scale Fisheries

The social-ecological dimensions of small-scale fisheries in Hawai'i and Baja Peninsula, Mexico encompass the natural and social 'systems-to-be-governed' (Kooiman and Bavinck 2013). Together, they shape governing interactions. In this section we describe the diversity (in gears, communities, and species targeted), scale (of communities, markets, of the resource and the fishery ecosystem), and their complexity and dynamics in both geographies.

Hawai'i

Coastal nearshore fisheries in Hawai'i, as with other tropical systems, are multi-species, multi-gear fisheries that are typically data-poor (Dalzell 1996). In Hawai'i, nearshore fisheries are inclusive of fisheries that operate in coral reef and estuarine environments, as well as coastal embayments and deep-reef habitats. The species targeted in these fisheries are correspondingly diverse, with a primary focus on reef finfish and invertebrates, estuarine and coastal shellfish, algae/seaweed, and schooling coastal pelagic species (Table 11.1). Gear types are diverse as well – fishers targeting coral reef and estuarine species commonly use spears, poles, nets, and traps, whereas fishers targeting coastal pelagic species (e.g. *Selar crumenophthalmus* [bigeye scad, akule] and *Decapterus spp.* [mackerel scad, 'ōpelu]), primarily use nets and hook-and-line gears, sometimes deployed from boats (Table 11.1)

Table 11.1 Fishery characteristics in Hawai'i and Baja, Mexico, including primary gear types, species targeted and markets for fishery species

Gear type & description	Primary species targeted	Primary market
Hawai'i		
Coral reef fishery gears: spears, poles, nets, and traps	Coral reef finfish and invertebrates, estuarine and coastal shellfish, algae/seaweed	Primarily consumed in the household or given away for home consumption to friends and family; some is sold commercially in local commercial fish markets, grocery stores, and informally to other venues
Surround nets, throw nets, hook-and-line pole gears	Schooling coastal pelagic species (e.g. *Selar crumenophthalmus* [bigeye scad, akule] and *Decapterus spp.* [mackerel scad, 'ōpelu])	Consumed in the household, also with a significant commercial fishery for local fish markets, grocery stores, and other venues
Bottomfish fishery: multi-hook handlines and pole-and-line gears	Species complex of deep-reef demersal piscivores (primarily Lutjanids)	Local commercial markets, with some export out of Hawai'i
Traps and nets: aquarium/ornamental fishery	Coral reef finfish and invertebrates	International ornamental fish markets
Mexico		
Artisanal trawls	Shrimp	Primarily intended for export to US and other countries, also consumed nationally
Artisanal longlines	Pelagic sharks	Meat consumed locally; fins exported to Asian markets
Hook and line	Finfish (groupers/snappers, yellowtail, ocean whitefish)	Higher quality species (groupers, snappers) shipped to other states in Mexico or exported
Gillnets - bottom-set and buoyed	Benthic fish, coastal elasmobranchs, mullet	Most benthic fishes consumed regionally or nationally; coastal elasmobranch meat consumed locally, while fins exported to Asian market
Traps	Octopus, lobster, swimming crab, finfish	Lobster destined for primarily Asian markets; other species groups primarily consumed regionally
Hookah diving	Clams, penshell, scallops, abalone, sea cucumber, geoduck	Abalone, geoduck, sea cucumber exported to Asian markets; other species consumed locally; certain species of clams are primarily consumed within the household
Jig	Squid	Exported

(Kittinger 2013). There is also a boat-based fishery targeting a complex of deep-reef demersal piscivores (snappers and groupers, or "bottomfish"), which relies on a multi-hook handline gear (Table 11.1).

Hawai'i is extremely culturally diverse and a majority of its inhabitants hail from geographies with strong ties to fishing, e.g., Native Hawaiians, Japanese, Chinese, Americans, Portuguese, Filipino, Korean, Samoan, Tongan, and Micronesian. Some of these cultures brought their fishing traditions to Hawai'i and their techniques or technologies were adopted by those already living there. However, the multicultural society currently present in Hawai'i presents unique planning and governance challenges. First, new residents may be unaware of fishing rules or do not share the same values as those living there for many years or generations. In other words, Hawai'i residents new to the islands may place a higher value on fish consumption in the present at the expense of future stocks, while locals with deeper ties to Hawai'i may have a greater incentive to conserve fish for the future. Second, different cultures may exhibit different epistemologies or ways of knowing that may not be captured in western planning and management processes. Third, nearshore reef habitats have ecological limits, which may be exceeded by lack of fisher knowledge about those limits and enforcement of fishing regulations.

In Hawai'i, it is difficult to separate fishing activities into sectoral categories (e.g., recreational, commercial), which are commonly used in US fishery management to develop plans and management actions. Subsistence, recreational, commercial, and cultural fishing can all occur on the same fishing trip, and often the factors that determine whether a catch is sold for profit or consumed can vary depending on the fishery and motivations of fishers (Glazier 2006). Kittinger, Glazier, and colleagues have used the post-landings disposition and distribution of seafood, or "fish flow," as a way to distinguish between market segments in these diverse small-scale fisheries (Kittinger et al. 2015; Glazier et al. 2013; Vaughan and Vitousek 2013). For example, coral reef and estuarine fisheries are primarily non-commercial as most catch is kept for consumption within the household or given as part of customary exchange (Kittinger 2013), which maintains social ties among fishers and their social-kinship network (Severance et al. 2013).

Fishers that sell their catch report their sales through the State of Hawai'i's commercial marine landings database. Most of this fish stays within Hawai'i, where they are sold in local fish markets and grocery stores. In addition to reported sales, fishers also engage in 'informal' selling to offset costs or as an additional way to generate revenue. This 'informal' market comprises non-reported sales to friends, family, restaurants, fish markets, grocery stores and other venues. While it is unknown how much of the catch is directed toward this market segment, social survey research shows that it is not an insignificant amount (Kittinger et al. 2015). A few species are sold into markets that reach beyond Hawai'i, such as the aquarium fishery (which primarily targets coral reef finfish and invertebrates) and the bottomfish fishery, both of which are exported to external markets.

Some nearshore fisheries display seasonal variability, including fisheries for schooling coastal pelagics, which enter nearshore embayments to spawn typically in the late summer to fall. Similarly, seasonal fishing events are centered around other spawning events, such as for O'opu. These endemic gobies spawn in freshwater streams and rivers and the larvae hatch in coastal nearshore waters; juveniles then return to stream systems and are targeted on their up-river runs, which typically occur in the fall. Bottomfish can also display seasonal variations, which can affect effort among fishers. While some species display some seasonal variability, many species are available throughout the year and all of these fisheries are active year-round in these tropical environments.

The diversity, scale, complexity, and dynamics that typify nearshore fisheries in Hawai'i make goal setting difficult, and both targets and solutions ambiguous. With no license requirement for recreational or customary/traditional fishers, there are little data available to gauge fishing effort or craft rules to govern it effectively. Although resources are surveyed more often, it is difficult for managers or researchers to determine whether critical ecosystem thresholds have been crossed. With better social and ecological data, practitioners can begin to develop a clearer understanding of thresholds, feedbacks, and targets. With very little data and *de facto* open-access conditions, the systems-to-be-governed continue to constitute a 'wicked' problem.

Baja Peninsula, Mexico

Marine resource extraction is of primary economic and cultural importance along the peninsula of Baja California Sur, Mexico (Brusca et al. 2004; Lluch-Cota et al. 2007), and is characterized by a high degree of diversity (Cudney-Bueno and Turk-Boyer 1998; Erisman 2011). This is particularly true for small-scale fisheries, or *pesquerías ribereñas,* that rely on the extraction of myriad resources such as benthic and pelagic bonyfishes, elasmobranchs, molluscs, crustaceans, and other marine invertebrates using a wide variety of fishing gears, such as traps, hook and line, gillnets, artisanal longline, artisanal shrimp trawls, jigging and diving (Table 11.1) (Cudney-Bueno and Turk-Boyer 1998).

Due to the region's high diversity and productivity of marine resources, and importance of small-scale fishing, many communities along Baja's peninsula are characterized by a high degree of immigration from other regions in Mexico (Young 2001). Additionally, ephemeral, seasonal fishing camps are common (Cartamil et al. 2011; Ramirez-Amaro et al. 2013). In this context, a fishing "community" on the Baja Peninsula may be hard to define and delineate, and can be considered relatively heterogeneous in cultural and socio-economic form. The uneven distribution of fisheries rights (via fishing permits and concessions), and the tendency for fisheries rights to accumulate under wealthier and more powerful individuals (Cinti et al. 2009, 2010; Basurto et al. 2012) has further exacerbated socio-economic disparities in the region.

Despite the fact that small-scale fishing operations occur on very small fiberglass vessels (~8 m) with no more than a three-person crew, diverse market dynamics link these fisheries across multiple scales. Depending on the cultural norms and monetary values associated with different marine resources, species are used for subsistence and commercial purposes, and are distributed to local, regional, and international markets. Species that have marginal or no market value are consumed within the household, traded or redistributed to other households, or kept as bait. Typically, various species of finfish, mollusks, sharks, and rays are consumed locally and regionally, while lobster, abalone, geoduck, shrimp and shark fin often have direct ties to international markets.

Importantly, these fisheries are subject to extreme fluctuations in productivity due to El Niño Southern Oscillation-driven changes in oceanographic conditions (Collins et al. 2002; Pérez-Brunius et al. 2006). Due to this variability in the ocean environment, many fishers in the region distribute their risk by engaging in a variety of fisheries using different gears and traveling to different locations, and have multiple livelihoods. Changing biophysical conditions are not the only source of uncertainty and complexity in Baja small-scale fisheries; strong connections with international markets, and a high degree of volatility in local market prices, render these fisheries vulnerable to market forces. In recent years, variability and complexity in coastal communities have increased due to a confluence of factors including the recent economic crisis, increased conflict in Mexico associated with the narcotic trade, a subsequent halt in tourism, and changing oceanographic conditions such as seasonal reductions in dissolved oxygen and pH (Micheli et al. 2012).

Similar to Hawai'i, the extremely dynamic nature of these fisheries poses significant governance hurdles to overcome. With constant changes in catch composition due to strong seasonality and inter-annual shifts in species availability, the demographic, spatial, and temporal changes in fishing "communities", and rapidly fluctuating markets at different scales, the dynamism of these fisheries adds a great deal of complexity for governance. Coupled with limited governance resources at the state and local level, scientific monitoring, policy-making and enforcement are difficult tasks to achieve. As in the case of many small-scale fisheries throughout the world, effective governance must address food security and livelihoods in addition to ecological sustainability and economic efficiency. Thus Mexican agencies are faced with a particularly arduous task given constraints in resources and highly dynamic fisheries.

The Emergence of Co-management in Small-Scale Fisheries

Hawai'i

Since Hawai'i obtained statehood, coral reef fisheries have been managed by a chronically underfunded, centralized agency located in the urban state capital of Honolulu. For decades, many fishing communities on outer islands and rural areas became disenfranchised by the unresponsiveness of the state agency and geographic distance separating them from fisheries decision-making. Budgetary constraints facing the enforcement agency (The Department of Conservation and Resource

Enforcement) have limited monitoring and enforcement on several islands. This lax enforcement has resulted in *de facto* open-access conditions and resource decline of Hawai'i coral reef fisheries. Further exacerbating the open-access situation, technological advances have allowed fishers to enter previously inaccessible areas causing increased or displaced fishing effort. The conflict caused by resource decline and increased competition over scarcer resources coupled with a management agency unable to adequately monitor and enforce resource rules caused many communities to take enforcement into their own hands, resulting in disputes and conflict among fishers in many areas (Ayers and Kitinger 2014).

Conversely, past success of the *ahupua'a* management system (Kaneshiro et al. 2005; Kittinger et al. 2011) and a revival of traditional management practices (Friedlander and DeMartini 2002; Poepoe et al. 2003) led many communities to seek greater autonomy over the management of small-scale fisheries. As a result, many communities across Hawai'i lobbied the State of Hawai'i to gain legitimate management authority over their adjacent marine areas. The results of this lobbying and a window of opportunity opened the door for the legal pathway of co-management of Hawai'i's coral reef fisheries to emerge in 1994. Although co-management has taken several forms in Hawai'i, Community-Based Subsistence Fishing Areas arose as a novel idea that could combine traditional or local management practices devised by Native Hawaiians or Hawai'i communities and help ensure that food security needs were met. The Community-Based Subsistence Fishing Areas allowed communities to share management authority with the state and partner with them in devising rules tailored to local conditions.

Since the formal pathway for co-management was created in 1994, a 2-year pilot project has resulted in a single successful partnering with the state of Hawai'i to co-manage marine resources. However, the non-renewal of Hawai'i's only experiment, in the context of marine resource co-management, has not deterred other communities from pursuing institutional change. Increased community interest has coincided with an increase in marine NGOs in Hawai'i to meet the rising community demand for support and capacity-building for institutional change and co-management. Although co-management (via Community-Based Subsistence Fishing Areas) has endured its share of process- and capacity-related challenges since its inception, it remains a promising alternative to existing hierarchical management processes, which have been largely ineffective in curtailing resource decline over the past few decades. Co-management, via Community-Based Subsistence Fishing Areas, provides a mechanism for tailoring resource rules to local conditions, injecting greater citizen participation in the decision-making process, and sharing management authority between resource users and the government.

Baja Peninsula, Mexico

The complex, dynamic, and diffuse nature of small-scale fisheries along the Baja Peninsula, coupled with limited governance resources and personnel in federal fisheries agencies, has rendered centralized governance efforts inefficient and ineffective. For example, high costs of generating relevant fisheries data, a general disregard for, and exclusion of, fishers during policy-making processes, and low government capacity to enforce these laws, has generally resulted in *de facto* open-access conditions. Commencing in the 1980s and gaining effort through the 1990s, the transformation towards neoliberal policies, represented a concerted effort by the state to address governance shortcomings in resource extraction economies. These policies, intended to bolster foreign investment opportunities and export markets, corresponded with a change in the Mexican constitution, undermining the foundation of communal property rights and governance (Vasquez-Leon 1999). Needless to say, these efforts by the state only exacerbated resource depletion and *de facto* open access conditions (Young 2001). As such, the Mexico government is often faulted for its inability to prevent the demise of coastal and marine resources (McGoodwin 1980; Vasquez-Leon 1999; Hastings and Fischer 2001; Young 2001; OECD 2006; Cinti et al. 2009).

At the same time, attempts by fishers themselves to self-govern their behavior and safeguard their resources may not be enough to deal with external drivers prevalent in the region, such as the encroachment of outside fishers, climate driven disturbances, or market volatility. For example, in some documented cases, self-governance, informal rule setting, and *de facto* no- fishing reserves declared by fishers themselves, were not enough to stifle unsustainable and illegal fishing practices. Without legal backing and support of these informal governance regimes, and supplementary enforcement provided by the federal government, communities are often unable to successfully keep roving bandits and other illegal fishing activity at bay, and are forced to abandon their communal contracts (Cudney-Bueno and Basurto 2009). The difficulty of governing Mexico's marine resources from just one institutional level provides a powerful impetus for moving towards co-management; sharing the costs of monitoring, decision-making, and enforcement through a co-management configuration may provide a viable alternative to current governance short-comings.

The perceived legitimacy of a movement towards co-management is particularly strong in fishing cooperatives in the state of Baja California Sur. Mutual trust between local resource users and central government authorities has evolved over many years in the region, commencing in times of resource scarcity and crisis in their lobster and abalone fisheries (Ponce-Díaz et al. 2009b). Realizing they did not have the necessary resources to adequately assess the status of the valuable but depleted lobster and abalone populations along the Vizcaino coast of Baja, the federal fisheries agency entered into an agreement with fishing cooperatives of the region: fishers would actively partake in collecting fisheries data, alongside government scientists, that would feed into decision-making processes and advise annual catch limits. In exchange for their active participation in stock evaluations, cooperatives would receive place-based, exclusive access to these resources for 20-year periods, with the potential for renewal. Importantly, enabling conditions for co-management, such as trust among scientists, fishers, and the central government, have emerged slowly after decades of collaboration in the state of Baja California Sur (Ponce-Díaz et al. 2009b).

Institutional Capacity: Hierarchical Decision-Making Processes and Policy

Successful co-management requires institutional capacity at various levels of the governing system, in addition to opportunities and pathways for institutional interplay across these levels (Young 2006). In general, scholars have identified three important levels of governance, although these levels are not always mutually exclusive. Meta-order (Kooiman and Bavinck 2005) or constitutional level (Ostrom 2005) governance usually sets forth the guiding principles and foundations, which both constrains and provides opportunities for subsequent levels of governance. Second-order (Kooiman and Bavinck 2005) or collective-choice level (Ostrom 2005) governance constitutes the main decision-making arena where policies are formed, guided by meta-order processes, and influencing first-order processes. Finally, first-order (Kooiman and Bavinck 2005), or operational level (Ostrom 2005) governance, is representative of day-to-day decisions carried out by individuals and groups within the constraints of these other governance levels. We will use these three levels of governance as a framework for discussing the institutional capacity for co-management in our two governing systems. We also evaluate the institutional interplay, or governing interactions that exist across these levels in formal and informal co-management arrangements.

Hawai'i

Meta-order Governance

Hawai'i is part of the United States and thus its fisheries management systems are broadly reflective of top-down approaches that characterize US fisheries management. The Public Trust Doctrine tends to be well established in state waters of the US and provides the foundational guidelines for how ocean resources are governed. The doctrine calls for governments to manage natural resources held in common (the trust body), in the best interest of the citizens of the state (beneficiaries of the trust). Although the US does not have a formal Federal Public Trust Doctrine (Turnipseed et al. 2009), it does informally adopt the Doctrine regarding the nation's authority to manage resources on behalf of the citizens. The Doctrine also protects people's rights to fishing, commerce, and navigation in states and informally in US waters (Turnipseed et al. 2009). However, Hawai'i is also a Polynesian archipelago, with a rich tradition associated with Native Hawaiian customary management, and aspects of these systems are embedded in the Hawaiian legal system as well as in the social mores and norms of fishers across the State (Mackenzie 1991). The nexus between US and indigenous Hawaiian systems are reflected in the State's legal system, which is underpinned by the State of Hawai'i constitution. Together the Public Trust Doctrine and the State of Hawai'i constitution protect everyone's right to fish, in addition to ensuring that resource rules do not privilege one group over another.

In the parlance of the interactive governance framework, the constitution provides the 'meta-level' order, providing the structural basis for rule-making and authority (Fig. 11.1a) (Kooiman and Bavinck 2013). The constitution defines the fundamental principles of governance and scopes of authority within the Hawai'i jurisdiction, and the current constitution amended in 1978 reflects the long evolution of constitutions starting with the Hawaiian monarchy.

Second-Order Governance

Up until recently, Hawai'i coral reef fisheries were exclusively managed by the centralized resource management agency (The Division of Aquatic Resources or DAR; Fig. 11.1a) situated on the State's most densely populated island of O'ahu. Although there are satellite offices on the other islands, management decisions were almost exclusively made by the main office in Honolulu. For decades, fisheries rules and regulations were required to be uniform across the state, despite significant differences in island geography, ecology, and population across the archipelago.

Even with a legal system that includes aspects of Native Hawaiian legal systems, the State of Hawai'i is organized primarily via hierarchical governance, following fisheries management institutions in the US. These institutions comprise the 'second level' order in the interactive governance framework, representing the organizations and institutions that interact within the milieu of the existing legal and constitutional framework to structure fisheries governance in this system (Fig. 11.1a). In Hawai'i, the policies and approaches applied in fisheries management are thus mediated by both the 'meta-level' legal systems and the 'second-level' institutional arrangements within the State. The Department of Land and Natural Resources is the primary agency responsible for fisheries management in Hawai'i (Fig. 11.1a). Within the Department, the Division of Aquatic Resources manages all marine and freshwater resources in the State, with programmatic focal areas in commercial fisheries and resource enhancement; aquatic resources protection, habitat enhancement, and education; and recreational fisheries.

First-Order Governance

At the functional level, the Division of Aquatic Resources administrates several programs that comprise its primary management activities and responsibilities for nearshore fisheries. Some of the primary fishing-related programs include: (1) Licensing and permitting programs, such as the commercial fishing licensing and bottomfish fishing vessel registration programs; (2) Spatial management measures such as Fishery Replenishment Areas, Marine Life Conservation Districts, and other designations that have specific rules and regulations; and (3) Fishing rules and regulations, including size limits for species, closed seasons, and gear restrictions. The Division of Aquatic Resources also works with other agencies within the Department of Land and Natural Resources such as the Division of Conservation

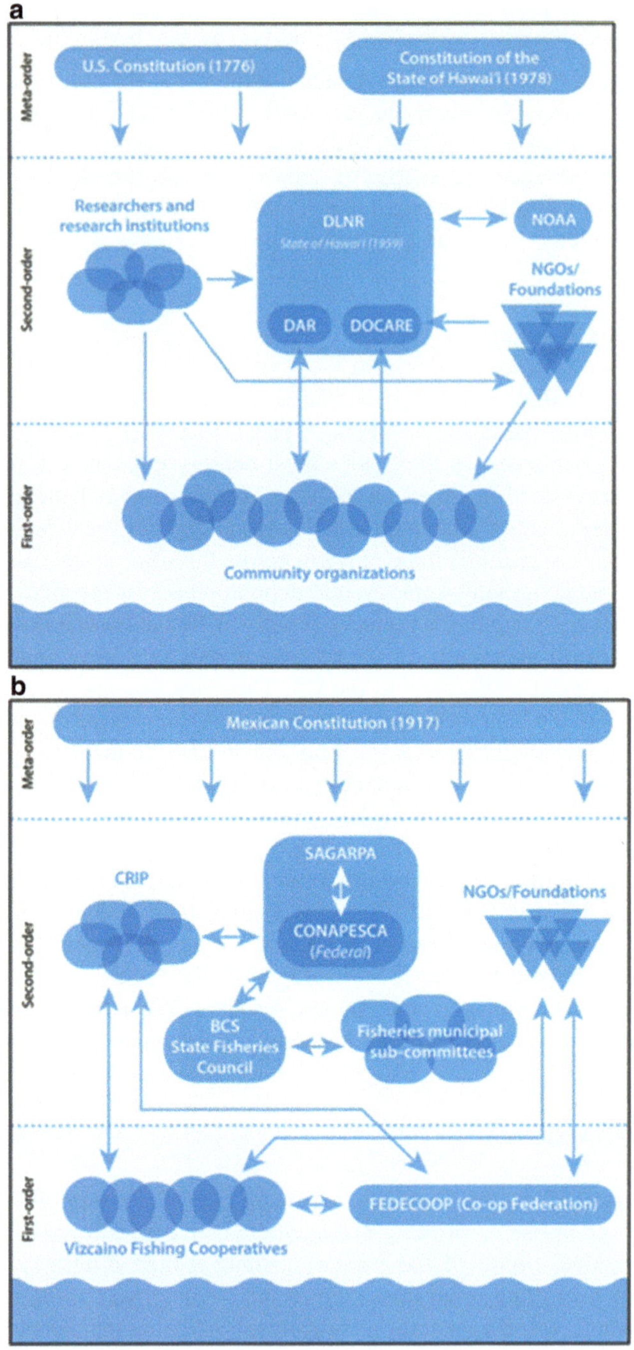

Fig. 11.1 Governance levels (meta-, second-, and first-order levels) as described by Kooiman and Bavinck (2005). For Hawai'i (**a**) and Baja California Sur (**b**), respectively. Acronyms: Department of Land and Natural Resources (DLNR), Division of Aquatic Resources (DAR), Division of Conservation and Resources Enforcement (DOCARE), National Oceanic and Atmospheric Administration (NOAA), Secretariat for Agriculture, Livestock, Rural Development, Fisheries and Food (SAGARPA), National Fisheries Commission (CONAPESCA), Regional Fisheries Science Institute (CRIP), Baja California Sur (BCS) (Figure credit: Jerker Lokrantz)

and Resources Enforcement, which is responsible for fisheries enforcement in the State. In the interactive governance framework, these programs comprise the 'first order' structures that affect how human actors and institutions interact in fisheries management systems to affect outcomes.

With regards to 'governing interactions,' or the "mutually influencing relations between two or more actors or entities in a governance setting," (Kooiman and Bavinck 2013) the Division of Aquatic Resources and the Department of Land and Natural Resources can work collaboratively with community organizations in a variety of ways. First, the Makai Watch program was developed to allow active participation of local communities to serve as the 'eyes and ears' for the State's Division of Conservation and Resources Enforcement, strengthening enforcement capacity at the local level. This program was incubated through two pilot sites in the late 1990s and was formalized in 2005, and is implemented in several sites around the State by active Makai Watch volunteers, NGO partners, and community organizations. Another mechanism for collaboration exists in the Community-Based Subsistence Fishery Area designation, the primary means through which co-management occurs. Community-Based Subsistence Fishery Areas allow communities to craft rules and management plans for their areas, pending state approval. Despite the fact that there are multiple opportunities for organized interests to subvert the process of participatory rulemaking, these co-management areas and associated rule-making mechanisms continue to be a promising policy instrument for managing small-scale fisheries in Hawai'i.

Baja Peninsula, Mexico

Meta-order Governance

The Mexican Constitution sets forth the guiding principles and values to which all Mexican fisheries law must adhere, and contains important language setting the legal and moral framework for fisheries co-management (Fig. 11.1b). Article 27 of the constitution clearly states that the Nation owns land and water resources, and articulates the Nation's duty of transferring resource use or property rights to individuals or groups of individuals (Ponce-Díaz et al. 2009a). Although unrelated to fisheries, the Mexican Constitution contains an important clause, which enabled informal co-management arrangements to materialize in Baja California Sur even before fisheries co-management legislation was in place. The Code for Civil Procedure states that voluntary agreements between or among parties must be considered contractual (Ponce-Díaz et al. 2009b). As such, in the early 2000s informal fisheries "sub-committees" emerged in every municipality of Baja California Sur, convening constituents from municipal, state, and federal governance levels, and giving local resource users more clout in decision-making processes (Ponce-Díaz et al. 2009b). Thus, the Mexican Constitution is integral in "meta-order" governance (Kooiman and Bavinck 2005) by providing the guiding principles in formal and informal fisheries co-management arrangements.

Second-Order Governance

There are numerous agencies, organizations, and laws, or second-order governance institutions (Kooiman and Bavinck 2005), pertinent to fisheries co-management in Mexico, guided by the meta-order principles stated above (Fig. 11.1b). The central decision-making body, the federal fisheries agency (CONAPESCA), is housed within the ministry for agriculture, livestock, rural development, fisheries and food (SAGARPA), which ultimately oversees all fisheries policy implementation around Mexico. One of CONAPESCA's primary responsibilities is the distribution of fishing rights to commercial fishers for designated species through permits, or less commonly, concessions. Permits are issued for 2–5 year periods, and dictate the species or species groups, gear types, seasons, and geographic delineation, whereby groups of fishers, or individual fishers, are allowed to fish. However, the most common permit "escama", gives fishers the authority to fish essentially *any* species of finfish found in the region, and the authorized geographic delineation spans an entire municipality. Therefore, fishing rights are often overlapping and not exclusive, creating conflict among fishers and communities (Cinti et al. 2009). Concessions are a much stronger form of property rights, issued by the government to fishing cooperatives every 20 years, with exclusive place-based rights for the capture of a particular species. INAPESCA, the federal agency responsible for fisheries science, also sits as a commission under SAGARPA's jurisdiction, and works closely with regional fisheries science centers (CRIP) to integrate science into policy.

First-Order Governance

First-order governing "takes place wherever people, and their organizations, interact in order to solve societal problems and create new opportunities" (Kooiman and Bavinck 2005) given the constraints and opportunities imposed by meta- and second-order governance as described above. In the context of Mexican small-scale fisheries, first-order governance occurs among individual fishers and fisher organizations called cooperatives. To date, nine fishing cooperatives from the Vizcaino region of Baja California Sur have successfully participated in co-management. These cooperatives were granted long-term concessions from the federal government to extract spiny lobster and abalone, and have emerged as highly organized and sustainable fisheries, as recognized by the Marine Stewardship Council. The Vizcaino cooperatives adhere to stringent rules devised by the fishers themselves, have implemented several no-take marine reserves, actively participate in scientific monitoring, have highly effective internal rotating enforcement bodies, and exhibit strong vertical integration in the global market (Ponce Diaz et al. 2009b; McCay et al. 2014). The nine Vizcaino cooperatives are incorporated at a higher level by FEDECOOP, a federation providing processing, marketing, scientific evaluation, coordination, and interest representation to all nine member cooperatives (Fig. 11.1b). As such, we assert that the distribution of governance power, in the case of FEDECOOP, is relatively equitable across users and government agencies, approaching the most common conception of co-management – cooperative (Jentoft 1989; Sen and Nielsen 1996; Ponce Diaz et al. 2009b).

Recent efforts have been made to decentralize the decision-making processes pertinent to fisheries, and increase communication and coordination (governing interactions) between the 'governing system' and 'system-to-be-governed'. In 2001, the CONAPESCA office was moved from Mexico City to Mazatlán, Sinaloa, in an effort to bring decision-making processes directly to the regions affected by these decisions. Then in 2007, important legislation was passed in the General Law of Sustainable Fishing, setting the stage for re-allocating governance power among federal, state, and municipal bodies on matters including permit and license application, the development of fisheries management plans, and in monitoring and enforcement (Ponce Diaz et al. 2009a, b). The 2007 General Law of Sustainable Fishing became an important legal platform for the creation of formal state and municipal level fisheries councils, intended to represent local and regional interests in decision-making processes, and the National Fisheries Council, intended to provide support, coordination, and continuity across governance levels and sectors (Ponce Diaz et al. 2009b). Although the state of Baja California Sur had already instituted informal fisheries "sub-committees" at the municipal level in the early 2000s, the 2007 law provided legal backing for the existing sub-committees.

Evaluating Co-management Success

Although the pathway for co-management in Hawai'i has existed for approximately 20 years, little summative progress has been made during this time. To date, the primary experiment for co-management involves one 2-year pilot project that was not renewed because both sides were unhappy with the relationship. Within Community-Based Subsistence Fishing Areas, communities can submit rules and management plans, pending approval by the state. However, draft rules for these special co-managed areas must go through the same administrative rule-making process as any other rule change. This process provides citizens with up to four opportunities to comment and express their concerns, but has also provided more opportunity for organized interests to subvert the rule-making process. Without rule approvals from the state resource management agency, the two permanently designated co-management areas are co-management areas on paper only. Since the state retains so much authority and power in the co-management relationship and minimal rights are held by local-level resource user groups, communities are left with very little governance power and autonomy. After 20 years of fits and starts, Hawai'i is still early in its transition to co-management due to: a lack of capacity – both at the state and community levels; a complicated administrative rulemaking process; organized interests opposed to local autonomy and place-based rulemaking; and an organizational culture at resource management agency that is accustomed to and trained for distant, centralized management. To ameliorate this, the State of Hawai'i and communities could commit to a few co-management areas for several years in different areas and rigorously evaluate them to assess their effectiveness relative to hierarchical management. If the co-management areas are effective, then additional areas could be added and if not, both communities and the State of Hawai'i could learn from their mistakes.

In Baja, despite relative success in the co-management regime developed between FEDECOOP and municipal, state, and federal levels of governance, this nonetheless, is a rare example of successful co-management in Mexican fisheries. Scholars have suggested that the failure in power redistribution in fisheries governance may be attributed to several factors. First, State Fisheries Councils do not have sufficient legal status, as they were not created by parliamentary law, do not have sufficient personnel, have limited responsibility and lack any long-term organizational structure (OECD 2006; Ponce Diaz et al. 2009a). Second, permitting and licensing still only occurs in federal offices, state offices only act as point of interception for permit applications (Ponce Diaz et al. 2009a). Third, movement of federal fisheries offices to Mazatlán from Mexico City in 2001 has not decentralized decision-making power, only changed geographic location (OECD 2006). Attempts at increased stakeholder involvement, starting in the early 1990s, has also failed due to the fact that meetings are held in Mexico City or Mazatlán, and small-scale fishers opinions are often not considered at these meetings (OECD 2006). As such, co-management in the vast majority of fisheries in Mexico can be characterized as consultative, whereby mechanisms exist for governments to consult with users, but most decisions are essentially taken by the government (Jentoft 1989; Sen and Nielsen 1996) with the exception of the FEDECOOP cooperatives described above.

Conclusions and Ways Forward

Given disparate histories and cultures, and resulting institutional arrangements, the trajectory towards co-management is slightly divergent between these two geographies. First, the nature of the small-scale fisheries in Hawai'i is primarily subsistence-based, while in Baja it is commercially oriented with a focus on highly lucrative exports (abalone and lobster). Second, coastal communities in Hawai'i have a particularly strong history and culture surrounding marine resource extraction, dating back centuries. Thus, there is still knowledge and recognition of historical use and practices, generating strong political will at the community level to return to the locally-based governance (consistent with the traditional *ahupua'a* system). However, the transition to co-management has been relatively recent. On the Baja Peninsula, however, traditional cultures first inhabiting the peninsula have long-since disappeared, and current coastal communities are recently settled representing decades of marine resource extraction on the peninsula, relative to Hawai'i's hundreds of years of experience. However, co-management (in its various forms) has been in place in the region for several decades, much longer than in Hawai'i. While developed world contexts such as Hawai'i are making some forays into collaborative governance, clearly co-management comes second in the first world. In Baja, the decades-old movement towards co-management was associated with the perceived difficulty of governing at one institutional level alone (either state or community), resource scarcity and crisis, and subsequent decades of collaboration among scientists, fishers, and practitioners.

Despite the inherent differences between the two geographies, through this comparative case study of co-managed small-scale fisheries in Hawai'i and Baja, considerable similarities are apparent across the developed/developing nation dichotomy. First, small-scale fisheries in both case studies are characterized by diversity, scale, complexity, and dynamism. Ironically, these characteristics are both critical to the success of these small-scale fisheries, while simultaneously contributing to the 'wickedness' of the governance problems (Jentoft and Chuenpagdee 2009). For example, the ability to target multiple species using multiple gear types is critical for sustaining livelihoods, but poses significant obstacles for monitoring, assessment, and evaluation of management effectiveness. Similarly, across both geographies, misinformed policies, meager state resources, and disenfranchisement of fishers have led to *de facto* open-access conditions, eroding historical governance regimes. Critical to the transformation towards co-management, Hawai'i and Mexico both have strong constitutional backing at the meta-level. And while small-scale fisheries from both geographies are situated within the context of highly centralized decision-making, their exemplar co-management cases both occur in states that enjoy more autonomy than other states in their respective federations. Despite the difference in temporal scale of contemporary fisheries governance between the two geographies, both have a history and culture of interactive governance and collective action, important for a transition back to decentralized approaches. Finally, despite the impetus and desire for small-scale fisheries in these geographies to move towards co-management, both have experienced a long and sometimes stagnant process. While Hawai'i may not be exemplar of developed nation small-scale fisheries, this comparison nonetheless has demonstrated remarkable parallels between these seemingly disparate geographies.

The bumpy, disjointed transition to co-management governance exhibited thus far in Hawai'i and Baja fisheries illustrates the importance of being patient and having realistic time horizons during governance transitions. The transition towards co-management needs to be thought of as a constantly-evolving process, not as an outcome. If short-term struggles and challenges characterize the process, failure is not necessarily destined (Chuenpagdee and Jentoft 2007). Rather, co-management should be thought of as a non-linear, iterative process requiring political will, understanding, and capacity at both central government and local resource user levels (Chuenpagdee and Jentoft 2007). Of equal importance is consideration of local context, as co-management is certainly not a panacea, nor is it immune to unintended consequences and failures. The struggles evidenced during this transition highlight the need to employ formative (process) and intermediate criteria (e.g. transaction/transformation costs) in addition to summative (outcome) measures. Too often, studies overlook the 'creation story' and do not examine transaction costs, and the genesis of self-organization, consensus-building, and collective action processes (Berkes 2010). A commitment must also be made to measure baselines and the summative effects of management interventions through monitoring and evaluation. These co-management experiences demonstrate the importance of long-term commitment. More planning-oriented staff and commitment are needed at the state level, including more flexibility and adaptive rulemaking. The commitment aspect should not only apply to communities and governments, but also to NGOs and grant-making organizations. Co-management is a process that doesn't happen overnight.

However, the groundswell of interest demonstrates the demand that exists for greater community-level involvement and autonomy in governance. Furthermore, small-scale fisheries in both geographies have previously been characterized by low governability from a single institutional level alone. For example, diversity, cross-scale phenomena, and dynamism all characterize the 'systems-to-be-governed,' necessitating increased coordination across levels of governance. Likewise, it is difficult for governments (or resource users) to carry out governance tasks by themselves due to limited resources, lack of compliance, and a suite of other factors. Thus, improved governability may be achieved by increasing governing interactions between the governing system and the 'system-to-be-governed'. In sum, co-management offers a tangible way forward in overcoming "wicked" governance problems in small-scale fisheries across the developing and developed nation divide. However, practitioners, researchers, and funders must proceed cautiously, proactively learning from mistakes, and tailoring projects to local circumstances.

References

Armitage, D. R., Plummer, R., Berkes, F., Arthur, R. I., Charles, A. T., Davidson-Hunt, I. J., Diduck, A. P., Doubleday, N., Johnson, D. S., Marsche, M., McConney, P., Pinkerton, E., & Wollenberg, E. (2009). Adaptive co-management for social–ecological complexity. *Frontiers in Ecology and the Environment, 7*, 95–102. doi:10.1890/070089.

Ayers, A. L., & Kittinger, J. N. (2014). Emergence of co-management governance for Hawai'i coral reef fisheries. *Global Environmental Change, 28*, 251–262. http://doi.org/10.1016/j.gloenvcha.2014.07.006

Basurto, X. (2005). How locally designed access and use controls can prevent the tragedy of the commons in a Mexican small-scale fishing community. *Society and Natural Resources, 18*, 643–659.

Basurto, X., & Speer, J. (2012). Structuring the calibration of qualitative data as sets for qualitative comparative analysis (QCA). *Field Methods, 24*, 155–174. doi:10.1177/1525822X11433998.

Basurto, X., Cinti, A., Bourillón, L., Rojo, M., Torre, J., & Weaver, A. H. (2012). The emergence of access controls in small-scale fishing commons: A comparative analysis of individual licenses and common property-rights in two Mexican communities. *Human Ecololgy, 40*, 597–609. doi:10.1007/s10745-012-9508-1.

Berkes, F. (2009). Evolution of co-management: Role of knowledge generation, bridging organizations and social learning. *Journal of Environmental Management, 90*, 1692–1702. doi:10.1016/j.jenvman.2008.12.001.

Berkes, F. (2010). Devolution of environment and resources governance: Trends and future. *Environmental Conservation, 37*, 489–500. doi:10.1017/S037689291000072X.

Brusca, R. C., Findley, L. T., Hastings, P. A., Hendrickx, M. E., Torre Cosio, J., & van der Heiden, A. M. (2004). Macrofaunal diversity in the Gulf of California. In J. L. E. Cartron, G. Ceballos, & R. Felger (Eds.), *Biodiversity, ecosystem and conservation in northwestern Mexico* (pp. 179–203). New York: Oxford University Press.

Cartamil, D., Santana-Morales, O., Escobedo-Olvera, M., Kacev, D., Castillo-Geniz, L., Graham, J. B., Rubin, R. D., & Sosa-Nishizaki, O. (2011). The artisanal elasmobranch fishery of the Pacific coast of Baja California, Mexico. *Fisheries Research, 108*, 393–403. doi:10.1016/j.fishres.2011.01.020.

Castro, A. P., & Nielsen, E. (2001). Indigenous people and co-management: Implications for conflict management. *Environmental Science & Policy, 4*, 229–239. doi:10.1016/S1462-9011(01)00022-3.

Chuenpagdee, R., & Jentoft, S. (2007). Step zero for fisheries co-management: What precedes implementation. *Marine Policy, 31*, 657–668. doi:10.1016/j.marpol.2007.03.013.

Cinner, J. E., Daw, T. M., McClanahan, T. R., Muthiga, N., Abunge, C., Hamed, S., et al. (2012). Transitions toward co-management: The process of marine resource management devolution in three east African countries. *Global Environmental Change, 22*, 651–658. doi:10.1016/j.gloenvcha.2012.03.002.

Cinti, A., Shaw, W. W., Cudney-Bueno, R., & Rojo, M. (2009). The unintended consequences of formal fisheries policies: Social disparities and resource overuse in a major fishing community in the Gulf of California, Mexico. *Marine Policy, 34*, 1322–1334.

Cinti, A., Shaw, W., & Torre, J. (2010). Insights from the users to improve fisheries performance: Fishers' knowledge and attitudes on fisheries policies in Bahía de Kino, Gulf of California, Mexico. *Marine Policy, 34*, 1322–1334. doi:10.1016/j.marpol.2010.06.005.

Collins, C., Castro, C., Asanuma, H., Rago, T. A., Han, S. K., Durazo, R., & Chavez, F. P. (2002). Changes in the hydrography of central California waters associated with the 1997–98 El Niño. *Progress in Oceanography, 54*, 129–147.

Cudney-Bueno, R., & Basurto, X. (2009). Lack of cross-scale linkages reduces robustness of community-based fisheries management. *PloS One, 4*, e6253.

Cudney-Bueno, R., & Turk-Boyer, P. J. (1998). *Pescando Entre Mareas del Alto Golfo de California: Una Guia Sobre la Pesca Artesanal, Su Gente y Sus Propuestas de Manejo. Centro Intercultural de Estudios de Desiertos y Océanos. Sonora, México.*

Dalzell, P., Adams, T., & Polunin, N. (1996). Coastal fisheries in the Pacific Islands. *Oceanography and Marine Biology: An Annual Review, 34*, 395–531.

Erisman, B. E., Paredes, G. A., Plomozo-Lugo, T., Cota-Nieto, J. J., Hastings, P. A., & Aburto-Oropeza, O. (2011). Spatial structure of commercial marine fisheries in Northwest Mexico. *ICES Journal of Marine Science, 68*, 564–571. doi:10.1093/icesjms/fsq179.

Espinoza-Tenorio, A., Espejel, I., & Wolff, M. (2011). Capacity building to achieve sustainable fisheries management in Mexico. *Ocean and Coastal Management, 54*, 731–741. doi:10.1016/j.ocecoaman.2011.07.001.

Friedlander, A. M., & DeMartini, E. E. (2002). Contrasts in density, size, and biomass of reef fishes between the northwestern and the main Hawaiian islands: The effects of fishing down apex predators. *Marine Ecology Progress Series, 230*, 253–264.

Friedlander, A. M., Shackeroff, J. M., & Kittinger, J. N. (2013). Customary marine resource knowledge and use in contemporary Hawai'i 1. *Pacific Science, 67*, 441–460. doi:10.2984/67.3.10.

Gelcich, S., Edwards-Jones, G., Kaiser, M. J., & Castilla, J. C. (2006). Co-management policy can reduce resilience in traditionally managed marine ecosystems. *Ecosystems, 9*, 951–966. doi:10.1007/s10021-005-0007-8.

Gelcich, S., Hughes, T. P., Olsson, P., Folke, C., Defeo, O., Fernández, M., Foale, S., Gunderson, L. H., Rodríguez-Sickert, C., Scheffer, M., Steneck, R. S., & Castilla, J. C. (2010). Navigating transformations in governance of Chilean marine coastal resources. *Proceedings of the National Academy of Sciences of the United States of America, 107*, 16794–16799. doi:10.1073/pnas.1012021107.

Glazier, E. W. (2006). *Hawaiian fishermen*. Belmont: Wadsworth Publishing Company.

Glazier, E. W., Carothers, C., Milne, N., & Iwamoto, M. (2013). Seafood and society on O'ahu in the main Hawaiian islands. *Pacific Science, 67*, 345–359.

Hastings, R. M., & Fischer, D. W. (2001). Management priorities for Magdalena Bay, Baja California, Mexico. *Journal of Coastal Conservation, 7*, 193–202. doi:10.1007/BF02742481.

Jentoft, S. (1989). Fisheries co-management: Delegating government responsibility to fishermen's organizations. *Marine Policy, 13*, 137–154.

Jentoft, S. (2005). Fisheries co-management as empowerment. *Marine Policy, 29*, 1–7. doi:10.1016/j.marpol.2004.01.003.

Jentoft, S., & Chuenpagdee, R. (2009). Fisheries and coastal governance as a wicked problem. *Marine Policy, 33*, 553–560. doi:10.1016/j.marpol.2008.12.002.

Jentoft, S., McCay, B. J., & Wilson, D. C. (1998). Social theory and fisheries co-management. *Marine Policy, 22*, 423–436. doi:10.1016/S0308-597X(97)00040-7.

Kaneshiro, K. Y., Chinn, P., Duin, K. N., Hood, A. P., Maly, K., & Wilcox, B. A. (2005). Hawaii's mountain-to-sea ecosystems: Social–ecological microcosms for sustainability science and practice. *EcoHealth, 2*, 349–360.

Kittinger, J. N. (2013). Participatory fishing community assessments to support coral reef fisheries co-management. *Pacific Science, 67*, 361–3381.

Kittinger, J. N., Koike, H., Stamoulis, K. A., Teneva, L. T., Kittinger, D. S., Oleson, K. L. L., et al. (2015). From reef to table: Social and ecological factors affecting coral reef fisheries, artisanal seafood supply chains, and seafood security. *PLoS One*. doi:10.1371/journal.pone.0123856

Kittinger, J. N., Pandolfi, J. M., Blodgett, J. H., Hunt, T. L., Jiang, H., Maly, K., McClenachan, L. E., Schultz, J. K., & Wilcox, B. A. (2011). Historical reconstruction reveals recovery in Hawaiian coral reefs. *PloS One, 6*, e25460.

Kooiman, J., & Bavinck, M. (2005). The governance perspective. In J. Kooiman, M. Bavinck, S. Jentoft, & R. Pullin (Eds.), *Fish for life: Interactive governance for fisheries*. Amsterdam: Amsterdam University Press.

Kooiman, J., & Bavinck, M. (2013). Theorizing governability – The interactive governance perspective. In M. Bavinck, R. Chuenpagdee, S. Jentoft, & J. Kooiman (Eds.), *Governability of fisheries and aquaculture: Theory and applications* (MARE publication series 7). Dordrecht: Springer.

Lluch-Cota, S. E., Aragon-Noriega, E. A., Arreguin-Sanchez, F., Aurioles-Gamboa, D., Bautista-Romero, J. J., Brusca, R. C., et al. (2007). The Gulf of California: Review of ecosystem status and sustainability challenges. *Progress in Oceanography, 73*, 1–26.

Mackenzie, M. K. (1991). *Native Hawaiian rights handbook*. Honolulu: Native Hawaiian Legal Corporation, Office of Hawaiian Affairs, Distributed by University of Hawai'i Press.

Martin, K. S. (2006). The impact of "community" on fisheries management in the US Northeast. *Geoforum, 37*, 169–184. doi:10.1016/j.geoforum.2005.05.004.

McCay, B. J., Micheli, F., Ponce-Díaz, G., Murray, G., Shester, G., Ramirez-Sanchez, S., et al. (2014). Cooperatives, concessions, and co-management on the Pacific coast of Mexico. *Marine Policy, 44*, 49–59. doi:10.1016/j.marpol.2013.08.001.

McClenachan, L., & Kittinger, J. N. (2013). Multicentury trends and the sustainability of coral reef fisheries in Hawai'i and Florida. *Fish and Fisheries, 14*, 239–255. doi:10.1111/j.1467-2979.2012.00465.x.

McGoodwin, J. R. (1980). Mexico's marginal inshore Pacific fishing cooperatives. *Anthropological Quarterly, 53*, 39–47.

Micheli, F., Saenz-Arroyo, A., Greenley, A., Vazquez, L., Espinoza Montes, J. A., Rossetto, M., & De Leo, G. A. (2012). Evidence that marine reserves enhance resilience to climatic impacts. *PloS One, 7*, e40832. doi:10.1371/journal.pone.0040832.

OECD. (2006). *Agriculture and fisheries policies in Mexico*. Paris: OECD Publishing.

Ostrom, E. (2005). *Understanding institutional diversity*. Princeton: Princeton University Press.

Peckham, S. H., Maldonado Diaz, D., Walli, A., Ruiz, G., Crowder, L. B., & Nichols, W. J. (2007). Small-scale fisheries bycatch jeopardizes endangered Pacific loggerhead turtles. *PloS One, 2*, e1041. doi:10.1371/journal.pone.0001041.

Pérez-Brunius, P., López, M., & Pineda, J. (2006). Hydrographic conditions near the coast of northwestern Baja California: 1997–2004. *Continental Shelf Research, 26*, 885–901. doi:10.1016/j.csr.2006.01.017.

Poepoe, K., Bartram, P. K., & Friedlander, A. M. (2003). The use of traditional Hawaiian knowledge in the contemporary management of marine resources. In N. Haggan, C. Brignall, & L. Wood (Eds.), *Putting fishers' knowledge to work: Conference proceedings* (pp. 328–339). Vancouver: Fisheries Centre, University of British Columbia, Canada.

Ponce-Díaz, G., Arregín-Sánchez, F., Díaz-de León, A., & Alvarez Torres, P. (2009a). Promotion and management of marine fisheries in Mexico. In G. Winter (Ed.), *Towards sustainable fisheries law: A comparative analysis* (pp. 233–252). Gland: IUCN.

Ponce-Díaz, G., Weisman, W., & McCay, B. (2009b). Co-responsabilidad y participación en el manejo de pesquerías en México: lecciones de Baja California Sur. *Pesca Y Conserv, 1*, 1–9.

Ramirez-Amaro, S. R., Cartamil, D., Galvan-Magaña, F., Gonzalez-Barba, G., Graham, J. B., Carrera-Fernandez, M., Escobar-Sanchez, O., Sosa-Nishizaki, O., & Rochin-Alamillo, A. (2013). The artisanal elasmobranch fishery of the Pacific coast of Baja California Sur, Mexico, management implications. *Sciencia Marina, 77*, 473–487. doi:10.3989/scimar.03817.05A.

Saenz-Arroyo, A., Roberts, C. M., Torre, J., Carino-Olvera, M., & Enriquez-Adrade, R. R. (2005). Rapidly shifting environmental baselines among fishers of the Gulf of California. *Proceedings of the Royal Society of London, 272*, 1957–1962.

Sagarin, R. D., Gilly, W. F., Baxter, C. H., Burnett, N., & Christensen, J. (2008). Remembering the Gulf: Changes to the marine communities of the Sea of Cortez since the Steinbeck and Ricketts expedition of 1940. *Frontiers in Ecology and the Environment, 6*, 372–379. doi:10.1890/070067.

Sala, E., Aburto-Oropeza, O., Reza, M., Paredes, G., & Lopez-Lemus, L. G. (2004). Fishing down coastal food webs in the Gulf of California. *Fisheries, 29*, 19–25.

Sen, S., & Nielsen, J. R. (1996). Fisheries co-management: A comparative analysis. *Marine Policy, 20*, 405–418.

Severance, C., Franco, R., Hamnett, M., Anderson, C., & Aitaoto, F. (2013). Effort triggers, fish flow, and customary exchange in American Samoa and the Northern Marianas: Critical human dimensions of Western Pacific fisheries. *Pacific Science, 67*, 383–393.

Turnipseed, M., Roady, S. E., Sagarin, R., & Crowder, L. B. (2009). The silver anniversary of the United States' exclusive economic zone: Twenty-five years of ocean use and abuse, and the possibility of a blue water public trust doctrine. *Ecology Law Quarterly, 36*(1), 1–70.

Vasquez-Leon, M. (1999). Neoliberalism, environmentalism, and scientific knowledge: Redefining use rights in the Gulf of California fisheries. In J. M. Heyman (Ed.), *States illegal practices*. Oxford: Berg.

Vaughan, M. B., & Vitousek, P. M. (2013). Mahele: Sustaining communities through small-scale inshore fishery catch and sharing networks. *Pacific Science, 67*, 33.

Young, E. (2001). State intervention and abuse of the commons: Fisheries development in Baja California Sur, Mexico. *Annals of the Association of American Geographers, 91*, 283–306.

Young, O. (2006). Vertical interplay among scale-dependent environmental and resource regimes. *Ecology and Society, 11*(1), 27.

Chapter 12
Governability of Small-Scale Lobster Fisheries in the Wider Caribbean

Iris Monnereau and Patrick McConney

Abstract Lobster fisheries in the Wider Caribbean region offer an interesting case for governance analysis. From the onset, these fisheries have been exclusively developed for the export market and have generated considerable foreign exchange and extensive livelihood opportunities. While the development of the fishery in the region took place in a similar period (between the 1950s and 1960s) with a similar end market (mainly trade to the US), and the lobster species harvested is identical throughout the region, the governance modes employed in different countries can be quite diverse. This results in differences in exploitation of the resource, value chain of the fishery and well-being of the fisheries. However, these factors will also in turn influence governance and governability. This chapter will analyze the implications of different governing modes in three countries, Belize, Jamaica and Nicaragua, on small-scale lobster fisheries. Specifically, it looks at the commonalties and variances in the governance system and system to be governed of lobster fisheries in the three countries, as well as the developments that underlie these differences and similarities. The chapter shows that the diversity in fisheries call for particularistic governing systems, and also that their diversity is actually the result of different governing modes. The governance mode and diversity of system-to-be-governed are linked by interactive relationships, and understanding the bi-directional interactions between them is crucial in order to improve governability and the wellbeing of fishers and by extension the wider society. Governability assessment of the three fisheries shows that the co-governance governance mode of Belize, resulting in a fair system-to-be-governed and governing system is most appropriate.

Keywords Fisheries governance • Governability • Lobsters • Caribbean

I. Monnereau (✉) • P. McConney
CERMES, University of the West Indies, Cave Hill, Barbados
e-mail: iris.monnereau@cavehill.uwi.edu; patrick.mcconney@cavehill.uwi.edu

S. Jentoft, R. Chuenpagdee (eds.), *Interactive Governance for Small-Scale Fisheries*, MARE Publication Series 13, DOI 10.1007/978-3-319-17034-3_12

Introduction

A lobster dinner in the United States (US) is a considered a luxury. The growing demand for lobster in the US since the beginning of the twentieth century is in line with global developments whereby consumers increasingly consume "high value foods" (Acheson 1997; Delgado et al. 2003; Monnereau 2012). The increase in demand initially stimulated increased output by the US lobster industry, but as demand grew, particularly during the latter half of the century, it stimulated lobster fisheries worldwide. The development of the lobster fishery should be seen in light of more general global trends in rising fish consumption, production, and trade. The industrialization of the world's oceans (Bavinck 2011), also known as the *blue revolution*, took place in little over a century, but with the most rapid fishery development taking place since the 1950s. This was the result of the combined effects of increasing populations and wealth, new technological possibilities, freezer facilities, urbanization, dietary changes, and the expansion of seafood markets (Bailey 1987; Hersoug 2005; Bavinck 2011). As a consequence, international fish trade rose from USD 6.1 billion in 1980 to a record value of USD 129.8 billion in 2011 (FAO 2010, 2014). The development of the Caribbean spiny lobster (*Panulirus argus)* fishery should be seen in the light of these global development trends.

The increasing demand and high unit price for lobster on the international market, combined with new harvest and postharvest technology, resulted in lobster fishery development throughout the world, including in places where previously no commercial lobster fishery had existed. In the Caribbean, lobster landings were small until the international demand from the US stimulated export-oriented commercial lobster fisheries throughout the region since the 1950s. Fishers increasingly built wooden lobster traps or used scuba gear to benefit from these new economic opportunities. Large US freezer ships went down to collect lobster tails, connecting the Caribbean fishers with American consumers (Vega 1978; Vilas 1989). The increase in volume and value of lobster from the Wider Caribbean region at the international level reflected the growing openness to, and integration into, international trade of many fisheries (Monnereau 2012). The lobster fishery in the Wider Caribbean currently provides widespread livelihood opportunities and generates extensive foreign exchange and tax income for several governments (Monnereau 2012; Chakalall et al. 2007). The annual value of the lobster fishery in the region is approximately USD 500 million (Chakalall et al. 2007) making it the most important fishery in the region in terms of value. The fishery represents an important source of employment and foreign exchange earnings for a large number of countries in the region (FAO 2007).

Some scholars have argued that the global fish trade involving developing countries has not necessarily translated into increased human development (Kurien 2004, 2005; Hersoug et al. 2005; Béné et al. 2010). Fishers might not actually reap the benefits of the global fish trade and trade can negatively affect food security, local economies, and livelihood options for the poor (Ruddle 2008). The lobster fishery in the Wider Caribbean region provides an interesting comparative case as it has, from its onset, focused nearly exclusively on the export market and has developed only since the 1950s yet many striking differences between the fisheries can be found. One of these differences relates to the level of overexploitation of the resource. While in Belize it is fully exploited, Jamaica and Nicaragua's lobster fisheries are both overexploited (FAO 2007). Marine fisheries in general, and particularly export-based fisheries in developing countries, are principally vulnerable to overexploitation (Thorpe and Bennet 2004). The spiny lobster in the Wider Caribbean is heavily fished throughout much of its range, being overexploited in six countries and either fully exploited or stable in others (FAO 2007). The underlying reasons for the overexploitation can be diverse and vary between countries, including, *inter alia*: high level illegal fishing; open access of the fishery; and lack of adequate management policies in place or implementation thereof. One could argue that even though the fisheries crisis in one country might resemble that in another country, when looking at the details, they are different. The circumstances in which they occur are diverse as well as the viewpoints on rules and regulations and stakeholder involvement to solve the problem of overexploitation.

In this chapter we compare the governability of lobster fisheries of Belize, Jamaica and Nicaragua using the governability assessment framework with emphasis on a value chain perspective. Governability can be defined as "the overall quality for governance" (Kooiman et al. 2008). This quality is assumed to be situated partly in the system-to-be-governed which relates to the fish chain and fishing communities, the governing system (the institutions and organizations that have a steering role in small-scale fisheries), and in the governing interactions (i.e. how the governing system and system-to-be-governed are related). Whether a fisheries system is more or less governable depends on the inherent traits and constructed capabilities of all three systems (Chuenpagdee and Jentoft 2013). We will therefore examine the system-to-be-governed, governing system and governing interactions for each of the three countries. When examining the system-to-be-governed we will focus mostly on the social aspects of the system-to-be-governed.

The choice of the three countries is based on the argument that each has developed a significantly different lobster governance mode ranging from co-governance to hierarchical governance. The overall result of the interaction between market parties, state, and civil society in each country can be seen as a distinct governance mode (Kooiman et al. 2008). These national fisheries governance modes in the different countries in the region can be expected to develop out of existing interactions and transactions. They are embedded in larger societal structures and are the result of dissimilar historical trajectories, the varying importance of the lobster fishery as an economic sector, geographical factors and the interactions among civil society, state, and market actors. In *Fish for Life,* Kooiman et al. (2005) distinguish three ideal types of governance modes of interaction: self-governing, co-governing, and hierarchical interaction. They argue that all societies demonstrate—and require—mixes of these three modes. This chapter examines the commonalties and variances in the system-to-be-governed, governing system and the developments that underlie these differences and similarities. We argue that these differences have explicit implications for small-scale lobster fisheries in each country.

Methodology

This work is based on fieldwork by the first author in Belize, Jamaica, and Nicaragua. The comparative case-study approach is valuable for examining the relationship between contextual factors and a specific research topic (Yin 2003). During fieldwork in the three countries, informal interviews as well as unstructured and semi-structured in-depth interviews were conducted, questionnaires administered, and participant observation carried out. The same author also participated in state and non-state policy meetings. The fieldwork was carried out between 2006 and 2009 across the three Caribbean countries and additional interviews were conducted with importers in the US and Europe. In each of the selected countries, there were different types of fishing groups both in terms of scale (small-scale and industrial) and gear (diving and trapping). This chapter is based on 23 observation days at sea, 118 interviews, 88 wellbeing surveys and other communication. All data presented in this chapter are published in Monnereau (2012) (Fig. 12.1).

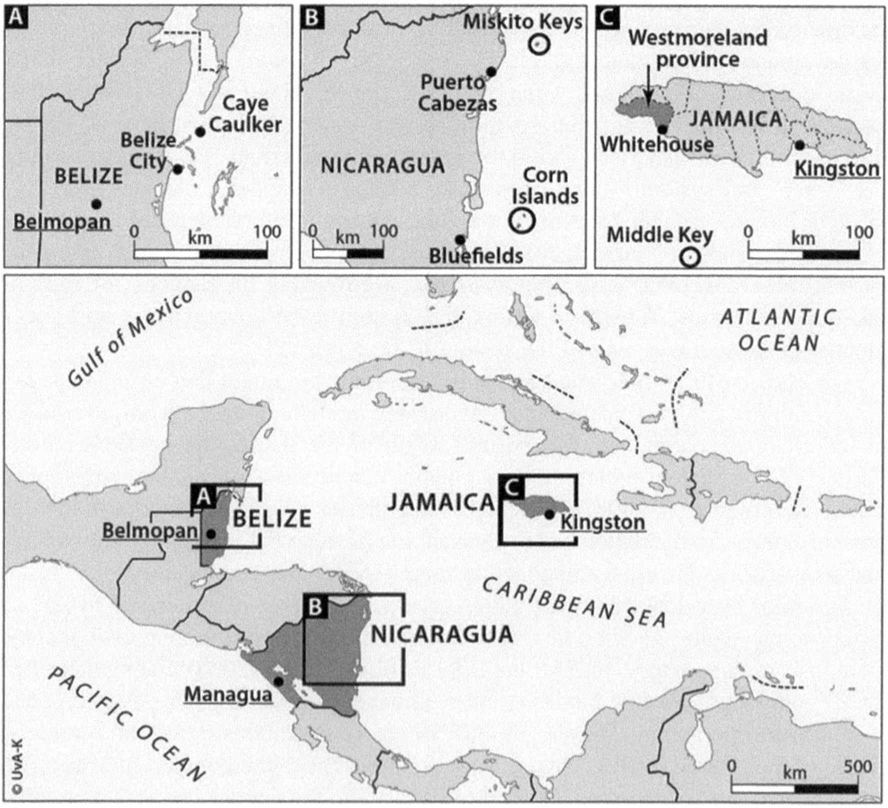

Fig. 12.1 Research countries and locations in Belize, Jamaica and Nicaragua

System-to-Be-Governed

Social ecological systems such as fisheries can be characterized by complexity, diversity, dynamics and differences in scale (Bavinck et al. 2005; Kooiman et al. 2005). Diversity is about heterogeneity of system elements (Bavinck et al. 2005; Jentoft and Chuenpagdee 2009). Fishers can be small-scale, intermediate scale or industrial fishers. The post-harvest link can extend to local, national or international markets or a combination thereof. Within the ecological system the diversity, complexity and dynamics can be found within, *inter alia*, the complex life-cycle of the spiny lobster and the fact it's a transboundary and shared resource. The spiny lobster is found in the Western Atlantic, from Bermuda and North Carolina's eastern US coast, down to Rio de Janeiro, Brazil. Throughout its different life stages lobsters move through different sea habitats and consequently have a complex life cycle requiring three distinct habitats: open ocean (larvae); a shallow, vegetated coastal zone (juveniles); and coral reef (adults) (Butler et al. 1997). As lobster is a shared resource by virtue of its planktonic larval dispersal developments, the destruction of nursery grounds in one country can potentially affect the prospective harvest of the neighboring countries (Ehrhardt et al. 2011). Although we acknowledge that diversity, complexity and dynamics can be found in the ecological system, in this chapter we focus further purely on the human aspects of the lobster fishery.

Wider Caribbean lobster fisheries share a number of common features but also exhibit significant diversity. The number of lobster fishers varies from a little over 2,000 in Belize, over 15,000 in Nicaragua to 20,000 in Jamaica (Monnereau 2012). The three countries all harvest the same species of spiny lobster and in each country one can find trap fishers and divers. Yet, the technique and scale differ by country. In Belize there are only small-scale fishers. Small-scale refers to smaller vessels with the added connotation of low levels of technology and capital investment per fisher. In Belize there are both small-scale trap fishers as well as divers. Trap fishers fish with wooden traps, while divers are free-lung divers. In Jamaica there are mostly small-scale fishers with a small industrial fleet. The industrial fleet uses wooden traps and small-scale fishers mostly chicken-wire traps. Divers can be free-lung, use SCUBA-equipment or hookah. Hookah refers to divers using an umbilical cord for air supplied from the surface by a compressor on board a boat. Nicaraguan lobster fishers are the most diverse with small-scale (wooden) trap fishers, small-scale divers (only SCUBA), industrial divers (SCUBA) and industrial trap fishers (wooden traps) (see Table 12.1).

These different techniques and scale-levels demand a different set of investments, skills and lead to different levels of remuneration. A free-lung diver in Belize, for example, is an independent fisher who leaves for 8–9 days at a time on a sailing dory. His investments include a dugout canoe, mask, fins, long pole with metal hook for lobsters and a spearfishing gun for fin-fish. He pays his captain 4 lbs. of lobster each day or the equivalent thereof on a daily basis but sells his catch to a middlemen or processing plant (one of two fishing cooperatives) of his liking. Fishing grounds in Belize are located off the coast in shallow waters. Trap fishers

Table 12.1 Diversity of lobster fishers in Belize, Jamaica and Nicaragua illustrated by technique and scale

Features of diversity	Belize	Jamaica	Nicaragua
Population[a]	331,900	2,715,000	6,080,478
Coastline in km[b]	386	1022	910
Number of fishers	2,026	20,000	15,720
Volume exports (lbs. of tails per year)	533,315	700,000	1,100,000
Type of fishing	Small-scale	Small-scale and industrial	Small-scale and industrial
Small-scale trappers	X	X	X
Small-scale divers	X	X	X
Industrial trappers		X	X
Industrial divers			X
Per cent of total catch by fishing type	100 % small-scale	60 % small-scale and 40 % industrial	50 % small-scale 50 % industrial
Input/output structure	Mostly export market (part also for tourist market)	Only part is destined for export	Nearly all exported

Source: Adapted from Monnereau (2012), unless noted
[a]Worldbank average 2010–2014
[b]CIA factobook. https://www.cia.gov/library/publications/the-world-factbook/fields/2060.html

live in close proximity to their fishing grounds while divers have to travel 1–2 days to reach their fishing grounds.

A diver in Nicaragua uses SCUBA tanks which he most commonly rents from a particular middleman he is indebted to. He is officially, however, an independent worker who goes out to sea on day trips together with a 'bubbleman' (someone who follows his air bubbles from the boat) and a captain. A Nicaraguan industrial diver on the other hand goes out for 21 days with 24 other divers and 25 helpers on a 'mother vessel'. He uses up to 16 SCUBA tanks a day. He works independently during the day from a dugout canoe fisher and pays his helper and the captain a fixed price. Both small-scale and industrial divers in Nicaragua face very insecure working circumstances and often face problems with 'decompression illnesses', also known as 'the bends' (Ehrhardt 2005, 2006).

In Jamaica one can find small-scale fishers who are day fishers fishing in inshore areas. A large percentage, nonetheless, fish in the most productive fishing grounds of Jamaica; the Pedro Bank. The Pedro Bank is one of the largest offshore banks in the Caribbean Basin (Espeut 2006). Large concentrations of lobsters are found on the Pedro Bank, which accounts for about 60 % of the total lobster landings. Although it is the best fishing ground in Jamaica, it is 80 km from the mainland, and thus very remote for fishers who reside on the two tiny inhabited atolls on the bank. Fishers are mostly multi-species small-scale fishers who make use of traps, SCUBA equipment, hookah or free lung diving.

In all three countries intermediaries, processing plants, and international importers play an important role in the fishery yet they vary in professionalism, magnitude

Table 12.2 Features of the system-to-be-governed (SG) in Belize, Jamaica and Nicaragua

Features of SG	Belize	Jamaica	Nicaragua
Entry barrier intermediaries	Medium level entry barrier; acting as such is illegal so one cannot set up a formal organization to act as a middleman but widely practiced	Medium level entry barrier; middlemen require capital to bind fishers and build lobster inventory	Intermediaries entry level varies greatly between "bucket ladies" and official intermediaries; official intermediaries requiring medium-high investments
Entry barrier processing plants	High entry barrier: there are only two exporting processors allowed (the fishing cooperatives)	Capital- intensive yet still often small-scale processing so medium level entry barrier	Entry barrier extremely high (large capital investments, know-how)
Export rules and practices	Processing in hands of fishing cooperatives	Processing in hands of many small (some illegal) exporters; only few larger official exporters	Approximately ten large processing plants
Quality standards and practices	Quality is good although one case of salmonella has recently been detected	Quality is poor	Quality is excellent
Dependency	Intermediaries mostly independent; one cooperative has strong ties with international importers; other cooperative has weak and constantly changing ties	Fishers highly dependent on intermediaries; processing plants are small with weak ties with international US importers	Unofficial intermediaries weak ties; official intermediaries strong ties; processing plant strong ties with international US importers

Source: Adapted from Monnereau (2012)

of the enterprises, levels of investment, as well as interrelationships. There are differences in entry barriers within the different groups as well. Entry barriers refer to the difficulty of new competitors to enter the trade and can be related to capital investment or technological barriers, and cultural, social or geographical barriers.

The differences between intermediaries, processing plants, and international importers relate, *inter alia,* to the complexity of fisheries. Complexity in fisheries arises in relation to the multiple linkages that occur within the chain and between fishery and non-fishery activities (Bavinck et al. 2005). In this chapter we focus on the linkages within the chain (Table 12.2).

In Belize, the principal value chain actors are fishers and fishing cooperatives. Fishing cooperatives are the only agents able to export seafood, and fishers are formally not allowed to sell through intermediaries. The cooperatives are owned by the fishers and all benefits flow back to the fishers at the end of the fiscal year by means of a second payment. There are therefore no external commercial parties involved. Although intermediaries are officially illegal in Belize, sale of lobsters through intermediaries does occur. In Jamaica, the chain is characterized by a multitude of actors between fishers and the end consumer. The latter can be located in another country (served via intermediaries, processing plants, and international importers), or in the local tourist industry. Lobster fishers in Jamaica frequently make use of intermediaries, as they are often based very far from the market and require help in obtaining fuel, credit, food, and even drinking water. There are a few large processing plants but only a limited amount is designated for exports. The large proportion ends up in the hospitality industry. Traders travel as far as 15 h by boat to pick up the lobster in Jamaica's most productive fishing grounds far out from the mainland. Fishers depend on the middlemen for fresh drinking water and fuel which are scarce on these atolls.

In the Nicaraguan lobster fishery, the fishers generally make use of intermediaries who transfer the product to processing plants. These intermediaries differ in entry-level barriers, and scale of operation. They differ in scale and size as well and vary from 'bucket-ladies' to official middlemen. Bucket ladies buy a bucket of lobster a day to sell to processing plants and have no investments whereas official middlemen run shops and supply credit for traps, ice and fuel. Intermediaries have a strong influence over fishers as the latter are dependent on them for fuel, ice and gear, however, fishers also benefit from the relationship in terms of credit for procuring this in times of need.

In Jamaica, there are many smaller processing plants, some of which are legal some illegal. As the quality of the lobster in this country is inferior, the barrier to entry into processing is lower than in Nicaragua. This is also the case because they are smaller-scale and thus require less capital and technological know-how, although they do export directly to the US. In Nicaragua, the entry barrier for processing plants is very high due to the high level of capital and technological investment as they are larger processing plants.

Dynamics in fisheries derive from the multiplicity of linkages within the chain, as well as the uncertainty due to unpredictable external factors ranging from environmental effects on fish stocks to global markets (Bavinck et al. 2005). System-to-be-governed, governing system and the governing interaction are variable over time. Most lobster produced in the three countries is sold on the international market, except for Jamaica where a large part is also destined for the national tourist market. Household consumption of lobster is low across the three countries (Monnereau 2012). Surveys conducted among lobster fishers in the three countries (Monnereau 2012) led to the conclusion that on average 81 %[1] of the fishers preferred eating other seafood over lobster (although in terms of value it was the most esteemed marine species to fishers). Fishers eat lobster, but these are mostly illegal sized ones they cannot sell in the market. Lobsters also do not have the same cultural appreciation as in many countries in the North such as in the US or European countries. However, lobster is highly valued by fishers as a commodity and desired by tourists and foreigners.

Fishers, intermediaries, and processing plants are all inserted in global value chains that connect them "upwards" with foreign buyers. The international export of lobster to the US and Europe requires strict handling procedures and food safety

[1] N=88.

standards to be in place in all countries' processing plants. As quality is not easily discernible, trust plays an important role in exporters' relationships with lobster importers. These relationships are reinforced by credit advances. Although all processing plants require high investments as a result of the mandatory compliance with Hazard Analysis Critical Control Point (HACCP) guidelines, there are significant quality differences between the three countries. These result from variations in the handling of the product before processing. Quality is highest in Nicaragua, followed by Belize, then Jamaica.

Governing System

Lobster is a transboundary resource that is facing severe challenges with overexploitation throughout the region. Addressing this requires regional cooperation. Several Regional Fisheries Bodies (RFB) have become involved in lobster fisheries governance, notably the Western Central Atlantic Fisheries Commission (WECAFC), the Caribbean Regional Fisheries Mechanism (CRFM) and the Organization of Fishing and Aquaculture in Central America (OSPESCA). However, none of these advisory RFBs has a management decision-making mandate, so they only indirectly influence national policy. National actors and arrangements are the most directly influential when considering lobster fisheries governance in Belize, Jamaica, and Nicaragua. Yet, both national and regional governance is often overshadowed by the extra-regional influence of the export markets (especially the US) communicated primarily through the value chain.

Our data, presented in the section on governing systems, will also show that diversity not only calls for particularistic governing systems; their diversity is also actually the result of different governing approaches. We consider the lobster fisheries governing system in the three countries by examining their development orientations, the orientation of states towards their fishing sectors, market-state relations and stakeholder representation. Lobster resources in the Wider Caribbean are facing a number of serious management problems. The open access nature of the fisheries, the high level of illegal, unregulated and unreported fishing, and declining stocks are the main problems. To counter these problems, all three countries have similar management regulations and policies (Table 12.3). All three countries have implemented a tailweight limit, a size limit, a closed season, and law prohibiting harvest of berried females and moulting lobsters. Only Belize, however, has prohibited the use of SCUBA gear for commercial fishing purposes as well as the presence of an industrial fleet. Fishers need licenses to enter the fishery in all countries yet only in the case of Nicaraguan industrial boats have the number of licenses been limited (since 2009). Each country has a minimum size, closed season and ban on catching egg-bearing females. However, the length of the closed season will vary and in Belize the minimum size is smaller than in Nicaragua and Jamaica. As part of the catch is sold to and consumed by the hospitality industry in Belize and Jamaica it is more difficult to ban the use of lobster during the closed season.

Table 12.3 Management regulations in place in lobster fishery in Belize, Jamaica and Nicaragua

Management regulations	Belize	Jamaica	Nicaragua
Size limit (tail weight in ounces)	4.2	5	5
Size limit (length in cm carapace)	7.62	7.62	7.60
Closed season	1 April–30 June (3 months)	1 March–30 May (3 months)	1 March–30 June (4 months)
Berried females prohibition law	Yes	Yes	Yes
Molting lobsters prohibition law	Yes	Yes	Yes
SCUBA prohibition	Yes	No	No
Limit to # licenses	No	No	Yes (since 2009)
Industrial fleet (number vessels)	No	Yes (4)	Yes (78)
Gear regulations	Using scuba gear is prohibited	Using scuba gear is allowed; hookah as well	Using scuba gear allowed

Source: Adapted from Monnereau (2012)

In Belize the pivotal involvement of fishing cooperatives in fisheries management since the 1960s is undisputed. The state has been pro-developmental, committed, and supportive of the small-scale fishers' initiative in the 1960s to organize as cooperatives. The cooperatives act as intermediaries between fishers and government. From the early 1960s, the government has granted exclusive rights over lobster export (and all other seafood) to fisheries cooperatives. Only two fishing cooperatives are allowed to export seafood products, and as fishers are owners of the fishing cooperatives, no commercial market parties are involved, and all benefits derived from the fishery flow back to the fishers. The two exporting cooperatives are located in Belize City; other smaller cooperatives are closely connected to one of these two cooperatives. Since 1965, requests from foreign firms to harvest, process, or export fish have been rejected. This protective measure is aimed at securing the profits of the fishery for the fishers, as no large commercial intermediaries are present to skim the profits. High export earnings have strengthened the cooperatives economically, thus translating into political strength and a determination to protect the privilege of the monopoly over export that they enjoy. The fact the commercial use of SCUBA gear in fisheries in Belize is prohibited has resulted in high levels of job safety for the fishers. Free-lung diving is enabled partly, however, by geographical factors that have left Belize with a very extensive reef with shallow waters (Table 12.4).

In Belize the management laws and regulations on closed seasons, limits to weight and length, and the catching of berried females are generally well enforced. Although illegal, unregulated and unreported fishing exists, in Belize it is lower in comparison to the other two countries. Fishers themselves believe only 9 % of fishers fish during the closed season, and the percentage of fishers believed to catch

Table 12.4 Features of the governing system of the lobster fisheries in Belize, Jamaica and Nicaragua

Features of governing system	Belize	Jamaica	Nicaragua
Stakeholder representation *Fishery Advisory Board (FAB)*	Cooperatives are represented on FAB	Industry not represented on FAB; cooperatives not represented	Industry (processor) interests represented on FAB; cooperatives weakly represented
Stakeholder representation (Non-Governmental Organizations (NGOs) and Marine Protected Areas (MPA)	State-NGO cooperation exists; NGOs own and manage MPAs; state-NGO relation formalized	NGOs and tourist market parties active in MPA management; state-NGO cooperation beginning and formalized	State-NGO cooperation limited; few MPAs exist
State-market relations in governing system	Involvement of market parties limited; cooperatives only parties with exporting license	Market parties play an important role in the fishery, but very limited role in fisheries governance; processors not organized	Important role market parties; processing plants well organized, able to influence decision making
Enforcement	Relative high enforcement	Relative low enforcement	Relative low enforcement
Weight limits/closed season	Yes/medium success	Yes/low success	Yes/low success
Access	Open access, but exclusive rights for domestic small-scale fishers	Open access, but limited for industrial fleet	Open access, but limited access for industrial fleet

Source: Adapted from Monnereau (2012)

berried lobster is the lowest of the three as well. In relation to the involvement of NGOs, Belize recognized its limited means and ability to manage its natural resources early on, and so established and formalized the aid provided by national and international NGOs mostly in relation to MPAs. In addition, the full array of stakeholders involved is represented on the FAB, including the fisheries cooperatives and NGOs. The FAB is a strong and powerful group of parties influencing decision-making on lobster fishery issues in Belize.

In Jamaica, the state initially attempted to create and support strong fishing cooperatives from the start of the fishery in the 1950s. Yet, due to organizational problems, many of these cooperatives have not been very successful and have not evolved into organizations with either market or decision-making powers. The government has maintained a special focus on the small-scale fleet, and few industrial boats have been licensed. Historically, fishers have also had subsidized fuel, mesh wire, engines, and boats. The regulations on closed season, weight, and size limits have had limited success. Illegal, unregulated and unreported fishing is considered to be high and data collection poor. Fishers themselves judge that approximately 50 % of the fishers fish for lobster during the closed season, 39 % fish for undersized lobster, and 33 % catch berried females.[2] These numbers are high and show the lack of management enforcement and control. The interests of small-scale fishers are poorly represented in decision making, while the subsidies by means of tax exemptions and cheap fuel were curtailed in the 1990s, when the government ended the subsidies and services to the fishing industry. All decisions concerning the fishery are taken by the government, although users have been involved in the process, and an FAB has recently been established. This FAB has failed to involve all stakeholders and market parties and fishing cooperatives were not represented. The role of NGOs and market parties in management of MPAs has been increasing in Jamaica. These relations can be seen as public-private partnerships and are beginning to become formalized.

In Nicaragua the lobster fishery developed rapidly beginning in the 1950s. By the end of the 1970s there were 100 industrial ships in operation in the region. Foreign capital, mainly from the US, was invested in boats and processing plants. The civil war in the 1980s, however, put an end to the fishery for a decade. After the recovery period for the lobster resource during these years of civil war with little fishing effort lobster harvests were very high. After 1990, the fishery once again gained prominence as an economic activity on the Caribbean coast of Nicaragua and over 70 boats were licensed. Since the initiation of the lobster fishery in Nicaragua in the 1950s, the large processing plants and industrial fleet owners (which to some extent are often one and the same) have been very influential in lobster fisheries governance. They are represented well in the FAB while other civil society organizations such as NGOs or fishing cooperatives are weakly represented. The fishing cooperatives that have developed are small, and have neither market powers nor decision-making powers. The fishery is highly overexploited, with extremely high levels of illegal, unregulated and unreported fishing (Ehrhardt 2005, 2006). The author argues

[2] N = 84.

that 60 % of all landed lobster consists of undersized lobster that has not yet reached an adult stage. Both secondary literature and interviews with fisheries managers and experts indicate that catching berried females and undersized lobsters, and fishing during the closed season are commonplace in Nicaragua. In Nicaragua, fishers in the surveys indicated they were of the opinion nearly 40 % of the fishers would catch berried females, slightly fewer, 31 %, they believed would catch undersized lobster, whereas in their opinion nearly 40 % would continue catching lobster throughout the closed season.[3] Monitoring and enforcement is low and corruption is prevalent. No MPA on the Caribbean coast in Nicaragua has yet been established, although the process has been initiated (Gonzalez and Jentoft 2010). NGOs are, however, currently slowly initiating cooperation with the Nicaraguan state and the lobster industry to improve the sustainability of the resource (Gonzalez and Jentoft 2010).

Governance Modes and Governing Interactions

In capture fisheries, the three modes of governance all influence governability. In Belize we have seen a *co-governance* mode where societal parties join hands with a common purpose in mind. Co-governance implies the use of organized forms of interaction for governing purposes (Kooiman et al. 2008). *Hierarchical governance*, such as can be found in Nicaragua and partly in Jamaica, is characterized by the interactions between a state and its citizens. It is a top-down style of interaction, expressing itself in policies and laws. In Table 12.5 we present more details on the governance modes and orientation of the state. In this section we discuss the governance modes of Belize, Jamaica and Nicaragua based on our findings in the system-to-be-governed and governing system presented previous.

In Belize the state has been supportive of the small-scale fishers since the 1960s. The cooperatives are secure financially which has translated into political strength. NGOs have become influential in fisheries governance in Belize as well, mainly through the establishment and protection of MPAs. The state in Belize has formalized the aid provided by national and international NGOs. In addition, the full array of stakeholders involved is represented on the FAB, including the cooperatives and NGOs. The FAB is a strong and powerful group of parties influencing decision-making on lobster fishery issues in Belize. We therefore conclude the Belizean lobster fishery governance style is one of co-governance. The development trajectory of the lobster fishery in Belize and the decisions that have been made in favor of the fishing cooperatives and prohibition of SCUBA gear and an industrial fleet has been very positive for small-scale fishers.

The governance style in Jamaica is more difficult to pinpoint and can be regarded as defective co-governance, later developed into hierarchical governance. In Jamaica, the state has attempted to create and support strong fishing cooperatives

[3] N=30.

Table 12.5 Governance modes and orientation of the state towards fishing sector

Features of governing system	Belize	Jamaica	Nicaragua
Governance mode	Co-governance	[Defective] co-governance/ hierarchical governance	Hierarchical governance
Development orientation of the state	Social democratic	Neo-liberal	Neo-liberal
Dominant domestic groups	Middle-class elite and foreign investors	Middle-class elite and foreign investors	Small traditional elite, oligarchic but some foreign investors
Orientation of the state towards fishing sector	Protecting interests of small-scale fishers and cooperative system	Initial focus on small-scale fishers and cooperatives failed. Majority of fishers small-scale	Focus on development of the fishery and foreign exchange; processing plants large influence; large industrial fleet
Impacts of small-scale fishers	High remuneration, safe working conditions, economic alternatives (for trappers), large influence on decision-making	Medium remuneration, medium safe working conditions, few economic alternatives, low influence on decision making	Low to medium remuneration, unsafe working conditions (for divers), economic alternatives limited to narco-trafficking, low influence on decision-making

Source: Adapted from Monnereau (2012)

from the start of the fishery in the 1950s. Yet, the cooperatives have not been very successful and have not evolved into organizations with either market or decision-making power. The government has maintained a special focus on the small-scale fleet, however, and few industrial boats have been licensed (only four). The state has also subsidized fuel, mesh wire, engines, and boats for fishers. Public-private partnerships between market actors such as hotel chains and the state, have recently developed and are being formalized. Although the co-governance mode has to date been unsuccessful, the state is making attempts to move again in the direction of co-governance. SCUBA gear is not prohibited but the level of use is much lower than in Nicaragua and as a consequence accidents are not as frequent. Small-scale fishers are however often poorly represented in decision-making.

In Nicaragua the governance style can be regarded as hierarchical. From the onset it has focused on development of the fishery and specifically the large industrial fleet. However, the top-down approach from the state is not unilateral. The state has had difficulties in withstanding the pressure from interest groups, such as the industrial fleet owners and processing plant owners, as they are both politically and economically very powerful. Fishing cooperatives or NGOs have been much less successful in influencing decision-making.

Discussion and Conclusion

The three countries show distinct 'qualities for governance' (i.e. qualities relate to characteristics in the system-to-be-governed such as wellbeing of fishers and dependency relations as well as the governing system relating to stakeholder interactions). This difference is partly situated in their differences in the system-to-be-governed, the governing system, and in the governing interactions (i.e. how the governing system and system-to-be-governed are related). This chapter has shown that the system-to-be-governed in lobster fisheries in the Wider Caribbean show clear commonalities and differences. While lobsters are mostly produced for a similar end market in the US and all fishers either use traps or dive for lobster our data also shows great heterogeneity. Belize only has small-scale fishers while both Jamaica and Nicaragua also have an industrial fleet. In Jamaica and Nicaragua SCUBA equipment is used while in Belize this is prohibited. The lobster chain also shows a great variety and types of intermediaries as well as processing plants and the dependencies and level of influence to influence the lobster chain. The dynamics of the chain in turn influence the governing system. Due to the fact there are few successfully functioning fishing cooperatives of small-scale fishers in Nicaragua and there are such a large number of different fishing fleets results in low prices. Processing plants on the other hand are very influential in the chain and have a large influence over other actors in the chain. In Jamaica the absence of many large-scale processing plants leaves a void for intermediaries while cooperatives to influence prices have been less successful. In Belize fishers are well organized in cooperatives and officially no middlemen in the lobster chain exist. Yet, in reality they are present.

However, as they are partly illegal they do not influence management of the fishery.

This chapter has shown distinct governance modes across the three countries and the implications thereof for small-scale fishers. There is no doubt that the Belizean lobster fishery is the most harmonious of the three, and closest to the ideal type of co-governance. Co-governance impacts small-scale fishers most positively in comparison to the hierarchical types of governance in Jamaica and Nicaragua. This study has also shown, however, that the characterization of hierarchical governance by Kooiman et al. (2005, 2008) leaves little room for the nuanced types of governance systems described in our findings. When analyzing the modes of co-governance and hierarchical governance in our case studies, one can see many differences with regard to the role of the state, NGOs, and markets. Market interests in Jamaica are relatively absent from the decision-making process, whereas in Nicaragua they are heavily involved in fisheries governance. In Nicaragua, however, NGOs are virtually absent in the decision-making process, whereas in Jamaica they are becoming increasingly important. Although Nicaragua's governance mode involves cooperation between the state and powerful market parties, we have also seen that the state has limited powers and that the term 'hierarchical state' does not do justice to the current state of affairs in Nicaragua. Market party influence over decision-making in the lobster fishery has thus been disproportionately large in comparison to civil society actors.

The boundaries between the state and market in governance are fluid. The three concepts of governance developed by Kooiman et al. (2005) do not adequately describe the governance system in Nicaragua where the state has a rather limited steering ability, the market is very powerful and fishing groups have very little decision making power. One could argue that this type of state reflects a 'soft state' in line with Myrdal (Myrdal 1970). A 'soft state' reflects that there are laws and regulations regarding conservation but there is a deficiency in law observance and enforcement and public officials are known to collude with powerful persons and groups to circumvent legal constraints on market transactions. Rotberg (2004) makes a distinction between weak states and strong states. States may be inherently weak according to him because of geographical, physical, or fundamental economic constraints; or they may be basically strong, but temporarily weak because of internal antagonisms, management failures, greed, despotism, or external attacks. Evans distinguishes between developmental and predatory states (1989). One could also argue that the state is perhaps more perversely strategic rather than 'soft' and deliberately turns a blind eye to illegal practices. Rather than considered just 'soft' they could be seen as 'strategically soft'. This chapter does not attempt to analyze all different types of state involvement in fisheries.

Interactive governance refers to societal interactions between actors in the public sphere and is opposed to 'private' activities. However, in lobster fisheries governance in the Caribbean, both in system-to-be-governed and governing system, the role of 'private' activities is crucial. The state's role in lobster fisheries governance is both heavily influenced by private actors, or lack thereof. Incorporating work that has been done on the Global Value Chain (GVC) approach (e.g. Gereffi 1994; Gereffi and Korzeniewicz 1994; Gereffi et al. 2005; Bair 2008; Gibbon et al. 2008) in the current governance framework is therefore crucial. In this approach the role of private actors along the value chain and governance are given a much more central role and provide pathways for analyzing the role of market parties within the system-to-be-governed. The GVC approach helps understand how global industries and consequent value chains are organized by examining the structure and dynamics of different actors involved in a given industry. In today's globalized economy with very complex and dynamic interactions, such as the case with fisheries, the GVC methodology is a useful tool to trace the shifting patterns of global production, link geographically dispersed activities and actors of a single industry such as the lobster fishery, and determine the roles they play in developed and developing countries alike. When analyzing governance modes of fisheries in a national context we thus argue that the varied role of market parties in system-to-be-governed and governing system should be taken into account more than is currently the case.

Kooiman et al. (2008) highlight the influence of power relationships within governance interactions. They also state that socio-political cultural traditions are of importance in enabling or restricting governance interactions. This chapter has been a modest attempt to show how different socio-political traditions within the same region regarding a fishery for a single target species will influence the system-to-be-governed and governing system and the small-scale lobster fisheries as a result. The chapter has shown that the diversity in fisheries call for particularistic governing systems, and also that their diversity is actually the result of different governing modes. Different interactions in the three countries will influence the structure of the value chain, such as the presence of an industrial fleet, use of harmful fishing equipment, and the type and extent of stakeholder involvement in governance. Rather than diversity in fisheries calling for a particular mode of governance, our results show that in fact this diversity is often the result of choices made within a country's particular governance mode. Governability assessment of the three fisheries shows that the co-governance governance mode of Belize, resulting in a fair system-to-be-governed and governing system is most appropriate. Jamaica and Nicaragua's lobster fisheries governability can be enhanced we argue by improving and creating better opportunities for the involvement of fishers in the governance system, enhancing the benefit and power sharing along the lobster value chain and diminishing the role of very powerful market parties.

References

Acheson, J. M. (1997). The politics of managing the Maine lobster industry: 1860 to present. *Human Ecology, 25*(1), 3–7.

Bailey, C. (1987). The political economy of fisheries development in the Third World. *Agriculture and Human Values*, 35–48.

Bair, J. (2008). Analysing global economic organization: Embedded networks and global chains compared. *Economy and Society, 37*(3), 339–364. Retrieved May 25, 2014, from http://www.informaworld.com/openurl?genre=article&doi=10.1080/03085140802172664&magic=crossref||D404A21C5BB053405B1A640AFFD44AE3

Bavinck, M. (2011). The mega-engineering of ocean fisheries: A century of expansion and rapidly closing frontiers. In S. Brunn (Ed.), *Engineering earth: The impacts of mega-engineering projects* (pp. 257–273). Dordrecht: Kluwer Press.

Bavinck, M., Chuenpagdee, R., Diallom, M., van der Heijden, P., Kooiman, J., Mahon, R., & Williams, S. (2005). *Interactive fisheries governance: A guide to better practice*. Amsterdam: Centre for Maritime Research.

Béné, C., Hersoug, B., & Allison, E. H. (2010). Not by rent alone: Analysing the pro-poor functions of small-scale fisheries in developing countries. *Development Policy Review, 28*(3), 325–358. doi:http://doi.wiley.com/10.1111/j.1467-7679.2010.00486.x

Butler, M., Herrnkind, W., & Hunt, J. (1997). Factors affecting the recruitment of juvenile Caribbean spiny lobsters dwelling in macroalgae. *Bulletin of Marine Science, 61*, 3–19.

Chakalall, B., Mahon, R., McConney, P., Nurse, L., & Oderson, D. (2007). Governance of fisheries and other living marine resources in the wider Caribbean. *Fisheries Research, 87*, 92–99.

Chuenpagdee, R., & Jentoft, S. (2013). Concerns and problems in fisheries and aquaculture: Exploring governability. In M. Bavinck, R. Chuenpagdee, S. Jentoft, & J. Kooiman (Eds.), *Governability of fisheries and aquaculture: Theory and applications* (MARE publication series 7, pp. 33–44). Dordrecht: Springer. doi:10.1007/978-94-007-61-7-0_4.

Delgado, C., Wada, N., Rosegrant, M. W., Meijer, S., & Ahmed, M. (2003). *Fish to 2020: Supply and demand in changing global markets*. Penang: Food Policy Research Institute/WorldFish Center.

Ehrhardt, N. (2005). *Assessment of the economic impact of illegal undersized landings of spiny lobsters, Panulirus argus, in the Nicaraguan fishery*. Danish Agency for International Development (DANIDA) Final Report to the Ministry of Development, Industry, and Commerce (38 p.). Government of Nicaragua.

Ehrhardt, N. (2006, April). *Integrated study of the spiny lobster fishery in the Atlantic coast of Nicaragua with special emphasis on the issue of diving*. Danish Agency for International Development (DANIDA) Final Report to the Ministry of Development, Industry, and Commerce. Managua: Government of Nicaragua.

Ehrhardt, N., Puga, R., & Butler, M. (2011). Implications of the ecosystem approach to fisheries management in large ecosystems. The case of the Caribbean spiny lobster. In L. Fanning, R. Mahon, & P. McConney (Eds.), *Towards marine ecosystem-based management in the Wider Caribbean*. Amsterdam: Amsterdam University Press.

Espeut, P. (2006). *The wild frontier: Living and fishing on the Pedro Cays of Jamaica: A socio-economic assessment*. Jamaica: Kingston.

Evans, P. (1989). Predatory, developmental and other state apparatuses: A comparative political economy perspective on the third world state. *Sociological Forum, 4*(4), 561–587.

FAO. (2007, September 19–29). *Report of the fifth workshop on management of the Caribbean Lobster Fisheries in the WECAFC area. Merida, Yucatan, Mexico* (FAO Fisheries Report No 826, 55pp.). Rome.

FAO. (2010). *State of the world's fisheries and aquaculture 2008*. Rome: FAO.

FAO. (2014). *The state of world fisheries and aquaculture 2014*. Rome: FAO.

Gereffi, G. (1994). The organization of buyer-driven global commodity chains: How U.S. retailers shape overseas production networks. In G. Gereffi & M. Korzeniewic (Eds.), *Commodity chains and global capitalism* (pp. 95–122). Westport: Praeger.

Gereffi, G., & Korzeniewicz, M. (Eds.). (1994). *Commodity chains and global capitalism*. Westport: Praeger.

Gereffi, G., Humphrey, J., & Sturgeon, T. (2005). The governance of global value chains. *Review of International Political Economy, 12*(1), 78–104. Retrieved May 23, 2014, from http://www.tandfonline.com/doi/abs/10.1080/09692290500049805

Gibbon, P., Bair, J., & Ponte, S. (2008). Governing global value chains: An introduction. *Economy and Society, 37*(3), 315–338. Retrieved May 25, 2014, from http://www.informaworld.com/openurl?genre=article&doi=10.1080/03085140802172656&magic=crossref||D404A21C5BB053405B1A640AFFD44AE3

Gonzalez, C., & Jentoft, S. (2010). MPA in labor: Securing the Pearl Cays of Nicaragua. *Environmental Management, 47*(4), 617–629.

Hersoug, B. (2005). Exporting fish, importing institutions- fisheries development in the Third World. In B. Hersoug, S. Jentoft, & P. Degnbol (Eds.), *Fisheries development: The institutional challenge* (pp. 21–92). Delft: Eburon.

Hersoug, B., Jentoft, S., & Degnbol, P. (Eds.). (2005). *Fisheries development: The institutional challenge*. Delft: Eburon.

Jentoft, S., & Chuenpagdee, R. (2009). Fisheries and coastal governance as a wicked problem. *Marine Policy, 33*(4), 553–560. Retrieved May 30, 2014, from http://linkinghub.elsevier.com/retrieve/pii/S0308597X08001917

Kooiman, J., Bavinck, M., Jentoft, S., & Pullin, R. (Eds.). (2005). *Fish for life: Interactive governance for fisheries*. Amsterdam: Amsterdam University Press.

Kooiman, J., Bavinck, M., Chuenpagdee, R., Mahon, R., & Pullin, R. (2008). Interactive governance and governability: An introduction. *The Journal of Transdisciplinary Environmental Studies, 7*(1), 1–11.

Kurien, K. (2004). *Fish trade for the people: Towards understanding the relationship between international fish trade and food security*. Trivandrum: Centre for Development Studies.

Kurien, J. (2005). *Responsible fish trade and food security* (FAO Fisheries Technical Paper 456). FAO: Rome.

Monnereau, I. (2012). *The red gold rush: The impact of governance styles on value chains and the well-being of lobster fishers in the Wider Caribbean*. Unpublished PhD thesis, University of Amsterdam, the Netherlands.

Myrdal, G. (1970). *The challenge of world poverty*. Middlesex: Penguin Books.

Rotberg, R. (2004). The failure and collapse of nation-states: Breakdown, prevention and repair. In R. Rotberg (Ed.), *When states fail: Causes and consequences* (pp. 1–49). Princeton: Princeton University Press.

Ruddle, K. (2008). Reconsidering the contribution of fisheries to society and millennium development goals. In K. Tsukamoto, T. Kawamura, T. Takeuchi, T. D. Beard Jr., M. J. Kaiser (Eds.), *Fisheries for global welfare and environment* (pp. 399–411). 5th World Fisheries Congress. Tokyo: Terrapub.

Thorpe, A., & Bennet, E. (2004). Market-driven international fish supply chains: The case of Nile perch from Africa's Lake Victoria. *International Food and Agribusiness Management Review, 7*(4), 40–57.

Vega, S. (1978). The development of spiny lobster fishing in Belize, 1920–1977. *Belizean Studies, 3*, 1–6.

Vilas, C. (1989). *State, class, and ethnicity in Nicaragua: Capitalist modernization and revolutionary change on the Atlantic coast*. Boulder: Lynne Rienner Publishers.

Yin, R. (2003). *Case study research: Design and methods* (3rd ed.). Thousand Oaks: Sage.

Part IV
Rights and Justice Concerns – Securing Access

Introduction

The sustainability of small-scale fisheries is inevitably about rights: rights of access and tenure, and more broadly human rights. The latter is stressed in the 2014 Voluntary Guidelines for Small-Scale Fisheries developed by FAO and endorsed by member states. Fisheries rights should therefore not be perceived of as a governance tool only, but also part of a wider concern for justice where small-scale fisheries have considerable interests at stake, regardless of where they are situated. A focus on justice requires addressing access to resources and food security, both of which are closely connected. In *Chap. 13*, Moenieba Isaacs focuses on South Africa and its snoek and west coast rock lobster small-scale fisheries, with particular emphasis on how these fisheries play a role in the food system. In South Africa, the irony is that the diet of the poor and many fishers catching high quality fish consists mostly of cheap processed and industrialized food, not very healthy due to its fat, salt and sugar content. In recent years a new small-scale fisheries policy has been introduced to improve this situation. Her chapter assesses the governance of the food system and the fisheries.

Chapter 14 by Alyne Elizabeth Delaney is a case study of coastal Japan, where cooperatives have traditionally been, and still remain, important for fisheries governance. In fact, Japan stands out as an example of successful co-governance and collective property rights. Major change has however occurred in the last two decades as the cooperative sector has undergone a merger process at the regional level, shifting power from local to regional organizations. This, the author argues, has led to the disempowerment of small-scale fisheries and local communities, reducing governability. In *Chap. 15*, Derek Johnson and Sölmundur Karl Pálsson analyze the small-scale fishery of Gimli, Lake Winnipeg, Canada. They argue that despite ample evidence that governance of the fishery is a wicked problem, particularly in terms of how past governance interventions shape current governability; the fishery's institutions have actually successfully navigated that wickedness over the past 40 years. The two central institutions involved have been a surprising combination of a state marketing board and a carefully designed individual transferable quota (ITQ) system. While fishers express varying degrees of irritation with the governing system, on balance it still retains considerable support among them for various reasons, including the economic and ecological stability it has brought.

Staying in the same hemisphere in *Chap. 16*, Adam Soliman also has something to say about quota systems involving transferability. His case study is situated in Alaska and focuses on a community rights-based system developed for the halibut and sablefish fishery as a way of defending the interests and opportunities of small-scale operators, including newcomers, and supporting the local community, who could easily have become disadvantaged by a dis-embedded ITQ system which has been introduced in many countries of the north. The author argues that this governing system is enhancing governability while addressing larger concerns such as those related to community survival in remote areas. In the chapter written by Jeppe Høst about Denmark's ITQ system (*Chap. 17*), we get a less optimistic description of what has followed in its wake. The chapter describes the changes that have occurred as a consequence of privatization and transferability, one of them being the concentration of quota rights to fewer, large-scale operators with privileged access to legal advisors and financial capital. On the other side, the small-scale sector has experienced major decline despite several safeguard policies, resulting in the marginalization of social objectives and reduced governability. Despite that, the ITQ system has also led to a counter movement where small-scale fishers have organized at community level to become more effective in the quota market.

Chapter 13 The Governability of Small-Scale Fisheries Food System in South Africa – The Case of *S*noek and West Coast Rock Lobster

Moenieba Isaacs

Abstract Poor people's consumption of good quality nutrition from fish protein is compromised by the high demand for high quality fish protein from wealthy consumers in the developed world (particularly United States, Europe and Japan). In South Africa, the diet of the poor, and many of those catching high quality fish, consists mostly of cheap processed and industrialized food, rich in fat, salt and sugar. A new small-scale fisheries policy has been developed to help rectify the problem. This policy is based on a human rights approach to fisheries governance and with an aim to promote social and economic justice for small-scale fishers and small-scale communities in South Africa. A key aspect of this policy is to protect livelihoods and promote food security through allocating fishing rights to community entities. How the right to livelihoods translates into the right to food in small-scale fishing communities is a major question, however. This chapter examines this issue by looking at the food system of two important small-scale species – *Thyritesatun* (snoek) for "nutritious" consumption and sale and *Jasus lalandii* (west coast rock lobster; WCRL) for "luxury" consumption. The governance of the food system and the challenges for the governability of the snoek and WCRL small-scale fisheries are critically assessed.

Keywords Small-scale fisheries • Governability • Food security • Food system • Livelihoods • Supply chain • South Africa

Introduction

South Africa faces a 'double burden' of malnutrition, both under-nutrition and obesity, which is a characteristic of a growing number of Middle Income Countries (Monteiro et al. 2004). The double burden is related to South Africa's extreme levels

M. Isaacs (✉)
Institute for Poverty, Land and Agrarian Studies (PLAAS), University of Western Cape, Cape Town, South Africa
e-mail: misaacs@uwc.ac.za

S. Jentoft, R. Chuenpagdee (eds.), *Interactive Governance for Small-Scale Fisheries*, MARE Publication Series 13, DOI 10.1007/978-3-319-17034-3_13

of inequality and the country's large poor population and is worsened by rapid change in the South African food system. Many poor people are increasingly distanced from self-sufficiency as they abandon local dietary foods and increase their dependence on processed and industrialized foods. The changes in dietary practices have impacts on food quality, variety and public health. Is there a role for fisheries in starting to address this burden? Fish in the human diet is essential to secure human health needs as it contains sufficient protein containing all essential amino acids, lipids with essential omega3 fatty acids, vitamins and minerals. It is crucial that fish protein of sustainably harvested species like snoek forms part of the diet of poor and small-scale fishing communities.

This study focuses on two important small-scale species on the west coast of South Africa. *Thyritesatun* (Snoek) is a low value but important source of protein, essential vitamins and minerals for many poor households in the Western Cape Province. It is also a well-known species in the South African consumer market. A significant amount of this species is imported from New Zealand (where it is known as barracouta) and sold as local "Cape snoek" by retailers. *Jasus lalandii* (West Coast Rock Lobster-WCRL) is an inshore, high value species and mainly processed and marketed internationally by established fishing companies. The WCRL is an important source of cash income for fishers. The snoek species is optimally utilized whilst the WCRL is over-exploited. In both fisheries, the fishers do only part of the harvesting and none of the processing and marketing. Market access and a proper cold chain are key for many fishers to improve their livelihoods and food security. The traditional food system of snoek and the modern food system of WCRL contribute to improving livelihoods and food security of small-scale fishers in Ocean View, which is a local community on the Cape where small-scale fisheries are an important industry and where my research is located. The case study describes the supply chain of snoek and its contribution to food (nutritional consumption). Also, the supply chain of WCRL and its contribution to the cash economy is discussed. Other fish species that are regularly consumed are Cape bream, mullet, hake, sardines and pilchards (in tins). Fresh sardines are not consumed locally and mainly used as bait. During the survey undertaken for this study it became apparent that there were many schools of sardines in the Bay that fishers could harvest, but all of it was frozen for crayfish bait. Chicken is still the cheapest form of protein as it is regularly available – 48 % of the interviewees stated that they get the protein mainly from chicken while 43 % said snoek is their main and also preferred form of protein.

The global food system and the length of the value chain intensify the complexity of scale (global/local), the geographic area, the temporal nature, and the diverse markets that are part of both fisheries. Whether fisheries governance will work or not in providing food security and quality depends on the characteristics (diversity, complexity, dynamics and scale) of the natural and social systems related to the fishery that is being governed. It also depends on the capacity/capability of the 'governing system' to address challenges in the food security in the food system, the quality of food, and the type of interactions (Kooiman et al. 2005). This chapter argues that the governance of small-scale fisheries needs to include the entire food system to fully understand the food security needs of the poor and in particular as the governability of the snoek and WCRL species. The study also examines the appropriateness of the present governing system, particularly the use of individual transferable quotas (ITQs) and its impact on small-scale fishers.

As mentioned above, the concept of governability for this case study is applied to the supply chain of two important small-scale species (WCRL and snoek) in Ocean View. The inputs and the interactions with buyers and sellers of both species are discussed and analyzed. The key question addressed in this chapter is how the governance system copes with issues of social and economic justice within the food system, and the extent to which the governance system is able to allocate rights that will secure the right to food (own consumption) and the right to livelihood (fish for cash).

Governance of the Food System

The World Food Summit of 1996 defined food security as existing "when all people at all times have access to sufficient, safe, nutritious food to maintain a healthy and active life" – i.e. access includes availability and affordability of food as well as how food is used. Food security is a complex issue with multiple environmental, social, political and economic determinants. A comprehensive and holistic analysis of how the current organization of food production, processing, distribution and consumption contributes to food security requires broadening the concept of a "food system" beyond only those activities. A host of other economic, social, and environmental drivers affect food security as well, and the interactions among these drivers, activities and outcomes are complex. A broader definition of food systems therefore includes: the interactions between and within bio-geophysical and human environments, which determine a set of activities; the activities themselves (from production through to consumption); and outcomes of the activities (contributions to food security, environmental security, and social welfare) (Ericksen 2008). Kneafsey et al. (2013) link questions of food security to the food system by arguing for economic access (affordability), food quality (nutrition, safety, taste, social acceptability), and a better understanding of the consumption and purchasing practices of food amongst the global rich.

The demand of high quality food at low cost is a key issue in the food system (Freidberg 2003). Here, quality goes beyond food safety by connecting relations of production and consumption with the development of transnational food commodity chains. Definitions of quality in each commodity chain are about interactions between the physical characteristics of the fish, food practices, and production strategies. The different aspects of each of these become important in individual commodity chains (Mansfield 2003). Key to this debate is how it impacts on the poor or those who produce and catch the food. According to Kurien (2004, 5) "the economic attribute that varies most with regard to fishery products is price. The price ratio of bluefin tuna to anchovies can be of the order of 200:1. However, both can be delicacies: the former for the well-to-do Japanese, the latter for the poor Sri Lankan. Bluefin tuna is fish for "luxury consumption" and anchovies are fish for "nutritional consumption". In the aggregate, their contribution to direct food security – fish as food – is vastly different."

The food system is globalized with the integration of national economies into the global economy through trade and investment rules and privatization, and supported by technological advances. On the other hand, localization is a process that reverses the trend of globalization by discriminating in favor of the local (Hines 2000, 4; Hinrichs 2003). This chapter argues for more integrated approaches to the globalization of high value species (WCRL) and localization of fish for nutritional consumption species (snoek). For example, if fishers market their fish globally for 'luxury consumption', it could increase their income in the form of cash. Snoek, on the other hand, is a significant protein in the lives of many poor working class households in Cape Town and hence fishers should continue to supply the local market for local consumption (localization of nutritional important species). However, researchers (Henson et al. 2000; Henson and Mitullah 2004; Mansfield 2003; Gibbon and Ponte 2005; Kambewa et al. 2008; HLPE 2014) warn us that the food quality standards imposed by international trade institutions and importing countries are excluding small-scale fishers from benefitting from the high value fish trade.

A food system approach takes holistic view of food governance, including all processes and infrastructure involved in feeding a population, such as growing, harvesting, processing, packaging, transporting, marketing, consumption, and disposal of food and food-related items (Ericksen 2008). According to Marsden (2013) the food system requires reflexive governance to respond to the dominant public-privately regulated food system as the latter lacks the capacity to respond to the top-down (quantity of food), landscape pressures. Reflexive governance should include the range of actors and establish more context dependent and spatially embedded food systems, innovations and niches. Economic and non-economic institutions are shaped by social and political institutions but also place-based forms of governance that allow actors and organizations to reconsider their current practices (Marsden 2013). More networked forms of governance become reflexive where flexible actor arrangements complement a more static and compartmentalized working of the state department (Jessop 2003). It requires multi-level governance to achieve the desired social, policy and knowledge integration required (Feindt 2010). One may need national and international policy structures and frameworks and also an active form of regional and local governance to occupy new policy spheres and spaces over time and scales. Regarding the value-chains of high value and low value species the governance approach needs temporal, cross-scale, cross-departmental policy integration, and more embedded, multi-level and multi-tiered governance. The interactive governance perspective (Kooiman 2008, 4) brings together a conceptual framework – "where governing bodies in the fish chain are in touch with the operational parts of that chain in order to ensure that the concerns of the fish chain become part of the governing efforts." This forms part of a fish chain analysis where one can situate the notion of food security and the food system in the natural environment, society (social and political), and the production systems (pre-harvest, harvest and post-harvest). This is undertaken here for two important small-scale species in South Africa (Kooiman et al. 2005).

The interactive governance framework analyzes the food system through a human rights-based approach with an emphasis on social justice. The right to livelihoods and food security is linked to the right to healthy food. Key to the class action case for small-scale fisheries in South Africa is their struggle to be recognized and to practice their livelihoods to ensure food security. Kneafsey et al. (2013) argue that the human rights-based approach to food through the right to food, namely localized food will improve governance at international, national and local levels, creating a legal framework for small-scale producers to ensure access to land and water. The concept of food sovereignty is broadly conceived as the rights of people to protect local production, maintain nutritious diets and regulate market access. The individuals who produce, distribute and consume food should beat the center of decisions pertaining to food systems and policies as opposed to the current global food system where corporations and market networks dominate. A localized food system is situated with social meanings, ecological realities, cultural identities and a knowledge system (Via Campesina 1996, 2007; Ibarra et al. 2011).

The importance of ensuring economic access at a household level to healthy, nutritious food for poor people and clearly defining what the role of government should be is a governability issue, which will be investigated in this chapter with reference to snoek and WCRL in the context of the Ocean View community.

Small Scale Fisheries Governance

International instruments and commitments, such as the Voluntary Guidelines on Responsible Governance of Tenure of Land, Fisheries and Forests in the context of National Food Security (*Tenure Guidelines*) and the Voluntary Guidelines on the Progressive Realization of the Right to Adequate Food in the Context of National Food Security (*Right to Food Guidelines*) endorse the important role of small-scale fisheries in providing food security. More specifically, the UNs Special Rapporteur on Right to Food makes "the explicit link between right to food and rights of those who produce it, to fair access to resources such as fish and water" (Masifundise Development Trust, Samudra, July 2013). Food sovereignty, while a necessary condition, is not sufficient for the right to food to be realized. In 1996, South Africa enshrined the right to food in the Constitution (Chapter 2, Section 27) and in the small-scale fisheries policy. However, it is an open question whether South African authorities have the will to make small-scale fishers more food secure and less vulnerable. This will depend on the governability of small-scale fisheries – the quality of representation of local committees, the nature of the rights allocation process, and the relationship between trawlers and small-scale fishers. The social and economic justice dimensions of these instruments are addressed in the new small-scale fisheries policy in South Africa which has created an action space for fishers to participate in a formalized value chain with a development agenda that is concerned with poverty alleviation, food security, access to financial capital, and subsidies. In the new small-scale policy, the state will make budgets available through national, provincial and local governments to strengthen the capacity, training and skills of local community entities and cooperatives.

The post-apartheid South African Government has relied on individual transferable quota-based (ITQ-based) fisheries management as part of a range of reforms designed to broaden access to fisheries, particularly for marginalized groups like artisanal and small-scale fishers. ITQs were introduced in the late 1980s in South Africa and in many other countries as a mechanism for economic rationalization that functioned by adapting fishing capacity to resource availability. In theory, ITQs are *de facto* property access rights or privileges and are primarily concerned with promoting economic efficiency rather than conservation, community welfare or equity (Copes and Charles 2004; McCay 2004; Sumaila 2010). The established fishing companies were required to increase their race and gender component by partnering with Black Economic Empowerment (BEE) companies if they wished to maintain their quotas. The fisheries department argued that Black Economic Empowerment would fit into the Government's broader macro-economic policy of reducing poverty, the rationale being that ITQs and Black Economic Empowerment in established fishing companies would provide secure, good quality jobs based on the Government's minimum wage regulatory framework and that these benefits would 'trickle down' to vulnerable fishing communities. The new Government was seeking to formulate a fisheries policy that would address popular expectations for a more equitable redistribution of access rights, while at the same time maintaining an internationally competitive fishing industry. Transformation created a space for many new entrants to access fishing rights to achieve equity, without the necessary infrastructure, financial capital, and business skills to manage the quota. These new entrants had no option but to enter into catching, processing and marketing agreements with large industrial companies. The consequence of restructuring was the concentration of rights in the hands of a few rights holders (Isaacs 2011).

Governance of small-scale fisheries in South Africa is situated within this wider transformation of fishing rights in South Africa and starts with a story of struggle and resistance to the long-term rights-allocations in the form of individual transferable quotas (ITQs). This resistance has come in the form of litigations and advocacy (e.g. as marches to Parliament) and has been aligned with national and international civil society organizations. In 2004, the Artisanal Fishers Association, Masifundise and the Legal Resources Center, with support from academics, launched a class action suit against the Minister of the Department of Environmental Affairs and Tourism (DEAT). The case, "Kenneth George and Others vs. the Minister (2007)", used the Constitution (of 1996) and the Equality Act (2000) to litigate against the reform process (ITQ allocation of fishing rights) in light of its social and economic impacts. The allocation system opened the door to elites within communities, who captured the benefits (fishing rights) of participation at the expense of communities and the marginalized bona fide small-scale fishers who were actually supposed to benefit from the transformation (Isaacs 2011).

The South African government has formally recognized small-scale fisheries through a participatory policy process, adopted by Cabinet in 2012. The new small-scale fisheries policy promotes a human rights-based approach, food security, co-management, customary practices, and allocates multi-species (basket of rights) to community legal entities with a strong development agenda. The Fisheries Department has already amended the Marine Living Resources Act 18 of 1998 (Draft Marine Living Resources Amendment Bill 2013). The amended bill formally recognizes the livelihoods of small-scale fishers. In March 2014, the National Council of Provinces endorsed the changes and the bill is now ready for cabinet approval and the President's signature. This means implementation of the small-scale fisheries policy can start in 2014. The amendments enable the Fisheries Department to focus on livelihoods, food security, value-chains, local economic development and overall benefits to small-scale fishing communities.

Methodology

Fieldwork for the Ocean View fish chain analysis started in 2011. A short survey was conducted to determine the purchasing and consumption practices of fisher households and consumers. A total of 300 households in the community were surveyed, using ten local fieldworkers. Individual interviews were also conducted with fish buyers, boat owners, fishers, rights holders, organizational representatives and fisheries officials. Finally, focus group interviews with rights-holders and follow-up workshops with fishers were organized during the period 2011–2014.

As a researcher who grew up in the community of Ocean View, I knew of the cultural value and significance of snoek in the consumption practices of the local households. This can be seen in the way they prepare the fish, but also through preservation techniques like drying, salting, smoking, and wrapping the fish in brown paper and storing it in oak barrels, also known as "moortjies".

I participated in the development of the new small-scale fisheries policies in the Western Cape Province of South Africa, and attended community meetings in which policies were discussed.

Results

Case Study

Ocean View is a colored township in the Western Cape (Fig. 13.1), established by the Apartheid state in 1968 as a result of the Group Areas Act, when all colored people were forcibly removed from Simonstown, Noordhoek, and Redhill. The estimated population is around 40,000. Of the 300 people I interviews, 32 % are males and 68 % females. Seventy-six percent of the respondents have high school

Fig. 13.1 Ocean View location (Isaacs 2013)

education, and 69 % are aged between 30 and 60 years. Above 40 % are unemployed and 29 % are on social welfare such as child pension and disability grants. Poor families stay in government housing or live in a shanty in someone's back yard, whilst the well-off live in owned or bonded properties with large grounds, drive a car and send children to quality schools that were formerly for white children only (Isaacs 2013).

Rights Allocations

The snoek fishery is part of the linefishery, which DAFF manages in terms of Total Allowable Effort (TAE), DAFF has allocated traditional line-fish permits on a long-term basis since 2006. The permits are based on Catch Per Unit Effort (CPUE), which permits fishers to catch as much as the carrying capacity of their vessel per day. Nationally 450 line-fish permits were allocated in 2006 for a period of 8 years, until December 2013. In addition 1,000 interim relief permits were allocated as part of the out of court settlement with the claimants who opposed the ITQ system. In Ocean View, 7 line-fishing permits have been allocated; one of them to a woman as part of the long-term rights allocation process to expire at the end of 2014.

Most of the women rights-holders are in the WCRL sector. This has, however, created gender tensions where the women act like "armchair fishers" and sub-contract fishermen to harvest their quota for a daily wage. The fishers feel they should have the rights, as they are the active fishers and not the women. There are a handful of women who fish their own quota and the small-scale fisheries policy aims to ensure that men and women get equal benefits from this sector. Women's economic and governance role (representation in institutions) is protected in the policy.

The fishers in Ocean View were divided into long-term rights (2005–2015) and interim rights holders for WCRL and linefish. There are also recreational rights-holders and poachers competing for WCRL. The WCRL allocation in Ocean View include 100 near shore permits due to expire at the end of 2015 with each permit allocated between 400 and 603 kg each (altogether 40 t) and the 104 interim rights holders allocated 138 kg each (altogether 14.5 t). In Ocean View the total is 54.5 t with an estimated value around $110,000.

The majority of the fishers (104) operate on interim relief permits, which are valid until the implementation of the new small-scale fisheries policy in the 2014–2015 fishing season. The permits allow them to catch 60 snoek, yellowtail or Cape bream per day, or 420 per week (this can be in one catch). In 2014, interim relief permits were given out for 8 consecutive years, but these rights are considered insecure, as they need to be renewed annually. Some fishers are also fishing on recreational permits, which permit only ten snoek per day. The type of allocation (interim permits, traditional linefish permits, or recreational permits) depends on who (permit holder, skipper) owns the infrastructure (boats, gear, transport). Fourteen employed forty-two local fishers to catch over 10 t WCRL in the 2013–2014 seasons. The catching rights are owned by many armchair fishers of Mamre, Atlantis, and Khayelitsha about 50–100 km from Ocean View. The fish factory sends the freezer vehicle to the landing sites, where fish is weighed and then transported factory immediately. The fish factory then electronically transfers the money to the interim rights-holders.

Small-Scale Fishers Representation

Organizational representation in Ocean View is often erratic and based on family lineages. Organizations are created and recreated to benefit from current regulations and creating one organization will continue to be difficult. Isaacs (2013) states that rights holders in Ocean View do not align with national organizations such as the South African Linefish Association, or national community-based organizations such as the Artisanal Fishers Association, established in 1992, and Coastal Links (established through the NGO, Masifundise, in 2004). Although some fishers were part of the class action case (see above), they split from the group drafting the small-scale fisheries policy. Some members continue to attend workshops and meetings scheduled by community organizations. However, many fishers are left out and hence do not get key information on small-scale policy and fishing rights allocation. The nature and structure of organizations along the coast have been (and may well now again be) captured by the elite to serve their needs, often in the process excluding poor fishers who are without much agency (Isaacs 2006).

Currently there are six or seven local community organizations representing fishers, but these do not communicate well with one another. Fishers claim that the representatives of local organizations only capture the benefits for themselves and their families. Isaacs (2003) found that information from the fisheries department (DAFF) does not reach fishers – leaders engage with the department in meetings but do not share the information with the fishers.

Some of the fishers residing in Ocean View were founding members of the Artisanal Fishers Association, but split in 2009 and re-joined a branch of the original 1992 fishers' organization – the Ocean View/Witsands Artisanal Fishers Association (OVWAFA). More recently-formed organizations in Ocean View include the Lighthouse Fishers Forum (LFF), established in 2009, the Association of Deep South Traditional Artisanal and Subsistence Fisheries Sectors (TASFS), established in 2004, the Deep South Fishers Collective Alliance (DSFCA), established in 2009, and the Deep South Fisheries Local Co-Management Committee (DSFLCC), established in 2011. The Rasta community (which follows the Rastafarian spiritual movement) has also claimed fishing rights (see Isaacs 2013).

In 2010, many existing rights holders in linefish, WCRL, and abalone were insecure with the development and implementation of the new small-scale policy. The Ocean View Community Fishing Forum (OVCFF), representing existing rights holders, were in a dilemma as existing rights were allocated individually whereas the new small-scale policy promoted collective rights.

Fish for Cash: West Coast Rock Lobster

The *rock lobster fishery* encompasses West and South Coast species. The WCRL fishery is commercially exploited on the West Coast of South Africa. Larger vessels use lobster traps while small-scale vessels use hoop nets. Rock lobsters are found at depths less than 80 m. Unsustainable catches over long periods, especially in the 1980s, and an increase in export trade resulted in increased fishing pressure and declines in catches and TAC. With state intervention in stock rebuilding and the introduction of Operational Management Plans in 1997 the resource was stabilized (Johnston and Butterworth 2005; Hauck 2008).

The TAC for the WCRL 2013–2014 fishing season (15 November to 21 April) has been set at 2,167.06 t (about 250 t less than the previous season). The catch apportioned to the commercial offshore sub-sector is set at 1,356.56 t and for the commercial near shore sub-sector at 451 t. The TAC apportioned for the subsistence (small-scale/interim relief) sub-sector is set at 276 t (138 kg per fisher for 1,782 fishers). The recreational fishing season has been restricted to 26 days with a bag limit of four WCRLs per person.

WCRL is rarely consumed by fishers as they prefer to sell to local agents for 12–20$ per kg or to the local elite for 3–4 $ each. The local rights holders (long term and interim relief) all sell their WCRL to local agents who sell to fishing companies who then export the catches internationally. There is a strong demand locally for WCRL during festive seasons – weddings and religious festivals (also see Hauck's 2008) work in Hout Bay. In Ocean View, WCRL are sold to marketing agents working for established fishing companies and then processed and exported internationally. This season (2013–2014) the fisheries department came to an agreement with marketing agents and established companies that the WCRL of interim relief with be sold for R 200 per kg ($20). The agents, having freezer vehicles at the landing site in Witsands, measure and assess the quality and then transport it to established companies who hold the lobster in tanks for export. Dead lobster is also bought at a reduced price and exported cooked and frozen.

The WCRL season is from October to May. During this period interim and long-term rights-holders, take their boats to either Kommetjie or Witsands where they launch them. A woman fisher said that she prefers Kommetjie as the water is calmer, making it easier to launch the boat. Most vessels in Ocean View are launched from Witsands launching site where the sea is a more treacherous due to the swell, requiring skills and experience to steer the boat through the waves. Fishers then take a trip north in the direction of Slangkop Lighthouse –known for the best quality lobster but often yield limited catches. The best WCRL catches are in the direction of Cape Point, but during the last WCRL fishing season (November 2013–May 2014) the sea temperature was higher than normal and hence impacted the quality of catch. The lobster can only be landed in Witsands and not Kommetjie which frustrates many fishers, as they would like to land their fish at Kommetjie. At the landing site there are separate monitors measuring the interim rights catch and long-term recreational permit catch. The African National Congress' veterans are also employed by the fisheries department to monitor overfishing and the catching of undersized lobster and lobster tails. These veterans are dressed in black and white camouflage clothes and target the poachers. They have the authority to check all interim relief vessels at landing sites. This is an example of militarizing marine resource management under the auspices of community development.

Local Demand

Cape Town is one of the most unequal towns in the world with the largest gap between rich and poor, hence the high demand locally for WCRL by the wealthy, especially for religious celebrations (Eid and Christmas), New Year celebrations, weddings and dinner parties. Hauck (2008) and van Sittert (1993) have written about the increasing local demand for WCRL and consequently increased informal trade.

The WCRL was a regulated sector that promoted export through the Crawfish Export Act of 1940. For over 60 years, a substantial local market grew over time but was often ignored by the authorities. Although the resource declined in the 1960s and 1970s, the informal and unregulated trade increased, as the economic alternatives on the west coast were limited (van Sittert 1994). Thus, opportunities to supply the local and export markets were taken up by informal fishers, who were increasingly regulated by government, and who had few other livelihood options (van Sittert 1994). A strong local demand for lobster by the catering industry, restaurants and wine farms provided an ongoing market for fishers who relied on the informal fishery to contribute to their household's income. WCRL is an important source of cash for many fisher households in Ocean View. This resource is also fished illegally and will have negative impacts on future sustainability and governability of the WCRL fishery.

Fish for Food: Snoek

Snoek is one of the most well-known fish species in South Africa, abundant in the Western Cape Province and with a strong market locally. At present, the Ocean View snoek fishery can be described as informal, which means that no proper cold chain is maintained, resulting in adverse sanitation impacts (Isaacs 2013).

Snoek Supply Chain

The snoek supply chain in Ocean View is short. It starts with the quota right-holder, the skipper and his crew who catch the fish, which is then auctioned to the langana the landing site. The fish is sold fresh to consumers directly off the "bakkie" on the day of harvesting. The fishers practice a 'gazat' system of catch distribution – 50 % of the fish goes to the boat owner for supplying the vessel, fuel, and bait and the other 50 % is distributed to the crew. Negotiations over the price of fish take place between skippers and langanas, but boat owners and fishers complain that the langanas collude to bid down prices. They know that 'cash is king' and many fishers are forced to sell at low prices to get some cash income. At the landing site in Kommetjie, snoek is sold to the langanas for \$1–3 per fish. Many langanasuse contract workers to sell the snoek to local consumers in Ocean View. The langanas do not use ice, as their goal is to sell all fish on the same day, either fresh or processed (salted, dried or smoked). Gutting and cleaning add an extra \$1 to the value and they then resell it for \$3–5 per snoek (\$7–9 if sold on credit). Dried snoek is sold at \$1–2 per piece snoek roe is sold at \$2–3 per kg, and smoked snoek costs \$1–2 per piece. In Ocean View there are two langanas selling fish for cash or credit to many poor households.

If the fish is not sold at the end of the day or if there is an oversupply, langanas and fishers will conserve it by salting, air drying or smoking it. Surplus snoek is also sold on credit to repay welfare payments. The langanas in Ocean View do not have fishing rights or boats, but this is not necessarily the case in the rest of Cape Town, where some langanas have fishing rights and are boat owners who contract workers to sell their fish. Most fishers have preferences for one of the langanas, who also provides them with cash advances ('voorskoot'). Although this creates a poverty trap of indebtedness for fishers and consumers, it also provides credit at those times when people need it most, since the fishers and poor households do not qualify for help from financial institutions. The same relationship exists between boat owner and crew. However, when the crew is in debt, they often do not return and move to another boat. The interdependencies between fishers, boat owners and sellers are evidence of the ways in which risk is managed and is reflected in the sharing system (see the 'gazat' system explained above).

Generally, those who buy snoek usually eat all of the snoek and nothing goes to waste. Respondents of the questionnaire were asked how they prepare snoek. The most popular preparation is *Langsous* (snoek boiled in water, salt, pepper and served with brown bread). Boiling snoek with water, potato, onions, and tomatoes is another common way of cooking it. Most people say they prefer to fry snoek with salt and pepper, lightly floured and served with tomato bredie. When they buy '*pap snoek*', known as spoilt, many make *frikkedel* – fish cakes with it. Smoking is also best when the snoek is partly spoilt or '*pap*'. Another popular meal is fish soup made of snoek heads, tomatoes, onions, and lots of chilies, green peppers, and potatoes. Snoek curry and breyani is not very popular but pickled snoek is very popular during Easter (Isaacs 2013).

Isaacs (2013) found that the purchasing, consumption and food use practices in Ocean View indicate that snoek is an important part of the diet of the poor and is situated within the culture and traditional practices of the community. The price of snoek is around 1$ per kg whilst the WCRL is sold for 20$ per kg, i.e. at 1:20 price ratio.

Discussion

Key Challenges to the Governability of WCRL and Snoek

Fisheries governance in South Africa has mainly focused on rights allocation, management systems and more recently development. The role of fisheries governance in ensuring food security is not prioritized not even at the national governmental level. Within an ITQ system of rights allocation the livelihoods of the individual rights holder, whether interim or long-term, is protected.

The interactive governance framework, in this chapter, is mainly applied to the food system of two important small-scale species that support the livelihoods of the poor of Ocean View. I am using one lens of interactive governance to make an argument for food democracy to be situated in food sovereignty and the right to food within a broader human rights based approach. Snoek is important in this context as it is central to the lives of fishers. The concept of food sovereignty speaks directly to the snoek species with regard to localization but also in terms of what Kurien (2004) calls "nutritional consumption". There is, however, a need to address the governability challenges that are based on the right to food (own consumption) as snoek is crucial for food security and nutrition in poor communities. This means, among other things, that there is a need to legislate against trawlers targeting snoek and preventing them from catching it as bycatch. Moreover, incorporating snoek into the centralized value chain will take a good source of protein and nutrition away from a poor community like Ocean View. One way, to ensure that snoek is allocated to small-scale fishers is to secure investment in infrastructure to improve the cold chain and encourage traditional preservation techniques such as salting, drying, smoking. Furthermore, the authorities should support local labeling for snoek caught off the west coast of South Africa and also insist on labeling imports from New Zealand as "baracouta".

Kurien (2004) further highlights how food security is indirectly ensured through livelihoods, job creation and responsible trade. In the case of WCRL, it can be sold for cash. The lessons from the WCRL supply chain clearly show that the fishers who are subcontracted to catch WCRL are not involved in processing and marketing agreements with agents of established companies. Fishers are only to a limited degree benefitting from profits being made in the WCRL sector. Hara (2014) argues that market access is part of the problem and more focus should be given to the structure and governance of the value chain. In the food system it is important that the governability of the market supports poor fishers. It is important to establish support structures to help the poor market their fish in order to increase the benefits of the market to them. The WCRL fishery also faces a few governability challenges. One such challenge is the monopoly that market agents employed by established companies to buy the catches seem to have.

Both fish for food and fish for cash can promote social and economic justice for small-scale fishers in Ocean View, but whether this happens depends on governability of these species and the food system.

Trawlers Harvesting Small-Scale Species

Snoek is caught as a targeted linefish species by small-scale fishers and as by-catch by company-owned trawlers, whereas wholesalers import barracouta (*Thyrsites atun)* from New Zealand, which is marketed under the label "snoek". Some trawl vessel skippers specifically target snoek. This has a negative impact on the availability of the resource for small-scale fishers and hence the protein intake of poor communities such as Ocean View (Isaacs 2013). Small-scale fishers using line fishing methods account for 64 % of the local catch of snoek while trawlers account for 36 % of catch. A considerable amount of imported barracouta from New Zealand is also sold as "Cape snoek" as highlighted above (Hara 2014) (Table 13.1).

In many countries around the world, large-scale trawling has been banned or severely restricted by the state at various junctures for different reasons. The list includes inter alia, Indonesia, Trinidad, Malaysia, Costa Rica, Brazil, Venezuela, Ecuador, Hong Kong, India and Senegal. In these countries the adverse ecological

Table 13.1 Snoek landings and imports in South Africa

Sources of snoek	2013
Linefish (informal market)	6,638,139
Deep sea hake trawl (commercial market)	3,650,270
Hake logline (commercial market)	3,491
Inshore trawl (commercial market)	709
New Zealand barracouta (imports)	4,441,370

Source: Isaacs (2013) and Hara (2014)

and socio-economic impacts inflicted on coastal small-scale fishing communities by large-scale trawling led to considerable physical violence and consequently to disruption of the normal livelihoods of small-scale fishing communities. This had adverse income effects and led to food insecurity.

Conclusion

How can we ensure that the most vulnerable people (fishers and their families) at the end of the supply chain are protected in the free market system? Kurien's (2004) report highlights the importance of trade of small-scale fisheries, whilst the recent HLPE (2014) report suggests that the commercialization of small-scale supply chains to international markets can have mixed impacts on food security and nutrition. In fisheries social-science literature the discussions on the ITQ-system overshadow the debates on the food system, food sovereignty and food security. Furthermore, the literature on small-scale producers shows that they are negatively incorporated into vertically integrated and centralized food value chains (Hara 2014).

The moral discourse promoted in international, regional and national legal instruments (new small-scale fisheries policy) is a necessary step in improving livelihoods and food security of coastal communities generally and in South Africa in particular. However, this is not sufficient. The governability of rights allocation processes and, more general fisheries management processes are presently at an all-time low. Moreover, aquaculture is still seen as the magic bullet for small-scale fishers, even though it has failed dismally in Africa to produce fish for food security and livelihoods. Politicians are raising expectations of the poor, vulnerable and marginalized fishers while actually creating opportunities for community elites to grab the rights of small-scale fishers. As a result, the poor suffer in terms of food security and nutrition. The new small-scale fisheries policy can reverse this situation and help address food security, sustainable livelihoods, and poverty alleviation concerns in fishing communities.

The national mandate of the Department of Agriculture, Fisheries and Forestry (2012) is to provide food security to the poor of South Africa. A recent report (HLPE 2014) stressed the important nutritional contribution of fish to the diet of poor people in terms of high quality protein, essential fatty acids, omega oils, and nutrients in vitamins if consumed two or three times a week. The poor, in other words, need to consume more fish, such as small pelagics and snoek. To what extent the South African fisheries authorities are prepared to protect vulnerable fishers from local market agents and prevent established companies from grabbing all the upstream benefits in the supply chain in order to provide good quality, affordable and nutritious fish protein to South Africa's poor remains to be seen.

Acknowledgements The author is grateful for the financial support from the National Research Foundation of South Africa under the South Africa-Tanzania collaborative research agreement. I would like to acknowledge the important contribution of the editors Ratana Chuenpagdee and Svein Jentoft for their valuable inputs. Further, I would like to thank external reviewers for their constructive contribution to this chapter.

References

Copes, P., & Charles, A. (2004). Socio-economics of individual transferable quotas and community-based fishery management. *Agricultural and Resource Economics Review, 33*, 171–181.

Department of Agriculture, Forestry and Fisheries. (2012). *2009/10 Performance review of fishing right holders – Overall report/summary: Limited commercial and full commercial rights holders*. Pretoria: Department of Agriculture, Forestry and Fisheries.

Ericksen, P. J. (2008). Conceptualizing food systems for global environmental change research. *Global Environmental Change, 18*(1), 234–245.

Feindt, P. (2010). Reflexive governance of global public goods: Multi-level and multi-referential governance in agriculture policy. In E. Brousseau, T. Dedeurwaerdere, & B. Siebenhüner (Eds.), *Reflexive governance for public goods*. Boston: MIT Press.

Fishing industry handbook: South Africa, Namibia and Mozambique. (2011). Thirty-ninth edition. South Africa: George Warman Publications.

Food and Agriculture Organization of the United Nations. (1996). *Rome declaration on world food security*. Rome: FAO.

Freidberg, S. E. (2003). Culture, conventions and colonial constructs of rurality in south–north horticultural trades. *Journal of Rural Studies, 19*(1), 97–109.

Gibbon, P., & Ponte, S. (2005). *Trading down: Africa, value chains and the global economy*. Philadelphia: Temple University Press.

Hara, M. (2014). Analysis of South African commercial traditional linefish snoek value chain. Running head: Improving fisher benefits in the snoek value chain. *Journal of Resource Economics, 29*(3), 279–299.

Hauck, M. (2008). Re-thinking small-scale fisheries compliance. *Marine Policy, 32*(4), 635–642.

Henson, S. J., & Mitullah, W. (2004). *Kenyan exports of Nile Perch: Impact of food safety standards on an export-oriented supply chain* (World Bank policy research working paper no. 3349). Washington, DC: World Bank.

Henson, S. J., Brouder, A. M., & Mitullah, W. (2000). Food safety requirements and food exports from developing countries: The case of fish exports from Kenya to the European Union. *American Journal of Agricultural Economics, 82*(5), 1159–1169.

Hines, C. (2000). *Localization: A global manifesto*. London: Earthscan.

Hinrichs, C. C. (2003). The practice and politics of food system localization. *Journal of Rural Studies, 19*(1), 33–45.

HLPE. (2014). *Sustainable fisheries and aquaculture for food security and nutrition*. A report by the High Panel of Experts on Food Security and Nutrition of the Committee on World Food Security, Rome.

Ibarra, J. T., Barreau, A., Del Campo, C., Camacho, C. I., Martin, G. J., & McCandles, S. R. (2011). When formal and market-based conservation mechanisms disrupt food sovereignty: Impacts of community conservation and payments for environmental services on an indigenous community of Oaxaca, Mexico. *International Forestry Review, 13*(3), 318–337.

Isaacs, M. (2003). *Understanding the social processes and politics of implementing anew fisheries policy, the marine living resources act 18 of 1998, in South Africa*. Unpublished doctoral thesis, University of Western Cape, South Africa.

Isaacs, M. (2006). Small-scale fisheries reform: Expectations, hopes and dreams of 'a better life for all'. *Marine Policy, 30*, 51–59.

Isaacs, M. (2011). Paradigm shift – From individual transferable quotas (ITQs) to collective allocations – A struggle for small-scale fishers in South Africa. *MAST, 10*(2), 63–84.

Isaacs, M. (2013). Small-scale fisheries governance and understanding the snoek (Thyrsites atun), supply chain in the Ocean View fishing community, Western Cape, South Africa. *Ecology and Society, 18*(4), 17. doi.org/10.5751/ES-05863-180417

Jessop, R. (2003). Governance and meta-governance: On reflexivity, requisite variety and irony. In H. P. Bang (Ed.), *Governance as social and political communications* (pp. 101–139). Manchester: Manchester University Press.

Johnston, S. J., & Butterworth, S. D. (2005). Evolution of operational management procedures for the South African West Coast rock lobster (Jasus lalandii) fishery. *New Zealand Journal of Marine and Freshwater Research, 39*, 687–702.

Kambewa, E., Ingenbleek, P., & Tilburg, A. (2008). Improving income positions of primary producers in international marketing channels: The Lake Victoria–EU Nile Perch case. *Journal of Macromarketing, 28*(1), 53–67.

Kenneth George and Others vs. the Minister Equality Court of South Africa. (2007). Court order. Cape Town, South Africa.

Kneafsey, M., Dowler, E., Lambie-Mumford, H., Inman, A., & Collier, R. (2013). Consumers and food security: Uncertain or empowered? *Journal of Rural Studies, 29*, 101–112.

Kooiman, J. (2008). Interactive governance and governability: An introduction. *The Journal of Transdisciplinary Environmental Studies, 7*(1), 1–11.

Kooiman, J., Bavinck, M., Jentoft, S., & Pullin, R. (Eds.). (2005). *Fish for life*. Amsterdam: Amsterdam University Press.

Kurien, J. (2004). *Responsible fish trade and food security* (Food and Agriculture Organization (FAO) technical paper, 456). Rome: Food and Agriculture Organization of the United Nations.

Mansfield, B. (2003). Fish, factory trawlers, and imitation crab: The nature of quality in the seafood industry. *Journal of Rural Studies, 19*(1), 9–21.

Marsden, T. (2013). From post-productionism to reflexive governance: Contested transitions in securing more sustainable food futures. *Journal of Rural Studies, 29*, 123–134.

Masifundise Development Trust. (2013). *Getting Rights Wrong: The Global Partnership for Oceans is blatantly ignoring the voices of the world's small-scale fisher peoples* (Samudra Report, No.65). Cape Town: Masifundise.

McCay, B. J. (2004). ITQs and community: An essay of environmental governance. *Agricultural and Resource Economics Review, 33*, 162–170.

Monteiro, C. A., Moura, E. C., Conde, W. L., & Popkin, B. M. (2004). Socioeconomic status and obesity in adult populations of developing countries: A review. *Bulletin of the World Health Organization, 82*(12), 940–946.

Republic of South Africa. (1996). Constitution.

Republic of South Africa (RSA). (2000). Equality act, Act No. 4.

Republic of South Africa (RSA). (2012). *Small-scale fisheries sector for South Africa*. Pretoria: Government Gazette, 20 June 2012. Vol. 564, No. 35455.

Sumaila, R. U. (2010). A cautionary note on individual transferable quotas. *Ecology and Society, 15*, 36.

Van Sittert, L. (1993). 'More in the breach than the observance': Crayfish conservation and capitalism. *Environmental History Review, 17*(4), 20–46.

Van Sittert, L. (1994). *'Red gold and black markets': The political economy of the illegal crayfish trade c.1890 – c.1990*. Cape Town: Department of History, University of Cape Town, South Africa.

Via Campesina. (1996). *The right to produce and access to land. Voice of the Turtle*. Available from: http://www.voiceoftheturtle.org/library/1996%20Declaration%20of%20Food%20-Sovereignty.pdf

Via Campesina. (2007). *Nyéléni declaration. Sélingué, Mali: Forum for food sovereignty*. Available from: http://www.foodandwaterwatch.org/world/global-trade/NyeleniDeclaration- en.pdf/view

Chapter 14
Japanese Fishing Cooperative Associations: Governance in an Era of Consolidation

Alyne Elizabeth Delaney

Abstract Using Miyagi Prefecture, Tohoku, Japan, as a case study, this chapter highlights the difficulties consolidation of fishing cooperative associations (FCAs) presents currently to small-scale fisheries governance in Japan. Historically, Japanese small-scale fisheries are known for their successful management through fisheries cooperatives and traditional local institutions. Significant change has taken place in the last 20 years, however, with the consolidation of local, port-level FCAs into prefecture level ones. A question is thus raised about the overall quality for governance, i.e. capacity and capability, in such large-scale mergers of local, port-level cooperatives into prefectural ones. In other words, is the fisheries system more or less governable with the new governance arrangement? Consolidating cooperatives may be economically rational; yet doing so disempowers local, small-scale fishers and does not, contrary to stated goals, provide better services to member fishers.

Keywords Fisheries cooperatives • Fishing territories • Japan • Mergers • Power

Introduction

This chapter investigates the Japanese fishing cooperative association (FCA) consolidation movement through an interactive governance lens, asking how the merging of FCAs has impacted the overall capacity and capability for good governance of small-scale fisheries.

The governance system for fisheries management in Japan involves a hierarchical, co-governance institutional framework.[1] For the small-scale coastal fisheries, the primary institutions of importance include prefectural level FCAs, Area Fishery Coordinating Committees, and local FCAs. Area Fishery Coordinating Committees are prefectural committees with 16 members: fishermen

[1] For a detailed description of the entire system see Makino and Matsuda (2005).

A.E. Delaney (✉)
Innovative Fisheries Management, Aalborg University, Aalborg, Denmark
e-mail: ad@ifm.aau.dk

S. Jentoft, R. Chuenpagdee (eds.), *Interactive Governance for Small-Scale Fisheries*, MARE Publication Series 13, DOI 10.1007/978-3-319-17034-3_14

(9), academic experts (4), and individuals with a public interest (usually local government officials) (Makino and Matsuda 2005). The Area Fishery Coordinating Committees are the committees which recommend rights and licenses. At the prefectural level, the FCA stipulate broad fishing regulations; local FCA regulations take these broader regulations as a starting point, and may make more detailed restrictions in their own local context.

Today, the local – port FCAs – is being left out due to the consolidation of cooperatives. The most extreme cases involve the formation of a single FCA for an entire prefecture. The ultimate goal of Japan Fisheries (JF; the national level) was to have only 250 FCAs remaining nationwide by 2008 (JF website). The objectives in the FCA Amalgamation Promotion Act (1998), were to make FCAs administratively, financially, and organizationally stronger in order to be able to provide better services to member fishers (Sato 2011).

Has the capacity for governance increased with FCA mergers? Have the mergers provided better services to member fishers? Do the locals' view of their responsibilities towards coastal resources and their expectations of the role of a fisheries cooperative conflict with the upper level's rationale for consolidations? In other words, what is the fit between elements across these two governance system levels? In answering these questions, this chapter focuses in particular on the scale issue of interactions and relationships between the local and prefectural levels in Miyagi Prefecture. The chapter also looks at how responsibility and power in managing fisheries resources has taken shape in this new era of consolidated fishing cooperatives.

In order to answer these questions, the chapter begins with a brief introduction to interactive governance, the history of consolidation of Japanese institutions, generally, as well as a discussion of power and scale in institutions. Following this, the chapter presents Japanese small-scale fisheries as background. With these sections as a basis, the chapter next focuses on interactive governance through the consolidation of fisheries cooperatives: the institutional levels, the consolidation process, the responsibilities of each institutional level, as well as perspectives on consolidation. The chapter ends with a discussion on these perspectives and what it means for small-scale fisheries governance in Japan.

Interactive Governance

Much has been written about the management and co-management of coastal resources (see e.g. Jentoft et al. 1998; Wilson et al. 2003; Pinkerton 2011) and the importance of fishing cooperatives for social and environmental sustainability (e.g. Pomeroy 1995; Armitage et al. 2009). The study of institutions, after all, can generate useful insights for governance of natural resources (Ostrom et al. 2002). Yet this chapter investigates not simply fisheries management and fisheries institutions, but looks at the issue of mergers of the Japanese FCAs through a governance perspective (Kooiman and Bavinck 2005). Interactive governance studies the interactive nature of contemporary society, reflecting upon the fact that "actors consult each other

[and] take into consideration what others do" (Jentoft and Chuenpagdee 2009, 554). Critically, in being interactive, governance includes "the whole of interactions taken to solve societal problems and to create societal opportunities, including the formulation and application of principles guiding those interactions and care for institutions that enable them" (Kooiman and Bavinck 2005, 17). Thus, governance is the shared, collective effort of a number of actors as well as organizations and institutions, including government, communities and civic organizations (Jentoft and Chuenpagdee 2009). Interactive governance recognizes that different modes (hierarchical, co-, self-) have different characteristics which may foster governance. It proposes that with conducive interactions, all modes can work effectively, especially if the conditions are right and if there is structural governability (Jentoft and Chuenpagdee 2009).

The social-ecological systems found in coastal resource governance are complex. Good coastal resource governance depends upon focusing upon the people targeting the resources (Gutierrez et al. 2011). A part of addressing such complexity is the need to understand elements of "meta-level governance" whereby values and principles, for example, are examined (Song et al. 2013). Interactive governance itself is founded on the idea of the existence of ethical values (Symes 2006). The understanding of the need for explicit details on, and attention to, these values and principles found in governance stems from acknowledgement that all levels of governance are value-laden (Kooiman 2003; Song et al. 2013). Consequently, it is "the normative and cognitive concerns of fishery stakeholders [that] underpin the overall governance process, guiding, shaping and inspiring decisions and actions" (Song et al. 2013, 168). People and actors, including institutions, are both constrained and enabled by their surroundings; they also have agency (Bavinck and Kooiman 2013).

Methodology

This chapter uses a case study approach, focusing primarily on the experiences of FCA members in Kuromatsu,[2] Miyagi Prefecture. Background literature on FCAs, Japanese fisheries policies, and consolidation-related literature were reviewed for elucidating the meta-level perspective; four semi-structured interviews were also carried out with Japanese fisheries researchers and local FCA representatives. Qualitative, empirical data on the local level were gathered using standard anthropological research methods of semi- and un-structured interviewing (Bernard 2011) focusing on the experiences of informants during two waves of consolidations (1999 and 2007) as well as current experiences. These interviews took place with fishers, FCA staff, and a politician in Kuromatsu in 2011 (10), 2013 (7), and 2014 (8). Key informant interviews were also conducted with two Japanese researchers and two FCA representatives over the telephone during the winter of 2013–2014.

[2] A pseudonym has been used so as to protect the identity of people given the political nature of current fisheries rights discussions in Miyagi Prefecture.

Historical Background: Consolidation of Japanese Institutions

Japanese society has a long history of consolidating institutions, beginning with the modern period in the 1870s. Large-scale mergers of villages, towns and cities took place, for example, in the 1880s and the 1950s (Rausch 2012). In recent years, the country has seen the amalgamation of cities, towns, and villages, of agricultural and fishing cooperatives. The reasons for such consolidations are numerous and varied, yet almost all have an economic impetus: mergers are expected to provide cost savings through economies of scale and the simplification of government bureaucracies and services (Rausch 2005).

In the 1980s and 1990s, the declining birth rate, aging population and financial difficulties that Japan was facing put pressure on the central government to merge municipalities again (Rausch 2012). In 2008, for example, 22 % of the Japanese population (127 million) was over 65 years of age (Makino 2011). It is estimated that by 2055, the population will fall to below 90 million with more than 40 % being over 65 (Makino 2011). At the same time, in the fisheries, production dropped from 12.8 (1984) to 5.4 (2009) million metric tonnes with the fishing population decreasing almost 50 % (1991 to 2009) (Sato 2011). With such a decline in working population foreseeable, FCA mergers were pushed for many years in advance.

Following the description of the FCA mergers, this chapter investigates the resultant change to fisheries governance due to consolidation and how they have affected the quality of governance of small-scale fisheries, focusing on the interactions and relationships between the local and prefectural levels. In order to do so, attention is paid to the aspect of scale among institutions. Fisheries – as natural systems – and their related governance and social systems, function on differing spatial and temporal scales. Matching the operational scales of these systems is thus an important aspect affecting governability (Bavinck and Kooiman 2013).

Institutional Scale and Power

Scale is an important characteristic of societal systems, along with diversity, complexity and dynamics (Kooiman and Bavinck 2005). While Gibson et al. (2000, 218) define scale as the "spatial, temporal, quantitative or analytical dimensions used to measure and study any phenomenon", from a governance perspective, scale also refers to the "time and space dimensions of systems to be governed as well as to governing systems" (Kooiman and Bavinck 2005, 14). Scale significantly impacts the way social institutions and systems function. Institutions and systems are a part of larger structures and events. For example, the local Kuromatsu perch fishery is nested within a wider Miyagi prefectural fishery, as well as within larger coastal dynamics. The governing system for such perch fisheries is also nested within larger administrative units that scale up to the national level. Occurrences in any one system at a particular level have consequences for all other levels.

Institutions must have communication and they must have authority (Wilson and Degnbol 2003). Societal systems are also inherently diverse, complex and dynamic (Bavinck and Kooiman 2013). In many instances, institutions cooperate and coordinate together, but they can also serve as arenas of struggle over definitions and interpretations (Wilson and Degnbol 2003) as well as over rights and responsibilities. Such diversity, complexity, and dynamism pose fundamental challenges to their governability at different scales (Bavinck and Kooiman 2013).

Any process of creation and maintenance of institutions involves challenges to power among social groups. As Wilson and Degnbol (2003) point out, groups will always push for interpretations of institutions, which reflect their own interests and their own need for control and power. Discussions of power as a trait of individuals (Wartenberg 1990) focus on how power comes to be controlled and steered by these individuals. The group which coordinates the action of the most people, by whatever means, wields the most power and has the greatest influence on how the institution will be defined and interpreted (Wilson and Degnbol 2003).

Scale has a demonstrable influence on how institutions operate. At the lowest level, known as first-order governing (Kooiman 2003), decision-making can be handled by face-to-face interactions where differences can be debated and compromises reached, such as in a local FCA management group which will work on discussions until consensus is reached (Short 1989).

It is anticipated and expected that at the higher institutional levels (Kooiman 2003), these processes of communication, the ones allowing for maximum sensitivity to factual truth and social values, will begin to break down. It is believed that the increased difficulty of maintaining uniformity and consistency in the way institutions function at higher scales will mean that increasingly coercive mechanisms will be needed for them to be governable (Wilson and Degnbol 2003).

The Japanese small-scale, co-governance system is one based upon local and true co-management agreements, where power is shared with government agencies and where proper conditions exist to assist conflict resolution and communication. It is also a system which includes knowledge generation and interactive learning, resulting in the possibility for adaptive co-management (Makino 2011).

Institutions are created and maintained over time; people witness and interpret behavior of other people in terms of the institution, and then base their own behavior on their own interpretation (Wilson and Jentoft 1998; Wilson and Degnbol 2003). The process of creating and maintaining institutions may involve contests of power among social groups.

Coordinated social action can be described as operational power (Wilson and Degnbol 2003). Power is something that is exercised and played out in social relationships (Jentoft 2007, citing World Bank 2004); it is a process and also "an indispensable component of fisheries and coastal management… [involving] considerable risks" (Jentoft 2007). Powerful institutions, such as the state or cooperatives may be included in these risks. The group which coordinates the action of the most people, by whatever means, directly or indirectly, from legal or monetary coercion to prestige and solidarity, wields the most power and has the greatest influence on how the institution will be defined and interpreted (Wilson 2003). There will always be winners and losers with these social arrangements, including in the fisheries, and with such institutions, "a political position on relations of power, conflict, and social justice" will always be expressed through the actions of the institutions (Jentoft 2007, 428). Yet, relations among the various actors may be made more equitable by power-sharing among the stakeholders (Pinkerton 1989; Jentoft 2007) in co-management institutions or interactive governance arrangements (Jentoft 2007).

Co-management, such as seen in this Japanese case, is defined as the sharing of such power; it is also empowerment (Jentoft 2005). Yet, in the face of FCA mergers and the (potential) taking-away of local level decision-making, they risk becoming dis-empowered and marginalized, for instance, in resource management decision-making.

Japanese Small-Scale Fisheries

Today, more than 186,000 fishers (87 %) work in coastal waters, often in small ports and in remote communities. As Makino (2011, 63–64) has pointed out, "due to the complexity of the system and its intensive nature, fisheries coordination and resource conservation cannot be implemented effectively in a top- down, command-and-control manner". One small section of coast may be divided into multiple types of rights, with hundreds of FCA members, and scores of different species. Thus small-scale, inshore, coastal fisheries are managed FCAs, branch cooperatives (*shisho*), and fisheries management organizations (FMOs) (Makino 2011).

The Japanese FCA system is an example of a successful small-scale, inshore fisheries co- and self-governance system (Makino and Matsuda 2005). The system has primarily been investigated through its co-management institutions, FCAs, and through the theoretical view of commons and common property theory (see, e.g., Ruddle and Akimichi 1984; Lim et al. 1995). The system is hierarchical in nature with local, prefectural, and national institutions. Uniquely, legal protection is provided to local-level institutions by Japan's Fisheries Law (proposed 1948, passed 1949).

In this law, the FCA has legally guaranteed, exclusive fishing rights to coastal areas (Ruddle 1987, 1989; Cordell 1989). These rights are based on historical precedent (Kalland 1995) and in most circumstances they limit overexploitation of the resources (Befu 1980; Makino 2011). In the Modernization period (1868–1901), Japan introduced a top-down, market-based rights system, which met with disastrous results through conflicts and over-fishing (Makino and Matsuda 2005); consequently, it was shelved and Japan returned to community and historically-based rights.

Today FCAs are locally, regionally and nationally linked organizations, which market products, supply gear, and work as credit unions. FCA membership entitles fishers to use rights to resources found within the territory of their local, community FCA while the FCA itself holds the right to the resource, as assigned to them by the prefectural level. Management of fishing territories ultimately takes place at the local level (Short 1989) by local committees within the local FCA.

Rights are assigned by the prefecture in 5–10 year periods to cooperatives, which manage these resources through their own committees in the local FCA. Decision-making is usually made through consensus and social pressure can often be high to ensure compliance (Matsuda and Kaneda 1984; Short 1989). The system allowed "management to operate on the appropriate scale and promotes local fishermen innovation, improving coastal fisheries for fishermen and their communities" and "enabled fishermen to incorporate local fishery knowledge and expertise into the management process" (McIlwain 2013, 10). Thus, in a pre-consolidation era, fit between the two governing system levels analyzed in this chapter was often found.

Japan Fisheries, JF Miyagi, and FCAs

Japanese FCAs have certain cooperative principles serving as their working guidelines. These principles include being an autonomous association of voluntary membership where fishers and cultivators unite in order to meet their economic, social, and cultural needs and aspirations through a democratically controlled, jointly owned enterprise (Sato 2011). Cooperatives must be "based on the values of self-help, self-responsibility, democracy, equality and solidarity" and must furthermore have members who believe in the "ethical values of honesty, openness, social responsibility and caring for others" (Sato 2011, 46). The principles consist of meta-level normative guidelines which cooperative members are expected to put into practice.

The Japanese FCA hierarchy includes Japan Fisheries (JF),[3] prefectural cooperatives (e.g., JF Miyagi), branch cooperatives (e.g. Kuromatsu FCA), and smaller, port-specific FCAs (e.g. Matsugahama). The national level has, along with the Ministry of Agriculture, Forestry, and Fisheries (MAFF), in the last two decades, pushed to have only one cooperative per prefecture. In 2013, there were 976 coastal FCAs (Japan Fisheries N.d.). At this time, three prefectures had only one FCA and six prefectures had a prefectural FCA plus "branch"[4] ones, including Miyagi (Sato 2011), which now has four branch FCAs plus JF Miyagi. Though the attempts to merge all cooperatives into prefectural level cooperatives may, by these numbers

[3] JF is the official name of Japan's national level fisheries cooperative association; also known in Japanese as *zengyoren*.

[4] Change over time makes the terminology confusing. Formerly "local" FCAs were at the port level and no longer officially exist in the nine prefectures with prefectural FCAs; though their fishing-ground management boards do. What are now branch cooperatives were formed from the consolidation of these local, port-level FCAs.

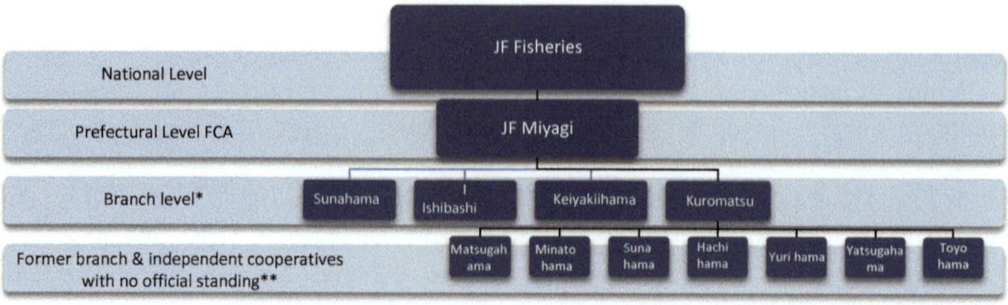

Fig. 14.1 Miyagi Prefecture FCA hierarchy. *Kuromatsu became a "branch" cooperative to JF Miyagi in 2007; branch names are pseudonyms; ** Until October 1999, cooperatives at this level were independent and worked directly with the prefectural level

(9), appear to have failed, more than 1,100 local cooperatives have merged with others, halving the total number of cooperatives (Fig. 14.1).

The Road to Consolidation

There are significant economic and demographic reasons for merging fisheries cooperatives, particularly the globalization of neoliberal thought – with its calls for having the market address social and economic difficulties (e.g. Mansfield 2004). With neo-liberalism there is a push towards an enhanced role of the private sector, which also includes the environment – even in the fisheries – where it is expected that private business and capital will, through market mechanisms, enable "more innovative and efficient environmental solutions" (Mansfield 2004, 313).

With the same global trend towards neo-liberalism, Japan is also experiencing greying of the population, with the resultant decline in active FCA membership becoming a serious issue. In 2010, for example, the elderly accounted for 23 % of the general Japanese population; in coastal communities, they made up 32 % (Fig. 14.2).

The situation in FCAs is even worse. In Ishikawa Prefecture, for example, fishers over 60 passed 50 % of the active fishing population in 2003 (Japan Fisheries 2009). Overall in Japan, fishermen over 65 make up one third of the fishing population (Popescu and Ogushi 2013). Tied to the greying of the fishing group is the loss of "youth groups" and women's groups in the FCAs. Youth groups are an important Japanese cultural tradition, and in the FCAs they served the important purpose of disseminating research, knowledge and innovation through cooperation with extension activities.

As the Japanese Ministry for Agriculture, Forestry, and Fisheries (MAFF) highlighted in a report (2011), one of the main trends in fisheries cooperatives is the

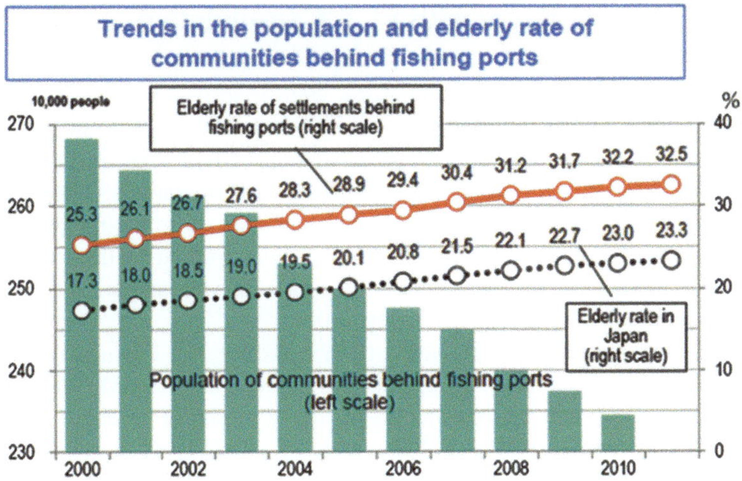

Fig. 14.2 Trends in the population and elderly rates in fishing port communities (Source: MAFF 2011b)

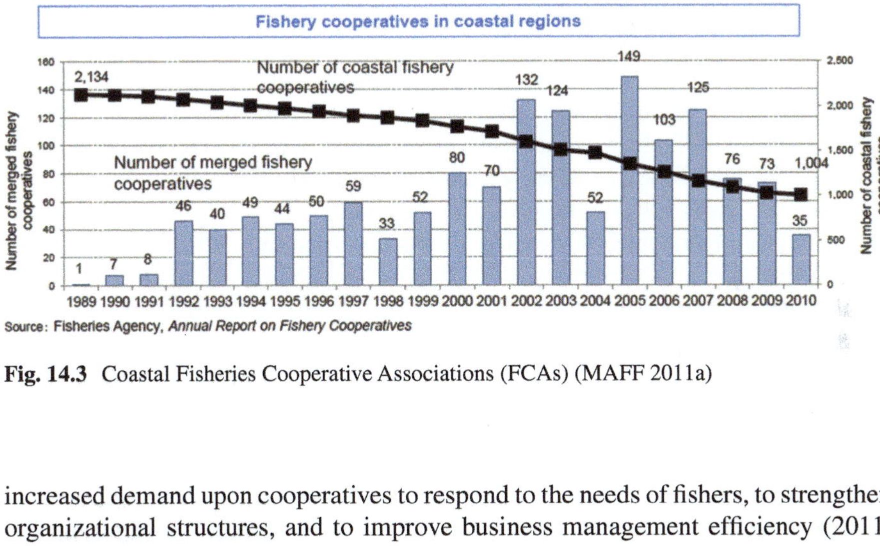

Fig. 14.3 Coastal Fisheries Cooperative Associations (FCAs) (MAFF 2011a)

increased demand upon cooperatives to respond to the needs of fishers, to strengthen organizational structures, and to improve business management efficiency (2011, 17). Tsuchie (2006) notes in a case study from Shimane Prefecture that mergers took place there due to a downward trend in production, fish value, and workers. With 70 % of coastal FCAs posting deficits (Tsuchie 2006), JF and MAFF decided change was needed. They believed consolidating FCAs would reduce administrative expenses, increase the financial health of reformed FCAs, and increase earnings from fish prices due to improved market integration and through new business development resulting from mergers (Fig. 14.3).

Kuromatsu's Experience with FCA Consolidation

In Kuromatsu, mergers among cooperatives were first discussed in 1953, but the differences among the cooperatives were too great. In 1987, however, a consolidation research committee was finally formed after the Miyagi Prefectural Fisheries Research Centre proposed rationalization of the FCAs' operations. After 6 years of consideration, the committee decided that consolidation might be possible. Consequently a FCA Rationalization Committee was formed by the town government with six more years of discussions and negotiations taking place before formal consolidation took place in 1999 (Delaney 2000).

Until FCA consolidation in October of 1999, the town of Kuromatsu had seven fishing cooperatives experiencing long-term decline in active, full-time members. *Nori*,[5] the most important marine species for the community (in economic terms), was harvested by the majority of FCA members. In 1972 there were 804 families[6] harvesting *nori*; in 1999, the year of the first FCA consolidation there were only 105 remaining. As time passed and membership numbers continued to decline, FCAs such as Kuromatsu FCA which were formed by these "branch" FCAs, themselves became branch FCAs of the main prefectural FCA, JF Miyagi, in 2007.

The consolidation progressed in steps: banking operations combined immediately; buying and selling by the end of 2000; selling of seaweed was to be put into two groups by the end of the 2010–2011 season; management of fishing grounds, the biggest sticking point in negotiations, was never merged and continues today as the prerogative of branch FCAs (Delaney 2000).

Officially, FCAs in Miyagi Prefecture merged into one prefectural FCA on April 1, 2007 with only four (including Kuromatsu) serving as branch FCAs. The fear held by many fishers and community members throughout the decades-long consolidation negotiation process was that if there is only one FCA per prefecture, then fewer cooperative officers control all of the fisheries rights, limiting locals' rights, benefits and responsibilities.

Local-Level Benefits and Rights

Despite the cooperatives' principles outlined above, the main benefit of FCA membership comes through access to resource rights. For most members this is the primary *raison d'être* for the cooperatives' existence. Other primary benefits may include savings and banking, marketing, equipment purchasing, and education.

In terms of resource rights, in coastal Japan, FCA members have rights to wild species harvest as well as cultivation of fish, seaweeds, and other maritime species (e.g. sea squirt, oysters). The type of environment in the local coastal waters, as well as historical uses, affect the type of rights awarded. *Nori* cultivation (aquaculture),

---

[5] *Porphyra* spp.; a seaweed most known in the West as a wrapper for sushi.

[6] In Japan, FCA membership is held by the household head (or by a widow of a household head).

for example, though begun in Kessennuma in the 1840s (Miyagi 1993) did not become widespread throughout Miyagi until the post-WWII era when new biological understandings made true cultivation, as opposed to harvesting of wild resources, possible. With the push for cultivation a change came in the rights available for members; "special demarcated rights" were awarded so that aquaculture could take place in areas not previously designated for cultivation.

Notions of equity and fairness in terms of resource rights allocations are based on local understandings in Japan's hierarchical society. Full members (*sei kumiaiin*) and junior members (*jun kumiaiin*) may have differences in rights. Also, the length of a household's membership may also impact resource distribution. For example, in Kuromastu, following the expansion of membership in one of the newer FCAs formed after WWII, the original members were allocated more seaweed cultivation space than newcomers; this changed in the 1980s after pressure for equality could no longer be withstood (Delaney 2003).

In addition to the now equal share of cultivation space provided to FCA members, there are other methods and means to ensure equity. Such means include conducting an annual lottery immediately prior to the new harvest season. This is done as the seabed and water quality vary from place to place. Consequently, a lottery provides a way to vary the space one receives ensuring fairness and parity. Such notions of equity and fairness can also be seen in how fish auction stalls are allocated (Bestor 2004) and even in the allocation of temporary housing following the 2011 Great East Japan earthquake and tsunami (personal communication, Japanese researcher).

Local Level Responsibilities

Initial fieldwork (1995–1996) uncovered the fairly common view that consolidation of the FCAs would "never take place". At that point in time, consolidation discussions had already been on-going for almost a decade. Many problems stood in the way: the issue of debts – some FCAs held debt while others had significant savings; the presidency- any president was assumed to put the needs of his local members over those from other communities; and particularly, the issue of fishing[7] grounds. The fishing ground issue was seen to be the main point of contention. In the end, in some areas, the issue was actually never overcome, but rather, as touched upon earlier, "by-passed" by having management councils (*rijikai*) continue operating today at the sub-branch level.

Why would the fishing grounds be such a problem for consolidation? Obviously it is an economic issue; access to resources is access to income. But there is also a deeper issue, that of taking care of the resources one received access to through one's ancestors. One thus has an obligation to take care of the fishing grounds, just as one's children will also do. A number of FCA members and wives commented on

[7] "Fishing grounds" (*gyoba*) is a general term which refers to any fisheries area, including wild capture and aquaculture.

how taking care of fishing grounds was their own personal responsibility (Delaney 2000, 2003). Such responsibility is connected to the household and related ancestor worship in their culture: At the heart of ancestor worship is a contract and a part of this contract is the obligation to maintain the household, something which is both literal and figurative (Parry 2014). In the end, the government changed the fisheries law so that "branch" FCAs could maintain control of their own fishing territories rather than have control go to the new consolidated FCA management bodies; "without that [change], many cooperatives would not have decided to merge" (personal communication, Japanese fisheries researcher).

By changing the fisheries law, the higher governance level acknowledged the "roadblock" local-level desires have thrown up in the path towards their hopes for FCA consolidations. By agreeing to consolidate their FCAs, local level members have shown their ability to adapt to a changing management environment. Local level members' insistence on fighting consolidation until they were allowed to maintain control of their own fishing grounds shows the importance of their view of responsibility towards fishing ground management and well as the importance of holding economically-valuable rights; it also provides a beautiful example of how their local level values actually guided and shaped decisions being made in the governance system (Song et al. 2013).

FCA Mergers and Governance Capabilities

If, as the ministry (MAFF) states, a continued importance of cooperatives is to respond to the needs of fishers, as well as to strengthen organizational structures and improve business management efficiency, how have they managed these goals? Can one actually balance all three? Or does one, e.g. business management efficiency, really serve as the primary focus at the expense of other goals? Such imbalance among competing goals, for example, can be seen in European fisheries where the Common Fisheries Policy was, theoretically, for environmental, economic, and social sustainability (Symes and Phillipson 2009), yet in practice the social aspect always came last, if at all.

Services Changes

The primary change for FCA members has been in the loss of local staff and services (see, for example, Table 14.1). When there were seven FCAs in the town most members could walk to the FCA office from their homes or local wharf because they were close by and work with staff with whom they had long-standing personal relationships. It was an easy trip to come in for help and service, important, for example, during the *nori* season when FCA members were processing their harvest until late in the evening and even into the night (often working 20 h days). Thus a local location was key for enabling easy access and service for members such as

Table 14.1 Pros and Cons of FCA consolidation in Miyagi for Kuromatsu branch FCA

Pros
Fishing grounds continue to be locally managed
Merged marketing
Fiscally sound cooperative
Cons
Limited Banking hours/days
Loss of local ATMs
Limited services (no office locally)
Difficult lines of communication
Limited direct-say
Decreased number of leadership positions

banking, which could be done in person, or purchasing supplies. These services were key for people living in rural and isolated coastal neighborhoods. Contacts with staff were also on a personal basis and served as important social connections. As one staff member ruefully noted to me, "it was easier in the old days, when one could simply ask directly."

Services such as marketing were also merged, for the most part to the benefit of all. Indeed this began even before official mergers with small groups forming organically on their own, years before official FCA consolidation. For example, in Kuromatsu, three FCAs grouped their harvests together for judging the quality of the harvest and sales, thereby increasing the prices they could receive from buyers.

Decision-making on the management of local resources by fishing ground committees was also made locally by officers of the local FCAs. In the early stages of the merger, these old FCAs, now new branch FCAs, still maintained their own boards to make such decisions. And these then met with the management board in the main FCA to report on the decisions. Following the 2007 mergers into Miyagi FCA, the boards continued working with the Kuromatsu branch FCA, but with fewer members per committee.

Perspectives on FCA Consolidations

With mergers, the local-level perspective can sometimes be overlooked in the work to address larger, overarching concerns of the second and meta-levels. Also, when a larger body is made up of many small parts, how does one decide where to place resources? With merged municipalities, for example, some areas of a town are given greater economic support than others. This is seen with tourism (Rausch 2012) and also seen after the 2011 earthquake and tsunami (Aldrich and Sawada 2014) with aid coming to a larger, merged municipality made up of many small, previously independent towns. The further the local level group is from the decision-making body, the more difficult it may be for the local-level to influence final decisions.

The consolidation of Japanese FCAs, despite it being an process on-going for over two decades, remains controversial and in dispute. The process has been "successful" in some cases, if looked at in economic terms; it has limited the need for the government to cover the debts of small fisheries cooperatives and increased the profitability of some. In 2005, for example, 378 cooperatives were 47 billion yen in debt (MAFF 2008). There is a view by some that since FCAs have experienced such a great decline in members, it is understandable that they have a deficit. There is also a fear that the mergers are a temporary solution to on-going economic decline as seen in Shimane Prefecture (Tsuchie 2006); initial savings were made with the reduction of personnel with mergers, but this was only a short-term improvement.

Yet many researchers believe FCA consolidations have been made solely with economic considerations, at the expense of the local and cultural view. Though some sub-branch FCAs have managed to retain their fishing ground management committees, fishing territory management is only one aspect of FCA responsibilities. This loss of the "local perspective" is similar with what is seen with Japan's latest round of municipal mergers (Rausch 2012).

As one Japanese maritime studies researcher noted, "Miyagi has been going through a lot of disputes over consolidation for the last a couple of decades and the whole post-3.11 debates are still heated" (personal communication 2013). Indeed, the 2011 tsunami strengthened attempts to change fisheries governance in the name of recovery. The governor of Miyagi Prefecture pushed to prioritize private business over FCA members for licenses in the name of "recovery" without consulting fishermen or the FCAs. The result was an opposition letter signed by 14,000 fishermen opposing the new plan (Wilhelm 2011).

There have been complaints by those working for consolidating the FCA structure that fishermen tend to be traditional and unwilling to change, similar to, what Song et al. (2013) noted, namely that people's values and principles are slow-changing and deeply ingrained. That is one explanation for the fight against FCA mergers and for the process taking as long as it has without meeting the goal of 250 FCAs (976 existed in 2013). This is also why proposals to allow outside (the FCA) businesses to receive fishing rights have also been fought over many years. For many fishermen, such proposed changes are the same as the FCA mergers, it means taking resource use decision-making and harvesting rights away from local people with whom responsibility over the resources resides.

Conclusion

Has the capacity for governance increased with FCA mergers? In the end, have the mergers provided better services to member fishers? The case study presented of Miyagi Prefecture provides a good first investigation, though further research is needed to see how this compares with other consolidation experiences in Japan. Though consolidation processes here appear to parallel consolidation elsewhere, the events of March 11, 2011 (earthquake and tsunami) may set it slightly apart.

In Kuromatsu, the answer to whether mergers provided better services to member fishers is mixed. On the one hand, the case provides a good example of how successful co-management can, especially with governmental and political support, add to the capacities and capabilities of local fishers and communities. The system was designed so that though the prefectural level officially assigns rights to resources, it is the members of the local level who actually manage the resources themselves and enforce the rules (both formally and informally). The FCAs are also linked closely with scientists and the prefectural fisheries research centers. Such close connection has enabled technological advances in *nori* and oyster aquaculture techniques, e.g. improving members' harvests and incomes.

Unfortunately, on the other hand, the case also shows how all-important economic considerations have penetrated Japanese fisheries at the expense of other concerns such as local traditions and cultural understandings of rights and responsibilities, as well improved services and interactions with FCA staff.

In focusing on the economic rationalities driving cooperative consolidations, Japanese fisheries are following the global trend of neoliberalism, which tends to overrule not only local culture and institutions, but also marginalize local input. Despite the economic rationality behind consolidations, local port communities, FCAs, and fishers argued against mergers for decades.

Japanese fishers do not blindly hold on to these rights, however. There are numerous instances of fishing rights being sold in order for coastal development to take place. In fact, land reclamation of coastal areas in Japan can only take place through the purchase of these rights (see, e.g. McKean 1981). Giving up the rights completely, versus giving them up to others to manage the resources, appear to be different matters to locals. Good stewardship of marine resources is thought to come from the holding of resource rights. In Japan, the holding of rights and good stewardship also connects to the importance of ancestor worship. At the heart of ancestor worship is a contract and a part of this contract is the obligation to maintain the household (Parry 2014). Fishing grounds, the access to which come through the household head, is also included within the concept of household responsibilities.

Finally, linked to local management of fishing grounds is the issue of power. Just as the formation of FCAs was empowering historically to small-scale coastal fishers, so is managing their own resources today. The taking away of management responsibilities and the actual dis-embedding of local social relations due to the separation between place and space (Giddens 1991) carries not only a risk that resources will not be managed properly (in the locals' view), but with it also comes the risk of locals becoming disenfranchised of their rights. In today's world of "economically rational" fisheries and fisheries management, this is a vital point. As we look around the globe, we must be careful while arguing for decisions to be made primarily keeping economic considerations in mind that we are not in fact, empowering others at the expense of the local, small-scale fishers who are the ones who will have to live with the consequences of governance failure.

References

Aldrich, D., & Sawada, Y. (2014). The physical and social determinants of mortality in the 3.11 tsunami. *Social Science & Medicine, 124*(2015), 66–75.

Armitage, D. R., Plummer, R., Berkes, F., Arthur, R. I., Charles, A. T., Davidson-Hunt, I. J., Diduck, A. P., Doubleday, N. C., Johnson, D. S., Marschke, M., McConney, P., Pinkerton, E. W., & Wollenberg, E. K. (2009). Adaptive co-management for social–ecological complexity. *Frontiers in Ecology and the Environment, 7*, 95–102.

Bavinck, M., & Kooiman, J. (2013). Theorizing governability – The interactive governance perspective. In R. Chuenpagdee, S. Jentoft, M. Bavinck, & J. Kooiman (Eds.), *Governability of fisheries and aquaculture*. Dordrecht: Springer.

Befu, H. (1980). Political ecology of fishing in Japan: Techno-environmental impact of industrialization in the Inland Sea. *Research in Economic Anthropology, 3*, 323–347.

Bernard, H. R. (2011). *Anthropological research methods: Qualitative and quantitative approaches* (5th ed.). Lanham: Alta Mira Press.

Bestor, T. (2004). *Tsukiji: The fish market at the center of the world*. Berkeley: University of California Press.

Cordell, J. (1989). *A sea of small boats*. Cambridge: Cultural Survival.

Delaney, A. E. (2000, June). *Consolidating Japanese fishing cooperatives for economic stability: Formal dilemmas and informal action*. Paper presented at the International Association for the Study of Common Property Biannual meeting, Bloomington, IN.

Delaney, A. E. (2003). *Setting nets on troubled waters: Environment, economics, and autonomy among nori cultivating households in a Japanese fishing cooperative*. Doctoral thesis. Department of Anthropology, University of Pittsburgh, USA.

Gibson, C. C., Ostrom, E., & Ahn, T. K. (2000). The concept of scale and the human dimensions of global change: A survey. *Ecological Economics, 32*, 217–239.

Giddens, A. (1991). *Modernity and self-identity. Self and society in the Late Modern Age*. Cambridge: Polity Press.

Gutierrez, N. L., Hilborn, R., & Defeo, O. (2011, February). Leadership, social capital and incentives promote successful fisheries. Letter. *Nature, 470*(7334), 386–389.

Japan Fisheries. (N.d.). *Japan fisheries (JF), Zengyoren homepage*. Retrieved on April 25, 2014, from http://www.zengyoren.or.jp/syokai/jf_eng2.html

Japan Fisheries. (2009). *Dissemination of activities: A guide. National Fisheries Improved Dissemination Staff Council*. Fisheries Agency. Tokyo: Department of Research Guidance. Retrieved April 25, 2014, from Japan Fisheries website: http://www.zengyoren.or.jp/syokai/jf_eng2.html

Jentoft, S. (2005). Fisheries co-management as empowerment. *Marine Policy, 29*(2005), 1–7.

Jentoft, S. (2007). *In the power of power: The understated aspect of fisheries and coastal management*. Human Organization. Winter 2007, 66(4), 426–437.

Jentoft, S., & Chuenpagdee, R. (2009). Fisheries and coastal governance as a wicked problem. *Marine Policy, 33*(4), 553–560.

Jentoft, S., McCay, B., & Wilson, D. (1998). Social theory and fisheries co-management. *Marine Policy, 22*(4–5), 423–426.

Kalland, A. (1995). *Fishing villages in Tokugawa Japan*. Honolulu: University of Hawaii Press.

Kooiman, J. (2003). *Governing as governance*. London: Sage.

Kooiman, J., & Bavinck, M. (2005). The governance perspective. In J. Kooiman, M. Bavinck, S. Jentoft, & R. Pullin (Eds.), *Fish for life: Interactive governance for fisheries* (pp. 11–25). Amsterdam: Amsterdam University Press.

Lim, C. P., Matsuda, Y., & Shigemi, Y. (1995). Co-management in marine fisheries: The Japanese experience. *Coastal Management, 23*, 195–221.

MAFF (Ministry of Agriculture, Forestry and Fisheries). (2008). *Gappeisokushin-hōkigen-go no aratanagyokyōtaisakunitsuite report*. Retrieved April 23, 2013 from, http://www.jfa.maff.go.jp/j/keiei/gyokyou/pdf/kaikaku2008.pdf

MAFF (Ministry of Agriculture, Forestry and Fisheries). (2011a). *"Statistics of FCAs", October, 2011. Fishery agency*. Tokyo: Government of Japan.

MAFF. (2011b). *Fisheries white paper*. Ministry of Agriculture, Forestry, and Fisheries, Fisheries Division. http://www.jfa.maff.go.jp/j/kikaku/wpaper/pdf/2011_jfa_wp.pdf. Accessed 11 July 2014.

Makino, M. (2011). *Fisheries management in Japan: Its institutional features and case studies* (Fish and fisheries series 34). London: Springer.

Makino, M., & Matsuda, H. (2005). Co-management in Japanese coastal fisheries: Institutional features and transaction costs. *Marine Policy, 29*, 441–450.

Mansfield, B. (2004). Neoliberalism in the oceans: "Rationalization," property rights, and the commons question. *Geoforum, 35*(3). Retrieved from http://www.sciencedirect.com.zorac.aub.aau.dk/science/journal/00167185313–326

Matsuda, Y., & Kaneda, Y. (1984). The seven greatest fisheries incidents in Japan. *Senri Ethnological Studies, 17*, 159–182.

McIlwain, K. (2013). *Catch shares in action: Japanese common fishing rights system*. Environmental Defense Fund.

McKean, M. (1981). *Environmental protest and citizen politics in Japan*. Berkeley: University of California Press.

Miyagi Prefectural Government. (1993). *Miyagi prefecture's traditional fishing methods IV: Cultivation (Nori)* (69 pp). Ishinomaki, Miyagi: Miyagi Fisheries Research Center.

Ostrom, E., Dietz, T., Dolsak, N., Stern, P. C., Stonich, S., & Weber, E. U. (Eds.). (2002). *The drama of the commons. Committee on the human dimensions of global change. National Research Council*. Washington, DC: National Academy Press.

Parry, L. R. (2014). Ghosts of the Tsunami. *London Review of Books, 36*(3), 13–17. Retrieved January 31, 2014, from http://www.lrb.co.uk/v36/n03/richard-lloydparry/ghosts-of-the-tsunami

Pinkerton, E. (Ed.). (1989). *Co-operative management of local fisheries: New directions for improved management and community development*. Vancouver: University of British Columbia Press.

Pinkerton, E. (Ed.). (2011). *Co-operative management of local fisheries: New directions for improved management and community development*. UBC Press.

Pomeroy, R. (1995). Community-based and co-management institutions for sustainable coastal fisheries management in Southeast Asia. *Ocean and Coastal Management, 27*(3), 143–162.

Popescu, I., & Ogushi, T. (2013). *Fisheries in Japan: Note*. Report requested by the Committee on Fisheries, European Parliament.

Rausch, A. (2005). Municipal mergers in rural Japan: Easy on the powerful, severe on the weak: Discussion paper. *Electronic Journal of Contemporary Japanese Studies*. Retrieved from http://www.japanesestudies.org.uk/discussionpapers/2005/Rausch.html

Rausch, A. S. (2012). A framework for Japan's new municipal reality: Assessing the Heisei gappei mergers. *Japan Forum, 24*(2), 185–204. doi:10.1080/09555803.2012.671845.

Ruddle, K. (1987). *Administration and conflict management in Japanese coastal fisheries*. Rome: FAO.

Ruddle, K. (1989). Solving the common-property dilemma: Village fisheries rights in Japanese coastal waters. In F. Berkes (Ed.), *Common property resources: Ecology and community-based sustainable development*. London: Belhaven Press.

Ruddle, K., & Akimichi, T. (1984). Introduction in maritime institutions in the Western Pacific. *Senri Ethnological Studies, 17*, 1–9.

Sato, M. (2011) *Major businesses (credit, marketing and supply) of fisheries cooperative Associations in Japan*. Annex 1 Information on FCAs in Japan (pp. 41–51). Retrieved March 19, 2014.

Short, K. M. (1989). Self-management of fishing rights by Japanese cooperative associations: A case study from Hokkaido. In J. Cordell (Ed.), *A sea of small boats*. Cambridge: Cultural Survival.

Song, A. M., Chuenpagdee, R., & Jentoft, S. (2013). Values, images, and principles: What they represent and how they may improve fisheries governance. *Marine Policy, 40*, 167–175.

Symes, D. (2006). Fisheries governance: A coming of age for fisheries social science? *Fisheries Research, 81*, 113–117.

Symes, D., & Phillipson, J. (2009). Whatever became of social objectives in fisheries policy? *Fisheries Research, 95*, 1–5.

Tsuchie, S. (2006). Managerial challenges with the merger of fishing cooperatives, 10 years after the merger of Ota City, Shimane Prefecture fisheries cooperatives. [English translation]. *Journal of the North Japan Fisheries Economics, 34*(3), 100–111.

Wartenberg, T. E. (1990). *The forms of power: From domination to transformation*. Philadelphia: Temple University Press.

Wilhelm, J. (2011). *Some preliminary thoughts on the Special Zones for Disaster Recovery in Fisheries (Tokku)*. Retrieved January 15, 2014, from http://www.wilhelm.jp/files/presentation/tokku_wilhelm.pdf

Wilson, D. C. (2003). Conflict and scale: A defense of community approaches in fisheries management. In D. C. Wilson, J. R. Nielsen, & P. Degnbol (Eds.), *The fisheries Co-management experience: Accomplishments, challenges and prospects* (pp. 193–209). Dordrecht: Kluwer Academic Publishers.

Wilson, D., & Degnbol, P. (2003, September). *Scale and aquatic resource management: Some thought experiments*. Paper presented at People and the Sea II – Conflicts, Threats and Opportunities. Centre for Maritime Research (MARE). Amsterdam, the Netherlands.

Wilson, D. C. & Jentoft, S. (1998). Structure, agency and embeddedness: Sociological approaches to fisheries management institutions. In D. Symes (Ed.), *Alternative management systems* (pp 63–72). Oxford: Blackwell Science.

Wilson, D. C., Degnbol, P., & Nielsen, J. R. (2003). *The fisheries co-management experience: Accomplishments, challenges and prospects* (Fish and fisheries series 26). Dordrecht: Kluwer Academic Publishers/Springer.

World Bank. (2004). *Good management practice in sustainable fisheries: Policy brief 12. Key concepts 1: Fisheries management systems and governance*. Retrieved March 5, 2014, from http://siteresources.worldbank.org/INTARD/Resources/335807-1323958783967/8321851-1323959552668/WBPolicyBrief12.PDF

Chapter 15
Governability and Its Discontents in the Fishery of Lake Winnipeg Since the Late 1960s: The View from Gimli

Derek Johnson and Sölmundur Karl Pálsson

Abstract A key contention of governability assessment is that fisheries governance problems are wicked: they cannot be addressed definitively through simple technical expedients nor can interventions to address them satisfy all parties equally. Several aspects of wickedness are evident in the case of the fishers operating from Gimli on Lake Winnipeg but in this chapter we emphasize the historical one. Specifically, our chapter looks at how past governance interventions shape present governability in fisheries. A severe social, economic, and ecological crisis resulting from decades of poorly controlled and exploitative resource extraction shook the Lake Winnipeg fishery in the 1960s. That crisis was averted through intervention by the Canadian state that had two central elements: a state marketing board and an individual transferable quota system. The intervention of the state led to a much more governable and sustainable fishery, though not without contradictions that are now leading to some expressions of dissatisfaction amongst fishers. Unlike in the 1960s, however, current institutional shortcomings are irritants rather than threats to the very existence of the fishery. The governability of the fishery remains high. In our chapter, we use a social wellbeing analytical perspective to illuminate fishers' mixed feelings about current institutional arrangements in the Lake Winnipeg fishery.

Keywords Individual transferable quotas • Social wellbeing • Lake Winnipeg • Gimli • Freshwater Fish Marketing Corporation • Governability • Supply management

Introduction

The case that we examine here is unusual because it is a good news story in the governability of small-scale fisheries. The fishery on Lake Winnipeg as practiced by fishers from Gimli in the late 1960s was a disaster. Catches had been plunging for a decade or more, fishers were fighting for the remaining available fish, and returns on

D. Johnson (✉) • S.K. Pálsson
Department of Anthropology, University of Manitoba, Winnipeg, Canada
e-mail: derek.Johnson@umanitoba.ca; palssons@cc.umanitoba.ca

S. Jentoft, R. Chuenpagdee (eds.), *Interactive Governance for Small-Scale Fisheries*, MARE Publication Series 13, DOI 10.1007/978-3-319-17034-3_15

the fishery were highly unequally skewed in favor of companies based in the United States. These traders used their economic muscle to maintain fishers in conditions of debt servitude and poverty. The Lake's fisheries governance arrangements, in short, were manifestly failing to achieve ecological sustainability and economic equity.

In relatively short order from 1969 to 1972, the federal and provincial levels of government in Manitoba, Canada coordinated their efforts to develop a new governing system to address the critical problems faced by the Lake Winnipeg fishery. They instituted a new hierarchical management apparatus that shifted control of the fishery away from merchants to the state. Central to the new arrangements were two new institutions: the Freshwater Fish Marketing Corporation (FFMC) that nationalized the purchase of fish, forcing private traders out of the industry, and what eventually became a limited individual transferable quota system that imposed effective controls on fishing effort. This unorthodox combination of institutional interventions successfully addressed the governability challenge of the earlier period and led to the subsequent 40 years of improving biological and economic conditions.

In contrast to the governability challenges of the 1960s, current concerns look relatively benign. Fishers in Gimli are increasingly disgruntled with pricing arrangements through the FFMC, they feel inadequately involved in the lake's governance, they recognize shortcomings in the quota system, and they worry for potential major but as yet unrealized ecological and governance threats. None of these issues are serious challenges to the governability of the fishery, but they are warnings of potential problems that need to be addressed.

We are keenly aware of the serious threats that ITQ based systems pose to fishing communities (McCay 1996, 2004; Copes and Charles 2004) so our interest in this chapter is to understand how, in the Lake Winnipeg case, the socially corrosive effects of ITQs appear to have been minimized. From the point of view of governability, we suggest that under specific conditions, ITQs can be one part of larger sets of governance interventions that may improve both the capacity to govern and quality of governance aspects of governability for small-scale fisheries. This is by no means a general endorsement of ITQs but rather just a reminder that governance solutions in fisheries have to be tailor made for each case (cf. Degnbol et al. 2006). In keeping with the idea that governability in fisheries is a wicked problem, our chapter also argues for a broader lesson about the importance of taking a historical perspective when assessing governability. For many fishers operating from Gimli, the memory of the ecological and economic crisis they, their parents, or grandparents experienced in the 1960s is still fresh and is an important part of the continued support for the distinctive governing system for the Lake Winnipeg fishery. History moves on, however, and the fishery's governors need to demonstrate responsiveness to current fisher concerns about pricing and inclusion.

Our argument is grounded methodologically in a social wellbeing approach. We use instruments from the social wellbeing toolkit to gauge satisfaction of fishers with current governance instruments and governing interactions. Our findings show the ways in which current governing interactions are perceived as legitimate and also are subject to dissatisfaction.

Social Wellbeing and the Assessment of Governability

For interactive governance theory, governability is an informed judgment about the 'capacity for governance' of a particular human system, or how well it addresses problems and creates opportunities (Kooiman and Bavinck 2013, 12). Governability can also be argued to imply a quality or evaluative dimension: a governing system's capacity for governance may be compromised if it is perceived to not meet locally important values, such as inclusion, equity, or transparency (Svein Jentoft: personal communication). From the governability perspective, human fisheries systems consist of three sub-systems: the system-to-be-governed, the governing system, and the system of governing interactions. According to Chuenpagdee and Jentoft (2013), these three intersecting systems are one axis in the assessment of governability of particular fisheries systems. They are set against ideal typical modes of governance from the interactive governance perspective: hierarchical, co-governance, and self-governance. Fisheries on Lake Winnipeg have been governed through a hierarchical structure, but with opportunities for consultation, since the major changes of the late 1960s so our analysis focuses on the hierarchical row of the governability assessment table (see Chap. 2 of this book). As the bulk of our data come from the fisher perspective, we view the governability of the fishery's hierarchical mode from the bottom-up. Our research did reveal some evidence of self-governance in, for example, norms around access to fishing grounds. As this, however, was not the focus of our research, as such institutions seemed minor in importance, and as they appeared not to be integrated into co-governance arrangements, we have chosen not to systematically assess these other modal dimensions of governability in the Gimli fishery.[1]

Our explanation of the Gimli case follows the sequence of fisheries sub-systems. We provide an overview of the main characteristics of each that integrates, in the governing system section, an overview of the historical development of the FFMC and the ITQ system. We emphasize how the fishery's current governability reflects the response to the 1960s crisis. In the discussion section, we take up issues of fit: how well the current hierarchical mode, and its two central institutions, appear to meet fisher expectations and values. In other words, we make a judgment about the degree to which governability is supported by a sense of legitimacy among Gimli fishers.

The evidence that we draw on for our assessment of the legitimacy of current arrangements was gathered using specific instruments from a toolkit for the assessment of social wellbeing in fisheries (Coulthard et al. N. d.). Coulthard et al.'s methods handbook builds on a large body of work that lays out a theoretical approach to understanding wellbeing in social relational, subjective, and material terms (WeD 2007; Gough and McGregor 2007; Coulthard et al. 2011; Weeratunge et al. 2014). As an analytical perspective, the social wellbeing perspective provides an insightful way of understanding the factors that people in particular places or

[1] We use the phrase Gimli fishery as a shorthand to refer to the collective effort of fishers who fish from Gimli harbour. While there are demographic differences amongst these fishers, they share sufficient social, historical and cultural connections that it makes sense to refer to them as a place-based fishing group.

situations find important to living well. As environmental characteristics, including governance arrangements and governability, figure large in the quality of life for small-scale fishers, there is a strong case for bringing the two perspectives together. Given limitations of space here, however, we do not elaborate theoretical connections between the approaches, but restrict ourselves to using social wellbeing as an entry point into fishers' subjectivities.

Within Coulthard et al.'s broader social wellbeing toolkit (N. d.), we draw particularly on the results we obtained from two data gathering tools: the person generated index and the governance relationship assessment index. The first of these is a means to identify and rank the factors that fishers identify as critical to their subjective wellbeing. The second also allows identification and ranking, plus indicators of satisfaction with the relationship, but in this case of relationships that fishers indicate as essential to their relational wellbeing. Pálsson undertook fieldwork using these tools and other methods during the fishing season of 2012/2013 and the full results of that work are published elsewhere (Pálsson 2014). The fieldwork was concentrated on fishers from the Municipality of Gimli, Manitoba and social wellbeing interviews were conducted with 20 fishers, of whom 18 use only open skiffs and two use open skiffs and enclosed boats for targeting white fish in the North Basin of Lake Winnipeg. Results were interpreted based on the larger ethnographic understanding of fishing from Gimli that Pálsson obtained. Pálsson also interviewed officers of the FFMC and the Provincial Department of Fisheries, retired officers, and other experts on the Lake Winnipeg fishery.

Social-Ecological System-to-Be-Governed: Historical and Contemporary Dimensions

Lake Winnipeg is the 10th largest freshwater lake in the world. It is 400 km long and the surface area of the lake is about 23,750 km^2 (Duguid and Brandson 2005). The lake is an ancient source of fish for the First Nations and metis groups of the region (Tough 1997; Russell 2000) but it was not until the arrival of a wave of Icelandic immigrants to the Gimli area in the 1870s that the commodity potential of the fishery really expanded (Þorsteinsson 1945; Kristjansson 1965; Houser 1986; Arngrímsson 1997). By the beginning of the 1880s many fishing companies, such as William Robinson & Co., Dominion Fish Company and Ewing and Fryer, began to set up working stations around the lake (Thompson 1976; Gerrard 1985; Tough 1997). These companies, many of which were contractors to larger American firms, rapidly established themselves as the dominant economic players on the lake (Thompson 1976; Gerrard 1985). Using their superior financial resources and market knowledge, they controlled the provision of inputs and the marketing of fish and were thus able to exercise considerable economic power over Icelandic First Nations, and metis fishers alike (McIvor 1966; Gerrard 1985). First Nations, metis and non-native fishers were concerned from early on that the companies were driving over-exploitation of the lake through using economic means to compel fishers to overfish (Mochoruk 1957; Tough 1984, 1997).

These concerns of fishers in the early years of the fishery had accentuated by the 1950s and 1960s. Catches for walleye and whitefish slumped, raising concerns about stock depletion (Mochoruk 1957; Lake Winnipeg Quota Review Force 2011). Whitefish catches dropped from 1,601,878 kg in 1950 to 341,950 kg in 1969 while walleye declined from 2,534,065 kg in 1950 to 922,761 kg in 1969 (Lake Winnipeg Task force: 123). Fishers became ever more indebted (McIvor 1966; Gislason et al. 1982; Gerrard 1985), as one older fisher reported:

> It was very difficult, very difficult. I know grown men at the end of the year that would fish for four or five months of the year and when they got to settle up and when they realized that they got nothing, they were crying! (Pálsson 2014, 54)

These conditions led to a whole scale reform of the fishery's governing system that will be outlined in the next section of the chapter on the governing system.

Currently, the fishing industry on Lake Winnipeg generates around $18.6 million annually and provides employment for almost 1,000 people (867 fishers and 173 hired hands) (Manitoba Conservation and Water Stewardship, Fisheries Branch 2012b). The lake is divided into three areas: South Basin (38 %), Channel area (24 %) and the North Basin (38 %).[2] The fishers along the shores of the lake have different ethnic backgrounds. The largest group is First Nations and metis, but there are also fishers of Ukrainian, Icelandic, and mixed descent. Our focus in this chapter is on the fishery conducted from the Municipality of Gimli (90 km North of Winnipeg, Manitoba) (see Fig. 15.1), where fishers are largely of Icelandic or mixed descent. Historically, the Gimli area has been considered as the most prominent fishing community of the lake, which is reflected in its predominance in statistics on quota holders. The communities along the shore of Lake Winnipeg are divided into 12 areas. Gimli belongs to area number two, as does the nearby town of Winnipeg Beach. There are 150 fishers located in area two, who hold 325 open skiff quotas or 22.6 % of the total open skiff quota entitlements. Fishers in area two have the largest share of the yearly total quota allowed, or 21.2 % (Manitoba Conservation and Water Stewardship, Fisheries Branch 2012c; Pálsson 2014).

The fishing year is divided into three seasons: the spring season (from 2 days after 80 % of walleye spawning completion has been reached until July 10th), fall season (September 1 to October 31) and the winter season (first ice to March 31). The spring and fall seasons are conducted on the open water where the fishers either use 20–25 ft open skiffs or 40–60 ft whitefish boats (limited in use to the spring season in the North Basin of the lake). The winter season is, however, completely different where the fishers harvest through the ice by using an auger attached on an old snow vehicle know as a Bombardier. The trend on Lake Winnipeg in recent years has been for fishers to use the open water seasons to fill their quotas as it is much more expensive to harvest during the winter season and not every fisher holds enough quota to fish year round (Manitoba Conservation and Water Stewardship 2011; Pálsson 2014).

The most important species at present is walleye [*Sander vitreus*] due to its price which has been almost been three times higher than other species (Lake Winnipeg Task Force 2011). Other species that are commercially important for the fishers are

---

[2] Figures in brackets refer to percentage of fishers in each area.

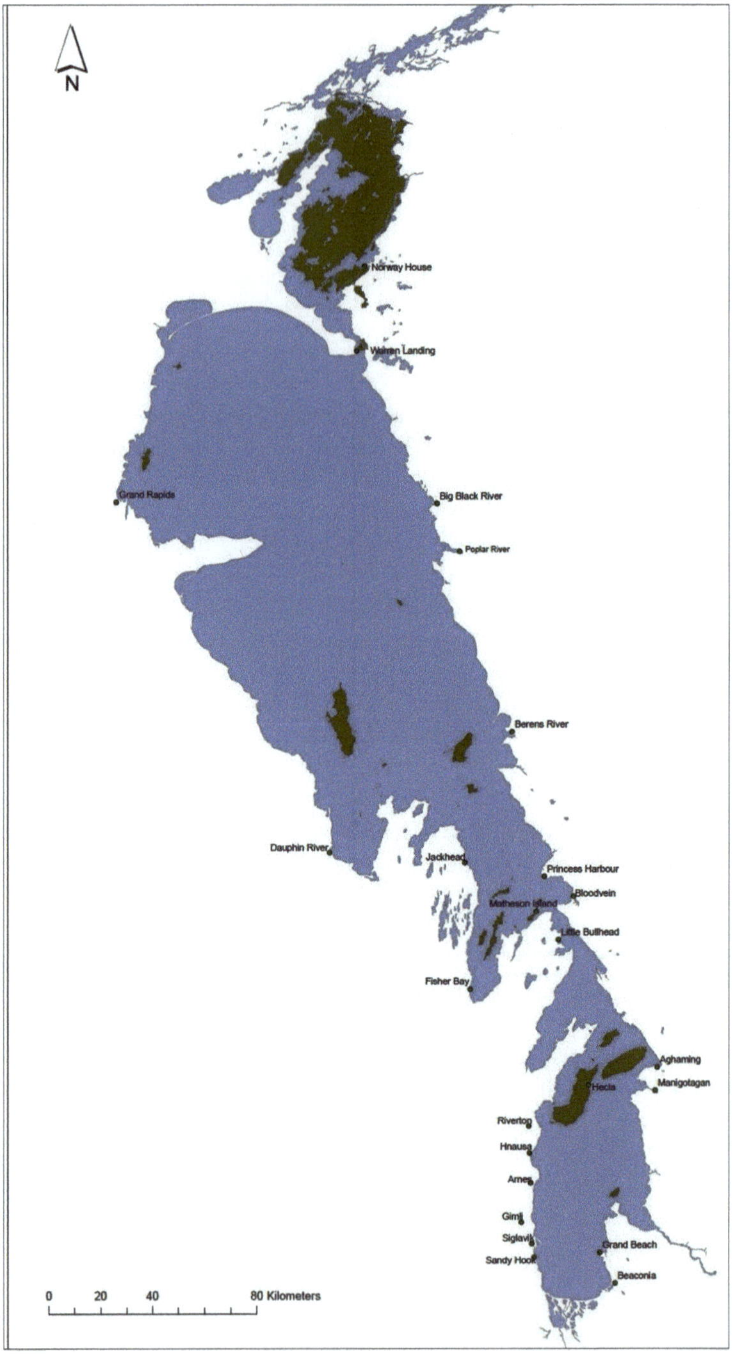

Fig. 15.1 Map of Lake Winnipeg (Source: Manitoba Conservation and Water Stewardship, Fisheries Branch 2012a)

lake whitefish [*Coregonus clupeaformis*], sauger [*Sander canadensis*], northern pike [*Esox lucius*], mullet [*Catostomus commersonii*], perch [*Perca flavescens*], goldeye [*Hiodon alosoides*], bass [*Morone chrysops*] and common carp [*Cyprinus carpio*]. Walleye, sauger and lake whitefish contribute most of the catch. In the 2009–2010 season, for example, they were 96 % of the lake's total catch (Manitoba Conservation and Water Stewardship 2011).

Governing System: Key Actors and Post-1960s Transformation

Final authority for decision-making and implementation in the Lake Winnipeg Fishery rests with the federal and provincial levels of government, which share responsibility for the governance of fisheries in Canada's inland waters (Manitoba Conservation and Water Stewardship 2009). These levels of government have been historically responsive to fisher concerns as voiced through formal and informal channels and legislation and practice does reflect the imprint of fisher contributions. Nonetheless, governance of the fishery is consultative rather than participatory or collaborative (Maclean 2010) and the concentration of decision-making power particularly with the Provincial Minister of Fisheries has made it, on balance, hierarchical in mode.

Following from the Natural Resources Transfer Agreement of 1930, the Federal Government has jurisdiction over conservation while the Provincial Government has the authority over the use and allocation of fish resources (Manitoba Conservation and Water Stewardship 2009). In practice, however, most of the responsibilities of the Federal Government, especially regarding day to day management, have been transferred to the Provincial level. In the case of Manitoba, the Manitoba Ministry of Conservation and Water Stewardship (MCW), Fisheries Branch, handles the day to day management of the lakes within the province, including Lake Winnipeg. The Interlake regional offices of the MCW Fisheries Branch are responsible for Lake Winnipeg. They manage fish stocks, set the amount of fish that can be taken out of the lake, set the dates when the fishing season begins and concludes, set regulations on mesh size and amount of gear and, finally, handle all the transfers of the quota.

Two organizations have been particularly important in representing fisher interests, the Lake Winnipeg Fisheries Management Advisory Board and the Manitoba Commercial Inland Fishers Federation. The Advisory Board was disbanded by its fisher representatives in 2007 but has been replaced by the Lake Winnipeg Fisheries Co-Management Board, another fisher organization with a much greater potential to move the governing system towards co-governance (Maclean 2010). As we did not conduct research on the formative activities of this body, however, we are unable to make any statements about the degree to which it appears to be succeeding in significantly strengthening fishers' power in decision-making.

The current governing system, the governance mode, and governing interactions of the fisheries of Lake Winnipeg reflect the dramatic intervention into the fishery that the Canadian state made in response to the crisis that the fishery faced in the 1960s. Most importantly, the central institutional planks of the current governing system, the Freshwater Fish Marketing Corporation and the individual transferable quota, or Quota Entitlement were an outcome of that historical turning point.

The crisis conditions in the Lake Winnipeg fishery in the 1960s, and similar conditions on other Canadian inland water bodies, led the Federal Government to establish the McIvor Commission to investigate the marketing of freshwater fish in Canada (Lamb 1975). Under the influence of long-standing and significant expressions of concern by fishers, channeled in the 1960s by the Manitoba Fishers' Association (the earlier name of the Manitoba Commercial Inland Fishers' Federation), the main finding of the commission supported fisher contentions of a highly unjust distribution of the returns of the fishery for the fishers (McIvor 1966). The committee suggested that the Federal Government establish a Freshwater Fish Marketing Corporation to serve as a single desk trader for freshwater fish in a few provinces in Canada (McIvor 1966). The Board would be similar in concept to the Canadian Wheat Board, which became in 1935 the sole agent for wheat (until 2012) in western Canada (Schmitz and Furtan 2000). The Federal Government enacted the committee's recommendations into law in 1969 when it established the crown corporation, the Freshwater Fish Marketing Corporation (FFMC) (Gislason et al. 1982; Gislason 1999).

The FFMC holds a monopoly on the purchase of fish and their export from the participating provinces. Fishers, however, can sell their product directly to customers locally and to restaurants, retailers and hotels if they apply for a special dealer license (Manitoba Conservation and Water Stewardship 1994; Pálsson 2014). The local market for fishers is not big, which means that most fishers on the lake sell the majority of their catch to FFMC, since the FFMC is required to buy their fish. The profits of the FFMC are returned to the fishers in form of a final payment (Pálsson 2014). Every spring, the FFMC gives the fishers the price for the fish, which is usually 75–85 % of projected market value for upcoming year. The price changes slightly before the winter season. Fishers get paid according to production and target species.

The establishment of the FFMC was a big step in changing the fisheries of Lake Winnipeg. The fishers suddenly had a guaranteed buyer for their fish, a buyer who had the mandate to maximize fisher incomes (Gislason 1999, 121). The ecology of the fishery remained a serious problem, however, since both the whitefish and walleye stocks were in a bad shape. Moreover, in 1970, a shock struck fishing communities when the lake closed to fishing, due to mercury pollution from a pulp mill discharge (Gislason 1999). While the closure of the lake brought hardship to fishers, it provided the provincial government a period of time to explore options for dealing with the overfishing problem (Gislason 1999). Building on consultations with fishers, the provincial government chose to implement a non-transferable individual quota system when the lake re-opened in 1972 (Gislason 1999). The system was supposed to eliminate the race for fish by guaranteeing the fisher a certain amount of fish. Gislason et al. (1982) argued that it was also "*intended to provide equalization of potential income* (145)".

The quota system in its initial non-transferable form had several significant drawbacks: initial quota allocations were too low to provide a reasonable income level, entry into the system was extremely restrictive, opportunities to expand fishing operations were limited, and the non-transferability of the quota shares meant that they could not be a store of value for their users (Gislason 1999, 123). Through the 1970s and 1980s, through the Lake Winnipeg Fisheries Advisory Board (Gislason 1999, 123) and the Manitoba Commercial Inland Fisheries Federation (Maclean 2010, 103), fishers therefore agitated for a less restrictive system where they could buy and sell quota (Manitoba Conservation and Water Stewardship 1989). In response, the Government of Manitoba made the quota system fully transferable. However, to prevent too much accumulation of quota, the government restricted the number of quota shares that a single fisher could own from four to six, depending on the community in which the fisher lives (Gislason 1999, 125).

At present, quota shares are fixed on a total allowable catch of three species for Lake Winnipeg of 6.52 million kilograms. The quota system is multi-species which only includes the three historically most important species: walleye, sauger and lake whitefish (Gislason et al. 1982; Gislason 1999). There are three types of quota entitlements. The first one is an open skiff quota entitlement that ranges from 2,270 to 7,940 kg (any combination of walleye, whitefish and sauger). Then there are 44 whitefish fleet quotas of approximately 15,880 kg, including an allowance to harvest 4,550 kg of either walleye or sauger (Pálsson 2014). The third type of quota entitlement is a quota of 650 kg for retired fishers who still want to fish (Coughlin 2006; Manitoba Conservation and Water Stewardship, Fisheries Branch 2012b; Pálsson 2014).

A third intervention to resolve the 1960s crisis of less long-lasting importance arose from the Manitoba Agriculture Credit Corporation (MACC) that had been established in 1958. The original mandate of the MACC was to facilitate acquisition of credit for Manitoba farmers, especially for younger ones. In 1970, the MACC opened the Fisheries Loan Program, which was an important step for securing finance for the fishers. The program enabled fishers to finance their purchase for equipment by applying for a "check off" system with the FFCM. This meant that the FFMC withheld 10–20 % from its payments to fishers who owed money to MACC (G.S. Gislason, personal communication).

At the time, the loan program was certainly vital for the fishers to finance their own operations. However, the loan program has much less contemporary relevance for fisher well-being. Gimli fishers now increasingly turn to the commercial banks for finance. As one fisher pointed out,

> The fishermen's loan program is not much used by us because there is a lot of money that we need that is not available through them. (Pálsson 2014, 148)

Governing Interactions: Successes and Fisher Concerns About Their Quality

Over the history of the fishery of Manitoba since the late nineteenth century, and particularly since the 1960s, there is considerable evidence that fishers and fisher representatives have been able to voice their views and concerns to government officials (Gislason 1999; Maclean 2010). The success of these interactions is indicated by the major institutional changes with the implementation of the FFMC and the quota system, and the transition of this latter to transferability in the 1980s. It is also indicated by less dramatic, but still significant modifications to governance arrangements, such as the introduction of the retirement quota and the maintenance of quota levels in the face of Manitoba Department of Fisheries wishes to reduce them (Maclean 2010). Again, however, as final decision-making power rests with the state and as there is also considerable evidence of the state overriding or ignoring fisher views (Maclean 2010), we hold that these interactions are framed within a hierarchical mode of governing.

Our research focused on the broader question of what constitutes wellbeing for fishers in Gimli but, as relations with the state are such an important part of their work and thus of their wellbeing, we found much of direct relevance to understanding governing interactions in the fishery. Our most directly pertinent results came from our governance relationship assessment survey where we asked fishers in Gimli what relationships were most important in influencing their fishing and fish marketing practices.

The importance of the two dominant fisheries governance institutions on Lake Winnipeg on fishers' lives is evident from the ranking in Table 15.1. It is clear from the table, also, that fishers are dissatisfied in their relationships with these institutions and that there is desire to change them. This desire for a changed relationship is particularly strong for the FFMC. Discussions with fishers about these findings revealed, however, that perceptions were not uniformly negative but, rather, that views were a mix of positive and negative perceptions of the institutions. Fishers felt that their relationships with the FFMC and the Provincial Government were problematic from the point of view of their material wellbeing, yet they valued how the institutions allowed them to meet important subjective goals.

Fishers were critical of the FFMC's inability to maintain their incomes and the increasing threat of concentration of the fishery through the quota system. The

Table 15.1 Governance relationship assessment results (N = 20)

Relationship (Ranked in order of influence)	(Percent) Fishers who cite this relationship as being important	Satisfaction with relationship scale 1–5	(Percent) Fishers who wish to change this relationship
FFMC	65	2.91	55
Provincial Government (proxy for quota system)	45	2.63	15
Family	35	5	0
Crew	25	4.75	0
Other fishers	25	4.4	0
Banks	10	3.5	0
Natural Resources Officers	10	1.5	5
Competing users	5	2	0
Customers	5	5	0
Federal Government	5	1	5

FFMC has a major impact on fishers' financial security as it buys all the fish that are offered to them. Fishers are increasingly frustrated with the prices that the FFMC pays, which they say have declined for the past 10 years:

> Yeah, it [FFMC] has provided a certain stability. The stability is however, not there when they are offering you 80 cents a pound for a fish, that 12 years ago used to be over $2 a pound. (Pálsson 2014: 136)

> It is so hard when you see the end price being so high, and then what we are getting. (Pálsson 2014: 137)

Fishers felt that declining prices were a direct result of decreasing efficiency and increased bureaucratization of the FFMC (Pálsson 2014). The fishers often pointed to reports in the media to support their arguments, including one claiming that the directors of the FFMC were making dubious expenditures on their business trips when marketing the fish (King 2012).

Concerns regarding the quota system have also emerged. Despite limitations on individual quota entitlement ownership, more successful fishers have found ways to indirectly control additional quota entitlements:

> It is almost like a monopoly on the quota system now. All the big guys have it and they are not going to give it up and they keep buying more and they are keeping the prices of quotas way up there and they are just controlling it. (Pálsson 2014, 149)

As is typical with individual transferable quota type arrangements, prices for quota have increased significantly, making it harder for new entrants to get into the fishing industry, especially in the South Basin of the lake. Earlier, the price for quota was once $1 or $2 CAD per pound but it is now difficult to find a price below $5 per pound (Pálsson 2014).

Despite fisher frustrations with the FFMC, some fishers are reluctant to say they want to see the FFMC go because it is a guaranteed buyer for their fish that they otherwise would not have:

> It [the FFMC] enhances [our ability]to live well. Absolutely, it provides certain stability. You know what is coming and what is being paid for (Pálsson 2014, 135)

> I know what I am going to get paid... I know when I start fishing for the season that I am going to get paid certain amount for certain grade of fish, so if I want to pursue that grade of fish, that I can and I know I will get paid for it. So, that's for me is the biggest thing with it and I also know that the cheques will never bounce from the FFMC, it is reliable, it is established (Pálsson 2014, 135)

Some fishers made similar claims about the quota system. As long as a fisher holds a quota, he or she still can make income from it. Essentially, by buying quota, the fisher is buying a job as it gives an entitlement to a guaranteed amount of fish and, through the FFMC, a guaranteed buyer: *"by buying quota, you are buying a job, and you know what you are getting...well a certain amount of money for your fish"* (Pálsson 2014, 110). Even when a fisher is too old or infirm to fulfill his or her yearly quota, the fisher can rent their quota to someone else: *"They are getting older and stuff and they don't want to sell and it is a system that works well. The younger guys can make a little more and the older guys can hold their quota longer"* (Pálsson 2014, 110).

In our interviews, fishers were more positive about the impacts of these institutions on their subjective wellbeing. As is typical of small-scale fisheries elsewhere in the world, fishers in Gimli are motivated to fish not just for the real material benefits of income, the relational benefits of their family based enterprises, but also for the way fishing supports subjective values that give them meaning as fishers. Fishers live from fishing, but also for fishing; they enjoy their work, and especially that it gives them the freedom of being their own bosses and planning their own time (Pálsson 2014). Some of the fishers we interviewed observed that both the FFMC and the quota system support their independence and freedom as compared to the earlier merchant-led fishing economy:

> You never made money, you were always broke, you were always owing money, the fishermen borrowed money for this and that equipment. The companies owned the nets, boats and everything. You were not independent, you were essentially run by them! (Pálsson 2014, 142)

> I think it is ours [the FFMC], it gives us a real stake in the fisheries that I think we wouldn't have it if the fish companies came back. There were fish companies here before; we would lose a lot of our independence. A lot of people think they will be free under the fish companies but I think it just will be the opposite. (Pálsson 2014, 142)

FFMC control of the market also frees fishers from having to spend time to market their product or find new buyers. This allows fishers to concentrate on catching fish and engaging in private marketing to the extent that they wish. They can use the spare time left with their families, the relationship that they rated the most important to them in the Governing Relationship Assessment survey. As one fisher emphasized:

> They help it [enhance the ability to live well], because we don't have to deal with the brokers and fish buyers. We don't have the time to sell our fish. When we fish, we fish til 8 or 9 pm, then we go out to the shed, and I just don't have the time, or the desire to hire someone else to do the work. That would take a lot of time from fishing and our family as well. (Pálsson 2014, 143)

Similar arguments were made about the quota system with regard to the independence it permits. The fishers pointed to their past dependence on the fish trading companies for capital. The quota system is designed to prevent that by restricting the quota to owner-operated boats. By this restriction, the fisher is the one who should make money from the fishery, as one fisher underlined: *"I don't have to give any percentage to companies. The quota is mine, and the money that I make from the quota is mine!"* (Pálsson 2014, 146)

Discussion

From the perspective of fishers in Gimli, the Lake Winnipeg fishery's governing system and governing interactions are cause for significant frustration but also provide acknowledged benefits. Fishers have concerns about the fishery's hierarchical mode of governance, but still recognize that the governing system has been frequently responsive in important ways to their needs since the major interventions at the turn of the 1970s. It is as yet unclear to us what the scope and effectiveness of the new Lake Winnipeg Fisheries Co-Management Board is, but it may be a means to address fisher concerns about governing interactions. Fishers also have significant, and what appear to be, growing concerns with the two major institutions that they interact with, the FFMC and the Provincial Government's individual transferable quota system. Here also, however, they see contributions, particularly when they view the institutions in comparison with the dire circumstances fishers faced prior to the 1970s. It appears to us that overall Gimli fishers do not favor abolishing these institutions but rather favor finding ways to make them more responsive and efficient.

From the Gimli fishery case, the most striking aspect that deserves further reflection is the fishery's distinctive institutional mix. It appears from a contemporary perspective as an unorthodox pairing: a vestigial social democratic marketing institution with an avant-garde neo-liberal rights-based institution. Importantly, however, the latter institution was conceived as a solution to the ecological problem at a time when the ideological application of ITQ systems in fisheries was still years away. While we are well aware of the social costs of ITQ systems elsewhere in the world, the Gimli case does give interesting pause for reflection. Its application there clearly has led to predictable ITQ effects of concentration and exclusion (cf. (McKay 2004; Pálsson 2006; Carothers 2008) but these appear to be more muted than elsewhere. This attenuation of impact is due in part to the restrictions on individual quota concentration – admittedly poorly enforced – but also because of the regulation on non-owner-operator control of quota endowments and the FFMC monopoly that effectively blocks vertical integration in the industry. These factors may also account for why, though fishers express concerns about the problem of intergenerational transfer of quotas due to their perception as a retirement store of capital for older fishers, this issue seems to be less acute than elsewhere.

There are several lessons of the ITQ system in combination with the FFMC. First, context is important. The Gimli case reminds us that ITQs must be understood in particular places; it suggests that the socially corrosive effects of ITQs may in some circumstances be tempered by other institutions, such as the FFMC and regulations such as the specific restrictions on concentration and owner operation. Second, despite its imperfections and fisher claims that the organization could be made more efficient, the FFMC shows that government marketing bodies can still achieve valuable social and economic goals and that they should thus remain part of the policy toolkit. Third, the history of the two institutions suggests lessons about the connection between legitimacy and governability. In the Gimli case, the FFMC and the ITQ system gained legitimacy because they addressed fundamental concerns for fishers about equity and ecological sustainability. That legitimacy is now increasingly in question for the material and relational reasons that we explain above which suggests that the FFMC and the Provincial Government need to find ways to address fishers' demands for a greater share of the income from the fishery and for greater inclusion in decision-making through the Co-Management Board. In the context of increasing global production of farmed white fleshed fish that competes directly with walleye and whitefish, addressing the former complaint may require awareness raising about market conditions as much as demonstrations of effort to increase the efficiency of the FFMC. The need for outreach and institutional adaptation is complicated, however, by the deep-seated ideological resistance to state intervention that has grown in Canada in recent years and which makes a dispassionate assessment of the FFMC difficult.

For governability theory more broadly, we identify several further lessons. First, the Gimli fishery case is evidence for governance as a wicked problem. The case shows that history is important: fisheries governability on Lake Winnipeg reflects the 'trace' of past interventions and experience (Jentoft and Chuenpagdee 2009). For many fishers in Gimli, current governance arrangements continue to mark a major improvement in contrast to the pre-1970s period. Nonetheless, a historical sensitivity also suggests that governability is not fixed, but rather varies over time and must be consciously maintained. Current fisher concerns about the quality of governing interactions and about the efficiency of the governing system need to be sensitively and meaningfully addressed. Second, methodologically, the governability assessment framework is highly ambitious and, certainly in the context of a single chapter, it is impossible to speak to all of its dimensions in any meaningful way. We thus drew selectively on it so as to be able to leave space for sufficient depth of empirical analysis. We focused on questions related to hierarchical governance and, even there, emphasized particular questions over others. Specifically, we used the social wellbeing approach as a tool to understand perceptions of the fishers, as part of the system-to-be-governed, of the governing system, and of governing interactions. With regard to the governing system, we sought to show how the contemporary government institutional matrix has been adjusted to fit, still imperfectly, fisher values of independence, fairness and also social relatedness. In the area of governing interactions, we looked in detail at the unique combination of the FFMC and the constrained ITQ system as institutions that are central in structuring interactions between fishers and the state. Both institutions do have a high degree of legitimacy, reflecting the fishery's history and the stability that they have provided, but there are also clear frustrations with them. We do think, even based on this incomplete application of the governability assessment framework, that we have enough evidence to show that the fishery of Gimli has had a high level of governability since the beginning of the 1970s. Prior to that point, Gimli fisher remember decades during which the capacity for governance of the fishery was low, as measured by its inability to grapple with fundamental problems of economic exploitation and ecological degradation. Fortunately, fisher frustrations were finally effectively acknowledged in the late 1960s by the Federal and Provincial governments which triggered a major overhaul of the fishery's governing system. Third, we also struggled with the mode of governance idea. While useful in providing reference ideal types, and while we did commit to the idea that the Lake Winnipeg fishery is governed hierarchically, the fishery's governing system has strong traces of co-governance that make it arguably an in-between mode of governance. Alternatively, the fishery can be seen as governed through a 'hierarchical mode with good governing interactions' (R. Chuenpagdee, personal communication). We will need to do more research before being able to judge which label makes more sense. Finally, we think the Gimli case illustrates the potential of marrying interactive governance theory with other approaches. In this case, we think social wellbeing provides a useful set of tools for understanding processes and perceptions of governability. Gimli fishers assess the governability of the fishery by the degree to which the governing system and governing interactions meet their material, relational and subjective needs. The pithy answer is they do, but imperfectly.

Conclusion

In the 1960s, the Lake Winnipeg fishery was in an ecological and social crisis. From the perspective of fishers in Gimli, this could also be seen as a crisis of governability, as the governing system was failing to address their fundamental material needs and their need to be heard. To their credit, the Federal and Provincial governments responded with major institutional interventions that dramatically enhanced the capacity of fishers to pursue their occupation and to live well. While the interventions could have come sooner – Gimli fishers had been frustrated with the governing system for decades – they have proven resilient over the subsequent 40 years and have shown themselves able to adapt to fisher concerns. Nonetheless, fishers in Gimli have mixed feelings about the FFMC and the ITQ system. They appreciate the stability and freedom that the institutions have brought, but they chafe at what they see as the inadequate income the FFMC allows them, the inequalities of the ITQ system, and the failure of the governing system to consistently and meaningfully involve them (cf. Maclean 2010). There are thus pressing governance needs that need to be addressed at present. The new Lake Winnipeg Co-Management Board may be the vehicle to do that, but it is as yet too early to say whether it will be a success or not. Should these governance problems be addressed in ways that reduce fisher concerns, it will be testament to the Gimli fishery continuing to exhibit high governability.

When combined with a social wellbeing methodology to bring out relative fisher satisfaction with governing relationships, the governability assessment framework provides a useful lens to structure an analysis of the distinctive and heterodox institutional arrangements in the Lake Winnipeg fishery. We emphasized its value particularly in relationship to a historical context that has been very important in shaping current perceptions of the fishery's governing system. The framework also provides the analytical perspective from which to understand how ITQs and hierarchical governance, an instrument and a governance mode not normally associated with the vitality and wellbeing of small-scale fisheries, can under particular conditions actually help sustain small-scale fisheries. The social wellbeing perspective makes it seems apparent that the significant institutional changes in the management of the Lake Winnipeg fishery in the early 1970s led to a major improvement in the fishery's governability, not only in terms of its capacity to be governed but also in terms of the quality of its governing interactions. Fishers have continued material and relational discontents with current governance arrangements, but they also admit a general, sometimes grudging, and satisfaction with them.

There remains considerably more work to be done to understand the governability of fisheries on Lake Winnipeg. In future research it will be necessary to consider much more fully the ecology of the lake. High, and perhaps unsustainable levels, of biological primary productivity in Lake Winnipeg have led to exceptionally high catch volumes. If catch levels, particularly of walleye, were to drop in future, it could be a trigger for latent discontent and lead to a threat to the governability of the fishery if the current institutional framework is increasingly seen by fishers as unresponsive to their needs. It is also necessary to expand the scope of our view of the lake. To date, we have looked at the lake's governability from the point of view of fishers from Gimli. Their relatively positive perspective on the Lake's governing system is not matched by communities elsewhere (Maclean 2010) and a fuller assessment of the governability of Lake Winnipeg's fishery will only come from a study that includes those communities, particularly those where First Nations and metis fishers dominate the population. We could also do more to bring out the voices of Lake Winnipeg's fishery managers who might provide a counterweight to fisher frustrations on governing interactions and the governing system. Even in Gimli, we would like to deepen our appreciation of the various views on the FFMC and the ITQ system. It is quite possible with further study that we would see differences in the community that are not currently apparent.

Acknowledgement We would like to acknowledge funding from the University of Manitoba Graduate Fellowship program and travel funding from the Social Science and Humanities Research Council of Canada funded project *Too Big To Ignore: Global Partnership for Small-scale Fisheries Research*. We also thank Mr. G.S. Gislason, an anonymous reviewer, Ratana Chuenpagdee, and Svein Jentoft for their constructive comments on earlier drafts of the chapter.

References

Arngrímsson, G. (1997). *Nýja Ísland: Saga of the journey to New Iceland*. Winnipeg: Turnstone Press.

Carothers, C. (2008). "Rationalized out": Discourses and realities of fisheries privatization in Kodiak, Alaska. In M. Lowe & C. Carothers (Eds.), *Enclosing the fisheries: People, places and power* (Fisheries society, symposium 68). Maryland: Bethesda.

Chuenpagdee, R., & Jentoft, S. (2013). Assessing governability – What's next? In M. Bavinck, R. Chuenpagdee, S. Jentoft, & J. Kooiman (Eds.), *Governability of fisheries and aquaculture: Theory and applications* (pp. 335–349). Dordrecht: Springer.

Copes, P., & Charles, A. (2004). Socioeconomics of individual transferable quotas and community-based fishery management. *Agricultural and Resource Economics Review, 33*(2), 171–181.

Coughlin, W. (2006, January 11). *An overview of Lake Winnipeg fisheries*. Presented at the 23rd annual Red River Basin land & water international conference, Winnipeg. Retrieved from http://www.redriverbasincommission.org/Conference/23rd_Proceedings/Coughlin.pdf

Coulthard, S., Johnson, D., & McGregor, A. (2011). Poverty, sustainability and human wellbeing: A social wellbeing approach to the global fisheries crisis. *Global Environmental Change, 21*, 453–463.

Coulthard, S., Amarasinghe, O., Koralgama, D., Paranamana, N., Sandaruwan, L., Britton, E., Bene, B. J., Pouw, N., Abunge, C., Mbatha, P., & Daw, T. (N.d.). *Exploring wellbeing in fishing communities: Methods handbook*. Northumbria University, UK.

Degnbol, P., et al. (2006). Painting the floor with a hammer: Technical fixes in fisheries management. *Marine Policy, 30*, 534–543.

Duguid, T., & Brandson, N. (2005). *Restoring the health of Lake Winnipeg – Canada's sixth great lake*. A report by the Lake Winnipeg Implementation Committee.

Gerrard, N. (1985). *A history of Icelandic river and Isafold settlements*. Aalborg: Saga Publication.

Gislason, G. S. (1999). From social thought to economic reality: The first 25 years of the Lake Winnipeg IQ management programme. In R. Shotton (Ed.), *Use of property rights in fisheries management*. Western Australia: Fremantle.

Gislason, G. S., MacMillan, J. A., & Craven, J. W. (1982). *The Manitoba commercial freshwater fishery: An economic analysis*. Winnipeg: University of Manitoba Press.

Gough, I. R., & McGregor, J. A. (2007). *Wellbeing in developing countries: From theory to research*. Cambridge: Cambridge University Press.

Houser, G. J. (1986). *Framfari: 1877 to 1880*. Winnipeg: Gimli Chapter Icelandic National League of North America.

Jentoft, S., & Chuenpagdee, R. (2009). Fisheries and coastal governance as a wicked problem. *Marine Policy, 33*(4), 553–560.

King, T. (2012, March 28). Crown corp's expenses fishy: Fishers. *Interlake Spectator*.

Kooiman, J., & Bavinck, M. (2013). Theorizing governability – The interactive governance perspective. In M. Bavinck, R. Chuenpagdee, S. Jentoft, & J. Kooiman (Eds.), *Governability of fisheries and aquaculture: Theory and applications* (pp. 9–30). Dordrecht: Springer.

Kristjansson, W. (1965). *Icelandic people in Manitoba: A Manitoba saga*. Winnipeg: Wallingford Press.

Lake Winnipeg Quota Review Task Force. (2011). *Technical assessment of the status, health and sustainable harvest levels of the Lake Winnipeg fisheries resources*. Manitoba: Manitoba Minister of Water Stewardship.

Lamb, C. W. (1975). An analysis of prices paid to fishermen before and after the establishment of a fishery cooperative. *Marine Fisheries Review, 37*(3), 36–38.

MacLean, J. (2010). *Fathoming Lake Winnipeg: The role of commercial fishers and their local knowledge in decision-making*. Master's thesis, Natural Resources Institute, University of Manitoba.

Manitoba Conservation and Water Stewardship. (1989). *Manitoba fisheries: Five year report to the legislature year ending March 31, 1989*. Winnipeg: Manitoba Water Stewardship.

Manitoba Conservation and Water Stewardship. (1994). *Manitoba fisheries: Five year report to the legislature year ending March 31, 1994*. Winnipeg: Manitoba Water Stewardship.

Manitoba Conservation and Water Stewardship. (2009). *Manitoba fisheries: Five year report to the legislature year ending March 31, 2009*. Winnipeg: Manitoba Water Stewardship.

Manitoba Conservation and Water Stewardship. (2011). *Annual report 2010–2011*. Winnipeg: Manitoba Water Stewardship.

Manitoba Conservation and Water Stewardship, Fisheries Branch. (2012a). Map of Lake Winnipeg. Received on April 16 2014 via email.

Manitoba Conservation and Water Stewardship, Fisheries Branch. (2012b). *Profile of Manitoba's commercial fishery*. Winnipeg: Manitoba Water Stewardship.

Manitoba Conservation and Water Stewardship, Fisheries Branch. (2012c). Table of Lake Winnipeg Quota – Overall production analysis (2012). Received on March 28th 2014 via email.

McCay, B. J. (1996). Social and ecological implications of ITQs: An overview. *Ocean and Coastal Management, 28*(1–3), 3–22.

McCay, B. J. (2004). ITQs and community: An essay on environmental governance. *Review of Agricultural and Resource Economics, 33*, 162–170.

McIvor, G. H. (1966). *Commissioner, Report of commission of inquiry into freshwater fish marketing*, Government of Canada.

Mochoruk, J. (1957). *Formidable heritage: Manitoba's North and the cost of development 1870–1930*. Winnipeg: University of Manitoba Press.
Pálsson, G. (2006). Nature and society in the age of postmodernity. In A. Biersack & J. B. Greenberg (Eds.), *Reimagining political ecology*. Durham/London: Duke University Press.
Pálsson, S. (2014). *Where the past meets the present: An assessment of the social and ecological determinants of well-being among Gimli Fishers*. Master's thesis. Thesis, Faculty of Graduate Studies, University of Manitoba.
Russell, F. (2000). *Mistehay Sakahegan, the Great Lake: The beauty and the treachery of Lake Winnipeg*. Winnipeg: Heartland Publications.
Schmitz, A., & Furtan, H. (2000). *The Canadian wheat board: Marketing in the new millennium*. Regina: Canadian Plains Research Center.
Thompson, S. O. (1976). *Riverton and the Icelandic settlement*. Riverton: Thordis Thompson.
Tough, F. (1984). The establishment of a commercial fishing industry and the demise of Native fisheries in Northern Manitoba. *The Canadian Journal of Native Studies, 2*, 303–319.
Tough, F. (1997). *As their natural resource fail: Native people and the economic history of Northern Manitoba, 1870–1930*. Vancouver: UBC Press.
Weeratunge, N., Béné, C., Siriwardane, R., Anthony, C., Johnson, D., Allison, E. H., Nayak, P. K., & Badjeck, M.-C. (2014). Small-scale fisheries through the wellbeing lens. *Fish and Fisheries, 15*, 255–279.
WeD. (2007). *Wellbeing in developing countries research group. Methods toolbox*. Retrieved from http://www.welldev.org.uk/research/methods-toobox/toolbox-intro.htm
Þorsteinsson, Þ. (1945). *Saga Íslendinga í Veturheimi*. Winnipeg: Columbia Press Ltd.

Chapter 16
Alaska's Community Quota Entities Program for Halibut and Sablefish: Between Governability Challenges and Opportunities

Adam Soliman

Abstract Individual Transferable Quotas (ITQs) can have severe negative impacts on small-scale fisheries. Such market-based management naturally favors large-scale participants, and cannot support small fishing operations or traditional fisheries without modification. One such modification designed to support small-scale fishers exists within the North Pacific Fishery Management Council's ITQ market off the coast of the state of Alaska in the form of Community Quota Entities (CQEs). CQEs obtain quota from the market and lease the quota to community members. CQEs allow for flexibility, and locally-tailored programs can directly involve fishers and fishing communities in the governing system. Financial and logistical barriers, however, have resulted in very limited quota share purchases under the program. Old Harbor, represented by the CQE Barnabas, Inc., was the first community to purchase halibut shares under the program, and it remains only one of two communities to do so as of 2013. Nonetheless, the community supports the program and its potential to provide opportunity to access the fisheries. Additionally, the Old Harbor program, governed by a community-based board of directors, has included a number of structures in the design of its CQE system to promote sustainable harvesting, self-enforcement, and inclusion of new and younger fishers. Despite substantial hurdles, the CQE model contains features that enhance governability. In Old Harbor's CQE system, the key governing interactions are focused on the core intent of CQEs to support small-scale, remote fisheries. These targeted governing interactions enhance governability by encouraging investment and engagement in local communities because they reflect the needs and concerns of those communities. Further development of affirmative action policies and supportive financing structures that reflect additional community needs could further enhance the governability of this system. Small-scale fisheries that face similar challenges under an ITQ or other market-based system—could benefit from similarly fine-tailored modifications to their existing governing systems.

A. Soliman (✉)
Director, The Fisheries Law Centre, Vancouver, Canada

Adjunct Professor, Faculty of Law, The University of British Columbia, Vancouver, Canada
e-mail: adamsoliman@me.com

S. Jentoft, R. Chuenpagdee (eds.), *Interactive Governance for Small-Scale Fisheries*, MARE Publication Series 13, DOI 10.1007/978-3-319-17034-3_16

Keywords CQE • Individual Transferable Quota (ITQ) • Fisheries law

Introduction: The Development of CQES in Alaska

In many fisheries jurisdictions, governing systems that make significant use of market-based policy instruments such as Individual Transferable Quotas (ITQs) have had severe negative effects on small-scale fisheries (Olson 2011; Soliman 2014a). Although ITQs have frequently been commended as an effective system of governance (Davis 1996), they have also faced heavy criticism for the unintended consequences that they create (Bromley 2009; Carothers and Chambers 2012; Soliman 2014b). One major criticism centers around the effects that ITQ programs have on the distribution of wealth and economic power within fisheries. These market pressures contribute to another major shortcoming of the ITQ system: the significant effect it has upon communities that rely on fishing.

These impacts on small-scale fishing communities are not surprising. ITQ schemes use market forces as a management mechanism, and market forces tend to favor participants who are more "efficient" in strictly economic and financial terms; this efficiency is strongly correlated with size. Although small-scale fisheries may in fact be "efficient" at sea, they may, for example, be more reluctant to invest. Underlying the commitment to ITQs is a basic assumption that under an ITQ program, fishery resources are distributed efficiently as free market forces work to eliminate overfishing and overcapitalization (Munro 2009; Soliman 2014a). With this redistribution, the market for ITQs encourages less efficient fishers to sell their quota shares and leave the fishery (Copes 1986). The theoretical end result is a more efficient fishery with fewer fishers.

Unfortunately, this type of fishery will naturally benefit larger and more efficient fishers that crowd out small-scale fishers, redistributing the economic benefits of the fisheries to industry (Bromley and Macinko 2008; Soliman 2014a). Scholars have even recently demonstrated the ways in which privatization dramatically alters the basic human-marine relationships that define fishing communities (Carothers and Chambers 2012). ITQ regimes in particular have been criticized for causing negative social impacts, such as the concentration and monopolization of quota shares in large commercial entities (Bromley and Macinko 2008; Soliman 2014a). In the groundfish fishery off the coast of Alaska, privatization has resulted in empirical end results that include consolidation and concentration of the fishery, a reduction in crew employment, increase in crew income, and changes to traditional and indigenous labor and community patterns (Olson 2011). These results have tended to disadvantage Alaska's small, remote, indigenous fishing communities in the transition to privatization (Carothers 2011).

Theoretical definitions of efficiency, it is important to note, ignore these externalities as well as the destructive impacts that operations can have on the environment and the social impact of fisheries collapse (Wingard 2000; Lowe and Carothers 2008; Soliman 2014a). As such, regulators began looking for ways to mitigate these harms. In response to these concerns, some jurisdictions have modified their ITQ-based governing systems by adding exceptions and structures intended to support remote communities, smaller fishing operations, and/or traditional and aboriginal fisheries.

Off the coast of Alaska, an example of one such "modified" supportive governing system was established by the North Pacific Fishery Management Council in the form of Community Quota Entities (CQEs) (Langdon and Springer 2007). This model emerged in a heavily exploited fishery, and the economic efficiencies of such a large-scale fishery serve to exacerbate the exclusion and marginalization of Alaska Natives there (National Research Council 1999). However, the market-based ITQ fisheries for halibut and sablefish for the Gulf of Alaska also include governing provisions that enable non-profit CQEs to obtain quota and lease it to community residents. The CQE program provides an opportunity to analyze whether small-scale fisheries can survive and thrive despite existing market-based governing systems when modified with locally-tailored programs designed to directly involve fishers and fishing communities in the governing system. To accurately assess the system's governability, however, it is critical to first understand the background of CQE programs and the conditions in Alaska's fishery.

The Alaska Native Claims Settlement Act, passed in 1971, resulted from an extended effort by Alaska Natives to obtain compensation for their ancestral lands (National Research Council 1999). The Act provided $96.2 billion and 44 million acres of land to Alaska Natives. The Alaska Native Claims Settlement Act created 12 regional corporations and more than 200 village corporations, and these for-profit entities received the Alaska Native Claims Settlement Act proceeds (National Research Council 1999). Alaska Natives became the shareholders. Despite the optimism that benefits would accrue following this distribution, some communities lacked necessary managerial capacity and oversight, and as a result monies flowed away from some Alaskan communities (National Research Council 1999). Some The Alaska Native Claims Settlement Act corporations experienced more success, however, and well-trained, responsive managers and community communication and participation played a key role in that success (National Research Council 1999).

Against this backdrop, the North Pacific Fishery Management Council established a Community Development Quota (CDQ) program in 1992 to allocate a portion of the fish harvest to Alaska Native villages and village coalitions. Like CQEs, the goal of CDQs is to enhance the economic and social development of small-scale fishing communities (National Research Council 1999). Under a CDQ program, a portion of the Total Allowable Catch is allocated to a community, and rather than allocating a total amount of fish, the quota reflects a percentage or share of the annual Total Allowable Catch established by the North Pacific Fishery Management Council (National Research Council 1999). The program places responsibility over a specific allocation of biomass within the Bering Sea fishery to residents of coastal Alaskan villages; this allocation seeks, in part, to mitigate against the limited opportunities of indigenous people to participate in the development of commercial fisheries over the past century (National Research Council 1999).

The CDQ quota is divided among six community organizations that are comprised of alliances of the Alaska Native Claims Settlement Act certified villages near the Bering Sea (National Research Council 1999). The organizations receive royalties from the shares and those royalties are used for community development (National Research Council 1999). Though the programs have local interests at heart, some critics have criticized the CDQ program for analogizing community development to fisheries development (National Research Council 1999); these criticisms may in turn apply to the CQE program.

The CQE program was a modification of the Individual Fishing Quota (IFQ) regime that had already been established for groundfish in the Gulf of Alaska (Stoll and Holliday 2014). The CQE program allows specially-designated, eligible communities in the Gulf of Alaska to establish non-profit entities to purchase and hold quota shares for mandatory lease to, and use by, community residents (NOAA N.d.; Stoll and Holliday 2014). The CQE program's intent is to improve the economic conditions of remote coastal communities by enabling them to purchase and use quota shares (State of Alaska Department of Commerce N.d.). Functionally, CDQ and CQE programs are almost identical: communities create local non-profit organizations who allocate quota shares to small-scale, community fishers. CQEs, like CDQs, represent a modification to the ITQ system. However, the government gives quota shares to CDQs for free under ANCSA, while those associated with CQEs have to be purchased (Richmond 2013).

The CQE program was proposed by the Gulf of Alaska Coastal Community Coalition in 2000 and authorized by the North Pacific Fishery Management Council in 2002 (Langdon and Springer 2007). The North Pacific Fishery Management Council oversees fisheries management in Alaska. The National Marine Fisheries Service promulgated regulations for the CQE program in 2004 using its authority under the Magnuson-Stevens Act, and in conjunction with the implementation of those regulations, the Council stated that "[a]llowing qualifying communities to purchase halibut and sablefish quota share for use by community residents will help minimize adverse economic impacts on these small, remote, coastal communities in Southeast and South central Alaska, and help provide for the sustained participation of these communities in the halibut and sablefish IFQ fisheries." (U.S. Office of the Federal Register 2004). In 2005, the Alaska State Legislature authorized a loan program for eligible organization to purchase quota shares (Langdon and Springer 2007; Department of Commerce 2014). The program was authorized to meet National Standard 8 of the Magnuson-Stevens Act to "take into account the importance of fishery resources to fishing communities in order to (A) provide for the sustained participation of such communities, and (B) to the extent practicable, minimize adverse economic impacts on such communities." (Magnuson-Stevens Act). In conference proceedings examining the implementation of CQEs, Langdon and Springer (2007) suggest that the program has not yet met this standard.

Since Alaska's CQE program became effective, only two CQEs have purchased quota shares through the program. Old Harbor, Alaska, represented by Barnabas, Inc., was the first community to purchase shares. Old Harbor's CQE has created a distribution system that is meant to be equitable, accountable, and achieve the community's goal of providing opportunities for both long-established residents and new entrants (NOAA N.d.). Even though only two communities are fully active through the purchase of halibut shares, 21 of the 42 eligible communities have completed the process of forming a CQE and obtaining National Marine Fisheries Service approval. This means that half of the eligible communities have invested substantial time and resources in preparing to participate in the program (Langdon and Springer 2007; NOAA 2013; Soliman 2014c).

The lack of quota purchases by additional communities can primarily be categorized as: (1) barriers to purchasing and (2) program-related restrictions (NOAA N.d.; Langdon and Springer 2007). These barriers are not legal restrictions; they arise from the high price and limited availability of quota shares. With respect to financing, Alaska amended its loan program in 2004 in anticipation of the CQE program, making CQE organizations eligible to obtain state-provided loans for the purchase of quota shares. The special loan terms that exist for CQEs under Alaska's state managed finance program include: a 35 % down payment or the collateral equivalent, a 15-year repayment schedule, and an interest rate of prime plus 2 % (Langdon and Springer 2007; Langdon 2008). In 2014, the interest rate for CQEs was 5.25 % (Department of Commerce 2014). To date, however, these terms have been far more prohibitive than assistive for CQEs pursuing sustainable financial status. This is in large part because loan terms are neither favorable, nor conducive to the purchase of a quota share. In addition to the large down payment and a lack of longer-term repayment options, borrowers are responsible for all of the direct costs of financing including initial surveys and inspections, appraisals, and title insurance (Langdon and Springer 2007). Despite these limitations, loans are still offered to these community entities as a form of aid built into the design of the CQE program.

Despite the low level of participation in the program, the model has strong potential to support small-scale fisheries on a broader scale. Small-scale fishers have demonstrated their support of CQEs, even in the planning stages of their involvement. More than half of eligible communities have invested substantial time and money preparing to participate (Langdon and Springer 2007). The City of Craig Planning Department, for example, has produced an economic analysis and business plan for implementation of the CQE program. In this business plan, the city planners state: "The City of Craig is being presented an opportunity to increase quota share for halibut and sablefish by participating in this program. Participation in this program will result in increased quota share for the local area which will have direct economic benefits to local fishermen, fish buyer, and processors and will result in additional income for services that support the fishing industry." (Templin 2004). Langdon and Springer (2007) have summarized interview responses from community members who, despite their frustration with financing barriers, express the potential for the program to provide earnings opportunities for community members, especially among youth. If the program was failing to promote coastal community interests, public support for CQEs would presumably decline.

This being said, frustration in these communities continues as income from fishing remains scarce. Excerpts from communications with Larsen Bay and Chenega

Bay fishers suggest that the CQE program creates a strong ideal for these communities to strive for, but this ideal is, in reality, much more difficult to obtain (Langdon and Springer 2007). One main concern expressed by fishers is the current lack of actual entry points within the CQE regime for new and young fishers to support a CQE and make it viable in the long term (Langdon and Springer 2007). As many communities still rely upon subsistence living (National Research Council 1999), the complexity required of solutions is even greater. While the CQE presents a variety of opportunities for these communities, it is far from perfect.

With such a limited sample size, it is immensely difficult to determine the governability of CQEs in general, especially in light of the variability of small-scale fishing communities and the "hybrid" nature of the governing system. As such, the relative of success of CQEs in supporting small-scale fishers is not yet clear. Nonetheless, the early successes of Alaska's CQE in including local community members in the decision making and governing systems, and the resulting openness to the program among fishing communities suggest that these components of the CQE's program design could have wider applicability, particularly in developing countries with similar communities, circumstances, and challenges. The successes of Old Harbor's working CQE, along with its limitations, provide a test model for a system with enhanced governability in small-scale fisheries under market-based regimes, as described below.

Governability Theories and Application

Traditionally, the term "governance" was applied to governments and the functions that they perform, but scholars now define the term much more broadly. "Governance" principles are considered to exist not only within governments, but also within private ventures, communities, the general public, universities, etc. (Kooiman et al. 2005). Governance is perceived as the collective, aggregate and integrated process of all these governing actors, which "can be more or less organized and routine, rarely harmonious but typically interactive" (Chuenpagdee and Jentoft 2009).

The viability of small-scale fisheries governed under a CQE model like that of Old Harbor may be more thoroughly understood through the application of the governance analytical framework. The governance analytical framework uses the Interactive Governance Approach to model and distinguish systems within three governance modes: self-governance, co-governance and hierarchical governance (Kooiman et al. 2005). "Governability" is the key quality that the governance analytical framework distinguishes. This quality is assumed to be situated partly in the system to be governed (e.g., fisheries and communities based on fisheries); partly in the governing system, (e.g., the institutions and organizations that direct and control fisheries); and partly in the governing interactions (i.e., how the governing system and the system to be governed are linked and communicate) (Bavinck et al. 2005).

The inherent traits and constructed capabilities of all three systems are what make a fishery more or less governable. To enhance governability, the governing system must be proficient, the system to be governed must be open and willing, and the tools the governing system uses—the governing interactions—must be effective. The governance analytical framework provides a means to analyze governance of fisheries, and fishery management tools—such as CQEs—by identifying important variables that help us to understand how and why governance implementations do or do not achieve the desired outcomes (Bavinck et al. 2005; Kooiman et al. 2005). Long-term sustainment of the resource, profitability, and the social well-being and justice of communities and stakeholders within the fishery reflect the desired outcomes in this approach (Kooiman et al. 2005; Bavinck et al. 2013).

In this new line of analysis, it is possible to have more governance but less government. In other words, solutions to governance issues may not consist of government action alone, but may also involve other sectors of society, including civil society organizations. These solutions may develop values, principles, and goals through a social exchange between the public and private "governors," involving stakeholders themselves in the governance of a given system (Bavinck et al. 2005; Kooiman et al. 2005). In Alaska, for example, the governance for CDQs includes the formation of boards of directors comprised of community members (National Research Council 1999). Each group, under the guidance of the state and applicable state and federal laws, has developed its own approach and tailored its program through a local interpretation of needs (National Research Council 1999). Where, as in Alaska, federal and state regulations lack of specificity as to how communities must organize or implement quotas, substantial variability in the groups arises (National Research Council 1999), but at least in the existing CDQ models there, stakeholder involvement is an integral part of the governing system. CQEs similarly involve stakeholders in the governing system and this involvement enhances the governability of the system. Involving stakeholders in governance is often seen as a way of increasing governability of fisheries and coastal systems (Chuenpagdee and Jentoft 2009) despite the fact that there are always transaction costs involved in doing so.

From a theoretical perspective, the CQE system—like many forms of fisheries governance—fails to fit easily into any one mode of governance. Instead, the management of small-scale fisheries through CQEs creates a system that reflects a "hybrid" of the three governance modes: hierarchical governance, where the state has all the power; co-governance, where the state and the individuals cooperate and govern together; and self-governance, where individual autonomy is at its strongest and governmental power at its weakest. CQE governance in Alaska includes components that fit into each of these three categories. Overarching federal and state statutes, regulations, and regional fisheries management councils comprise a government hierarchy; the community- and citizen-based board of directors cooperates with the government to implement laws and programs in a form of co-governance; and the CQE community has substantial autonomy in the design of how to implement, organize, and enforce its quotas.

As this case study discusses, several key issues arise in the assessing the governability and success of CQEs, and as these issues are analyzed, the observation that the current system in Alaska does not fit easily into one governance category becomes clear. This case study will examine five questions to assess the governability of CQEs: (1) How well does the program support the overall management objectives of sustaining the resource in the long term and enabling fishers to be profitable? (2) What is the legal nature and force of the rights created by the primarily ITQ-based governing system? (3) What is the ability or capacity of the program to support Alaska's small-scale fisheries adequately? (4) In what ways might the program be enhanced (e.g., by granting territorial rights) to support small-scale fisheries more effectively? (5) Could CQEs or other similar modifications in other communities, particularly in the South, support small-scale fisheries struggling under ITQ regimes or other market-based systems?

The Governability of CQEs

CQE programs are potentially important from a governance perspective because they may help to meet the criticisms of community-based programs within larger market-based systems by providing a better model for co-operative management and co-governance with community stakeholders (Deacon 2007; Costello 2012). Fisheries management can be made more effective by involving local interests in the decision making process (Wilen 1985). The CQE provides a vehicle for this higher level of local stakeholder involvement. In theory, a successful CQE can purchase additional quota shares, growing the commercial presence and fishing industry of the CQE, and yielding positive outcomes for small-scale fishers. As the presence of the CQE grows within the community—the system to be governed—the community's involvement and investment in that system should also grow, ultimately yielding a more responsive governing system. These factors would theoretically result in a more governable system.

From the governability perspective, the CQE program is in part an effort to promote co-governance and self-governance. The system to be governed includes the community small-scale fisheries as well as the ITQ fishery as a whole. The CQE's authority to allocate quota shares to community residents is a combination of co-governance and self-governance. The co-governance aspect of this is that National Marine Fisheries Service retains authority to approve a CQE for a village or collection of villages; without National Marine Fisheries Service approval, a CQE cannot begin to hold or allocate quota shares. Self-governance exists once a village resident requests quota shares from their CQE; at that point, it is the CQE's decision whether or not to allocate the shares to that resident. This decision is completely independent of National Marine Fisheries Service; the CQE is given the sole authority to make the initial allocation.

As shown above, this shift to local self-governance creates an unusual situation: while the majority of fishing operations are regulated at the national or state-wide level through a hierarchical governance model, CQEs are governed at the local level, at least in terms of distribution and, in part, enforcement. CQEs do submit plans to the federal government for approval, but these plans are designed to allow for significant autonomy, making CQEs appear more like co-governance than self-governance. But while the government does have the power of approval, as in a hierarchical system of governance, the implementation of these CQEs is made so local—through accountability procedures, the requirement that beneficiaries be local, and overall local ownership or management—that the system as a whole looks more like a self-governance model. There is no real continuing oversight. Once a CQE is created and approved, judicial review is minimal and deferential. There are no strict reporting requirements or other monitoring options of the effectiveness of the programs. This broad autonomy coupled with the heavy oversight at conception of the CQE makes these programs hard to fit into any one governance structure, particularly when placed alongside hierarchical ITQ regimes. CQEs are, in fact, a unique situation. This uniqueness, however, reflects a program carefully tailored to meet local needs, and it is this very characteristic which gives the CQE model the potential to result in highly governable systems.

CQEs in Practice

The CQE program is intended to promote the welfare and economic vitality of remote coastal fishing communities but also to shift the responsibility of management of these resources to the local level. The non-profit entities must be incorporated under the laws of the state of Alaska. These CQEs are then able to purchase quota share and lease it to their residents in any manner that the CQE deems as efficient and necessary, provided that the lessee meets the residency requirement and is listed on the National Marine Fisheries Service registry of persons eligible to receive and fish IFQ. A resident is eligible if he/she has lived in the eligible community for 12 consecutive months, and has claimed residency only in the community for those 12 months. If an applicant is not on the National Marine Fisheries Service registry (and some young fishers are not), then guiding that applicant through the registration process is the CQE's responsibility. These rules keep CQE effects local. This intensive local involvement promotes a governing system that is more responsive to the needs of small-scale fishers, and encourages fishers to be more involved in the system as interested stakeholders.

The decision making body of the CQE program is the non-profit entities' boards of directors. This local authority allows the CQE to employ collaborative and cooperative governance to satisfy the program's stated mandates to increase fishing activity, promote economic growth, and support community members carrying on fishing traditions for remote coastal communities (NOAA N.d.). These goals are enshrined in both the non-profit's founding documents, and in its proposals to the National Marine Fisheries Service for approval of CQE formation. These proposals often include statements about how the CQE will remain accountable to the local population it is designed to serve, and National Marine Fisheries Service approval is based, in part, on these representations. As a result, accountability to the population is built into the program's systems of governance, devolved to the local communities.

Beyond this, the CQE program does not mandate any uniform standard specifying how CQEs make quota allocation decisions within their communities. Each CQE is free to consider what their community's needs are at the time of the decision. Beyond civil actions that can be taken against non-profits, or attacks on the general application and approval process at the federal level, there is no appeal process that can be used to contest the decision of the board regarding whether or not quota share will be allocated to a particular resident. This further highlights the self-governance aspects of these corporations, as they are largely insulated from day-to-day government control.

With only two CQEs fully participating in the market to provide data, it is difficult to gain a clear picture on exactly how these systems would work in broader application. An examination of one of these CQEs, however, can yield lessons for other communities as to which components of the program work well and which do not. The first CQE to actively lease quota was Cape Barnabas Inc., the CQE for the community of Old Harbor. Cape Barnabas has developed a reasonably transparent and rigorous policy for making equitable quota share allocation decisions. Cape Barnabas allocates shares using a point system that gives preference to those with experience, equipment, investment, and who employ community residents on their vessels. Quota is leased in quantities that are large enough to be economically viable. Cape Barnabas also sets aside 20 % of the halibut quota for inexperienced community residents with limited resources (e.g., small boats). This entry level quota pool is open to applicants who own less than a set amount of IFQ, and who have no immediate family members owning more than the designated amount. To equitably allocate quota shares, the CQE conducts community outreach and employs clearly articulated standards for applicant selection (Sea Grant Alaska 2009). This community outreach includes education and providing other information members of their own community.

Cape Barnabas' selection process begins with the CQE decision-making board reviewing the initial application for eligibility under the residency and registration requirements. If the applicant is eligible, he or she is ranked based on the CQE's scoring criteria. Because Cape Barnabas identifies entry-level fishers as a target group, the board considers these fishers separately from the general quota pool. The other scoring criterion consists of what the community's needs are at the time of the decision making process. For example, the CQE sets aside 20 % of leases for small-boat fishers seeking to "get their foot in the door," and limits the number of leases per household to two, ensuring that the benefits of the program are spread throughout the community (Sea Grant 2009). Community needs are also reflected in the awarding of points for the hiring of local residents as crew members (Sea Grant 2009). The application also includes an affidavit signed by the applicant, which expresses the applicants' affirmation of eligibility, commitment to pay crewmembers and the CQE in compliance with CQE requirements, and agreement to return unreached quota to the CQE by a specified date (Sea Grant Alaska 2009).

Cape Barnabas also requires an applicant to its program to submit a fishing plan specifying when the applicant plans to fish; with whom the applicant will fish; who will work in the crew; and who will buy the fish (Sea Grant Alaska 2009). Cape Barnabas set this requirement in order to ensure that each lessee has a plan that will work when applied in "real world," thus ensuring that it will support the economic viability of their community. After it has paid off the initial loan that was required to fund its creation, the corporation also plans to reduce the lease rates that it charges to fishers to the level needed to cover only its operating overhead. More net revenue will then be retained by the fishers, to the benefit of the community in which they live. As the CQE is a not-for-profit entity; it cannot retain earnings and all monies must flow back to the community (Sea Grant Alaska 2009).

Sustainability of Fisheries Under CQEs

As noted above, the CQE program is designed to promote community viability in remote coastal communities. Halibut and sablefish are very important to Alaska's coastal communities, having created over 10,000 jobs in the past 10 years (State of Alaska Department of Commerce N.d.). These jobs exist in processing plants, vessel equipment, supply, and repair, as well as on the fishing vessels themselves. The presence of a viable fishery creates this multitude of supporting industries as an inherent part of it.

But does this program meet its goals of resource sustainability? The objective of promoting sustainability of the fishery is met because the CQE program is part of a larger plan (here, the IFQ regime) which, like other Total Allowable Catch regimes, utilizes a set a quantitative limitation on the amount of fish that can be taken from the sea (State of Alaska Department of Commerce N.d.). This sustainability objective is also supported by the scoring system that Cape Barnabas has implemented. The scoring system takes into account not just the type of fisher that a community resident is but also the type of household that he/she comes from. For instance, someone in the 15–25 age group receives more points than someone over the age of 55. The younger fishers have at least a theoretical incentive to fish sustainably because they will need the fish stock to be viable for many years to come. Sustainability of the stock is promoted further by the deductions that a CQE can make from their scoring system as a penalty for overfishing. These penalties reduce the chance that someone who fishes too much will be allocated quota shares. All of these factors benefit the community, and align with the goal of fishery sustainability for both CQE members and the fishing community as a whole (Sea Grant Alaska 2009).

Effective cooperation within the community can in turn increase the degree of cooperation, and therefore the quality and effectiveness of co-governance, between the larger entity (in this case, the CQE) and the resource manager. In a non-CQE case, Munro et al. (2013) showed how effective cooperation among Canadian ITQ holders led fishers to go far beyond normal cooperation actions. Typically, cooperation is only accepting and conforming to the regulator's requirements; in this case, fishers were pro-actively recommending that the regulator implement a conservation initiative. Because the CQE governing system is community-based and, in part, community enforced, evidence of such cooperation-driven conservation may emerge over time as well.

Steps Toward the Future: Can We Live with Property Rights?

In fisheries, ITQ schemes can be thought of as private property regimes to the extent that a fisher who holds a share of the Total Allowable Catch has an exclusive right to harvest that part of the annual catch of fish (Macinko and Bromley 2004). It is this authorization (rather than the fish) that can be sold, leased, and dealt with by the holder. Property rights therefore may exist with regard to the quota, which are the exclusive right to harvest (Stewart 2004).

This right is limited to the proprietary interest, however, and does not amount to a full legal property right (Macinko and Bromley 2004; Soliman 2014b). In a bid to protect the natural resource at stake, nations restrict and limit this entitlement. Should the natural resource continue to be depleted, then states generally wish to be able to revoke or reduce these harvest rights. In order to avoid lengthy procedures and allow for swift management, such restrictions are usually laid down in law, and the ITQ is framed in such a way as to avoid recognizing it as "property" in a legal sense. Some laws may speak of 'quotas,' while others describe the rights as 'quotas allocated to a license,' and yet others confer 'entitlements' under 'licenses or permits' (Garcia et al. 2003).

Thus, 'full' property rights are never granted, and the rights that do exist may be ambiguous and open to interpretation. The effort to avoid recognition as "property" is the reason why quotas are usually granted for free: Governments generally want the ITQ to be seen as a restricted license, permit or quota—that is, as a proprietary interest that does not amount to full ownership—and fear that the payment of a fee for such entitlement could give rise to monetary claims in case of later expropriation, revocation or reduction of that entitlement. There also appears to be a correlation between the presence of legislation requiring compensation for deprivation of property, and the desire to avoid giving property right status to ITQs (Leal 2000; Soliman 2014b).

The issue regarding property rights versus a license is one that receives inconsistent treatment by the courts. It is this inconsistent treatment that suggests that ITQs are not full property rights and gives rise to anxiety about the situation surrounding them. In Canada, for example, courts are likely to find ITQs to be a property right within private conflicts, but when the issue is between an individual and the government they are regularly construed as mere privileges (Soliman 2014b). Indeed, in *Malcolm v. Canada* (2013), a court in British Columbia determined that aggregate shares could be reduced without compensation.

In the US, legislation explicitly provides that ITQs do not create any kind of property right, and that no compensation is payable in case of revocation of the rights (Magnuson-Stevens Act). US legislation describes ITQs explicitly as permits only. This system is summarized by Stewart as being based on the concepts of the people's ownership of the resource, and the government's sovereign right to conserve and manage it (Stewart 2004). Many have argued that the right granted by the US is more of a 'revocable privilege' than any kind of property right (Arnason 1999; Leal 2000). Communal small-scale fisheries, in particular, clearly do not hold full communal rights because CQEs allow the purchase and sale of quota shares (Ziff 1996). These quota shares are held by the community, but then leased to individuals.

Regardless of this inconsistency in the judiciary and between different nations, it is clear that the uncertainty about the property status of fishing rights in ITQs and CQEs influences the perspectives (and perhaps the participation) of fishers and fishing communities of the effectiveness of these systems. As described more fully below, the application of some form of discrete territorial rights might balance this uncertainty and promote greater openness toward the governing system and willingness to participate in that system, enhancing overall governability.

Governing Interactions in CQE

Despite existing components in Old Harbor's system that support small-scale fishers, the fact that after a decade since the program's inception it remains only one of two CQEs leasing quota shares in Alaska despite the interest of other eligible communities reveals opportunities to further enhance system governability. Specifically, governing interactions that play a key role in governability in this context include more supportive finance structures, affirmative action policies and territorial rights.

Alaska's CQE system offers the state run loan program for fishers purchasing quota shares. As discussed above, however, this financial support does not bear out in reality but instead represents a substantial—even prohibitive—barrier to entry. Financing options represent a key governing interaction that profoundly influences the governability of the CQE system, and this is clear despite the limited sample size of functioning CQEs. Within Old Harbor, entrants cannot access the fishery if they cannot access funds. Outside of Old Harbor, financial barriers simply block additional CQEs from purchasing quota shares. Community members have voiced their frustration with this reality (Langdon and Springer 2007).

More carefully tailored finance options that better met the needs of Alaska's native communities could reduce financial barriers and result in new entrants to the market. Langdon and Springer (2007) have suggested apprentice programs, award quotas such as those in the CDQ program, 5- to 7-year grace periods, and grants in lieu of loans as financing options that would alleviate financial barriers. Additional options might include lengthier loan repayment options with lower incremental payments, or grace periods for years with low catch. The critical characteristic of any changes to the system must be, however, that they're carefully tailored to the needs of that community. Careful tailoring of the critical governing interaction of financing carries the potential to profoundly impact governability.

Similarly, affirmative action policies that support women fishers and younger fishers would reduce entrance barriers for those groups. A system supporting the entry of new fishers to the market would in turn promote greater community engagement. With more communities voicing their concerns and investing in the system, governability of CQEs will be enhanced.

Some communities investing in preparing to purchase shares under a CQE have built in such mechanisms, such as allocating half of the quota for new, younger entrants (Langdon and Springer 2007). Old Harbor identifies entry-level fishers as a target group, so the board considers these fishers separately from the general quota pool. Old Harbor also has set aside 20 % of leases for small-boat fishers to reduce barriers to entry.

Governability could also be enhanced by the addition of a governing interaction that establishes discrete territorial rights. A territorial right that would be useful in this context would provide a fisher or group of fishers with a long-term property right in a distinct area, along with the exclusive right to fish that area (Wyman 2008). Territorial rights, then, do not include full property rights, but rather one or a few sticks from the "bundle of rights," depending on the system's particular design. For example, the territorial right would likely include the right to exclude along with the right of use and perhaps the right of control; it would likely not include the property right of disposition or alienability. It is important to note that the key quality that would allow such a governing interaction to enhance governability is the instrument's capacity to be adapted to the needs of a particular community. Thus, the size, scope, season—all the details of such a discrete property right—would depend upon the wants and needs of the fishing community as well as the biology and other characteristics of the given fishery itself.

Some scholars have argued that a major shortcoming of the ITQ system is the lack of any long-term right to access the resource in question and this difference may have impacts on incentives to conserve the resource (Wyman 2008; Soliman 2014b). Providing a right, rather than a mere privilege, changes the inherent calculus of the individual actors and provides long term incentives to manage the fishery. Such rights could even be granted to a group made up of fishers who actively coordinate fishing efforts through a cooperative or community, such as Alaska's CQEs, further supporting those conservation incentives. In addition, discrete territorial rights would yield an additional layer of self-governance, promoting the monitoring and enforcement of the system. Such discrete rights, then, could enhance the investment of fishers and the fishing community in the governing system, enhancing governability.

There is little doubt that National Marine Fisheries Service has the authority to implement a management regime that includes territorial rights though the various fishery councils. The ever-growing field of the public trust doctrine strongly suggests that the federal government may divest fishing rights into exclusive hands of a few as long as such divestment: (1) promotes the public interest and (2) does not impair other related public rights (*Illinois Central* 1982; Lynch 2007). The Magnuson-Stevens Act specifically allows for the establishment of "limited entry programs" so long as National Marine Fisheries Service takes into account: (1) present participation in the fishery; (2) historical fishing practices in and dependence on the fishery; (3) the economics of the fishery; (4) the cultural and social framework of the fishery; and (5) "other relevant considerations." Therefore, it would be entirely possible to implement territorial rights within the Alaskan CQE program (Magnuson-Stevens Fishery Conservation and Management Act, 16 U.S.C. § 1853 1996).

While financing options and affirmative action represent governing interactions that respond to problems in the governing system, territorial rights reflect the needs of the system to be governed, i.e., the needs of the fishery to remain sustainable. At the same time, territorial rights would serve to meet the needs of fishers as well, providing a measure of exclusivity that would enhance their investment in the system by providing a more concrete category of property rights. Similarly, although financing and affirmative action primarily represent ways to engage the fishing community, the suggested modifications would have ancillary benefits to the fishery as well. For example, younger fishers would theoretically have more incentive to conserve since they could remain in the fishery for a longer period of time. Similarly, reducing financial barriers for new entrants could yield conservation benefits if more of the quota shares are used by fishers with smaller boats and less destructive gear. Each of these three governing interactions carries with it the potential to enhance the governability of CQEs in Alaska. Moreover, based on the challenges Alaska has faced in implementing the program to date, these governing interactions should be closely examined in any broader application of CQEs or similar modifications.

CQE as an Option to Support Small-Scale Fisheries in Developing Nations that Are Eager to Implement ITQ

Having seen some benefits of ITQ systems in the developed world, as well as the drawbacks, there is a broader question: would small-scale communities in developing countries facing the challenges of an ITQ regime benefit from having this type of hybrid governing system? Where a market-based catch share program exists or where such a program is likely to be imposed, a community-based program like a CQE provides a model that seeks to protect the rights of small-scale fishers.

The differences between developed and developing countries will likely be more significant than the differences among Alaska's CQE eligible communities. Developing countries may have different management capacities (Rieser 1999), or different logistical or financial capabilities (Clark 1985; Pauly et al. 2002). Even within the Alaskan experience with CQE's, the financial requirements (Langdon and Springer 2007) and the logistical limitations (Langdon 2008) pose serious problems in regards to the success of the CQE program. Even the specific challenges which faced Old Harbor pose a lesson related to the intensive difficulty which faces any organization attempting to establish their own CQE within the Alaskan system (Langdon 2008). Old Harbor's success depended as much on the ability to engage in proper networking, having the right contacts, and other "soft power" style options as much as it did on the CQE program that was developed. For example, having a CQE applicant prepare a written plan for the use of quota shares will increase the likelihood of success of that applicant because the applicant will have input from the decision makers on whether his/her plan is "workable." This may be very difficult for newly formed CQEs. Old Harbor had an advantage from the start because it contained the same staff from its Native Corporation as the CQE itself (Melissa Berns, personal communication). This meant that the staff was highly organized from the start and that they had experience developing and managing an investment portfolio prior to forming the CQE. This will likely not be the case for most developing CQEs where organization and experience may not be readily available. Some have noted, however, that the co-governance of the management process, as has occurred in the CDQ and CQE systems, could lead to the co-management of the resource in question (National Research Council 1999).

It has also been suggested (in Soliman 2014a) that the ITQ system is ultimately unable to engage the entirety of a developing society. The result is that the communities that require the jobs provided by fisheries are not fully developed by the ITQ system. The presumption of economic efficiency ignores the fact that many individuals go out of business in the process. Just as in Alaska, however, the key to CQE success is in its capacity to adjust and respond to different community needs.

Thus, if the basic principles of the Alaskan CQE program were implemented in another country's existing ITQ regime, it would have the potential to enable the poorer regions of a country remain economically viable. Enhancement of viability could emerge from the mandatory planning and application process that the CQE program has developed. Moreover, the flexibility built into the program allows the CQE to adjust to community needs throughout that process. For example, one of the unfortunate effects of ITQ regimes, the concentration of quota shares into the hands of only a few entities, can be alleviated if a certain percentage is retracted from the market place and set aside for small-scale fisheries and traditional fisheries (Soliman 2014a). CQEs also possess the potential to disperse an increased level of money to the communities that they are a part of (Berkes 1989). Sustainable livelihoods are integral for a successful development program in general, and fisheries are no exception (Worm et al. 2009).

There are many additional factors that change the governability analysis of a CQE program in the developing world, and each factor must be considered in the design of the system and any modifications to it—and ultimately those factors will weigh into a case-by-case investigation of whether that particular locally-tailored CQE or the CQE model itself is a good match for the system to be governed. The capacity to self-govern or co-govern—or to convince a government to give up some of its hierarchical governance—is very different in developing countries. The system's capacity to provide oversight, minimize corruption, monitor programs, and enforce regulation all must be assessed to determine whether a CQE will be a governable system in a given country. Despite these challenges—both in implementing the system itself and in assessing which system design best matches a country's particular situation—this author is not ready to completely discount the benefits of trying a CQE system in the South, starting with indigenous populations. It is possible that through these efforts, overall governability on one extreme or the other may compromise to find a middle ground.

Conclusion

Alaska's CQE program is an important enhancement intended to protect and enhance the viability of small-scale fisheries in that state. The program attempts to respond to the negative societal impacts that ITQ regimes typically have on smaller remote coastal communities. This program intends to make some quota shares available to remote communities by allocating a percentage to qualified communities, which then can lease it to community residents.

In the State of Alaska, in the first community to actively acquire quota shares through its CQE, the beneficial impact has been felt in the system governed—not only within the CQE community, but also in the fishery as a whole. By promoting a consistent dialogue between the community residents, the community managers, and the citizens of Alaska, the CQE program at Old Harbor has built a more inclusive decision making structure, as evidenced by the community decision making board of directors formed by members of CQE communities. Old Harbor was able to recruit the staff from its Native Corporation as the staff for its CQE. This enabled Old Harbor to begin the CQE process with an experienced and familiar management.

Despite existing components in Old Harbor's system that support small-scale fishers, the fact that it remains only one of two CQEs that have purchased quota shares in Alaska as of 2013—despite the interest of other eligible communities—reveals opportunities to further enhance governability within the CQE framework. Specifically, governing interactions that play a key role in governability in this context include more supportive finance structures and affirmative action policies. More carefully tailored finance options that better met the needs of Alaska's native communities could reduce financial barriers and result in new entrants to the market. Similarly, affirmative action policies that support women fishers and younger fishers would reduce entrance barriers for those groups. A system supporting the entry of new fishers to the market would in turn promote greater community engagement. With more communities voicing their concerns and investing in the system, governability of CQEs will be enhanced.

As the program develops and barriers to purchasing and entry are alleviated, the program has every indication of meeting its goal of supporting small-scale fisheries in remote regions. The governability of these unique entities, and of fisheries that use these entities, defies easy definition. CQEs bear characteristics of self-governance, co-governance and hierarchical governance. And yet, the sample size of CQEs is very small; broader patterns will be revealed if the program continues and expands. CQEs do represent an attempt to broaden access to fishing outside of ITQs. Governed by localities and individual stakeholders, the model embraces empowerment of the community and devolution of power from the government. The CQE program supports a peer-enforcement structure, for example, with the goal of sustainability and stability. Although there is little competition for quota shares among community residents at this time (Melissa Berns, personal communication), if such competition develops, peer enforcement will apply penalties for violations of the program such as overfishing. These penalties are built into the scoring system for determining who receives the quota shares from Old Harbor's CQE. Competing residents would therefore have an incentive to report violations in order to improve their chances of receiving a quota share allocation. Whether or not these programs work, it is likely that the input and localized management will continue to have a profound effect on the lives of fishers.

Although the exact contours of the analytical governance framework applicable to these "hybrid" entities have not yet been sharpened, involving the stakeholders in the decision-making over their access to subsistence resources yields enhanced governability. Implementing community-based programs like the Alaskan CQE program can change the way developing countries manage their resources and develop more just property rights structures. Further, community-based management systems have the potential to provide developing countries with opportunities to build more governable systems and in turn develop more powerful tools to stop the overexploitation of their resources and to help remote communities stay economically viable.

References

Alaska Native Claims Settlement Act, 43 U.S.C. § 1601 (1971).

Arnason, R. (1999). Property rights as a means of economic organization. In R. Shotton (Ed.), *Use of property rights in fisheries management: Proceedings of the Fishrights 99 conference* (pp. 14–25). Rome: FAO.

Bavinck, M., Chuenpagdee, R., Diallom, M., van der Heijden, P., Kooiman, J., Mahon, R., & Williams, S. (2005). *Interactive fisheries governance: A guide to better practice*. Amsterdam: Centre for Maritime Research.

Bavinck, M., Chuenpagdee, R., Jentoft, S., & Kooiman, J. (Eds.). (2013). *Governability of fisheries and aquaculture: Theory and applications* (MARE publication series 7). Dordrecht: Springer. doi:10.1007/978-94-007-61-7-0_4.

Berkes, F. (1989). *Common property resources: Ecology and community-based sustainable development*. London/New York: Belhaven Press.

Bromley, D. W. (2009). Abdicating responsibility: The deceits of fisheries policy. *Fisheries, 34*, 280–302.

Bromley, D. W., & Macinko, S. (2008). *Rethinking fisheries policy*. Juneau: Department of Fish and Game.

Carothers, C. (2011). Equity and access to fishing rights: Exploring the community quota program in the Gulf of Alaska. *Human Organization, 70*(3), 213–223.

Carothers, C., & Chambers, C. (2012). Fisheries privatization and the remaking of fishery systems. *Environment and Society: Advances in Research, 3*, 39–59.

Chuenpagdee, R., & Jentoft, S. (2009). Governability assessment for fisheries and coastal systems: A reality check. *Human Ecology, 37*, 109–120.

Clark, C. W. (1985). *Bioeconomic modelling and fisheries management*. New York: Wiley.

Copes, P. (1986). A critical review of the individual quota as a device in fisheries management. *Land Economics, 62*, 278.

Costello, C. (2012). Introduction to the symposium on rights-based fisheries management. *Review of Environmental Economics and Policy, 6*, 212–216.

Davis, R. (1996). Individually transferable quotas and the Magnuson Act: Creating economic efficiency in our nation's fisheries. *Dickinson Journal of Environmental Law and Policy, 1*, 267–314.

Deacon, R. (2007). The efficiency gains from fully delineating rights in an ITQ fishery. *Marine Resource Economics, 22*, 347–362.
Department of Commerce, Community & Economic Development, Alaska. (2014). *Financing section: Interest rates*. Retrieved on November 16, 2014, from http://commerce.state.ak.us/dnn/ded/FIN/InterestRates.aspx
Garcia, S. M., Zerbi, A., Aliaume, C., Do Chi, T., & Lasserre, G. (2003). *The ecosystem approach to fisheries: Issues, terminology, principles, institutional foundations, implementation and outlook* (FAO fisheries technical paper 443). Rome: FAO.
Illinois Central Railroad Co. v. Illinois, 146 U.S. 387, 453 (1982).
Kooiman, J., Bavinck, M., Jentoft, S., & Pullin, R. (Eds.). (2005). *Fish for life: Interactive governance for fisheries*. Amsterdam: Amsterdam University Press.
Langdon, S. (2008). The community quota program in the Gulf of Alaska: A vehicle for Alaska native village sustainability? In M. E. Lowe & C. Carothers (Eds.), *Enclosing the fisheries: People, places, and power* (Symposium 68, pp. 55–74). Bethesda: American Fisheries Society.
Langdon, S., & Springer, E. (2007). Gulf of Alaska community quota program: Status and issues. In P. Cullenberg (Ed.), *Alaska's fishing communities: Harvesting the future: Conference proceedings*. Fairbanks: Alaska Sea Grant College Program, University of Alaska Fairbanks.
Leal, D. (2000). *Homesteading the Oceans: The case for property rights in U.S. fisheries* (PERC policy series, issue number PS-19). Bozeman: PERC.
Lowe, M. E., & Carothers, C. (Eds.). (2008). *Enclosing the fisheries: People, places, and power* (Symposium 68). Bethesda: American Fisheries Society.
Lynch, K. (2007). Application of the public trust doctrine to modern fishery management regimes. *New York University Environmental Law Journal, 15*, 285.
Macinko, S., & Bromley, D. (2004). Property and fisheries for the twenty-first century: Seeking coherence. *Vermont Law Review, 28*, 623.
Magnuson-Stevens Fishery Conservation and Management Act, 16 U.S.C. § 1853 (1996).
Malcolm v. Canada, Fisheries and Oceans. FC 363 (2013).
Munro, G. (2009, August). *The efficiency of rights based fisheries management schemes and the quest for resiliency: An overview*. Keynote address at Conference on Efficient Fisheries Management. Fishing Rights and Flexibility, Reykjavik, Iceland.
Munro, G., Turris, B., Kronbak, L., Lindroos, M., & Sumaila, R. U. (2013). *Catch share schemes, the theory of dynamic coalition games, and the groundfish trawl fishery of British Columbia*. Paper presented for the NAAFE Forum, Petersburg, Florida.
National Research Council (U.S.), Committee to Review the Community Development Quota Program. (1999). *The community development quota program in Alaska*. Washington, DC: National Academy Press.
NOAA. (N.d.). *Community quota and license programs and community quota entities*. Alaska Regional Office.
NOAA Fisheries Service (NMFS). (2013). *Report on holding of individual fishing quota (IFQ) by residents of selected Gulf of Alaska fishing communities 1995–2012*. Retrieved November 14, 2014, from http://alaskafisheries.noaa.gov/ram/reports/ifq_community_holdings_95-12.pdf
Olson, J. (2011). Understanding and contextualizing social impacts from the privatization of fisheries: An overview. *Ocean and Coastal Management, 54*, 353–363.
Pauly, D., Christensen, V., Guénette, S., Pitcher, T. J., Sumaila, U. R., Walters, C. J., Watson, R., & Zeller, D. (2002). Towards sustainability in world fisheries. *Nature, 418*, 689–695.
Richmond, L. (2013). Incorporating indigenous rights and environmental justice into fishery management: Comparing policy challenges and potentials from Alaska and Hawai'i. *Environmental Management, 52*(5), 1071–1084.
Rieser, A. (1999). Prescriptions for the commons: Environmental scholarship and the fishing quotas debate. *Harvard Environmental Law Review, 23*, 393.
Sea Grant Alaska. (2009). *Community quota entities: Workshop proceedings*. Retrieved on November 14, 2014, from http://www.npfmc.org/wp-content/PDFdocuments/rural_outreach/CQEworkshopProceedings09.pdf

Soliman, A. (2014a). Using individual transferable quotas (ITQs) to achieve social policy objectives: A proposed intervention. *Marine Policy, 45*, 76–81.
Soliman, A. (2014b). Do private property rights promote sustainability? Examining individual transferable quotas in fisheries. *Seattle Journal of Environmental Law, 4*, 245.
Soliman, A. (2014c). Achieving sustainability through community based fisheries management schemes: Legal and constitutional analysis. *Georgetown International Environmental Law Review, 26*, 273.
State of Alaska Department of Commerce. (N.d.). Community quota entity program (CQE). Retrieved November 14, 2014, from http://www.commerce.state.ak.us/bsc/CDQ/cqe/background_cqe.htm
Stewart, C. (2004). *Legislating for property rights in fisheries* (FAO legislative study 83). Rome: FAO.
Stoll, J. S., & Holliday, M. C. (2014). *The design and use of fishing community and regional fishery association entities in limited access privilege programs*. U.S. Department of Commerce, NOAA. NOAA Technical Memorandum NMFS-F/SPO-138.
Templin, B. (2004). *Community quota entity (CQE) program economic analysis and business plan for the City of Craig*. Alaska: City of Craig. Retrieved November 14, 2014, from http://www.craigak.com/documents/CQE_Plan.pdf
U.S. Office of the Federal Register. (2004). Fisheries of the exclusive economic zone off Alaska; individual fishing quota program, community purchase—Final rule. *Federal Register, 69*(84), 23681–23694. Retrieved on April 30, 2004.
Wilen, J. (1985). Towards a theory of the regulated fishery. *Marine Resource Economics, 1*, 369–388.
Wingard, J. D. (2000). Community transferable quotas: Internalizing externalities and minimizing social impacts of fisheries management. *Human Organization, 59*, 48–57.
Worm, B., Hilborn, R., Baum, J. K., Branch, T. A., Collie, J. S., Costello, C., Fogarty, M. J., Fulton, E. A., Hutchings, J. A., Jennings, S., Jensen, O. P., Lotze, H. K., Mace, P. M., McClanahan, T. R., Minto, C., Palumbi, S. R., Parma, A. M., Ricard, D., Rosenberg, A. A., Watson, R., & Zeller, D. S. (2009). Rebuilding global fisheries. *Science, 325*, 578–585.
Wyman, K. (2008). The property rights challenge in marine fisheries. *Arizona Law Review, 50*, 511.
Ziff, B. (1996). *Principles of property law*. Toronto: Carswell.

Chapter 17
Governing Through Markets: Societal Objectives, Private Property Rights and Small-Scale Fisheries in Denmark

Jeppe Høst

Abstract This chapter explores the relationship between small-scale fisheries and market-based governance systems such as individual transferable quotas. The analysis is prompted by the wide-ranging introduction of private and transferable *Vessel Quota Shares* in the Danish demersal fisheries. The chapter documents both quantitative and qualitative changes in the fisheries due the introduction of market mechanisms. Among these changes has been a concentration of resource ownership, changes in the social organization favoring larger operations with better access to legal advisors and financial capital. Consequently, operators in small-scale fisheries are facing challenges, when acting on the quota market, and in general the small-scale sector is in serious and structural decline. The chapter discusses characteristics of the Danish fishing fleet in relation to instruments and actions available for a governing system *governing through markets*. The specific instruments put in place to protect and safeguard small-scale fisheries are analyzed and discussed. As part of this, the chapter asks how the introduction of market mechanisms changes and transforms the overall governability of the small-scale fisheries. The chapter argues that social objectives have been marginalized and that this development reduces the overall governability of the sector. Moreover, participation of operators in the distribution of fishing quotas has increased through joint quota companies, guilds and through the flexible and dynamic, but individual, every-day market transactions.

Keywords Privatization • Social objectives • Market governance • ITQs • Policy design

J. Høst (✉)
Department of Ethnology, SAXO-Institute, University of Copenhagen, Copenhagen, Denmark
e-mail: jeppeh@hum.ku.dk

S. Jentoft, R. Chuenpagdee (eds.), *Interactive Governance for Small-Scale Fisheries*, MARE Publication Series 13, DOI 10.1007/978-3-319-17034-3_17

Introduction

The purpose of this chapter is to discuss the relationship between small-scale fisheries and market-based governance principles such as individual transferable quotas, sometimes also termed rights or wealth based fisheries management. Based on a case study from Denmark, the chapter examines how the introduction of private property rights has changed the nature of the *system-to-be-governed*, and in particular the *interactions* between the *governing system* and the *system-to-be-governed* (Kooiman et al. 2005; Bavinck et al. 2013). Using such an interactive governance approach it is possible to contrast the different principles of meta-governance and examine the change from a common quota management system to the new market-based governance. The development can be described as a movement from state-centric and vertical management to interactive governance using markets as governance instruments (Torfing et al. 2012). In other words, the chapter documents how the distribution of fishing opportunities through markets has changed the social and material dynamics of the fleet, and discusses how this has altered the governability of the sector, with a particular focus on small-scale fisheries (see further Høst 2015). The chapter begins with a short historical introduction to the quota management system in Denmark, and then explains the transition to market-based fisheries governance. Following this introduction, the chapter examines the overall design of market-based governance and focuses in particular on two central governance instruments and policy objectives, namely regulations governing concentration of quota shares, and a specific measure designed to protect and promote small-scale fisheries. Looking at the interactions and discussions around these governance instruments and interactive processes, the overall governability is discussed in relation to societal concerns and objectives. Anti-concentration rules and safeguard measures were intended to guard and promote near-shore fisheries but have, due to its failures, been subject to many public discussions and concerns. The interactions and discussions around these instruments reveal the strengths and weaknesses of market-based fisheries management.

Growth and Decline, TACs and Quotas

In this section I give a short historical introduction to shifting governance systems of North European fisheries in general and Danish fisheries in particular. I briefly outline the different images, instruments and actions involved in fisheries governance in these regions as well as characterize the different consequences of each meta-governance style for the different orders of governance – from the meta-governance level to the day-to-day handling of the governance interactions (Bavinck et al. 2013). This enables me to contrast the governance of a common national quota with the new regime based on market mechanisms. From the 1950s and onwards northern European fishing capacity grew dramatically with the introduction of, among other things, new efficient gear, electronic fish finders, larger vessels and engines (Søndergaard 2004; Karlsdóttir 2005). In Denmark alone, catches increased more than tenfold from 166,000 tonnes in 1945 to almost 2 million tonnes in 1980 (Holm 2002). The new industrial reduction fisheries were responsible for a substantial part of the growth, but the increase was imminent in all corners of the industry. Consequently and in response to more or less uncontrolled international competition (especially in the North Atlantic and the North Sea) a range of new governance instruments and institutions were introduced at the international level (Rozwadowski 2002; Karlsdóttir 2005). This international governance process took off in the 1970s, and its exact historical timing and introduction reflected the dramatic expansion of national Exclusive Economic Zones (EEZs) (Nandan 1987). In most areas, EEZs were to be expanded from 6 to 200 nautical miles. For Europe in particular, this challenged the seasonal mobility and patterns of fishing activities (Rozwadowski 2002; Karlsdóttir 2005). What were previously open waters of the North Sea were to be entirely divided into zones belonging to United Kingdom, Norway, Sweden, Denmark, Germany, the Netherlands, Belgium and France. Neither fish nor fishing activities, of course, adhered to the planned zones and borders. Danish fishers saw zoning as a disaster. Therefore, as a response to the coming expansion of the EEZs, there was a search for alternative solutions with regard to the complex fisheries issue.

An alternative to EEZs was to split the North Sea fisheries into a number of national quotas between surrounding North Sea countries. These national quotas could be based on a new scientific tool, the estimation of Total Allowable Catches (TACs). Thus, from the mid-1970s a range of species in the North Sea were split between countries through national fishing quotas estimated from historical catches. In contrast to the principle of exclusive zones, quotas allowed each country to carry on its fishery based on national quota shares. The reasoning behind fishing quotas in northern Europe was thus influenced by geo-political transformations as much as concern over marine resources. For the purpose of this chapter, the important change to note was a shift from input regulation through the management of gears, seasons and catch areas to governance, or rather management, through output regulation (TACs and national quotas). Historically, it was also a shift from a liberal and growth oriented strategy through subsidies to a management based on limited fishing opportunities. In a rather tacit manner, this shift also altered the object of governance. While the implementation of TACs and national quotas solved the puzzle of how to avoid the negative implications of EEZs, it also changed the principles and dynamics of governance. These changes will be examined in the following section.

Governance Interactions

The estimation of Total Allowable Catches (TACs) represented the state-of-the-art of scientific knowledge and modeling of fish mortality, and the new management approach put advanced marine biology at the heart of fisheries management (Karlsdóttir 2005). The result was a more hierarchical governance structure with the state defining and implementing quota limits in alliance with international scientific institutions. Hierarchical structures, in other words, placed scientific advice and knowledge above the every-day experiences of fishers. Fish stocks were computed by scientists and decided upon through political negotiations and only then caught by fishers. The interactions between fishers and the character of the governing system had changed. Instead of having governmental support to pursue the best and largest catches around the world, fishers now had to comply with quota limits, licensing and control systems. In addition fishers were now competing over a limited quota, and not as previously over wild fish resources. On the other side, these new instruments put an end to decades of international governance discussions on how to manage international fisheries (Rozwadowski 2002; Karlsdóttir 2005). However, quota management also changed the very nature of the system-to-be-governed. The biological governability of marine resources was enhanced, as it established scientific advice and fixed distribution keys between countries as the central points in management of fish stocks (Rozwadowski 2002; Karlsdóttir 2005). But new challenges around resource distribution also appeared.

In Denmark, the first national quotas were introduced for North Sea herring in 1974 and quotas for other species soon followed.[1] As mentioned above, this increased the stability and biological governability of the North Sea fisheries, at a moment when many species were in decline due to increased fishing pressure (Karlsdóttir 2005). However, the cardinal problems of national quotas were soon exposed as the national quota had to be split in some way between resource users. This distribution was complex because of the many different fishing techniques, seasons, vessels sizes and so on, but nonetheless had immediate economic implications for fishers. Already within the first months of the new quota system, gillnetters raised concerns. The gillnetters needed time to undertake their fishing activities, and they accused trawlers of catching too much of the overall quota in the first months of the fishery. This and similar concerns emerged under the national quota management. The lower the TAC and national quota, the more problematic and vocal were the discussions and protests. The problem of how to distribute limited resources between operators appeared almost unsolvable and revealed new types of conflicts. Most noticeable was the conflict between gear types as mentioned above and the conflicts between fishers of a particular region and mobile fishers visiting their fishing grounds. Over the years these conflicts, which were a direct result of the governance instrument, led to the common national quota being split into monthly or weeklong periods. Later, criteria such as vessel length and fishing activity were added to the allocation procedure. The image that guided these instruments was an image of *free and equal* access between operators. This image was a legacy from the previous management system (mainly on gear) and liberal culture in the producer organizations. This meant that resources were to be free of cost and that the different licenses and regulations should apply equally to all. This principle was however

[1] This and the following sections are based on a reading of the weekly magazine for fishers, "Dansk Fiskeritidende" between 1970 and 1980, supplemented with personal interviews.

very difficult to implement in practice, and policies resulted in fragmentation of the fleet into different interest groups based on vessel size or mobility (Søndergaard 2008; Vedsmand et al. 1996). In addition, attempts to solve one conflict in a particular area often resulted in shifting the problem to another fishing area. Thus, interactions between the governing system and the system-to-be-governed developed in complex and path dependent manners.

Day-to-Day Activities

The quota system meant new rhythms for operators on a day-to-day basis and they were affected by the introduction of new registration and monitoring systems. In order to be allocated a quota, vessel owners had to sign up for a specific fishery for a 2-week period. The quota system, therefore, was an important agent in producing the movements between fishing areas and in setting the seasons. In some cases, fishers could, if they caught their quota within the first week, sign up for a new fishery in the second week. With the possibility of another quota allocation within the same period, fishers often chose to go out despite at times poor weather, jeopardizing their safety and making their work conditions much more difficult, while leaving smaller operators frustrated in the harbor. The result was larger landings, lower prices and higher costs of operation. Operators invested in larger vessels while municipalities and producer organizations argued for larger harbors in order to host the new vessels. Built into the management system and governance principles was an incentive for a high volume fishery.

At the same time, for the management institutions involved in the administrative day-to-day management, the common quota system increased the complexity and insecurity. Feedback from catch registrations had to be integrated into dynamic models, and new quota sizes had to be re-calculated on a weekly basis. On a meta-governance level, the system also gave rise to concern. Evidently the overall economic performance was not improved by the rigorous system and governance principle. As a result the administration as well as politicians had to tackle protests and complaints over quota sizes from discontented fishers and criticisms from economic advisors (Andersen and Andersen 2000; Frost and Løkkegaard 2001; Frost et al. 2005; Tarbensen 2012). Despite licenses and scrapping schemes, there was no real solution to the overcapacity in the sector and its continuous growth. Simultaneously, there were many and often large reductions in quotas from one year to the next, indicating that in the 1990s neither Danish fishers nor fish resources were doing well (Flaaten 2013). Small near-shore fisheries co-existed with large-scale fisheries and mid-sized trawlers within the same regulatory system, which caused conflicts between gear types, vessel sizes as well as between mobile vessels and those only fishing in one region. The governance instruments, even though they were guided by the principle of free and equal access to all operators, also came with incentives for increasing volume and mobility favoring larger operators.

Towards a Market-Based Quota System

The management of the common national quota was a difficult and complex policy issue. The rationing of quotas in shorter periods created a dichotomy between operators on one side and administration and scientific advisors on the other side. In addition, the governing system was poorly equipped for the unpredictability of wild capture fisheries; every change in national quotas came as a shock in the sector, and resulted in new economic pressure. Thus, in the 1980s and 1990s radical alternatives were discussed amongst the producer organizations and within the administration. As part of this discussion, privatization of quotas had been debated, as this would create a more stable division of fishing opportunities. However, a majority of the fishers were against privatization for ideological reasons: the private ownership of a common resource was against the tradition of the share-based organization in fishing operations. Giving vessel owners private ownership over fish resources would create a dividing line between them and their crew, a line that would only grow with time, as new generations of young fishers would enter the sector without ownership of quotas. This was against the dominant *image* of governance at the time, which advocated *free and equal access* between fishers. On a practical level, fishers saw privatization as a loss of flexibility. This was of grave concern since it was often flexibility that assured fishers that they could survive the ever-persistent seasonal and environmental fluctuations of the ocean. In case of a decline in one resource, fishers could shift to another catch area or target another species. With fixed quota shares they feared this would not be as easy anymore. However, due to persistent economic problems in the state-administered system and rising discontent with the growing number of regulations and governance instruments, the idea of privatization slowly won ground among operators and politicians. Inside the governing institutions, market mechanisms were increasingly promoted and presented as the only proper solution to "the fisheries problem", which was, in other words, the political distribution of fishing opportunities (Søndergaard 2008).

Market Based Fisheries Management

From 2003 onwards, transferable individual quotas were introduced with regard to a number of pelagic species. The most notable case was that of the large-scale herring fisheries which introduced individual and transferable quotas in 2003. Based on the herring experience as well as others, private and transferable quotas were finally introduced for demersal species in 2007 through the Vessel Quota Share System. With this change market mechanisms became central in distributing fishing quotas. In a market-based governance system the total allowable catch is split between operators in individual, private and transferable quota shares. Quota shares are, in other words, given to operators as private property. The quota share thus represents the right to harvest a certain percentage of the annual catch of a specific species. The innovative element is that quota shares can be traded and leased between operators which, according to some economic literature, results in fishing opportunities being given to most efficient resource users (Grafton 1995, 1996, 1999; Leal 2002; Mansfield 2004; Arnason 2005; Dupont et al. 2005; Sanchirico and Kroetz 2010). Whether this is correct or not is disputed and subject to ongoing discussions (See for examples McCay 1995; Eythorsson 1996; Helgason and Pálsson 1997, 1998; Arnason 2005; Andresen and Højrup 2008; Andersen and Lauridsen 2010; Einarsson 2011; Olson 2011).

In recent years, and in many governing institutions, as in the case of the EU Commission on fisheries, *governing through markets* has taken centre-stage as an image of governance (Helgason and Pálsson 1997; Hersoug 2005; European Union 2009; Schou 2010; European Commission on Fisheries 2012). Seen from a meta-governance perspective, the distinctive feature of market-based governance is that the complex and dynamic distribution of fishing opportunities is relocated from the state to individuals or companies interacting through a market. Thus, it constitutes a shift from common resource management to a private property regime. It should be noted that market-based management systems presuppose a rather advanced level of environmental knowledge (TACs) in addition to well-implemented monitoring and control systems. Although private ownership of fish resources, or ownership of catch areas has been common throughout history, the introduction of private property rights as a modern principle, image and instrument of governance has a relatively short history. Academically, the potential benefits of private property were the object of discussions in the 1950s and onwards (Gordon 1954; Scott 1955). These academic discussions were responses to concerns over overcapacity and quota distribution problems shaped by the rapid development of ocean capture fisheries in the twentieth century. However, the first actual modern governance instrument based market mechanisms were introduced in the 1980s in countries like Iceland, New Zealand and Canada (see Johnson and Pálsson this volume). In the following sections I will take a closer look at the Vessel Quota Share (VQS) system, as an example of market-based and interactive fisheries governance.

Policy Setup

It is well documented that individual and transferable quotas can lead to concentration of ownership and introduce the risks of "slipper skippers" and "tenant-fishers" (McCay 1995; Helgason and Pálsson 1998; Copes 2004; Olson 2011). Concentration of ownership is partly the intention, as private and transferable quotas are often introduced to promote structural adaptation by reducing overcapacity. But in the governance of small-scale fisheries, concentration of ownership is one of the crucial aspects, since the ownership of resources are most often concentrated in large-scale operations. In Denmark the term "small-scale fisheries" is not used as a category in the management of fisheries. Instead, "near-shore", or more literally correct "coastal fisheries", are used to describe the heterogeneous fleet of smaller operators that mainly fish from one port and most often return home every day.

As in other countries, coastal fishers in Denmark feared transferable quotas, and the responsible Ministry of Food and Agriculture tried to address this concern in the actual policy design. Firstly, this was done through the overall design and secondly via a set of specific regulations on quota ownership and a safeguard measure to protect coastal fisheries. Below, I will first discuss the overall policy design and then turn to the maximum ownership regulation and coastal fisher safeguard.

In the overall design, quota shares were attached to vessels and not to individuals (Ministry of Food and Agriculture 2005, 2006). This, it was believed, would slow down the trade. Quotas would only be for sale when a vessel was for sale, and not broken up in portions and sold back and forth. Moreover, in the initial design, quota allocations were calculated on the basis of a vessel's catch-history over the three preceding years and thus represented the diverse activities of fishing vessels targeting different species in different catch areas. With this approach it was argued that ownership was given to the actual users. In order to buy more quota, operators had to buy a whole vessel with the quota attached to it, and then later scrap the excess vessel. Since vessel overcapacity was perceived as the main problem in the sector, this arrangement would contribute to a decrease in tonnage and in the number of vessels. In the imagined scenario, an initial concentration of vessels would be followed by a more stable situation, where a lower number of vessels would have a more reasonable amount of quota to run their operations.

However, to make the system flexible, a few exceptions to the principles described above were added. These exceptions allowed, for example, owners to sell off minor parts of the quota (<25 %) and detach quotas from a vessel in order to specialize their operations, or get rid of unwanted quota shares from an acquisition (which would reflect another vessel's historic activities). In addition, quota pools and leasing between operators was introduced and enabled further flexibility in day-to-day operations. Quota pools were to be almost self-governing with regard to the balance between quota shares in the quota pool and annual outtake by the members. A number of quota pools were established: some local, others regional and one more or less national in its scope. Inside these quota pools, operators could lease quotas back and forth through online bidding or with the help of the quota manager. In addition to this basic design, an important supplement was the specific policy instruments aimed at limiting concentration and promoting (or rather protecting) coastal fishers from market mechanisms. These will be further discussed in the sections below.

Maximum Limit of Ownership

As mentioned above, one of the main concerns, raised by both the sector and by politicians, was the need to avoid an unwanted concentration of resource ownership in too few hands and to prevent absentee ownership.[2] This was indeed a complex

[2] Legally, a fishing vessel can be owned in Denmark by either a commercial fisher with "A" status, or by a company of which two-thirds is held by commercial fishers with "A" status. A commercial fisher with "A" status is defined as a person who: (1) has Danish citizenship or who has been living in Denmark for at least two years; (2) has been occupied with commercial fishing with "B" status (having an income from commercial fishing) for at least 12 months; and (3) has at least 60 % of his income from commercial fishing.

policy issue as the basic mechanism in a market-based distribution system is the reduction of fishing capacity, by allowing operators to buy each other out of business (Schou 2010). The solution was to set a maximum limit on the number of vessels one person or company could acquire. The limit, agreed in coordination with the producer organization, *Danmarks Fiskeriforening*, was set at four. This meant that only quota shares from four other vessels could be *acquired* and *transferred* to the original vessel. It is important to note the double condition, *acquire and transfer*, since it later proved to be one of several loopholes in the regulation. The maximum limit on ownership only came into being legally just before the promulgation of the legislation in January 2007, and thus it was decided that those acquisitions made prior to that date (and prior to deciding the maximum limit) would not count in relation to the limit. A handful of operators had already been buying vessels from the moment the political decision was made in 2005, and these were accordingly not included in the regulation. With a maximum number set on acquisitions, the system was designed to allow for the desired structural adaptation while at the same time ensuring social objectives with regard to the number of operators, geographical distribution and absentee ownership. However, as I will show later, the "four vessel rule" was weak both in its formulation and actual application. To promote the small-scale sector the maximum ownership regulation was supplemented with the coastal fisher safeguard.

Coastal Fisher Safeguard

The *coastal fishery* in Denmark is generally appreciated for creating jobs and for its use of environmentally sound fishing practices. In addition to this, there are the cultural and educational values that come with it and contribute to tourism and local identity. It is however difficult to define any precise category on national basis that grasps the diverse characteristics of the coastal fisheries, and the qualities described above can be both discussed and disputed. However, when the VQS system was implemented, a specific coastal fisher safeguard was designed in order to protect and promote coastal fisheries. The safeguard measure was based on vessels less than 17 m in length and with 80 % of fishing trips lasting less than 72 h. Vessels that adhered to these principles could choose to sign up for a 3-year period and each year receive extra allocations of cod and sole, two of the most important demersal species. If an operator signed up for the coastal fisher safeguard it would not be possible to sell quota shares to operators outside the scheme. The safeguard measure was made optional, that is, operators were not forced into it. Those who did sign up were only obligated for a 3-year period and after that they could choose to renew the membership or not. It was also decided that the safeguard was to be evaluated by a committee after the first 3 years.

The above sections have introduced the policy design with its specific measures and basic elements. In the following sections, I shall review the development of both quota ownership and policy design after implementation of the VQS system.

Policy in Motion

After the actual implementation of the VQS system in January 2007,[3] quota trading quickly took off between operators and through shipbrokers. The so-called race for quotas made prices rise many times as quotas soon became a main trading commodity, often comprising 70–80 % of a vessel's trading value. Since cod is regarded as the most important demersal fishery, the investments in cod VQS captures the main trends in quota trading as a whole between 2007 and 2011.

The Danish demersal fishery is divided into five catch areas. In all five catch areas the number of operators (having cod VQS) has, as expected, decreased. The area that shows the largest degree of concentration is the Eastern Baltic. In 2011, 10 vessels caught more than 50 % of the cod and 25 vessels accounted for 75 % of the catch. While one company holds 8 % of the quota, the median share is 0.24 %, which indicates a rather unequal distribution between quota holders. In the 2007 initial allocation (based on the 3 year catch history), 50 % of quota shares were distributed among at least 35 operators.[4]

In other words, the quota shares are concentrated in fewer hands and with greater difference between the largest holders and the average holder. The total number of VQS owners in the Eastern Baltic declined from 221 in 2007 to 150 in 2011. Only 109 vessels took part in actual fishing. In general, the largest operators have increased their share, while the overall number of operators has reduced. Regional ownership has also changed and ownership has been concentrated in much fewer harbors. For example, in the Eastern Baltic a significant amount of quota shares are now *owned* by fishers operating from the other end of the country. For these operators the investments in quotas has been used to expand their businesses and fish the whole winter season in the Baltic Sea, which combined with their other VQS has allowed them to reach full annual production.

The numbers indicate a clear concentration of ownership and fishing activities. It is doubtful whether the degree of concentration of activities on these vessels was in line with the intentions of the "four vessel rule". This could be explored through a closer look at the vessel trade. However, more importantly, there has been a growing recognition that the real problem was not the actual concentration, but that the maximum rule was an inadequate instrument to govern the concentration issue. There were several reasons for this. In formulation it was unclear, for example, if each of the four vessels allowed in the maximum rule could, in theory, already consist of quota from four other vessels linked together by the previous owner. Each of these could in turn comprise of quota from four other vessels and so on. This might sound counter intuitive for some, but could technically be arranged rather easily as a way to get around

[3]The following analysis is based on official data from the Ministry of Food, Agriculture and Fisheries (VQS sizes and catch registrations in 2007 and 2011) and the EU fleet register (for gear types in 2007 and 2011). Where the data has been ambiguous (in particular concerning gear types), I have induced corrections based on my personal observations and interviews.

[4]A large portion (16 %) was allocated to so-called "Flying Dutchmen", vessels existing only on paper as licenses and catch history. This could be due to wreckage, but in this case it also represents the early quota trade between 2005 and 2007.

the maximum regulation. An actual and illustrative example of this is a vessel being sold back and forth as part of a quota trade: In the final deal, the vessel held fishing rights for more than four million euro (O'Riordan 2012), equaling at the time, 7 % of the total Danish quota for plaice in the North Sea, plus almost 10 % of plaice in the Skagerrak catch area. The high quota shares are even more remarkable considering that the vessel was made of fiberglass, less than 5 m in length, and had neither an engine nor wheelhouse. Its function was simply to hold and accumulate quota, while the paperwork and financial transactions were dealt with. It is unclear if the vessel ever moved or touched water. Likewise, the exception allowing 25 % of quota shares to be separated and sold from a vessel could be used – through repetitive trades – to split and trade VQS. If an operator, for example, bought a vessel and immediately sold 25 % back to the original owner, the end result would in fact be equal to the first owner selling of 75 %. In other words, through the use of accountants, legal advisors and by the loose interpretation of the legal text the regulation was slowly eroded. Similarly, the double condition in the regulation, defining that only a maximum of four vessels could be *acquired and transferred*, was just as weak. In effect, it meant not a limit on acquisitions but only a limit on how many of the acquisitions could be *transferred permanently* to the original vessel. In other words, a company or person could buy as many vessels as possible, much more than four vessels, and use the leasing system to annually de facto transfer them to the main vessel. This loophole effectively made the "four vessel rule" obsolete (Ministry of Food and Agriculture 2012).

Revised Concentration Rules and Public Debate

The regulation governing concentration did not work. Instead it created a fleet of quota holding vessels, and as an indirect result of this, the quality of data on vessels was weakened. Increasingly, the fleet data represented the quota holding structure more than the active fishing fleet. Quota trading, such as the above, indicates that some operators treated vessels under the VQS system only as holders of investments and movements of capital. As shown above, the "four vessel rule" was weak in its concrete application due to its flawed design, but what can be questioned is why the administration was not able to deliver in terms of the intentions of the regulations. I will try to reflect on this from a governability perspective later in the discussion.

As the failures of the maximum regulation became increasingly clear, concern was raised in the public media, by NGOs and among politicians. In 2012, the Minister of Food and Agriculture responded with a new regulation on maximum ownership (Ministry of Food and Agriculture 2012). The new approach was based on maximum percentages as opposed to the "four vessel rule". Maximum percentages were a more transparent measure to evaluate concentration, as it set a maximum on the amount one vessel and individual person could hold. Thus the new instrument should in theory have improved governability for the governing system. However, with maximum percentages being between 5 and 10 %, the new regulation reflected the flaws of the earlier rules. This is the case because percentages between 5 and 10 means that marine resources could only be harvested by 10–20 vessels.

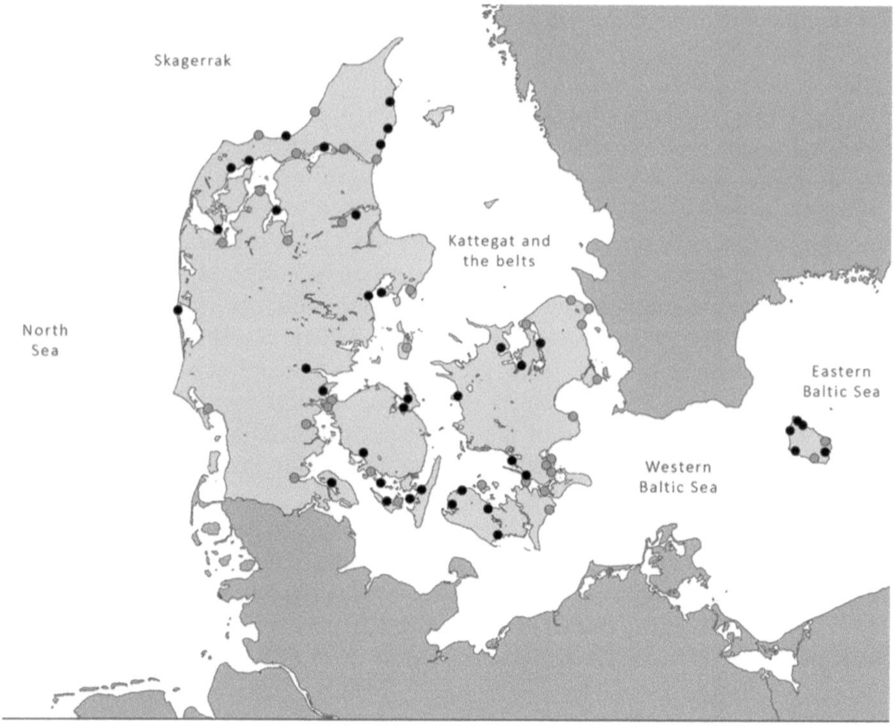

Fig. 17.1 Map showing decline in the number of harbors with commercial fishing vessels between 2005 and 2012. *Black dots* are harbors which no longer have any commercial fishing vessels, but had in 2005. *Grey dots* are harbors with half or less the number of vessels as compared to 2005

Seen from a small-scale fisheries perspective, the 2007 regulation and the inventive ways to work around it favored operators that could comprehend the legal and administrative side of the trade. In effect, this gave a much larger role to accountants and legal advisors working together with fishing companies, banks and investors. On the other hand, self-employed coastal fishers were less eager to engage in risky investments in quota and use of external capital. With the initial weak formulation of the maximum ownership rule the road was paved for a substantial concentration of fishing rights, which was to have a severe and irreversible impact on the coastal fishery (see Fig. 17.1).

Evaluation of the Coastal Fisher Safeguard

As mentioned above a coastal fisher safeguard was introduced as part of the VQS system and was to be evaluated after 3 years. This evaluation was very critical:

> The Coastal Fisheries Committee firmly believes that the coastal fishing scheme [safeguard] was made with the intent to preserve and develop the Danish coastal fisheries, which

> the current scheme does not seem to help. Based on the analysis conducted and based on many conversations with coastal fishermen around the country the Coastal Fisheries Committee assesses that coastal fisheries in Denmark are virtually being phased out. (Living Sea 2012 [2009], 3)

One of several problems identified in the report was the allocation of extra quotas. These extra allocations were distributed relative to the amount of VQS already held by the vessel owners. Technically, this meant that the largest quota holders also received the largest extra shares. In that way there were, according to the evaluation, a number of operators who received large amounts of the supposed coastal fisher quota and a majority who received tiny amounts (Living Sea 2012). Even though these vessels, receiving large extra amounts, were less than 17 m long there was a sentiment, both in the sector and among civil society, that these operators were not proper "coastal fishers". The evaluation painted a rather negative picture of the future for Danish coastal fisheries, which accordingly were "virtually being phased out". An important reason for this negative picture was the combination of individual market behavior and the optional character of the coastal fishery safeguard. Quota shares were simply sold at higher prices outside the coastal fisher safeguard, and the safeguard did little to promote coastal fisheries. This created an incentive for potential sellers to exit the safeguard measure. But it also created the possibility for a "coastal vessel" to buy quota shares inside the safeguard and then after the 3-year period not renew the status as a "coastal" vessel. In this way the "coastal fleet" was not safeguarded in the long run.

The critical evaluation did give rise to some debate, but regardless of the debate the coastal fisher safeguard was anyway prolonged for another 3 years without any substantial changes to it. In the subsequent evaluation, 3 years later in 2012, the committee that was made up of coastal fishers could not agree on a joint statement, and as a consequence a wider board of stakeholders, including civil society, was established to come up with alternatives. This led to some concrete changes to the safeguard through the inclusion of further species, the differentiation between gear types and shortening the definition of fishing trips to 48 h. Whether this is enough to turn the tide of Danish coastal fisheries is doubtful, as the changes did not overturn the optional character of the safeguard. Thus, even though the definition and rewards of the coastal fisher safeguard was improved, operators intending to sell their VQS would still be able to do so to large-scale operators. Ownership and access to marine resources, in other words, will continue to move from small to large-scale operators and be concentrated in the hands of fewer operators. Above, I have outlined and discussed two specific governance instruments that were designed to govern the relation between small and large-scale fisheries. The maximum ownership regulations, as well as the coastal fisher safeguard, illustrate the potential as well as limits of market-based fisheries governance. In the following section, I will focus on the general governance implications of market-based instruments for small-scale fisheries. At the heart of this matter lies an unanswered question, that is, whether and how market-based fisheries governance qualifies as interactive governance?

Discussion and Conclusions

During the last 50 years the Danish state has governed its fisheries using three distinct governance principles. Each of these has produced different governance interactions and relations between the governing system and the system-to-be-governed. Years of liberal growth managed through simple input regulation ended in the 1970s with the introduction of national quotas and TACs. Quotas and TACs in turn opened a new chapter of governance, as the political and economic distribution of the limited fishing opportunities became a core governance problem. A new vertical and state-centric mode of management was instituted, and in this regime scientific advisors and state administrators gained importance while the day-to-day knowledge of fishers was marginalized. The system was an improvement in terms of the biological governability of fish stocks, but was problematic in other ways. The interactions between administrators and operators were conflict-ridden and complicated, and perhaps most importantly, the overall economic performance was unsatisfactory due to persistent overcapacity in the sector. Gradually economic performance became the core governance problem. From 2003 onwards the national quota management was replaced with a governance system of private and transferable quotas. With this shift, the Danish state chose a market-based system to handle the dynamic distribution of fishing opportunities between operators. In this regime the economic distribution of resources between operators were to be self-governing through market mechanisms. In contrast, in the previous governance regimes, distribution was based on the principle of equality and competition between operators. Equality guided the concrete design of governance instruments and interactions, although it became increasingly difficult to administer. The introduction of private quotas and market mechanisms was thus a break from the *state guaranteed free and equal access*, and thus a radical shift in governance approach. This new *image* of governance was influenced by economic performance and less focused on social equality. Based on methodological individualism, the central actor or target of market-governance was not a crew of share-based fishers, but an operator able to invest and organize efficient and large-scale production. Here we find the root of the challenge for small-scale fisheries in market-based fisheries governance. Small-scale operators seek economic independence, but are faced with large investments in quota. In consequence, small-scale operators refrain from buying or rely on leasing of quotas instead of investing permanently. As the Danish case illustrates this process can quickly change the social landscape and composition of the fleet, resulting in large-scale quota owning and specialized operations on the one hand, and a diverse category of quota leasers and small-holders on the other hand (Byskov 2010; Høst 2015). The share based, independent and self-employed fishing practice was in principle linked to the free and equal access of ocean resources (Højrup and Schriewer 2012), and with a change to market-based governance, the economic independence of small-scale operators had to be guaranteed through governance instruments, that is, through safeguards and regulations governing the market. In this respect, the Danish case presented in this chapter, is hardly a best practice example. There are, however, one or two points to be learned from the Danish case and the interactive processes examined in this chapter.

Neither the maximum regulation nor the coastal fisheries safeguard were enough to hinder an undesired economic and social restructuring of the Danish fisheries. The question that remains today is whether this failure was due to poor policy design or, instead, indicates that market-based fisheries governance is close to impossible to frame and control and therefore should not be implemented for small-scale fisheries? In this case the use of markets for governance has had an almost irreversible negative impact on Danish small-scale fisheries. According to Torfing et al. the use of (quasi-)markets in governance "require a relatively tight top-down meta-governance where goals and economic conditions are clearly defined in advance and outcomes are evaluated with reference to pre-given standards [...]" (Torfing 2012, 202). In the Danish case presented above the goals and conditions were not clearly translated from political decisions to concrete policies. The political and societal intentions invoked at the dawn of privatization were never fully carried out in terms of concrete policy formulations. Further, political goals have not been administered and monitored properly nor evaluated soon enough. After the first critical evaluation of the coastal fishery safeguard no actions were taken, and only after 6 years were small changes made to the policy design. These changes were not made in reference to the initial standards and goals, but rather reflected the undesired transformation of the sector. Thus, the democratic foundation of governance has been questionable. This is true not only with regard to policy implementation, but also in relation to broader civil society concerns. On the other side, democratic norms and moral in the fisheries sector itself have not been strong enough to cater for potential loopholes and legal short cuts. It seems reasonable to conclude that in general, policies introducing the market-based governance in fisheries will have to deal with significant pressure and creative efforts of profiteering in the system-to-be-governed. In this case, with loose interpretations and control of the initial policy setup, the ground was set for moral depreciation among operators. This is critical, as the introduction of market mechanisms for the governance of fishing has increased the potential for self-governance. But as the case in this chapter has shown self-governance needs a strong policy framework as well as democratic foundation.

Ultimately, there was a contradiction between societal objectives such as ecosystem health, job creation and spread of value chains and individual maximizing market behavior. It has not been the purpose of this chapter to speculate about why the administration did not deliver a clear enough policy, or why operators did not adhere to a greater moral discipline. It can be asked, however, how the introduction of market mechanisms in the distribution of fishing opportunities has improved governability? Market-based governance limits the involvement of society at large. Thus for society in a broad understanding it decreases governability. It institutes private properties and ignites an irreversible transformation process. While participation and economic governability have increased for some operators, concerns regarding social equity and ecosystem wellbeing remain.

This conflict illustrates the difficulties in using markets in interactive fisheries governance. Quota markets as a governance method shares assumptions and ideological reasoning with interactive governance instruments such as New Public

Management. However, in New Public Management a public authority contracts or auctions public services to private entities. These kinds of quasi-markets differ from the type of markets involved in market-based fisheries management on several accounts (Torfing 2012). First of all, in a quota market, the contracts are not made between state and private entities, but between multiple private entities in an ever-changing quota market. What was a public governance matter, namely the distribution of fishing opportunities, are now taken care of by a new private configuration of fishing companies, banks, investors, legal advisors and accountants. Thus the Danish case illustrates that in this process the initial policy creation is crucial. With private transfers and bank credits involved, it is extremely difficult to change and reverse policy failures. In that respect interactive governance and market-based governance is a contradiction. This is a point worth noting as market-based fisheries governance is promoted worldwide as a quick fix to overcapacity problems. If transferable quota shares should be recommended at all for small-scale fisheries, it should be as part of a "slow" process where policy regulation is fully anchored in the sector.

In conclusion, the Danish case presented in this chapter offers several learning points in terms of the relationship between small-scale fisheries as a *system-to-be-governed* and transferable quota shares as a *governing system*. Seen in the context of interactive governance (Kooiman et al. 2005), the recommendations for market-based governance of small-scale fisheries derived from the analysis in this chapter are as follows. Firstly, governance should be based on strong and clearly defined policies derived from elaborate stakeholder consultations. Secondly, segmentation of the fleet and quota markets should be based on strict and non-optional schemes. Thirdly, proper and potent monitoring and evaluation processes should be designed along with institutions to govern and change the system. Finally, transparent and reliable mechanisms for stakeholder involvement should be introduced so that stakeholders are given proper tools and power with regard to self-governing market mechanisms. The Danish vessel quota share system has none of the above, and hence the future development of Danish coastal fisheries is in a precarious and vulnerable situation, despite being supported by both government and civil society.

References

Andersen, M., & Andersen, K. (2000). *Fiskere om fisk og fiskeri: essensen af 77 samtaler med danske kystfiskere om økologisk fiskeri: afsluttende rapport om en landsdækkende interviewundersøgelse blandt danske kystfiskere*. Grenå: Landsforeningen Levende Hav.

Andersen, K., & Lauridsen, H. (2010). From fishing to speculation. *SAMUDRA, 56*, 21–25.

Andresen, J., & Højrup, T. (2008). The tragedy of enclosure. The battle for maritime resources and life-modes in Europe. *Ethnologia Europaea, 38*(1), 29–41.

Arnason, R. (2005). Property rights in fisheries: Iceland's experience with ITQs. *Reviews in Fish Biology and Fisheries, 15*(3), 243–264. doi:10.1007/s11160-005-5139-6.

Bavinck, M., Ratana, C., Jentoft, S., & Kooiman, J. (2013). *Governability of fisheries and aquaculture: Theory and applications*. Dordrecht/New York: Springer.

Byskov, S. (2010). *Fiskeriet der forsvandt: eksempler fra et dansk fiskerierhverv i opbrud 1990–2008*. Esbjerg: Fiskeri- og Søfartsmuseet.

Copes, P. C. A. (2004). Socioeconomics of individual transferable quotas and community-based fishery management. *Agricultural and Resource Economics Review, 33*, 171–181.

Dupont, D. P., Fox, K. J., Gordon, D. V., & Grafton, R. Q. (2005). Profit and price effects of multi-species individual transferable quotas. *Journal of Agricultural Economics, 56*(1), 31.

Einarsson, N. (2011). *Culture, conflict and crises in the Icelandic fisheries: An anthropological study of people, policy and marine resources in the North Atlantic Arctic*. Uppsala: Acta Universitatis Upsaliensis.

European Commission on Fisheries. (2012). *Transferable fishing concessions*. Brussels: European Union.

European Union. (2009). *Commission green paper on the reform of the common fisheries policy*. Fisheries Commission.

Eythorsson, E. (1996). Coastal communities and ITQ management. *Sociologia Ruralis, 36*(2), 212–223.

Flaaten, O. (2013). Institutional quality and catch performance of fishing nations. *Marine Policy, 38*(2), 267–276. doi:http://dx.doi.org/10.1016/j.marpol.2012.06.002

Frost, H., & Løkkegaard, J. (2001). Individuelle omsættelige kvoter – Kort belysning af vigtige spørgsmål for danske fiskeri. In Fødevareøkonomisk Institut (Ed.), Copenhagen: Statens Jordbrugs- og Fiskeriøkonomiske Institut.

Frost, H., Løkkegaard, J., & Andersen, J. (2005). *Forvaltning af det danske konsumfiskeri*. Kbh.: Fødevareøkonomisk Institut.

Gordon, H. S. (1954). The economic theory of a common-property resource: The fishery. *The Journal of Political Economy, 62*(2), 124–142.

Grafton, R. Q. (1995). Rent capture in a rights-based fishery. *Journal of Environmental Economics and Management, 28*(1), 48–67. doi:http://dx.doi.org/10.1006/jeem.1995.1004

Grafton, R. Q. (1996). Individual transferable quotas: Theory and practice. *Fisheries Reviews in Fish Biology and Fisheries, 6*(1), 5–20.

Grafton, R. Q. (1999). *Private property and economic efficiency: A study of a common-pool resource*. Dunedin: University of Otago.

Helgason, A., & Pálsson, G. (1997). Contested commodities: The moral landscape of modernist regimes. *Journal of the Royal Anthropological Institute: Incorporating "Man", 3*, 451–471.

Helgason, A., & Pálsson, G. (1998). Cash for quotas: Disputes over the legitimacy of an economic model of fishing in Iceland. In D. Miller & J. G. Carrier (Eds.), *Virtualism: A new political economy*. New York: Oxford.

Hersoug, B. (2005). *Closing the commons: Norwegian fisheries from open access to private property*. Delft: Eburon.

Højrup, T., & Schriewer, K. (2012). *European fisheries at a tipping point* (Estudios Europeos, no 1). Murcia: edit.um.

Holm, P. (2002). Sjæk'len, årbog for Fiskeri- og Søfartsmuseet, Saltvandsakvariet i Esbjerg. Sjæk'len, årbog for Fiskeri- og Søfartsmuseet, Saltvandsakvariet i Esbjerg.

Høst, J. (2015). *Market-based fisheries management – Private fish and captains of finance* (Mare series). New York: Springer.

Karlsdóttir, H. M. (2005). *Fishing on common grounds: The consequences of unregulated fisheries of North Sea herring in the postwar period*. Göteborg: Ekonomisk-Historiska Institutionen vid Göteborgs Universitet.

Kooiman, J., Jentoft, S., Pullin, R., & Bavinck, M. (2005). *Fish for life: Interactive governance for fisheries*. Amsterdam: Amsterdam University Press.

Leal, D. R. (2002). *Fencing the fishery a primer on ending the race for fish*. Bozeman: PERC.

Living Sea. (2012). *Evaluation of coastal fishing scheme*, translated from Danish.

Mansfield, B. (2004). Neoliberalism in the oceans: "Rationalization", property rights, and the commons question. *Geoforum, 35*(3), 313–326.

McCay, B. J. (1995). Social and ecological implications of ITQs: An overview. *Ocean and Coastal Management, 28*(1), 3–22.

Ministry of Food and Agriculture. (2005). *New regulation* (pp. 1–5). Copenhagen: Ministry of Food and Agriculture.
Ministry of Food and Agriculture. (2006). *New regulation* – Press release. Copenhagen: Ministry of Food and Agriculture.
Ministry of Food and Agriculture. (2012). *New concentration rules* – Press release. Copenhagen: Ministry of Food and Agriculture.
Nandan, S. N. (1987). The exclusive economic zone: A historical perspective. In *The law and the sea: Essays in memory of Jean Carroz* (Vol. 2012, pp. 171–188). New York: FAO.
O'Riordan, B. (2012). *Transferable fishing concessions: Plastic dinghies, ITQ's and wild speculation*. Retrieved from http://www.inshore-ireland.com/index.php?option=com_content&task=view&id=1010&Itemid=1542012
Olson, J. (2011). Understanding and contextualizing social impacts from the privatization of fisheries: An overview. *Ocean and Coastal Management, 54*(5), 353–363. doi:10.1016/j.ocecoaman.2011.02.002.
Rozwadowski, H. M. (2002). *The sea knows no boundaries: A century of marine science under ICES*. Copenhagen: International Council for the Exploration of the Sea in Association with University of Washington Press, Seattle.
Sanchirico, J. N., & Kroetz, K. (2010). *Economic insights into the costs of design restrictions in ITQ programs*. Resources for the future.
Schou, M. (2010). Sharing the wealth. *SAMUDRA, 55*, 18–23.
Scott, A. (1955). The fishery: The objectives of sole ownership. *The Journal of Political Economy, 63*(2), 116–124.
Søndergaard, M. K. (2004). *Teknologisk udvikling i dansk fiskeri 1945–2000* (Fiskeri- og Søfartsmuseets studieserie; nr. 16). Esbjerg: Fiskeri- og Søfartsmuseet.
Søndergaard, M. K. (2008). *Dansk fiskeri c. 1945–2005: teknologi, udvikling og forestillinger om kontrol*. PhD, Syddansk Universitet, Esbjerg.
Tarbensen, K. (2012). *For alle Danmarks fiskere: Danmarks Fiskeriforening 125 år, 1887–2012*. Fredericia: Danmarks Fiskeriforening.
Torfing, J. (2012). *Interactive governance: Advancing the paradigm*. Oxford/New York: Oxford University Press.
Torfing, J., Peters, B. G., Pierre, J., & Sørensen, E. (2012). *Interactive governance: Advancing the paradigm*. Oxford: Oxford University Press.
Vedsmand, T., Friis, P., & Raakjær, N. J. (1996). *The Danish fishing industry – Structure, policy formulation and control of Danish fisheries*. Roskilde: Roskilde Universitetscenter, Institut for Geografi og Internationale Udviklingsstudier.

Part V
Marine Protected Areas – Securing Space

Introduction

In this section, the focus is on another popular management tool which is spreading around the world. As with transferable quota rights, marine protected areas (MPAs) have potentially negative effects on small-scale fisheries in so far as they exclude people from accessing their fishing grounds. But MPAs may also make a positive contribution to small-scale fisheries provided that tenure rights of small-scale fisheries people are respected and secured, and if they have a say in the way MPAs are designed and operated. People in small-scale fisheries of course have governability concerns with regard to the long term sustainability of their marine ecosystem, but they also face serious threats to food security and livelihoods.

Chapter 18, by Leopoldo Cavaleri Gerhardinger, Fabio de Castro, and Cristiana Simão Seixas, is a case study from southern Brazil where MPAs have been introduced as a way of scaling up small-scale fisheries governance, albeit with mixed results. The authors highlight the challenges that actors are facing. While the MPA has yet to realize its stated goals, it has created a window of opportunity for co-governance reform, which the authors argue is an achievement in itself. They express the view that governability must be seen as a continuous learning process mediated by changes in structures, values, and interactions. In *Chap. 19*, by Merle Sowman, the governability of small-scale fisheries in South Africa and how they are affected by MPAs is the issue of concern. Sowman argues that MPAs have largely disregarded the socio-cultural rights and livelihood needs of people living in small-scale fishing communities within or in the vicinity of MPAs. She identifies the existence of a "mismatch" between the state-centric and natural science-based approaches and the realities that small-scale fishing people face where MPAs have been instituted. She also argues that a more people-centred approach is urgently needed; something that almost 20 years of democracy has not been able to deliver as yet.

Chapter 20, by Alice Joan G. Ferrer, takes us to the Taklong Island National Marine Reserve in the Philippines. She contends that the hierarchical mode of governing, instituted by the Department of Environment and Natural Resources, has overall not been responsive to small-scale fisheries challenges within the MPA. The problem has been lack of interaction, poor information and weak representation of local small-scale fishers, which have led to low governability. This has been a problem from the very beginning of the MPA, namely from "step zero" onwards, which has resulted in the MPA becoming a non-functional system unable to deliver on its sustainability goals. José J. Pascual-Fernández, Inés Chinea-Mederos and Raquel De la Cruz-Modino take us to the Canary Islands of Spain in *Chap. 21,* where several MPAs have also been introduced and others proposed. Their focus is on how MPAs as a social institution rather than a technical measure are interfering with established use rights. In some of the proposed MPAs small-scale fishers are challenged by a much larger and increasingly powerful recreational fishing sector. Thus, the control over governance that small-scale fishers used to have, through the customary fisher organizations, *cofradías*, is increasingly challenged by other stakeholders. Involving other stakeholders in the MPA decision-making process maybe both a challenge and an opportunity for governability, as MPAs promote new patterns of interaction between groups, according to the authors. Collaboration between stakeholders will not be easy given the absence of a history of collaboration across sectoral boundaries. For collaboration to improve, new institutional formations that enable more constructive interaction are called for.

Estelle Jones, Heidi Schuttenberg, Tim Gray and Selina Stead report on a number of MPA case studies in Thailand in *Chap. 22*. The loss of coastal-marine biodiversity and the over-exploitation of coastal fisheries, due partly to the destruction of mangrove forests, have called for governance initiatives that in some cases have been successful while in others less so. What seems to matter is the degree to which the implementation of MPAs has been negotiated in a way that has yielded legitimacy, trust and cooperation in governance interactions, thus enabling effective conflict resolution and the imposition of sanctions for non-compliance, resulting in improved governability. Although hierarchical governing modes should not be ruled out, the authors hold that effective coastal governance may be more likely by supporting the development of a patchwork of smaller community-based self-governance or co-governance initiatives.

Chapter 18
Scaling-up Small-Scale Fisheries Governability Through Marine Protected Areas in Southern Brazil

Leopoldo Cavaleri Gerhardinger, Fabio de Castro, and Cristiana Simão Seixas

Abstract This chapter investigates governing interactions at the Baleia Franca Environmental Protection Area (Santa Catarina state, South Brazil) as an example of new opportunities and challenges to scale-up small-scale fisheries governability through Marine Protected Areas (MPAs). Previous studies on MPAs in Brazil highlight the innovative aspects of these governing systems such as their well-functioning, active, and progressive management councils. We describe the increasing response of the governing system to fisheries issues that are largely aligned with governance paradigms of collaboration and social learning. Despite all efforts and some notable accomplishments in responsiveness and performance, we point out the challenges related to the mismatch between the governing system and the systems-to-be-governed that hinders fishers' political agency and limits small-scale fisheries governability at broader territorial levels. We identify and analyse the wicked problems faced by actors engaged in processes of transformation in coastal-marine governance and provide suggestions for improving governability.

Keywords Coastal Governance • Participation • Leadership • Brazil • Conservation Unit • Innovative Institutional Arrangements • Fisheries Management

Introduction

Since the 1960s, fishers and scientists have witnessed an accelerating fisheries crisis in Brazil. The crisis has been characterized by a rapid erosion of the diversity of ecological knowledge, practices, and identities amongst small-scale fishers, known

L.C. Gerhardinger (✉) • C.S. Seixas
Environmental Studies and Research Centre, University of Campinas, Campinas, Brazil
e-mail: leocavaleri@gmail.com; csseixas@unicamp.br

F. de Castro
Centre for Latin American Research and Documentation,
University of Amsterdam, Amsterdam, The Netherlands
e-mail: f.decastro@cedla.nl

S. Jentoft, R. Chuenpagdee (eds.), *Interactive Governance for Small-Scale Fisheries*, MARE Publication Series 13, DOI 10.1007/978-3-319-17034-3_18

in Brazil as artisanal fishers (Diegues 1983; Dias Neto and Marrul Filho 2003; Vasconcelos et al. 2007; Rebouças et al. 2006; Gerhardinger et al. 2009) as well as a rapid decrease in abundance, richness and diversity of marine communities and species, particularly those of interest to fisheries (Castello 2010). However, recent political and institutional changes in Brazil have opened opportunities for new interactions in order to address this crisis. Decentralization of some decision-making processes, participatory mechanisms, and the creation of new Marine Protected Areas (MPAs) are ingredients that have redefined the 'rules of the game' in which the governability of small-scale fisheries are expected to improve (*e.g.*, Cordell 2006). Fisheries governance through MPAs is frequently proclaimed as an important strategy in Brazil (MMA 2013). This chapter will investigate the changes, obstacles, and opportunities associated with this governance scheme in the governability of small-scale fisheries, using a case study of a large-scale governing system named '*Baleia Franca* [Southern Right Whale] *Environmental Protection Area*' (BF-EPA).

The BF-EPA is part of a national protected areas governance system called National System of Conservation Units (SNUC, its Portuguese acronym), under the aegis of the Ministry of Environment. SNUC encompasses 12 categories of protected areas divided into two main groups – sustainable use areas that allow consumptive use (*e.g.*, human residency, customary activities, managed extractive activities) and full protection sites allowing only non-consumptive use (*e.g.*, research, tourism). The institutional architecture and participatory mechanisms between both groups are very different. The former is based on more socially inclusive governance and geared towards reconciling economic and environmental goals. The latter is based on more top-down governance and restrictive of human intervention. EPAs are sustainable use protected areas, and are usually extensive areas including both public and private land, and crossing territorial jurisdictions and governmental institutions at federal, state, and municipal levels. Their social (and environmental) complexity creates major governance challenges.

In a recent study on MPAs in Brazil, Gerhardinger et al. (2009, 2011) noted that the BF-EPA had a particular governing approach when dealing with institutional, bureaucratic, and financial challenges shared by other MPAs in the country. This approach included active engagement of BF-EPA staff in partnerships with local actors. These local constituencies were mandated high levels of decision-making power and autonomy in the process of establishing the BF-EPA Management Council (hereafter BF-EPA Council) in 2005. This council is often treated in the literature as an innovative institutional space for integrating policies and actors in social learning through collaborative governing modes (Macedo 2008; Macedo et al. 2013). Further official recognition of this innovation came in 2012, when the BF-EPA was designated by the Ministry of Environment as a pilot-project to support the reformulation of national guidelines for elaboration of management plans of protected areas. If well developed, this process can potentially influence governability of all protected areas in the country. However, though expectations for innovation within this particular governing system are high, both in academic and policy terms, fisheries governability remains arguably poor at the EPA territorial level.

The self-proclaimed innovative nature of such interactions, as will be shown, offers us an intriguing case through which to analyse the challenges and opportunities of dealing with the social-ecological fisheries crisis. The implementation of the BF-EPA generated expectations for increased governability through augmented performance of the governing system. It also raised expectation over its capabilities to respond to fisheries problems and enable opportunities for sustainable territorial development of the sector. The improvement in governability of the (fisheries-related) social system was also explicitly desired, *e.g.* through increased responsiveness of self-governing modes as an outcome of participative and social learning incentives offered to fishing actors (Macedo et al. 2013). Ultimately, the BF-EPA Council was expected to alleviate the mismatch between institutional and ecological systems by scaling-up fisheries governability through the operation of a problem-solving platform dealing with issues at EPA territorial-level.

This chapter will first provide a general description of the fisheries natural and social systems and respective governing system. This will be followed by a description and analysis of the main fisheries governability issues emerging at the BF-EPA in the past decade. Finally, we discuss the major territorial-level governability challenges and distil the insights and lessons offered by this case study. We focus our analysis on the conduciveness of governing interactions at the interface between the social and the governing system.

The description and analysis of governing interactions is based on in-depth semi-structured interviews and participant observation in the central-southern coastal area of Santa Catarina state in 2007–2008 (Gerhardinger et al. 2009, 2011; Macedo et al. 2013) and 2011–2012. In the later period, semi-structured interviews were carried out with eight key individuals, identified through peer recommendations of BF-EPA Council members (the identity of the interviewees was protected due to pre-interview shared agreement), belonging to the BF-EPA governance system. Three non-structured interviews were also conducted to include complimentary perspectives from different sectors (*e.g.*, State agents, resource users (including fishers), academics and environmentalists). Furthermore, observational data were recorded in nine BF-EPA Council meetings and numerous informal encounters. We have also thoroughly analysed official minutes of 30 BF-EPA Council meetings from 2005 to 2012. Governability assessment follows the step-wise approach synthesized by Chuenpagdee and Jentoft (2013). The interactive governance approach is used in the description and analysis of the system-to-be-governed (natural and social systems), the governing system (BF-EPA Council), and governing interactions regarding fisheries issues.

Systems-to-Be-Governed

BF-EPA encompasses a high diversity of coastal-marine ecosystems. It lies in a regional transition zone and includes several ecosystems such as bays, estuaries, sandy beaches, mangroves, sandy dunes, rocky shores, salt marshes and coastal lagoons and lakes. Biological productivity in this region is generally higher than in other tropical Brazilian coastal ecosystems, partly due to small resurgence currents in some locations. There are strong seasonal variations under the influence of sub-Antarctic Atlantic waters, continental freshwater discharge (winter), and predominance of subtropical waters of the Brazilian current (summer). As a result, the area is characterized as an ecotone with particularly high biodiversity due to the presence of both tropical and temperate marine communities (Floeter et al. 2007). A recent marine ichthyologic richness study reported 203 species of marine fish in an area immediately northwards of the BF-EPA border (Bertoncini et al. in prep.).

Small-scale fisheries at BF-EPA dates back to the Portuguese period in the eighteenth-century, when fishers and farmers from the Azores and Madeira islands migrated to the south of Brazil (Lago 1961). Small-scale fisheries and small-scale agriculture predominated until the 1960s, when national development policy encouraged rapid economic transformation (Borges 2008). This resulted in urban expansion, demographic growth, market integration, expansion of tourism and industrial development (Diegues 1983; Câmara 2001; Polette and Vieira 2009).

The fishery system was also transformed from the 1960s as a result of policies prioritizing industrial fisheries (Diegues 1983; Capellesso and Cazella 2011; Oliveira and Silva 2012). As a result, small-scale fisheries have declined and the local economy has shifted to services such as tourism, ports, and other sectors (Diegues 1999; Filardi 2007; Vasconcelos et al. 2007). In addition, several families rely on pensions and unemployment benefits, particularly during seasonal fishing bans on certain species (Capellesso and Cazella 2011).

The fishery system in the region encompasses a number of different fisheries ranging from small to large-scale industrial fisheries. Industrial fisheries are predominantly based on bottom trawling, seining, long-line, rods with live baits (tuna), and passive gears such as bottom or surface-set gillnets and traps. Small-scale fisheries combine several types of small vessels (*e.g.,* small purse-seining and undecked boats, aluminium speedboats, canoes, and closed cabin boats), and a large range of fishing gears. Gomes (2012) has identified 22 fishing gears used at sea and in lagoons, such as gillnets (used for seining or passive fishing), nets used for bottom trawling, cast nets, hand-lines, rods, and long-lines. The diversity of the fishing systems is reflected in the catch composition. Gomes (2012) has identified 62 folk fish species belonging to 37 scientific species captured in the BF-EPA territory. In coastal lagoons, summer shrimps are the main targeted resource, followed by crabs and finfish such as mullets (Seixas 2002; Filardi 2007).

The small-scale fishery systems of BF-EPA are fundamentally dynamic due to their coupling to the natural system. Although they occur throughout the year, activities are amplified during the winter, following the dynamics of main fishing migratory resources (*e.g.*, mullet *Mugil liza* between May–July and bluefish *Pomatomus saltatrix* after July).

Fish is sold in local markets or to related industries, frequently through middlemen (Filardi 2007). Conflicts between small-scale and industrial fisheries are diffuse and chronic, although fishing actors move between industrial and small-scale fisheries – *i.e.*, small-scale fishers sometimes become crew members in industrial fisheries (Filardi 2007; Saraiva 2010; Oliveira and Silva 2012). Conflicts amongst small-scale fishers are also common due to the impact of different fishing gears or due to disputes over certain fish resources (Rodrigues 2011). Some of the main problems mentioned by small-scale fishers are lack of enforcement, corruption, institutional misfit and other public policy flaws (Filardi 2007; Rodrigues 2011), as well as conflicts with other coastal marine actors.

Governing System

Over the past decade, several authors have outlined promising ongoing incipient coastal governance initiatives in the central-south coast of Santa Catarina state. Seixas and Berkes (2003), for instance, describe historical changes and tensions in governance modes in one of many coastal lagoons in the region (Ibiraquera lagoon). Rebouças et al. (2006) propose actions for participatory and integrated management of small-scale fisheries at a broader territorial level. These emerging initiatives were seen as part of a relatively new *territorial ecologization dynamics* (*sensus* Rebouças et al. 2006) that valued cultural patrimony while seeking opportunities for economic and political inclusion of traditional small-scale fishing and agricultural communities (*e.g.* Rodrigues 2011). These interaction patterns were part of the process of designing and implementing new governing instruments for *sustainable territorial development* (Cerdan et al. 2011) within or surrounding the boundaries of the broader governance structure of the BF-EPA.

Designated through Federal Decree IBAMA N°14 in 2000, the BF-EPA encompasses 1,561 km^2 along 130 km of coastline (Fig. 18.1). This region supports nearly 800,000 people in nine municipalities in the south-central coast of Santa Catarina state (South Brazil). The statutory objective for governance of the BF-EPA is framed around the protection of the Southern Right Whale (*Eubalaena australis*), as stated in the regulation and planning of territorial occupation and use of the regions' coastal and marine ecosystems.[1]

The BF-EPA is under the jurisdiction of different governmental agencies. The Ministry of Fisheries and Aquaculture, created in 2009, is generally in charge of fisheries and aquaculture issues. However, the mandate to oversee fisheries issues within SNUC protected areas has been disputed and since 2011 the responsibility has been assigned to the Ministry of Environment. Finally, municipal and state level agencies may also engage in fisheries issues.

Two governing instruments are crucial for the implementation of protected areas in Brazil – Management Councils and Management Plans. The BF-EPA Management Council was created between 2004 and 2006 through a bottom-up process of

[1] "...[to] *protect, in Brazilian waters, the Southern Right Whale (Eubalaena australis), organize and guarantee the rational use of regional natural resources, organize the occupation and use of water and land, organize recreational and touristic use, activities of research and the traffic of boats and airplanes.*" (Federal Decree IBAMA, 2000 N° 14, Art.1; our translation).

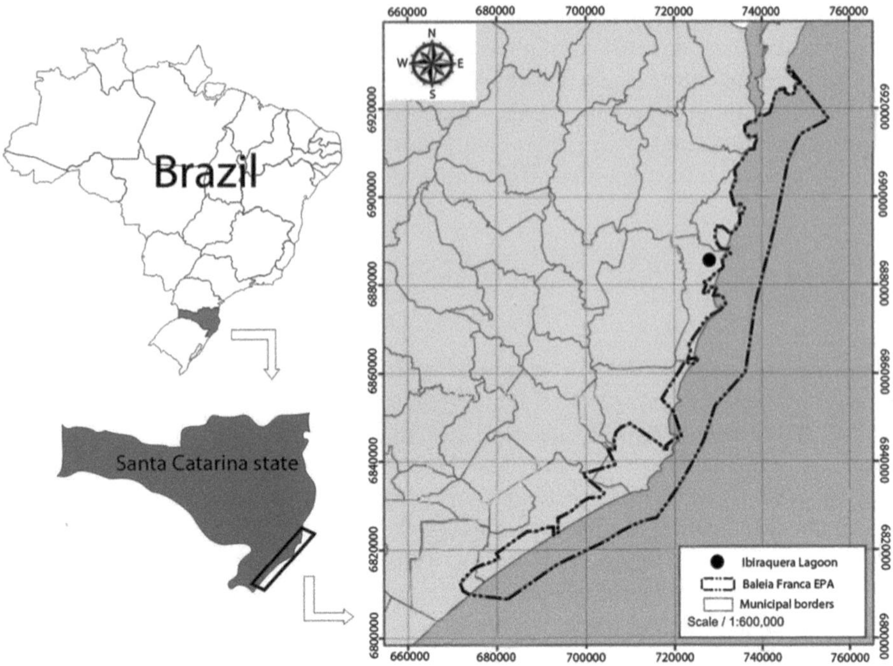

Fig. 18.1 Baleia Franca Environmental Protection Area (After Macedo et al. 2013)

multi-stakeholder institutional interaction (Fig. 18.2). The statutory roles of the Management Council are: (a) to stimulate the participation of different actors in the elaboration, implementation and review of management plans; (b) to facilitate the multiple uses of the area; and (c) to formulate guidelines for actions to integrate, optimize and make compatible the livelihoods of local populations with the site's conservation objectives. The Management Council is composed of 42 elected members (see Gerhardinger 2014), equitably distributed across three social groups – public organizations, resource users (*e.g.*, small-scale fishers, tourists, mining companies, port services providers), and environmental organizations. The Council is also supported by Working Groups organized around topical governing issues, five Technical Chambers, and an Executive Committee, whose role is to facilitate meetings and serve as the secretariat (Fig. 18.2).

In 2012, most of the 42 Management Council members, despite differences over specific issues (*e.g.*, mining development *vs* lagoon conservation) were aware of the need to work collectively, and therefore took the opportunity to pro-actively reshape undesirable structures in the governing system. For example, SNUC requires a Management Council for EPA but does not specify its role (consultative or deliberative). Although most government managers consider their role as consultative, the

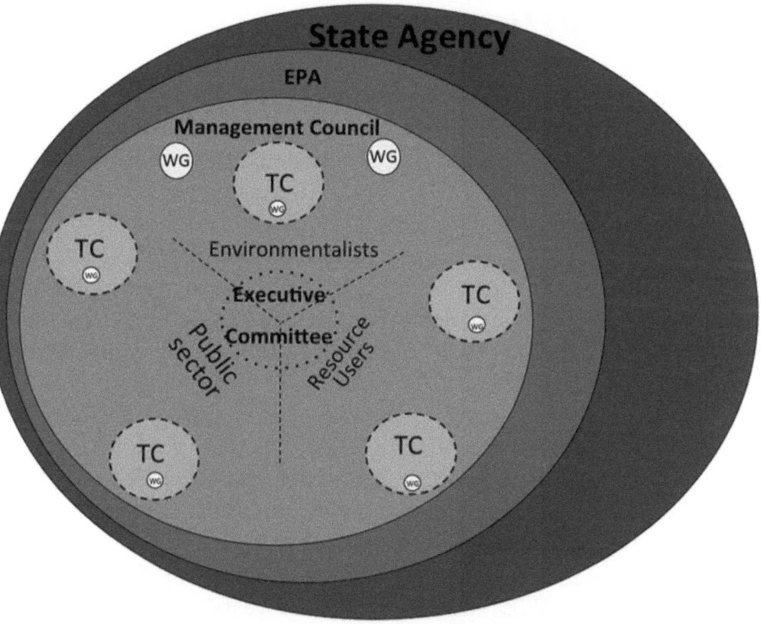

Fig. 18.2 Institutional architecture of the Baleia Franca Environmental Protection Area (EPA) Management Council, including its respective socio-political support entities (*e.g.*, *TC* technical chamber, *WG* working group). There are currently (as of 2010) five Technical Chambers operating under the following themes: Biodiversity Management; Territorial Management; Protection and Monitoring; Sustainable Economic Activities; Southern-right Whale Conservation

Management Council continuously sought to establish a deliberative role by operating according to a self-designed institutional structure and decision-making procedures.

The Council has also been involved in promoting a bottom-up management plan (see Macedo et al. 2013; NEXUCs' 2012). Such a plan was led by key people from communities, universities, and NGOs and was largely enabled by the head of the BF-EPA. The approach was locally referred to as a '*transgressive approach*'[2] and was recognized by the Protected Areas Federal Agency (ICMBio for its Portuguese acronym) as a *pilot project* to inspire possible reforms in the federal guidelines. Considering that ICMBio is currently responsible for approximately 10 % of Brazilian territory, the potential agency of BF-EPA Council members in transforming the governing system has been (and remains) high.

The following section will explore governability through a description and analysis of the main small-scale fisheries issues related to the responsiveness of coastal and marine governing systems particularly how they related to the operation of the BF-EPA Council between 2005 and 2012.

[2] For a detailed analysis of the '*transgressive approach*' see Gerhardinger (2014).

Scaling Up Fisheries Governability

Over the last decade, societal response to fisheries problems in the south-central coast of the State of Santa Catarina has been primarily mediated through governing arrangements for the implementation of the BF-EPA. However, in its initial phase (2000–2003), the capacity of the BF-EPA to govern small-scale fisheries was limited because the structure to foster fisheries governance was not yet in place.

This initial phase was characterized by a top-down bureaucratic process that resulted in the creation of a 'paper park'. Upon the arrival of a new park manager in 2003, the process became more inclusive. This process coincided with political changes at the national level with the election of a left-leaning national government (Hochstetler 2008). Grounded in a progressive discourse of social justice, several activists were invited to support governmental agencies and develop collaborative initiatives with civil society organizations (Hochstetler and Keck 2007). Leadership has played a major role in bridging the gap between bureaucratic and informal systems, creating space for identification and mobilization of constituencies to establish legitimate participatory mechanisms. As described above, the creation of the BF-EPA Management Council is a case in point. It has become quite responsive to a plethora of fisheries issues since 2005. Moreover, over the years the Council has also supported in various ways (politically and/or technically) the self-organization of complimentary fisheries-related governing systems in the central zone of its borders.

We first focus on the claim made for and the negotiation process involved in getting a Protected Area constituted for the aquatic system of the Ibiraquera Lagoon (Fig. 18.1) – adjacent to BF-EPA. Although the final decision has not yet been made, the Council has been actively supportive of a *Marine Extractive Reserve* claimed by the local stakeholders (Vivacqua 2012). Second, the Council played an important role in providing a platform for discussions about the seasonal opening of the sandbar between the sea and the Ibiraquera lagoon. Customary practices regulating the seasonal opening of the lagoon mouth have become a source of conflict between local fishers and other users (*e.g.*, tourism, water sports) in the last few decades (Seixas and Berkes 2003; Berkes and Seixas 2005). Fishers and tourists disagree on the criteria to be used for opening the lagoon mouth. In 2010 the *Ibiraquera Lagoon Mouth Opening Management Committee* was created in order to coordinate public and private interests' vis-à-vis the seasonal opening. Discussions were held about the conflict and possible alternative ways forward within the BF-EPA Council. The Fisheries Technical Chamber (FTC) in particular played an important role. Nowadays, decisions of this new committee are based on a set of agreed principles and criteria for problem-solving. Local knowledge is obtained from three local experts (skilled fishers). Once direct intervention (removal by trucks) of the sand barrier is needed to re-establish water inflow into the lagoon, our informants claim that a more conventional governing approach would require a bureaucratic and costly environmental licensing process. Several BF-EPA Management Council actors were directly engaged or supportive of this largely novel governing mechanism in Brazil.

Between 2008 and 2012, three initiatives/projects relevant to creating opportunities for small-scale fisheries under the label of '*territorial development*' were implemented with an interface with the BF-EPA: *Sustainable Territorial Development*

Project (2008 onwards); *Territorial Laboratory* (2009–2010); and *Southern Santa Catarina Territory* (2009 onwards). The two former projects were led by universities and the Santa Catarina State Rural Development Agency. The latter focused on aquaculture opportunities in coastal lagoons in the South of the BF-EPA, and was part of the national *Fisheries and Aquaculture Territorial Development Policy* of the Ministry of Fisheries and Aquaculture. These projects were aimed at fostering economic incentives that are sensitive to endogenous characteristics of institutionalized units of territory lying fully or partly within the borders of BF-EPA. However, according to our informants, although many Council members have taken part in these different projects and the need for integration amongst them repeatedly stressed, the capacity of the BF-EPA Management Council has been limited.

In 2009, the BF-EPA Management Council representatives supported an ambitious small-scale fisheries monitoring initiative to upscale fisheries governance along the coast of Santa Catarina state. The program was initiated by the Rural Development Agency of the State of Santa Catarina (an active representative of BF-EPA Management Council) with technical support from UNIVALI (Universidade do Vale do Itajaí) and financial support from the Ministry of Fisheries and Aquaculture. The main goal was to implement a monitoring program amongst 237 fishing communities comprising 1,500 fishers in 33 coastal cities. The proposal aimed to engage fishers in participatory monitoring of fish harvests, oceanographic parameters, structure and dynamics of a diverse fishing fleet and gears throughout the coastal seascape. However, the project was discontinued in 2011 due to a number of reasons. Firstly, the regional agency CEPSUL (Traducao do CEPSUL esta estranha Cheque no Google se ha exemplos. Minha sugestao seria: CEPSUL (Center for Marine Biodiversity Research and Conservation of the Southern Region) (Southern Region Marine Biodiversity Research and Conservation Center of the Ministry of Environment) and managers of all the MPAs along the coast of Santa Catarina State had limited involvement. Second, fisher organizations avoided the project, as they were not pleased by it. Third, local partnerships and funding were disrupted (Foppa et al. 2011).

Another initiative worth mentioning was the creation of the Fisheries Technical Chamber (FTC) in 2007. In contrast to the former institutional arrangement of issue-specific Working Groups, the FTC mandated on all local and regional-level challenges in small-scale fisheries under the BF-EPA Council umbrella. The creation of this forum was key to linking fishers' grassroots organizations with state bureaucracies. Between 2007 and 2010, the FTC had the direct support of an external consultant/facilitator to co-design and implement a FTC-Action Plan (Rodrigues 2011). The consultant's hybrid position as an autonomous United Nations Environmental Programme consultant and as representative of the BF-EPA staff led to increased participation of fishers in the Council. The FTC-Action Plan included the implementation of capacity building courses on fisheries management, an agenda for the elaboration of a local fishing management instrument called *Fishing Accords*[3] in order to tackle fisheries conflicts, and an agenda for the elaboration of a

[3] Fishing accords are defined in Brazil as '*...the body of specific measures derived from consensual treaties amongst diverse fishing resource users and management agency in a geographically defined area*' (IBAMA IN Nº 29/2002).

Participative Fisheries Management Plan for the BF-EPA. In early 2010, the FTC had been effective in proposing new regulations for the mullet (*Mugil liza*) fishery, a socioeconomically relevant activity in the region, and in mediating conflicts between small-scale fishers and recreational spear-fishers. Despite the fact that the governance system had become more responsive, the FTC Action Plan was poorly implemented. In fact, by 2010 all co-designed Action Plans of the BF-EPA Council had failed because it went beyond the Council's capabilities and/or mandate. The FTC was finally discontinued during a Council regimental reform in late 2010, and fisheries issues were transferred to a new Biodiversity Management Technical Chamber (Fig. 18.2).

The end of the FTC, the discontinuation of the consultancy work to foster the participation of fishers in the Council, and emerging fishing conflicts drove the decline of the fisheries governability in this period (Macedo et al. 2013).

The situation improved in 2011 when fisheries issues assumed importance again and the BF EPA actively collaborated with fishers who were now led by State authorities. Despite tensions between BF-EPA authorities and some fishers from the southern part of the territory, a series of capacity building and assessment workshops were carried out and a preliminary collaborative working agenda was proposed for co-designing a *Participative Fisheries Management Plan* in 2012. This plan, however, was put on hold as it would become a chapter of the EPA Management Plan which was still under design.

Fishers' participation remained low at the Council until mid-2012, when the '*Artisanal Fishers Movement of Santa Catarina Coast*' emerged in the region. A former Council member was particularly influential in this process and led the movement, together with other small-scale fishers. This grassroots movement, frustrated by the severe depletion of the fisheries and transformations in small-scale fisher identities, organized around the ban of industrial fisheries inside the BF-EPA. Although an initiative of approximately only 30 fishers, this bottom-up mobilization has potential in terms of new opportunities for interactions between the social and the governing systems. Fishers' participation in the Council, however, as suggested by one informant, remains one of the main challenges of this governing system:

> *What have we done wrong? We have invested so much in fisheries education* [pt: formação; capacity building]. *But where are the fishers?"* (BF-EPA Management Council member)

Trimble et al. (2014) investigated the reasons behind fisher's non-participation in meetings with government staff, including marine protected area managers, in the southeaster coast of Brazil. They concluded that (i) the timing of the meetings were often not nor were fishers properly invited to the meetings; (ii) the meetings were carried out by government staff and were often biased *i.e.*, not respecting different sources of knowledge or fostering consensus building; (iii) there was a lack of transparency and (iv) no clear objectives, procedures and intended outcomes of meetings contributed to fisher non-participation in such decision-making arenas. Next we will explore the patterns found at BF-EPA, some of which coincide with those listed above.

Fishers' Interaction with the Governing System

Accounting for the challenges and opportunities for fishers' to participate in governance is key to improving governability. However, participation is a costly activity, as it requires time and motivation, and needs to be prioritized. Therefore, in order to enable the participation of fishers, a good match between the governing system and the social and natural systems is required. In the case of the BF-EPA, we identified several mismatches between these systems that are related to the level and quality of participation of small-scale fishers in the Council. We will explore in particular the structural mismatches between the fishery system and the governing system, with a particular focus on contrasting or alternative governing images in interactions between fishers and other actors.

The dynamics of the fishery system and the natural system are closely connected. For example, during the mullet (*Mugil liza*) fisheries (May–July) and the bluefish (*Pomatomus saltatrix)* and drummers (*Micropogonias furnieri*) fisheries thereafter, participation of fishers in governing interactions can be challenging. Likewise, during the summer (Dec–Feb), many fishers work in tourism. Therefore, despite the innovations observed in the Council, meetings and workshops across the year should account for patterns in the dynamics of natural systems (e.g., seasonal migration/availability of main fish resources). However, this governing system often follows bureaucratic and political schedules, which usually falls to a low priority in the fishers' schedule. Alternatively, fishers rely on non-fishers to represent them *(e.g.*, environmental institutions often speak for fishers in Council meetings) or fishers' representatives that do not engage in fishing activities themselves and/or do not properly articulate the interest of small-scale fishers. Interestingly, some of the Council members blame the lack of fishers' participation on their '*passiveness*', or because they are subject to '*paternalism*', or '*clientelism*',[4] rooted in historical social relations in the region, as illustrated by the quotations below:

> *In fisheries I believe there is such an apathy... it is historical I think to be dependent on government, on the Fishers' Unions [Colonia de Pesca in Portuguese]. These Unions operate a lot with such a political exchange of favours, with the* defesos [compensation during fishing bans] *more the Fishers' Union. They live upon that pattern; the more people become dependent, the more the Colony receives. Archaic politics but continues to be valid...* (Council member)

> *...people living along the coast in the littoral are very dependent on the cycles of nature... 'There is fish, great we have fish! There is no fish, it is because of God's will'... and thus you live as you can and wait for things to get better. On the one hand it is interesting to learn from these traditional communities, the recovery of the sacred, improved connection to nature and understanding of natural cycles. But there is this apathy. These are not entrepreneurial communities.* (Council member)

[4] Social relations between "patrons" (rich, powerful and influential elites) and "clients" (poorest and powerless) in which the former provides jobs, protection, infrastructure, and other benefits in exchange of votes and various forms of loyalty (Johnson 2010; Basurto et al. 2013).

The boundaries among the diverse fishing modalities are source of another mismatch that has direct consequences for the responsiveness of the governing systems. Fishers' social system is intrinsically diverse. They continuously move through a spectrum of modalities between independent small-scale fisher and hired job in industrial fishing boats (Filardi 2007). This mobility is asymmetrical, however, as they shift from an autonomous production system to an economically dependent employment system. Nonetheless, fishers often make use of this divide strategically. On the one hand, their role permeates their position between that of autonomous small-scale fishers and employed industrial fishers; on the other hand, they emphasize their distinctive position as small-scale fishers in order to position themselves and claim their rights in particular context of internal and external conflicts. Oliveira and Silva (2012) argue that the '*cristalization*' of these two fishing categories in the bureaucratic system is reflected in recent efforts to build an objective juridical-political language for fisheries management in the country.[5]

A third mismatch is related to images of the system-to-be-governed across stakeholders. EPAs are the most heterogeneous category of protected areas in Brazil, comprising a diverse range of stakeholders that are generally characterized by highly asymmetric power relations. Therefore, the way the socio-environmental challenges are problematized and how solutions are proposed will depend on how governing images are shaped and how they interplay in governing interactions. In highly asymmetrical power structures, some governing images tend to prevail over others. The fact that the BF-EPA is named after a flagship species indicates that the governing image for this territory emphasises a very particular set of interactions with the natural system, in contrast to the complexity of the 'statutorily-defined' system-to-be-governed.

This image not only influences the representation of the BF-EPA among local fishers as a territory of the whale but also influences how fishers perceive of themselves in this governing system. Our analysis, as well as those of Bueloni (2012), Gomes (2012), and Palhares (2013), has indicated severe communicative obstacles in the application of the image of the Southern-Right Whale as a denominator for a new territory expected to be co-designed. For instance, Gomes (2012) points out that fishers often relate the BF-EPA to the whale itself or to local NGOs, as explained by one Council member:

> *We perceived a confusion. When they* [fishers] *complained about the 'APA'* [EPA – Environmental Protection Area], *it was more about the Southern-Right Whale Project and about the Southern-Right Whale Institute* [marine conservation Non-Governmental Organizations], *because the actions carried out by these institutions were more intensive on the beach, particularly with the fishers.* (Council member)

The BF-EPA mandate focused on a single species makes the Southern-Right Whale a key 'agent' in the mainstream governing image, a trend observed in other parts of Latin America as well (Few and Tortorici 2013). When discussing and proposing fisheries regulations, the BF-EPA Council has engaged in statutory statements that presuppose the customary rights of traditional and small-scale fishers, evoking livelihood security and autonomy. Thus, although the problematization of

[5] Brazilian Fisheries Code – Law N° 11.959, June/2009.

socio-environmental challenges at BF-EPA Council meetings goes beyond human-whale interactions, the governing image of whale protection conflicts with fishers' images of small-scale fishing protection.

Therefore, governability is hampered because the governing system misplaces fishers' political ecology and agency patterns. The current governing system presupposes the unnecessary and costly need to adapt local ecological knowledge. This is not simply requiring too much of the fishery social system, but perhaps is also inherently counterproductive. Ultimately, the governing system is about governing humans-in-ecosystems at a defined terrestrial-marine borderline. In other words, the BF-EPAs' territorial governance mandate is much broader than the whale-focused image communicated to all stakeholders through the EPA name. We therefore suggest that when communicating about the BF-EPA, all actors would benefit from the usage of alternative/complimentary images synthesizing broader biogeography/ecosystems – even with absolutely no formal changes in statutory governance mandate.

Finally, the way solutions and opportunities are explored among local fishers and other actors represent another mismatch in the governing system. Formalized and institutionalized interactions carried out under rigid bureaucratic structures hinder fishers' participation, not only because of their limited experience in this realm, but also due to their subordinated position towards other groups. This structural problem is often overlooked or interpreted as being a result of fishers' limited knowledge of formal institutions, as suggested by one of our informants:

> *The fisherman understands very little about institutions* [pt: institucionalidades], *they seldom understand their own... that the Fishers' Union, the association or the assembly/guild* (key-person of BF-EPA)

It is interesting to see the contradiction emerge from participatory procedures implemented under different images of governing systems. On the one hand, fishers are continuously called upon for improved citizenship and participation through engagement in social learning processes. On the other hand, they are simultaneously pressed to 'learn' how to perform in a bureaucratic institutional arrangement, which does not translate into an immediate increase in decision-making power or even fishing power. In the case of the BF-EPA, we argue that the increasing frustration with the limited results of efforts to build organizational capacity among fishers raises questions about the method of political inclusion through top-down training systems. We are thus pressed to remain critical and cautious of compulsory training schemes that are deliberately bound to institutional building processes for environmental governance.

Final Remarks

Governability of small-scale fisheries is often characterized by highs and lows due to complex social and biophysical features. Small-scale fisheries are embedded in broader economic, political and social processes where power relations play a key role. BF-EPAs are a territorial representation of this heterogeneous socioecological context where dynamic and emergent processes are highly influenced by multiple factors at different scales. As a result, the up-scaling of small-scale fisheries in marine protected areas is faced with several dilemmas. This case study reveals three key factors influencing fisheries governability in Brazil: mismatches between the social and governing system affecting fishers' political agency (limited participation), institutional instability and leadership.

Limited Participation

Over a decade, the BF-EPA governing system has been transformed and has created new governing interactions through participatory mechanisms. Increased governability of fisheries was enabled by a participative and inclusive step-zero process to designate and activate a BF-EPA Council. This new governing instrument was scaled-up to improve the fit between the fisheries social system and systems-to-be-governed at the EPA territorial level (nine coastal municipalities). Governability thus increased substantially with the collective capacity of actors to elaborate more sophisticated images of problems and opportunities in fisheries. In some cases, this process is reflected in more sustainable use of fisheries such as in the case of Ibiraquera lagoon. Nevertheless, the potential role of BF-EPA in fostering sustainable territorial development still lags behind expectations, and many informants argued that things would only improve through the design of the long-awaited participatory EPA Management Plan.

The limited participation of small-scale fishers, however, reflects power asymmetries related to various mismatches outlined in governing interactions held at the interface between social and governing systems. Also, even though fisheries has been an important theme recurrently dealt with by the Council, it has often been regarded as a secondary issue because of competing demands to implement a very broad territorial governance mandate that focuses on whale conservation. Substantial efforts to increase fishers' participation were placed on formal training and inclusion in the EPA Management Council structure. However, due to the interplay of the outlined meta-order (image) mismatches, small-scale fishers showed low level of responsiveness. Limited participation is also often associated with perceptions of fishers' behaviour (*e.g.*, passiveness or clientelism) or misbehaviour (*e.g.*, predominance of corrupt or self-interested leaders).

The economic permeability between small-scale and industrial fisheries further illustrates fishers' agency patterns. While such a distinction is necessary for small-scale fishers to be empowered in such participatory forums, they have somehow to balance their priorities between a crystallized artisanal fisher category or stay as a marginalized category with more economic flexibility. We have thus also noticed an increased tension emerged from the polarization between industrial vs small-scale fishers due to the institutionalization of these two categories in the Brazilian fisheries legislation.

We have thus described several symptoms partly associated with a wicked problem. This problem can be framed as follows: while the identity of the fisheries social system reflects the dynamics of the natural system in order to prosper and be viable, the governing system frequently assumes its own identity and governmentality over the fisheries social system, to which formalism and bureaucratic dynamics is unnatural. Small-scale fisheries, on the other hand, presuppose stability in natural systems provisions and dynamics, and a rich and diverse set of communicative typologies mirroring the natural system. We suggest that improved governability should emerge from alternative ways of dealing with this wicked problem of mismatch in systems' identities. However, for this to happen, mutual learning and adaptation of both fishers' political ecology and/or governing system is necessary. We do not intend to disregard the important role of a formal learning and capacity building process, nor the very significant initiatives undertaken by BF-EPA actors. Nonetheless, we contend that the learning process must emerge spontaneously from the interactions between actors of both governing and social systems. We suggest that the required learning, representational, and political activation patterns shall emerge from gradual, experiential, and predominantly informal and reflexive interactions between actors with agency in the interface between these systems.

Institutional Instability

Power asymmetries in the governing system influence not only the level of participation of fishers but also fisheries governability. These asymmetries result in an unstable governing system and limitations in institutional capacities.

Despite increased efforts at inclusiveness and innovation, Brazil is a recent democracy characterized by historical legacies of inequality and dependency, with limited institutional capacity and social organizations. We have shown that BF-EPA actors have been actively trying to identify and transform an untenable governing system in which Management Plans in protected areas have been criticized for their overly technical, diagnostic/normative-oriented, top-down and ultimately inoperative nature. In the last 10 years protected areas have been downgraded, downsized, and reclassified in Brazil (Bernard et al. 2014). As a result stakeholders deal with complex interrelated governability challenges or wicked problems related to limited institutional capacities and strong influences from informal and formal politics.

During the first operational phase of the BF-EPA Council (2006–2009), other subsidiary governing instruments and interactions were activated to improve governability. For example, the FTC was created and designed to match the participation and learning demands of fisheries governing systems. However, the subsequent shutting down of the FTC in 2010, despite its improving performance, combined with the failure to implement the participatory Action Plan in subsequent years, indicates serious limitations to governability. Since 2010, progress in fisheries governability in the BF-EPA has declined due to the breakdown of formal initiatives (second order interactions *sensu* Kooiman et al. 2005) concerning fishing issues (*e.g.*, fishing accords and participative fisheries management plan). It was only in 2012 that a strategic re-orientation took place in the BF-EPA, with full acknowledgement that fishing issues should now be nested as a special chapter in the participatory Management Plan. Institutional instabilities and consequent strategic adaptations in governing MPAs has had a direct effect on fisheries governability. Thus, it is clear that the challenges in fisheries governability are not only an outcome of poor BF-EPA performance but also due to the external influences of national and regional politics.

Leadership

While power relations have affected levels of participation and institutional functioning, key actors enjoying particular attributes and resources have helped enable improvements in governing interactions. Such actors have mobilized their resources in order to facilitate social interactions, knowledge exchange and production, and conflict resolution. Moreover, they have the ability to bridge different institutional arrangements and reconcile different perceptions. A few actors have played this role at different moments, fostering the development of the BF-EPA Council and the engagement of small-scale fishers in the process. The shift of the BF-EPA from a consultative to deliberative body, a crucial transformative process desired by most Council members, was triggered by the BF-EPA chief in charge and supported by other actors such as researchers, environmental organizations and other government authorities.

The involvement of small-scale fishers in EPA governance was enabled by an external consultant. His hybrid institutional background enabled him to facilitate interactions between hierarchical and self-governing modes of governance.

Key actors involved with the BF-EPA Management Council have been able to seek new opportunities through influence of individuals (leadership) and new institutional (formal and informal) mechanisms. Outcomes, however, have taken different forms at different times. Leadership has been key to minimizing power asymmetries in participatory initiatives, and in seeking new opportunities for institutions to upscale fisheries governability to the territorial level under unstable institutional conditions.

In sum, the BF-EPA case illustrates the complexity of small-scale fisheries in coastal-marine governance processes in Brazil. While BF-EPA actors have yet to achieve desired outcomes, they have done a great job in opening up a window of opportunity for broad-level governing systems reform. While many actors are still frustrated and some have given up along the way, the formulation of a collaborative and adaptive Management Plan for an entire coastal-marine territory is an achievement in itself. Therefore, fisheries governability must be seen as a continuous learning process mediated by changes in structure, values, and interactions. Despite the several challenges faced to date, the lessons learnt in the process of developing the BF-EPA Council so far provide a positive outlook for the future. We thus hope the ideas outlined in this chapter contribute to the hypothesis that they are now inspiring larger transformative systems change, thus responsible for what may (or hopefully) be their most fruitful journey for scaling-up coastal-marine governability in Brazil.

References

Basurto, X., Bennett, A., Weaver, A. H., Van Dyck, S. R., & Aceves-Bueno, J. S. (2013). Cooperative and noncooperative strategies for small-scale fisheries' self-governance in the globalization era: Implications for conservation. *Ecology and Society, 18*(4), 38. doi:dx.doi.org/10.5751/ES-05673-180438

Berkes, F., & Seixas, C. S. (2005). Building resilience in lagoon social–ecological systems: A local-level perspective. *Ecosystems, 8*, 967–974.

Bernard, E., Penna, L. A., & Raújo, E. (2014). Downgrading, downsizing, degazettement, and reclassification of protected areas in Brazil. *Conservation Biology: The Journal of the Society for Conservation Biology, 28*, 1–12. doi:10.1111/cobi.12298.

Borges, J. C. L. (2008). A Pesca em Santa Catarina/Brasil: Regime Jurídico da Estrutura Administrativa entre 1912 à 1989. In *II Encontro de Economia Catarinense: Área Temática: Economia Rural e Agricultura Familiar* (pp. 114–124).

Bueloni, F. S. (2012). *Mudanças Temporais na Utilização da Baleia Franca, Eubalaena australis, pelas comunidades locais dos Municípios de Imbituba e Garopaba, litoral de Santa Catarina, Brasil.* Dissertação de Mestrado em Ecologia (Ciências Biológicas), Universidade Federal de Santa Catarina, Florianópolis, 119p.

Câmara, M. R. (2001). *O turismo no litoral de Santa Catarina: tensões, conflitos e reorganização espacial.* Dissertação (Mestrado em Geografia), Universidade Federal de Santa Catarina, Florianópolis, 215p.

Capellesso, A. J., & Cazella, A. A. (2011). Pesca artesanal entre crise Econômica e problemas socioambientais: Estudo de caso nos municípios de Garopaba e Imbituba (SC). *Ambiente and Sociedade, 14*, 15–33.

Castello, J. P. (2010). O futuro da pesca da aquicultura marinha no Brasil: A pesca costeira. *Ciência e Cultura, 62*(3), 32–35.

Cerdan, C., Freire Vieira, P., Policarpo, M., Vivacqua, M., Capellesso, A., Castro Rodrigues, H., Martinel, B., Cordeiro, E., Lesage, A., Meynard, F., Pedrosa, A., Adriano, J., Leonel, M., & Ferreira, M. (2011). *Desenvolvimento territorial sustentável na zona costeira do Estado de Santa Catarina Brasil* (Documento de Trabajo N° 87, 61p). Programa Dinámicas Territoriales Rurales. Rimisp, Santiago, Chile.

Chuenpagdee, R., & Jentoft, S. (2013). Assessing governability – What's next. In M. Bavinck, R. Chuenpagdee, S. Jentoft, & J. Kooiman (Eds.), *Governability of fisheries and aquaculture: Theory and applications* (pp. 9–30). Dordrecht/New York: Springer. doi:10.1007/978-94-007-6107-0.

Cordell, J. (2006). Brazil: Dynamics and challenges of marine protected area development and coastal protection. In *Scaling up marine management: The role of marine protected areas* (pp. 58–77). Washington, DC: The World Bank.

Dias Neto, J., & Marrul Filho, S. (2003). *Síntese da situação da pesca extrativa marinha no Brasil.* Brasília: IBAMA.

Diegues, A. C. S. (1983). *Pescadores, camponeses e trabalhadores do mar* (287p). São Paulo: Ática.

Diegues, A. C. S. (1999). A Sócio-Antropologia das Comunidades Marítimos no Brasil. *Revista Etnográfica, III*(2), 361–375.

Few, M., & Tortorici, Z. (Eds.). (2013). *Centering animals in latin American history.* Duke University Press.

Filardi, A. C. L. (2007). *Diagnóstico da Pesca Artesanal do Município de Garopaba (SC): Potencialidades e Obstáculos para a Gestão Adaptativa para o Ecodesenvolvimento* (p. 257). Universidade Federal de Santa Catarina.

Floeter, S. R., Rocha, L. A., Robertson, D. R., Joyeux, J. C., Smith-Vaniz, W. F., Wirtz, P., Edwards, A. J., Barreiros, J. P., Ferreira, C. E. L., Gasparini, J. L., Brito, A., Falcón, J. M., Bowen, B. W., & Bernard, G. (2007). Atlantic Reef fish biogeography and evolution. *Journal of Biogeography, 35*, 22–47. doi:10.1111/j.1365-2699.2007.01790.x.

Foppa, C. C., Bonatti, S., Medeiros, R. P., & Borgonha, M. (2011). Monitoramento Participativo da Pesca Artesanal Marinha do Estado de Santa Catarina: Desafios Sociopolíticos. In V Simpósio Brasileiro de Oceanografia: Oceanografia e Políticas Públicas, Santos (pp. 1–5).

Gerhardinger, L. C. (2014). *Innovation, governability and agency of key-individuals at Southernright Whale Environmental Protection Area (Santa Catarina state, Brazil)*. Doctoral thesis, Interdisciplinary Programme on Environment and Society, Campinas State University, Núcleo de Estudos e Pesquisas Ambientais.

Gerhardinger, L. C., Godoy, E. A. S., & Jones, P. (2009). Local ecological knowledge and the management of marine protected areas in Brazil. *Ocean and Coastal Management, 52*(3–4), 154–165. doi:10.1016/j.ocecoaman.2008.12.007.

Gerhardinger, L. C., Borgonha, M., & Bertoncini, A. A. (2010). *Memórias do Mar: Biodiversidade, Conservação e Cultura no Litoral Brasileiro* (1st ed.). Caravelas: Ecomares.

Gerhardinger, L. C., Godoy, E. A., Jones, P. J., Sales, G., & Ferreira, B. P. (2011). Marine protected dramas: The flaws of the Brazilian national system of marine protected areas. *Environmental Management, 47*(4), 630–643. doi:10.1007/s00267-010-9554-7.

Gomes, G. O. (2012). *O Conhecimento Local sobre Mudanças nos Estoques Pesqueiros na APA da Baleia Franca, no Litoral Sul de Santa Catarina*. Universidade Federal de Santa Catarina.

Hochstetler, K. (2008). Organized civil society in Lula's Brazil. In P. R. Kingstone & T. J. Power (Eds.), *Democratic Brazil revisited* (pp. 33–53). Pittsburgh: University of Pittsburgh Press.

Hochstetler, K., & Keck, M. E. (2007). *Greening Brazil: Environmental activism in state and society*. Durham: Duke University Press.

Johnson, D. S. (2010). Institutional adaptation as a governability problem in fisheries: Patron–client relations in the Junagadh fishery, India. *Fish and Fisheries, 11*(3), 264–277.

Kooiman, J., Bavinck, M., Jentoft, S., & Pullin, R. (Eds.). (2005). *Fish for life: Interactive governance for fisheries*. Amsterdam: University of Amsterdam Press.

Lago, P. F. A. (1961). Contribuição geográfica ao estudo da pesca em Santa Catarina. *Revista Brasileira de Geografia: Separata. Rio de Janeiro, 1*, 121–215.

Macedo, H. S. (2008). *Processos Participativos na gestão de áreas protegidas: estudo de caso em unidades de conservação de uso sustentável da zona costeira do Sul do Brasil*. Dissertação de Mestrado – Universidade Federal de Santa Catarina, Centro de Filosofia e Ciências Humanas. Programa de Pos-Graduação em Sociologia Politica. Florianópolis.

Macedo, H., Vivacqua, M., Rodrigues, H. C.-L., & Gerhardinger, L. C. (2013). Governing wide coastal-marine protected territories: A governance analysis of the Baleia Franca Environmental Protection Area in South Brazil. *Marine Policy, 41*, 118–125.

Ministério do Meio Ambiente – MMA. (2013). *Cadastro Nacional de Unidades de Conservação*. Retrieved June 13, 2014, from http://www.mma.gov.br/areas-protegidas/cadastro-nacional-de-ucs

NEXUCs'. (2012). *Unidades de Conservação do Brasil: o caminho da Gestão para Resultados* (536p). São Carlos: RiMa Editora.

Oliveira, O. M. B. A., & Silva, V. L. (2012). O Processo de Industrialização do Setor Pesqueiro e a Desestruturação da Pesca Artesanal no Brasil a Partir do Código de Pesca de 1967. *Seqüência, 33*, 329–357. doi:http://dx.doi.org/10.5007/2177-7055.2012v33n65p329

Palhares, R. (2013). *As Relações entre Unidades de Conservação Federais Marinho-Costeiras do Estado de Santa Catarina e Populações Humanas*. Tese de Mestrado em Ecologia. Universidade Federal de Santa Catarina, 148p.

Polette, M., & Vieira, P. F. (2009). The strides and gaps in Brazilian integrated coastal zone management: An undercover evaluation of the scientific community perceptions and actions. *Ocean Yearbook, 23*, 670–685.

Rebouças, G. N. M., Filardi, A. C. L., & Vieira, P. F. H. (2006). Gestão Integrada e Participativa da Pesca Artesanal: Potencialidades e Obstáculos no Litoral do Estado de Santa Catarina. *Ambiente and Sociedade, 9*, 83–104.

Rodrigues, H. C. L. (2011). Relatório Final referente ao Produto 3 para a implementação das atividades previstas no edital n° 07/2010, contrato n° 050/2010 CEADES/MPA – Ciclo de Apoio a Gestão Compartilhada, no Território Sul Catarinense, 85 pp.

Saraiva, S. (2010). Duelo entre Pescadores e Atuneiros em Garopaba. *O Popular*, 1–2.

Seixas, C. (2002). *Social-ecological dynamics in a management system: Investigating a coastal lagoon fishery in Southern Brazil*. Tese de Doutorado em Natural Resources and Environmental Management. University of Manitoba, U. M., Canada.

Seixas, C. C., & Berkes, F. (2003). Learning from fishers: Incorporating local knowledge in policy design and assessment. In P. H. F. Vieira (Ed.), *Conservação da diversidade biológica e cultural em zonas costeiras: enfoques e experiências na América Latina e no Caribe* (pp. 333–371). Florianópolis: APED Editora.

Trimble, M., Araujo, L. G., & Seixas, C. S. (2014). One party does not tango! Fishers' non-participation as a barrier to co-management in paraty, Brazil. *Ocean and Coastal Management, 92*, 9–18.

Vasconcelos, M., Diegues, A. C. S. A., & Sales, R. R. (2007). Limites e possibilidades na gestão da pesca artesanal costeira. In A. L. Costa (Org.), *Nas Redes da Pesca Artesanal* (pp. 15–83). Brasília: IBAMA – MMA.

Vivacqua, M. (2012). *Dilemas da Conservação e Desenvolvimento na Gestão Compartilhada da Pesca Artesanal: Conflitos e Sinergias nos Processos de Criação de Reservas Extrativistas Marinho-Costeiras em Santa Catarina. Tese de Doutorado em Sociologia Política.* Florianópolis: Universidade Federal de Santa Catarina.

Chapter 19
Governability Challenges Facing Small-Scale Fishers Living Adjacent to Marine Protected Areas in South Africa

Merle Sowman

Abstract The governability of small-scale fisheries located adjacent to Marine Protected Areas (MPAs) in South Africa has increasingly come under scrutiny as communities, social science researchers, NGOs and human rights activists challenge current governance approaches that disregard the socio-cultural rights and livelihood needs of fishing communities living within or adjacent to MPAs. Drawing on research conducted in seven case studies in South Africa, this chapter explores the current mismatch between the realities facing fishing communities impacted by MPAs and the state-centric and natural science-based approach to governance adopted by South Africa's fisheries management and conservation authorities. This approach to MPA governance persists despite a suite of policy reforms and political rhetoric that indicates the embrace of a more people-centred approach to natural resource governance. The key focus of this chapter is to gain a deeper understanding as to why this mismatch persists despite almost 20 years of democracy and policy reforms. While the devastating impact of South Africa's political history is evident in all cases, other factors that inhibit meaningful change and formation of robust governance systems, are highlighted. These include the persistence of a natural-science paradigm; the divergent principles, values, worldviews and images amongst governance actors; institutional shortcomings; failure to recognize and respect local and customary forms of governance; and the lack of attention to implementation mechanisms that are informed by all governance actors.

Keywords Small-scale fisheries • Marine protected areas • Governability • Poverty • Fishing communities • People-centred approach • South Africa

---

M. Sowman (✉)
Department of Environmental and Geographical Science, University of Cape Town, Cape Town, South Africa
e-mail: merle.sowman@uct.ac.za

S. Jentoft, R. Chuenpagdee (eds.), *Interactive Governance for Small-Scale Fisheries*, MARE Publication Series 13, DOI 10.1007/978-3-319-17034-3_19

Introduction

The transition to democracy in 1994 in South Africa, catalyzed new forms of governance that sought to address past injustices and improve the socio-economic conditions of poor and marginalized communities. The law reform process and institutional restructuring that took place permeated all sectors of society including fisheries and conservation management. Expectations amongst small-scale fishers were high that rights to resources would be restored or re-allocated and that legal provisions for this sector would ensure its protection, development and wise management. However, despite a progressive Constitution and a suite of natural resource management policies and laws that require respect for human rights, restitution and equitable access to resources, coastal fishing communities living in or adjacent to "no-take" Marine Protected Areas (MPAs) in South Africa, continued to be denied access to traditional fishing grounds.

South Africa, like many countries across the world, is employing MPAs as a key tool for biodiversity conservation, habitat protection and more recently, fisheries management (Helvey 2004; Laffoley 2008; DEA 2012). MPAs are considered by marine scientists in South Africa to be the backbone of its marine conservation strategy (Lemm and Attwood 2003), and an increasingly important management tool for rebuilding depleted fish stocks (Kerwath et al. 2013). Despite the ongoing scientific debates surrounding the value of MPAs as a tool for improving fisheries management and fisheries productivity in particular (Hilborn et al. 2004; Kolding 2006), South Africa has embarked on an ambitious programme to expand the coastal and marine area under protection and establish a representative network of MPAs (DEA 2012; Sink et al. 2012). This is in response to South Africa's commitment to a host of international multilateral agreements including the Convention on Biological Diversity and its various programmes of work, the Johannesburg Plan of Implementation (2002) and the targets set at the World Parks congress in Durban in 2003. Over the past 15 years, South Africa has developed an impressive suite of policies, laws and strategies[1] to give effect to these commitments. Currently, in South Africa, approximately 23.2 % of the coastal zone is under some level of protection (refer Fig. 19.1) of which 9.1 % comprises 'no-take' MPAs[2] (DEA 2012). Recently South Africa has declared its first off-shore MPA, in waters surrounding Prince Edward Islands, an area covering approximately 180,000 km^2. The National Protected Areas Expansion Strategy (RSA 2008) has set ambitious targets for MPAs, aiming to have 25 % of the coastal zone declared MPAs by 2028 of which 15 % of the coastal area would be declared "no-take" zones. However, while plans

[1] These include the Living Marine Resources Act of 1998, National Environmental Management (NEM) Act of 1998, NEM: Protected Areas Act of 2003 and its amendment of 2014, NEM: Biodiversity Act of 2004, as well as various Biodiversity Assessments (2004 and 2008) and the National Protected Area Expansion Strategy (RSA 2008).

[2] "No-take" MPAs refer specifically to MPAs where no fishing is permitted. Other MPAs may allow fishing of certain species and may also have zones where certain activities (eg boating, fishing, only passive recreation) are allowed.

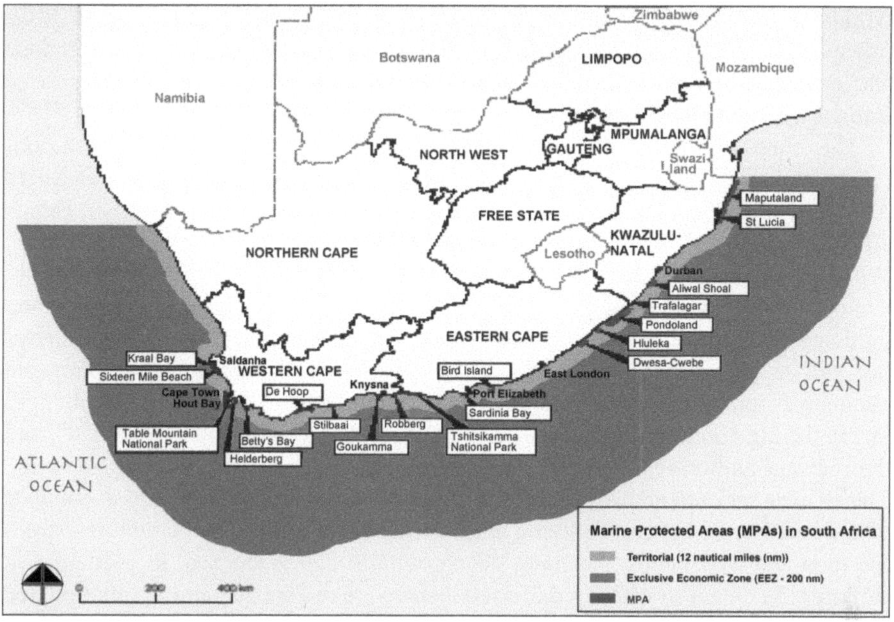

Fig. 19.1 Marine protected areas in South Africa

to expand South Africa's MPA network is high on the conservation agenda, the focus is shifting to enhance protection of the offshore marine environment (Sink et al. 2012).

Over the past century, many coastal fishing communities have been affected by the establishment and management of MPAs (Sunde and Isaacs 2008; Sowman et al. 2011; Masifundise Development Trust (MDT) 2013). Increasingly, the governability of small-scale fisheries systems in such contexts, has come under scrutiny as fishing communities, social science researchers, NGOs and human rights activists challenge current governance approaches that disregard the socio-cultural rights and livelihood needs of these communities (Sunde and Isaacs 2008; Sowman et al. 2011, 2014a, b, c; Emdon 2013; Jackson et al. 2013; MDT 2013; Sunde 2013, 2014; Sunde et al. 2013; Williams 2013). Pressure on government from these groups to review and revise the regulations restricting access to MPAs has been met with opposition from members of the marine science and conservation community who are concerned about establishing a precedent and argue instead that lack of alternative livelihoods is the key challenge (Minutes of MPA Forum meetings, 2012–2013; expert witness for ECPTA in the case of State vs D Gongqose 2012). Lack of access to MPAs has led to growing discontent amongst fisher communities who argue for restoration of rights in terms of culture and custom and for meeting food and livelihood needs. In recent years, communities have approached the Human Rights Commission, marched to Parliament and launched legal challenges against the state to demand their rights to resources (K George and others v Minister of Environmental

Affairs and Tourism 2005 and 2010; Sunde et al. 2013; State vs D Gongqose 2012; Sowman et al. 2014a, b, c; Sunde 2014). Twenty years after democracy, governance and governability of small-scale fisheries in the context of MPAs appears to be fraught with difficulties.

This chapter is based on information obtained from case study research conducted in seven small-scale fishing communities[3] living adjacent to MPAs in South Africa (or areas being considered for MPA status) (Sunde and Isaacs 2008; Hauck 2009; Sowman et al. 2011, 2014c; de Greef 2013; Emdon 2013; Jackson et al. 2013; Williams 2013; Sunde 2013, 2014; Sunde et al. 2013). Researchers in all sites conducted household surveys, key informant interviews, focus groups meetings and participated in local meetings and activities. This research also draws on information and insights gained from participation in various fisheries and MPA policy and planning forums over the past 5 years. It explores the realities of and challenges facing fishing communities impacted by MPAs – "the system-to-be-governed", and the relevant authorities' approach to MPA governance – "the governing system". The chapter seeks to address questions of governance and governability in contexts where small-scale fishers are living adjacent to MPAs and rely on natural resources for food and livelihoods and have strong cultural ties to the sea. In particular, it explores the extent to which small-scale fishers are recognized and accommodated in planning and decision-making processes relevant to MPAs. Further, it examines the fit between policy rhetoric and practice after 20 years of democracy. This chapter seeks to answer these questions by drawing on the theoretical foundations of the Interactive Governance Approach (Kooiman et al. 2005; Bavinck et al. 2013) and using the Interactive Governance framework outlined by Jentoft (2013). In particular, it explores how principles, values, worldviews and images influence governability at every level of governance.

Exploring Governability of Small-Scale Fisheries in the Context of MPAs

The Nature of Governance Interactions

The departure point for an assessment of small-scale fisheries governance in the context of MPAs in South Africa is that the mode of governance relevant to coastal resources is largely hierarchical. Yet, within this hierarchical system principles such as participation, co-management, access to information, respect for local and traditional knowledge, are required by law. This section focuses on understanding the nature of the governance interactions between the governing system and the system-to-be-governed.

[3] Case studies include fishing communities at Olifants Estuary, Langebaan lagoon, Hangberg in Hout Bay, Tsitsikamma, Dwesa-Cwebe, Hluleka and Kosi Bay.

The Interactive Governance Approach (Bavinck et al. 2013) is mainly concerned with understanding the exchanges, communications, collaboration, knowledge production and decision-making processes that take place between actors involved in the process of governance. The Interactive Governance Approach is premised on a set of normative principles that regard inclusive participation, and deliberation amongst civil society actors and joint problem solving as expressions of democracy and thus desirable (Kooiman and Bavinck 2013). It presumes that problem identification, and development of objectives, plans and solutions should be deliberated upon by the different governance actors and jointly agreed to. While hierarchical governance suggests a top down approach and expects government to play a leading role, the Interactive Governance Approach suggests that high levels of participation, information sharing and joint problem solving should still guide planning, management and decision-making. Thus the nature of the interactions between the governing system and the system-to-be-governed, irrespective of the mode of governance, will have a bearing on the governability of the system. Figure 19.2 provides a conceptual framework of the interactive governance approach that is used to guide the analysis of case material in this chapter.

In the case of small-scale fishing communities living adjacent to MPA's in South Africa, *de facto* governance is largely state-driven, top-down, regulatory and dominated by natural science (Sunde and Isaacs 2008; Sowman et al. 2011; Hushlak

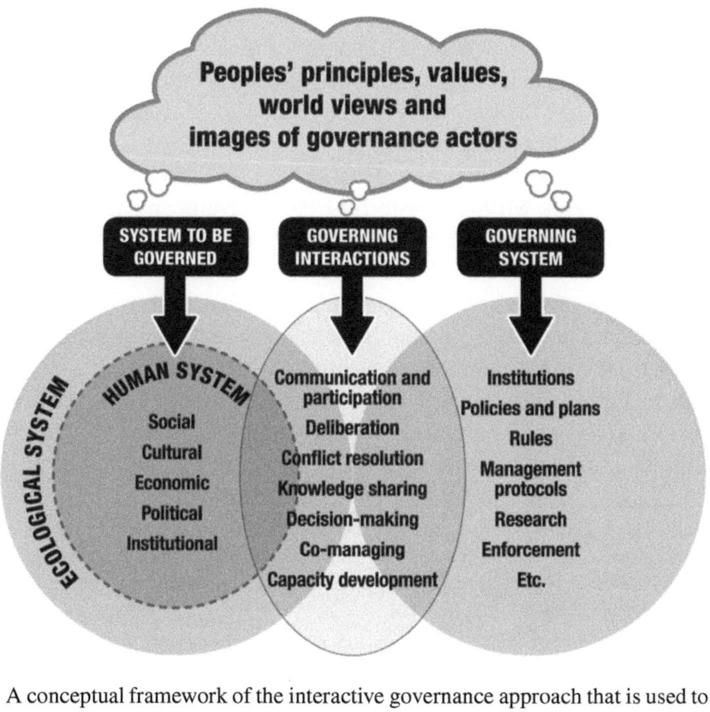

Fig. 19.2 A conceptual framework of the interactive governance approach that is used to guide the analysis of case material in this chapter

2012). This is surprising in view of the commitments South Africa has made to a host of international multi-lateral agreements relevant to MPAs that call for recognition of the rights of indigenous peoples, greater participation in decision-making and the sharing of benefits from protected areas. The promulgation of the Policy for the Small-scale Fisheries Sector in South Africa in June 2012 (DAFF 2012) confirmed South Africa's commitment to the protection and development of small-scale fishers including recognition of their customary rights and systems in so far as they are consistent with the Bill of Rights.

Yet, the nature of governance interactions between small-scale fishers and other governance actors in the context of MPAs is not participatory and deliberative, nor are the principles of free and informed consent, access to information, accountability, transparency, recognition of local and indigenous knowledge, respect for customary systems of governance and law, upheld in most of these interactions. In fact, in many MPAs, the nature of interaction between key governing actors – namely government officials, local fishing communities, and other stakeholders, is largely conflictual. There is significant evidence from newspaper articles, research reports, minutes of meetings, legal papers and magistrate court records that conflict in MPAs, especially between local fishing communities and conservation officials and agencies, is prevalent (Newspaper articles various (2009–2014); K George and others v Minister of Environmental Affairs and Tourism 2005 and 2010; Sunde and Isaacs 2008; Hauck 2009; Mbatha 2011; Sowman et al. 2011, 2014a, b, c; de Greef 2013; State v D. Gongqose and two others 2012; Emdon 2013; Sunde 2013, 2014; Sunde et al. 2013). These conflicts take various forms including protests outside conservation agency offices, "protest" fishing,[4] marches to parliament, arrests and fines or imprisonment, letters and petitions to the relevant government Ministers, court cases, as well as physical harassment. Examples of such conflicts and governance interactions are provided below.

The Hangberg fishing community in Hout Bay, who live adjacent to the Table Mountain National Park MPA (refer to Fig. 19.1) have been harvesting west coast rock lobster (WCRL) in waters adjacent to the settlement since the nineteenth century (van Sittert 1994). Since the MPA was established as a no-take lobster sanctuary in 1934, and reinforced by the proclamation of the Table Mountain National Park MPA in 2004, the community has been actively excluded from their traditional fishing grounds. After the democratic elections in 1994, the community expected government to grant some level of access to these waters. However, this has not been the case and has resulted in increased frustration amongst the traditional fishers of Hangberg and fuelled further "illegal"[5] fishing activities. Fishers say they are

[4] There are various examples where fisheries have openly defied the state and fished in protected areas. For example in 2007, approximately 70 armed local fishers including both those involved in recreational and subsistence fishing, entered the Tsitsikamma MPA in the southern Cape and fired gunshots in front of the conservation officers.

[5] The term "illegal" has been placed in quotation marks as fishers are claiming rights to these waters where they have historically fished. In cases where small-scale traditional fishers have continued to practice their fishing traditions, the term "informal" fishing is used in preference to "Illegal" fishing.

tired of waiting for government to authorize access to traditional fishers. Furthermore, in recent years, interim relief permits have been allocated to community members many of whom have no historical links to fishing in the area (fisher leader, personal communication, March 2013). A recent discussion with a group of fishers indicated that some fishers in Hangberg no longer care about conservation because they see others plundering resources in their waters and argue they are more concerned with their immediate requirements for food and income (fisher, personal communication, March 2013). Recent research has revealed a significant increase in informal and illegal harvesting of resources within the MPA some of which can be attributed to the frustrations of traditional fishers and their desire to assert their rights to resources but also out of necessity to support their families (Hauck 2009; de Greef 2013). Although there have been a few meetings between conservation agencies and fishers over the past 5 years, there has been little progress with respect to addressing the historic rights of these traditional fishers. Instead government has issued interim relief permits[6] to some traditional fishers to harvest resources beyond the MPA boundaries and high levels of informal and illegal fishing have continued within the MPA.

While conservation agencies do in some cases consult the public and local fishing communities regarding development of management plans, or to obtain input regarding certain proposals, these consultations are not meaningful as fishers concerns and perspectives have seldom led to changes in MPA governance. For example, in the case of a proposal to prohibit fishing in the Olifants Estuary in 2008, the concerns raised by local fishers at a stakeholder meeting were ignored, despite the fact that they have been fishing in these waters for nearly 100 years, are dependent on the fishery for food and livelihoods (Sowman 2009).

Objections by the fishers and their social partners[7] to the proposed MPA outlined in a draft management plan (Anchor Environmental Consulting 2008) resulted in a lengthy process of consultation and negotiation between government, their consultants, local fishers and their social partners. Willingness by government to discuss the grievances of the fishers was largely facilitated by an intervention by the Legal Resources Centre, a public interest legal NGO who wrote a letter to the then Minister of Environmental Affairs and Tourism demanding that the rights and socio-economic needs of this fishing community be recognized and addressed. The fishers argued throughout these discussions that their customary, cultural and livelihood rights needed to be recognized and that the proposed no-take MPA was unconstitutional and served only the interests of conservation and other stakeholders. These

[6] Interim relief permits were the outcome of an Equality Court Ruling in 2007, to provide traditional fishers with legal access to marine resources under specified conditions, while a new small-scale fishing policy was being developed. These permits are still issued on an annual basis to traditional fishers that meet certain government criteria until such time as the new Small-scale Fisheries Policy (DAFF 2012) is implemented.

[7] The fishers of the Olifants Estuary have been working closely with social science researchers at the University of Cape Town, two NGOs namely, Masifundise Development Trust and the Legal Resources Centre, and a community-based organization Coastal Links, for several years to have their rights recognized and their livelihoods preserved.

arguments were supported by provisions in the draft and later final Small-scale Fisheries Policy (DAFF 2012) which requires recognition of customary practices, cultural and socio-economic rights of traditional fishers balanced with sustainable use principles. After several meetings and workshops with government managers and scientists over a 5 year period, there was finally support from government and other stakeholders to allow local fishers to continue fishing in the estuary. While the fishers supported the calls for strengthening protection of threatened resources and habitats, they felt that they should not bear the burden of conservation efforts (Jackson et al. 2013). However, they agreed that the existing protected area at the mouth of the river, an area of approximately 1 km in extent, should be retained and formalized and that they should play a role in its monitoring and management (Jackson et al. 2013).

While the state has committed to co-management in policies and legislation relevant to natural resource management and fisheries management in particular, and has developed guidelines to implement such an approach,[8] evidence from various research projects suggests that co-management in coastal fishing communities is largely state-driven and involves communities in so far as it serves government's agenda (Hauck and Sowman 2003; Schell 2011; Mbatha 2011; Hauck et al. 2014; Hansen et al. 2014).

Further, there is very little evidence that local and indigenous knowledge of fishers is effectively incorporated into MPA planning and management processes. In fact, such knowledge is considered by the majority of fisheries and conservation scientists encountered in this research not to be valid since from their perspective it must be verified by science to be of any value to management and decision-making (Minutes of various meetings, 2011–2014). For example, a workshop with marine scientists and fishers was intended to facilitate sharing of knowledge to generate input for management proposals for the Olifants estuary. At a meeting to prepare for this workshop, one of the main government fisheries scientists commenced the meeting by stating categorically that "only information that could be verified by science" would be considered (Minutes of meeting held at MDT 2011). This position set the tone for the workshop which was not conducive to building relationships, trust and mutual respect. The fishers' felt they needed to defend their information, scientists were cautious and skeptical about fishers' inputs and fishers questioned the basis for some of the scientific information presented (participant observation, Ebenhaeser 2011).

Research in the seven cases revealed that there are few attempts by scientists and conservation managers to recognize and incorporate fisher knowledge in MPA management. However, in Kwa-Zulu Natal, EKZN Wildlife has made efforts through their co-management committees and community monitoring systems, to involve local communities in information gathering and knowledge exchange. However, the

[8] The People and Parks initiative of the Department of Environmental Affairs is one such example although it has a strong terrestrial focus and has made considerable progress in engaging local communities and other stakeholders in planning and management decision-making. However, this approach is not evident in the MPAs investigated in this study.

focus has been on ecological monitoring, with less effort directed towards incorporating local and indigenous knowledge in management decisions. While there have been interventions (eg beach walks with local fishers, community surveys) to recognize and incorporate fisher knowledge in the process of reviewing zonation plans for Langebaan and Dwesa-Cwebe MPAs, these efforts have to date not resulted in changes to MPA management. Co-production of knowledge as envisaged in the Interactive Governance Approach is not taking place in South Africa's fisheries and MPA governance processes in any meaningful way.

The above review and discussion suggest ongoing challenges to meaningful engagement amongst actors (fishers, scientists, managers and other stakeholders) in MPA governance processes in South Africa. This is having a significant impact on the governability of such areas, with negative social consequences for small-scale fishers and long term implications for sustainability. These shortcomings exist largely due to mismatches that occur at every level of governance but most significantly at the level of meta-governance.

Governance Mismatches and Why They Persist

This section explores some of the mismatches that exist at different levels of governance and seeks to better understand why they persist. In many respects, it can be expected that if there are major mismatches at the level of principles, values, worldviews and images, it is likely that mismatches will occur at the other levels of governance (Kooiman and Bavinck 2013). Thus if there are fundamental differences at this level, it is likely that the institutions established and rules formulated to steer and regulate the system, will affect governance at different levels. Table 19.1 provides a summary of some of the key mismatches that occur at the different levels of governance. The discussion that follows will focus mainly on the mismatches that occur at the level of meta-governance and explore possible reasons for their persistence.

Divergent Principles, Values, Worldviews and Images

Fisheries and protected areas in South Africa cannot be understood without considering the historical, political, economic and social context in which these sectors operated. In the case of fisheries, small-scale fishers were actively excluded from participating in the fishing industry while in the case of MPAs forced removals, displacement of communities, dispossession of land, restricted access to resources were all key features of the historical development of terrestrial and marine protected areas (Brockington et al. 2008; Claassens and Cousins 2008; Sunde and Isaacs 2008; Sowman et al. 2011). In addition to political motivations, those steering governance firmly believed that state or private ownership of the marine commons was the only effective means of safeguarding resources and protecting biodiversity

Table 19.1 A summary of some of the key mismatches that occur at the different levels of governance

	Those steering governance	Actors in the system-to-be-governed
Meta governance level	**Principles and values:** Largely informed by neoliberal thinking, mainstream conservation discourses and principles such as sustainable use, conservation and protection of biodiversity. Management supported mainly by natural science. Participation of stakeholders as required	**Principles and value:** largely informed by culture, customary practices and governance systems. Respect and care for nature and each other. Resources will replenish themselves. Management informed by local knowledge and experience. Participation in decisions is fundamental
	Worldviews: State or private ownership of marine commons is required to ensure sustainability and protection of natural resources. Society has a responsibility to "care for nature". Nature must be controlled and managed by those with requisite knowledge	**Worldviews:** Ownership is vested in the community, rights of access derive from membership of the group and are relational. Local norms and rules dictate social obligations and resource responsibilities. Nature and society are inextricably linked, identity and culture are tied to people's relationship with natural resources and each other
	Epistemologies Driven by positivist and natural science based approaches. For knowledge to be used and valued it must be measurable, replicable and verified by science	**Epistemologies:** Based on experiential knowledge, long-standing practices and beliefs passed down through generations by cultural transmission. Knowledge is adaptive, holistic and context specific
	Images: Poor fishing communities degrade and deplete natural resources due to their poverty status unless government imposes and enforces rules. If left unchecked a "Tragedy of the commons" scenario is likely to be the outcome	**Images:** Government's conservation policies and practices have led to forced removals, restrictions on access to marine resources and harassment. Nature conservation efforts impact livelihoods and exacerbate poverty
Mismatches at the institutional level	Multitude of state institutions at national, provincial and local level with different responsibilities for SSFs and MPAs but little co-ordination	Confusion amongst communities and stakeholders regarding powers, roles and responsibilities of different state institutions
	No formal recognition of local and customary governance systems including customary laws.	Local and customary governance systems guide the behavior and decisions of those living under such systems including their interaction with natural resources.
	Abundant rhetoric about co-management with local fishing communities and other actors despite absence of functioning co-management institutions (except in KZN).	Absence of formal co-management structures (except in KZN) and mechanisms for interaction and decision-making amongst local governance actors.
	Resource rules made and enforced by the state. No recognition of local, customary and cultural norms and practices.	State rules perceived as illegitimate and often inappropriate. Local communities operate within the 'legal system' that suits their immediate needs.
	Enforcement performance measured in terms of number of fines issued, arrests made and sentences passed.	Conservation enforcement measures seen by many as harassment and a violation of human rights.

Day to day management	Monitoring of status of stocks and habitats undertaken by scientists and conservation officials.	Outsiders (scientists) have access to fishers' resources for scientific purposes – yet locals cannot access these resources.
	Focus on practical actions to achieve conservation and fisheries management objectives and targets (e.g. clearing alien vegetation, rehabilitating environments, restocking resources).	Objectives and targets largely unknown to local communities. Some conservation practices contrary to local values, beliefs and practices
	Managers and conservation officers respond directly to natural resource problems and events (beached whale, pollution spill) according to procedures and plans.	Local communities respond to problems and events according to local experience and/or customary systems and rules.
	Failure to comply with fisheries and MPA rules results in harassment, fines, confiscation of equipment, imprisonment.	Harvesting resources for food, livelihood and cultural purposes is a human right. Local people have rights to resources and do not perceive state rules as legitimate.

Based on Sunde and Isaacs (2008), Hauck (2009), Sowman (2009), Sowman et al. (2011, 2014c), Mbatha (2011), de Greef (2013), Emdon (2013), Jackson et al. (2013), Sunde (2013, 2014), Sunde et al. (2013), and Williams (2013)

(Hersoug 2002; Sunde et al. 2013). Fishing communities living in or adjacent to these protected areas were not consulted or compensated for loss of access to traditional fishing grounds.

The transition to democracy and the promulgation of a plethora of policies and laws signaled a new more people-centred approach to conservation and fisheries management. Yet, despite these new policies including a small-scale fisheries policy that requires redress and fulfillment of human rights, South Africa continues to embrace neoliberal ideology and consumer capitalism in its approach to natural resource management and conservation. The persistence of the neoliberal ideology articulated in South Africa's macro-economic policy published soon after the democratic elections and reinforced by the recently published National Development Plan (National Planning Commission (NPC) 2012), confirm government's bias towards privatization and faith in the markets to allocate resources efficiently and generate profits that will lead to "trickle down" benefits for the poor. The NPC is clear in its support of big industry and states "*Small-scale fisheries cannot be regarded as a way to boost employment. Capital-intensive industrial fisheries offer better salaries and better conditions of employment than small-scale low-capital fisheries. Reducing the rights allocated to industrial fisheries to award them small-scale operations simply cuts jobs*" (NPC 2012, 229). The significant contradictions that exist between South Africa's neoliberal macro-economic policy and the social justice imperatives underpinning South Africa's new natural resources legal regime, provide significant barriers to meaningful reform in the marine conservation arena.

For many small-scale fishing communities, people and nature are inextricably linked and degradation of the environment or loss of access to resources threatens their livelihoods and their cultural identity and group integrity (Wicomb and Smith 2011; Sunde 2013). In many communities, fishing practices are bound up with cultural beliefs and practices and a strong sense of sharing resources harvested, especially with those in need, prevails (Mbatha 2011; Sunde 2013, 2014; Sunde et al. 2013; Williams 2013). Local and customary governance systems have guided and in some contexts continue to guide resource allocation, use and management decisions although these systems are evolving in response to state imposed rules and other external factors (Kepe 1997; Sunde et al. 2013; Sunde 2013; Williams 2013; Sowman et al. 2014a, b). For coastal communities living adjacent to MPAs, they believe they are entitled to access and use marine resources for livelihood and cultural purposes as has been the practice of many previous generations. There is a belief that nature will replenish itself and traditional harvesting practices will not lead to collapse of stocks (Mbatha 2011; Williams 2013; Sunde 2014). Fishers recognize that they have a responsibility to care for nature but participation in management decisions is regarded as a basic social norm. Yet, 'no take' rules and restricting access have been imposed on many fishing communities living adjacent to MPAs without any consultation and with significant negative social consequences (Sunde and Isaacs 2008; Hauck 2009; Emdon 2013; de Greef 2013; Sowman et al. 2014c).

While there have been shifts in the conservation discourse in South Africa, in practice the philosophy and approach to fishing in MPAs remains consistent with the fortress conservation approach of the 1960s and 1970s. MPAs are considered sacrosanct and opening up of "no take" MPAs for fishing is considered by many marine scientists to be "the thin edge of the wedge" and an "erosion of the conservation estate" (Minutes of Meetings 2012–2014). In commenting on proposals to open up Dwesa-Cwebe MPA to limited local fishing, a marine scientist who has been involved in scientific research to inform MPA planning and management in the area for several years said the following *"It makes no sense in terms of South Africa's conservation planning programme which is currently being driven by highly reputable management authorities ... to further reduce this figure by opening up a protected area of high quality linefish habitat to fishing... such an action flies in the face of South Africa's international commitments to biodiversity conservation. There is a clear and urgent need to add to the extent of no-take network rather than to reduce it (Marine scientist, powerpoint presentation, 2011).* These strongly held beliefs regarding the importance of MPAs, especially no take MPAs, as a conservation and fisheries management tool, are supported by the images held by this group of actors regarding the degradation and overexploitation of resources that would result from opening up MPAs to local fishing communities. There is a firm conviction that such action would result in a "tragedy of the commons" and huge losses to society as a whole.

These beliefs are driven by a conviction that MPAs are a key tool for conservation and fisheries management. Marine ecologists and fisheries scientists in South Africa are at the forefront of endeavors to expand the network of MPAs in the country. Their work is largely informed by natural science although there is an increasing recognition that social science is needed to better inform MPA planning and management (Sowman et al. 2014d). However, there is still a high level of skepticism regarding the value and integrity of local and traditional knowledge to inform these processes. Thus there exist tremendous methodological and epistemological barriers to knowledge sharing and integration. The dominance of the natural science paradigm in conservation and fisheries management in South Africa has perpetuated the unequal power relations that have existed for decades between conservation and fisheries scientists on the one hand and local communities on the other. This represents a fundamental limitation to governability of small-scale fishers in the context of MPAs in South Africa.

The differences in principles, values, worldviews and images of various actors engaged in fisheries governance in the context of MPAs are profound and significantly affect the type of institutional set up, the design and application of rules, as well as the actions and decisions taken by different actors in day-to-day management (See Fig. 19.2). These mismatches also significantly influence the nature of governance interactions and the effectiveness of governance interventions identified to address problems. Simply putting in place mechanisms to improve communication, deliberation, co-ordination, exchange of information and conflict resolution – will not necessarily address the root causes leading to these governance challenges.

Institutional Shortcomings

The institutional shortcomings encountered in this research are not surprising given the mismatches found at the level of meta-governance. The multitude of state institutions involved in fisheries and MPA governance in South Africa, as well as the many policies and laws governing these institutions, has resulted in overlapping jurisdictions and a lack of clarity regarding roles and responsibilities of the various actors. To further confuse matters, the Constitution recognizes customary law as an independent and equal source of law as well as the authority of traditional leaders and has promulgated legislation[9] to this effect. Various statutes relevant to fisheries and protected area management contain provisions that promote co-management and the establishment of community managed conservation areas. Yet, state governance systems and day-to-day management practices do not reflect an acceptance of this legal pluralism. The lack of clarity has exacerbated the confusion amongst local fishing communities regarding government's intentions to implement the new fisheries and conservation rhetoric articulated in post-Apartheid policies, laws and strategies.

This is well illustrated by examining the underlying philosophy and approach to enforcement that persists in fisheries and MPA management in South Africa. In general terms, small-scale fisheries are considered to be a threat to protected areas due to "illegal" fishing. The response has been to increase enforcement efforts in MPAs. Enforcement performance is measured in terms of the number of arrests, fines and convictions. The higher the numbers the greater the success rate of the enforcement programme (Minutes, MPA forum held at Aliwal Shoal in 2014). There is no interrogation of the possible root causes of the ongoing "illegal" harvesting in protected areas and how these enclosures have affected local livelihoods, food security, cultural and religious expression.

Local fishing communities regard these state-imposed rules as illegitimate and in most cases do not understand the reasons for the establishment of the MPA. They view enforcement measures as harassment and a violation of their human rights (Emdon 2013; MDT 2013; Sunde 2013, 2014; Sowman et al. 2014c). Furthermore, for small-scale fishers the concept of a "no-take" protected areas to rebuild declining fish stocks, restore degraded habitats and enhance ecosystems for the greater public good, are difficult concepts to embrace, especially when other sectors (e.g. recreationists, commercial fisheries, mining and energy) are granted rights to exploit marine resources in or adjacent to their territories. The management plans, protocols and practices are focused on meeting conservation, ecological and fisheries management objectives, which largely fail to give attention to the historical, political, socio-economic and cultural context of the local fishing communities in these coastal areas.

Failure to recognize and respect local and customary forms of marine use and governance is a further reason for mismatches at the institutional level (Sowman

[9] For example the Traditional Governance and Leadership framework Act of 2003 contains provisions to enable delegation of authority over natural resources to traditional leaders.

et al. 2014a, b; Sunde et al. 2013). This situation persists despite the fact that over 50 % of Africans in South Africa live according to some form of African customary law (Mnisi 2007 in Sunde et al. 2013), and the South African Constitution recognizes customary law and institutions in so far as they are consistent with the Bill of Rights. Furthermore, South Africa has committed to several multi-lateral agreements that require respect for the rights of indigenous peoples and local communities and their involvement in decision-making. While the new Small-scale Fisheries Policy (DAFF 2012) includes principles and provisions to give effect to these rights (Sowman et al. 2014a) it remains silent on how these principles will be applied in MPAs. Except in the case of Kosi Bay, fisheries and conservation scientists do not recognize the existence of customary systems and are reluctant to negotiate changes to current access and use rules in MPAs.

Recent research is enhancing understanding of the complex local and customary marine governance systems in South Africa (Hauck 2009; Wicomb and Smith 2011; Ferris 2013; Sunde et al. 2013; Sunde 2013, 2014; Williams 2013) and is forcing recognition of what Borrini-Feyerband (2008) refers to as "other governance types" including self-governance. Increasing calls by local communities to have their customary, cultural and socio-economic rights recognized, is likely to require a review and reconsideration of access and use rights in many of South Africa's MPAs.

Lack of Attention to Policy Implementation Processes

South Africa has undergone a major law reform process and is considered to have one of the most progressive Constitutions and suite of natural resource management laws in the world (Hauck and Sowman 2003). However, while much emphasis has been placed on improving participation in policy formulation, and setting in place institutions, processes and standards to promote transformation, insufficient attention has been given to the design of these institutions, and the practical procedures and human resource requirements to give effect these policies and laws. Furthermore, implementation typically falls back to government which is hampered by lack of resources, limited capacity, weak leadership, high staff turnovers as well as lack of political will to ensure effective implementation of laws (Sowman et al. 2014a, b). In many cases, mechanisms for civil society actors to provide input and direction to the implementation phase are lacking. The contradictions inherent in policy frameworks governing economic development and natural resource management in South Africa further undermine the implementation process as these will be open to interpretation by the department mandated to implement the policies.

The importance of monitoring and learning that takes place in the process of policy implementation is well documented (Ojha et al. 2013). The insights and lessons gleaned from such processes are vital to enable the adaptation of policies and plans once understanding of practical implementation and challenges become apparent. Such an approach requires the ongoing involvement of civil society and other governance actors to ensure accountability and that policies and management plans are appropriate to different local contexts. The involvement of all governance actors in the process of implementation – review – reflection and adaptation is critical in order to enhance governability of small-scale fishers living in and adjacent to MPAs in South Africa.

Conclusion

This research has explored the nature of governance interactions that take place amongst actors involved in small-scale fisheries governance in the context of MPAs in South Africa. It has also sought to understand the mismatches that exist in the process of governance, why they persist and what the implications are for governability. The assumption of the Interactive Governance Approach that different governance actors will find a way to work collaboratively, deliberate on complex problems and seek to resolve conflicts and give up some of their identity and power in the process (Bavinck et al. 2013) is not borne out in the cases examined in this research. Whilst it is recognized that significant time is required to overcome the devastating impact of South Africa's political history, there are several other factors that inhibit governability.

Foremost amongst these factors are the divergent values, principles, worldviews and images of different actors and the persistence of a natural-science dominated paradigm in achieving conservation and management of small-scale fisheries living adjacent to MPAs. This paradigm is the "powerhouse" of fisheries and MPA management in South Africa. The disjunctures at the meta-governance level reverberate throughout the governance system leading to institutional set ups that are not effective and respected by all stakeholders, and interpretations that are not always consistent with Constitutional imperatives. This in turn leads to management actions and decisions that on a practical daily basis impact heavily on especially poor and marginalized fishing communities. The absence of a set of shared principles and common vision in fisheries and MPA management in South Africa means that the framing of problems, the interpretation of policy and law, the approaches to planning and management as well as the resolution of problems, are often contested. Fisheries and conservation managers are committed to rebuilding fisheries stocks and maintaining and expanding the conservation estate while civil society and their social partners are seeking redress and demanding that human rights be respected first, and are thus approaching governance from very different philosophical, ontological and epistemological positions. Song et al. (2013) have suggested that governance challenges could be reduced if the values, images, and principles of stakeholders are made explicit, understood, and articulated in policy and decision-making processes. This of course is desirable but is unlikely to happen without firstly acknowledging that fundamental mismatches do exist, agreeing on the root causes and then being willing to explore why they persist and how they can be reconciled. This requires political will and leadership as well as wise facilitation to encourage engagement on these fundamental meta-governance mismatches. It also requires ongoing research in order to provide robust evidence for the root causes underlying governability challenges in small-scale fisheries and MPAs in South Africa, ideas for addressing challenges and showcasing examples where governance interactions and processes have led to more equitable and sustainable outcomes.

References

Anchor Environmental Consulting. (2008). *Olifants Estuary management plan* (draft). Cape Town: University of Cape Town.

Bavinck, M., Chuenpagdee, R., Jentoft, S., & Kooiman, J. (Eds.). (2013). *Governability of fisheries and aquaculture: Theory and applications* (MARE publication series, Vol. 7). Dordrecht/New York: Springer.

Borrini-Feyerabend, G. (2008). *Governance as key for effective and equitable protected area systems*. IUCN Commission on Environmental, Economic and Social Policy (CEESP), Briefing Note 8.

Brockington, D., Duffy, R., & Igoe, J. (2008). *Nature unbound. Conservation, capitalism and the future of protected areas*. London: Earthscan.

Claassens, A., & Cousins, B. (Eds.). (2008). *Land, power and custom: Controversies generated by south Africa's communal land rights act*. Athens: Ohio University Press.

DAFF (Department of Agriculture, Forestry and Fisheries). (2012). *Policy for the small-scale fisheries sector in South Africa*. Pretoria: DAFF.

de Greef, K. (2013). *The booming illegal abalone fishery in Hangberg: Tough lessons for small-scale fisheries governance in South Africa*. Minor Dissertation towards a Master of Science in Conservation Biology, University of Cape Town, South Africa.

DEA (Department of Environmental Affairs). (2012). *South African environmental outlook, 2012: A report on the state of environment, Chapter 5: Oceans and coast, Draft 1*. Pretoria: DEA.

Emdon, L. (2013). *Gender, livelihoods and conservation in Hluleka, Mpondoland c. 1920 to the present: Land, forests and marine resources*. Master's thesis, University of Cape Town, South Africa.

Ferris, L. (2013). A customary right to fish when fish is sparse. Managing conflicting claims between customary rights and environmental rights. *Potchefstroom Law Journal, 16*(5), 556–614.

Hansen, M., Ramasar, V., & Buchanan, K. (2014). Localizing global environmental governance norms: Implications for justice. In M. Sowman & R. Wynberg (Eds.), *Governance for justice and environmental sustainability*. New York: Earthscan from Routledge.

Hauck, M. (2009). *Rethinking small-scale fisheries compliance: From criminal justice to social justice*. Doctoral thesis, University of Cape Town, South Africa.

Hauck, M., & Sowman, M. (Eds.). (2003). *Waves of change: Coastal and fisheries co-management in South Africa*. Cape Town: University of Cape Town Press.

Hauck, M., Mbatha, P., & Raemaekers, S. (2014). Enhancing benefits to small-scale fishers along the east coast of south Africa. In R. Wynberg & M. Hauck (Eds.), *Sharing benefits from the coast: Rights, resources and livelihoods*. Cape Town: UCT Press.

Helvey, M. (2004). Seeking consensus on designing marine protected areas: Keeping the fishing community engaged. *Coastal Management, 32*, 173–190.

Hersoug, B. (Ed.). (2002). *Fishing in a sea of sharks. Reconstruction and development in the South African fishing industry*. Delft: Eburon.

Hilborn, R., Stokes, K., Maguire, J.-J., Smith, T., Botsford, L. W., Mangel, M., Orensanz, J., Parma, A., Rice, J., Bell, J., Cochrane, K. L., Garcia, S., Hall, S. J., Kirkwood, G. P., Sainsbury, K., Stefansson, G., & Walters, C. (2004). When can marine reserves improve fisheries management? *Ocean and Coastal Management, 47*, 197–205.

Hushlak, A. (2012). *Integrating traditional ecological knowledge in South Africa's small-scale fisheries: The Olifants Estuary gillnet fishery*. Master's thesis, University of Cape Town, South Africa.

Jackson, S., Sowman, M., & Cox, J. (2013). *Fishers' proposals for fishery management in the Olifants Estuary*. Unpublished report, Environmental Evaluation Unit, University of Cape Town and Masifundise Development Trust.

Jentoft, S. (2013). *TBTI working group 6: Governing the governance*. Concept note prepared by Jentoft, S. in collaboration with Chuenpagdee. R. Unpublished document.

Kenneth George and others v The Minister of Environment and Tourism (EC 1/2005 and 2010).

Kepe, T. (1997). *Environmental entitlements in Mkambati: Livelihoods, social institutions and environmental change on the wild coast of the Eastern Cape*. Bellville: Programme for Land and Agrarian Studies, University of Western Cape.

Kerwath, S. E., Winker, H., Götz, A., & Attwood, C. G. (2013). Marine protected area improves yield without disadvantaging fishers. *Nature Communications, 4*, 2347.

Kolding, J. (2006). *MPAs in relation to fisheries – What are the biological and fish stock implications?* Paper presented at the Norwegian Fisheries Forum, 24–26 October 2006.

Kooiman, J., & Bavinck, M. (2013). Theorizing governability – The interactive governance perspective. In M. Bavinck, R. Chuenpagdee, S. Jentoft, & J. Kooiman (Eds.), *Governability of fisheries and aquaculture: Theory and applications* (MARE publication series, Vol. 7). Dordrecht/New York: Springer.

Kooiman, J., Bavinck, M., Jentoft, S., & Pullin, R. (Eds.). (2005). *Fish for life: Interactive governance for fisheries*. Amsterdam: Amsterdam University Press.

Laffoley, D. (2008). *Towards networks of marine protected areas. The MPA plan of action for IUCN's World Commission on Protected Areas*. Gland: IUCN WCPA.

Lemm, S., & Attwood, C. (2003). *State of marine protected area management in South Africa*. Unpublished report, WWF, South Africa.

Masifundise Development Trust (MDT). (2013, September). *FishersNet, Pan African Issue* (Vol. 1). Mowbray: Masifundise Development Trust.

Mbatha, P. N. (2011). *Sharing benefits from coastal resources with rural communities in South Africa: The influence of institutional arrangements*. Master's thesis, University of Cape Town, South Africa.

NPC (National Planning Commission). (2012). *National development plan: Vision for 2030*. Pretoria: National Planning Commission.

Ojha, H. R., Hall, A., & Sulaiman, R. (Eds.). (2013). *Adaptive collaborative approaches in natural resource governance: Rethinking participation, learning and innovation*. London: Routledge.

Republic of South Africa (RSA). (2008). Department of Environmental Affairs and Tourism(DEAT). *National biodiversity expansion strategy*. Pretoria: DEAT.

Schell, N. (2011). *Small-scale fisheries in South Africa. Stakeholders' understandings and perceptions of co-management in South Africa*. Master's thesis, University of Cape Town, South Africa.

Sink, K., Holness, S., Harris, L., Majiedt, P., Atkinson, L., Robinson, T., Kirkman, S., Hutchings, L., Leslie, R., Lamberth, S., Kerwath, S., von der Heyden, S., Lombard, A., Attwood, C., Branch, G., Fairweather, T., Taljaard, S., Weerts, S., Cowley, P., Awad, A., Halpern, B., Grantham, H., & Wolf, T. (2012). *National biodiversity assessment 2011: Technical report. Volume 4: Marine and coastal component*. Pretoria: South African National Biodiversity Institute.

Song, A. M., Chuenpagdee, R., & Jentoft, S. (2013). Values, images, and principles: What they represent and how they may improve fisheries governance. *Marine Policy, 40*, 167–175.

Sowman, M. (2009). An evolving partnership: Collaboration between 'experts' and a net-fishery. *Gateways: International Journal of Community Research and Engagement, 2*, 119–143.

Sowman, M., Hauck, M., van Sittert, L., & Sunde, J. (2011). Marine protected area management in South Africa: New policies – Old paradigms. *Environmental Management, 47*(4), 573–583.

Sowman, M., Sunde, J., Raemaekers, R., & Schultz, O. (2014a). Fishing for equality: Policy for poverty alleviation for South Africa's small-scale fisheries. *Marine Policy, 46*, 31–42.

Sowman, M., Raemaekers, S., & Sunde, J. (2014b). Shifting gear: A new governance framework for small-scale fishers in South Africa. In M. Sowman & R. Wynberg (Eds.), *Governance for justice and environmental sustainability*. New York: Earthscan/Routledge.

Sowman, M., Rajagopalan, R., Sharma, C., & Sunde, J. (2014c). Making space for small-scale fishing communities: The use of spatial management measures in fisheries management and marine conservation. In S. M. Garcia, T. Rice, & A. T. Charles (Eds.), *Governance for fisheries and marine conservation: Interactions and co-evolution*. Chichester/Hoboken: Wiley-Blackwell.

Sowman, M., Raemaekers, S., & Sunde, J. (2014d). *Guidelines for integrating human dimensions into marine protected area (MPA) planning and management*. Prepared by the Environmental Evaluation Unit, University of Cape Town for WWF-SA.

State versus Gongqose plus two others. (2012). (E382.10). Unreported.

Sunde, J. (2013). *Living customary law along the South African coastline: Securing the rights of small-scale fishing communities*. Unpublished Report, Legal Resources Centre, Cape Town.

Sunde, J. (2014). *Customary governance and expressions of living customary law at Dwesa-Cwebe: Contributions to small-scale fisheries governance in South Africa*. Doctoral thesis, University of Cape Town, South Africa.

Sunde, J., & Isaacs, M. (2008). *Marine conservation and coastal communities: Who carries the costs? A study of marine protected areas and their impact on traditional small-scale fisher communities in South Africa*. Document prepared for the International Collective in Support of Fishworkers.

Sunde, J., Sowman, M., Smith, H., & Wicomb, W. (2013). Emerging proposals for governance of tenure in small-scale fisheries in South Africa. *Land Tenure Journal, 1*(13), 117–146. Food and Agriculture Organization of the United Nations, Rome.

Van Sittert, L. (1994). *Red, gold and black markets: The political economy of the illegal crayfish trade c.1890-1990*. Department of History, University of Cape Town, South Africa.

Wicomb, W., & Smith, H. (2011). Customary communities as 'peoples' and their customary tenure as 'culture': What to do with the Endorois decision. *African Journal of Human Rights Law, 11*, 422–446.

Williams, S. (2013). *Beyond rights: Developing a conceptual framework for understanding access to coastal resources at Ebenhaeser and Covie, Western Cape, South Africa*. Doctoral thesis, University of Cape Town.

Chapter 20
Assessing Governability of Small-Scale Fisheries in Taklong Island National Marine Reserve in the Philippines

Alice Joan G. Ferrer

Abstract This chapter identifies and describes the challenges in small-scale fisheries governance in the context of Taklong Island National Marine Reserve in Guimaras Province, Philippines. Results show that diversity (diverse species of flora and fauna; many fishers highly dependent on the fishery resources and other users), complexity (relationship between key fishery habitats and fish catch; low cooperation and compliance to laws and regulations and conflicts within and among stakeholder groups), dynamism (decline in the quality of the resources overtime; low mobility), and scale (relatively small-sized marine reserve with stakeholders from local to national levels) characterize the natural and social systems in the marine reserve, which lend themselves to low governability. The hierarchical governing mode with the Department of Environment and Natural Resources heading the interim Management Board has not been responsive to governance challenges posed by small-scale fisheries in the reserve. The lack of representativeness and inadequate exchange of information have led to poor quality interaction between the systems. Barriers have been created since "step zero" contributing to low governability. The case illustrates that a "non-functional marine reserve" is detrimental to the sustainability of coastal resources and wellbeing of the people.

Keywords Marine reserve • Protected area • Taklong Island

Introduction

The Philippines is an archipelago of more than 7,000 islands, 2,200,000 km^2 of territorial waters, and 36,000 km of coastline. About one-half of the country's population (close to 95 million according to the 2010 Census of Population) live in coastal areas. Coastal resources in the country provide essential food, livelihood, and income to majority of the people.

A.J.G. Ferrer (✉)
Division of Social Sciences, University of the Philippines Visayas, Miagao, Iloilo, Philippines
e-mail: agferrer@upv.edu.ph

S. Jentoft, R. Chuenpagdee (eds.), *Interactive Governance for Small-Scale Fisheries*, MARE Publication Series 13, DOI 10.1007/978-3-319-17034-3_20

The small-scale fisheries account for large employment and contribute a significant share to fishery output. In 2012, an estimated 1.5 million small-scale fishers (those using fishing vessels of three gross tonnes or less, or those not using boats) contributed about 26 % to total fishery output and 33 % to total fishery value (Bureau of Fisheries and Aquatic Resources 2014). Thus, the future of the fisheries in the country depends to a large extent on how well small-scale fisheries will fare in the coming decades. One important determining factor is the state of fishery resources and the ways by which they are being currently conserved and managed.

While the country is considered to have the most diverse marine ecosystem in the world and prioritizes marine conservation and protection (Carpenter and Springer 2005), it faces many challenges. Described as "being in turbulent seas", the country's marine resources were characterized as having "depleted fishery resources, degraded coastal environment and critical fisheries habitat, low catches/incomes and dissipated resource rents, physical losses and/or reduced value of catches" (Luna et al. 2004, 355). Several factors contributing to this poor state are improper post-harvest practices and inefficient marketing, inequitable distribution of benefits of resource use, inter-sectoral and intra-sectoral conflicts, poverty among small-scale fishers, and inadequate systems and structures for fisheries management. These threats remain, and the burden lies most heavily on the small-scale fishers who are highly dependent on fishery resources and identified as the poorest people in the country (Philippine Statistics Authority – National Statistical Coordination Board 2012).

In the 1970s, marine protected areas (MPAs) were introduced in the country as a fisheries conservation and management tool (White et al. 2006a). As described by the World Conservation Union (IUCN 1994, 7; Kelleher 1999, xvii), a protected area is "an area of land and/or sea especially dedicated to the protection of biological diversity, and of natural and associated cultural resources, and managed through legal or other effective means". The country is known to have the most number of MPAs and the largest number of community-based MPAs in the world (Pollnac et al. 2001; Weeks et al. 2010). In 2008, more than 985 MPAs covering an area of almost 15,000 km^2 were established in the country (Weeks et al. 2010). The most recent MPA database (as of July 2014) documents the presence of 1,800 MPAs (Cabral et al. 2014).

While there is evidence to suggest that MPAs in the country and elsewhere have positive biological impacts (Dalby and Sorensen 2002; Gell and Roberts 2003; Russ et al. 2003, 2004; Adan 2004; Alcala et al. 2004; De Guzman 2004; Hilborn et al. 2004; White et al. 2006b; Razon et al. 2012), their establishment can also have negative livelihood impacts if small-scale fishers are denied access to areas where they have fished, pushing them to go to other open areas (Javier 2003; Bennett and Dearden 2014). Thus, resistance from fishers, and conflicts between and among different resource users, are expected.

Experience with MPAs in the country points to management and sustainability in particular as enduring challenges. For instance, an inventory of marine reserves in the Visayas region shows that only 34 % of the 564 marine reserves were functional (Alcala et al. 2008). The poor performance of MPAs is often associated

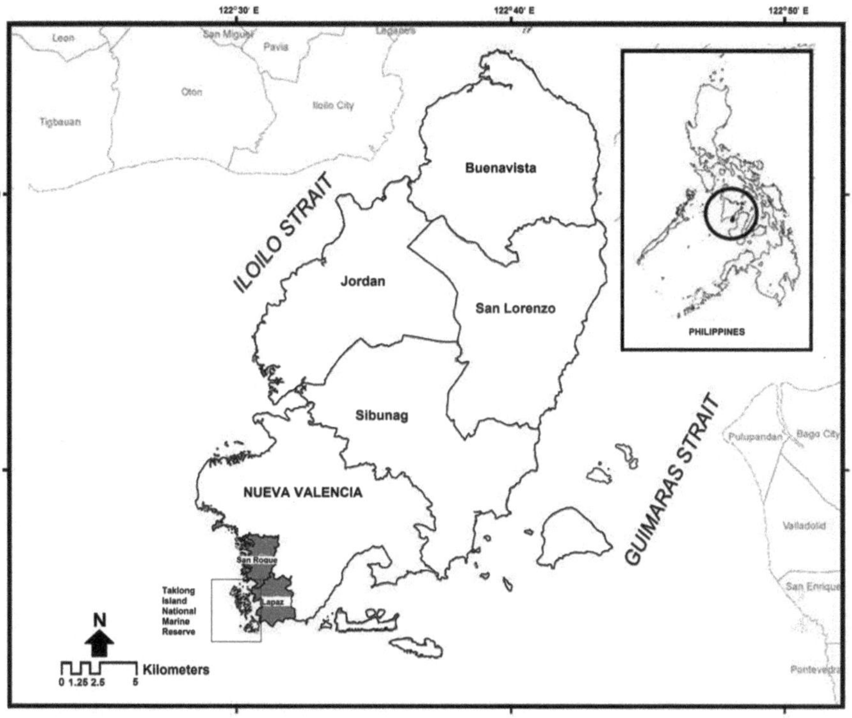

Fig. 20.1 Location of Taklong Island national marine reserve

with the failure of the governing system, which may include local governments, communities, non-government organizations or the national government (De Guzman 2004). Not considered in such analysis and conclusions is that the nature of the system in which MPAs operate may be inherently dynamic, complex, and diverse (Chuenpagdee and Jentoft 2009). These characteristics put limits on what can be realistically achieved.

Using a governability assessment framework (Kooiman et al. 2005, 2008; Chuenpagdee et al. 2008; Chuenpagdee 2011; Bavinck et al. 2013), this chapter offers a way of explaining the performance of systems using the Taklong Island National Marine Reserve (TINMR) in Guimaras Province, Philippines as an illustration (Fig. 20.1). TINMR[1] is a protected area of 11.4345 km^2 (84 % of which are water, while the rest is land). It encompasses 46 islets including the Taklong and Tandog Islands of barangays[2] Lapaz and San Roque of the municipality of Nueva Valencia (Table 20.1).

[1]Information was drawn mainly from PAWCZMS' 2011 Highlights of Accomplishments; TINMR: Brief Profile and Historical Background by DENR PENRO-Guimaras.

[2]A barangay is the basic political unit in the Philippines.

Table 20.1 Basic information on Taklong Island national marine reserve

Location[a]	10.4 -10.43333 °North; 122.48333-122.51917 °East
Municipality/city	Municipality of Nueva Valencia
Province	Guimaras
Area	1,143.5 ha (183 ha land and 960.45 ha water)
No. of years as marine reserve	24 years
Habitat[a]	Coral reef (fringing), mangroves, sea grass, algal bed
Fish biomass/density[a]	Very low (5.3 kg/500 m)
Coral cover[a]	Poor (25 %)
Surrounding barangays	Lapaz and San Roque
Population in the surrounding barangays	970 households; 2006 individuals as of 2011
No. of households into municipal fishing (est) in the area	364 households
Established by	Department of Environment and Natural Resources
Legal foundation	Presidential Proclamation No. 525 dated February 8, 1990; Designated as initial component of the national integrated protected area system under the 1992 National Integrated Protected Areas System Act
Management board	Interim Protected Area Management Board, headed by the Regional Director of the Regional Office VI (Western Visayas) of the Department of Environment and Natural Resources
Superintendent	Personnel from Department of Environment and Natural Resources-Provincial Environment and Natural Resources Office (DENR-PENRO) in Guimaras Province
Law enforcement group	None
Infrastructure	UP Visayas Marine Biological Station; Eco-Park by the Department of Environment and Natural Resources VI (Western Visayas)
Office	No physical office
Integrated protected area plan	None
Funding source	Department of Environment and Natural Resources
Monitoring and evaluation	None
Major activities in the area	Fishing (using legal and illegal gears and methods), research, recreation (swimming, boating, snorkeling, diving)
Current management issues and problems	Fishing using illegal gears (e.g., fine-meshed net) or methods (e.g., blast fishing, use of poison); continued illegal activities such as cutting of mangroves, encroachment of the commercial fishers, and collection of plant species and bird eggs

[a]From Alcala et al. (2008); Other data from Key Informant Interviews and Focus Group Discussions and DENR-PENRO (n.d., 2012)

In 1990, the Taklong Island National Marine Reserve (TINMR) was established under Presidential Proclamation No. 525 "for the benefit and enjoyment of the Filipino people under the administration of the Department of Environment and Natural Resources." With the passage of the National Integrated Protected Areas System (NIPAS) Act of 1992, TINMR became an "initial component" of the integrated protected area system in the country. The final declaration of TINMR under the National Integrated Protected Areas System Act of 1992 requires a DENR study and recommendation for a Congressional action. More than two decades later, TINMR has still not been declared a protected area under the integrated protected area system in the country.

In 2008, TINMR was described as "non-functional" (Alcala et al. 2008) amidst the continuing degradation of the marine resources. Meanwhile small-scale fishers, who are highly dependent on resources in TINMR, continue to suffer in poverty, face food and nutrition insecurity, and lack of opportunities for livelihood and income diversification.

The hierarchical mode of governance in TINMR, with the interim Management Board headed by the Department of Environment and Natural Resources, has not been responsive to the needs and challenges in TINMR. The low governability in TINMR can be traced back to the pre-implementation stage when local communities were not consulted before TINMR area was declared a MPA. Local participation, moreover, has remained low during the implementation years of the MPA.

This chapter examines the relationship between small-scale fisheries and TINMR, and how the inherent characteristics of the natural and social systems may affect governability. Insights obtained from the analysis about factors enhancing or limiting governance can help form realistic or attainable expectations and help identify areas where governance can be improved. The specific questions raised include: What are the characteristics of the area that warranted its declaration as a national marine reserve? How do the governing institutions of TINMR perform? What are the apparent implications of the hierarchical mode of governance on governability of the reserve and local small-scale fisheries? What limitations in governability are present?

In what follows, the systems-to-be-governed, which consist of the TINMR ecosystem and its resources and the social system (the users and other stakeholders), the governing system (the steering mechanisms), and the interaction between the two systems that may give rise to governance challenges are examined. The chapter ends with a discussion and conclusion.

The data for this chapter came from a Stakeholders Analysis in 2013 for TINMR funded by the GIZ[3] Environment and Rural Development Program-Integrated Coastal Management in the Philippines. The data were collected using mixed methods such as key informant interviews, focus group discussions, and a review of secondary sources of information. Data were collected from 1 April 2013 to 30 May 2013. Twenty-one key informants from the local government units (municipality of Nueva

[3] Deutsche Gesellschaft fur Internationale Zusammenarbeit – Environment and Rural Development Program.

Valencia and the province of Guimaras), DENR regional and provincial offices, and academia were interviewed. Thirteen focus group discussions were organized, involving 43 adult male small-scale fishers, 45 adult female fisher folks, 18 youth, and eight DENR personnel. A validation meeting was conducted on 22 July 2013 in barangays San Roque and Lapaz in the municipality of Nueva Valencia, Guimaras Province.

Systems-to-Be-Governed

The Natural System

TINMR (11.4345 km^2) is small in area relative to Sagay National Marine Reserve (320 km^2), which is the other nationally-designated MPA in Western Visayas region. The two largest no-take MPAs in the country are Tubbataha Reef National Park (968 km^2) and Apo Reef Natural Park (275 km^2) (Weeks et al. 2010). The natural system in TINMR is characterized by high diversity, complexity, and dynamism, and low scale (Table 20.2). TINMR is a diverse natural habitat[4] for endemic birds, reptiles, and marine fishes. It is also the nesting and breeding grounds for endangered sea turtles and Tabon birds (Tabon scrubfowl, Megapodius cumingii) (DENR-PENRO N.d.). TINMR also has reef associated fishes (198 species in 32 families), sea grasses (20 ha of 7 species), soft and hard corals (100 ha of coral reefs, 107 species in 50 genera belonging to 13 families of scleractinian corals), seaweeds (at least 77 species), and diverse mangroves (209 ha with 25 true mangrove species).

Fish catch in TINMR is dependent on the conditions of mangroves, sea grasses, and coral reefs. The protection and conservation of these habitats are crucial to the long-term sustainability of the fisheries in the reserve. However, the diversity of species of flora and fauna will continue to attract different stakeholder groups representing different interests who engage in different activities in the reserve.

Resources in TINMR have degraded over time. In the 1970s, coastal marine resources were said to be rich by key informants. Soft corals were as big as a basin and mangroves lined the coasts. The open access regime and increasing population in the area led to the degradation of resources in the 1980s. Activities with negative impacts on the resource include illegal fishing (blast fishing, cyanide or other use of poison in fishing, encroachment of commercial fishers from the neighboring island province of Negros, and use of compressors in fishing so fishers can stay longer underwater), illegal cutting of trees (forest trees and mangroves) for charcoal, fuel, and building materials, and other destructive activities (gleaning in mangrove areas, and collection of plant species and bird eggs). These activities continued despite TINMR's status as a marine reserve. In August 2006, bunker fuel from an oil tanker that sank nearby polluted TINMR and affected the coastal ecosystem and households along the coasts.

[4] Information was drawn mainly from PAWCZMS' 2011 Highlights of Accomplishments; TINMR: Brief Profile and Historical Background by DENR PENRO-Guimaras.

Social System

The social system of TINMR is characterized by high diversity and complexity, low dynamism, and high scale (Table 20.2). The high diversity and complexity are attributed to the presence of many stakeholders with varied interests and activities in TINMR. The major stakeholders are the residents (including the fishers) and officials in the affected barangays of San Roque and Lapaz, the local governments (municipal and province), national government agencies such as the Department of Environment and Natural Resources (DENR) and the Bureau of Fisheries and Aquatic Resources (BFAR), academia and researchers, tourists (local and foreign), and organizations with activities in the area.

The number of stakeholder groups grew slowly over the years, up until the oil spill in 2006. TINMR was in the national and international news as a result of the oil spill. Since that time, tourism and research activities have increased in the area.

The residents of barangays San Roque and Lapaz that border TINMR are mostly fishers. They consider TINMR as their traditional fishing ground. According to key informants, about 80 % of the 634 households in barangay Lapaz and 336 households in barangay San Roque rely on small-scale fishing within TINMR for food and as a source of income.

Fishers use small boats and passive gears (gillnet, traps, hook and line). Their gears are designed for shallow waters, which make TINMR an ideal fishing ground

Table 20.2 Governability of Taklong Island national marine reserve under the hierarchical mode of governance

	Systems-to-be-governed	
Attributes	Natural system	Social system
Diversity	High (abundant and diverse species of flora and fauna and endemic species)	High (many fishers highly dependent on the fishery; many other users – tourists, researchers)
Complexity	High (fish, mangroves, sea grasses, corals)	High (low cooperation, conflicts within and among stakeholder groups)
Dynamism	Low to medium (decline in the quality of the resources due to many stakeholders with different interests conducting various activities in the area	Low (most barangay people are native to the area and not mobile despite limited income opportunities in the area)
Scale	Low to medium (11.4345 km^2 endemic birds, reptiles, and marine fishes as well as endangered species)	High (major stakeholders include those from the barangay to those at the national level)
Governing system	"Control and administration" placed under the Department of Environment and Natural Resources; Interim Protected Area Management Board dormant for years and recently revived; Very limited resources (funds, people, equipment) for proper management; inadequate enforcement of fishery rules and regulations	
Governing interactions	Poor quality of interactions; mostly meetings ("to inform" of decisions of the interim management board). Fishers and local residents claim no participation in the management; poor compliance to rules and regulations	
Outcome	Low governability	

Table 20.3 Basic information about small-scale fishers from Barangays San Roque and Lapaz

No. of fishers with information	43
Age (mean)	50.02
No. of years in school (mean)	7.38
Married (percent)	86.05
No. of years as fisher (mean)	32.21
No. of gears used (mean)	2.2
Gears used	Fish pen, hook and line, longline, gillnet, fish coral
Fish caught	Striped mackerel, Ponyfish, Silver-Biddy, Yellowtail scad, Tooth ponyfish, Short mackerel, Threadfin bream, Freckled goatfish, Brushtooth lizardfish
With other sources of income (percent)	81.40
Other sources of income	Carpentry, fishing crew, driving, farming, gamefowl breeder, laborer, electrician, livestock raising

Note: Data are from Focus Group discussions with 43 fishers in barangays Lapaz and San Roque

for them. Gears for deep-sea fishing are also too expensive for most fishers who fish primarily for food. During the typhoon season, a number of islands in TINMR provide shelter for fishers when strong winds prevent them from fishing.

Peak fishing season is from November to May while the lean season is from June to October. Fishers fish 6 days a week during peak season and 4 days a week during the lean months. Fishing activity, which used to be male-dominated, has become a husband and wife activity. This enables the husband and wife to take home all the catch and income from fishing, unlike when non-members of the family not the wife join the husband in fishing. When they are not in school, children also help their fathers fish or their mothers glean. Selling of catch is done by the fishers themselves or their wives and is primarily confined to the barangay and the adjacent barangays.

The focus group discussions with 43 fishers revealed that fishers have relatively limited formal education, have spent more than half of their lives in fishing, and have few livelihood opportunities (Table 20.3). Livelihoods are largely dependent on coastal resources. Closing TINMR to any fishing activity would therefore negatively affect many fishers.

Small-scale fishers in adjacent coastal barangays of Tando and Lucmayan, and the land-locked barangay of Salvacion are also dependent on fishery resources in TINMR. Key informants from Tando reported that 90 % of the households in the barangay are fishers using gillnet and long lines. In barangay Lucmayan, according to key informants, 90 % of the 518 households in 2012 (or 25 % of the total households of 2,170 of the barangay) are small-scale fishers from Sitio[5] Dungca-an. For Lucmayan fishers, the islets in TINMR serve as transient areas for resting before going farther offshore to fish. Although Salvacion is a non-coastal barangay, 458 fishers fish in TINMR.

[5]A sitio is a smaller geographic area within a barangay.

During the pre-implementation stage of TINMR in the late 1980s, no local organizations were present in the barangay. At that time, the residents were widely dispersed physically and socially. The absence of any local organization meant a lack of local community participation prior to the proclamation of TINMR. A key informant from the DENR confirmed that no consultations happened and that they did not happen as they were not required at that time.

The first two people's organizations, Lapaz Fisherfolk Aquatic Resources and Mangrove Management Association Inc. and San Roque Coastal Environment Program Association Inc., were founded in the early 1990s, (after the proclamation of TINMR) under the Coastal Environment Program of the Department of Environment and Natural Resources. When the program ended in the latter half of the 1990s, the organizations became inactive until their revival in 2009 with the infusion of new funds from DENR for a mangrove rehabilitation project. In 2013, the Lapaz Fisherfolk Aquatic Resources and Mangrove Management Association Inc. and San Roque Coastal Environment Program Association Inc. had 57 and 46 members, respectively. The members were residents of the barangays, mostly women engaged in mangrove planting and maintenance work. Members are represented in the interim Management Board of TINMR.

The fishers in barangays Lapaz and San Roque, however, do not identify themselves with these organizations. They claimed that the organizations are "not for and by the fishers" but rather paid work for those who want mangrove rehabilitation.

The Lapaz Small Fisherfolk Association was founded in 1996 by the Provincial Office for Agricultural Services but became inactive soon after due to problems with management. The Association was reorganized in 2012 by fishers themselves in response to the need for collective action to confront concerns regarding TINMR. Similarly, the San Roque Fishers Association was established in February 2013 because livelihood assistance and funds could only be given through a barangay-level fisher organization.

Conflict occurs between fishers of barangays Lapaz and San Roque and fishers from others barangays. During the focus group discussions, fishers from Lapaz and San Roque identified fishers from other barangays and transient fishers (from Negros Occidental and municipalities in the province of Guimaras) as users of illegal and destructive fishing methods. Also, they questioned the entry of "outsiders" in TINMR for tourism and research purposes while they were restricted from entering the area, especially the area near the marine biological station.

TINMR is increasingly becoming a recreational area. Tourism activities include mostly island hopping, swimming, and diving/snorkeling. Three resorts opened around 2009. A local tourism association with 100 members (mostly women in the barangay) also came together in 2009 under the Community-Based Rural Tourism Program of the Province of Guimaras. In 2012, TINMR received 1,475 visitors: 792 were local visitors (from Guimaras and adjacent provinces in the region), 653 were national visitors, and 30 were foreign nationals (DENR- PENRO 2013).

Significant research has also been conducted in TINMR. The University of the Philippines Visayas (UPV) has conducted research in TINMR since the 1960s. UPV's involvement in TINMR started with field biology classes. In the 1970s, UPV established a marine biology station on a portion of donated land within TINMR. In the late 1980s, UPV supported DENR in its move to declare the area a marine reserve. More researches were conducted in 2006–2013 under the Oil Spill Response Program of UPV.

Stakeholders can be classified into two groups in terms of their support of TINMR's inclusion in the integrated protected area systems network. The first group is generally supportive of it. DENR and the Bureau of Fisheries and Aquatic Resources see the TINMR area as an important natural resource in need of protection. Local government units think that TINMR being declared a protected area will boost tourism in the area. Local peoples' organizations (Lapaz Fisherfolk Aquatic Resources and Mangrove Management Association Inc. and San Roque Coastal Environment Program Association Inc.) see the declaration of TINMR as a protected area resulting in more work related to mangrove rehabilitation and protection and consequently better pay.

Ironically, the "presumed" primary beneficiaries do not support TINMR and have resisted its inclusion as a protected area. These beneficiaries include local residents, the barangay council, and the local fishers' organizations (the San Roque Fishers Association and the Lapaz Small Fisherfolk Association) of barangays Lapaz (the Lapaz Small Fisherfolk Association) and San Roque (the San Roque Fishers Association) as well as those in other neighboring barangays. While the fishers are aware of the importance of protection and conservation of resources, this is offset by their greater concern that a protected area will impact adversely on their fishing and livelihoods more generally.

Governing System

The governing system of TINMR has not been responsive to governance challenges (see Table 20.2). At first, there was confusion as to who has jurisdiction over TINMR. The 1991 Local Government Code bestowed on the local governments to manage their territorial waters, something affirmed in the 1998 Fisheries Code. TINMR is in the municipal waters (within 15 km from the shoreline) of the municipality of Nueva Valencia and yet DENR has control and administrative powers over it. There was also a misunderstanding that DENR had the sole responsibility to manage TINMR resulting in limited attention given to the participation of local governments (provincial and municipal).

Today, TINMR is still not part of the national integrated protected area system as required by the 1992 National Integrated Protected Areas System Act. The Act classified all protected areas that had been legally declared prior to the Act's enactment to be part of the protected area system. The DENR has not yet submitted documents to support a recommendation to Congress for a legislative enactment establishing TINMR under the integrated protected area system. One important requirement that is missing is a management plan for TINMR. In 2005, there was an attempt to start with the paper work for such a management plan. However, this proved challenging with documents travelling back and forth, getting lost, and eventually only resurfacing in 2013. A key informant reported that officials of one barangay within TINMR did not sign documents supporting the final declaration of TINMR as part of the national integrated protected area system.

As a national marine reserve, the management of TINMR rests with the interim Protected Area Management Board with the head of the regional office of DENR as chair. Based on the Implementing Rules and Regulations of the 1992 National Integrated Protected Areas System Act, the composition of the Management Board is finalized once a law is passed placing TINMR under the integrated protected area system. Currently, the interim Management Board has 13 members, six of whom are from the two directly affected barangays of Lapaz and San Roque (Punong Barangays, chairs of peoples' organizations Lapaz Fisherfolk Aquatic Resources and Mangrove Management Association Inc. and San Roque Coastal Environment Program Association Inc., and chairs of the Barangay Fisheries and Aquatic Management Council). The rest include members from the Bureau of Fisheries and Aquatic Resources-Region VI, Provincial Planning and Development Office of Guimaras Province, as well as the local chief executives of the province of Guimaras and the municipality of Nueva Valencia, and a member from academia. In 2011, the interim Management Board created the Technical Working Group (PAMB Resolution No. 1 Series of 2011) to address issues pertaining to fish coral operations, the tour route for TINMR visitors, installation of signages, the fee system and the General Management Plan for TINMR (DENR PENRO-Guimaras 2012). It was also at this time, after many years of inactivity since its creation in the late 1990s, when the interim Management Board became "active" (based on the number of meetings conducted). In 2012, capacity building support was given to the interim PAMB by GIZ-EnRD Program-ICM. Based on the focus group discussions and interviews with key informants from the barangays, it was apparent that local people perceive the interim Management Board as "all meetings, no action", "selective in documenting meetings by recording things that support their interests", and, "lengthy meetings with minimal output." The fishers resented the decision of the interim Management Board to declare "no fishing activity" in TINMR in November 2012 to be effective on 1 January 2013. Although the barangay council and the peoples' organizations have representatives in the interim Management Board, they claimed that the discussion was dominated by technical people, namely representatives from government agencies, academia, and local government.

The execution of approved plans and programs for TINMR, including law enforcement, is the duty of Protected Area Superintendent. However, being a superintendent is not a fulltime job but rather one of many functions of the provincial environment staff. No support staff is assigned to the TINMR superintendent.

Key informant interviews and focus group discussions participants said that the superintendent was normally absent in TINMR. This situation enables fishers to continue fishing within TINMR but not without constant fear of getting caught. Fishers acknowledge that other local people may report them to the superintendent who visits on occasion. Such a situation has resulted in suspicion and subtle conflict among local people.

A major issue affecting the governability of TINMR is insufficient funding. According to a key informant from DENR, the regular budget for the maintenance and operating expenses (MOOE) of TINMR in 2013 was P157,000 (US$ 3,536), which is used for travel expenses of the Superintendent and also meetings of the interim Management Board. DENR has presence in TINMR when there are project funds available. During the years 1993–1996, DENR staff were deployed in TINMR under the Coastal Environment Program. By 1996, only the assigned superintendent could be seen in TINMR. The presence of DENR was again felt in TINMR during the oil spill in 2006. Using compensation funds, the DENR established an eco-park in the reserve area. The key informants and focus group discussions participants felt that the eco-park structures that were built were unnecessary and mostly non-functional.

Moreover, there are no other instruments to support the implementation of TINMR. No law enforcement group has been created to protect TINMR. Markers to indicate physical boundaries and to guide fishers so as they do not encroach into the marine reserve are absent. The buoys placed earlier were reported stolen or destroyed by the local fishers although they denied this, claiming that the damage in particular was due to strong waves and winds.

Interactions Between the Systems-to-Be-Governed and the Governing System

The poor interaction between local fishers, local residents and various governing actors in charge of TINMR is responsible for making the system, as a whole, less governable (Table 20.2). For one, the interests of the local people were not considered at the outset due to inadequate consultations with them regarding the establishment of TINMR. Thus, it was most likely the case that local specificities were not adequately considered in decisions about TINMR. Possibly this is the reason why the concept of a marine reserve is not clear to fishers. Also, no information has been shared with fishers for many years. In other words, people's views about the marine reserve do not match with the stated goals of the marine reserve.

Further, fishers and other local residents reported that they neither participate in the management of TINMR nor in any decision making pertaining to it. The local residents claimed that they are not aware of what is going on in TINMR. Mechanisms are limited through which information and knowledge flow to the stakeholders, particularly to the local people. Meetings between the people and the interim Management Board are not common, and, when they happen, are mostly to inform people of decisions and not to exchange ideas and solicit options to solve problems. That is, no formal venues for feedback exist. For instance, in a meeting where the ban on fishing in TINMR was announced, many fishers were not able to attend because they did not know about the meeting or found it too costly in terms of time and transportation costs. This is despite having representatives from the barangay and of the people's organization in the interim Management Board. Community meetings, albeit infrequent in recent years, were held to inform the people of the interim Management Board's decisions rather than to consult them on decisions to be made.

Discussion

Marine resources within TINMR are rich and diverse but decades of poor protection from destructive fishing and other anthropogenic pressure within the marine reserve have resulted in degradation. TINMR is an example of a "non-functional marine reserve" where the establishment of a protected area has not guaranteed actual protection of the natural and social systems within it because of poor implementation of regulatory measures.

Social factors largely determine the long-term sustainability of MPAs (Christie 2004). Community support has been found to be the main factor playing a significant role in the success of other MPAs in the country (Pollnac et al. 2001; De Guzman 2004; White et al. 2006a; Christie and White 2007; Razon et al. 2012). Leaving the fishing communities out of the implementation process can result in the building of barriers between the fishery managers and the fishing communities (Nielsen et al. 2004) leading to poor governing interactions. In the case of TINMR, community support and cooperation is the critical missing factor, and the local communities feel alienated from the resource they consider their own. They feel that they are displaced from their traditional spaces by other stakeholder groups such as tourists and researchers. Thus, conflict is created or intensified between and among stakeholder groups (Nielsen et al. 2004; Christie and White 2007).

The case of TINMR highlights the importance of involving local communities from the preparatory stage onwards while establishing protected areas. It is at this stage that issues and concerns may emerge (Chuenpagdee and Jentoft 2007; Jentoft et al. 2007). In the case of TINMR, the problem of limited governability started in the pre-implementation stage due to no local consultations and information dissemination and education, a fault that remained during implementation as well. Chuenpagdee and Jentoft (2007) argue that the "step zero" of MPAs under co-management regimes is as important as the implementation stage and hence requires champions and facilitators to suggest what can be realistically achieved. Further, Chuenpagdee et al. (2013) highlight the significance of "step zero", using examples of MPAs in Spain and Mexico. They found that the proposal stage of MPAs, when things are still fluid, cannot be divorced from power struggles and politics. Different stakeholders are at that stage promoting their own interests and securing their positions. The same situation holds true for TINMR.

The case of TINMR also illustrates that a hierarchical mode of governance (with DENR at the helm of management of TINMR) can fail to protect resources and promote the social and economic welfare of people affected by the reserve. TINMR is beset by very limited financial resources, which is also the biggest concern in most MPAs in the country (Lowry et al. 2009; Maypa et al. 2012). The necessary instruments (i.e. management plan, guidelines) and other resources (personnel and equipment) that support the implementation of the marine reserve are either absent or inadequate.

Earlier, Kuperan et al. (1998) have shown that transaction costs associated with central government management is significantly lower at the beginning than was the case with co-management marine protected areas in San Salvador, Philippines. The reverse was true during the implementation period 1988–1996. Community members are more likely to comply with rules and regulations developed by the community as a whole as opposed to regulations imposed by an external regulatory authority. This is true in case of TINMR where there is low compliance to the rules and regulations set by the interim PAMB.

Meanwhile, Javier (2003) found that DENR (highly centralized and bureaucratic) cannot provide all the necessary and timely support to enable MPA enforcers to effectively manage the resource; thus efforts must come from the local governments. This was supported by La Viña et al. (2010) who found that, under the 1992 National Integrated Protected Areas System Act, management decisions are subject to the bureaucratic hierarchy of the DENR and the cumbersome procedures of national budgets and funding. They recommended decentralization of the management of protected areas given that local governments are autonomous, and the management actions and funding are decided locally and quickly. Both studies (Javier 2003; La Viña et al. 2010) recommended that for small areas, community-based resource management should be established. For larger jurisdictional areas, co-management styles may be more effective as there is a bigger pool of resources available for different protected area management activities. According to La Viña et al. (2010, 40), DENR is "generally agreeable to the disestablishment of small coastal or marine designated protected areas, leaving the National Integrated Protected Areas System Act more applicable to globally significant marine ecosystems such as the Tubbataha Reef." Given this recommendation, it is necessary to relook at the management of TINMR and the hierarchical mode of governance. Such a mode of governance appears unsuitable given the fact that local small-scale fishers are highly dependent on the fishery resources but yet remain poor and simultaneously the quality of the resource has declined. The natural resource system of TINMR is not very extensive and can be put entirely under the control of local communities and fisheries institutions. There is a need to improve mechanisms to endure genuine representation of local people and to prevent the systems-to-be-governed and the governing system to be alienated from one another. Also, to increase support of the fishers of the marine reserve is the need to introduce successful alternative income generating projects, which can lower dependence on the ecosystem and can improve MPA performance (Pollnac et al. 2001; Gjertsen 2005).

Conclusion

The case of TINMR illustrates low governability as well as the pitfalls of decision-making to establish protected areas that does not involve participation of the local community from "step zero" until the implementation stage. It also reminds us about the dangers of having "non-functional marine reserves" that instead of protecting the resource and the people dependent on the resource does the exact opposite.

The factors that led to the failure of TINMR being able to reverse the decline of the quality of resources and promote the well-being of the fishers and other local people can be found in the systems-to-be-governed, the governing system, and the interaction of these systems. Each system is beset with governance challenges that affect their governability. The richly diverse natural system has many users with differing interests engaged in different activities within TINMR. Without effective law enforcement, the natural system cannot recover. Small-scale fishers are highly dependent on the fishery resources that exist within the reserve. Their small boats and gear types limit their ability to go offshore or to other fishing grounds. Other sources of income are also limited in the area. Meanwhile, the governing system is inadequate to deal with the challenges and the hierarchical mode of governance fails because of lack of genuine community support. From the pre-implementation to implementation stages, there is poor interaction between the interim Management Board and local residents and other stakeholders, leading to low governability.

These limitations should be seen as opportunities that can guide changes needed in TINMR in the future. These changes involve providing venues for consultation and mechanisms for improving the exchange of ideas and information among stakeholders so as to improve policies. There is a need to identify the appropriate organizational structure to manage TINMR and ensure that the probability of failure in governance is minimized, if not eliminated. Transferring the management of TINMR to the local government and the people in the barangay is likely to considerably increase governability.

Acknowledgement The author thanks the Deutsche Gesellschaft fur Internationale Zusammenarbeit –Environment and Rural Development Program Integrated Coastal Management for the funds; Terence Dacles, Isaac Abello, and Carlo Custodio for the support; Joey Pedrajas and Jinky Hopanda for the efficient assistance; Ratana Chuenpagdee, Svein Jentoft, and the anonymous referees for helpful comments; and the Key Informants and Focus Group Discussion participants for the time and information shared.

References

Adan, W. R. (2004). Multiple small-scale marine sanctuaries in municipal waters: The Magsaysay example. In *Department of Agriculture-Bureau of Fisheries and Aquatic Resources (DA-BFAR) In turbulent seas: The status of Philippine marine fisheries* (pp. 232–236). Cebu City: Coastal Resource Management Project.

Alcala, A. C., Russ, G. R., & Maypa, A. P. (2004). Evidence for fishery enhancement effects of marine reserves in central Philippines. In *Department of Agriculture-Bureau of Fisheries and Aquatic Resources (DA-BFAR) In turbulent seas: The status of Philippine marine fisheries* (pp. 215–218). Cebu City: Coastal Resource Management Project.

Alcala, A., Bucol, C., & Nillos-Kleiven, P. (2008). *Directory of marine reserves in the Visayas, Philippines*. Foundation for the Philippines Environment and Silliman University-Angelo King Center for Research and Environment Management (SUAKCREM), Dumaguete City, Philippines.

Bavinck, M., Chuenpagdee, R., Jentoft, S., & Kooiman, J. (Eds.). (2013). *Governability of fisheries and aquaculture: Theory and applications* (MARE publication series, p. 7). Dordrecht: Springer. doi:10.1007/978-94-007-61-7-0_4.

Bennett, N. J., & Dearden, P. (2014). Why local people do not support conservation: Community perceptions of marine protected area livelihood impacts, governance and management in Thailand. *Marine Policy, 44*, 107–116.

Bureau of Fisheries and Aquatic Resources. (2014). *Fish contribution to the economy*. Retrieved March 15, 2014, from http://www.bfar.da.gov.ph/profile?id=18#post

Cabral, R. B., Aliño, P. M., Balingit, A. C. M., Alis, C. M., Arceo, H. O., Nañola, C. L., Jr., Geronimo, R. C., & Partners, M. S. N. (2014). The Philippine marine protected area (MPA) database. *Philippine Science Letters, 7*(2), 300–308.

Carpenter, K. E., & Springer, V. G. (2005). The center of the center of marine shore fish biodiversity: The Philippine Islands. *Environmental Biology of Fishes, 72*, 467–480.

Christie, P. (2004). Marine protected areas as biological successes and social failures in Southeast Asia. *American Fisheries Society Symposium, 42*, 155–164.

Christie, P., & White, A. T. (2007). Best practices for improved governance of coral reef marine protected areas. *Coral Reefs, 26*(4), 1047–1056.

Chuenpagdee, R. (2011). Interactive governance for marine conservation: An illustration. *Bulletin of Marine Science, 87*(2), 197–211. doi:10.5343/bms.2010.106.

Chuenpagdee, R., & Jentoft, S. (2007). Step zero for fisheries co-management: What precedes implementation. *Marine Policy, 31*, 657–668.

Chuenpagdee, R., & Jentoft, S. (2009). Governability assessment for fisheries and coastal systems: A reality check. *Human Ecology, 37*(1), 109–120.

Chuenpagdee, R., Kooiman, J., & Pullin, R. (2008). Assessing governability in capture fisheries, aquaculture and coastal zones. *Journal of Transdisciplinary Environmental Studies, 7*(1), 1–20.

Chuenpagdee, R., Pascual-Fernandez, J. J., Szelianszky, E., Alegret, J. L., Fraga, J., & Jentoft, S. (2013). Marine protected areas: Re-thinking their inception. *Marine Policy, 39*, 234–240.

Dalby, J., & Sorensen, T. K. (2002). *Coral reef resource management in the Philippines: With focus on marine protected areas as a management tool*. Master's thesis, University of Copenhagen, Botanical Institute, Department of Physical Ecology.

De Guzman, A. B. (2004). *A fishery in transition: Impact of a community marine reserve on a coastal fishery in northern Mindanao, Philippines* (EEPSEA research report no. 2004-RR6). Singapore: EEPSEA.

Department of Environment and Natural Resources – Provincial Environment Office (Guimaras) (DENR-PENRO). (2012). *Taklong Island National Marine Reserve: Annual report 2011*. Guimaras: Philippines.

Department of Environment and Natural Resources – Provincial Environment Office (Guimaras) (DENR-PENRO). (2013). *Taklong Island National Marine Reserve: Annual report 2012*. Guimaras: Philippines.

Department of Environment and Natural Resources – Provincial Environment Office (Guimaras) (DENR-PENRO). (N.d.). *Taklong Island National Marine Reserve (TINMR): Brief profile and historical background.*

Gell, F. R., & Roberts, C. M. (2003). Benefits beyond boundaries: The fishery effects of marine reserves. *Trends in Ecology & Evolution, 18*(9), 448–455.

Gjertsen, H. (2005). Can habitat protection lead to improvements in human well-being? Evidence from marine protected areas in the Philippines. *World Development, 33*(2), 199–217.

Hilborn, R., Stokes, K., Maguire, J. J., Smith, T., Botsford, L. W. J., Orensanz, M. M., Parma, A., Rice, J., Bell, J., Cochrane, K. L., Garcia, S., Hall, S. J., Kirkwood, G. P., Sainsbury, K., Stefansson, G., & Walters, K. (2004). When can marine reserves improve fisheries management. *Ocean and Coastal Management, 47*, 197–205.

IUCN. (1994). *Guidelines for protected area management categories. CNPPA with the assistance of WCMC*. Gland/Cambridge: IUCN.

Javier, M. E. P. (2003). *Do institutions affect the performance of marine protected areas? Evidences from the Philippines* (EEPSEA research report no. 2003-RR5). Singapore: EEPSEA.

Jentoft, S., van Son, T., & Bjørkan, M. (2007). Marine protected areas: A governance system analysis. *Human Ecology, 35*, 611–622.

Kelleher, G. (1999). *Guidelines for marine protected areas*. Gland/Cambridge: IUCN.

Kooiman, J., Bavinck, M., Jentoft, S., & Pullin, R. (Eds.). (2005). *Fish for life: Interactive governance for fisheries*. Amsterdam: Amsterdam University Press.

Kooiman, J., Bavinck, M., Chuenpagdee, R., Mahon, R., Pullin, R., & Salagrama, V. (2008). Governability of aquatic resources: Conceptual exploration and applications. *Journal of Transdisciplinary Environmental Studies, 7*(Special Issue), 1.

Kuperan, K., Abdullah, N. M. R., Pomeroy, R. S., Genio, E., & Salamanca, A. (1998, June). *Measuring transaction costs of fisheries co-management*. Paper presented at the Seventh Biennial Conference of the International Association for the Study of Common Property (IASCP).

La Viña, A. G. M., Kho, J. L., & Caleda, M. J. (2010). *Legal framework for protected areas: Philippines*. IUCN-EPLP No. 8.

Lowry, G. K., White, A. T., & Christie, P. (2009). Scaling up to networks of marine protected areas in the Philippines: Biophysical, legal, institutional, and social considerations. *Coastal Management, 37*(3–4), 274–290.

Luna, C. Z., Silvestre, G. T., Green, S. J., Carreon, M. F., III, & White, A. T. (2004). Profiling the status of Philippine marine fisheries: A general introduction and overview. In *Department of Agriculture-Bureau of Fisheries and Aquatic Resources (DA-BFAR) In turbulent seas: The status of Philippine marine fisheries* (pp. 3–11). Cebu City: Coastal Resource Management Project.

Maypa, A. P., White, A. T., Cañares, E., Martinez, R., Eisma-Osorio, R. L., Aliño, P., & Apistar, D. (2012). Marine protected area management effectiveness: Progress and lessons in the Philippines. *Coastal Management, 40*(5), 510–524. doi:10.1080/08920753.2012.709465.

Nielsen, J. R., Degnbol, P., Viswanathan, K. K., Ahmed, M., Hara, M., & Abdullah, N. M. R. (2004). Fisheries co-management—An institutional innovation? Lessons from South East Asia and Southern Africa. *Marine Policy, 28*(2), 151–160.

Philippine Statistics Authority – National Statistical Coordination Board. (2012). *Fishermen still the poorest sector in 2009*. Retrieved March 15, 2014, from http://www.nscb.gov.ph/pressreleases/2012/PR-201206-SS2-01_pov2009.asp

Pollnac, R. B., Crawford, B. R., & Gorospe, M. L. G. (2001). Discovering factors that influence the success of community-based marine protected areas in the Visayas, Philippines. *Ocean and Coastal Management, 44*, 683–710.

Razon, B. C., Liao, L. M., & Nakagoshi, N. (2012). Success and failure of marine protected area management affecting the fish catch by adjacent fishermen in Sarangani Bay, Mindanao, Philippines. *South Pacific Studies, 33*(1), 1–23.

Russ, G. R., Alcala, A. C., & Maypa, A. P. (2003). Spillover from marine reserves: The case of Naso vlamingii at Apo Island, the Philippines. *Marine Ecology Progress Series, 264*, 15–20.

Russ, G. R., Alcala, A. C., Maypa, A. P., Calumpong, H. P., & White, A. T. (2004). Marine reserve benefits local fisheries. *Ecological Applications, 14*(2), 597–606.

Weeks, R., Russ, G. R., Alcala, A. C., & White, A. T. (2010). Effectiveness of marine protected areas in the Philippines for biodiversity conservation. *Conservation Biology, 24*(2), 531–540.

White, A. T., Aliño, P. M., & Meneses, A. T. (2006a). *Creating and managing marine protected areas in the Philippines*. Retrieved March 15, 2014, from http://oneocean.org/download/db_files/creating_and_managing_mpas.pdf

White, A. T., Gomez, E., Alcala, A. C., & Russ, G. (2006b). Evolution and lessons from fisheries and coastal management in the Philippines. In T. McClanahan & J. C. Castilla (Eds.), *Fisheries management: Progress towards sustainability*. Retrieved on 15 March 2014, at http://oneocean.org/download/db_files/Chap5.Phil.06.Book.pdf

Chapter 21
Marine Protected Areas, Small-Scale Commercial Versus Recreational Fishers: Governability Challenges in the Canary Islands, Spain

José J. Pascual-Fernández, Inés Chinea-Mederos, and Raquel De la Cruz-Modino

Abstract Marine protected areas (MPAs) are promoted as an effective model for the management of marine areas worldwide. They are not only a technical management measure but also a social institution that interacts with existing use rights. In the Canary Islands, several marine reserves have already been created, while others have been proposed. Some of the already created protected areas were promoted and supported by small-scale fisher organizations. Newly proposed areas are to be backed by different institutions and small-scale fishers. For small-scale fishers marine reserves have some advantages in terms of co-governance and increased involvement in rule making and surveillance. However, increasingly, other stakeholders like recreational fishers are demanding inclusion in the governing process. It is recreational fishers who are usually the most unsupportive of MPAs and thus pose governability challenges. Involving them, therefore, in discussion about MPAs may help improve governability although it will require institution building on their side. We conclude that MPAs' inception processes are both a challenge and an opportunity for governability, as they promote new patterns of interactions between stakeholder groups.

Keywords MPAs • *Cofradías* • Governability • Step zero • Marine reserves • National Parks • Recreational fishing • Small-scale fisheries • El Hierro • Tenerife

---

This article is based on research conducted under the project "Governance challenges for sustainable small-scale fisheries: creating synergies with marine conservation and tourism" (GOBAMP II, CSO2013-45773-R, financed by Ministry of Economy and Competitiveness of Spain), with the support of the Vice-rectorate of Research and International Relations of the University of La Laguna, Tenerife. We acknowledge the collaboration of the Network of Marine Reserves of the General Secretary of Fisheries, Ministry of Agriculture, Food and Environment of Spain.

J.J. Pascual-Fernández (✉) • I. Chinea-Mederos • R. De la Cruz-Modino
Institute of Social and Political Sciences, Universidad de La Laguna, Tenerife, Spain
e-mail: jpascual@ull.es; ichinea@ull.es; rmodino@ull.es

S. Jentoft, R. Chuenpagdee (eds.), *Interactive Governance for Small-Scale Fisheries*, MARE Publication Series 13, DOI 10.1007/978-3-319-17034-3_21

Introduction

Marine Protected Areas (MPAs) have been promoted as an effective marine resource management tool to address environmental issues because they lend themselves well to an ecosystem approach. However, MPAs are not simple technical fixes, which can be designed based on biological criteria alone. Rather, social, economic, institutional, and territorial aspects need to be incorporated in the design and implementation of MPAs. From an interactive governance theory perspective, MPAs are complex systems that are meant not only to preserve marine environments, but also "to solve societal problems and create societal opportunities" (Kooiman et al. 2005, 17). Hence to improve governability, MPAs need to pay significant attention to stakeholder involvement, institution building and conflict resolution (Chuenpagdee and Jentoft 2007; Chuenpagdee et al. 2013).

MPAs can be as diverse as their goals (Jentoft et al. 2011), but one common goal is to conserve marine and coastal ecosystems. A MPA implementation often has impacts on local communities because it transforms conventional fisheries management systems (Thorpe et al. 2011), introduces or enhances tourism activities, and alters customary access to marine resources and decision-making processes related to the governance of the marine areas involved. When MPAs are conceived primarily for conservation and are consequently imposed on local communities, it is common for conflicts to occur. These conflicts arise mostly because of different perspectives on the state of the marine environment and what can and should be done about it (Jentoft et al. 2012). Studies show that when these differences are addressed in the design and implementation of MPAs, success can be achieved. This is especially the case when local stakeholders, particularly fishing communities, are involved and lead these processes (Pascual Fernández and De la Cruz Modino 2008, 2011; Kalikoski and Vasconcellos 2008).

In Spain, national and regional governments legislate on fisheries and the protection of the marine environment. There are MPAs with different designations, objectives, and legal frameworks, as there are different administrations involved in their enactment. Some of these MPAs place emphasis on the needs of small-scale fishers,[1] while others may be more focused on strictly conservation goals or other supra-local goals such as tourism amongst others (De la Cruz Modino and Pascual-Fernández 2013). Currently, in Spain there are 14 MPAs designated as *Reservas marinas de interés pesquero* (Marine Reserves with Fishing Interest, MRs)[2]; three of these have been established in the Spanish region of the Canary Islands, where our case studies are located. The level of involvement of small-scale fishers, through their organizations, called *cofradías*, varies in each case. In general, however, the

[1] European Union usually defines small-scale as being fishing activities carried out by boats less than 12 m in length and not using towed gear (Council Regulation (EC) No 1198/2006 on the EFF) (Macfadyen et al. 2011).

[2] The state has declared 10 Marine Reserves with this designation (see http://goo.gl/A5jF2O, accessed 16 July 2014), some of them established and managed jointly with regional governments, and the Valencian and Galician regional governments has two more each.

cofradías are local non-profit public corporations, representing the interests of the whole fishing sector (Pascual-Fernández 1999), with a deeply-rooted history (Erkoreka Gervasio 1991; Alegret 1996).

While the involvement of local stakeholders in the creation of MPAs is deemed necessary, who these local stakeholders are needs to be defined in each case (Chuenpagdee et al. 2013; Pomeroy et al. 2005, 2007). The degree of local involvement or influence in the creation of protected areas depends on the characteristics of stakeholders and the nature of interactions between different actors in the area. Following Mitchell et al. (1997) and Mikalsen and Jentoft (2001), the most relevant traits of all stakeholders involved are legitimacy, urgency, and power. These three factors are not distributed evenly among the different user groups in the coastal zone.

In the Canary Islands, small-scale fisheries are regarded as legitimate as they have a long history attached to them in coastal areas. Small-scale fisheries are not only important to fisher livelihoods but also to other actors such as restaurants, fish-shops, and residents. While anglers, with their recreational activities, also have a history in the area, their legitimacy may be weaker because they are less dependent on the fisheries for livelihoods. In relation to urgency, again small-scale fishers depend on fishing not only for economic reasons, but also cultural ones. As mentioned above, recreational fishers fish for leisure and hence urgency of access is lower for them. With respect to power, although traditionally *cofradías* have always had a major influence on fishery and coastal zone policies, today recreational fishers and other coastal stakeholders, who are demanding to be heard, appear to have more power than before (Heck et al. 2011; Hattam et al. 2014). The demands of new actors have added complexity to governing interactions, and create governability challenges. Recreational fishers in general are not members of associations and hence the demands of those who speak as their leaders do not necessarily represent the voice of recreational fishers as a whole, reducing their legitimacy. As of December 2013, there were a total of 98,134 recreational fishers (according to the number of licenses issued) in the archipelago (Source: Canarian Government). Data from our research pertaining to Tenerife suggests that only 4 % of them were associated with any organization.

Recreational fishers have significant power due to their numbers and potential political influence. Although they can be dormant stakeholders (Mitchell et al. 1997), if the conservation agenda and governance of small-scale fisheries compromises their situation, they are likely to become active. In developed countries, the increasing number of recreational fishers has placed them in the spotlight, presenting governance challenges and uncertainties (Hickley et al. 1998; Pitcher and Hollingworth 2002; Pawson et al. 2008). It is necessary, therefore, to further analyze the impact of recreational fisheries on society, ecology and governance (ICES 2013).

Governing systems in coastal areas face challenges to cope with the presence of new actors more generally, not only recreational fisher groups but also conservation organizations, scuba divers, tourist operators and scientists, without marginalizing small-scale fishers. This chapter examines how small-scale and recreational fishers interact in a shared space and discusses whether these interactions contribute to making marine reserves more or less governable. We focus on two marine reserves on different islands, one in Tenerife, not yet implemented, and another in El Hierro. We begin by presenting a general description of the study areas and the current status of fisheries management in the archipelago, including conservation efforts and the development of recreational fisheries, and conclude with a discussion on how all this impacts upon governance.

Our research on marine reserves in the Canary Islands began in 2001 in El Hierro and 2004 in Tenerife. In general, we employed a mixed method approach, which was mostly qualitative and longitudinal, including interviews, participant observation, and surveys using questionnaires. Archival work and a literature review were also undertaken. In the study of recreational fishers in Tenerife, we developed field-sampling methods (more than 100 sampling points on the island) to conduct on-site interviews (more than 1,000), and phone surveys (500 interviews every 3 months of the year). In El Hierro, qualitative methodologies were used to examine interactions between small-scale fishers and recreational fishers. Finally, key informant interviews were conducted with fishers, the local population, recreational users, scientists and administrators to obtain their opinions about MPAs and gather information about stakeholder roles in the governance of the area.

El Hierro and Tenerife Islands

The Canary Archipelago comprises seven islands and several islets with a total coastline of about 1,500 km. Marine ecosystems surrounding the Archipelago can be described as sub-tropical, characterized by high biodiversity but low volumes of particular species. The environment is under increasing pressure due to transformations along the shoreline caused by construction and coastal infrastructure, climatic change, the emission of contaminants, and fishing amongst others. The natural side of the system-to-be-governed is, in other words, fragile.

There are many fishing ports and beach areas around the coast of each island. In El Hierro (nearly 11,000 inhabitants in 2013), most small-scale fisheries activities are based in the harbour of La Restinga. La Restinga is the home of the only *cofradía* in El Hierro. In Tenerife (almost one million inhabitants), there are ten *cofradías* and small-scale fisheries are carried out from a large number of harbours. On both islands, the majority of fishing boats are small-scale, i.e. less than twelve meters in length, and trawling is completely banned. In total, there are approximately 1,000 small-scale fishers in the archipelago earning their living from fishing.[3]

The government's role in managing the islands' coastal zone was, for a long time, very limited, with only a few specific regulations in place. The creation of autonomous regions in Spain led to the transfer of responsibilities for fishing in

[3]Total employment in fisheries and aquaculture in the Canary Islands is 1,662 persons as of September 2014 (Source: Canarian Institute of Statistics). There is no official data about employment in the fleet of less than 12 m but our current estimate is around 1,000 fishers.

internal waters to the Canarian Government, while the central government maintained control over territorial waters.[4] The distribution of responsibilities between the central administration and regional government led to fragmentation of legislation and control of fishing activities in the archipelago, making things more complicated. The first-sale of fish has been regulated since 2002, but the level of compliance is far from perfect, as the points of first sale are few and not available for much of the day in many areas. Furthermore, according to fishers interviewed, there is a market for furtive fishing especially to restaurants, which in some cases results in undercutting small-scale fisheries and reducing their market share. The economic crisis in Spain from 2008 onwards resulted in high unemployment in the Canary Islands, making poaching a real option for the unemployed, and even for those who just wanted extra income.

Meanwhile, recreational fisheries in the archipelago have increased. To fish recreationally a license must be obtained. As of December 2013, there were 98,134 licenses in operation (Department of Fisheries, Government of the Canary Islands), with the vast majority belonging to anglers who fish from shore or boat, and to a lesser extent, spear fishers. It is very difficult to obtain precise data about boat anglers, as they share the same licenses as those who fish from the shore (Pascual Fernández et al. 2012b). In our previous research on recreational fishing in Tenerife, we found it easy to obtain data for shore-based anglers but not for spear-fishers or boat anglers (Pascual Fernández et al. 2012a, b). In other words, understanding recreational fisheries is more complicated than understanding the small-scale fisheries as data is often incomplete and less reliable. The percentage of recreational fishers, moreover, is highly variable across islands. For example, recreational fishers constitute 4.3 % of total population in Tenerife and 19 % in El Hierro (Source: Canarian Government).

In this context, concerns about the overall situation vis-à-vis fishing resources and the marine environment in general are increasing. One suggested way to improve the health of marine ecosystems has been through the creation of marine reserves in the Canary Islands (Bacallado et al. 1989). As of now, three marine reserves have been established, one around the island of La Graciosa, another in the south west of El Hierro, and a third one in the south west of La Palma, in 1995, 1996 and 2001, respectively. The El Hierro MR was sought by the local *cofradía*, while in La Graciosa local fishers did not support the creation of a MR (Chuenpagdee et al. 2013; De la Cruz Modino and Pascual-Fernández 2013). As Chuenpagdee et al. (2013) and Jentoft et al. (2012) argue, the level of support for MPAs depends on whether stakeholders' views were taken into account or not. The benefits of

[4]Article 148 of the 1978 Spanish Constitution specifies that regional governments have the capacity to legislate and manage maritime areas (Suárez de Vivero and Frieyro de Lara 1994); hence, fisheries responsibilities are shared in Spain. Both national and regional governments legislate on the protection of the marine environment under their jurisdiction. For regional governments, this relates to internal waters, as those situated between capes and specified by the state. As Suárez de Vivero et al. affirms: "This division of competences also affects territorial distribution: the Central Administration have exclusive competences over the Territorial Seas (TS) and the Exclusive Economic Zone (EEZ) – where most national fishing areas are located – whereas the regional governments restrict their action to Internal Waters (IW)" (1997, 199).

marine reserves, as perceived by small-scale fishers, were mainly linked to improved catches, increased surveillance, and limited entry of boat anglers, spear-fishers or small-scale fishers coming from other harbours. In some cases, local small-scale fishers were happy to accept additional restrictions on gears and the designation of no-take zone that MRs brought with them. The success of some of these reserves, especially in the case of El Hierro, raised hopes of creating new reserves in other areas of Spain (García Allut and Jesús 2009). Two MRs have been planned in Tenerife, the neighbouring island. A lesson learned from the Tenerife case is that planning needs to take into account the social side of the system-to-be-governed, the governing system and governing interactions. These elements, particularly those of the social system, are very different in El Hierro and Tenerife, as the number of fishers, small-scale and anglers, as well as other stakeholder groups involved, is far greater in the case of Tenerife. Understanding the diversity, complexity, dynamics and scale of these systems is necessary in order to enhance marine reserves governability.

Projects for Marine Reserves of Fishing Interest in Tenerife

Tenerife is the largest of the Canary Islands and is characterized by a mild climate year-round. Its natural beauty, European culture and reputation for being safe has made it a popular tourist destination since the 1970s. In 2013, 4,770,660 tourists visited Tenerife (Source: ISTAC). Population growth and tourism development, along with their infrastructural and housing demands, have impacted the marine environment and the coastline. Tourism competes for space, workforce and capital with small-scale fishing. The same is true of aquaculture or recreational fishing.

The creation of marine reserves in Tenerife was proposed based on the findings of scientific studies carried out in the 1970s and 1980s (Bacallado et al. 1989). From 2004 onwards, the Island Government (*Cabildo Insular de Tenerife*[5]) began to promote their creation, obtaining funding from an external EU related project.[6] Feasibility studies were carried out by research institutions in two of the areas proposed by previous studies, Teno and Anaga. *Cofradías* were involved from the beginning in the planning process, together with scientists and the *Cabildo*. While the *Cabildo* has no legal authority to establish MRs, it may contribute to their design and promote their implementation with the regional and national governments who

[5] The *Cabildo Insular* is a local institution of government in the Canary Islands, created from 1912, There are seven *Cabildos* one for each of the Islands, and they have the capacity to regulate some areas like tourism or environment.

[6] Project PARQMAR, financed by INTERREG IIIB call of proposals, focused on the marine protected areas of Azores, Madeira and the Canary Islands.

can legislate.[7] Meetings with local communities and civil society organizations were held in many places around Tenerife in order to present and discuss proposals. The degree of participation and support for the initiative varied, with low involvement from people not related directly to small-scale fisheries or conservation groups. Small-scale fishers in some communities rejected the idea initially because they feared that the MRs would severely compromise their activities. In order to explain what an MR could imply for small-scale fisheries, the voices from fishers of El Hierro (where the most successful marine reserve in the Canaries is located) were useful. Fishers participating in the discussion became keener after hearing the El Hierro experience, and how an MR could reduce the pressure from poachers by increasing the surveillance in the area. Yet, the MR process took several years until some consensus as to where they could be established and what would be permitted in them was achieved in 2008.

Regional and national fisheries administrations, notably those with legal capacity to enact an MR, were not really involved in the initial process. Only a civil servant from the regional administration participated occasionally, but that too without the authority to make any commitment. Recreational fishers were consulted, but they did not participate effectively in the negotiations about the proposals. The problem with recreational fishers is that they are highly diverse and lacking in terms of a representative organization. However, when finally an agreement on MRs was being finalized with the Island Government (*Cabildo*), recreational fishers used mass media and their political influence to block it. Moreover, they gained a seat at the negotiating table. Finally, in 2010 a partial agreement, when the *Cabildo* supported the two MR proposals with some areas for recreational fishing, was decided upon. Not all the *cofradías* agreed with this final outcome, and expected further consultations with the Department of Fisheries.

The process changed the power dynamics within the governing system. The *Cabildo* assumed a management role vis-à-vis island fisheries, while the regional government did not collaborate actively because it saw this initiative as a way to undermine its legislative powers in interior waters. Although the national government has the power to create MRs in territorial waters, funding for new MRs in the country was halted due to the financial crisis in Spain. Funding for existing MRs, moreover, was reduced substantially, in some cases compromising protection and surveillance of the MRs as the coordination budget was greatly cut. The delay in the process meant that the proposal lost its momentum.

[7]The LOTRACA or Organic Law 11/1982, Supplementary Transfers to Canary Islands is a law passed on 10 August 1982 together with the Autonomy Statute of the Canary Islands. This law transfers to the Canary Islands some State government competencies, to offset the special circumstances of the Canary Islands, specifically insularity and remoteness. The capacity to regulate fishing in some areas or aquaculture was transferred with this law and Royal Decree 1938/1985. This has not been delegated to *Cabildos*.

Conflicts Between Small-Scale and Recreational Fisheries in El Hierro

The island of El Hierro is the smallest and least populated among the seven islands of the Canary Island archipelago, with only 268.71 km^2 of land-surface and 10,979 inhabitants in 2013.[8] The village of La Restinga, the main settlement linked to the marine reserve, is located in the southwest of the Island and was for a long time mainly dependent on small-scale fisheries. However, in the last few decades alternative livelihood opportunities related to tourism and other services have emerged (De la Cruz Modino and Pascual-Fernández 2013).

The maritime area close to La Restinga is named the *Sea of Calms*, an important fishing ground where fishing activity has been developing and was improved with the creation of a local *cofradía* and the construction of a fishing harbour. Nowadays, there are 29 active fishers and 31 boats, characterized by their small size (most of them around 7–8 m length, some larger 10–12 m), suitable for a multiple fishing gear strategy with a diverse range of target species. In addition to the *cofradía*, the cooperative *Pescarestinga* is another key governing institution, created by local fishers to recover control of the marketing of catches.

Sea of Calms is a fragile ecosystem with a high diversity of species, exceptional underwater visibility, and a warm sea surface temperature, making it potentially a great tourism destination, especially for scuba divers. Divers come year round as well as visitors from other islands mainly in the summer and on public holidays. Local people, including many fishing families, manage the majority of tourism services such as lodging, restaurants and bars, resulting in increased household incomes.

Fishing and tourism development generate concerns in the area in terms of ecosystem degradation. Local fishers have always been aware of the fragility of their environment. Fishers have instituted their own conservation rules (e.g., banning fishing gears such as fish pots, trammels and longlines) and have advocated for the creation of a marine reserve in the *Sea of Calms* to protect benthic resources. Supported by the national and regional fisheries administrations and scientists of the University of La Laguna (in Tenerife), a design and establishment process for the creation of a marine reserve began with active participation from the *cofradía*.[9] However, no other stakeholders were involved in the consultation process and in the actual design. As a result scuba diving and angling were banned in the no-take zone (angling in the buffer zone too) and recreational fishing by boats was prohibited. Initially, these restrictions led to conflicts with dive centres, which were feeling the pinch. However, it turned out that the area has attracted more tourists, thus more than doubling the number of diving centres in the area.

[8] The permanent population slightly exceeds two million inhabitants in the Archipelago and more than ten million tourists visit the Canaries every year; see http://goo.gl/tcSY6i, accessed July 11 2013.

[9] See De la Cruz Modino, Pascual Fernandez (2013) and Jentoft et al. (2012).

We can consider El Hierro MR as an example of a co-governance system, where fishers, scientists and government cooperate to support the sustainable development of small-scale fisheries and conservation of marine resources. In September–October 2011, an underwater volcano erupted in the middle of the MR, which seriously affected marine life, fishing and other marine activities in nearby areas. The *cofradía,* supported by Canarian universities and the Marine Reserve management, established a temporary ban on fishing in the most affected area from October 2011 to January 2013 for pelagic species, and to March 2013 for demersal species. The national and regional governments, at the request of the *cofradía*, supported the fishing ban for 6 months and provided funding to small-scale fishers[10] so that they could endure the lack of income. The regulation permitted boat fishing for pelagic species and recreational fishing from the shore or by boat in certain places.

The temporary ban in El Hierro MPA was not well received by recreational fishers or by business associations. In particular, they did not approve of the initial national decree (Order AAA/1990/2012, September 13) that limited angling from the shore in many areas of the Island. One of their arguments was that limiting recreational fishing could reduce the number of tourists on the island at a time of economic crisis. In this particular instance, the new president of the Cabildo sided with the recreational sector breaking a long tradition of cooperation with professional fishers. Demonstrations took place in the main town on the Island, consequently resulting in some restrictions being modified.

Conflicts between recreational fisher groups and small-scale fishers became acrimonious. However, not all recreational fishers were united. Those who were part of the Canarian Association for Responsible Fishers supported conservation measures, hence not alienating themselves from small-scale fishers. Other recreational fishers who are not members of this association exerted pressures in multiple ways on local politicians. Social networks, local radios and local web-based newspapers played a role in exacerbating the conflict. In this open-access, online format, people have opportunities of making anonymous comments on the issues, viewable to all. Some of these comments have created anger, as they were perceived as unfair and insulting by both sides. The leader of this informal group of recreational fishers clashed with the president of the *cofradía*, and any negotiation between them became impossible.

The ban ended in March 2013, but recreational fishers and scuba divers continue to demand inclusion in decision-making related to fisheries and ecosystem management. Such inclusion is critical given the possibility of establishing a Marine National Park. A proposal to do so is currently being considered by the national government, the *Cabildo* (Island government), municipalities, *cofradía* and a variety of organizations and civil society in El Hierro. The proposal is being opposed by the

[10] ORDER 26 September 2012 established a temporary closed season on the island of El Hierro (BOC, October 1st 2012 http://goo.gl/nIsOaZ). Order AAA/1990/2012, 13 September established a temporary closed season on fishing activities around the island of El Hierro, BOE September 21, 2012 (http://goo.gl/43vjeS), Order AAA/2788/2012, 21 December modified Order AAA/1990/2012, BOE 27 December 2012 (http://goo.gl/QS7gLZ).

same recreational fishers group because it feels that a National Park will lead to some restrictions on recreational fisheries. The president of the *Cabildo* has been critical of the project, urging no further restrictions on recreational fisheries. The Canarian Association for Responsible Fishers, on the other hand, has given its initial support for the initiative, as they feel the restrictions on recreational fisheries would be minimal and benefit the whole Island in terms of tourism development. Small-scale fishers would not be greatly impacted by the regulation as they have already imposed gear restrictions on themselves some years ago. Legislation pertaining to National Parks allows small-scale fishing inside the protected area, with specific restrictions, as happens in the National Park of the Atlantic Islands of Galicia and the National Park of Cabrera Archipelago in the Balearic Islands, the only two existing with marine areas in Spain. The *cofradía* started to support the creation of an MPA when they realized it would be a good model for preserving small-scale fisheries and promoting conservation.

Conflicts are to be expected. This is an ongoing and dynamic step-zero process (Chuenpagdee et al. 2013), with no detailed road map or proposal on the table at the moment, and with competing images held by different stakeholders of what the Park can bring (Jentoft et al. 2012). As the process evolves, knowledge about the Marine National Park becomes more precise for all stakeholders, reducing the *image* gap between them. A National Park implies a very different governance structure from the Marine Reserve; governing bodies of the former require the involvement of more stakeholder groups by law, which creates some uncertainties. However, this more inclusive structure may improve the interaction between stakeholders groups who are currently disconnected and suspicious of each other.

Discussion

In recent years, fisheries policy in Europe has moved towards greater involvement of stakeholders in regulatory decision-making, notwithstanding country differences in terms of power sharing mechanisms and the distribution of government responsibilities among agents (Mikalsen and Jentoft 2008). In Spain, the structure of fisheries management involves national and regional governments, both with some capacity to govern and legislate in specific areas. Small-scale fishers have long been integrated in the governing process in coastal areas through the *cofradías*. What has happened in many instances is that the diversity of uses and stakeholders has increased, diminishing the relative importance of small-scale fishers. Recreational fishers, scuba divers, snorkelers, surfers and tourists are all demanding a say in coastal affairs. Each of these stakeholder groups has a variety of relationships with other groups, most importantly with small-scale fishers. These relationships can be mutually beneficial or conflictive but certainly add complexity to the system. Furthermore, interaction and exchanges inside each of these groups have soared, as a consequence of new opportunities facilitated by Internet.

With respect to scale issues, the situation is very different in Tenerife and El Hierro. El Hierro's total population is a just under 11,000 while that of Tenerife is around one million. In El Hierro, everyone knows each other. This is not necessarily an advantage as conflicts or confrontations amongst friends and acquaintances may evoke strong feelings and reactions. In Tenerife, the number of recreational fishers is more than threefold the total population of El Hierro. El Hierro should, therefore, be more governable than Tenerife, despite the current conflict. Yet, it is clear that the governability of fisheries in both places is lower than before because of the significant transformations that have taken place in coastal activities (Chuenpagdee and Jentoft 2013).

Changes in social systems frequently create chaotic situations. In our two case study areas, the system-to-be-governed changed relatively quickly, as new stakeholders entered the arena. Apart from recreational fishers, there are surfers, scuba divers or tourism operators, all making demands that need to be taken into account. However, various actors within the governing system do not necessarily react to the changing situation in similar ways. The Fisheries Department at the national level (Secretaría General de Pesca), like the Department of Fisheries at the regional government level, has taken time to adjust to new stakeholder claims. Integrating new stakeholder groups is not an easy task, as the majority of them are not well organized. Recreational fishers are usually not considered by many, including government, as definitive stakeholders, as they lack a degree of legitimacy and urgency. However, the fact that they are poorly organized (as is also the case with scuba diving operators) does not hinder their capacity to influence media or politicians. In both El Hierro and Tenerife, leaders have been chosen keeping in mind their closeness to the media, local institutions and local politicians. Similarly, the media and social networking sites have been used to mobilize local people in relation to their demands. They have been able to compensate weak organizations (like in Tenerife) or the lack of them for some groups (in the case of El Hierro), with the wide use of these media tools to put pressure on institutions.

The challenges faced can be highlighted by reflecting upon the deliberations that took place when a proposal for a Fisheries Law in the Canary Islands in 2009 was put forward. The draft proposal was presented to *Cofradías*, and received inputs from them. The proposal included restrictions on recreational fishers, which led them to protest because they felt they had not been included in the process. The law was finally blocked. The governability of the system suffered, and the governing system had difficulties coping with the new situation. Transforming a conflictive relationship to one of collaboration constitutes a governability challenge, but is also a wicked problem (Chuenpagdee and Jentoft 2013; Jentoft and Chuenpagdee 2009), as the demand for more inclusive governance often involves groups that are poorly organized. In other words, while there is a need for being more inclusive, there is no clear way forward to achieve this goal. To create organizations involves time, effort and/or money. For most recreational fishers, investing any of the above is not worth it as the benefit they will attain is limited or nonexistent (Wade 1987). In El Hierro, however, the previously disorganized recreational fishers have built an association so as to participate in the discussion process about the Marine National Park.

Theories of collective action help explain why only recreational fishermen are organized, usually just a few of them, when something threatens their *status quo* (Wade 1987).

Cofradías traditionally have cooperated with the government in fisheries management in various ways. For example, most of current gear restrictions in the Canary Islands are a result of cofradías requesting such restrictions. Hence, this can be considered a form of co-governance (Kooiman and Bavinck 2005), as *cofradías* are not only consulted but also influence policy (Pascual-Fernández 1999).

The case of the Marine Reserve of El Hierro is illustrative of the benefits of co-governance and how co-governance can avoid many of the shortcomings of hierarchical management (Hind et al. 2010; Hogg et al. 2013). However, co-governance remains vulnerable. The government, in some instances, may decide to act on its own, hierarchically, and impose regulations on fishers. Another possibility is that the European Union imposes decisions on Spain and local *cofradías* have little say. The reverse is also possible. *Cofradías* may act on their own, in the name of self-governance, and for instance increase minimum landing sizes without the backing of the government, but supported by the compliance of local fishers and the assessment of scientists (Brito Hernández et al. 2013). This has happened in La Restinga in the case of an emblematic species, the parrotfish (*Sparisoma cretense*), and some others. The challenges in making co-governance work are greater given the diversity of stakeholders.

Whereas the development of the Marine Reserve in El Hierro in the 1990s did not result in confrontation between small-scale and recreational fishers, the new projects in Tenerife and El Hierro have led to such confrontation because of the necessity of being more inclusive, precisely with stakeholders related to recreational activities, like fishers or scuba divers. In marine reserves in Spain, local small-scale fishers and particularly *cofradías* are included in the 'governing system', and in the design and consultation process. In some cases, as in la Restinga, small-scale fishers have been the promoters of the reserve. They can affect or be affected by marine reserves, as the theory of stakeholder identification and salience postulates (Mitchell et al. 1997; Mikalsen and Jentoft 2001). This is because they possess three key attributes: power, legitimacy and urgency. The situation is very different for recreational fishers, as they have often not been considered as legitimate stakeholders. Moreover, their lack of organization has prevented them historically from participating. It is, therefore, more challenging to involve them now and to deal with the complex interactions between different stakeholder groups. In Tenerife, discussions about the projects for marine reserves have created an opportunity for small-scale and some recreational fishers to interact. The process helped both groups approach each others' positions, and has slowly favored a convergence of images about the main challenges related to fisheries arena. For instance, the same recreational fisheries leader that initially opposed the marine reserves is now participating in the Fisheries Local Action Groups[11] in Tenerife at the level of their managing board,

[11] These are associations formed from partnerships between fisheries actors and other local private and public stakeholders, who managed the Axis 4 funds of the EMFF dedicated to achieving the sustainable development of fishing areas. Source: https://webgate.ec.europa.eu/fpfis/cms/farnet

where small-scale fishers have the majority of votes. The interactions between some recreational fishers associations and small-scale fishers have improved in Tenerife. The situation is not stable, as the leaders of recreational fishers can be challenged at any time, because the associations they represent have only a small number of members. Though the current co-governance arrangements are dynamic and inherently unstable, they also offer an opportunity for more inclusive governance.

Conclusions

Increasing diversity of coastal activities, including that of recreational fishing, has resulted in a new governing scenario in many western countries. This new scenario is characterized by an increase in diversity, complexity and dynamics within the system due to more interactions between stakeholder groups. Different scale issues, moreover, are involved. In countries like Spain, sophisticated systems of co-governance were developed, creating or sustaining organizations like *cofradías* to cope with the dynamic nature of these areas. However, economic development has resulted in significant changes to the coast and the inclusion of new stakeholders such as recreational fishers and tourism operators. This requires a restructuring of the interactions between user groups, and a transformation of the governing system in order to cope with new governability challenges. Such a transformation is needed because the capacity of institutions and the governing system as a whole has been unable to adapt. Bureaucratic inertia makes changes a slow process as there are many who might resist change.

The existing scenario constitutes a wicked problem (Jentoft and Chuenpagdee 2009), as it is connected with societal changes on a large scale and is virtually impossible to solve definitively. Governing interactions must facilitate information flow and partnerships between stakeholders. They should also allow for conflicting images of the coast to be discussed openly. This requires a process of alignment between stakeholder groups that may be slow and difficult. Although facilitating governing interactions between these stakeholders groups is not easy, it is possible to find opportunities for collaboration. Stakeholder groups do not need to be in conflict with one another (Chuenpagdee and Jentoft 2013). Small-scale fishers, recreational fishers, scuba divers and tourism operators in El Hierro or Tenerife can cooperate and collectively govern the coast.

Step zero processes of MPAs constitutes a governability challenge, but in some cases may also facilitate cooperation (Chuenpagdee and Jentoft 2007; Chuenpagdee et al. 2013). During these periods, given that none of the actors are in control of the situation, a new dynamic of interactions between some of these groups may appear, as has happened in the case of Tenerife or may occur in El Hierro. Such processes constitute an opportunity to revise relationships between stakeholders, reach new agreements and create new rules. There is no guarantee that this is going to happen, but the opportunity is always present. Sometimes external actors can be the catalyst for change, and in MPA step zero processes these external actors are usually present.

MPAs define a set of access rights, uses and users. While the step-zero process to define access rights, uses and users may result in conflicts, it can also be an opportunity to develop new agreements and reach synergies between stakeholder groups.

References

Alegret, J. L. (1996). Ancient institutions confronting change: The Catalan fishermen's confradies. In K. Crean & D. Symes (Eds.), *Fisheries management in crisis* (pp. 92–98). Oxford: Fishing New Books/Blackwell.

Bacallado, J. J., Cruz, T., Brito, A., Barquín, J., & Carrillo, M. (1989). *Reservas marinas de Canarias*. Canarias: Consejería de Agricultura y Pesca de Canarias Secretaría General Técnica.

Brito Hernández, A., Hernández Pérez, J. C., Clemente Martín, M. S., & Dorta Morales, C. (2013). *Estado actual de la biodiversidad y los recursos marinos costeros en la Reserva Marina de La Restinga-Mar de las Calmas (El Hierro)*. La Laguna: Universidad de La Laguna- Viceconsejería de Pesca y Aguas del Gobierno de Canarias. Unpublished.

Chuenpagdee, R., & Jentoft, S. (2007). Step zero for fisheries co-management: What precedes implementation. *Marine Policy, 31*(6), 657–668.

Chuenpagdee, R., & Jentoft, S. (2013). Assessing governability – what's next. In M. Bavinck, R. Chuenpagdee, S. Jentoft, & J. Kooiman (Eds.), *Governability of fisheries and aquaculture: Theory and applications* (pp. 335–349). Dordrecht: Springer.

Chuenpagdee, R., Pascual-Fernández, J. J., Szeliánszky, E., Luis Alegret, J., Fraga, J., & Jentoft, S. (2013). Marine protected areas: Re-thinking their inception. *Marine Policy, 39*(0), 234–240.

De la Cruz Modino, R., & Pascual-Fernández, J. J. (2013). Marine protected areas in the Canary Islands: Improving their governability. In M. Bavinck, R. Chuenpagdee, S. Jentoft, & J. Kooiman (Eds.), *Governability of fisheries and aquaculture: Theory and applications* (pp. 219–240). Dordrecht: Springer.

Erkoreka Gervasio, J. (1991). *Análisis histórico institucional de las cofradías de mareantes del país vasco*. Vitoria: Gobierno Vasco.

García Allut, A., & Jesús, A. (2009). Becoming proactive agents. *Samudra Report, 53*, 15–18.

Hattam, C. E., Mangi, S. C., Gall, S. C., & Rodwell, L. D. (2014). Social impacts of a temperate fisheries closure: Understanding stakeholders' views. *Marine Policy, 45*(0), 269–278.

Heck, N., Dearden, P., McDonald, A., & Carver, S. (2011). Stakeholder opinions on the assessment of MPA effectiveness and their interests to participate at Pacific Rim National Park Reserve, Canada. *Environmental Management, 47*(4), 603–616.

Hickley, P., & Tompkins, H. (1998). *Recreational fisheries: Social, economic, and management aspects*. Oxford: Fishing News Books.

Hind, E. J., Hiponia, M. C., & Gray, T. S. (2010). From community-based to centralised national management–A wrong turning for the governance of the marine protected area in Apo Island, Philippines? *Marine Policy, 34*(1), 54–62.

Hogg, K., Noguera-Méndez, P., Semitiel-García, M., & Giménez-Casalduero, M. (2013). Marine protected area governance: Prospects for co-management in the European Mediterranean. *Advances in Oceanography and Limnology, 4*(2), 241–259.

ICES. (2013). *Report of the ICES Working Group on Recreational Fisheries Surveys 2013 (WGRFS), 22–26 April 2013, Esporles, Spain*. Denmark. ICES CM 2013/ACOM:23. Retrieved September 13, 2014, from http://goo.gl/6HazRA

Jentoft, S., & Chuenpagdee, R. (2009). Fisheries and coastal governance as a wicked problem. *Marine Policy, 33*(4), 553–560.

Jentoft, S., Chuenpagdee, R., & Pascual-Fernandez, J. J. (2011). What are MPAs for: On goal formation and displacement. *Ocean and Coastal Management, 54*, 75–83.

Jentoft, S., Pascual-Fernandez, J., De la Cruz Modino, R., Gonzalez-Ramallal, M., & Chuenpagdee, R. (2012). What stakeholders think about marine protected areas: Case studies from Spain. *Human Ecology, 40*(2), 185–197.

Kalikoski, D., & Vasconcellos, M. (2008). Marine protected areas and reconciling fisheries with conservation: Insights from the common property theory. In J. L. Nielsen, J. J. Dodson, K. Friedland, T. R. Hamon, J. Musick, & E. Verspoor (Eds.), *Reconciling fisheries with conservation. Proceedings of the Fourth World Fisheries Congress* (pp. 1211–1219). Bethesda: American Fisheries Society.

Kooiman, J., & Bavinck, M. (2005). The governance perspective. In J. Kooiman, M. Bavinck, S. Jentoft, & R. Pullin (Eds.), *Fish for life: Interactive governance for fisheries* (pp. 11–24). Amsterdam: Amsterdam University Press.

Kooiman, J., Bavinck, M., Jentoft, S., & Pullin, R. (Eds.). (2005). *Fish for life: Interactive governance for fisheries*. Amsterdam: Amsterdam University Press.

Macfadyen, G., Salz, P., & Cappell, R. (2011). *Characteristics of small-scale coastal fisheries in Europe* (IP/B/PECH/IC/2010-158 PE 460.059). Brussels: Policy Department Structural and Cohesion Policies, European Parliament, Fisheries.

Mikalsen, K. H., & Jentoft, S. (2001). From user-groups to stakeholders? The public interest in fisheries management. *Marine Policy, 25*(4), 281–292.

Mikalsen, K. H., & Jentoft, S. (2008). Participatory practices in fisheries across Europe: Making stakeholders more responsible. *Marine Policy, 32*(2), 169–177.

Mitchell, R. K., Agle, B. R., & Wood, D. J. (1997). Toward a theory of stakeholder identification and salience: Defining the principle of who and what really counts. *Academy of Management Review, 22*(4), 853–886.

Pascual-Fernández, J. (1999). Participative management of artisanal fisheries in the Canary Islands. In D. Symes (Ed.), *Southern waters: Issues of management and practice* (pp. 66–77). London: Blackwell's Science/Fishing New Books.

Pascual Fernández, J. J., & De la Cruz Modino, R. (2008). Los espacios marinos protegidos en España: ¿nuevas formas institucionales para las estrategias de apropiación? In O. Beltrán Costa, J. Pascual Fernández, & I. Vaccaro (Eds.), *Patrimonialización de la naturaleza: el marco social de las políticas ambientales* (pp. 199–221). Donosti: Ankulegi Antropologia Elkartea.

Pascual-Fernandez, J. J., & De la Cruz Modino, R. (2011). Conflicting gears, contested territories: MPAs as a solution? In R. Chuenpagdee (Ed.), *World small-scale fisheries contemporary visions* (pp. 205–220). Delft: Eburon.

Pascual Fernández, J. J., Chinea Mederos, I., Santana Talavera, A., Martín-Sosa Rodríguez, P., Moreira Gregori, P. E., & Rodríguez Darias, A. J. (2012a). *Análisis de los resultados finales y elaboración de conclusiones sobre los resultados de las encuestas presenciales y de la encuesta telefónica sobre pesca recreativa a la población de la isla de Tenerife* (Proyecto GESMAR MAC/2/C68). Santa Cruz de Tenerife: Instituto Universitario de CC. Políticas y Sociales (Univ. de La Laguna) – Cabildo de Tenerife. Unpublished.

Pascual Fernández, J. J., Chinea Mederos, I., Santana Talavera, A., Martín-Sosa Rodríguez, P., Rodríguez Darias, A. J., & Moreira Gregori, P. E. (2012b). *La pesca recreativa en Tenerife y su regulación*. Santa Cruz de Tenerife: Cabildo Insular de Tenerife.

Pawson, M. G., Glenn, H., & Padda, G. (2008). The definition of marine recreational fishing in Europe. *Marine Policy, 32*(3), 339–350.

Pitcher, T. J., & Hollingworth, C. E. (Eds.). (2002). *Recreational fisheries: Ecological, economic and social evaluation*. Oxford: Blackwell.

Pomeroy, R. S., Watson, L. M., Parks, J. E., & Cid, G. A. (2005). How is your MPA doing? A methodology for evaluating the management effectiveness of marine protected areas. *Ocean and Coastal Management, 48*(7–8), 485–502.

Pomeroy, R. S., Mascia, M. B., & Pollnac, R. B. (2007). Marine protected areas: The social dimension. In *Report and documentation of the expert workshop on marine protected areas and fisheries management: Review of issues and considerations. Rome, 12–14 June 2006.* FAO Fisheries Report No. 825 (pp. 149–181). Rome: Food and Agriculture Organization of the United Nations.

Suárez de Vivero, J. L., & Frieyro de Lara, M. (1994). Spanish marine policy – Role of marine protected areas. *Marine Policy, 18*(4), 345–352.

Suárez de Vivero, J. L., Frieyro de Lara, M., & Jurado Estevez, J. (1997). Decentralization, regionalization and co-management: A critical view on the viability of the alternative management models for fisheries in Spain. *Marine Policy, 21*(3), 197–206.

Thorpe, A., Bavinck, M., & Coulthard, S. (2011). Tracking the debate around marine protected areas: Key issues and the BEG framework. *Environmental Management, 47*(4), 546–563.

Wade, R. (1987). The management of common property resources – Collective action as an alternative to privatisation or state-regulation. *Cambridge Journal of Economics, 11*(2), 95–106.

Chapter 22
The Governability of Mangrove Ecosystems in Thailand: Comparative Successes of Different Governance Models

Estelle Jones, Heidi Schuttenberg, Tim Gray, and Selina Stead

Abstract The continuing loss of coastal-marine biodiversity and the over-exploitation of coastal fisheries in many parts of the world fuel debates about what governance arrangements can turn the tide on these declines. Using the governability framework, this chapter investigates the comparative strengths and weaknesses of self-governance and co-governance arrangements, by interrogating six case studies from a mangrove region in Thailand. Our examination of three successful and three unsuccessful case studies, identifies the characteristics that distinguish cases which were successfully able to continue long-term implementation of negotiated resource management arrangements: (1) trust and cooperation in governance interactions, which enables conflict resolution and informal sanctioning; and (2) stakeholders' perceptions that the solution was fair and legitimate. Our analysis also challenges three widely accepted "good practices" for managing natural resources: the need for extensive community engagement in designing solutions; the requirement for formal recognition of community-designed solutions; and the desirability of large inputs of funding from external donors. Our data showed that whilst these three good practices are desirable, they are not necessarily required for successful initiatives. We also offer observations about the influence of diversity, complexity, dynamics, and scale on governance outcomes.

Keywords Governability • Mangroves • Self-governance • Marine protected areas • Case studies

---

E. Jones (✉) • S. Stead
School of Marine Science and Technology, Newcastle University, Newcastle upon Tyne, UK
e-mail: estelle.jones@talk21.com; selina.stead@ncl.ac.uk

H. Schuttenberg
U.S. Agency for International Development, Washington, D.C., USA

University of Aberdeen, Aberdeen, Scotland, UK
e-mail: heidi.schuttenberg@gmail.com

T. Gray
Department of Politics, Newcastle University, Newcastle upon Tyne, UK
e-mail: tim.gray@ncl.ac.uk

S. Jentoft, R. Chuenpagdee (eds.), *Interactive Governance for Small-Scale Fisheries*, MARE Publication Series 13, DOI 10.1007/978-3-319-17034-3_22

Introduction

The continuing large-scale loss of coastal-marine biodiversity and over-exploitation of coastal fisheries (UNEP 2006) fuel debates about what type of governance arrangements can turn the tide on these declines. Discourse on coastal governance occurring in a range of disciplines is increasingly critical of the ability of top-down hierarchical management regimes to maintain the sustainable use and conservation of coastal systems (Chambers 1983; Pomeroy 1995; Barrett et al. 2001; McLeod and Leslie 2009). However, alternative arrangements focused on either self-governance or co-governance have also faced challenging issues. For example, self-governance initiatives that are able to maintain sustainability by harnessing community self-interest often occur at smaller spatial scales that are generally inadequate for making meaningful contributions to biodiversity conservation (Roe et al. 2000). Conversely, co-governance arrangements that negotiate complex power-sharing agreements at broader, biologically-relevant spatial scales, raise questions about their long-term sustainability given the larger financial and conflict-resolution resources they require (Christie et al. 2009b; Schuttenberg 2011).

In this chapter we investigate the comparative strengths and weaknesses of self-governance and co-governance arrangements, by investigating six case studies from a mangrove region on the west coast of Thailand, which is home to substantial marine biodiversity and diverse small-scale fishing communities. The case studies vary in their governability characteristics and their outcomes. Their "Systems-to-be-Governed" (SG) vary in the problems that motivated management action, their spatial scale, and the level of community support for negotiated solutions. Their "Governing Interactions" (GI) vary in the actor group initiating management action, as well as the level of consultation and the level of trust or conflict evident in negotiating resource management rules. Their "Governing Systems" (GS) vary in the specific resource use rules in place, and the extent to which on-going implementation is established through formal or informal sanctioning capacities. Finally, their outcomes vary in terms of whether negotiated management solutions have achieved sustained implementation, whether fisheries benefits have been realized, and whether biodiversity has been conserved or enhanced. By examining patterns in these variables, we:

(a) identify the governability characteristics associated with different successful outcomes, and;
(b) test key hypotheses (identified below) about the comparative strengths of these governance models and the conditions required for their success.

Crafting Effective Coastal Governance Arrangements: The State of the Debate

This book considers questions around the governability of small-scale fisheries involving three governance modes: hierarchical governance, co-governance (also referred to as collaborative governance), and self-governance (Jentoft and

Chuenpagdee, this volume). Here we provide a brief overview of relevant theories and arguments related to these forms of governance, distilling common themes from the literature about their comparative strengths and weaknesses, as well as the governability factors that are commonly considered essential to their ability to deliver different outcomes. Through this discussion we identify key hypotheses that guide our interrogation of the six Thai case studies presented in the subsequent sections.

Hierarchical governance is a “state-centric”, top-down mode of governance, focusing on legality, bureaucracy, and command and control (Gray 2005). It was the most common form of environmental management and conservation in temperate, western countries throughout the nineteenth and twentieth centuries, and was exported to the developing world through colonialism and international aid projects (Christie and White 2007). However the appropriateness of transplanting these governance systems into contexts with very different social and institutional contexts was questioned both in terms of their fairness to local people (Brechin et al. 2003) and their ability to achieve the conservation and development objectives, which they often claimed as their goals (Roe et al. 2000). Over the last 20 years, the reaction against hierarchical approaches to governance – particularly in developing country contexts where formal institutions are often underfunded and weak – has been so strong as to be characterized as revolutionary (Phillips 2003). The System-to-be-Governed demanded a more flexible governing system.

One of the earliest responses to the perceived short-comings of hierarchical governance was to devolve the responsibility and decision-making around natural resource management from distant state-led institutions to the local or community level (Chambers 1983). By empowering communities to self-govern, this “bottom-up” approach is based on the idea that communities will sustainably manage their natural resources into perpetuity when they have the right, ability, and incentive to do so (Pomeroy 1995). Community-based arrangements are predicted to be successful when they are focused on community needs and interests (Horwich and Lyon 2007); when communities have the skills to resolve resource use conflicts (Pretty 2003); when rules can be easily enforced by the community because they are highly visible and operate at a manageable spatial scale (Schuttenberg 2001); and when communities are more homogeneous in character (Pollnac 1998). Additionally, communities should have clear, formalized tenure or rights to the natural resources they will manage (Ostrom et al. 1999). This formalization removes the “tragedy of the commons,” gives communities confidence they will receive long-term harvesting rewards for short-term sacrifices or restrictions on resource use, and legitimizes their efforts to enforce resource restrictions with users or “poachers” from external communities.

As efforts to understand and foster community-based self-governance regimes expanded, a number of assertions were made about the strengths and weaknesses of this governance model. Proponents argued that it resulted in fairer resource use arrangements (Western and Wright 1994), which were sustained through time in the absence of strong formal institutions (White and Vogt 2000), and provided tangible benefits to communities (Kiss 1999). Conversely, critics asserted that reported increases in natural resource harvests mainly resulted from reallocating resources away from external communities (McShane and Newby 2004), that self-governed initiatives were too small to provide meaningful protections to biodiversity (Roe et al. 2000), and that the implementation of community-based initiatives often unraveled in the face of internal conflict (Agrawal and Gibson 1999) or external pressures (Berkes et al. 2006).

Collaborative or co-governance was, in part, a response to the perceived weaknesses of both hierarchical and community-based self-governance (Christie et al. 2009a). In the coastal arena, the goals and process of collaborative governance are described extensively in the integrated coastal management and ecosystem-based management literatures. These theories for governing systems aim to overcome weak formal institutions by establishing broad, ecosystem-scale resource-use rules that meet the needs of a wide range of stakeholders. Requirements for co-governance include institutional arrangements that are both vertically and horizontally integrated (Cicin-Sain and Knecht 1998), so that there is consistency in the formal rules of local, state, and national authorities (vertical integration) and that potential cross-sectoral interests have been reconciled (horizontal integration). Also, governing systems are more likely to be successful when they are focused on ecologically-relevant spatial scales (McLeod and Leslie 2009), when all stakeholders have been meaningfully engaged and genuinely agree with the negotiated solution (CBD Ad Hoc Technical Expert Group 2005), and when there are transparent conflict resolution mechanisms (Christie et al. 2009a).

The co-governance emphasis on ecologically-meaningful scales could enable it to compensate for the perceived weakness in small community-based self-governance arrangements. However, working at these larger scales brings with it the necessity to accommodate the needs and aspirations of diverse stakeholder communities, as well as to integrate with complex formal and informal institutional arrangements (Christie et al. 2009a, b). Some have observed that this additional complexity makes it difficult to negotiate a mutually satisfying solution and to sustain its implementation, amidst changing social, political, and ecological dynamics (Schuttenberg 2011). In the rest of this chapter we bear in mind these competing theoretical assumptions when comparing the experiences of our six Thai case studies.

Approach of this Chapter

The following section describes six Thai case studies in terms of their governability characteristics (SG, GS, GI) and their outcomes (section "Case studies"). We examine patterns in these variables to identify the governability characteristics associated with successful outcomes (section "Distinguishing characteristics of "successful" case studies") making use of the theoretical assumptions about the governance models discussed above (section "Patterns that challenge accepted 'good practices'"), and investigate the comparative strengths and weaknesses of self- and co-governance models (section "Comparative successes of the two post-hierarchical governance paradigms"). In section "Conclusion" we conclude with a reflection on whether the evidence presented here suggests that community-based self-governance and co-governance represent trade-offs between implementing negotiated resource management solutions and governing over spatial scales that can effectively conserve biological diversity.

The six case study case studies are derived from community group interviews and one-to-one key informant interviews conducted between January and June 2011. Community group interviews were conducted in eight fishing communities ($n = \sim 200$ attendees total) and explored the communities' involvement in management initiatives and their responses to local crises. Key informant interviews were conducted with 24 academics, national and regional government officers and non-government organizations (NGOs). Abbreviated notes were made on flip charts during the community interviews and later translated and recordings from key informant interviews were transcribed. Both were entered into NVivo and analyzed by descriptive coding which formed the basis for emerging themes, patterns of behavior and approaches to management. This study also measured the ecological effect of three of the MPAs discussed below (Jones 2014).

Case Studies

The six case studies occur in the same geographic area and have broad similarities in their System-to-be-Governed and their Governing Systems (section "Similarities in the case studies"). We first describe these similarities to contextualize the case studies; we then describe each case study, highlighting the differences in their governability characteristics and outcomes (section "Variation in the case studies").

Similarities in the Case Studies

System-to-Be-Governed

Ranong and Phang-nga are two provinces on the west coast of Thailand just south of the border with Myanmar (Fig. 22.1). This System-to-be-Governed is dominated on one side by coastal mountains covered in rainforest and plantations, and on the other by one of the country's largest mangrove forests. In the marine environment, sea grass beds, muddy and sandy shores and, to a lesser extent, fringing coral reefs contribute to some of the world's most diverse ecosystems (True and Plathong 2010). Sandwiched between these coastal mountains and mangrove forests are towns and villages supplied by one main road running the entire length of the coastline.

In the region there are two larger towns (population: Kapoe 6,800 and Kuraburi, 12,000) and many small villages scattered along the main route. At the north is Laemson National Park, the biggest state-managed MPA in this area, which runs about 40 km south on the mainland to a sheltered area by Koh Phra Thong Island. Two large mangrove estuaries are located at each end of this area which support rich

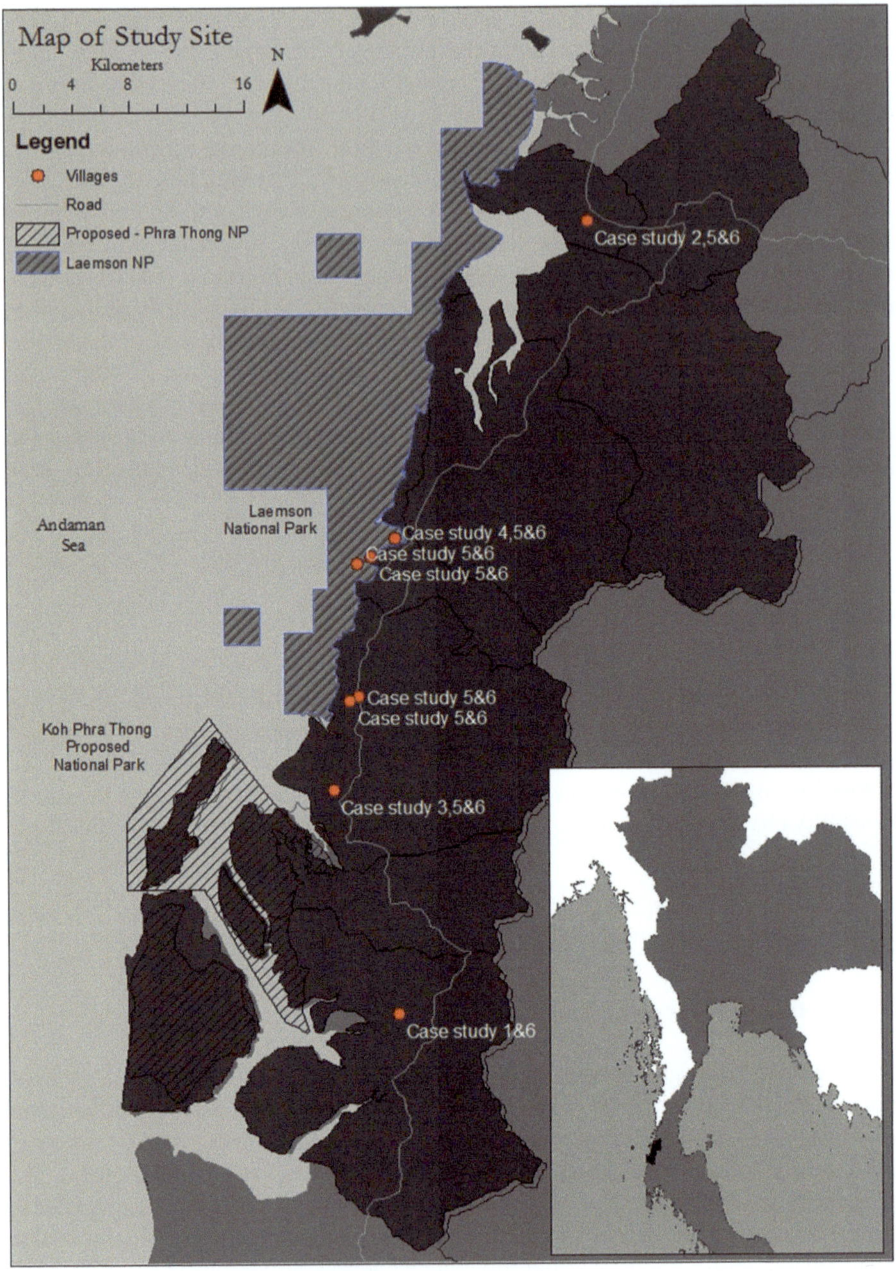

Fig. 22.1 Map of study site and insert of location in Thailand

biodiversity and are important fishing grounds. Eight fishing communities located on these estuaries have been involved in one or more marine management initiatives, the majority MPAs. These villages are populated predominantly by Thai Muslims, who practice fishing as their primary occupation and small-scale agriculture (oil and rubber plantations) as their secondary occupation. Large-scale plantations and aquaculture are also important economic activities here, and have contributed in the past to mangrove deforestation, though to a lesser extent than in other regions of Thailand. Commercial logging was outlawed in 1989 and all mangroves concessions expired by 2002. Illegal logging of mangroves trees is considered rare (per. comm. key informant, 2012), while replantation projects, undertaken in the revered King of Thailand's name, are frequent (NACA 2007; PhuketTV News 2011). The threat from logging has, however, moved upstream as high rubber and palm oil prices encourage conversion of the mountain slopes (per. comm. key informant, 2012). Moreover, the growing demand for seafood is placing increasing pressure on the range of marine taxa living in this productive ecosystem.

Along this coastline, small-scale fishing ('Small-scale fisheries') has been practiced for centuries, targeting a diverse range of species, including pelagic and demersal fish, squid and cuttlefish, shrimp, shellfish, jellyfish and crab. Today Small-Scale Fisheries is dominated by crab (Portunidae) fishing, with over 6 species commercially harvested using gillnets and box traps. What makes this fishery so important to this area is the value-added processing, whereby the crabs are either: (1) cooked, shelled and the meat packaged for sale to the big cities; (2) larger individual are transported, live, in tanks direct to market; or, (3) kept until they moult and sold as soft-shell crab to a highly lucrative market. What is socially important about this processing, especially the first, is that they employ local women and children who work from home or after school, contributing to community income and social cohesion through flexible working that can be more inclusive of the wider community. However, crab stocks along with other marine taxa are believed to be declining and local people have a growing awareness of the need for stock and habitat protection (per. comm. key informant, 2012).

Governing System

The Governing System comprises both the formal structures in place and the informal practices that take place. In Thailand, natural resource management was traditionally under the control of the state and highly centralized. In 1997 the 'People Constitution' was a major step away from this centralized system, whereby the role of 'citizen participation' was enshrined and delivered through a decentralization system:

> The State shall promote and encourage public participation in the preservation, maintenance and balanced exploitation of natural resources and biological diversity and in the promotion, maintenance and protection of the quality of the environment. (Anon 1997: Article 79)

This reform was followed in 1999 by the National Government Decentralization and Reform Act which aimed to replace what was perceived as an ineffective centralized system with an efficient, integrated and participatory system of governance and to resolve problems of overlapping and competing jurisdictions within government agencies. Thailand has three tiers of administration: national (often termed simply as 'Bangkok'); provincial (*Changwat*); and district (*Amphoe* and *Tambon*), and the 1997 Constitution aimed to devolve control away from Bangkok to the districts.

This devolved management is complex, but officially at least, it is administered through two offices, the *Amphoe* office and the *Tambon* Administration Organisation (TAO) (Fig. 22.2). At the *Amphoe* level, the district officer or *nai amphoe* oversees civil laws, policing, tax collection, district registers, vital statistics, and local elections. The district officer also has access to the *Gam nan* and *Puu Yai Baan* (see below) and convenes monthly meetings to inform them of government policies and instructs them on the implementation of these policies. As the chief magistrate, the district officer is also responsible for resolving land disputes and many villagers refer their disputes to him rather than to the court. The district officer reports directly to the provincial governor.

> …the provincial governor and the *nai amphoe* are to direct and oversee the operation of TAO to ensure that it functions within the framework of law and follows the rules laid down by the officialdom. (Rajchagool 1999, 39)

At the sub-district or *Tambon* level, there is one official office, the *Or-Bor-Tor*, but also a second group which is the old system of *Gam nan* and *Puu Yai Bann*. The

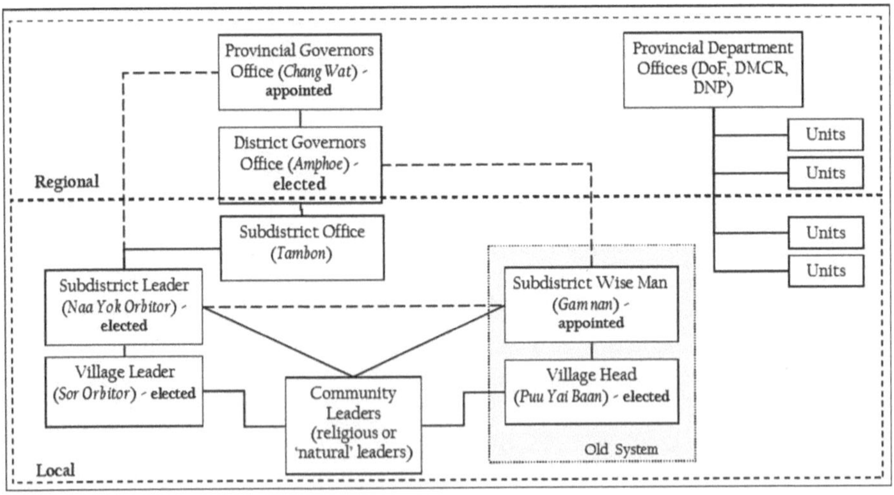

Fig. 22.2 Main institutions involved in the management and conservation of the marine and coastal area. *Solid lines* represent official chain of commands defined in legislation; *broken lines* represent unofficial chain of commands remaining from old systems of administration. The *grey box* represents old systems which still exist and hold political sway (created from key informant interviews)

Puu Yai Baan is the village head and the *Gam nan* is elected from a cluster of *Puu Yai Baans*, (between 6 and 10 villages) to lead the group. Initially elected for life, both parties are now elected every 5 years. The new Tambon Administration Organisation (TAO) Act B.E. 2537 (1994) gave powers to the *Or-Bor-Tor* which formally replaced the traditional system, though in practice the *Gam nan* and *Puu Yai Baan* still carry political weight (Rajchagool 1999). The TAO has the same structure as the old system, with the *Sor Or-bor-tor* acting as the village head and the *Naa Yok Or-bor-tor* as the regional head, but they are elected every 4 years, can only run for two terms, and are officers of the government.

There are three government agencies who are directly involved (there are a number of departments indirectly involved) in the management of coastal and marine resources: the Department of Fisheries (DoF), the Department of National Parks (DNP) and the Department of Marine and Coastal Resources (DMCR). Head offices of all these departments are based in Bangkok; regional units exist at the provincial and district level for the DoF and DMCR; and each national park has a team from the DNP. Staff in these departments tend to interact directly with each other at a similar level but receive their instructions, strategies and policies from above.

Variation in the Case Studies

Below we describe six case studies in which resource use rules were negotiated in response to perceived declines in fishery catches or ecosystem condition. A summary of the key features of each case study are presented for comparison in Table 22.1.

Case Study 1 The first case study involves a community's response to a decline in crab catch on which they are dependent for food and sale to market. After initial consultation within the village, this community responded by: firstly, establishing a conservation group; secondly, by dividing an area of mangrove into two zones – one for utilisation and the other for conservation and; thirdly by rearing and releasing young crab in the hope of improving stocks. This response was closely linked to the community's needs, and to the community's capacity to view its problems from different angles which generated a variety of solutions (SG). Obtaining full participation from villagers took time and awareness-raising. Once achieved, this led to the establishment of a gear limitation zone, which restricted gears perceived as 'modern' or non-traditional, and only traditional traps and mullet nets were permitted (GS).

Once support from within the village was established and management options were identified, villagers patrolled the area and confiscated illegal gear, mainly from people outside the village. This enforcement of the rules on 'external' fishers caused conflict with neighbouring villages, and wider support was sought from district officers to enforce this 'community-established' conservation area; enforcement is difficult due to intimidation and the fact that external fishers are not actually breaking

Table 22.1 Overview and comparison of key features in each case study

		Self-governance		Co-governance			
	Feature	Case study one	Case study two	Case study three	Case study four	Case study five	Case study six
System to be governed	Catalyzing problem	Low crab catches	Low crab catches	Habitat degradation	Conflict between SSF gears	Commercial fisheries encroachment	Mangrove deforestation
	Scale of solution	3 km^2	2 km^2	3 km^2	1,500 m out along 65 km of coast	315 km^2	~400 km^2
	Community support for solution	Yes – want to look at impacts of pollution on the MPA	Yes – want to expand in other villages	No – internal gear conflict waiting for outside help	Mixed – fatigued but need a solution	Mixed – de-prioritized after the tsunami, but support now re-emerging	Yes – protective ethos toward mangroves
Governing interactions	Solution initiated by	Community	Community	Conservation NGO	Communities	National park (NP)	NP, expert advisors and local elites
	Consultation in developing solution	Yes – within the village	Yes – between two villages and with local marine expert	Yes – between NGO, government (DMCR) and some groups in two communities	Yes – regional meeting with government officers and resource users	Yes – between NP, village leaders, and local fishers	Limited – village leaders inform community of rules to follow
	Trust and cooperation in interactions	Yes – willingness to take part and enforce	Yes – willingness to take part and enforce	No – internal conflict and mistrust in village around objectives	Mixed – village conflict but willing to engage	n/a	Yes – most villagers had concerns about mangrove deforestation

Governing system	Solution found/rules created in area	Yes – zonation with conservation, utilization and gear limitation zones	Yes – community no-take zone	Yes – seasonal no-take zone	Yes – no purse-seining zone beach to 1,500 m offshore	Yes – Sea guarding scheme with SSF in exchange for legitimate access to fisheries with approved gear	Yes – Mangrove cutting limited to personal use and requires 1:3 replanting. Need approval from committee
	Formal sanctioning capacity	No – no capacity in structure to recognize community spatial restrictions	No – no capacity in structure to recognize community spatial restrictions	No – no capacity in structure to recognize community spatial restrictions	No – no capacity in structure to recognize community spatial restrictions	Partial – National Park Act and informal agreement with communities	Yes – violators would be prosecuted under the Forest or National Park Acts
	Informal sanctioning capacity	Yes – social sanction and gear confiscation back by village leaders	Yes – social sanction and gear confiscation back by village leaders	No – spatial restrictions not recognized by some fishers	No – no social sanctioning to enforce spatial restrictions	No – SSF report violations & National Park implements enforcement	Yes – regular reporting, sometime used as a vendetta
Outcomes	Solution upheld	Yes – running for around 9 years	Yes – running for around 4 years	Mixed – still designated but poaching common	No – 6 months only	No – stopped after tsunami	Yes – people report violation & NP enforces based on these reports
	Fisheries success?	Yes	Yes	No	No data	No data	No data
	Biodiversity success?	No	No	No	No data	No data	No data

the law, but flouting locally agreed initiatives which impacts on management effectiveness (GI). Official government support was not forthcoming, however, the villagers' efforts appear successful for meeting fisheries objectives (Jones 2014) for improving the crab stocks and increasing Catch Per Unit Effort (CPUE).

Case Study 2 The second case study is tackling the same issue as the first, a perceived crisis in crab stocks (SG). As before there was self-diagnosis of a problem, but in this case external advice was taken to find a solution and a no-take marine protected area (MPA) was established (GS). Defining the boundaries was however challenging, conflict arose with a neighboring village and compliance was low. Through community meetings a compromise on the size of the protected area was achieved and a workable solution negotiated (GI). Whilst obtaining 100 % acceptance of the no-take zone by all villagers is an ongoing task, the community are undertaking enforcement themselves by monitoring, confiscating gear and reporting violations to village heads. This case is perceived as successful and independent assessments show the MPA is having an ecological effect (Jones 2014), but the management initiative appears unstable as it is exposed to the caprices of the local elites.

Case Study 3 The third case study describes an initiative co-developed by an international NGO and a community at a small, community-level scale. Management discussions were initiated by the NGO in response to a perceived need for species and habitat protection after degradation from the 2004 Asian Tsunami (SG). The NGO consulted with the community to gauge willingness and potential local benefits (e.g., ecotourism and crab nursery ground protection) were presented as an incentive. Subsequently a seasonal protected area was established with the aim of prohibiting fishing within its boundaries for 3 months each year, from February to April (GS). However not all resource users were supportive and conflicts arose. An internal gear conflict between crab trappers and crab netters, perceptions of limited consultation, and inadequate conflict resolution mechanism led to low levels of compliance and little support for its continuation (GI). The community felt formal recognition and government engagement would help, but it was not forthcoming. Independent assessment of the ecological effect of this management initiative found that neither fisheries nor biodiversity objectives were met (Jones 2014).

Case Study 4 The fourth case study illustrates a co-governed approach for tackling a gear conflict occurring between small-scale purse seiners and other small-scale fisheries. This approach was born out of post-tsunami funding which had some negative, unintended consequences by allowing fishers to up-grade their gear, and enabled small-scale purse seines to arrive in the village. This resulted in villagers fishing with purse seines close to the shore and impacting on other target species and taking many juvenile fish which caused conflict with other fisheries (SG). Advice and support were sourced from outside the villages with village heads, fisheries officers, district governors and the national park to negotiate a solution with the owners of the purse seines through a group meeting in the village. A spatial restriction was agreed, in which the new gear group would not operate within 1,500 m of the shore (GS); this agreement was upheld for 6 months and then collapsed due to the purse seiners displacement from productive fishing grounds. It appears the displaced fishers were not willing to continue with the negotiated rules, as they felt disadvantaged once fishing patterns became clear (GI). This contention continues with some villagers lobbying local officials for action.

Case Study 5 Our fifth case study is a state-led, co-governance arrangement motivated by a desire to strengthen enforcement of a large marine national park against commercial fishers (SG). The National Park identified a solution to this widely-acknowledged problem by reaching out to a network of fishing communities to act as national park guards to improve the park's enforcement capacity. In this trade-off local fishers were permitted fishing rights in exchange for improving the park's effectiveness against large fishing vessels. Because this arrangement violates Thailand's National Park Act (1961), special agreement was obtained for this project and a viable system was implemented based on enhanced cooperation between the park and local small-scale fisheries (GS). The 2004 Asian tsuanmi, however, undermined the initiative both physically and socially. The interviewees describe some unexpected consequences of aid relief, claiming that it changed community dynamics – reducing trust between groups and eroding community expectations of self-reliance. Additionally, the impacts of the tsunami necessitated a practical shift in national park priorities, away from enforcement of commercial fisheries, toward rebuilding the area (GI). Interviewees further eluded to an interest in restarting this co-governance arrangement, although we note that it has gone unimplemented for almost a decade.

Case Study 6 The final case study is a large-scale, state-led co-governance project to tackle mangrove deforestation and coastal protection (SG). Rules were created by a national park and the multi-stakeholder mangrove conservation committee, with limited direct consultation with resource users. These simple rules allowed some local use, but protection of the mangroves from large-scale clearance was prohibited (GS). This initiative has been successful in its goals of prohibiting large-scale mangrove clearing. Success is attributed to the allowance for local use (SG), awareness of the important role that mangroves play in coastal protection (GS) and project endorsement by the King of Thailand. Key informants reported that the king's endorsement incentivized compliance because people felt they have betrayed the popular king if they did not report violations (GI).

Patterns in Governability Characteristics and Outcomes

The studies presented above describe six case studies in which rules and regulations were negotiated and implemented to address perceived problems within mangrove habitat and fisheries in Southern Thailand. Of the six case studies, three are successful in that they continue to be actively enforced (Case Studies 1, 2, 6); additionally, two of these were found to increase the abundance and size of crab species targeted by fisheries compared to control sites (Case Studies 1, 2) (Jones 2014). Here we examine the governability characteristics associated with these successful case studies (section "Distinguishing characteristics of "successful" case studies"). We then highlight evidence from this study which appears to challenge three "good practices" that are broadly accepted within the coastal governance literature (section "Patterns that challenge accepted 'good practices'"). We close with an examination of the comparative successes of the governance paradigms used in this study (section "Comparative successes of the two post-hierarchical governance paradigms").

Distinguishing Characteristics of "Successful" Case Studies

Analysis of the data presented in Table 22.1 suggests two characteristics that distinguish the successful case studies from the case studies in which negotiated solutions fell apart: high social cohesion in governance interactions, and a perception of the solution's fairness and legitimacy.

Cooperation in Resolving Conflicts and Informal Sanctioning

Two distinguishing features of the successful case studies were high social cohesion and an ability to cooperate in governing interactions, which appears to have supported the success of these case studies in three ways: (a) finding an agreeable solution; (b) overcoming a lack of formal sanctioning capacity; and (c) achieving on-going enforcement. The high social cohesion evident in Case Studies 1 and 2 seems to contribute to the ability of these communities to negotiate and re-negotiate different solutions, in contrast to Case Studies 3 and 4. In part, the cohesion witnessed in the successful case studies would seem to result from the greater homogeneity of these communities. Compared to the unsuccessful case studies, Case Studies 1 and 2 were able to find solutions that met the needs for most people within the community; in contrast the solutions negotiated in Case Studies 3 and 4 broke down under perceptions that different segments of the community were being unfairly advantaged.

Part of the theorized reason that community-based, self-governing MPAs are able to succeed is their greater ability to enforce rules, without the need for external enforcement capacities which are often unavailable in low resource environments (GS) (Barrett et al. 2005). Strong social cohesion is one of the attributes associated with this greater enforcement ability, although theories acknowledge that it can either support or deter enforcement: strong community cohesion can confer an ability to enforce shared rules, or community members may shy away from enforcing rules in order to minimize conflict among community members (Halpern 2005) (GI). In the successful case studies, the high social cohesion appears to support on-going enforcement through informal sanctioning capacities, whereas in Case Study 3 low social cohesion and high internal conflict appears to be one of the factors that impedes enforcement of the MPA. Further, the ability to continue enforcing the negotiated solution in Case Studies 1 and 2, despite the absence of formal recognition of the MPAs' rules, appears to result, in a large part, from the high levels of cohesion seen in these communities.

Perception of Solution's Fairness and Legitimacy

The case studies presented here suggest that successful implementation of negotiated solutions occurred in cases where communities perceived the solutions to be fair and legitimate. In Case Studies 1 and 2, a perception of fairness seemed to confer legitimacy to the solution, even in the absence of formal government recognition. In contrast, the solution negotiated in Case Study 4 was rejected by a key sub-group within the community because they perceived it as benefiting others in the community while disadvantaging themselves. In Case Study 3, a solution which would seem to impact all community members in a similar way was perceived as illegitimate by a key sub-group within the community because they were not consulted in the solution's development (GI). Interestingly, Case Study 6 was perceived as fair and legitimate by a broad range of stakeholders over a large spatial area despite quite limited stakeholder engagement in its development. This case study's targeted exclusion of only commercial use of mangrove resources and its undiscriminating applicability to all people also created a solution that is broadly perceived as fair and enforceable, strengthening its legitimacy. In this case we further hypothesize that legitimacy was also conferred by the support of Thailand's revered King, whilst the end of logging concessions in 2002, investments in mangrove replantation, awareness raising by the national park and the 2004 Tsunami would all have added to local peoples' perception of the importance of mangroves.

Patterns that Challenge Accepted 'Good Practices'

We found little evidence of three widely-accepted conditions of successful community-based management

Community Engagement in Negotiating Solutions

The success of Case Study 6 in the absence of meaningful and representative community engagement is a fascinating positive deviant within the conservation literature. Effective community engagement is almost universally viewed as a prerequisite for effective governance of coastal systems (Ferse et al. 2010). It is seen as essential for achieving fair outcomes, fostering ownership, and realizing active, sustained implementation (Rudd 2000; Crawford et al. 2006; Pomeroy et al. 2007; Christie and Pollnac 2011). Yet, Case Study 6 appears to have achieved all these attributes of effective coastal governance through a top-down approach to the design and implementation of the program (GI). While we agree that thoughtful, well-implemented community engagement is very likely to enhance the success of coastal governance initiatives, Case Study 6 provides an interesting example that it is possible to establish the legitimacy of a program in its absence if other factors are present. In this case, the crucial factors were an existing appreciation of the importance of mangroves; indiscriminate rules which applied to all; the presence of a

Protected Area Committee which allowed some engagement of local elites; a concession that allowed local use, while only banning large-scale deforestation; and cultural reverence for the King.

Formal Institutionalization of Solutions

Another long-standing hypothesis in the coastal governance literature is the need for negotiated solutions to be validated through formal laws and other forms of state recognition (Cicin-Sain and Knecht 1998). This recognition is considered particularly critical in the case of community-led initiatives, which are viewed as more vulnerable to being overwhelmed by outside influences than other forms of governance (Schuttenberg 2001). We see Case Studies 1 and 2 as a potential challenge to this theoretical assumption, in that these case studies have been continuously implemented for a significant length of time producing fishery benefits, in the absence of formal recognition or sanctioning capacity. Further we believe there is a legitimate question about whether more formal, active state involvement would strengthen these governance arrangements (GS). It is possible that the commonly-held theory of state validation strengthening and enabling the expansion of these cases would hold. However, an alternative possibility also exists that increased state engagement could erode community ownership and self-reliance, as occurred with a community sanctuary in Apo, Philippines (Hind et al. 2010). Some types of government investment could also create conflicts within the community if one sub-group was seen to benefit more than another. Thus while we believe that sensitively crafted state validation would be helpful in sustaining and expanding these case studies, we concurrently assert that the effectiveness of these cases – in its absence – raises a flag that efforts to formalize these initiatives must be careful to preserve the characteristics that have supported their success thus far.

Role of Donor Support

A third widely-held belief within the fields of coastal governance and conservation is that limited funding or lack of resources is a major impediment to achieving sustained positive outcomes on meaningful spatial scales (Sale 2008). While we agree that insufficient resources can cause governance programs to fail, the case studies presented here illustrate that there are also potential pitfalls of high resourcing levels. Case Studies 4 and 5 are cases in which high levels of donor funding created conflicts, fuelled over-exploitation of coastal resources (Case Study 4), and undermined community self-reliance (Case Study 5). These examples suggest that where financial resources are concerned, there can be too much of a good thing. We assert that a more nuanced way of describing the role of financial resources in enabling successful coastal governance is that the resources must match a program's needs, being neither much lower nor much higher than what can be constructively applied to the situation at hand (SG).

Comparative Successes of the Two Post-hierarchical Governance Paradigms

Community-Based Self-Governance

The case studies presented here are consistent with the theorized strengths and weaknesses of community-based self-governance described in section "Introduction". Case Studies 1 and 2 were both successful in terms of sustaining implementation and enhancing fisheries; both were unsuccessful in enhancing biodiversity, which was believed to be related to the size of the MPAs which were very small (2–3 km^2).

Co-governance

Similarly the co-governance case studies presented reflected the issues discussed in section "Introduction", principally the difficulty of realizing the promise of these more complicated governance arrangements. The more heterogeneous character of the stakeholders involved in these initiatives makes it harder to negotiate solutions that are able to achieve multiple objectives (Case Study 3) and requires higher levels of facilitation and conflict resolution (Case Study 4). While Case Studies 4 and 5 are operating at spatial scales that have the capacity to produce meaningful biodiversity outcomes, the greater difficulty in achieving sustained implementation means this potential benefit remains elusive. In contrast, the one successful co-governance case study, Case Study 6, focused on a very narrow objective and, in practice, operated in a similar way to a hierarchical governance regime. Its success is a fascinating example within our sample of case studies because many aspects of its design and implementation run counter to major trends in contemporary conservation – such as recommendations for early community engagement – even though some engagement was facilitated through the Protected Area Committee, expectations for collaborative development of solutions, and an increasing lack of confidence that top-down initiatives can succeed.

Conclusion

On the System-to-be-Governed, this study has found that it is easier to craft governance solutions when the initiative's goals are more narrowly focused and the system-to-be-governed is more socially homogeneous, with high social cohesion, and operating at smaller spatial scale. On Governing Interactions, this study has found that stakeholders' perceptions of a governance initiative's relevance, fairness and legitimacy are the most important determinants of whether it continues to be implemented and enforced. On Governing Systems, this study has found that elements of hierarchical governance approaches can contribute to solving small-scale fisheries problems when they are well-designed and the formal institutions involved are respected. However, efforts to realize effective coastal governance at large, ecologically-meaningful spatial scales may achieve greater success by supporting the development of a patchwork of smaller community-based self-governance or co-governance initiatives.

References

Agrawal, A., & Gibson, C. C. (1999). Enchantment and disenchantment: The role of community in natural resource conservation. *World Development, 27*(4), 629–649.

Anon. (1997). *Constitution of the Kingdom of Thailand 1997*. Retrieved from http://www.asianlii.org/th/legis/const/1997/1.html#C001

Barrett, C. B., Brandon, K., Gibson, C., & Gjertsen, H. (2001). Conserving tropical biodiversity amid weak institutions. *Bioscience, 51*(6), 497–502.

Barrett, C. B., Lee, D. R., & Mcpeak, J. G. (2005). Institutional arrangements for rural poverty reduction and resource conservation. *World Development, 33*(2), 193–197.

Berkes, F., Hughes, T., Steneck, R. S., Wilson, J. A., Bellwood, D. R., Crona, B., & Worm, B. (2006). Globalization, roving bandits, and marine resources. *Science, 311*, 1557–1558.

Brechin, S. R., Wilshusen, P. R., Fortwangler, C. L., & West, P. C. (Eds.). (2003). *Contested nature: Promoting international biodiversity conservation with social justice in the twenty-first century*. Albany: SUNY Press.

CBD Ad Hoc Technical Expert Group. (2005). *Strategies for overcoming obstacles to implementation of integrated marine and coastal area management (IMCAM)*. Montreal.

Chambers, R. (1983). *Rural development: Putting the last first*. Essex: Longman Scientific & Technical.

Christie, P., & Pollnac, R. B. (2011). Socio-economy – Social dynamics of scaling-up marine protected area declaration and management. In J. Claudet (Ed.), *Marine protected areas: A multidisciplinary approach* (pp. 121–140). Cambridge: Cambridge University Press.

Christie, P., & White, A. (2007). Best practices for improved governance of coral reef marine protected areas. *Coral Reefs, 26*, 1047–1056.

Christie, P., Pollnac, R., Fluharty, D. L., Hixon, M. A., Lowry, K., Mahon, R., & Eisma-Osorio, R. L. (2009a). Tropical marine EBM feasibility: A synthesis of case studies and comparative analyses. *Coastal Management, 37*(3–4), 374–385.

Christie, P., Pollnac, R., Oracion, E. G., Sabonsolin, A., Diaz, R., & Pietri, D. (2009b). Back to basics: An empirical study demonstrating the importance of local-level dynamics for the success of tropical marine ecosystem-based management. *Coastal Management, 37*(3–4), 349–373.

Cicin-Sain, B., & Knecht, R. W. (1998). *Integrated coastal and ocean management: Concepts and practices*. Washington, DC: Island Press.

Crawford, B. R., Korompis, M. K., & Pollnac, R. B. (2006). Factors influencing progress in establishing community-based marine protected areas in Indonesia. *Coastal Management, 34*, 39–64.

Ferse, S. C. A., Mánez Costa, M., Schwerdtner Mánez, K., Adhuri, D. S., & Glaser, M. (2010). Allies, not aliens – Increasing the role of local communities in MPA implementation. *Environmental Conservation, 37*(1), 23–34. doi:10.1017/S0376892910000172.

Gray, T. S. (2005). Theorising about participatory fisheries governance. In T. S. Gray (Ed.), *Participatory in fisheries governance* (pp. 1–366). Dordrecht: Springer.

Halpern, D. (2005). *Social capital*. Cambridge: Polity.

Hind, E. J., Hiponia, M. C., & Gray, T. S. (2010). From community-based to centralised national management—A wrong turning for the governance of the marine protected area in Apo Island, Philippines? *Marine Policy, 34*(1), 54–62.
Horwich, R. H., & Lyon, J. (2007). Community conservation: Practitioners' answer to critics. *Oryx, 41*(3), 376–385.
Jones, E. V. (2014). *Does community management still need the State? Comparative analysis of marine protected area governance in Thailand*. UK: Newcastle University.
Kiss, A. (1999). *Making community-based conservation work*. Paper presented at the Society for Conservation Biology Annual Meeting, College Park, MD.
McLeod, K. L., & Leslie, H. (Eds.). (2009). *Ecosystem-based management for the oceans*. Washington, DC: Island Press.
McShane, T. O., & Newby, S. A. (2004). Expecting the unattainable: The assumptions behind ICDPs. In T. O. McShane & M. P. Wells (Eds.), *Getting biodiversity projects to work* (pp. 49–76). New York: Columbia University Press.
NACA. (2007). *Koh Yao Noi mangrove replanting in celebration of 80th birthday of His Majesty the King of Thailand*. Retrieved November 4, 2013, from http://www.enaca.org/modules/news/article.php?article_id=1692
Ostrom, E., Burger, J., Field, C., Norgaard, R., & Policansky, D. (1999). Revisiting the commons: Local lessons, global challenges. *Science, 284*(5412), 278.
Phillips, A. (2003). Turning ideas on their head: The new paradigm for protected areas. *The George Wright Forum, 20*(2), 8–32.
PhuketTV News. (2011). *Mangrove planting for our Majesty of the King of Thailand*. Retrieved from http://www.youtube.com/watch?v=HhzclzL8LRs
Pollnac, R. (1998). *Rapid assessment of management parameters for coral reefs*. Narragansett: Coastal Resources Center, University of Rhode Island.
Pomeroy, R. (1995). Community-based and co-management institutions for sustainable coastal fisheries management in Southeast Asia. *Ocean and Coastal Management, 27*(3), 143–162.
Pomeroy, R. S., Mascia, M. B., & Pollnac, R. B. (2007). Marine protected areas, the social dimension. *Report and documentation of the expert workshop on marine protected areas and fisheries management: Review of issues and considerations. FAO fisheries report* (pp. 149–182). Rome: FAO.
Pretty, J. (2003). Social capital and the collective management of resources. *Science, 302*, 1912–1914.
Rajchagool, C. (1999). Tambon administration organisation: Are the people in the dramatis personae or the audience? In *Transport and communications bulletin for Asia and the Pacific no.69* (pp. 29–58). New York: Economic and Social Commission for Asia and the Pacific/United Nations.
Roe, D., Mayers, J., Grieg-Gran, M., Kothari, A., Fabricius, C., & Hughes, R. (2000). *Evaluating Eden: Exploring the myths and realities of community-based wildlife management*. London: International Institute for Environment and Development.
Rudd, M. A. (2000). Live long and prosper: Collective action, social capital and social vision. *Ecological Economics, 34*(1), 131–144.
Sale, P. F. (2008). Management of coral reefs: Where we have gone wrong and what we can do about it. *Marine Pollution Bulletin, 56*(5), 805–809. doi:10.1016/j.marpolbul.2008.04.009.
Schuttenberg, H. Z. (2001). *A national perspective on community-based marine sanctuaries: Opportunities and challenges for establishing a network of community sanctuaries in Indonesia*. Master's thesis, University of Rhode Island, Kingston, Rhode Island.
Schuttenberg, H. Z. (2011). *Understanding and strengthening effective coral reef governance: A map & compass to guide strategic change in southeast Asia*. Doctoral thesis, James Cook University.

True, J., & Plathong, S. (2010). *Nomination document of The Andaman bioregion of Thailand for UNESCO World Heritage nomination*. Bangkok: Department of National Parks.

UNEP. (2006). *Marine and coastal ecosystems and human well-being: A synthesis report based on the findings of the Millennium Ecosystem Assessment* (p. 76). Nairobi: United Nations Environment Programme.

Western, D., & Wright, R. M. (Eds.). (1994). *Natural connections: Perspectives in community-based conservation*. Washington, DC: Island Press.

White, A., & Vogt, H. P. (2000). Philippine coral reefs under threat: Lessons learned after 25 years of community-based reef conservation. *Marine Pollution Bulletin, 40*(6), 537–550.

Part VI
Cross-Boundary Governance – Fostering Interactions

Introduction

Small-scale fisheries and their governance cannot be isolated from their external natural, social and political environment. Small-scale fisheries may be a system in themselves, but they are typically a system within a complex system that operates at various scales. Thus, governability challenges and solutions for small-scale fisheries are not always to be found internal to the fisheries governance system; cross-sector linkages must also be explored.

In *Chap. 23*, Andrew M. Song and Ratana Chuenpagdee explore the decline of the small-scale fishery of Newfoundland and Labrador in Canada and how it is linked to, and has been formed by, institutions and policies at provincial, national, and transatlantic scales. The chapter explores "meta-governing" principles embedded in these policy areas, such as equity and efficiency, which need to be considered for improving the governability of small-scale fisheries. Meta-governance, they argue, is also crucial to determine the general course the fishing industry should take in the province as a whole and what role small-scale fisheries should play. Maria Hadjimichael in *Chap. 24* provides insights into the complex situation that small-scale fishers in Cyprus find themselves in. The pattern is rather similar to the situation elsewhere: over-exploitation of marine resources, hierarchical governance failure as can be seen in poor enforcement and compliance, and ecosystem alteration due to climate change and invasive species. Existing power relations and struggles among different stakeholders must take part of the blame. The author submits that for governability to be enhanced new institutions must be created in order to strengthen the governance role of civil society at the community level.

The role of small-scale fishers in fisheries governance is also discussed by Silvia Salas, Julia Fraga, Jorge Euan and Ratana Chuenpagdee in *Chap. 25* in their case study from Yucatan, Mexico. The structure of the problem in Mexico is strikingly similar to that of Cyprus small-scale fisheries. Conservation efforts have been far from successful largely due to poor interaction among local small-scale fishers, despite the fact that they share fishing grounds and are in proximity of two nested

MPAs. The authors are concerned with how to facilitate better interaction between small-scale fishers in order to foster cooperation within and between communities and to overcome the conflicts that are hampering the governability of small-scale fisheries. Their study illustrates how images of MPAs influence stakeholders' perception, and that such images may substantially differ between neighboring communities. In *Chap. 26* Olivier Randin investigates the recent (2012) decision of the International Court of Justice regarding the marine territorial boundary dispute between Colombia and Nicaragua. Situating his empirical study in the San Andrés, Providencia and Santa Catalina Archipelago of Colombia, his focus is on how small-scale fishers have been victimized by decision at scales way beyond their control. In particular, he highlights how these fishers in Colombia are unable to access their traditional fishing grounds as they are now under Nicaraguan control. The chapter calls for an analysis of scales beyond the local community in order to grasp all the complexity, diversity and dynamics of interactions at multiple levels. The chapter also provides lessons about small-scale fisheries governance and governability when spatial scale is central. Joeri Scholtens, in *Chap. 27*, discusses a case with many similar traits to that of the Nicaragua/Colombia case, albeit from an area far away, namely the Palk Bay between India and Sri Lanka. Again, the livelihoods of small-scale fisheries communities are subject to big politics and disrupted by processes beyond local influence. In this case, small-scale fishers in Sri Lanka, particularly in the northern Tamil area, are suffering from the intrusion of Indian trawler fishers. The chapter discusses a number of factors affecting the governability of the fisheries system, such as the mismatch between the scale of governance and the scale of the problem, institutional fragmentation, power imbalances, and political interferences. The author posits that while co-governance might be the solution of the problem, constructive collaborative relationships are difficult, if not impossible, to create in practice given the local context.

Chapter 23
A Principle-Based Analysis of Multilevel Policy Areas on Inshore Fisheries in Newfoundland and Labrador, Canada

Andrew M. Song and Ratana Chuenpagdee

Abstract Small-scale fisheries in Newfoundland and Labrador have been shaped by multiple institutions and policies occurring at various scales. Some of these institutions and policies present opportunities, while others pose threats to the sustainability and the viability of this inshore fishery, a sector already facing a grim decline. This chapter analyzes contemporary policy areas at three different scales (provincial, national and transatlantic) and assesses their likely effects on the future of the small-scale fisheries and the rural communities who depend on it. Utilizing the governability perspective focusing on 'meta-level' governance, the governing principles embedded in these policy areas are given particular attention in the analysis. The discussion generates a set of questions that need to be considered in moving forward with the governance of the small-scale inshore sector in Newfoundland and Labrador. We argue that these inquiries are also crucial for charting the overall direction of the fishing industry in the province as a whole.

Keywords Governance principle • Fisheries policy • CETA • Direct fish sales • Fleet self-rationalization • Inshore fisheries • Newfoundland and Labrador

Introduction

The collapse of the cod population and the ensuing moratorium in the early 1990s was widely cited as a landmark event that signaled the downturn of the inshore fishery in Newfoundland and Labrador (Palmer and Sinclair 1997; Harris 1998; Davis 2014). The closure of the cod fishery resulted in a large reduction in the number of inshore ground fish licenses, particularly for those operating small boats less than

A.M. Song (✉) • R. Chuenpagdee
Department of Geography, Memorial University, St. John's, Canada
e-mail: amsong@mun.ca; ratanac@mun.ca

S. Jentoft, R. Chuenpagdee (eds.), *Interactive Governance for Small-Scale Fisheries*, MARE Publication Series 13, DOI 10.1007/978-3-319-17034-3_23

35 ft in length.[1] The number of registered boats in this category in Newfoundland and Labrador fell from 13,587 in 1992 to 7,640 in 2002, with much of this decline attributed to the removal of part time fishers as well as older fishers who were persuaded to retire from the industry (Shrank 2005; DFO 2014). The moratorium also had an impact on tens of thousands of community members across the province, who indirectly relied on the fishery for their jobs and income (Milich 1999; Hamilton and Bulter 2001; Davis 2014).

The fishing industry of Newfoundland and Labrador has since been transformed into a shellfish-based fishery, primarily targeting shrimp and snow crab, benefiting from the increased stock abundance. Shellfish have become a lucrative harvesting alternative to groundfish, generating over $400 million in landed value in 2013, and now comprising the majority of the economic value in the industry (e.g., 79.9 % of total landed value in 2013) (DFA 2014; May 2014). In the snow crab fishery, the mid-sized inshore vessels (between 35 and 65 ft in length) called 'longliners' enjoyed unprecedented prosperity early on through "supplementary" crab licenses issued beginning in 1986. It was not until 1995 that small-boat fishers also gained access to crab permits (Davis and Korneski 2012). These temporary quotas, albeit much smaller than the allocations of the larger boats, were eligible to small-scale inshore fishers who were able to maintain and defend the 'core' status,[2] and were later turned into permanent licenses (Davis and Korneski 2012). In 1997, a quota of 5,895 tonnes (out of a total allowable catch of 43,000 tonnes) was assigned to and utilized by over 1,800 inshore vessels less than 35 ft in length (Roy 1997; Shrank 2005). As such, Davis and Korneski (2012) note that the crab fishery has emerged over time as a vital contributor of income for many small-scale independent fishers across the province.[3] In the northern shrimp fishery, 300 boats in the mid-inshore range[4] were first allocated a quota in 1997 despite strong opposition from existing offshore license holders. This inshore quota was granted in addition to special allocations of quota distributed to several aboriginal groups and community organizations through the 1990s and 2000s (Foley et al. 2013). Similar to the effect of the crab fishery, Foley et al. (2013) state that the remarkable growth in the northern shrimp

[1] According to the federal policy on commercial fishing licenses for the Atlantic region, vessels less than 65 ft Length Over All are considered inshore fleets. However, in most contemporary policy and industry discourse in Newfoundland and Labrador, these inshore enterprises are further divided into an "inshore sector" that fishes in inshore waters (with vessel length under 35 ft) and a "nearshore sector" that operates in nearshore waters (with vessels between 35 and 65 ft). While this paper focuses on the inshore fisheries in general, we give a particular emphasis on the smaller scale fishers, operating in inshore waters with vessel length less than 35 ft. Following a policy change in 2007, the vessel length limit of this category was extended from 35 to 40 ft in order to reflect the trends of targeting species found further offshore (Gov. of NL, April 12, 2007a).

[2] The distinction of "core group" of fishers was created in 1996 as part of the Fisheries and Oceans Canada's Licensing Policy reform. A core fisher status is granted when the following criteria are met: be the head of an enterprise, hold key licenses, have an attachment to and be dependent on the fishery (DFO 1996).

[3] In total, the small-scale inshore sector in 2013 is estimated to comprise 2,493 enterprises and contains roughly 6,000 registered fishing vessels under 40 ft in length (DFO 2014; May 2014).

[4] Boats between 45 and 65 ft in length in this case.

industry has played a crucial role in alleviating the impact of the 1992 cod moratorium on the inshore harvesters, processors and processing workers, especially along the northeast coast of the province. For instance, over 50 % of northern shrimp catch has been landed in coastal communities by inshore owner-operators in each year since 1998 (Foley et al. 2013).

Presently, inshore fisheries in Newfoundland and Labrador are facing an uncertain future. Their economic viability largely depends on the catch of shrimp and crab, which had experienced remarkable growth in the last two decades. However, stagnancy in the production volumes of these key shellfish species in recent years has raised concern for the sector (Clift and Cooper 2014). Moreover, current speculation on the recovery of cod, once regarded as the bedrock of the inshore fishery and rural society in Newfoundland, has generated considerable anxiety and opposing opinions among fishers and other related fishery actors (Davis 2014, also see Khan and Chuenpagdee 2013; Mather 2013). The rebuilt cod stock would be expected to somehow influence the abundance of shellfish species, although much of the science on species interaction and ecosystem complexity remains largely equivocal (Lilly et al. 2000). While some argue that the inshore fishery is still the most important vehicle for advancing rural economic development (see Foley et al. 2013), what the future holds for this sector in these uncertain and changing environments is difficult to predict. Doug May, Professor of Economics at Memorial University, in a presentation to an industry-research audience, has proclaimed that fishers who survive the changes will do well financially (May 2014). Such a conclusion needs to be questioned, however, because it is important to look not only at who survives, but also why they survive while others struggle. Given that the sustainability of coastal regions in Newfoundland and Labrador has long been influenced by changes in fisheries policies (Sinclair 1985; Foley et al. 2013), understanding policies and their potential consequences is imperative for the future governance of the fishery and fishery-based communities.

Similar to other Canadian provinces, multiple regulative institutions are involved in managing the fishery in Newfoundland and Labrador. At the federal level, Fisheries and Oceans Canada (DFO) oversees fisheries harvest, and administers stock assessment, fleet regulation and quota management among other things. It also manages different groups of ocean users through the Oceans Act. Governance of the post-harvest activities, e.g., processing, marketing and trade of landed fish and shellfish, is under the mandate of the provincial government through the Department of Fisheries and Aquaculture (DFA). Supranational economic institutions in the form of multi- or bi-lateral trade agreements, such as the North American Free Trade Agreement (NAFTA), and intergovernmental management bodies, such as the Northwest Atlantic Fisheries Organization (NAFO), also play a role in the provincial fishery, primarily with regard to processing requirements and market access vis-à-vis foreign producers. They are also involved in regulating resource extraction in the continental shelf areas that lie outside Canada's Exclusive Economic Zone. At the local level, the Fish, Food and Allied Workers

(FFAW)[5] is a trade union that represents the wide-ranging interests of harvesters and processing workers operating in the province. Also, the Association of Seafood Producers as a processing industry lobby group exerts considerable influence at this scale. While governance at different and multiple levels is required, a high degree of coordination is necessary in order to avoid fragmentation and disconcerted effort in this largely hierarchical governing mode. Hence, striving for an institutional "fit" among these multi-level regulations and policies is expected to be of crucial importance for the governability of both inshore and offshore fisheries.

One way to "assess" policy coherence is through understanding the governing principles embedded in different policies (see Chap. 35 in this book). Principles represent an overarching vision for policy implementation and therefore support and guide the nitty-gritty of the provisions contained within. According to interactive governance theory (Kooiman et al. 2005), principles are part of the 'meta-level' governance elements that fundamentally influence governability (Chuenpagdee and Jentoft 2013; Song et al. 2013). They also contribute to the normative dimension of institutions, that is, they help define goals and objectives and also designate appropriate ways to pursue them through activation of required norms (Chuenpagdee and Song 2012). Debates about policy issues concerning small-scale fisheries in Newfoundland and Labrador can be explored in these "principle-based" terms to make sense of the broad array of intersecting institutions and gauge their impacts on the inshore fishery.[6] Such analysis sheds light on the possible implementation of these policies, and, thus, on the governability of the fisheries system.

This chapter focuses on three current and emerging policy issues that have generated wide discussion in the provincial fishery, namely direct sales of fish by harvesters, fleet self-rationalization, and the Canada-European Union free trade agreement (CETA). They span three different scales of institutions, i.e., provincial, federal and transatlantic, respectively. The study aims to identify and compare the guiding principles of these policy areas to search for consistency or disparity among them. For instance, we ask what the overarching principles are that can be inferred or extracted in each case, and how they shape the fishery. Also, to serve as a reference point for the analysis of fisheries-based principles, we draw upon two encompassing national and international guidelines deliberated in the Canadian Parliament and at the Food and Agriculture Organization of the United Nations (FAO) (i.e., the Fisheries Act and the Voluntary Guidelines for Securing Sustainable Small-Scale Fisheries, respectively). The study relies on a review of a wide array of documents such as published papers, grey literature including research reports and industry magazines, online newspaper articles, government legislation, and news releases. Based on this examination, we discuss issues and potential outcomes that these

[5] FFAW is now fully known as FFAW/Unifor. Unifor was created by the 2013 merger of the Canadian Auto Workers, of which FFAW was a member, and the Communications, Energy and Paperworkers Union of Canada.

[6] A recent policy paper on the future of small- and medium-scale fisheries and coastal communities in Newfoundland and Labrador also stresses the importance of principles and heavily relies on them to formulate policy recommendations (see Neis and Ommer 2014).

policy areas may bring to the inshore fishery and pose questions that are relevant for their future as well as the overall fishing industry in Newfoundland and Labrador. We conclude with some remarks about the governability of small-scale fisheries in Newfoundland and Labrador in the context of the current and prospective institutional set-up.

Understanding Three Fisheries Policy Areas at Issue and Their Principles

Direct Sales

In simple terms, the issue of direct sales of fish concerns who can sell the fish caught in coastal waters and to whom. Two pieces of provincial legislation and associated regulations have provided the legal basis for the sale and purchase of fish and fish products in Newfoundland and Labrador – the Fish Inspection Act and the Food and Drug Act (Eric Dunne Consulting Initiatives 2010). The first regulates the management of the fish processing and buying sectors and the overall marketing of fish and fish products, while the second specifies the standards and practices of various establishments in which seafood is prepared and/or sold to consumers. Characteristic of a hierarchical governance setup, these are the central regulative mechanisms through which the processing and marketing chains are structured for all fisheries in the province.

According to the current regulatory arrangements, a fish harvester operating within the inshore fishery in Newfoundland and Labrador can only sell his/her catch directly within the province and also only to a licensed fish buyer or processor. This implies that consumers are not permitted to purchase fish directly from a fish harvester at a wharf. Also, with the exception of the sea urchin fishery, which in recent years received a series of ministerial exemptions allowing harvesters to directly sell whole unprocessed sea urchins to out-of-province buyers (Pisces Consulting Limited 2014), there is no permanent direct sale of fish to other locales unless first processed by in-province processors. These 'minimum processing requirements', legislated by the Fish Inspection Act, specify the minimum levels of processing activity for all species that must be carried out in the province before being shipped out.

Nevertheless, direct sale of fish (typically cod) to consumers has been a common occurrence in the province. This happens not only because the complete prohibition of direct sales is largely unenforceable, but also because of a "Personal Use" catch category, provisioned in the Dockside Monitoring Program. Under this federal regulation, a portion of landings not sold to a licensed fish buyer or processor can be used for consumption or disposition directly by the harvester. In a DFA-commissioned report, Eric Dunne Consulting Initiatives (2010) speculates that in some areas especially large population centres such as the Avalon, Burin and Bonavista peninsulas, some of the fish under this category become a "cash crop" especially for small-scale fish harvesters, whose sales create additional income outside of their formal fishery earnings. The report also estimated that at least 1,180 of the 2,313 harvesters declaring some Personal Use landings appeared to have engaged in some level of direct sales in 2010 (see Eric Dunne Consulting Initiatives 2010).

The issue of Personal Use landings, and the ensuing direct sales, represents a loophole and a grey area in post-harvest policy. While the DFO has allowed personal possession and disposition of fish, they have no legal means to prevent direct sales because what happens next to the landed fish (whether consumed, sold, processed or given away) is a matter of provincial jurisdiction. In the meantime, DFA cannot do much aside from enforcing the existing regulation. The total ban of fisher-to-consumer direct sales is a unique arrangement observed in Newfoundland and Labrador. Other fish producing provinces in Canada, such as Nova Scotia, New Brunswick and British Columbia, all permit some degree of direct sales by harvesters, which typically involve a requirement of minimized pre-sale processing by harvesters (e.g., limited only to heading and gutting) and a sale only to individual customers or licensed food establishments who are clearly purchasing for personal/domestic consumption and not for re-sale (Eric Dunne Consulting Initiatives 2010).

Opinions on the current post-harvest policy have been markedly divided. Fish harvesters generally feel that total prohibition is too restrictive and would stifle their entrepreneurial spirit, as it deprives them of an opportunity to link up with buyers and processors outside the province (*The Navigator Magazine* 2014). Seafood consumers and the restaurant sector are also in favor of relaxing the rules to increase the availability of high quality fresh seafood at local commercial outlets, and to legitimize a longstanding and familiar practice of acquiring fish directly from fishers, especially in rural areas (Murphy and Neis 2011). On the other hand, governments are concerned that direct sales will inevitably introduce less controllable fish handling processes, leading to a greater health risk to seafood consumers and potentially tainting the image of Newfoundland and Labrador seafood. The fish processing industry also strongly resists any change in the policy citing a shortage of raw material for the processing plants, which already exists and will be exacerbated by increased direct sales (Eric Dunne Consulting Initiatives 2010; Smith et al. 2014).

It can be argued that the current ban on direct sale may have been founded upon accountability (for food safety), controllability and risk-aversion principles. The Fish Inspection Act and the Food and Drug Act and the subsidiary regulations are, first and foremost, geared towards minimizing health risks and maintaining government accountability for, and control of, any food safety-related incidents. Hence, the crux of government effort has been regulating and enforcing the post-harvest processes of buying, handling, storing, grading, processing, transporting and marketing in a manner that reduces risks to food-borne diseases and protects public health. Another set of principles that underlie the current provincial restrictions on direct sales (to customers and outside buyers) is arguably a socially-based one. Consistent with social principles such as equitable distribution of resource wealth and social sustainability, it is generally understood that the minimum processing requirements have played an important role in protecting local processing employment in out port communities and optimizing economic value for the rural parts of the province.

In recent years, however, there have been several conspicuous (and controversial) cases in which calls to relax the constraints on direct sales have been made by the harvesting industry and exemptions to direct sales granted. For instance, as mentioned earlier, an initial securing of outside buyers was permitted in the sea urchin fishery in 2010. Subsequently, shipping of whole unprocessed sea urchins was also allowed in 2013 (Pisces Consulting Limited 2014). In 2012, DFA granted Ocean Choice International, one of the most vertically integrated seafood companies in the province, a 5-year exemption to the minimum processing requirement of yellowtail founder, allowing the company to export up to 75 % of the species unprocessed, with the remaining fish processed by local workers at its plant in Fortune. In principle, such changes clearly represent an erosion of social principles that ensure wider community benefits in the form of rural employment in favor of economic ones that advocate trade liberalization and financial viability of specific industry players. In addition, the fact that DFA commissioned a study to examine the direct sales issue implies that pressure from harvesters to ease the restrictions was likely high.[7] Most recently, the newly appointed Minister of Fisheries and Aquaculture, Vaughn Granter, spoke positively about the issue in a radio interview:

> We need to be able to look at some of the things that the previous ministers looked at, you know, us being able to allow harvesters in the cod [fishery] to sell directly to local consumers and local restaurants… and I believe harvesters have been asking for this… And also I had some discussions with the fisheries department about selling catch to buyers outside of the province. I understand that the union is in favor of that, and we need to look at that as a pilot project. (CBC, October 6, 2014)

In line with the ongoing development that signals a greater infusion of market-oriented principles, DFA has shown interest in instituting fish auctions as a way to promote fisher involvement in the sale process. Fish auctions are assumed to be more efficient in price setting in fisheries as well as a stronger link for harvesters to the marketplace (Gov. of NL, April 12, 2007b). The objectives of establishing a fish auction include "improving operational efficiency at both the harvesting and processing sectors and to provide a market driven economic incentive to reduce seasonal landing gluts" in addition to raising product quality (Gov. of NL, April 12, 2007b). This particular initiative would also hint at the underlying principles of efficiency and reliance on market forces.

Fleet Self-Rationalization

The fishing industry in Newfoundland and Labrador has long been preoccupied with the issue of overcapacity. It is a persistent and pressing topic raised by many fishery-related actors across the board. A discussion paper on fishing industry renewal jointly prepared by the federal and provincial governments in 2006

[7] The report from this 2010 study, which finally appeared on the DFA website in September 2014, recommended that the province partially allow direct sales (see Eric Dunne Consulting Initiatives 2010).

considered it "the most critical underlying structural challenge" facing the industry (Gov. of NL and Gov. of Canada 2006, 1). Government-led initiatives for reducing capacity date back to the late 1960s with the 1969 report of the Royal Commission on the Economic Prospects of Newfoundland and Labrador, which requested a reduction in the number of fishers (Shrank 2005). Subsequent recommendations pointed to a similar conclusion. For example, the report of the 1982 Kirby Task Force on the Atlantic Fisheries and the 1993 report of the Federal Task Force on Income and Adjustments in the Atlantic Fishery (i.e., the Cashin report) both called for measures that would relieve overdependence on the fishery, with the latter recommending a 40–50 % reduction in both harvesting and processing capacity (Shrank 2005, 411). In response to these suggestions, a series of "adjustment" programs leading up to and following the cod moratorium was introduced by the federal government to assist fishers out of the cod fishery. With the implementation of the 1989 Atlantic Fisheries Adjustment Program, the 1992 Northern Cod Adjustment and Recovery Program, and the Atlantic Groundfish Strategy, which ended in 1998, the federal government spent nearly three billion dollars, mostly in Newfoundland, for the provision of income support to affected fishers and fish plant workers (Shrank 2005). This process worked in tandem with the combined effort of the provincial government and the FFAW in developing a professional certification program for the inshore sector. The professionalization of harvesters contributed towards reducing the number of fishers by eliminating marginal or part-time fishers, who were deemed to hold little attachment to the industry (Clarke 2003).

Despite achieving some success in the years that followed the cod moratorium, it has been suggested that the problem of overcapacity has been acutely revived, led by increased capacity in the crab and shrimp fishery. The prevailing sentiment, persuasively advocated by economists, is that there are too many inshore fishers with inadequately small quotas sharing the increasingly scarce amount of shellfish resources. Shrank (2005) and Clift and Cooper (2014), for instance, have both argued that this is a sure sign of excess capacity and that resolving this perennial issue is a crucial and an imminent task.

The main thrust of the federal government's response came in 2007 with the announcement of the outcomes of the federal-provincial Fishing Industry Renewal Initiative. Propelled by the general consensus observed in the premier's meeting as well as during the consultations with industry stakeholders such as harvesters and plant workers and also with wider groups including community organizations and non-industry associations, the initiative was heralded as providing a policy that will "fundamentally change the Newfoundland and Labrador fishery" (Gov. of NL, April 12, 2007c). Subsequently, it introduced measures that would lead to capacity reduction in the harvesting (mainly inshore and nearshore fleets under 65 ft) and processing sectors. The new initiative relies on the mechanisms of voluntary fleet rationalization by fishers belonging to the Independent Core category,[8] which would

[8] Independent Core is the category assigned to the head of a Core enterprise who is not party to certain "controlling trust agreements" with a person, corporation or other entity with respect to the licences issued in his or her name. The purpose of the creation of this category is to protect inshore

allow two enterprises being combined into one at the removal of the other enterprise along with an associated vessel and species licenses from the system (Gov. of NL, April 12, 2007d). One study estimates that the implementation of this policy has resulted in the cumulative exit of 400 fishing enterprises or 5 % of the total fleet (Policy Brief 1 2012). Although at the moment no single licensed fisher is permitted to combine more than two enterprises on a single vessel, Davis (2014) states that there exists increased pressure to allow for further consolidation on a single enterprise.

More recently, conceived to be a continuation of the Renewal initiative, another policy recommendation was brought forward at the provincial level. In July 2009, DFA, the Association of Seafood Producers and FFAW signed a Memorandum of Understanding (MOU) to formulate strategies for industry restructuring and address the issues of overcapacity and unviability. The final report released in February 2011 painted a rather grim picture of the industry by suggesting that only "between one-third and two-thirds of the fish harvesting operations currently operating in NL are viable" and further highlighted the particular unviability of inshore vessels under 40 ft in length (Report of the Independent Chair 2011). Consequently, it recommended reducing the size of the inshore fleet by 30–80 %, with the highest required reductions concentrated on the northeast and west coasts of Newfoundland as well as southern Labrador. To the dismay of the proponents of the Memorandum of Understanding, the recommendations were swiftly rejected by the provincial government on the grounds that too much focus was laid on rationalization instead of restructuring, and that the suggestions were formulated without an adequate discussion about a long-term vision for the industry (Walsh 2011).

As controversial as these policy suggestions seem, the reactions of several groups also varied significantly. Despite the official government claim that the consultation process was inclusive and genuine, outport communities felt that they were very much ignored throughout the largely top-down Memorandum of Understanding process (Walsh 2011). Furthermore, for inshore fishers in rural areas, rationalization essentially means a mass layoff, which would lead to an eradication of small-scale fleets in favor of larger, corporate fishing enterprises and even a decimation of rural communities (The Telegram, March 30, 2009; Policy Brief 1 2012; Smith et al. 2014). On the contrary, the governments, both federal and provincial, maintain that industry renewal and restructuring "represents the best opportunity for the fishery to remain the primary economic driver for many rural areas" and rationalization is the key mechanism to restructuring the fishery (Gov. of NL and Gov. of Canada 2006, 7). While the provincial government decided against implementing the far-reaching measures recommended in the Memorandum of Understanding, it nevertheless remains a strong supporter of rationalization through enterprise combining. The FFAW also appears to be an advocate of rationalization, claiming that "rationalizing

fishers' control over the decision to request a "transfer" of licenses they hold. The implication is that anyone who is in a controlling agreement will not have their licence renewed. This measure thus aims to further strengthen the independence of the inshore fleet (under 65 ft in length) protected by the Owner-operator and Fleet Separation policies in effect since the 1970s (DFO 2007).

means not only downsizing but also other ways to improve profitability in the industry" (Policy Brief 1 2012, 5). Hence, it produces a win-win solution that will allow for longer and more stable fishing seasons and improved incomes for those who remain in the fishery (Policy Brief 1 2012, The Western Star, March 4, 2011). Other actors have expressed an even stronger conviction towards rationalization. For instance, some economists, as well as large-scale processors and companies engaged in the offshore sector, have been frustrated at the slowness of the current pace of rationalization and have urged for more aggressive strategies of implementing fleet rationalization to achieve capacity reduction and ensure competitiveness in the international market (Clift and Cooper 2014; The Navigator Magazine 2012). Their assertion seems to converge on the understanding that rationalization of the harvesting sector is already ongoing in spite of limited government intervention, and a full-scale rationalization is inevitable for the industry to stabilize into the future. Therefore, they argue that it is imperative to plan ahead and make this transformation as orderly and efficient as possible.

Several governing principles that are fundamental to the debates on rationalization can be described as follows. It is perhaps intuitive to imagine that concerns with overcapacity and the policy impetus towards rationalization stems from an adherence to the principle of economic efficiency, viability and competitiveness. The Discussion Paper on Fishing Industry Renewal prepared by DFO and DFA stated that "we recognize the crucial role of the fishing industry within the province's economic and social structure, particularly in rural areas, but efficiency and competitiveness must be given greater emphasis." (Gov. of NL and Gov. of Canada 2006, 2). Similarly, the resultant policy announcement in 2007 indicated that the "Fishing Industry Renewal initiative should result in a sustainable, economically viable, internationally competitive, and regionally-balanced industry" (Gov. of NL, April 12, 2007e). In the final report of the Memorandum of Understanding process (Report of the Independent Chair 2011), the utmost importance of making the harvesting sector economically viable was also clearly demonstrated:

> In the inshore sector, the level of rationalization that is required to allow fish harvesters to achieve a relatively reasonable average income is significant. In the areas most dramatically affected by the shift to shellfish and the decline of the groundfish sector, as high as 80 percent of the current capacity must be rationalized in order for those who remain to achieve a *viable income*… In all other inshore fleets, significantly less (40 percent) rationalization is required in order to allow those harvesters who remain to achieve the desired enterprise *viability target*. This could be achievable and feasible through an enterprise combining process. (p. 61, emphasis added)

Other principles can also be discerned from the further perusal of the fleet self-rationalization policy. Government has so far opted for an approach that allows individual fishers to voluntarily apply for fleet combining. As federal Fisheries Minister Loyola Hearn, at the time of announcing the self-rationalization policy in 2007 put it: "This is about choice. These changes allow those who want to invest in their enterprise an opportunity to grow and enable those who wish to leave the industry an opportunity to leave on their own terms. Today, we have given fishers greater flexibility than ever before in determining their future and the ability to make decisions that allow them to adjust to ever changing market conditions" (Gov. of NL, April 12, 2007c). This explanation draws close to the tenets of rational choice theory, in which economic decisions are left to individuals to voluntarily choose what is best for them based on self-interest and cost-benefit calculation. According to this logic, achieving the financial viability of an independent fishing enterprise would be wholly attributed to fisher's prudent decisions. In the same way, fishers' voluntary decisions would be solely blamed in the case of insolvency.[9] Basing the governance of the small-scale fishing sector on these principles, the former provincial Fisheries and Aquaculture Minister Tom Rideout has stated: "It is clear that we all envision a rational, self-sustaining and stable industry that can engage and attract younger workers, but at the same time can allow those having a long term attachment to the industry an opportunity to continue their careers to the extent possible over the next five to ten years" (Gov. of NL, April 12, 2007c).

Aside from those principles that underlie the prevailing insistence on rationalization, it may be possible to gain a further nuanced understanding of this policy if we consider other principles that may have been (deliberately or inadvertently) neglected in the promotion of the policy. In particular, the principle of equity appears to be weakly articulated. There is an apprehension among small boat fishers and concerned citizens elsewhere that rationalization in the way it is currently set up favors larger inshore enterprises with bigger vessels as well as offshore fleets to the detriment of most people living in rural fishing communities and smaller boats. As the equity considerations are generally not taken into account in rationalization policies and programs, a disproportionate degree of capacity reduction may occur in the small-scale sector which may leave the future of the community and the local inshore fishery threatened and uncertain (Policy Brief 1 2012; The Telegram, March 30, 2009). With the possibility of wealth being unduly concentrated in those who get to remain in the fishery, the importance of considering how different groups will be affected by the self-rationalization policy (i.e., who wins and who loses) could be more meaningfully integrated through the activation of the equity principle.

Canada-EU Comprehensive Economic and Trade Agreement (CETA)

Newfoundland and Labrador's fishery has always been built on the export economy. In fact, from the sixteenth century onwards, the Newfoundland economy as a whole greatly depended on foreign markets, both for exports of fish (salted or frozen) and imports of food, supplies and manufactured goods. Lacking a significant domestic market, cod trade to countries in Europe and in the Caribbean remained crucial in the sustenance of the island economy. By the 1950s, the United States became the major importer of Newfoundland cod. More recently, the market has diversified

[9] For more on this 'making of self-managing fishers' in the Newfoundland and Labrador fishery context, see Bavington (2011) chapter 5.

with China, the United Kingdom, Russia and Vietnam becoming other major destinations of provincial seafood (DFA 2014).

The recently agreed Canada-European Union Comprehensive Economic and Trade Agreement (CETA) represents the biggest emerging trade mechanism through which the Newfoundland and Labrador fish and seafood products can be better connected to other export markets – in this case, the European Union (EU). When ratified in 2016, 98 % of EU tariffs will be reduced to zero and the import of seafood into this lucrative market will be 100 % duty-free within 7 years, providing the provincial fishing industry a significant competitive advantage over producers from other countries that do not have a free trade agreement with the EU. For instance, the two most important species for the provincial fishery, shrimp and snow crab, will be duty-free immediately upon CETA coming into effect.[10] The agreement will also eliminate "end-use" restrictions on provincial seafood products in the EU, meaning that fish products can be packaged and branded as Canadian products, rather than shipped away as raw material for European plants. This move is generally expected to create new opportunities for provincial seafood producers with respect to secondary processing, brand building and marketing strategies to deliver high-quality, premium products to EU markets (Gov. of NL, October 29, 2013; The Telegram, October 19, 2013). In order to gain unrestricted access to the EU market, however, the provincial government was required to make a concession by relinquishing minimum processing requirements for fish exports destined for the EU.

Some stakeholders, most notably the Progressive Conservative government, who participated in the trade negotiations as the ruling provincial party, and FFAW, have shown great optimism about the potential benefits of this agreement to the provincial fishery. The outgoing president of FFAW, Earl McCurdy stated "We've operated for years at a tremendous disadvantage to, say, Norway, Iceland and other competitors in terms of getting access to that important market… This, for the most part, removes those disadvantages and gives us a fighting chance" (The Telegram, October 19, 2013). Similarly, the provincial government estimated that CETA implementation will immediately save the industry $25 million due to the elimination of tariffs on fish and seafood in addition to opening up new opportunities worth over $100 million, which together will add over a billion dollars of value to the fishery (Gov. of NL, October 29, 2013). The concern over the waiving of the minimum processing requirement was also cautiously brushed aside by those who believe that CETA is a boon by insisting that it represents a relatively small concession to make. McCurdy again explained that "Quite frankly, Europe can't compete with us anyway. Their energy costs are higher. Their wage costs are as high or higher, and they're not nearly as close to the raw material as we are" (The Telegram, October 19, 2013).

The most vocal criticism of this trade deal in Newfoundland and Labrador has sprung from members of the opposition party. For instance, Jim Bennett, a member of the House of Assembly and the provincial Liberal Party, succinctly called CETA

[10] Current duties on frozen shrimp, cooked and peeled shrimp, and snow crab are 12 %, 20 %, and up to a rate of 8 %, respectively (Gov. of Canada, September 26, 2014).

"a complete sell-out of the small processing sector and small rural communities" (The Telegram, October 19, 2013). Scott Sinclair of the Canadian Centre for Policy Alternatives also asserted that CETA negotiations and the subsequent reduction of foreign trade barriers must not result in the erosion of independence for fishers and coastal communities. He cited the case in 2012 in which the provincial government rejected an "unprecedented request" by Ocean Choice International for a permanent exemption from minimum processing requirements for yellowtail flounder and redfish exports on the basis that there is a need to "ensure the long-term security of resources for the benefit of future generations of Newfoundlanders and Labradorians" (Sinclair 2013, 28).[11] These divergent opinions are still at the speculative stage, as we wait for the implementation of CETA and learn more about how it will impact the fishery and especially the small-scale fishing fleet in practice.

The main principles that underpin a trade policy such as CETA can be reasonably thought of as being of economic nature. The doctrine of free trade and greater economic integration holds sway in its rationale and in what it tries to accomplish. Interestingly, the preamble of the Consolidated CETA Text states a far wider basis for the workings of CETA. In addition to creating an expanded secure market through the reduction or elimination of barriers to trade and investment, it purports to support the involved parties' commitment to promote sustainable development in all economic, social and environmental dimensions. It encourages enterprises to respect principles of corporate social responsibility and further recognizes the rights and legitimacy of national governments to achieve various "non-economic" policies such as public health, environment, public morals and protection of cultural diversity (Gov. of Canada, September 26, 2014). Sinclair (2013), however, has contradicted this official position by contending that "the basic principles of trade and investment treaties and fisheries regulation are like oil and water; they do not mix. Particularly in the areas of supporting the inshore fishery and coastal communities, the potential conflicts between trade and investment treaty rules and Canadian fisheries regulations are numerous and profound" (Sinclair 2013, 22). More specifically, he argues that many, if not all, national and provincial fisheries policies are contrary to the principle of national treatment and non-discrimination provisions of trade and investment treaties. In other words, from a trade treaty perspective, these fisheries policies, including those traditionally considered commonsensical (e.g., Canadian fishing licenses are restricted to Canadians only, or majority Canadian-owned corporations), constitute discrimination based on nationality or local origin (Sinclair 2013). Moreover, the principles of adjacency and historical dependence, which have been used as the basis of fisheries resource allocation in Canada, thereby favoring fishers who are located near the resources or those who have a long history of resource use (Foley et al. 2013), are both in clear conflict with the national treatment and non-discrimination provisions (Sinclair 2013). Likewise, co-governance principles promoting local participation and control, such as subsidiarity, are at odds with the fundamental tenets of free trade. Hence, we can see that the principles underlying CETA may clash with those that support inshore, community-embedded,

[11] A temporary exemption was, however, approved, as mentioned in the previous section.

small-scale fisheries, which was in fact witnessed with respect to the waiving of minimum processing requirements. To this end, Sinclair (2013, 6) has argued that "for reasons of fairness and equity, these forms of positive discrimination are both desirable and morally compelling".

Discussion

In this analysis, we have identified some of the key governing principles that operate within the three selected policy areas affecting small-scale fisheries, i.e., direct sales, fleet self-rationalization and CETA. As these policy issues are implicated in multiple institutional scales and have been garnering much attention and debate in- and outside the province, they have the potential to extensively reshape the provincial fishery, and the small-scale inshore sector in particular. As posited by the interactive governance perspective, the divergence of 'images' related to these policy areas and the mismatches in the principles underlying them are potential sources of ungovernability. In order to situate these policy areas in a broader context, we include two other relevant national and international guidelines, namely the Fisheries Act (more specifically, the Bill C-45 which was introduced to amend the Fisheries Act) (Parliament of Canada 2007) and the recently adopted Voluntary Guidelines for Securing Sustainable Small-Scale Fisheries (FAO 2014) in the analysis. While the principles of the selected policy issues are inferred from government texts and related documents, those of the referenced policy guidelines are extracted from an explicit mention in the official documents (Table 23.1).

The issue of governability rests upon how the principles gleaned from these policy areas fares with each other, where they are situated in relation to the overarching guidance of national and international principles on small-scale fisheries, and what possible outcomes may arise from their implementation. First, a comparison of the principles represented in the three policy areas indicates a general adherence to (or a shift towards, at least, in the case of direct sales) economy-oriented principles. In particular, market-based efficiency and financial viability appear to be the central themes that drive these initiatives. Additionally, in the case of fleet rationalization, rationality focused on individual choice and self-determination accompanies the economic principles. We also note that legislations overseeing the sale of landed fish are predominantly based on the idea of safeguarding public health as well as the protection of rural employment and community integrity.

While the impetus carried forward in these policy areas points towards an efficiency-based and trade-driven future for the fishery, their comparison to the high-level guidelines helps reveal what is perhaps left out or under threat. The two overarching documents exhibit a considerable overlap in the principles they endorse. They both promote the conservation of fish and fish habitats through an ecosystem-based approach, a precautionary approach and sustainable development. Another major common feature relates to social concerns such as equity, equality, gender considerations and respect for cultures and aboriginal rights. There are also some

Table 23.1 List of governing principles characterizing the three policy areas and two overarching guidelines as they pertain to small-scale fisheries in Newfoundland and Labrador

Thematic categories	Direct fish sale	Fleet self-rationalization	CETA	C-45[a] (The Fisheries Act)	SSF guidelines
Economic	Efficiency (fish auction)	Efficiency	'Free trade principles' such as national treatment	Economic viability	Economic viability
	Market force reliance (fish auction)	Financial viability	Economic sustainability	Secure access and allocation stability	Economic sustainability
	Financial viability (direct sale)				
	Trade liberalization (direct sale)	Competitiveness			
		Rationality			
		Self-choice/independence			
Ecological			Environmental sustainability	Sustainable development	Environmental sustainability
				Ecosystem approach	Ecosystem approach
				Precautionary approach	Precautionary approach
Socio-cultural	Equitable distribution of resource wealth (min. processing requirement) – *under threat*	Equity – *neglected*	Corporate responsibility	Equity, fairness	Equity and equality (also gender)
	Social sustainability (min. processing requirement) – *under threat*		Social sustainability	Traditional knowledge	Social responsibility

(continued)

Table 23.1 (continued)

Thematic categories	Direct fish sale	Fleet self-rationalization	CETA	C-45[a] (The Fisheries Act)	SSF guidelines
				Public access	Fishing community participation
				Respect for aboriginal and treaty rights	Non-discrimination
					Respect of cultures
					Human rights and dignity
					Social viability
Management	Accountability		Legitimacy of national governments in regulating domestic affairs	Public participation	Accountability
	Controllability		Adjacency – *in conflict*	Adjacency	Transparency
	Risk-aversion		Historical dependence – *in conflict*	Historical use	Rule of law (rule-based approach)
				Rule compliance	
				Scientific information	
				'Best use' principle	
				Inter-jurisdictional cooperation	

[a]An Act respecting the sustainable development of Canada's seacoast and inland fisheries (for modernizing the Fisheries Act)

management-specific ones including participation and rule compliance that appear in both texts, in addition to economic viability. From this, two major incongruities emerged between the principles of the three policy areas and those advocated in the high-level documents. First, there is a general tendency among the three policy areas to underplay, or to shift away from, the social and cultural agenda deemed important in the high-level guidelines. The self-rationalization policy has so far been disproportionately abrasive and unkind to the inshore sector, to the detriment of rural communities. When it comes to CETA and direct sales, the elimination of minimum processing requirements would likely produce a similar reaction among those more conscious of the equity issues. In addition, specific to the situation that prohibits direct sales, small-boat fishers have raised a further equity-based claim that larger players in the fishery receive a favorable treatment, which is unfair to the small-scale sector. For instance, while big processing companies with their own factory trawlers have been given special exemptions to sell fish they catch unprocessed to outside buyers including overseas, small-scale fish harvesters and small processors are banned from carrying out such out-of-province sales. A fisher is quoted saying: "[It] seems like the cartel [of fish processors] has so much power over our government that it legislated that I can't sell or give away a DFO monitored fish on the dock to my friends and neighbours" (*The Navigator Magazine* 2014; also see Smith et al. 2014).

Secondly, the conflict between the principles entrenched in the free trade doctrine such as international competitiveness and national treatment and those that promote the integrity of domestic fisheries such as adjacency, historical dependence and community empowerment requires further attention. While the two major guidelines clearly account for the latter, strong political emphasis put on free trade by the current Canadian government with a number of bilateral and regional trade agreements presently under negotiation (e.g., the Trans-Pacific Partnership) raises a question as to how these two streams of principles can be reconciled. Balancing of these seemingly incompatible principles, therefore, poses a critical issue for fisheries governance generally and the future of the inshore fishery specifically. As noted by Sinclair (2013), there is growing pressure to forgo the principles that help retain geographical distinctiveness and strengthen domestic fisheries in return for those that advance greater global economic integration. If such a trend does become a reality, policies that aim to empower small-scale fisheries are likely to be squeezed out along with coastal rural communities, whose robustness in many ways hinges on the vibrancy of the small-scale sector.

The question then arises: what is the vision for the inshore fishery in Newfoundland and Labrador? Is it a future that primarily focuses on the financial aspect of fishing enterprises and processing plants (i.e., increased profits and viability), or is it something that also gives serious weight to the more intangible social, cultural and stewardship benefits of small-scale fisheries and rural communities? There have already been numerous calls across disciplines to develop a collective vision for the fishery. On the one hand, the provincial government was hoping to see greater discussion on the ultimate vision for a restructured fishery when it rejected the Memorandum of Understanding recommendations in

2011. Concerned economists have also indicated a need for a future strategic reorientation in the province's fishing industry, especially given the fading economic importance of the fishery in the province in light of the rising dominance of the oil and gas industry (Clift and Cooper 2014). On the other hand, a group of inshore fishers in Change Islands, a part of rural coastal Newfoundland, also decried the lack of vision as they "felt that each new policy or regulation is tagged onto previous ones without any vision of the future of the fishery and fishing communities" (Policy Brief 1 2012, 6).

We argue that the vision-setting starts with a careful consideration, and selection, of governing principles. They would need to be evaluated against and compared with each other for compatibility and consulted with higher-level principles for relevance. Only a genuine and open-minded negotiation among various interested parties would achieve this value-laden task. Although 'hard choices' between the seemingly incommensurable principles as well as a forceful overcoming of power imbalances among different groups are likely required at times, a "principle-based" process calls for a mode of governance that is interactive and interdependent rather than a hierarchical one, which is more suited for a top-down transfer of governing visions (Kooiman and Jentoft 2009). In this respect, community-based organizations, such as the Fogo Island Co-op (see Foley et al. 2013) and FFAW can presumably play a key role in organizing and mobilizing small-scale fishers locally as well as province-wide. A shift towards co-governance has already been underway since the cod moratorium, with the various levels of government announcing greater emphasis on integrated management and public involvement (Khan and Chuenpagdee 2013). Although transformative changes have yet to occur, such moves represent a crucial platform from which mutual conversations about principles can spring forward and proliferate. Once accepted by all actors, the principles would need to be made explicit to the public and reviewed and judged in light of newly emerging policies and changing circumstances. Having no vision for the fishery would mean having no principles to openly speak of. With an emphasis on a set of guiding principles, we can expect to create a normative basis for setting a coherent direction for fisheries policy.

As such, working with principles as part of 'governing interactions' represents a crucial step towards ensuring and enhancing the governability of fisheries. Small-scale fisheries around the world are continuously being integrated into the global economy, and they also face increasingly uncertain environmental conditions. This likely requires an application of multiple policies situated at various institutional scales to tackle the intensifying complexity. We submit that a careful examination of governance principles should be an important part of fisheries analysis and we encourage their ongoing exploration to help make sense of the current struggles.

Concluding Remarks

The Newfoundland and Labrador fishery appears to be at a critical juncture. Bolstered by the prevailing trends in demographics and market-driven self-rationalization, it is often described that the fishery is "moving away from a "social" fishing industry to a more competitive, market-based one" (May 2014). This tendency is also reflected in the three major policy areas, as our analysis of their underlying governing principles shows. Yet, with growing talk of cod recovery and continued discussion of ways of regenerating the rural economy, the fishery appears to have reached a crossroad. If the province wants to have coastal rural communities thrive in a manner that is socially inclusive, economically modest and culturally vibrant, the importance of the inshore fishery cannot be overstated. The inshore fleet, especially the small-scale sector, has been, and will continue to be, the lynch-pin of the social fabric of coastal rural Newfoundland and Labrador. To do that, however, the province will have to make a serious effort to better align fisheries policies with the set of principles that are currently under-appreciated and under increasing threat. On the other hand, if the province is inclined to solely concentrate on the financial viability of individual players in the industry and the associated market-based dimensions, the present policy trend would likely continue its course, resulting in a situation where larger boats that earn larger profit are promoted. With this however, we will almost certainly be forced to contend with the accelerated demise of small-scale fisheries and many rural communities in Newfoundland and Labrador.

Acknowledgments We thank two anonymous reviewers for their insightful feedback on an earlier draft of this chapter. If any errors remain, they are the sole responsibility of the authors.

References

Bavington, D. (2011). *Managed annihilation: An unnatural history of the Newfoundland cod collapse*. Vancouver: UBC Press.

Canadian Broadcasting Corporation (CBC). (2014). *The Fisheries Broadcast, Monday October 6, 2014*. Retrieved from http://podcast.cbc.ca/mp3/podcasts/nlfisheries_20141006_90774.mp3

Chuenpagdee, R., & Jentoft, S. (2013). Assessing governability – What's next. In M. Bavinck, R. Chuenpagdee, S. Jentoft, & J. Kooiman (Eds.), *Governability of fisheries and aquaculture: Theory and applications* (pp. 335–349). Dordrecht: Springer.

Chuenpagdee, R., & Song, A. M. (2012). Institutional thinking in fisheries governance: Broadening perspectives. *Current Opinion in Environmental Sustainability, 4*, 309–315.

Clarke, M. (2003). The professionalization of inshore fishers. In R. Byron (Ed.), *Retrenchment and regeneration in rural Newfoundland* (pp. 134–157). Toronto: University of Toronto Press.

Clift, T., & Cooper, T. (2014). The three R's. Rationalization, retrenchment, reorientation: The Newfoundland and Labrador Fishery after the MOU. *Newfoundland Quarterly, 107*(2), 37–40.

Davis, R. (2014). A cod forsaken place? Fishing in an altered state in Newfoundland. *Anthropological Quarterly, 87*(3), 727–758.

Davis, R., & Korneski, K. (2012). In a pinch: Snow crab and the politics of crisis in Newfoundland. *Labour/Le Travail, 69*(1), 119–145.

Department of Fisheries and Aquaculture (DFA). (2014). *Seafood industry year in review 2013*. St. John's: DFA.

Eric Dunne Consulting Initiatives. (2010). *Report of the review of the regulations and policy for direct fish sales in Newfoundland and Labrador*. St. John's: Eric Dunne Consulting Initiatives. Retrieved from http://www.fishaq.gov.nl.ca/publications/Direct_Fish_Sales_NL_2010%20.pdf

Fisheries and Oceans Canada (DFO). (1996). *Commercial fisheries licensing policy for eastern Canada – 1996*. Retrieved from http://www.dfo-mpo.gc.ca/fm-gp/policies-politiques/licences-permis/index-eng.htm

Fisheries and Oceans Canada (DFO). (2007). *Policy for preserving the independence of the inshore fleet in Canada's Atlantic fisheries*. Retrieved from http://www.dfo-mpo.gc.ca/fm-gp/initiatives/piifcaf-pifpcca/piifcaf-policy-politique-pifpcca-eng.htm

Fisheries and Oceans Canada (DFO). (2014). *Commercial fisheries*. Retrieved from http://www.dfo-mpo.gc.ca/stats/commercial/licences-permis/licences-permis-atl-eng.htm

Foley, P., Mather, C., & Neis, B. (2013). *Fisheries allocation policies and regional development: Successes from the Newfoundland and Labrador shrimp fishery*. St. John's: The Harris Centre, Memorial University. Retrieved from http://www.mun.ca/harriscentre/reports/arf/2011/11-12-ARF-Final-Mather.pdf

Food and Agricultural Organization (FAO). (2014). *Guidelines site voluntary guidelines for securing sustainable small-scale fisheries in the context of food security and poverty eradication*. Advance copy as endorsed by the 31st session of the FAO committee on fisheries (Rome, 9–13/06/2014).

Government of Canada. (2014, September 26). *Consolidated CETA text*. Retrieved from http://www.international.gc.ca/trade-agreements-accords-commerciaux/agr-acc/ceta-aecg/text-texte/toc-tdm.aspx?lang=eng

Government of Newfoundland and Labrador. (2007a, April 12). *Canada-Newfoundland and Labrador fishing industry renewal changes to DFO vessel replacement policy*. Retrieved from http://www.releases.gov.nl.ca/releases/2007/fishaq/0412bkfed2.htm

Government of Newfoundland and Labrador. (2007b, April 12). *Backgrounder. Newfoundland and Labrador fishing industry renewal strategy. Element: Implementation of fish auctions*. http://www.releases.gov.nl.ca/releases/2007/fishaq/0412bk6.htm

Government of Newfoundland and Labrador. (2007c, April 12). *Renewing the Newfoundland and Labrador fishing industry*. St. John's: Government of Newfoundland and Labrador. http://www.releases.gov.nl.ca/releases/2007/fishaq/0412n03.htm

Government of Newfoundland and Labrador. (2007d, April 12). *Canada-Newfoundland and Labrador fishing industry renewal: Fleet self-rationalization – Enterprise combining*. Retrieved from http://www.releases.gov.nl.ca/releases/2007/fishaq/0412bkfed1.htm

Government of Newfoundland and Labrador. (2007e, April 12). *Backgrounder. Newfoundland and Labrador fishing industry renewal: The process*. Retrieved from http://www.releases.gov.nl.ca/releases/2007/fishaq/0412bk1.htm

Government of Newfoundland and Labrador. (2013, October 29). *Fishing industry enters new era of opportunity with $400 million fund*. Retrieved from http://www.releases.gov.nl.ca/releases/2013/exec/1029n03.html

Government of Newfoundland and Labrador and Government of Canada. (2006). *Canada-Newfoundland and Labrador fishing industry renewal: A discussion paper*. Retrieved from http://www.fishaq.gov.nl.ca/industry_renewal/fs114_5_2006e.pdf

Hamilton, L. C., & Bulter, M. J. (2001). Outport adaptations: Social indicators through Newfoundland's cod crisis. *Human Ecology Review, 8*(2), 1–11.

Harris, M. (1998). *Lament for an ocean: The collapse of the Atlantic cod fishery*. Toronto: McClelland and Stewart.

Khan, A. S., & Chuenpagdee, R. (2013). An interactive governance and fish chain approach to fisheries rebuilding: A case study of the northern Gulf cod in eastern Canada. *AMBIO, 43*, 600–613.

Kooiman, J., & Jentoft, S. (2009). Meta-governance: Values, norms and principles, and the making of hard choices. *Public Administration, 87*(4), 818–836.

Kooiman, J., Bavinck, M., Jentoft, S., & Pullin, R. S. V. (Eds.). (2005). *Fish for life: Interactive governance for fisheries*. Amsterdam: Amsterdam University Press.

Lilly, G. R., Parsons, D. G., & Kulka, D. W. (2000). Was the increase in shrimp biomass on the northeast Newfoundland shelf a consequence of a release in predation pressure from cod? *Journal of Northwest Atlantic Fishery Science, 27*, 45–61.

Mather, C. (2013). From cod to shellfish and back again? The new resource geography and Newfoundland's fish economy. *Applied Geography, 45*, 402–409.

May, D. (2014). *People – Demographics and productivity in fishing industry*. Retrieved from http://www.mun.ca/care/CCFI_Workshop_Presentation.pptx

Milich, L. (1999). Resource management versus sustainable livelihoods: The collapse of the Newfoundland cod fishery. *Society and Natural Resources, 12*, 625–642.

Murphy, I., & Neis, B. (2011). *Navigating the legislative requirements for fisheries-tourism initiatives in Newfoundland and Labrador*. St. John's: Community University Research for Recovery Alliance (CURRA). Retrieved from http://www.curra.ca/documents/TCR_Fisheries_Tourism_Regulations_Report_Feb_14_Final_to_TCR_revised.pdf

Neis, B., & Ommer, R. (2014). *Moving forward: Building economically, socially, and ecologically resilient fisheries and coastal communities*. St. John's: Community University Research for Recovery Alliance (CURRA). Retrieved from http://www.curra.ca/documents/Revised_CURRA%20Policy%20Paper%20April%2016%202014%20to%20the%20printer%20%281%29.pdf

Palmer, C. T., & Sinclair, P. R. (1997). *When the fish are gone: Ecological disaster and fishers in northwest Newfoundland*. Halifax: Fernwood Publishing.

Parliament of Canada. (2007). *Bill C-45: An Act respecting the sustainable development of Canada's seacoast and inland fisheries*. Retrieved from http://www.parl.gc.ca/About/Parliament/LegislativeSummaries/bills_ls.asp?ls=c45&Parl=39&Ses=1

Pisces Consulting Limited. (2014). *Sea urchin fishery review*. Paradise: Pisces Consulting Limited. Retrieved from http://www.fishaq.gov.nl.ca/publications/Sea_Urchin_Report.pdf

Policy Brief 1. (2012). *Rationalization of the fishing industry: The case of Change Islands, NL*. Retrieved from http://gracilis.carleton.ca/tgeog/downloads/Using%20Local%20Knowledge,%20PB%201%20%28June%202012%29.pdf

Report of the Independent Chair: MOU Steering Committee. (2011). *Newfoundland and Labrador fishing industry rationalization and restructuring*. Retrieved from http://www.fishaq.gov.nl.ca/publications/mou.pdf

Roy, N. (1997, September 12–13). *The Newfoundland fishery: A descriptive analysis*. Prepared for a symposium on the efficiency of North Atlantic Fisheries. Reykjavík, Iceland.

Shrank, W. E. (2005). The Newfoundland fishery: Ten years after the moratorium. *Marine Policy, 29*, 407–420.

Sinclair, P. R. (1985). *From traps to draggers: Domestic commodity production northwest Newfoundland: 1850–1982*. St. John's: ISER Books.

Sinclair, S. (2013). *Globalization, trade treaties and the future of the Atlantic Canadian fisheries*. Ottawa: Canadian Centre for Policy Alternatives.

Smith, D. A., Vodden, K., Woodrow, M., Khan, A., & Furst, B. (2014). The last generation? Perspectives of inshore fish harvesters from Change Islands, Newfoundland. *The Canadian Geographer, 58*(1), 95–109.

Song, A. M., Chuenpagdee, R., & Jentoft, S. (2013). Values, images, and principles: What they represent and how they may improve fisheries governance. *Marine Policy, 40*, 167–175.

The Navigator Magazine. (2012). Bill Barry – Analyst and prophet? Processing and magnate talks politics, business and the fishery of the future, *15*(7), pp. 12–14.

The Navigator Magazine. (2014). An argument for inter-provincial free trade and free enterprise, *17*(9), pp. 12–13.
The Telegram. (2009, March 30). *Nothing rational about fisheries rationalization*. Retrieved from http://www.thetelegram.com/Letters-to-the-editor/2009-03-30/article-1458787/Nothing-rational-about-fisheries-rationalization/1
The Telegram. (2013, October 19). *CETA a big opportunity for fishery: McCurdy*. Retrieved from http://www.thetelegram.com/Business/2013-10-19/article-3434084/CETA-a-big-opportunity-for-fishery%3A-McCurdy/1
The Western Star. (2011, March 4). *FFAW president says fleet rationalization necessary*. Retrieved from http://www.thewesternstar.com/Business/2011-03-04/article-2299471/FFAW-president-says-fleet-rationalization-necessary/1
Walsh, D. (2011). What restructuring? Whose rationalization? Newfoundland and Labrador's memorandum of understanding on its fishing industry. In R. Chuenpagdee (Ed.), *World-small-scale fisheries: Contemporary visions* (pp. 81–97). Delft: Eburon.

Chapter 24
Power Dynamics and Community Failure in the Small-Scale Fisheries Sector in Cyprus

Maria Hadjimichael

Abstract Despite being an island state, fisheries have never been a major contributor to GDP in Cyprus. In social and economic terms, however, the island's extensive coastline is indicative of the importance of fishing for (but not confined to) coastal communities. Cyprus has been a member of the European Union (EU) since 2004 and hence has had to comply with the EU's Common Fisheries Policy and specifically with the Mediterranean Regulation. Fisheries resources suffer, in spite of a number of regulatory measures being in place, due to (i) overexploitation caused by the activities of the professional, recreational and illegal fishers and also weak enforcement of existing regulations, and (ii) ecosystem shifts due to factors such as climate change and the mushrooming of invasive species. Using the Interactive Governance Approach to supplement author's empirical data from different studies, this chapter aims to understand the governability (quality of governance) of the Cypriot small-scale fisheries sector and the failure of the fisheries management system to sustain fish stocks and more generally secure the small-scale fisheries sector. Given that in Cyprus fisheries are under a hierarchical mode of governance, failure to govern should as a first step be analyzed in terms of how power is exercised in the interaction between the state and its citizens, or at least the relevant user groups. Power may both enhance and distort governability, depending on who administers it and how it is used. The chapter concludes that existing power relations, interactions and struggles among different stakeholders partly lie at the root of the problem, but that problems are further exacerbated by an economic system which focuses on individual utility-maximization on the one hand and authorities, with a 'Divide and Conquer' approach on the other hand. New institutions need to be created and policies improved in order to strengthen civil society institutions at the community level. It is also important that fishing communities are involved in discussions and empowered so that they acknowledge their role and envision the way towards governable fisheries.

M. Hadjimichael (✉)
Innovative Fisheries Management, Aalborg University, Aalborg, Denmark
e-mail: mmh@ifm.aau.dk; maria.m.hadjimichael@gmail.com

S. Jentoft, R. Chuenpagdee (eds.), *Interactive Governance for Small-Scale Fisheries*, MARE Publication Series 13, DOI 10.1007/978-3-319-17034-3_24

Keywords Cyprus • Small-scale fisheries • Mediterranean • Power dynamics

Introduction

> *In a political system where nearly every adult may vote but where knowledge, wealth, social position, access to officials, and other resources are unequally distributed, who actually governs?* (Robert A. Dahl. 1961. Who governs? Democracy and Power in an American City)

Fisheries resource governance, particularly when it comes to small-scale fisheries, can be community, state or market-driven and nested in existing political, legal, and economic environments. This embeddedness raises questions about how markets, states, and other external or internal factors affect the capacities of communities and user-groups to respond to environmental change (McCay and Jentoft 1998). It has also been argued that the cause of resource degradation is institutional and that if the right rules and governance structures were put into place, natural resources would have been used wisely and conservation goals met (Acheson 2006). Jentoft (2007) has claimed that fisheries and coastal management ultimately rests on power: the power to design, decide, enforce, and implement regulatory decisions. That can be particularly true in systems governed under a hierarchical mode of governance, such as that in Cyprus. Through a case study of the governance of the Cyprus' small-scale fisheries, this chapter describes the fisheries system with focus on power relations and assesses their impact on governability.

McCay and Jentoft (1998) argue that overfishing may well be a sign of 'community failure' rather than Hardin's (1968) Tragedy of the Commons. Similarly, Harvey (2012) argues that the problem in Hardin's discourse is not the commons per se, but the failure of individualized private property rights to fulfill common interests in the way they are supposed to do. He suggests the need to focus on individual ownership of the resource and individual utility-maximizing behavior as the problem. While the state is often projected in opposition to individuals, looking only at the interests of the totality or even a specific class or a group, at the same time the state's power is both an individualizing and a totalizing form of power (Foucault 1982). Foucault (1982) argues that the strong dependence of an individual on the state is linked to a type of individualization (Foucault 1982) that would exaggerate one's drive for individual rather than collective gain. Therefore, in hierarchical systems, where actors are not encouraged to work collectively, the individual's link to the state is exacerbated, often resulting in the individual becoming disconnected from the community.

There are always winners and losers when it comes to fisheries management as decisions and constraints will "restrict stakeholders' degrees of freedom (including those of managers), and the relations stakeholders have to each other" (Jentoft 2007). Additionally, as management is about choices, it is recognized that trade-offs as to how a fishery is managed is a matter of values and political philosophies, which may differ among groups and individuals (Larkin 1977). This is true in all governance modes identified in the Interactive Governance Approach, but more so in those fisheries managed under a hierarchical mode of governance (the other two modes being self- and co-governance modes). Marx (1867) pointed out in the Capital, Volume 1 that "between equal rights force decides" in the context of a struggle between collective capital, i.e., the class of capitalists and collective labor, i.e., the working-class regarding working hours. Harvey (2012) has extended Marx's concept to include different struggles such as that for the urban commons. I would like to extend this analysis and argue that it (Marx's statement) is true for the commons in general.

Conger and Kanungo (1988) discuss the constructs of power and control and relate it to two different ways to view empowerment: (i) as a relational construct, describing the perceived power of an actor or organizational unit over others, or (ii) as a motivational construct, meaning that power is used to enable individuals to strengthen their self-determination or self-efficacy. The latter is the most commonly used definition of empowerment and the one used by the World Bank, according to which "central to this process are actions which both build individual and collective assets, and improve the efficiency and fairness of the organizational and institutional context which govern the use of these assets".[1] In order for individuals and communities to feel empowered, something that will consequently allow for the re-building of a community and perhaps the improvement in resource conditions, one must understand power relations and dynamics and their potential impacts on governability.

This chapter will use the Interactive Governance Approach to analyze and understand the governability (quality of governance) of the Cypriot small-scale fisheries sector, focusing mainly on power dynamics over the last decade in an attempt to understand the failure of the fisheries management system to sustain fish stocks and to secure the small-scale fisheries sector. Cyprus fisheries are governed under a hierarchical mode of governance. Therefore, failure to govern should as a first step be analyzed in terms of how powers exercised, between the state and its citizens, or at least the relevant user groups. Power may both enhance and distort governability, depending on who administers it and how it is used. The chapter is organized according to the framework for assessing governability steps described by Chuenpagdee and Jentoft (2013). First, the nature of the data used is presented. This is followed by a country profile with specific reference to the fisheries sector. The next section presents the case study results and assesses governability in Cyprus' small-scale fisheries by examining the characteristics of the natural and social systems-to-be-governed and decision-making in the governing system. Finally, the discussion attempts to bring all these elements together and considers how power can work to improve governability.

---

[1] http://web.worldbank.org/WBSITE/EXTERNAL/TOPICS/EXTPOVERTY/EXTEMPOWERMENT/0,,contentMDK:20244572~pagePK:210058~piPK:210062~theSitePK:486411,00.html

Nature of the Data

Data and supporting case studies presented in this chapter are an amalgamation of field work (of which some findings are already published in peer reviewed journals) carried out by the author in Cyprus using different methodologies. The author's experience as an active member of civil society, through her engagement with a local marine-related environmental Non-Governmental Organization (e-NGO) and continuous communication with local fishers, was also an important source of understanding and framing of the data used in the chapter. Three types of methodologies were used in this study: interviews, the analysis of media/newspaper articles, and the use of other secondary data from peer reviewed and grey literature such as relevant reports from the Cypriot authorities and from the European Commission and the FAO.

Specifically, data from Hadjimichael et al. (2013a, b) were collected during a 2009 field visit in Cyprus. The two studies are inter-connected as both are part of the same study exploring the 'Human dimension of the European fisheries governance'[2] (Hadjimichael 2010). Hadjimichael et al. (2013a) includes survey data using a conjoint analysis method to identify fishers' opinions with respect to their most and least preferred regulatory obligations and how they impact on their income. Hadjimichael et al. (2013b) focuses on fishers' problems and challenges that were repeatedly mentioned during the semi-structured interviews. Findings from these interviews complimented the survey data. Problems were further explored using secondary data such as past research literature, national reports, relevant national and European regulations. Follow-up interviews with governmental fisheries experts were also conducted. Interview data from Hadjimichael et al. (2014) that investigated the governance of marine aquaculture in Cyprus by exploring the contradictions that exist among stakeholders and the political forces at work are also used.[3]

Country Background

Cyprus, officially the Republic of Cyprus (RoC), is an island state in the Eastern Mediterranean Sea. The Republic of Cyprus only has control over the Southern part of the island (dark green color in Fig. 24.1). The island was partitioned after the Turkish invasion in 1974, which was a response to a military coup backed by the Greek military junta. The Southern part of the island is recognized internationally, whereas the Northern part (light green in Fig. 24.1) is under the control of the Turkish Republic of Northern Cyprus, a state only recognized by Turkey. The Republic of Cyprus is the third largest and third most populous island in the Mediterranean and

[2] 47 interviews with small-scale fishers.

[3] Three interviews with fishers' representatives and two interviews with coastal community representatives.

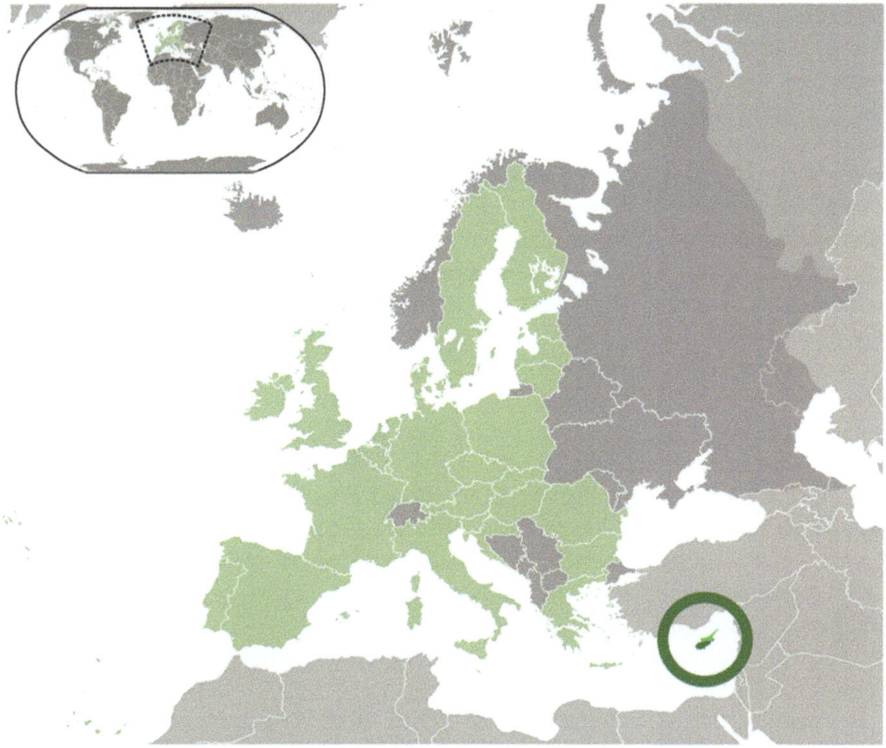

Fig. 24.1 The Republic of Cyprus (*in the dark green circle*) with respect to the European Union (*in washed out green*). The island of Cyprus is partitioned and the southern part, the Republic of Cyprus, is part of the European Union (*dark green*). The northern part, the Turkish Republic of Northern Cyprus (*light green*), is only recognized by Turkey (Source: Wikipedia)

has been a member of the European Union since 2004. The information provided pertaining to the period after 1974 is relevant only for the area the Republic of Cyprus has control over.

Brief History of Fisheries in Cyprus

Despite its relatively small contribution (less than 0.3 %) to the Gross National Income (GNI), the fisheries sector is important to the local economy of several coastal areas since it generates income and work opportunities, contributing to the social and economic welfare of the residents of these areas (Marttin et al. 2006). Recreational fishing is also a significant activity both in terms of number of persons involved and fishing effort (MOA 2012).

Fishing in Cyprus was exclusively carried out from small sailing or rowing boats using trammel nets and longlines until the Second World War. The first trawlers were built locally during the Second World War. In 1952 trawlers operating within the territorial waters of Cyprus were limited by law to ten and for conservation purposes were prohibited from trawling during the summer months. After 1960, when Cyprus' status changed from a British colony to an independent state, the trawler fleet numbered between 10 and 12 vessels before reducing to eight in 1979. The inshore fleet was also mechanized.

In 1964, the Department of Fisheries and Marine Research (DFMR) was formed within the Ministry of Agriculture and Natural Resources, as the governmental body responsible for the fisheries sector. By 1974 the fishing fleet, buoyed on by government assistance and programmes, numbered 12 trawlers and about 350 small wooden boats of 5–9 m overall length with marine diesel engines (forming the small-scale fisheries). About 600 people were involved in this fishery. Fishing was mainly with trammel nets, which catch a great variety of species, bottom-set gill-nets during certain seasons mainly for bogue (*Boops boops*) and picarel (*Spicara* spp.), and bottom-set longlines. This fishery was spread along the north and east coasts of Cyprus, an area mostly sheltered from the prevailing westerly winds and in which small coves and calmer waters encouraged the evolution of artisanal skills and traditions (Garcia and Demetropoulos 1986).

The Turkish invasion in 1974, and the subsequent partition of the island, had an important impact on the fishing industry in a couple of ways relevant to the governability discussion: (i) the loss of control of 60 % of the continental shelf and 45 % of the coastline (all valuable fishing grounds) by the Government of the Republic of Cyprus, and (iii) 300 fishers becoming refugees and moving to the South of the island. The Government assisted the 300 fishers through loans and grants, technical assistance and the building of new fishing shelters so that they could resume fishing in the non-occupied waters (Demetropoulos 1985; Hadjistephanou and Vassiliades 2004). The changes that occurred in 1974, nonetheless, led to the heavy exploitation of resources in the areas that remained under the control of the RoC (Garcia and Demetropoulos 1986) and also in the interdependence of fishers and the administration.

Assessing Governability in Cyprus' Small-Scale Fisheries

The governance of small-scale fisheries in Cyprus is certainly a wicked problem, namely "*a problem that is difficult to define and delineate from other and bigger problems*" (Jentoft and Chenpagdee 2009). The 1974 partitioning of the island, for example, led to an increase in the fishing effort in the south coast of the island. Governability challenges in the Cypriot small-scale fisheries relate to a whole range of issues such as overfishing, economic viability of the sector, conflicts over access to space and resource and power struggles (between different métiers but also between fishers and the authorities). Fish stocks around Cyprus' waters are not in a healthy state and catches have been declining since the 1990s. Dysfunctional and unformulated management by responsible authorities has exacerbated the situation. The length of the nets that can legally be set in the water by small-scale fisheries is greater than two million meters,[4] which is approximately seven times the coastline of the RoC.[5]

Hadjimichael et al. (2013b) has identified that the small-scale fisheries sector's vulnerability has increased due to: (i) the invasion of the lagocephalos (rabbit fish), known as silver tripe blassop (*Lagocephalus scleratus*) from the Red Sea and (ii) resource use conflicts with the recreational sector. The recreational sector, which includes a number of sub-sectors itself, is a very complex one. Small-scale fishers and recreational fishers in June 2013 also collectively opposed the issuance of licenses to purse seiners. Purse seiners in Cyprus were decommissioned using European Fisheries Fund. However the author found, during her 2009 interviews in Cyprus, that inshore fishers were expecting the introduction of new licenses in June 2013, shortly after the centre-right Democratic Rally (DISI) political party assumed power, which was what actually happened.

Natural System-to-Be-Governed

The Mediterranean basin is one of the most diverse and stable Large Marine Ecosystems in terms of species groupings and their share of the total catch (EOE 2008). Its complexity is also high as there are a number of specific environmental characteristics in the Mediterranean that influence fisheries in the region. For example, the narrowness of the continental shelf leads to a situation in which a substantial part of fishing activities are carried out close to the coast, coincident with the highest biodiversity of bottom fish (European Commission 2002). The constrained shelf area means that there is a high degree of competition for space among fishers from different fleets and different sectors (professional and recreational). Additionally, despite the region's low productivity status, the fisheries in the Mediterranean have demonstrated a surprising resilience to fishing compared with other areas of the Atlantic (Lleonart 2005).

According to the DFMR, the stocks of five important demersal marine fish species have been evaluated on a yearly basis based on regular samplings and on length and age distributions, but also on statistical data collected through the fishery statistical system that is based on landings (Marttin et al. 2006). Fisheries production declined in 1974 and 1975, most likely due to the political unrest during that time and to the lack of control by the Cypriot government over some fishing grounds. There has been a steady increase, on the other hand, from 1976 onwards. From 1986 to 1996, production rose again to above 2,500 metric tonnes per year. Since 1994, however, there has been a steady decline of the fisheries production, resulting

[4] Calculated using the regulations valid in 2012.

[5] Which is 291.283 km in length.

in a production of less than 1,750 tonnes in 2003 and 1,109 tonnes in 2012 (Marttin et al. 2006; DFMR 2012). According to official statistics, assessments based on the 2005–2010 national surveys have identified that the four main demersal species (bogue, picarel, surmulet [*Mullus surmuletus*], and red mulet [*M. barbatus*]) in Cypriot waters are being fished beyond sustainable levels and that the stocks are in a poor state. The General Fisheries Commission for the Mediterranean (GFCM) therefore suggested a decrease in fishing effort from the inshore fleet and the trawling fleet who exploit these resources (DFMR 2012). Picarel and red mullet particularly showed more than a 50 % decrease in their biomass since 2010 (DFMR 2012). The picarel fishery is of particular relevance for the discussion regarding the conflict between the inshore fishers and the purse seiners discussed later.

The growing population of an invasive species, referred to by the locals as lagocephalos (rabbit fish), has been a major issue for Cypriot small-scale fisheries as it has adversely impacted the dynamics of the natural system-to-be-governed (Hadjimichael et al. 2013b). The silverstripe blassop is widespread in the Indo-Pacific region and migrated from the Red Sea into the Mediterranean through the Suez Canal. It is known as a "lessepsian migrant". With climate change and the warming of the waters around Cyprus, this lessepian migrant dominates the waters of Cyprus (DFMR 2009). According to DFMR, Lagocephalos was first reported in Cyprus in 2000, although it has become more common in catches since 2004. Measuring up to 100 cm in length and 7 kg in weight, Lagocephalos can cause serious damage to the catch and fishing gear of fishers, using its powerful jaws. Additionally, it has no commercial value due to the presence of tetrodotoxin in its tissues.[6] A total of eight species of "lessepsian migrants" were identified during a national research survey, which was part of the "International Bottom Trawl Survey in the Mediterranean" known as MEDITS completed in July 2011. Of these two were rabbit fish species (*L. sceleratus* and *L. suezensis*) (DFMR 2011). The DFMR believes that the implementation of the action plan for the reduction of the number of Lagocephalos, which took place between May and August 2012, had positive results for the control and decline of their population (DFMR 2012).

Social System-to-Be-Governed

The dynamic nature of the Cypriot fisheries' social system-to-be-governed in terms of the composition of stakeholders has been outlined earlier. Currently, the Cyprus capture fishery consists of the small-scale fisheries, the trawl fishery and the multipurpose/polyvalent fishery. Additionally, as of June 2013, two purse seiners have received a license to operate in Cypriot waters. The multipurpose fishery (polyvalent gears) consists of boats of about 16 m overall length that use long lines in the waters of the Republic of Cyprus and in international waters in the Eastern Mediterranean. Adding to this diversity, recreational fishing in Cyprus has economic,

[6]A neurotoxin that can be a source of poison with a high fatality risk.

social and cultural importance. The DFMR issues approximately 4,500 licenses for recreational fishers, while it is estimated that hundreds of others fish with rods and lines without a license (DFMR 2012). Sport fishers need a license for the use of a boat with nets and long lines, for using spear-guns and for fishing with nets-without boats. According to official statistics, the recreational fishery captures about 15 % of the total Cypriot catch though this is not yet reflected in the Fishery Statistics, most likely because the DFMR has only recently focused on this fishery (DFMR 2008).

Small-scale fisheries consists of the majority of the Cypriot fleet in terms of the number of vessels and is composed primarily of small fishing vessels (less than 12 m) that use seasonally deployed passive gear. The number of effort days for this fleet varies from season to season and weather dependent. Depending on the time of the year, these small vessels target different species. Landings are mainly composed of *Spicara* spp. (mostly *S. smaris*), Boops boops, Mullus barbatus, *M. surmuletus*, *Pagellus erythrinus* and cephalopods (*Octopus vulgaris*, *Eledone moschata*, *Loligo vulgaris* and *Sepia officinalis*). The fleet also lands relatively large quantities of *Diplodus* spp., *Sparisoma cretense* and *Siganus* spp.

In 2010,[7] the then 500 active vessels of this category (the small-scale fisheries sector) formed approximately 94 % of the professional Cypriot fleet. These vessels engaged in seasonal fishing and were authorized to use trammel nets, anchored gill-nets, set longlines, pots and traps. Despite the large numbers of vessels, it accounted for just under half (1,628 GT) and 60 % (22,231 kW) of the tonnage and engine power respectively. In terms of employment, 87 % of the Cypriot fishers or 640 people work in this sector. This sector produces approximately 64 % of the weight of landings that comprise 69 % of the value of total Cypriot fishery landings, worth around seven million Euros in 2010. The fleet segment operated at a loss that year with a negative Gross Value Added[8] (GVA) of 8.5 million and a net loss of 11.8 million due to issues like limited fisheries resources and the increase in fuel prices.

Governing System

In order to unravel the overall quality of governance in the Cypriot small-scale fisheries we must first understand the different elements that affect the governability, depending on the specific mode of governance. As already mentioned, the Cypriot

[7] Data from Anderson et al. (2012) collected under the frameworks of the Data Collection Regulation (DCR). cf Council Regulation (European Commission (EC) No 1542/2000) and the data collection framework (DCF), cf. Council Regulation (European Commission (EC) No 199/2008).

[8] Signifies the added value the activity contributes to the national economy. The indicator may provide information on the socio-economic importance of the fishery, as economically important stocks are represented by high revenueswhile the associated costs are a measure of the level of effort applied in the fishery (Anderson et al. 2012).

small-scale fisheries are governed under a hierarchical mode of governance with authority being in the hands of the DFMR of the Ministry of Agriculture, Natural Resources and Environment. Government policy in the fisheries resources sector aims towards "*the sustainable and balanced management of the resources, the increased contribution of fisheries in the domestic production of fish and in the improvement of professional fishers' work conditions*".[9]

According to the DFMR, the management of fisheries resources is achieved through fisheries research, the collection of fishery data and their utilization for the exploitation of the Cyprus fisheries stock, and for the development of the Government's fisheries policy. Cyprus however does not have a national research institute, thus research is realized either by foreign researchers or by persons working within the fisheries administration who often do not have a proper research frame. DFMR staff collects data in accordance with the Community Data Collection Framework as well as taking part in the MEDITS survey. The activities of the DFMR include the development and management of fisheries and aquaculture, including the collection and analysis of relevant data, and the promotion of programmes for fishers (especially since Cyprus accession in the EU) such as the construction of fishing shelters and enforcement of relevant legislation.

Within the Cypriot Fisheries Law, the country's fishing fleet is divided into three fleet segments: the small-scale fisheries vessels (with a length of 6–12 m), the polyvalent (or longliners) (with a length of 12–24 m) and the bottom trawlers (with a length of 21–27 m). These vessels are categorized depending on their type of license: those fishing in the territorial waters of Cyprus and those that fish in international waters (eastern and central Mediterranean). Small-scale fisheries are further divided into three categories. Category A and B are for professional fishers for whom fishing is their main (category A) or part-time (category B) occupation,[10] and category C who, according to the Fisheries Law, are "*professional fishers who are only allowed to fish during weekend and public holidays*". In this chapter, the term small-scale fisheries does not include category C fishers.

Since Cyprus' accession to the EU, there has been an effort to harmonize national and Community (meaning the European Union) legislation to ensure compliance with the Common Fisheries Policy of the EU. The Common Fisheries Policy provides the umbrella framework for top-down control of fisheries resources at the EU level with Member States being required to transpose Community to national legislation. Mediterranean EU member states specifically are expected to implement Council Regulation (EC) No 1967/2006 concerning management measures for the sustainable exploitation of fishery resources in the Mediterranean Sea. There are a number of management measures applicable to the small-scale fisheries sector, drafted in the form of either National or Community legislation which include: (i) restrictive access to the fisheries by limiting the number of licenses issued for each fleet segment, (ii) effort control through restrictions on the use of fishing gears

[9] http://www.moa.gov.cy/moa/dfmr/dfmr.nsf/DMLAli_en/DMLAli_en?OpenDocument

[10] According to Cyprus Fisheries Law, Category B fishers must go to sea at least 120 days each year and land at least 1,000 kg of fish each year.

(quantities, soaking time, depth of deployment and distance offshore) and regulations in terms of fishing capacity (using scraping schemes, engine power restrictions, capping the fleet vessel register), (iii) market restrictions that define a set of minimum landing sizes, (iv) technical conservation measures for minimum mesh sizes, and (v) seasonal and area closures.

Access to the European Fisheries Fund and, the EU's assistance *"to the fishing industry and coastal communities to help them adapt to changing conditions in the sector and become economically resilient and ecologically sustainable"*[11] gave fishers the financial buffer to either exit or improve their profession. The most important funding schemes that supported a reduction in the fishing effort are schemes for the scrapping of vessels of the polyvalent and trawling fleet and schemes for providing a premium to small-scale fishers for "*activities that encourage the reduction of fishing effort on a voluntary basis*", similar to a buy-back program (DFMR 2012).

Decision-Making Structures

The power of decision-making lies with three main actors/institutions: the Director of the DFMR, the responsible Minister (Minister of Agriculture, Natural Resources and the Environment) and the Board of Ministers and Parliament (the House of Representatives of the RoC). The President of the RoC also has the authority to intervene.

According to Cypriot Fisheries Law, the Ministerial Board has the power with a decree to submit a regulation in the official newspaper of the RoC. These decrees can pertain to area and gear restrictions, fishing licenses and enforcement. The Parliament can cancel or amend these regulations within a time period of 20 days but the President of the RoC can accepts these amendments or not. Finally, the Director of the DFMR has the power to allocate licenses to professional and recreational fishers.

The DFMR is the institution dealing with day to day fisheries issues. The top-down approach of the responsible authorities is often criticized by fishers who feel powerless to change the situation within the sector. Despite opportunities for small-scale fisheries, recreational fishers (except for Category C) and e-NGOs to form an alliance an over reliance on the DFMR by these various groups has led to conflicts between them and the continuance of a top-down approach. A recent example of this top-down approach was the public consultation regarding the amendment of the Cyprus Fisheries Law that took place in October 2012 and involved an array of different fisheries stakeholders such as professional fishers, different groups of recreational fishers and relevant e-NGOs. The minutes of the meeting give a different view to what the stakeholders said took place. A newspaper article written by a representative of the recreational fishers present at the consultation described it as

[11] http://ec.europa.eu/fisheries/cfp/eff/index_en.htm

'not so public': "*it was clear that the intention was that the public consultation was not to be so public after all and to not include so much consultation*" (Aristidou 2012). Aristidou also used examples from his experience with the DFMR in general, and the public consultation specifically, to describe the policy of the DFMR and its Director in particular as one of 'Divide and Conquer'. This complex interaction among the different actors has created a dependency of fishers (mainly the professional ones) on the authorities, and has thus not allowed for the creation of collaboration among fishers (professional and recreational). The author had similar experiences with the Director of the DFMR between 2012 and his retirement in September 2013 whilst campaigning against issues such as overfishing and illegal shark fishing in Cyprus. It has often been suggested that public consultations and stakeholder inclusion methods in general are a way to legitimize and re-enforce existing hierarchies rather than change power structures (see Gray and Hatchard 2003 for such a critique of the Common Fisheries Policy). This appears to be the case also in Cypriot fisheries.

Conflicts and Power Relations

This section focuses on conflicts that revolve around issues of power in Cypriot fisheries in the past 10 years in an attempt to understand how power works in marginalizing small-scale fisheries. This does not mean that prior to that conflicts were not present and that power was absent. However, during the late 1970s and early 1980s, inshore fishers had the support of both the Government and the Cypriot Parliament. The move, for example, to close fishing for trawlers around October1982 was evidence of the political backing the Government and Parliament gave to the approximately 700 small-scale fisheries families at the expense of the 70 or so families in the trawler fishery, a large proportion of which were foreigners on contract as crew members (Garcia and Demetropoulos 1986).

Power relations in Cypriot fisheries have always been a complex and dynamic issue. Fishers (recreational, inshore, trawlers, polyvalent or purse seiners) have never held power as such but were in Foucault's words its "subjects". Power vis-à-vis fisheries in Cyprus resides in the following institutions in order of the least powerful to the most powerful: the DFMR (and particularly the Director of the DFMR who has the power to make final decisions, for example, in terms of distributing fishing licenses), the Minister of the Ministry of Agriculture, Natural Resources and the Environment which is the Ministry to which the DFMR belongs, the Cypriot Parliament, which is the decision-making body in charge of voting in new legislation and finally the President of the Republic who has the power to annul legislation passed by Parliament if (s)he considers it to be against the constitution.

The negative impact of recreational fisheries on small-scale fishers was raised on many occasions during interviews with small-scale fisheries who criticized recreational activities because they competed for the same space and resources. Specific issues raised were that (i) sometimes wealthy recreational fishers owned boats that were more powerful than those of small-scale, professional fishers, (ii) that there was a lack of restrictions on the activities of recreational fishers, and (iii) that recreational fishers sometimes illegally sold their catch to restaurants for a lower price than that of professional fishers.

However, recreational and small-scale fishers are not always in conflict; they agree that the goal of protecting the marine environment is not an attribute of a specific group, but of fishers as a whole. A representative from the recreational fishers said the following about conflicts: "*I see that those who are true in what they are doing, whether they fall under the professional or the recreational fishers' group, with all the connotation of the word and without hiding behind a title in order to support some interests, have the same aims and policies. My views are almost identical with many of the professional fishers, and we all agree that there are very bad professional and very bad recreational fishers. Most importantly however, there are a number of groups in between, who I call interest groups, who use their political power and contacts to do whatever they want*".

The introduction of a 2007 legislative measure which gave recreational fishers professional status (with the creation of category C) was an example often brought up by small-scale fisheries and some of the recreational fishers, as indicative of the power dynamics. According to this new measure (Number 132(I) of 2007), category C holders are allowed to use gill nets of restricted mesh and of a maximum size of 600 m during weekends and public holidays. The new measure was introduced to counter the introduction of EC Regulation 1967/06[12] that required recreational fishers to stop using gill nets (referred to as the gill net ban). According to many fishers, the introduction of Category C status was a mechanism used by decision-makers (Members of Parliament) to satisfy some influential and rich recreational fishers. The then president of the Association of Professional Fishers stated in the newspaper 'Cyprus-Mail': "*instead of the Cyprus Parliament approving the regulation (EC No 1967/2006) so that it can become law, after pressure from the masses of amateurs and in an attempt to attract votes, not only did it not approve it, but bypassed it and with a special legal amendment created a category C for fishers.*"[13] According to fishers, Category C holders were sometimes upgraded by the DFMR to a category A or B status (and sometimes category A or B fishers were downgraded to category C).

To further highlight the relevance and consequence of power in Cypriot small-scale fisheries, two cases that are indicative of the situation in Cyprus as a whole are briefly discussed. Firstly, in a study on the governance of marine aquaculture in Cyprus, conflicts were identified between the fishers and fishing communities who fish or live close to marine aquaculture farms (Hadjimichael et al. 2014). In one particular case, a strong accusation of nepotism was made regarding a farm that was according to the fishers adversely impacting their activities. A fishers' representative described protests of fishers and the coastal community who lived close to the

[12] Council Regulation (EC) No 1967/2006 of 21 December 2006 concerning management measures for the sustainable exploitation of fishery resources in the Mediterranean Sea.

[13] http://www.cyprus-mail.com/cyprus/parliament-accused-passing-fishing-laws-attract-votes

aquaculture farm that took place during the early 1990s. Their objections and protests, however, fell according to the fishers, on deaf ears, as the company that bought the farm in the 1990s had strong ties with a high level politician (then Minister). The second example is the recent case of the purse seiner licensing mentioned earlier. The announcement in favor of purse seiners did not come as a surprise to those who follow the politics of Cyprus' fisheries as the licensing was expected with the change of government.[14] Fishers' mobilizations, after the announcement pertaining to the licensing, led to the passing by the Cypriot Parliament of a new amendment to the Cyprus Fisheries Law, banning the purse seine in Cyprus waters. The then President of the RoC himself annulled the Parliament's decision on the grounds that such a decision was beyond their jurisdiction. The Parliament can take the issue to the Supreme Court but had not done so by the time of writing this chapter, even though meetings between the Parliament and the Minister have taken place.

It is important to note that at least two issues have come up regarding the licensing procedure of the purse seiners. Firstly, it appears that there has not been an official call for applicants interested in applying for a purse seine license and the application/approval procedural has not been transparent. Secondly, during a meeting with the relevant Parliamentary committee, the Minister suggested that the main catch of the purse seiner, which is the picarel, is in a healthy state[15] (something that data presented later challenges).

Discussion

Throughout this chapter, it has become clear that the low governability of the Cypriot fishing governance system lies at least partly in the complexity of power relations, interactions and struggles among different stakeholders. The current economic reality in Cyprus limits the possibilities for a quick solution to the Cypriot fisheries management problem, which is indeed a "wicked" one. Cyprus fisheries suffer from overfishing, weak enforcement and mismanagement. It is a problem embedded in issues created by an economic system that focuses on individual utility-maximization. Also, the concerned authorities through a 'Divide and Conquer' which attempts to create conflicts between professional fishers, recreational ones and e-NGOs, undermines the role of communities. The community is also "absent from the neo-classical model of the tragedy of the commons" (McCay and Jentoft 1998), and consequently the solutions offered do not focus on the community, even though community empowerment can be a key element of a wider strategy for improving both the ecological and the socio-economic aspects of

[14] A number of fishers also mentioned this during interviews conducted for the author's PhD research in 2009.

[15] The minutes from the meeting can be found here: http://www2.parliament.cy/parliamentgr/008_05f/008_05_4028.htm

fisheries management. The deteriorating state of marine resources in Cyprus requires action even though extraction (by both professional and recreational fishers) is unlikely to decrease with the current e economic situation in Cyprus. On the contrary, the situation will potentially worsen, and government cut backs will weaken the already weak patrolling and enforcement of fisheries legislation.

The dysfunctional relationship which exists between fishers and the Cypriot authorities challenges any attempts for collaboration between fishers and the DFMR or non-governmental research or other institutes that would be necessary to enhance the governability of small-scale fisheries on the island. There is a deep level of mistrust of authorities among small-scale fisheries, be it the DFMR or Cypriot Ministers of Parliament. This stems from the general feeling in Cyprus that nepotism and corruption is high. This feeling was aggravated by a series of actions that fishers considered unjust, most notably the creation of category C fishers and the licensing of purse seine fishing. The feelings towards the EU, however, were not as hostile as the majority of fishers believed that the country's accession to the EU had a positive impact on their profession for two main reasons: (i) fishers felt there was a higher body where they could challenge decisions taken by national decision-makers, and (ii) the European Fisheries Fund meant additional support for exit from the industry and improvement of fishing harbors etc.

A government that does not provide communities some roles in fisheries management loses an important opportunity, not only to support community viability but also to make management systems work more proficiently (Jentoft 2000). Co-management in fisheries, which is about the restructuring of relations and moving towards a more equal sharing of power among interested stakeholders (Raakjær Nielsen et al. 2004), has potential for improving governability. However, it is not on the political map of either government or other stakeholders. Moreover co-management neither removes conflicts or interests nor eliminates power games, but does provide a vehicle for handling them in a way that may bring small-scale fisheries out of the dire circumstances that they find themselves in.

In Cyprus, we can see an example of what Kooiman and Chuenpagdee (2005) describe as one typical of the global South, where the institutionalization of required responsibilities by the State leads to governing roles which are marginal, aimed at increasing catch, and based on self-, instead of common interest. Additionally, the notion of the state being the responsible institution for fisheries governance has created an open access regime where it is very difficult to not only control who fishes, but also monitor how many fish are extracted. I would agree with McCay and Jentoft (1998) and suggest that the overfishing problem in Cyprus is due to community failure, embedded in the historical and political backdrop of the island. On the one hand, individual small-scale fishers have a strong dependence on the state, partly because of financial aid expectations, and partly because of the country's history of poor collective decision-making. Earning a livelihood is the primary concern for fishers and with dwindling resources, any financial aid (subsidies), albeit small, is important. Regarding history, there are no studies to suggest the existence of Cypriot fishing communities with strong social bonds, perhaps explainable by the fact that Cyprus' history is one of continuous colonization. The current postcolonial condition persistent in Cyprus is largely determined by the idea of Europe and the desire to be recognized and confirmed as a modern European society (Argyrou 2010). With Europeanization, the established neoclassical economics paradigm became the norm, creating individuals having to choose between being rational individuals against being good communitarians.

Governance is what those in power want it to be and true emancipation of those who are ruled can only happen if the ruling class is willing to delegate power (Jentoft et al. 2005). Otherwise one cannot talk about democracy in a political system where knowledge, wealth, social position and access to officials are unequally distributed (Dahl 1961). A possible solution would be for the fishers, in collaboration with scientists and responsible authorities to follow the Spanish example of cofradias, or the Chilean example of caletas, both of which are community fishing cooperatives, where fishers register in a specific cofradia or caleta and have a number of responsibilities and consequently benefits too. In Chile, for example, small-scale artisanal fishers are restricted to operate within the coastal area of the caleta where they are registered. A survey of participating fishers in a number of Chile's caletas highlighted four social benefits because of this type of co-management: (i) improved rapport between fishers and the state, (ii) greater awareness among fishers about ecology and the benefits of management, (iii) cooperation between fishers and scientists, and (iv) unity between fishers with the principal benefit being fishers' newfound "consciousness" of the value of management – within the context of co-management and territorial use rights (Schumann 2007). Such structures should not be idealized but one must appreciate the role of both the concerned authorities and the fishers in making this type of regime successful. Moreover, in such structures the role of the administration needs to be decisive, with investments for example in training and improving organizations as well as in building the social capital of these institutions (Frangoudes et al. 2008).

Public/stakeholder consultations in Cyprus are still at a premature phase. Such consultations are held back due to the unwillingness of the responsible authorities to hand over part of their control. Generally, most civil society actors in Cyprus choose to conduct their advocacy activities through 'clientelistic' relations with political parties. Those who are outside of this system are not taken seriously. While consultation jargon has been introduced officially, it is mostly to show that the government is complying with EU rules (Perry-Kessaris 2013). In practice, small-scale fisheries livelihood concerns are not articulated nor do such fishers have the know-how to take an unjust issue further either in terms of making an official complaint or in filing a court case. Eventually, small-scale fisheries become submissive to the system and learn to become receivers of financial aid, hence becoming pawns in the social hierarchy. Both small-scale fisheries and recreational fishers agree that the common goal of protecting the marine environment is not an attribute of a particular group but of all fishers. Thus, a community fisheries cooperative does not need to exclude recreational fishers. On the contrary, adapting such a cooperative to the local context is the only way to make it successful small-scale fisheries. In conclusion, this case study on the governance of Cyprus' small-scale fisheries has argued that low governability of the sector is a result of power dynamics between concerned authorities and fishers. Power in this case study is distorting governance. The DFMR, the Cyprus Parliament and specific government representatives (from Ministers to the President himself) influence the balance of power at sea in ways that more often than not adversely affect the ruled class. However, power does prevent people from doing what they wish to do, but also colonizes the mind and affects people's thinking and actions (Argyrou 2010). Therefore, even though it is important for the small-scale fisheries to demand more participation communities will find it difficult to imagine a different system because the top-down decision-making system has become so entrenched. To better grasp the issue of power and participation, further study is needed to understand the reasons for the involvement (or not) of Cypriot fishers in the decision making process, along with more detailed analysis of the different discourses and involvement of different actors in this process. Improving governability requires the improvement of the quality of governance by ensuring that the governing system has the capacity to govern both effectively and in a just way. This requires the creation of new institutions and the improvement of policies that will strengthen civil society institutions at the community level (Jentoft 2000). It is also important that fishing communities become involved in discussions and empowered so as they acknowledge their role and envision a way towards governable fisheries.

References

Acheson, J. M. (2006). Institutional failure in resource management. *Annual Review of Anthropology, 35*, 117–134.

Anderson, J., Carvalho, N., Contini, F., & Virtanen, J. (2012). The 2012 annual economic report on the EU fishing fleet (STECF-12-10). EUR 25425 EN-2012.

Argyrou, V. (2010). Independent Cyprus? Postcoloniality and the spectre of Europe. *Cyprus Review, 22*(2), 39–47.

Aristidou, A. (2012). Δημόσια διαβούλευση και το μαύρο μας το χάλι. Politis Newspaper, page 77.

Chuenpagdee, R., & Jentoft, S. (2013). Assessing governability – What's next. In M. Bavinck, R. Chuenpagdee, S. Jentoft, & J. Kooiman (Eds.), *Governability of fisheries and aquaculture* (pp. 335–349). Amsterdam: MARE Publication Series.

Conger, J. A., & Kanungo, R. N. (1988). The empowerment process: Integrating theory and practice. *Academy of Management Review, 13*(3), 471–482.

Dahl, R. (1961). *Who governs?: Democracy and power in an American city*. New Haven/London: Yale University Press.

Demetropoulos, A. (1985). *Cyprus fisheries*. Marine Policy, January.

DFMR. (2008). Annual report, Department of Fisheries and Marine Research, Cyprus. Retrieved November 5, 2013, from http://www.moa.gov.cy/moa/dfmr/dfmr.nsf/DMLreports_en/DMLreports_en?OpenDocument&Start=1&Count=1000&Expand=2

DFMR. (2009). Annual report, Department of Fisheries and Marine Research, Cyprus. Retrieved November 1, 2013, from http://www.moa.gov.cy/moa/dfmr/dfmr.nsf/All/6980F1E5B5322AE94225771300313D1A/$file/%CE%95%CE%BA%CE%B8%CE%B5%CF%83%CE%B709.doc?OpenElement

DFMR. (2011). Annual Report, Department of Fisheries and Marine Research, Cyprus. Retrieved November 5, 2013, from http://www.moa.gov.cy/moa/dfmr/dfmr.nsf/All/EB79A1DC0F8B450

A42257A3F0024A383/$file/%CE%95%CE%BA%CE%B8%CE%B5%CF%83%CE%B72011.doc?OpenElement

DFMR. (2012). Annual Report, Department of Fisheries and Marine Research, Cyprus. Retrieved November 11, 2013, from http://www.moa.gov.cy/moa/dfmr/dfmr.nsf/All/EBCC10B9461861D142257C1400337AE1/$file/%CE%95%CE%BA%CE%B8%CE%B5%CF%83%CE%B72012.doc?OpenElement

EOE. (2008). *Mediterranean Sea large marine ecosystem*. Retrieved November 11, 2013, from http://www.eoearth.org/view/article/154549

European Commission. (2002). Communication from the Commission to the Council and the European Parliament laying down a Community Action Plan for the conservation and sustainable exploitation of fisheries resources in the Mediterranean Sea under the Common Fisheries Policy. Com. 535 final. Brussels, (09-10-2002).

Foucault, M. (1982). The subject and power. *Critical Inquiry, 8*(4), 777–795.

Frangoudes, K., Marugán-Pintos, B., & Pascual-Fernández, J. J. (2008). From open access to co-governance and conservation: The case of women shellfish collectors in Galicia (Spain). *Marine Policy, 32*(2), 223–232.

Garcia, S., & Demetropoulos, A. (1986). *Management of Cyprus fisheries*. FAO Fisheries Technical Paper 250, 40.

Gray, T., & Hatchard, J. (2003). The 2002 reform of the Common Fisheries Policy's system of governance – Rhetoric or reality? *Marine Policy, 27*(6), 545–554.

Hadjimichael, M. (2010). *The human dimension of the European fisheries governance*. Doctoral thesis, Bangor University, Bangor.

Hadjimichael, M., Kaiser, M. J., & Edwards-Jones, G. J. (2013a). The impact of regulatory obligations on fishers' income: Identifying perceptions using a market-testing tool. *Fisheries Research, 137*, 129–140.

Hadjimichael, M., Delaney, A., Kaiser, M. J., & Edwards-Jones, G. J. (2013b). How resilient are Europe's inshore fishing communities to change? Differences between the north and the south. *AMBIO, 42*, 1037–1046.

Hadjimichael, M., Bruggeman, A., & Lange, M. A. (2014). Tragedy of the few? A political ecology perspective of the right to the sea: The Cyprus marine aquaculture sector. *Marine Policy, 49*, 12–19.

Hadjistephanou, N., & Vassiliades, L. (2004). *The present status of fishery and information system in Cyprus*. GCP/INT/918/EC – TCP/INT/2904/TD-4.2. MedFisis Technical Document No. 4.2, 55.

Harvey, D. (2012). *Rebel cities: From the right to the city to the urban revolution*. London: Verso.

Jentoft, S. (2000). The community: A missing link of fisheries management. *Marine Policy, 24*(1), 53–60.

Jentoft, S. (2007). In the power of power: The understated aspect of fisheries and coastal management. *Human Organization, 66*(4), 426–437.

Jentoft, S., & Chuenpagdee, R. (2009). Fisheries and coastal governance as a wicked problem. *Marine Policy, 33*(4), 553–560.

Jentoft, S., Kooiman, J., & Chuenpagdee, R. (2005). National institutions. In J. Kooiman, M. Bavinck, S. Jentoft, & R. Pullin (Eds.), *Fish for life: Interactive governance for fisheries* (pp. 173–196). Amsterdam: University of Amsterdam Press.

Kooiman, J., & Chuenpagdee, R. (2005). Governance and governability. In J. Kooiman, M. Bavinck, S. Jentoft, & R. Pullin (Eds.), *Fish for life: Interactive governance for fisheries* (pp. 285–299). Amsterdam: University of Amsterdam Press.

Larkin, P. A. (1977). An epitaph for the concept of maximum sustainable yield. *Transactions of the American Fisheries Society, 106*(1), 1–11.

Lleonart, J. (2005). *Review of the state of world marine fishery resources* (pp. 49–64). FAO Fisheries Technical Paper 457. B5 Mediterranean and Black Sea.

Marttin, F., Barone, M., Bizsel, C., Fayed, S., Hadjistephanou, N., Krouma, I., Majdalani, S., Özdemir, A., Salem, A., & Vassiliades, L. (2006). *Brief introduction to the Eastern*

Mediterranean fisheries sector. GCP/INT/918/EC -TCP/INT/2904/TD-6. MedFisis Technical Document No. 6.

Marx, K. (1867). *Capital* (Vol. 1). New York: Penguin.

McCay, B., & Jentoft, S. (1998). Market or community failure? Critical perspectives on common property research. *Human Organization, 57*(1), 21–29.

MOA. (2012). *Ministry of agriculture and natural resources*. Annual Report for 2011. Nicosia. Retrieved November 5, 2013, from http://www.moa.gov.cy/moa/agriculture.nsf/All/F48D4924 45DD7ABEC2257A4C00283EBA/$file/Annual%20Rep-2011-english.pdf

Perry-Kersaris, A. (2013). Anemos-ity, Apatheia, Enthousiasmos: An economic sociology of law and wind farm development in Cyprus. *Journal of Law and Society, 40*(1), 68–91.

Raakjær, N. J., Degnbol, P., Viswanathan, K. K., Ahmed, M., Hara, M., & Raja Abdullah, N. M. (2004). Fisheries co-management – An institutional innovation? Lessons from South East Asia and Southern Africa. *Marine Policy, 28*(2), 151–160.

Schumann, S. (2007). Co-management and "consciousness": Fishers' assimilation of management principles in Chile. *Marine Policy, 31*(2), 101–111.

Chapter 25
Common Ground, Uncommon Vision: The Importance of Cooperation for Small-Scale Fisheries Governance

Silvia Salas, Julia Fraga, Jorge Euan, and Ratana Chuenpagdee

Abstract Like in many countries around the world, concerns about resource degradation due to high fishing intensity and use of illegal fishing gears have led to the creation of several protected areas in Mexico. Also as in other cases, these conservation efforts have not been very successful, especially in areas where boundaries are unclear; resource uses overlap, and enforcement weak. Under these circumstances, conflicts between users are likely to escalate, making the fisheries system and the protected areas ungovernable. As posited by interactive governance theory, how stakeholders interact depends partly on the inherent characteristics of the social system, including images that they have of each other, and of the governing system. Stakeholder interactions are also reflections of their willingness to cooperate with each other, which in turn affects the overall resource governability. We illustrate the importance of stakeholder cooperation for governability using a case study of two neighboring small-scale fishing communities, San Felipe and Dzilam de Bravo, on the Yucatan coast of Mexico. While sharing fishing grounds and two nested protected areas, fishers from these two communities had different images about what the protected areas were for, who benefited from them, and how they should be governed. The communities also differed in livelihood options, the level of internal organization, and in the mode of governance. Based on our findings obtained through participatory research, we discuss how to foster cooperation between small-scale fishers and promote co-governance in order to enhance resource governability in the area.

Keywords Cooperation • Participatory research • Protected areas • Resource governability • Small-scale fisheries • Mexico

S. Salas (✉) • J. Fraga • J. Euan
CINVESTAV, Merida, Mexico
e-mail: ssalas@mda.cinvestav.mx; jfraga@mda.cinvestav.mx; euan@mda.cinvestav.mx

R. Chuenpagdee
Department of Geography, Memorial University, St. John's, Canada
e-mail: ratanac@mun.ca

S. Jentoft, R. Chuenpagdee (eds.), *Interactive Governance for Small-Scale Fisheries*, MARE Publication Series 13, DOI 10.1007/978-3-319-17034-3_25

Introduction

Benefits of marine protected areas (MPAs) are often stated in terms of biodiversity improvements and "spill-over effects" (Guénette et al. 2000; Sale et al. 2005; Aguilar-Perera et al. 2006; Mora and Sale 2011) and, to a lesser extent, in terms of their economic contributions (Sumaila 1998; Hannesson 2001). Yet, one of the factors restricting the successful implementation of MPAs around the world is related to the social and institutional consequences of these instruments (Christie 2004; Hilborn et al. 2004; Jentoft et al. 2007; Mora and Sale 2011). As argued by several authors (White et al. 2002; Hilborn 2007; Berkes 2008; Jentoft and Chuenpagdee 2009), fisheries over-exploitation and environmental degradation in coastal areas have more to do with the socio-economic and institutional-political nature of the problems than with the resources themselves. For this reason, participation of fishers and communities in decision-making about the size, location and governance of MPAs is imperative to enhance fisheries governance, including MPA governability (Davy and Breton 2006; Berkes 2008; Charles and Wilson 2009; Jentoft et al. 2012). This is particularly important given that MPAs are likely to affect livelihoods and the viability of small-scale fisheries. The process can also be rather cumbersome when fishing communities are diverse, complex, and dynamic; thus a common vision or agreement on MPA goals may not be easily achieved (Hilborn et al. 2004; Jentoft et al. 2011).

Further complication can arise when participation of fishers from one community is not entirely independent from what goes on in neighboring communities, especially when they share common pool resources and also the protected areas (Fraga et al. 2006; Pajaro et al. 2010). In such instances, enhancing MPA governability is not only about improving fisher participation in the discussion about where MPAs should be situated, how big they should be, what activities should be allowed inside them, and who should make decisions. It is also about understanding how affected small-scale fishers interact in their own community and with others in nearby areas. These interactions reflect some inherent characteristics of the social system, including images that they have of each other and of the governing system, as well as of their capacity for self-organization and their willingness to cooperate (Gutiérrez et al. 2011; Jentoft et al. 2012; Ovando et al. 2013). We argue from the interactive governance perspective (Kooiman et al. 2005) that unless these interactions are well understood, in their own context, small-scale fisheries governability challenges will remain. Cooperation as a form of interaction is of specific interest in this chapter, given the peculiarity of the case study of two neighboring small-scale fishing communities, which share fishing grounds and protected areas, but not much else.

Situated in close proximity to each other, small-scale fishers of San Felipe and Dzilam de Bravo, on the Yucatan coast of Mexico, fish in the same nearshore waters. Both communities are located within the boundary of Dzilam de Bravo State Reserve (Fig. 25.1), declared through a top-down process by the state government in 1989 (Secretaria de Ecologia 2006). Concerns about the dwindling fisheries

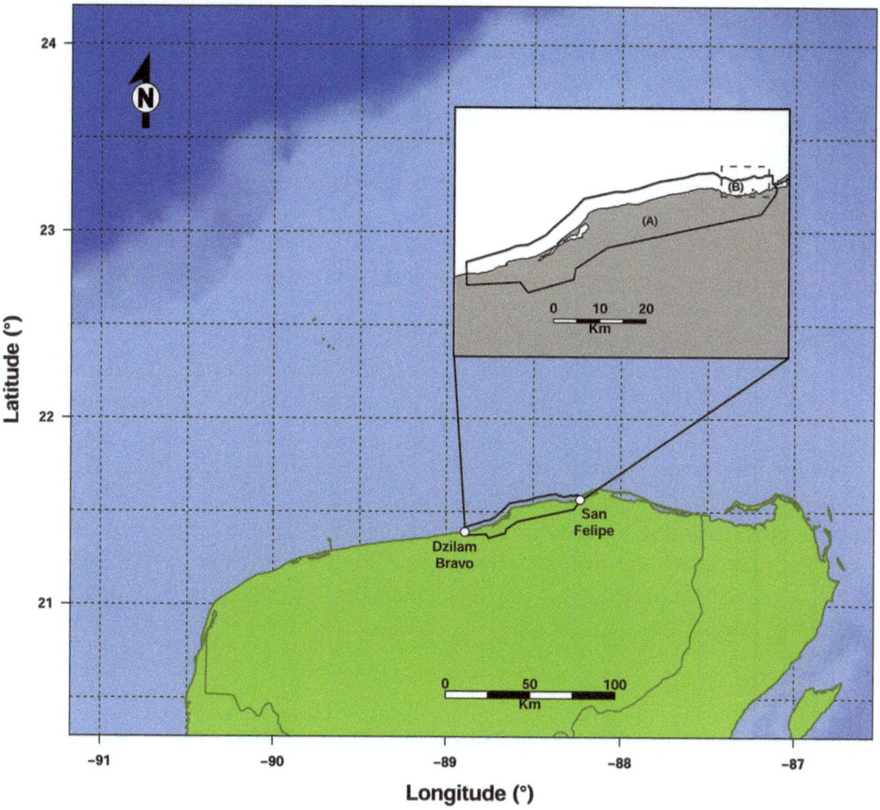

Fig. 25.1 Location of protected areas along the Yucatan coast, Mexico: (**A**) Dzilam de Bravo State Reserve and (**B**) Actam Chuleb Marine Reserve

resources and the 'paper park' status of the State Reserve drove the small-scale fishers of San Felipe to create their own marine reserve, Actam Chuleb, in 1995 (Chuenpagdee et al. 2004). Through their fishing cooperative and with support from the municipal government, they established their own rules and a local management committee to enforce them (Ayuntamiento Municipal de San Felipe 1999; Diario Oficial del Gobierno del Estado de Yucatán 2005; Aguilar et al. 2012). Rather than being praised for this conservation initiative, San Felipe small-scale fishers were reprimanded by state officials, especially as small-scale fishers of Dzilam de Bravo, whose access to fishing grounds were affected, lodged complaints. Since Actam Chuleb Marine Reserve is embedded within the boundary of the existing Dzilam de Bravo State Reserve, San Felipe fishers had no legal right to prevent others, like small-scale fishers from Dzilam de Bravo, from entering and fishing in Actam Chuleb. This situation with the MPAs adds another layer of complexity to the relationship between these two fisher groups, thus affecting the overall governability of small-scale fisheries in the area.

This chapter aims at understanding the complex relationship between small-scale fishers of San Felipe and Dzilam de Bravo and its effect on resource governability, taking into consideration the overlapping boundaries of the MPAs, their different origins, and how fishers perceive them. We ask what these two fisher groups have in common, what distinguish them from each other, and what factors and conditions determine their interactions. In accordance with meta-order governance posited by interactive governance theory (Kooiman and Jentoft 2009; Song et al. 2013), we further analyze the 'images' that small-scale fishers have about MPAs, based largely on their knowledge and perception about these protected areas and what they intend to do. This examination contributes to determining avenues to enhance cooperation between these two small-scale fishing communities. We argue that cooperation between small-scale fishers who share fishing grounds and protected areas is beneficial for the health of the ecosystem and for their livelihood viability.

In what follows, we present an overview of the two small-scale fishing communities and the historical development of their protected areas. Next, we briefly describe the various studies that we base our examination on, before presenting the key findings, and discussing the implications. We conclude with some recommendations about steps that can be taken towards enhancing cooperation and thus overall resource governability.

Background: The Communities and Their Protected Areas

The coastal area of Yucatan is rich in mangrove forests and submerged aquatic vegetation, which provide nursery and feeding grounds for many juvenile fish and crustaceans (Arceo-Carranza et al. 2010). These ecosystems support productive fisheries, of economic importance to small-scale fishers such as red grouper (*Epinephelus morio*), spiny lobster (*Panulirus argus*), and octopus (*Octopus maya*) (Pedroza and Salas 2011). Small-scale fisheries dominate the area and over time the sector has grown in capacity and efficiency as well as in number of fishers and boats, thus increasing competition for limited and dwindling resources (Fraga et al. 2008; Pedroza and Salas 2011). Rapid development in the coastal area—including the clear-cutting of mangrove forests for housing construction and burning of forested areas for pasture or agricultural land—has prompted several government initiatives to help protect the rich biodiversity of these areas and several others along the Yucatan coast (Chuenpagdee et al. 2002; Jesus and Euan-Avila 2008). Dzilam de Bravo State Reserve is one such example.

Established by the federal government in 1989 and designated as a RAMSAR site in 2000, the state reserve comprises about 69,000 ha (Secretaria de Ecologia 2006). It is situated at one end of the 'ring of cenotes' (sink holes), a unique hydrological system formed by the impact of a large meteor in ancient times. The reserve encompasses a range of coastal habitats, including submerged aquatic vegetation, inter-tidal zones, coastal dunes, and forests. High biological diversity characterizes this area. It contains 38 species of fishes, eight species of amphibians, and 148 species of birds, attracting a large amount of tourists each year (Secretaria de Ecologia 2006). Many of the endemic fish species are native to the cenotes and many marine species are of high commercial value.

Dzilam de Bravo State Reserve falls within the municipal boundary of Dzilam de Bravo and San Felipe (Fig. 25.1). Dzilam de Bravo, the larger of the two municipalities, is the third largest fishing community along the Yucatan coast, the majority of which are of small-scale. The labor force in the fisheries has been rising steadily, from 58 % of people dedicated to fishing in 1989 to 76 % by 2004 and close to 90 % by 2010 (INEGI 2014). The increase in fishing population is also due to fishers coming here from nearby communities during the octopus season (August to December), or for the sea cucumber season (November-February), thus putting heavy pressure on fisheries resources in the area (Salas et al. 2011). Many small-scale fishers also engage in supplementary income-generating activities, such as aquaculture and tourism. Mostly, efforts to develop alternative livelihoods are household-based or promoted by small groups of people. Limited interactions, both among small-scale fishers and with government agencies, make it difficult to coordinate actions.

Closely connected by sea, but rather far in travel distance (about 70 km by road) from Dzilam de Bravo, is San Felipe. As shown in Table 25.1, although smaller in size, San Felipe has some similar socio-economic characteristics as Dzilam de Bravo. While activities such as ranching and tourism provide additional income for the small-scale fishers (Aguilar et al. 2012), fishing is still considered the main economic activity in San Felipe, engaging about 90 % of the population.

Concerns about ecosystem health and livelihoods due to rapid and unplanned coastal development have prompted small-scale fishers of San Felipe to take action. With support from the municipality and local officials, they worked with the only fishing cooperative in the community to create the Actam Chuleb Marine Reserve mainly to promote conservation (Chuenpagdee et al. 2002; Fraga et al. 2006; Jesus

Table 25.1 Socio-economic characteristics of the fishing communities

	Dzilam de Bravo	San Felipe
Population (2010)	2,463	1,839
Female population (%)	48	49.0
Male population (%)	52	51.0
Illiterate people above 6 years old (%)	1.4	6.6
People without sanitary services (%)	4.9	4.3
People without power (%)	4.7	3.9
People without drinkable water (%)	25.0	3.9
People without medical services (%)	45	59.0
People involved in fishing (2010) (%)	90.0	65.2
Other activities	Commerce, ranching, aquaculture, tourism	Tourism, ranching

Sources: INEGI 2005; Datos económicos, demográficos y sociales, Gobierno del Estado deYucatan

and Euan-Avila 2008). The reserve covers an area of about 30 km^2 from the edge of the mangrove forests to about 2 km from shore. It encompasses dense nesting areas of several types of birds, with rich and diverse marine habitats providing shelter for many commercially important species, including lobster and other crustaceans. It also includes a small island named "El Cerrito," which contains rich archeological Mayan artifacts of high cultural value (Andrews and Gallareta 2003; Secretaria de Ecología 2006). The fishing cooperative, in coordination with local authorities and the municipality, defined the reserve as a self-regulated area where only subsistence fishing using non-destructive gears such as hooks and lines was allowed (Bjørkan 2009; Aguilar et al. 2012).

Given their small size and homogenous characteristics, governability of this small-scale fishing community could be assumed to be high. However, seasonality creates resource variability that presents some challenges to small-scale fisheries governance. While the creation of Dzilam de Bravo State Reserve did not lead to any serious disagreements, problems and governability challenges arose with the community-driven establishment of the Actam Chuleb Marine Reserve. Through decentralization, resource management was under the authority of local governments, which, for the most part, work collaboratively with the municipalities and fishers cooperatives in making local level decisions. Nonetheless legal acknowledgement of this reserve was not achieved and control over external users was not enforceable as was wished by small-scale fishers of San Felipe. As a consequence, conflicts arose between these two small-scale fishing communities and a stalemate lasted until 2006, when the Actam Chuleb Marine Reserve was officially recognized and incorporated into a special "extractive zone" within Dzilam de Bravo State Reserve.

Data Sources

For this study, we draw from three main sources of data. The first was a survey conducted in these communities between 2001 and 2003, which solicited their knowledge of the protected areas, their perception of their importance, and their level of participation in the management of these reserves. A total of 175 respondents in San Felipe and 231 in Dzilam de Bravo were surveyed. These included small-scale fishers (both members of the fisheries cooperative and non-members), tourism-businesses, housewives, scientists and other residents.

Second, we organized community workshops to discuss the future of the protected areas, after the completion of the surveys. One workshop was held in each community while a third one was held in the town of San Felipe, which was aimed to promote interaction and foster collaboration among members from both communities. Sixty people attended the workshop in San Felipe, 45 in Dzilam de Bravo and 48 in the third workshop. In all cases, participants were mostly small-scale fishers, but government officials and scientists working in the areas also joined.

We conducted another survey in 2008, using a questionnaire with open-ended questions, to obtain an update on the conditions of the protected areas. This survey was done in response to some fishers reporting that the protected areas were not operating effectively. This time the objective was to evaluate people's perceptions regarding coastal resources and management strategies of the protected areas at that time, and assess their level of involvement in management. The study also aimed to pinpoint which factors would encourage compliance with conservation measures in the protected areas. The study targeted three groups of respondents, i.e. cooperative fishers, non-cooperative (or independent) fishers, and middlemen. A total of 99 people were interviewed in San Felipe (about half were cooperative fishers) and 172 in Dzilam de Bravo (two-thirds were cooperative fishers).

In addition to the more systematic survey-based approach discussed above, we went back to visit the two communities in 2013 and 2014 to have conservations with some of the key informants, including the leader of the fishers cooperative, tourist guides and hotel managers to follow-up on the marine resource situation and the outcomes of the protected areas.

Key Factors Affecting Governability

Perceptions and Knowledge About Coastal Resources and the Protected Areas

From the first survey we observed that in general residents of Dzilam de Bravo and San Felipe acknowledged the importance of coastal ecosystems, such as mangroves and rocky areas, in terms of their roles as habitats for birds, and nursery and refuge areas for economically important species, like lobsters and groupers. Consequently, they expressed concerns about clear-cutting of mangrove forests and fishing in shallow waters, which they said affected juvenile lobsters. They also expressed concern that development in the area had led to concerns of coastal pollution such as excessive garbage. On the other hand, impact from tourism activities (e.g., boating operated by local guides who took tourists to visit the spring waters and the cenotes) was of least concern to the communities. This could be because, in both cases, tourism offered supplementary income to fishing and was considered a viable alternative livelihood option. The Dzilam de Bravo community particularly expressed the need to protect the cenotes, recognizing the biodiversity and tourism values of these areas within the reserve system. Both communities talked about environmental education as a way to inform people and help reduce pollution as well as a way to promote conservation.

The high level of awareness about environmental issues and the shared understanding about the importance of conservation make small-scale fisheries highly governable, and provide a good foundation for collaboration. Unfortunately, the two communities differed markedly in terms of their knowledge about the protected areas. Results from the surveys showed that the majority of the people in San Felipe (86 %) were familiar with the existence of the two protected areas and acknowledged their importance. While 75 % of the respondents from Dzilam de Bravo knew about their state reserve, only 14 % had heard of the Actam Chuleb Marine Reserve. Most of the respondents in Dzilam de Bravo were not able to explain why and how the state reserve was established, or how it operated. Many thought its purpose was to provide protection to plants and animals, and the majority of small-scale fishers in Dzilam de Bravo viewed this area mainly as an opportunity to develop tourism activities, given the presence of spring water and the cenotes.

Respondents in the San Felipe community were generally more aware of the limits of the Actam Chuleb Marine Reserve. However, only two percent recognized that the Actam Chuleb was nested within the Dzilam de Bravo State Reserve. Further, they were unaware about the legislative boundary of the protected areas and whether they were under the authority of the municipality of Dzilam de Bravo or San Felipe, or both. The overlapping boundaries of the two protected areas created confusion and misunderstanding among small-scale fishers in both communities. It also had consequences in terms of the images that small-scale fishers had about resource ownership and their role and responsibility in stewardship. For instance, San Felipe small-scale fishers had previously been very proud of their reserve and were willing to make sacrifices when thinking that the reserve was theirs to protect. However, with the changing image of the reserve, challenges in the governance of small-scale fisheries are likely to multiply.

Partnership and Cooperation Between Communities

On the whole, fishers in Dzilam de Bravo felt alienated from the management of their own protected area, having received no direct benefit from it. Less than 30 % of the respondents reported having been involved in the management of the state reserve. The fact that the state government was in charge of the management of the reserve may have left small-scale fishers of Dzilam de Bravo with no motivation for engagement. Another reason identified by the respondents was the general lack of interest from the research community and non-governmental organizations in this community and its reserve. This contrasts starkly with San Felipe where the high level of awareness, participation and self-governance capacity among small-scale fishers was attributed, to some extent, to these external stimuli (Jesus and Euan-Avila 2008). Interestingly, about 10 % of the San Felipe respondents did not like the idea of receiving financial support from international agencies. The reason proffered for this response was a feeling that these agencies might have special interests and may wish to impose some conditions upon them, thus affecting the self-governing system they had long employed. The different levels of participation, and thus experience, in resource governance between the two communities make governance of small-scale fisheries in the area more complex and less governable when involvement of fishers from both places is required.

The spatial overlap between Actam Chuleb and the state reserve raised awareness that led to action being taken in San Felipe. Small-scale fishers, community members, and local officials came together to search for options to legalize the management plan they had developed for their marine reserve, after recognizing that they could not impose sanctions on outsiders (like those from Dzilam de Bravo). Cooperation with Dzilam de Bravo community members was deemed necessary to facilitate the legal process. However, there was no history of collaboration between the two communities, notwithstanding their proximity to each other. The joint community workshop held in 2003 was a small step in facilitating discussion. The workshop participants suggested revision of operating rules and enforcement, search for funding to support monitoring and surveillance, and development of mechanisms to generate collaborative interaction between community members, including establishing a new committee to promote community-driven initiatives. Since only a few people from Dzilam de Bravo attended the workshop, it is doubtful as to whether cooperation would be fostered. Unless other avenues to improve interaction were found, the governability of small-scale fisheries would suffer as a consequence.

The Changing Land- and Seascape

Data from the 2008 survey revealed that the number of small-scale fishers combining fishing with tourism had increased in both communities. Recreational fishing and fishing for previously under-utilized species such as crab were also more common. When asked to compare the conditions of the resources in 2008 with 5 years earlier (the first period of study), many respondents indicated the steady deterioration of fisheries resources in the area. This situation created governability challenges as many fishers in both communities indicated that they turned to the protected areas, either to fish illegally or to increase income by bringing tourists there. According to the key informants, tourism income was perceived as the main benefit derived from the reserves, also attracting external investors. Tension between community members surfaced as many felt that only a few people benefited from the reserves. In San Felipe, the largest hotel in town (Hotel San Felipe) had a clear advantage in tourism business, with their ability to attract foreign tourists with package tours, which included accommodation, food and recreational fishing in the reserve. In Dzilam de Bravo, the fisher cooperative was seen as granting tourism-operation permits to family members and relatives. Unlike in San Felipe, the Dzilam de Bravo fisher cooperative played little role in promoting the wellbeing of small-scale fishers and was not involved in the discussion about the MPAs. From the governability perspective, the change in fishing practices and the perceived inequality in the communities will likely result in making small-scale fisheries less governable.

Small-scale fishers also seemed to have less faith in the conservation value of the protected areas. While about 72 % of the 98 fishers interviewed said that the protected areas continued to provide benefits for local fisheries and the community in general, the other 28 % of the respondents felt that these protected areas did not meet conservation goals as they were not effective in controlling illegal activities. In fact, a key informant interviewed in 2013 revealed that small-scale fishers from San Felipe had collectively agreed to stop supporting enforcement efforts in Actam Chuleb and in fact started to utilize the resources they had taken care of for a long time. In his words, "*It is like taking our savings out of the bank instead of allowing others to take a free ride*". Such a negative downturn towards conservation is likely to raise another governability challenge in the area.

Another noticeable change in the dynamics of fisheries was the heightened role of women in fisheries, the local economy and resource governance. In 1996, women in San Felipe came together to establish their own cooperative that focused on small crabs used as bait for the octopus fishery (performed by small-scale fishermen). Since 2005, another group of women initiated production of handicrafts made out of shells and sold them to tourists in Dzilam de Bravo. In addition to the new social and economic dynamics that emerged because of increased involvement of women, it also resulted in changes in governance when women wanted to be involved in resource management and conservation. The fisherwomen, in particular, became very active and vocal in decision-making about fisheries, after garnering interest from the media, funding organizations and government. The new dynamic in the community brought about by the women's groups may contribute to making small-scale fisheries more governable since additional income generated by women may help lessen household reliance on fisheries resources. Considering that women are keen on conservation, their involvement in resource governance may contribute to making the MPAs more beneficial to small-scale fisheries than they have been in the past.

Discussion

In their study of the Philippines and Indonesia, Pollnac and Pomeroy (2005) state that socio-economic variables can vary by groups within a community and can define behavioral responses towards the use and management of resources. Cinner and Pollnac (2004), on the other hand, observe that perceptions about environmental resources and the wealth of community members define not only how resources can be used in a place, but also the values people place on them, which consequently influence their response towards conservation initiatives. In this study, high economic dependence on coastal resources plays a key role in how communities interact with each other and with the marine ecosystem. Concerns about mangrove forests and rocky bottoms (as refuges of important commercial species) expressed by small-scale fishers of Dzilam de Bravo and San Felipe reflect the level of importance that they place on both fishing livelihoods and conservation. Recent changed attitudes of San Felipe fishers towards the marine reserve, which are less concerned with current conservation initiatives, suggest that a deeper understanding of what underlie people's priorities and actions is required. This includes a thorough examination of internal and external factors and conditions that are either conducive to or prohibitive of individual conservation efforts and community cooperation. A proper analysis of the small-scale fisheries system using the governability assessment framework would help reveal what these factors and conditions may be.

Values and images can change over time, especially when induced by ecological, social and political changes. As suggested by Salas et al. (2011), resource degradation, variability in environmental conditions, and changes in management policies can modify people's behavior and attitudes, and hamper possible cooperation. The recent decline in fisheries resources in the whole region (Pedroza and Salas 2011; Salas et al. 2011) has created high uncertainty of resource availability, which has, in turn, increased the cost of resource extraction. In the study areas, recurrent red tides, reduction in the availability of fishing resources, and increase in the fishing population have induced small-scale fishers from both communities to extend their fishing activities around or inside the state reserve. With insufficient enforcement capability (e.g., only two officers undertaking multiple tasks in the reserve, including patrolling), the level of compliance is low. The situation is not unique as lack of personnel dedicated to the monitoring and enforcement of protected areas is common in Mexico. Cudney-Bueno et al. (2009), for instance, report a situation in San Jorge reserve in Baja California, Mexico where members of the local government have not been able to deter fishing in the reserve, and after various confrontations with interlopers, have agreed to allow fishing in the area so as to prevent others from free-riding.

A major change in the governing system in the area deserves special attention because of its potential effects on values and images of small-scale fishers towards the MPAs, and thus governability. After several years of self-enforced regulatory practices, small-scale fishers in San Felipe lost their autonomy in decision-making about Actam Chuleb Marine Reserve. Since its creation, San Felipe fishers and their cooperative had worked closely with the community and the municipality in defining the governance of the Actam Chuleb Marine Reserve. They had agreed on operational rules and elected a group of fishers to represent them in coordinating surveillance activities. In effect, they have expressed a desire for self-governance and exclusive rights to the area. As shown in our study, the realization that their efforts were not legally acceptable came as a surprise to San Felipe small-scale fishers. To rectify the situation, they initiated discussions with local governments and Dzilam de Bravo small-scale fishers in order to provide legal protection to the Actam Chuleb. The result may not be what they had wished for, however. The legislative change that occurred to officially incorporate Actam Chuleb Marine Reserve as a special extractive zone within the Dzilam de Bravo State Reserve has come at a price. The multi-stakeholder reserve committee, which San Felipe small-scale fishers themselves recommended, implies that they would no longer have sole authority over Actam Chuleb. Further, with the state government assuming responsibility over Actam Chuleb, the local government in San Felipe has discontinued support to the cooperative for surveillance activities. This, along with the recent change in the local ruling political party, has created a deep division within the organization, and the eventual break-up of the fishing cooperative. Many small-scale fishers operate independently while disbanded members have formed a new tourism cooperative. In sum, San Felipe has been transformed from a place of high social capital, good communication and cooperation among small-scale fishers to one of factions and self-interest. Such a dramatic change presents a major governability challenge for sustainable small-scale fisheries in the area.

Is it possible to restore social cohesion and self-governance that has been eroded due to social dynamics and governance changes in the MPAs? We suggested earlier that fostering cooperation between the two communities may help improve small-scale fisheries governance. This requires, however, an understanding of values and images that underlie small-scale fishers' behavior, as well as an alignment of these elements with those of the governing system (Jentoft et al. 2010; Song et al. 2013). Also, as suggested by Gatewood (1984), human sociability is a process of negotiation in which individuals cooperate and/or compete with one another while pursuing diverse goals. Hence, cooperation among stakeholders can occur only when there is a perception of mutual benefit for those involved in the process. Although small-scale fishers in Dzilam de Bravo perceived little benefit from the MPAs early on, they became more interested in conservation because of the growth in tourism development in the area. While San Felipe small-scale fishers may no longer place such a high value on their MPA, they also benefit from tourism income. The common interest of both fisher groups in the development of eco-tourism offers a potential common ground for cooperation. The Actam Chuleb Marine Reserve once upon a time had unified small-scale fishers when fisheries resources were in decline. Hence, it may be possible for them to unite again, this time also with their neighboring fishers who similarly depend on fisheries resources and whose cooperation is necessary in order to achieve sustainability of small-scale fisheries livelihoods in the area.

Governance interventions are required to promote cooperation between the two communities in combining fishing and low-impact tourism. As stated by Skaperdas and Syropoulos (1996) and Cudney-Bueno et al. (2009), people need to perceive the potential benefits of their engagement in cooperative actions in resource governance. While these benefits may not be obvious, there is a synergistic action where the total effect is greater than the sum of the independent actions (Guttman 1996; White et al. 2002; McConney and Baldeo 2007; Ovando et al. 2013). However, only under certain conditions would an individual be motivated to participate fully in a collective action (Gatewood 1984; Gray et al. 2012). It can be assumed that cooperation will take place when the results are perceived as mutually beneficial. Developing a common vision for cooperation in business development and in conservation may need to be accompanied with a set of incentives, in the short- and long-term. One of the first steps may be to create a multi-stakeholder committee to develop a sub-regional plan for promotion of fishing and eco-tourism in the area.

In the case of Dzilam de Bravo more work needs to be done as the ground to promote internal and external cooperation is weak. There are a few starting points that may provide further grounds for cooperation. For instance, a few respondents in this community did mention that they were concerned about the decline of the fisheries resources. Some small-scale fishers from Dzilam de Bravo also acknowledged the efforts of fishers from San Felipe in protecting their reserve. While Dzilam de Bravo fishers are not sure how they can get directly involved in the management of the state reserve, they may be keen to participate in discussions about tourism-fishing cooperation. Unfortunately, the recent opening of the sea cucumber fishery in Dzilam de Bravo may have complicated the situation since it has led to illegal fishing that even the state government cannot control. The problems with the sea cucumber fishery would need to be addressed before talk of cooperation can begin.

Cooperation between different levels of government is another key element that needs to be fostered in order to enhance governability. Enforcement problems are worsened by gaps or overlaps in regulations as well as the lack of communication between government officials. Several legal instruments are in place regarding conservation of protected areas in Mexico (see Jesus and Euan-Avila 2008; Garcia-Frapolli et al. 2009; Cudney-Bueno 2009 for details). However, there is a lack of mechanisms to coordinate efforts by various agencies and to evaluate their efficiency (Vidal and Capurro-Filograsso 2008; Salas et al. 2011). Further, communication between those who share responsibility in the management process is a major problem (Jesus and Euan-Avila 2008; Garcia-Frapolli et al. 2008). For instance, exchange of information among government officials (municipality, community members and the Secretaria de Ecologia (Ministry of Ecology)) has been poor and is worsening in our case study area since the recent change of local government in San Felipe. This political rift is a good example of how small-scale fisheries governability is affected not only by what is happening with the fisheries resources and the fishing communities alone, but also by what goes on in the governing system that may be beyond the control of small-scale fishers.

Conclusion

This chapter illustrates the complexity of governance when only fishing grounds and protected areas are in common, but not much else. The difference in the characteristics of the two communities, e.g., in terms of social capital, level of organization, and perceptions of users, means that opportunities and willingness to participate in resource governance and cooperate to promote alternative economic activities are uneven. A complete governability assessment (Chuenpagdee and Jentoft 2013), along with a 'step zero' study (Chuenpagdee and Jentoft 2007; Chuenpagdee et al. 2013), could help examine what would be required to provide a level playing field for both communities, as well as offer ideas about innovative mechanisms and governing interventions that would result in greater cooperation for resource governance. Through this process, small-scale fishers and other communities of San Felipe and Dzilam de Bravo may be able to work collaboratively in defining objectives for protected areas, and formulating fishing rules and regulations that recognize local user rights and self-governance traditions. This will also provide opportunities for the governments to consider an appropriate channel through which to improve interactions between small-scale fishers and government officials, fostering co-governance in the future.

In the meantime, alternative employment activities must be considered to provide supplementary income to fishers' families, to reduce pressure on resources, and as a way to cope with uncertainty due to resource variability. One of the main challenges that implementation of protected area regulations face is finding options for displaced people. Coercive actions generally fail to achieve desirable outcomes, and instead generate conflicts. Hence local communities need to understand the purpose of protected areas and agree on the goals. The shared interest between small-scale fishers of San Felipe and Dzilam de Bravo in combining fishing with low-impact tourism offers an opportunity to explore mechanisms that can help foster and strengthen cooperation. This has also to be in line with the rules and regulations promoted in co-governance of fisheries resources and protected areas.

After 5 years of showing little interest in Actam Chuleb Marine Reserve, in 2014, the National Commission of Protected Areas, together with different environmental organizations, started conducting community workshops. These workshops aim to "revive" interest and participation of small-scale fishers and other communities in the management of MPAs. Community members once again recognize the importance of self-governance. They want to take advantage of a recent policy that introduces the concept "Refugio Pesquero" (fishing refuge) in Mexican law, as it offers an opportunity to generate community-based actions with the support of government agencies. San Felipe has a unique opportunity to engage in an initiative that addresses conservation issues by combining low impact fishing and ecotourism activities. The local Actam Chuleb Civil Association has secured a 5-year agreement to be part of a state ecotourism network, as well as committed itself to being more involved in administration of issues external to the community as opposed to with only fishing related activities. The strong interest in ecotourism highlights the economic transition that has taken place in the community. It remains to be seen, however, whether involvement in ecotourism will enhance governability of the marine resources in the area and increase cooperation between the two communities.

Acknowledgments We thank the people of San Felipe and Dzilam de Bravo for their kind collaboration and warm welcome while we undertook this research. The project was funded by the International Development and Research Centre of Canada, in collaboration with CARICOM, Laval University and the International Ocean Institute—Canada. We thank Yvan Breton and Brian Davy for their support during the project. The research team members, namely, Miguel Cabrera, Hector Rodríguez, Karin Alvarez, Josefina Morales, Anna-Cristina Gavaldon, Laura Vidal, Martha Uc, Nidia Echeverría and Felipe Bobadilla were instrumental to this research. Juan Carlos Mijangos helped with the organization of workshops. Comments on an earlier version of the chapter by Yvan Breton and Maiken Bjørkan have been helpful.

References

Aguilar, W. J., Castro- Castillo, N., & Couoh Cab, J. (2012). El manejo del área marina y costera protegida Actam Chuleb y los beneficios económicos que genera a los usuarios del municipio de San Felipe, Yucatán, México. *Estudios Sociales, XX*(40), 127–153.

Aguilar–Perera, A., Scharer, M., & Valdés-Pizzini, M. (2006). Marine protected areas in Puerto Rico: Historical and current perspectives. *Ocean and Coastal Management, 49*, 961–975.

Andrews, A., & Gallareta, N. T. (2003). *Proyecto Isla Cerritos y Cupul, 1984–1988*. Retrieved from http://ww.ncf.edu/andrews/ICAP.html
Arceo–Carranza, D., Vega–Cendejas, M. E., Montero–Muñoz, J. L., & Hernández de Santillana, M. (2010). Influencia del hábitat en las asociaciones nictimerales de peces en una laguna costera tropical. *Revista mexicana de biodiversidad, 81*(3), 823–837.
Ayuntamiento Municipal de San Felipe. (1999). Documento de declaración de Reserva Municipal Actam Chuleb. San Felipe, Yucatán, México, 11 de enero.
Berkes, F. (2008). Evolution of co-management: Role of knowledge generation, bridging organizations and social learning. *Journal of Environmental Management, 90*(5), 692–1702.
Bjørkan, M. (2009). Putting MPAs to work: A Mexican case study on community empowerment. *MAST, 8*(1), 11–31.
Charles, A. T., & Wilson, L. (2009). Human dimensions of marine protected areas. *ICES Journal of Marine Science, 66*, 6–15.
Christie, P. (2004). *Marine protected areas as biological successes and social failures in Southeast Asia* (pp. 155–164). American Fisheries Society Symposium.
Chuenpagdee, R., & Jentoft, S. (2007). Step zero for fisheries co-management: What precedes implementation. *Marine Policy, 31*, 657–668.
Chuenpagdee, R., & Jentoft, S. (2013). Assessing governability: What's next? In M. Bavinck, R. Chuenpagdee, S. Jentoft, & J. Kooiman (Eds.), *Governability of fisheries and aquaculture: Theory and applications* (pp. 335–350). Dordrecht: Springer.
Chuenpagdee, R., Fraga, J., & Euan, J. I. (2002). Community's perspectives toward marine reserve: A case study of San Felipe, Yucatán, Mexico. *Coastal Management, 30*(2), 183–191.
Chuenpagdee, R., Fraga, J., & Euan, J. I. (2004). Progressing toward co-management through participatory research. *Society and Natural Resources, 17*, 147–161.
Chuenpagdee, R., Pascual-Fernández, J. J., Szeliánszky, E., Luis Alegret, J., Fraga, J., & Jentoft, S. (2013). Marine protected areas: Re-thinking their inception. *Marine Policy, 39*, 234–240. doi:10.1016/j.marpol.2012.10.016.
Cinner, J. E., & Pollnac, R. B. (2004). Poverty, perceptions and planning: Why socioeconomics matter in the management of Mexican reefs. *Ocean and Coastal Management, 47*, 479–493.
Cudney-Bueno, R., Bourillón, L., Sáenz-Arroyo, A., Torre-Cosío, J., Turk-Boyer, P., & Shaw, W. W. (2009). Governance and effects of Marine reserves in the Gulf of California. *Ocean and Coastal Management, 52*(3–4), 207–218.
Davy, B., & Breton, Y. (2006). Coastal management in the wider Caribbean: Resilience, adaptation and community diversity. In Y. Breton, D. Brown, B. Davy, M. Haughton, & L. Ovares (Eds.), *Coastal resource management in the wider Caribbean: Resilience, adaptation, and community diversity* (pp. 1–16). Ottawa: Ian Randall Publishers- IDRC.
Diario Oficial del Gobierno del Estado de Yucatán. (2005). Acuerdo número 71. Poder Ejecutivo. Mérida Yucatán, jueves 29 de diciembre.
Fraga, E. J., Salas, S., & Chuenpagdee, R. (2006). Manejo comunitario de una área natural protegida en Yucatán, México (Fase II). CINVESTAV-Mérida- CIID. 214 p.
Fraga, J., Salas, S., & Mexicano-Cíntora, G. (2008). La pesca en Yucatán: de la abundancia a la escasez, a la fragilidad de las estructuras institucionales. In J. Fraga, G. J. Villalobos, S. Doyon, & S. A. García (Eds.), *Descentralización y manejo ambiental. Gobernanza costera en México* (pp. 133–148). Plaza y Valdés, Mexico.
García-Frapollia, E., Ramos-Fernández, G., Galicia, E., & Serranod, A. (2008). The complex reality of biodiversity conservation through Natural Protected Area policy: Three cases from the Yucatan Peninsula, Mexico. *Land Use Policy, 26*(3), 715–722.
García-Frapollli, E., Ramos-Fernández, G., Galicia, E., & Serrano, A. (2009). The complex reality of biodiversity conservation through natural protected area policy: Three cases from the Yucatan Peninsula, Mexico. *Land Use Policy, 26*(3), 715–722.
Gatewood, J. (1984). Cooperation, competition, and synergy: Information-sharing groups among Southeast Alaskan salmon seiners. *American Ethnological Society, 11*(2), 350–370.
Gray, S., Shwom, R., & Jordan, R. (2012). Understanding factors that influence stakeholder trust of natural resource science and institutions. *Environmental Management, 49*, 663–674.

Guénette, S., Chuenpagdee, R., & Jones, R. (2000). Marine protected areas with emphasis on local communities and indigenous peoples: A review. *Fisheries Centre Research Report, 8*(1), 56.
Gutiérrez, N. L., Hilborn, R., & Defeo, O. (2011). Leadership, social capital and incentives promote successful fisheries. *Nature, 470*, 386–389.
Guttman, J. M. (1996). Rational actors, tit-for-tat, and the evolution of cooperation. *Journal Ecological Behaviour Organization, 29*, 27–56.
Hannesson, R. (2001). The economics of marine protected areas. In R. Sumaila, & J. Alder (Eds.), *Economics of protected areas* (pp. 85–92). Fisheries Centre Research Report, *9*(8).
Hilborn, R. (2007). Managing fisheries is managing people: What has been learned? *Issue Fish and Fisheries, 8*(4), 285–296.
Hilborn, R., Stokesb, K., Maguire, J. J., Smithd, T., Botsford, L. W., Mangel, M., Orensanz, J. L., Parma, A., Rice, J., Bell, J., Cochrane, K., Garcia, S., Hall, S. J., Kirkwood, G. P., Stefansson, S. K., Stefansson, G., & Walters, C. (2004). When can marine reserves improve fisheries management? *Ocean and Coastal Management, 47*, 197–205.
INEGI. (2014). *Mexico en cifras*. Retrieved from http://www3.inegi.org.mx/sistemas/mexicocifras/default.aspx?src=487&e=31
Jentoft, S., & Chuenpagdee, R. (2009). Fisheries and coastal governance as a wicked problem. *Marine Policy, 33*, 553–560.
Jentoft, S. C., van Son, T., & Bjørkan, M. (2007). Marine protected areas: A governance system analysis. *Human Ecology, 35*, 611–622.
Jentoft, S., Chuenpagdee, R., Bundy, A., & Mahon, R. (2010). Pyramids and roses: Alternative images for the governance of fisheries systems. *Marine Policy, 34*, 1315–1321.
Jentoft, S., Chuenpagdee, R., & Pascual-Fernandez, J. J. (2011). What are MPAs for: On goal formation and displacement. *Ocean and Coastal Management, 54*, 75–83.
Jentoft, S., Pascual-Fernandez, J., De la Cruz Modino, R., Gonzalez-Ramallal, M., & Chuenpagdee, R. (2012). What stakeholders think about marine protected areas: Case studies from Spain. *Human Ecology, 40*, 185–197.
Jesus, A., & Eúan-Avila J. I. (2008). *Monitoring submerged aquatic vegetation changes with multispectral imagery and landscape ecology* (Vol. 2, pp. 339–350). Eighth international symposium on GIS and computer mapping for coastal zone management. CoastGIS 07, Santander España, October 8–10.
Kooiman, J., & Jentoft, S. (2009). Meta-governance: Values, norms and principles, and the making of hard choices. *Public Administration, 87*, 818–836.
Kooiman, J., Bavinck, M., Jentoft, S., & Pullin, R. (2005). *Fish for life: Interactive governance for fisheries*. Amsterdam: Amsterdam University Press.
McConney, P., & Baldeo, R. (2007). Lessons in co-management from beach seine and lobster fisheries in Grenada. *Fisheries Research, 87*, 77–85.
Mora, C., & Sale, F. P. (2011). Ongoing global biodiversity loss and the need to move beyond protected areas: A review of the technical and practical shortcomings of protected areas on land and sea. *Marine Ecology Progress Series, 434*, 251–266.
Ovando, D., Deacon, R. T., Lester, S. E., Costello, C., VanLeuvan, T., McIlwain, K., Strauss, C. K., Arbuckle, M., Fujita, R., Gelcich, S., & Uchida, H. (2013). Conservation incentives and collective choices in cooperative fisheries. *Marine Policy, 37*, 132–140.
Pajaro, M., Mulrennan, M. E., Alderc, J., & Vincent, A. (2010). Developing MPA effectiveness indicators: Comparison within and across stakeholder groups and communities. *Coastal Management, 38*(2), 122–143.
Pedroza, C., & Salas, S. (2011). Responses of the fishing sector to transitional constraints: From reactive to proactive change, Yucatan fisheries in Mexico. *Marine Policy, 35*(1), 39–49.
Pollnac, R. B., & Pomeroy, R. (2005). Factors influencing sustainability of integrated coastal management projects in the Philippines and Indonesia. *Ocean and Coastal Management, 48*, 233–251.

Salas, S., Bjørkan, M., Bobadilla, F., & Cabrera, M. A. (2011). Addressing vulnerability: Coping strategies of fishing communities in Yucatan, Mexico. In S. Jentoft & A. Eide (Eds.), *Poverty mosaics: Realities and prospects in small-scale fisheries* (pp. 195–220). London: Springer, Netherlands.

Sale, P., Cowen, R. K., Danilowicz, B. S., Jones, J. P., Kritzer, J. P., Lindeman, K. C., Planes, S., Polunin, N., Russ, G., Sadovy, Y., & Steneck, R. S. (2005). Critical science gaps impede use of no-take fishery reserves. *TRENDS in Ecology and Evolution, 20*(2), 74–80.

Secretaria de Ecologia. (2006). *Programa de manejo Reserva Estatal de Dzilam.*1ª edición. Mexico. Retrieved from http://www.seduma.yucatan.gob.mx/areas-naturales/documentos/Pm_Dzilam_Docto.pdf

Skaperdas, S., & Syropoulos, C. (1996). Can the shadow of the future harm cooperation? *Journal of Economic Behavior & Organization, 29*, 355–372.

Song, A. M., Chuenpagdee, R., & Jentoft, S. (2013). Values, images, and principles: What they represent and how they may improve fisheries governance. *Marine Policy, 40*, 167–175.

Sumaila, U. R. (1998). Protected marine reserves as hedge against uncertainty: An economist's perspective. In T. J. Pitcher, D. Pauly, & P. Hart (Eds.), *Reinventing fisheries management* (pp. 303–309). Dordrecht: Kluwer.

Vidal, L., & Capurro-Filograsso, L. (2008). Quantitative analysis of natural resource regulations leading to coastal ecosystems sustainability: Mexico as a case study. *Journal of Coastal Research, 24*(4), 876–889.

White, A. T., Salamanca, A., & Courtney, C. A. (2002). Experience with marine protected area planning and management in the Philippines. *Coastal Management, 30*, 1–26.

Chapter 26
Small-Scale Fishers, Changing Borders: The Case of San Andrés Archipelago (Colombia) and the International Court of Justice

Olivier Randin

Abstract Small-scale fisheries are not a fixed scale in terms of governance and governability. While they may be small by their individual size, small-scale fisheries are strongly interconnected to higher levels of governance. This interconnectedness makes them vulnerable to external influences and shocks, which in turn affect their governability. This chapter calls for an analysis of scales beyond the local community in order to grasp all the complexity, diversity and dynamics of interactions at multiple levels. As an illustration, the case of small-scale fisheries in the San Andrés archipelago on the Caribbean coast of Colombia will be looked at. Due to the International Court of Justice decision over boundary dispute between Colombia and Nicaragua in 2012, small-scale fishers in this area lost access to their traditional fishing ground. The chapter examines consequences of this decision using the governability assessment framework and provides lessons about small-scale fisheries governance when spatial scale is a critical issue.

Keywords Colombia • San Andrés • International Court of Justice • Cross-level analysis

Introduction

Small-scale fisheries are complex and diverse systems, typically embedded in local communities (Jentoft and Eide 2011). However, they are far from being just "part of the landscape" (Pauly 2009; Chuenpagdee 2011) and are subject to international events challenging their existence. In many instances, small-scale fisheries face issues beyond their boundaries. Unfortunately, most studies of small-scale fisheries are limited to the local scale (Chuenpagdee 2011). As argued by Johnson et al. (2005), to

O. Randin (✉)
Department of Geography, Memorial University, St. John's, Canada
e-mail: ojr628@mun.ca

S. Jentoft, R. Chuenpagdee (eds.), *Interactive Governance for Small-Scale Fisheries*, MARE Publication Series 13, DOI 10.1007/978-3-319-17034-3_26

fully understand challenges related to small-scale fisheries, an analysis is needed at all levels and scales. Scale is defined as "the spatial, temporal, quantitative or analytical dimensions used by scientists to measures and study objects and processes".

Governance takes place at multiple levels (Gibson et al. 2000) and quality of interactions between these multiple levels contribute to making the system more or less governable (Kooiman and Bavinck 2013). According to the interactive governance framework (Kooiman et al. 2005), scale plays a critical role in determining governability. Here, governance is understood as "the whole of public as well as private interactions taken to solve societal problems and create societal opportunities. It includes the formulation and application of principles guiding those interactions and care for institutions that enable them" (Kooiman and Bavinck 2005, 17). Governability is defined as "the overall capacity for governance of any societal entity or system" (Kooiman et al. 2008, 3).

This chapter underlines the sensitivity of small-scale fisheries – situated at a local level – to higher level events (international level). The consequences of this sensitivity may be exacerbated if the State does not act as a mitigating factor. Governability could be improved if – when looking for options and opportunities – a cross-level analysis was performed (Cash et al. 2006). To support this argument, the case study of the Colombian archipelago of San Andrés, Providencia and Santa Catalina is scrutinized. In 2012, the International Court of Justice (ICJ) decided to transfer about 76,000 km^2 of sea from Colombian jurisdiction to Nicaragua's. This international event has had dire consequences for small-scale fishers ("Effects of the Ruling of the International Court of Justice (ICJ) in the Archipelago" 2014).

The chapter is divided as follows: the second part briefly presents the context and the consequences of the ICJ judgment. The third part describes the archipelago and its fisheries (the natural system-to-be-governed), the small-scale fishers (the social system-to-be-governed), the governing system and their interactions following the interactive governance framework. The fourth part is a discussion on how governability of small-scale fisheries is affected by interactions across levels. A conclusion will sum up the salient points of this case study and suggest directions in which to examine small-scale fisheries.

Information was obtained by reviewing secondary sources, namely grey literature and institutional reports (NGOs, government, international organizations and project reports). Information on small-scale fisheries in the archipelago came mainly from CORALINA's website (the regional agency in charge of environmental management and planning), former CORALINA employees and researchers affiliated with CORALINA. To acquire a better understanding of fishers' perspective a one month field visit in the archipelago was devised. The first aim of this visit was to talk to fishers about their life, concerns, hopes and their relationship with the national government as well as with regional institutions. The second aim was to obtain information from other governing institutions and stakeholders in charge of fisheries about the change of jurisdiction and its impacts on small-scale fisheries governance. A contact person in the archipelago introduced me to institutions' officials. Within fishers' communities, semi-structured interviews were performed. Fishers were contacted using a snowball technique.

Nicaragua vs. Colombia, ICJ Decision

In December 2001, Nicaragua filed proceedings to the ICJ (The Netherlands), questioning the legal validity of the 1928 Barcenas-Esguerra Treaty between Nicaragua and Colombia. Nicaragua considered that the treaty was not valid and asked for a decision on the title of the archipelago and maritime delimitations with Colombia (ICJ, Application, Instituting Proceedings 2001; Mantilla 2009). Nicaragua argued that it was entitled to an outer continental shelf and claimed a maritime border related to this oceanographic feature. This claim would have left the islands of San Andrés, Providencia and Santa Catalina enclaves in Nicaraguan water. The ICJ, however, in its verdict, did not encourage the creation of enclaves of inhabited islands and favored the continuity of Colombian territory. Its decision, which is final and without appeal, left islands, cays and part of the sea to Colombia, but transferred about 76,000 km^2 of sea from Colombian to Nicaraguan jurisdiction (ICJ, Judgment, November 19th 2012). With these new international borders, small-scale fishers have lost important traditional fishing grounds. The president of Colombia, Juan Manuel Santos, has refused to recognize the ICJ decision.

The Colombian president decided unilaterally to grant small-scale fishers 6 months subsidies (January–June 2013) to temporarily attenuate the direct effect of the decision. The Secretariat of Agriculture and Fisheries, the institution in charge of fisheries in the archipelago, was given the responsibility to distribute the subsidies to small-scale fishers. The Secretariat faced two concerns in creating a list of small-scale fishers. First, many fishers were more interested in acquiring a subsidy than in finding new fishing grounds. Second, people who never fished tried to take advantage of the program. To single out free-riders was a difficult task that has only been partly attained. The president of an association of small-scale fishers on Providencia states that before the subsidies there were about 200 small-scale fishers on the island but now, after the announce of subsidies, the official list has about 500 small-scale fishers.

The decision of the State to grant subsidies to small-scale fishers created thus tensions in communities and accentuated the complexity of interactions among small-scale fishers. Rules regarding the beneficiaries of the subsidies were set by the central government. But, according to small-scale fishers, information about these rules were not well transmitted by the government, and not well understood by small-scale fishers. In the end, some fishers received money, others did not and some subsidies went to non-fishers. Subsidies are thus a sensitive subject on Providencia and Santa Catalina. In interviews, representatives of regional institutions and fishers alike talked about the disruptive effects that these subsidies had within small-scale fishers' communities and on their relation with institutions. For the Secretariat of Agriculture and Fisheries and CORALINA, the ICJ decision and the national subsidies have jeopardized a whole collaborative and working process with small-scale fishers. The former consider that an indirect message has been sent to small-scale fishers: that they do not need to work to obtain money. These institutions regret a unilateral action following a top-down governance model when they want to foster a bottom-up approach to management. According to small-scale fishers, the State only created trouble and jealousy in the islands. They wished the central government would have paid closer attention to the needs of the archipelago and that the government would have included them earlier in the judicial debate at the ICJ.

The Archipelago of San Andrés, Providencia and Santa Catalina

The archipelago of San Andrés, Providencia and Santa Catalina (total land area: 57 km^2) is a Colombian territory in the Caribbean Sea. The archipelago lies in the southwestern part of the Caribbean Sea, about 720 km northwest of the Colombian coast and about 230 km east of the coast of Nicaragua (Fig. 26.1). It is composed of three inhabited islands: San Andrés (27 km^2), Providencia (17 km^2) and Santa Catalina (1 km^2). The archipelago has many uninhabited cays and banks among which the most important are, with regard to fisheries, Albuquerque cays, East-Southeast Cays, Roncador, Quitasueño, Serrana, Serranilla and Bajo Nuevo. The archipelago of San Andrés, Providencia and Santa Catalina had about 180,000 km^2 of Exclusive Economic Zone prior the ICJ decision (Mow 2006; CORALINA-INVEMAR 2012).

In 2010, the population of the archipelago was estimated at about 73,300 persons (DANE 2005). The archipelago faces overpopulation and strong immigration regulations have been introduced to limit a flux of immigrants from continental Colombia. These immigrants flee the civil conflict on the mainland or look for job opportunities. Many come illegally and the population is estimated to be closer to 80,000 persons, the majority of which live on the island of San Andrés. Providencia and Santa Catalina is the smallest municipality and has about 5,000 inhabitants. The official languages of the archipelago are Spanish and English. The archipelago has a native ethnic group, the Raizals, who speak Creole (Mow 2006). The economy of the archipelago relies primarily on tourism, trade and fisheries activities. Livelihoods of the Raizals depend heavily on natural resources: fisheries, agriculture and livestock. Therefore, fisheries play an important role as a source of income and food security as well being a part of a wider social and cultural identity (CORALINA-INVEMAR 2012).

In 1953, the State decided to develop a free port in the island of San Andrés to encourage the economic development of the archipelago. This decision gave the impetus for a migration of continental Colombians to the island, turning the Raizals into a minority in their own land and fostering social inequalities (Howard 1992). Today, fisheries and agriculture are insufficient to feed the large population of the archipelago and most resources are imported from continental Colombia. Nevertheless, small-scale fisheries and agriculture remain important livelihood assets and still belong to the culture and traditions of the archipelago.

To manage the natural resources of the archipelago, CORALINA favored an ecosystem-based approach and decided to create a multi-zone Marine Protected

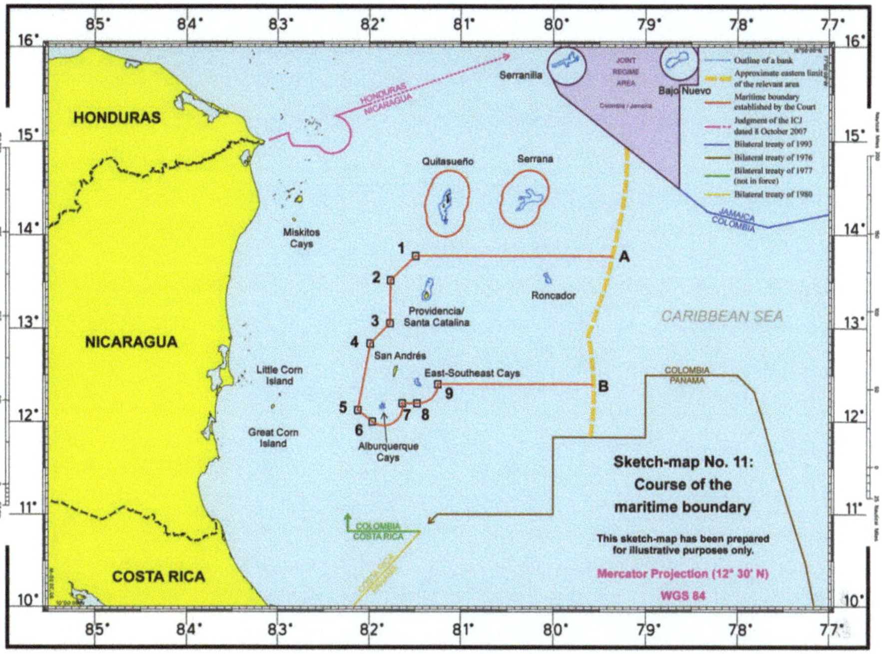

Fig. 26.1 New borders between Colombia and Nicaragua according to the ICJ decision (ICJ, *Nicaragua vs. Colombia*, Judgment, November 19th 2012, 89). Prior to ICJ decision: the 82nd meridian set the border between Nicaragua and Colombia. Post ICJ decision: Colombian waters are eastward of the *dotted line* and also within points A-1-2-3-4-5-6-7-8-9-B. The cays of Quitasueño and Serrana still belong to Colombia but are enclaves in Nicaraguan waters

Area (MPA); the Seaflower MPA. The MPA was created between 2000 and 2005. CORALINA fostered a participative management approach including stakeholders in the design process of the MPA. The first working sessions were tense and small-scale fishers, who enjoyed free and open-access to the sea until then, resented the presence of a managing institution that would impose itself on them (Gorricho and Rivera 2005). This open-ocean MPA (65,000 km^2) is divided into three administrative sections: the Southern Section with the coastal and marine areas of San Andrés island, the Central Section with the coastal and marine areas of Old Providence and Santa Catalina islands, the Northern Section with the uninhabited atolls of Roncador, Serrana and Quitasueño (CORALINA 2010; c.f. Fig. 26.2. of this chapter).

The Natural System-to-Be-Governed

Fisheries of the San Andrés archipelago are highly diverse, with about 65 species being fished (CORALINA-INVEMAR 2012). Among the main commercial species are queen conch (*Strombus gigas*), spiny lobster (*Panulirus argus*) and many

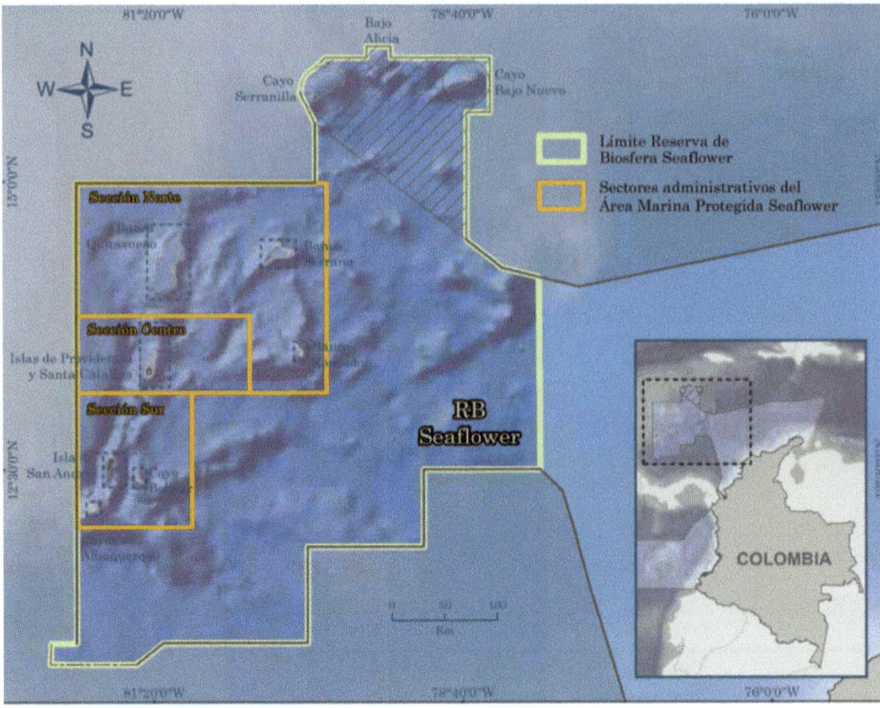

Fig. 26.2 Delimitation of the Seaflower MPA (CORALINA-INVEMAR 2012, 25), prior to the 2012 ICJ decision. The MPA is divided into three sections: the Northern Section (uninhabited), the Central Section (with Providencia and Santa Catalina), and the Southern Section (with San Andrés). The white line delimits the boundaries of the department of San Andrés, Providencia and Santa Catalina (prior ICJ decision) and also represents the boundaries of the *Seaflower Biosphere Reserve*, a designation by UNESCO to the department in 2000

species of finfish, such as snappers, e.g., yellowtail snapper (*Ocyrus chrysurus*), dog snapper (*Lutjanus jocu*), and groupers, e.g., goliath grouper, (*Epinephelus itajara*), nassau grouper (*Epinephelus striatus*) (Prada et al. 2004; CORALINA 2010; CORALINA-INVEMAR 2012). The number of species and habitats entails complex ecological relationships. But complexity of the natural system, though acknowledged, is poorly known: remoteness of the archipelago makes it hard and expensive to reach for scientific teams. The dynamics of the natural system are shaped by two important drivers: climate change and anthropogenic influences (CORALINA 2010). Monitoring of fish abundance and diversity has not been performed on a regular basis. However, decreasing landings of economically important species suggest a state of overfishing.

Most catches are large oceanic pelagic fish (36.2 %), followed by demersal fish (34.9 %). Catch of large oceanic pelagic species are represented by black bonito (*Thunnus Atlanticus*), shallow water bonito (*Katsuwonus pelamis*) and black marlin (*Makaira nigricans*). Main demersal species fished are yellowtail (*Ocyurus chrysurus*), ocean triggerfish (*Canthidermi ssufflamen*) and dog snapper (*Lutjnanus jocus*) (CORALINA-INVEMAR 2012). Finfish extraction in small-scale fisheries, between 2004 and 2007, was estimated at about 100–110 tonnes. Populations of groupers and snappers – species of commercial importance in the archipelago – have declined in several fishing areas (Prada et al. 2004).

While time series data and stock assessments for finfish are scarce and incomplete, signs of overfishing and abundance depletion can be observed in fishers' behavior. Small-scale fishers go further away from the islands to fish, changing their fishing habits from reef to pelagic fisheries. Moreover they engage in activities in all possible shelves, seamounts and banks. Catch Per Unit Effort is decreasing and an increase in juvenile catch has been observed (Prada et al. 2004; CORALINA-INVEMAR 2012). Commenting on abundance, a trap fisher on Providencia said in an interview:

> When I was young, I was a speargun fisher. I could fish for 3 or 4 hours and my boat was full. Today, speargun fishers need up to one day of fishing to sometimes get about 50 pounds of fish.

With regard to fishing boundaries, small-scale fishing mainly took place within the Seaflower MPA, prior to the ICJ decision. Small-scale fishers undertake activities in all possible reefs, shelves and go further away to pelagic regions to fish (Prada et al. 2004). With such a highly diverse, complex and dynamic natural system-to-be-governed governability is low.

The Social System-to-Be-Governed

The majority of small-scale fishers are Raizals. Their identity and cultural background can be traced back to Anglo-Saxon influences. They are, in their majority, Protestants and Creole speakers whereas continental Colombians speak Spanish and are catholic with a Hispanic cultural background (Mow 2006). Raizal small-scale fishers have a very strong cultural identification with the sea. A Raizal woman on Providencia said:

> The sea is part of our essence, it is who we are.

And to be a fisher is both a livelihood and cultural identity. In Santa Catalina, an old respected fisher said:

> I am a fisherman because it makes me free!

In 2012, the archipelago had a population 740 small-scale fishers and 178 boats: 538 small-scale fishers in San Andrés with 120 boats and 202 small-scale fishers in Providencia and Santa Catalina with 58 boats. Boats with the following features are considered small-scale: boats less than 12 m long with an outboard engine (14–150 horse power), with a capacity of less than 3 tonnes. The number of fishers aboard varies between 2 and 5 depending on the size of the boat (CORALINA-INVEMAR 2012). Small-scale fishers use mainly three techniques:

speargun fishing at lung capacity (scuba diving fishing is prohibited), trap fishing, hook/line and reel fishing. Net fishing is forbidden in the archipelago (CORALINA-INVEMAR 2012). Fishers on Providencia and Santa Catalina fish mainly with spearguns and hook and lines.

Small-scale fishers usually share their time between fishing and other activities such as agriculture, seasonal work in the local public administration and tourism. Those that are considered full-time fishers are few: on Providencia, a representative working for the Seaflower Keepers Project – a project fostering and supporting associations and cooperatives in the archipelago, amongst other things – estimated the number of full time fishers at about 90 for this island. He stressed that it is not possible to live solely on fishing. For instance, during the closed season of spiny lobster and queen conch, very often fishers have to find alternative livelihoods. Governing institutions are working on alternative livelihood projects: small-scale mariculture on Providencia and recreational fishing tourism (catch-and-release) on Providencia and San Andrés, for instance.

Complexity of the social system-to-be-governed is high due to two factors. The first relates to the relationship with the natural system: high population density in the archipelago, about 2,400 persons/km^2, puts much pressure on fisheries resources leading to overfishing (CORALINA 2010; Howard and Taylor 2010). With diminishing fishing resources, institutions set rules to protect stocks. But the need to live and earn an income sometimes leads fishers to fish illegally. The second factor relates to the relationship with governing institutions: communication and collaboration have proven difficult due to strained relationships between Raizals and the central government. These strained relationships have their roots in the history of the archipelago and the colonial behavior of national institutions since the establishment of the free port model in 1953 (Howard 1992; Mow et al. 2003).

Small-scale fishers' social interactions and dynamics in the archipelago are complicated and can be divided into interactions with actors external and internal to the archipelago. Regarding interactions with external actors, small-scale fishers have a strong sense of ownership of land and sea, and interactions with national institutions have been difficult since the implementation of the free port model in 1953. Nowadays, small-scale fishers have little trust for national institutions and native islanders are struggling to be recognized as culturally different from continental Colombia. Regarding interactions with internal actors, small-scale fishers describe themselves as individualistic and struggle to organize and act together. A lobster fisher, during an informal talk, said:

> In this island (Providencia and Santa Catalina), we (fishers) do not know how to organize. We are all one against the other. We are selfish! It is a small island, we should be capable of working together – but no. I don't know why we are so individualistic. In San Andrés, it's the same problem but even worse! There are the continental Colombians against islanders, but even so fishers are not capable to unite!

Fishing boundaries have recently been curtailed by the ICJ decision, adding pressure to small-scale fishers' social interactions and lifestyle. In other words, the natural system – its boundaries – has been reduced in size but the size of the social system remains unchanged. The overall governance quality of the social system-to-be-governed is rather low. Though diversity is low – the majority of small-scale fishers are of the same indigenous ethnic origin – which should facilitate governance, complexity and dynamics are high making the system less governable.

The Governing System

In Colombia, fisheries management follows a hierarchical, top-down governance mode. The Ministry of Agriculture and Rural Development is in charge of agriculture and fisheries. The National Authority for Mariculture and Fisheries (AUNAP) works under its authority. The AUNAP is in charge of implementing fisheries and aquaculture policies. In the archipelago, the Secretariat of Agriculture and Fisheries is responsible for the implementation of national policies.

The central State has legally acknowledged the geographic cultural and natural distinctiveness of the archipelago. In accordance with a decentralization process set up by the 1991 Colombian Constitution, a large degree of autonomy is granted to the archipelago to govern itself and manage its natural resources (Law 47 of 1993, art 1+5+23−25 in particular). To do so the archipelago has two specific institutions: CORALINA (Corporación para el Desarollo Sostenible del Archipiélago de San Andrés, Providencia y Santa Catalina), in charge of environmental related matters, and the Departmental Fishing Board (Junta Departamental de Pesca), a fisheries joint administrative and consultative committee.

CORALINA is a decentralized autonomous regional governmental agency in charge of the management of natural resources and environment. It also is in charge of the regional planning process pertaining to land and marine use. To fulfill its ecosystem-based management vision, CORALINA, local communities and stakeholders worked together to design the Seaflower MPA (Mow et al. 2003; Gorricho and Rivera 2005). The MPA (65,000 km^2) was constituted in 2005 as the 9th largest MPA in the world (Protect Planet Ocean 2010). However after the ICJ decision the area of the MPA has been reduced to half its original size. CORALINA fosters participatory management; involving small-scale fishers as much as possible in the decision-making process (CORALINA 2010).

CORALINA has a unique administrative role in Colombia as it can "enact and enforce environmental regulations on land and sea" (CORALINA 2010, 122). However, the responsibility of control and enforcement are duties of the maritime authorities (DIMAR) and the Colombian Navy. The priority of the Navy is to protect Colombian borders and fight drug-trafficking. Few resources are devolved to fisheries control and surveillance. Monitoring, control and surveillance are therefore deficient according to an interviewee at the Secretariat of Agriculture and Fisheries.

The second institution is the Departmental Fishing Board constituted in 2000. Its role, mainly consultative and informative, is to issue permits, manage small-scale fisheries areas and submit fisheries related suggestions to national authorities. The

Departmental Fishing Board is composed of one representative from: the archipelago's government, AUNAP, CORALINA (environmental regional institution), DIMAR (maritime authorities), the national university (for scientific advices), as well as one small-scale fishers' representative and one industrial fishers' representative (CORALINA-INVEMAR 2012).

With regard to complexity of the governing system, CORALINA and the Secretariat of Agriculture and Fisheries have their own individual institutional ways in which to deal with problems and offer solutions. These ways are sometimes difficult to match. However, the vision of an integrated, interactive and bottom-up oriented management is shared by both institutions. Both call for a Departmental Fishing Board with increased executive and legislative powers.

The relationship between these two institutions and central authorities at the national level is much more complex. National authorities, with a more vertical chain of command, are criticized by regional institutions (CORALINA and the Secretariat of Agriculture and Fisheries) and fishers alike for their lack of sensitivity to the specific challenges of the archipelago and for not implementing the vision of decentralization as inscribed in the 1991 Constitution.

An important issue that affects the institutional dynamics of fisheries management is the frequent changes in State policies regarding fisheries. From 1968 to 2008, the authority in charge of fisheries changed four times (Rueda et al. 2011, pp. 120–121):

> The last changes in the fishing authority (from INPA and INCODER to ICA) emphasize one of the biggest problems in the Colombian fishing industry: the changing institutional framework.

The last change to the institutional set-up occurred in 2012 when the AUNAP, under the Ministry of Agriculture and Rural Development, became the institution in charge of mariculture and fisheries. This lack of continuity in institutional structures impacts fisheries in terms of long term policies, data collection and liability of the governing system.

The geographical jurisdiction of CORALINA expands to the whole department. Its mandate encompasses issues of natural resource management and nature conservation to the livelihoods of communities benefiting from and affecting natural resources. At the regional level, governability is good due to strong relationships and interactions between institutions. However, the low trust level of small-scale fishers and their difficulty in organizing negatively affect regional governability. The main constraint on governance in the archipelago is the strained relationship between the national government and regional institutions.

Governing Interactions

Small-scale fishers are represented in several ways. On Providencia, there are three associations (Pesproislas, Asopescboth, Asocrab) and one cooperative (Fish and Farm Cooperative). The Secretariat of Agriculture and Fisheries encourages the participation of small-scale fishers in stock assessments. CORALINA fosters a participative approach to management. The Departmental Fishing Board includes a small-scale fisheries representative. This diversity should favor the integration of small-scale fishers in decision-making processes and interaction opportunities should reduce gaps in knowledge and interests. With regard to the dynamics of interaction, both CORALINA and the Secretariat of Agriculture and Fisheries express the need to promote participation that includes small-scale fishers. A representative of CORALINA underlined that, at the inception of the Seaflower MPA, more effort and care should have been put into socio-economic analyses and less on biological and ecological analyses.

Complexity of interactions is not solely found within governing institutions. It can also be found in the social system-to-be-governed and its capacity to receive and transmit information. As mentioned above, small-scale fishers recognize they have difficulties to work together. Means of communication and interaction exist but this does not mean that they are effectively utilized. Effectiveness depends on fishers' participation and their capacity to organize, which remains low.

With regard to the scale feature of the interactive governance framework, the State tends to impose its policies without taking sufficient steps to listen to ideas and propositions from its archipelagic region. Moreover, the national government appears to be less inclined to devolve power; limiting the range of action of CORALINA, the Secretariat of Agriculture and Fisheries and the Departmental Fishing Board to improve fisheries governance.

Discussion

To improve small-scale fisheries' governability, analysis should not just focus on the local level at which they operate. The analysis should be broadened to look at higher levels in order to understand the different interactions affecting, positively or negatively, small-scale fisheries. The ICJ decision is an opportunity to examine how the governance of small-scale fisheries is affected by multiple levels of interactions. In other words, it is a cross-level approach that casts light on small-scale fisheries' governability: a micro-macro framework approach where the micro level cannot be totally grasped without the macro level, and vice versa (Kooiman 2013).

Governability Issues

The ICJ decision is an external driver of change that reveals strengths and weaknesses of the governance in the archipelago. International stakeholders in this case were limited to three main actors: the State of Nicaragua, the State of Colombia and the ICJ. The conflict between the two States is a struggle for marine resources: fisheries, oil and maritime space. The tool used in this struggle for territory is a bilateral international treaty. Complexity lies in the scale at which one looks at the problem, the level at which the decision is taken to solve the problem and the level at which the consequence of the decision will have most effects. In this instance, governability depends on the ability of the governing systems to communicate with the system-to-be-governed.

The direct effects are felt by small-scale fishers who have lost their traditional fishing grounds. Conflicts are expected to increase with increasing numbers of fishers fishing in less fishing areas ("Effects of the Ruling of the International Court of Justice (ICJ) in the Archipelago" 2014). Governability is hindered, amongst other reasons, because of small-scale fishers' difficulty to organize and work as a group, and their low trust towards governing systems that reduce their desire to participate. Roots of this low trust are to be found in the colonial behavior of continental Colombians (Howard 1992).

Governing systems reacted differently according to their sensitivity. The Colombian president, in a unilateral decision, accorded subsidies in order to mitigate the consequences of loss of waters for small-scale fishers. However, the State, instead of acting as a mitigating actor receiving and absorbing the shock, (inadvertently) became an additional vector adding stress and tension to the social-system-to-be-governed and governing institutions of the archipelago. CORALINA and the Secretariat of Agriculture and Fisheries fostering a bottom-up approach would have favored a participatory scheme to find solutions. Governability has been undermined by the unilateral decision of the central government; a decision considered unfit by small-scale fishers and regional governing institutions alike.

The overall capacity for governance is reduced due to the existence of two different, overlapping, governing schemes: The national one has a hierarchical approach; the regional one fosters a participative and integrative approach. In addition, low governability has to be understood keeping in mind the historical and social context of the archipelago and the complex relationships and interactions between the islands and continental Colombia.

Interactions

Interactions at the Actor Level and at the Structural Level

Power plays an important part in governing interactions (Jentoft 2007). Interactions can be considered at the actor level in terms of "the willingness or ability of actors to participate" (Kooiman and Bavinck 2013, 20). These interactions can foster integration, learning, legitimacy, and accountability, but can also be politically naïve, inefficient and unrealistic (Kooiman and Bavinck 2013). At this horizontal level, actors (fishers) struggle to unite though they understand the many challenges the archipelago is facing.

Interactions can also be considered at the structural (or contextual) level, namely at interactions between the governing system and the system-to-be-governed

(Kooiman and Bavinck 2013). Looking at the vertical interactions of the present case study, fishers have low amounts of trust towards governing institutions (mainly national ones) and national institutions a low capacity to listen to and inform local communities.

Small-scale fishers have difficulties to express clearly their needs and wants with regard to stakeholders' participation even though associations and fora exist. On Providencia, a fisher stated:

> The biggest problem is to get together and work together.

The leader of a small-scale fisher association shared the same point of view:

> No one takes responsibility but everyone criticizes. I would picture the situation as follows: the members of the association throw stones at the leader and then hide behind the wall.

Small-scale fishers explain that this low quality of interaction is a consequence of a strong sense of individualism and few skills regarding collective functioning – resulting in scattered voices. In addition, small-scale fishers have low trust in governing institutions, reducing their willingness to interact.

Communication and interactions can also be difficult between institutions and small-scale fishers. The Deputy Director for Seas and Coasts at CORALINA, said:

> Today, many people see us as an entity that only regulates and sanctions.

And, talking about CORALINA, a fisher said:

> CORALINA makes me very angry: they forbid but they propose no alternatives.

The Deputy Director recognized that it is difficult to have a regional entity – CORALINA – that regulates and a national entity that gives subsidies directly to small-scale fishers without consultation with regional governing institutions. He underlined the necessity to work closer with other regional institutions as well as the community.

Interactions with the State

There is more to fisheries governance than management structures and institutional arrangements (Jentoft 2007). Tensions rise over power, and a governability analysis should be aware of power when it comes to interactions between systems. Colombia's Constitution aims at decentralization of power. However, the system is still very hierarchical and national authorities seem to have difficulties to share this power. For instance, the implementation of the Departmental Fishing Board faced resistance from central authorities. The case had to be brought to the Colombian Supreme Court (which ruled in favor of the Departmental Fishing Board and devolution of power). As Fox et al. (2005) summarize:

> Decentralization is essentially a process involving competition among competing vested interests.

In the present case, tensions arose between a regional system that leans on decentralization – in accordance with the Constitution – and national institutions still following a very hierarchical, vertical approach. Governability becomes less efficient because of conflicting governance systems.

Wicked problems can be a cause of low governability (Jentoft and Chuenpagdee 2009). But low governability could also be considered with regard to the wicked solutions implemented to solve a problem. A wicked solution could thus be broadly described as a technical solution having no clear time horizon, no clear purpose and with only a vague idea of how it will work. The Colombian State, with its subsidies, fostered a wicked solution developing more problems than offering solutions.

A speargun fisher said:

> These subsidies, it is really a mess! I don't understand why the government acted like this. They could have helped in other ways, with material or improved infrastructures for example.

An interviewee at the Secretariat of Agriculture and Fisheries commented on these subsidies:

> Instead of finding solutions or looking for other fishing areas, small-scale fishers are seeking these subsidies. Moreover, the Secretariat of Agriculture and Fisheries has ideas and plans to develop and to support fisheries in the archipelago but the central government did not ask for our opinion on the problem, nor if we had solutions to propose.

The Deputy Director for Seas and Coasts at CORALINA underlined that it has been a long process to establish positive interactions with small-scale fishers. The ICJ decision and State subsidies have jeopardized a whole working process with small-scale fishers. After the ICJ decision, small-scale fishers are less inclined to trust institutions (national or regional). They feel frustrated that all these years of protection and conservation of fisheries resources will benefit Nicaragua ("Effects of the Ruling of the International Court of Justice (ICJ) in the Archipelago" 2014) and feel that if they had been consulted earlier in the audition process, the ICJ decision might have been different. They resent the paternalistic attitude of the Presidency and the little knowledge it has of the archipelago. However, it is recognized that efforts are being made (Mow 2006, 3):

> The Colombian government has increasingly recognized that the management of the tiny remote oceanic islands is complex and that limitations are imposed by isolation, small land area, and distinct culture and social systems. However, they still need to understand that planning and management need to be based on linkages between biological systems – terrestrial, coastal and marine – and societal systems.

Governance is mainly about dealing with "highly contentious sociopolitical issues" (Kooiman 2013, 364) and governance interactions are about how to cope with unexpected events and deal with conflicts that are not likely to go away easily. According to CORALINA and the Secretariat of Agriculture and Fisheries, governance is negatively affected in two ways: a lack of contextual sensitivity from central authorities and a lack of devolved power to regional institutions. They consider that with effective power and proper means of communication and interaction, the effectiveness of governability should be increased in the archipelago.

In this context, governability is influenced by the capacity of the State to look at a region and recognize its specific problems as well as receive and transmit information. In other words, what is key is the capacity of national institutions to look at lower levels in the governing system. In order to enhance governability, institutions dealing with small-scale fisheries should: (1) look at levels above and below their own and (2) zoom-in (small-scale analysis) on the problem at the local level and zoom-out (large-scale analysis) to encompass other stakeholders at higher levels (Sayre 2009).

Levels and Scales

Analysis at one level only is insufficient to depict all the complexity and diversity of a system confronted with governability challenges. The ICJ decision, and its consequences for the archipelago's small-scale fisheries, highlights the necessity of a broader, more integrative perspective. The disruptive effect of this decision underlines the need to investigate the integrated nature of small-scale fisheries across levels and the ability to change scales of analysis – when the situation calls for it – when looking at small-scale fisheries.

Consequences of decisions taken at different scales have different meanings when observed from the international level and from the local level. What appears to be a logical, rational and equitable repartition of the sea between two States proves to be illogical, irrational and profoundly unfair in the eyes of the local population and its small-scale fishers. Both views are valid if observed independently. But if the two scales, international and local, are considered one against the other, there is a discrepancy between the two regarding the perception of a maritime boundary (Mantilla 2009). Complexity thus lies in the perception of scales when conceptualization of space and boundaries at higher levels – using a larger scale of analysis – are imposed on lower levels. Castro Gonzáles (2009) points out that seldom have the interests, hopes, and relationships with the marine system of the ones – fishers, sailors, traders – enjoying it, been taken into account. In sum, the scale at which the dispute over maritime borders took place did not take into account the scale where the strongest impacts would be felt.

The national level could become the interface between both. The magnitude of the impact of an international decision on the local level may depend much on the ability of the national government to mediate between the global and the local. In the present case, the State, instead of acting as an interface between the international level and the local level, hindered the creation of a space for problem-solving and opportunity creation. Its actions were, in fact, twice detrimental: (1) prior to the ICJ final judgment when it did not involve the local population in the formulation of the legal defense argumentation (Avella 2009) and (2) after the decision when it did not consult the population for solutions. The State did not consider governance as "political brokerage" (Chuenpagdee and Jentoft 2013, 346) and hence did not discuss issues and negotiate solutions, consequently reducing its governance capacity. This demonstrates the importance of having a governance system that is flexible and adaptable and where communication flows throughout all levels of the overall system.

Governing institutions should be able to interact in such ways that each institution adds to the knowledge of the other to solve problems or create opportunities. In this instance, the constructive sharing of information between levels could have played an important role.

The effects of international relations permeate through multiple levels from top to bottom and with consequences on governing systems as well as systems-to-be-governed. Governability of small-scale fisheries is thus affected by upper level decisions and events and should be considered within this complex "scalar flux" (Kidd and Shaw 2013, 184) interacting in "continued institutional turbulence" (Idem, 184). In this case study, "scalar flux" is understood as the inherent interactive feature of governance acting at multiple levels (and scales). Therefore, scale should not be considered as a hermetic boundary but as an analytical framework that leaves the definition of scale, and how it develops in a particular context, as an empirical question (Johnson 2006). Small-scale fisheries should in future be analyzed and integrated at higher levels and larger scales, far beyond the local level.

Conclusion

Images (Jentoft 2012) and narrative (Johnson 2006) of small-scale fisheries are sometimes too restrictive, too neat. Small-scale does not necessarily mean less diversity, less complexity and less dynamics and thus higher governability. The small-scale fisheries of the San Andrés, Providencia and Santa Catalina archipelago confirm the diversity, complexity and dynamics of small-scale fisheries as well as their integration in interactions extending far beyond the island community. And however remote these islands are, they are not isolated or out of reach of international events. The issue of scale and the risk of oversimplifications related to scale size should be acknowledged.

The San Andrés, Providencia and Santa Catalina archipelago faces several problems – international, institutional, societal and ecological – challenging a whole society and its marine ecosystem. This has a direct impact on small-scale fishers' daily life ("Effects of the Ruling of the International Court of Justice (ICJ) in the Archipelago" 2014). Measures taken by the national system are inadequate. They do not correspond to the needs and wants of regional institutions and small-scale fishers in the archipelago. In this instance, governability is affected by a low quality of communication and tensed power relations between stakeholders at different levels that are detrimental to interaction.

Berkes (2010, 494) points out that "focusing only at one level, whether local, national or international, is inadequate design for governance policy". A broader perspective is needed to grasp all the complexity and dynamics of fisheries governance. In this case study, the problem (international level) not only directly affected small-scale fishers (local level) but permeated from the international level through the national level and to the local level. Therefore, analysis of small-scale fisheries governance must take into account levels above and below the one that is directly studied (Sayre 2009) as small-scale fisheries' problems – and opportunities for solutions – can be embedded within multiple levels. In sum, governability of small-scale fisheries is influenced by interactions between governing institutions at different levels (international, national, regional), by interactions between institutions and small-scale fishers, and by interactions within the system-to-be-governed itself, namely small-scale fishers communities. Governability of small-scale fisheries could be improved by taking into consideration cross-level interactions. The analysis should consider integrating small-scale fishers beyond the local level and giving insights into how the national level and the international level impact the local. Second, governability analysis should give particular attention to the quality of interactions between the governing system and the system-to-be-governed. Indeed, the mere existence of channels of communication is not enough. Interactions can be considered to be a flux and hence the capacity and quality of emitting information is as important as the capacity and quality to receive information. Thus, the quality of governability depends, amongst other things, on the quality of interaction skills.

Acknowledgments I would like to express my gratitude to Professor Svein Jentoft and Professor Ratana Chuenpagdee for giving me the opportunity to write a chapter about this archipelago. I would like to thank the International Fisheries Management master's program and the Faculty of Biosciences, Fisheries and Economics of the UiT Arctic University of Norway (Tromsø, Norway) for partially funding this master's thesis research. Also thanks are due to two anonymous reviewers for their wise and very helpful comments. I would also like to offer my special thanks to June Marie Mow who made this journey to the archipelago so fascinating. Thanks to all interviewees for their time and shared knowledge. And last but not least, this research would not have been possible without fishers' participation: thank you for your time and generosity.

References

Avella, F. (2009). Bases Geohistóricas del Caribe Insular. In S. Mantilla (Ed.), *Fronterasen el Caribe: La disputa Colombo nicaragüense por San Andrés*, Providencia y Santa Catalina, Cuardernos del Caribe No 12, Publicación Especial, San Andrés Isla, Colombia.

Berkes, F. (2010). Devolution of environment and resources governance: Trends and future. *Environmental Conservation, 37*(4), 489–500.

Cash, D. W., Neil Adger, W., Berkes, F., Garden, P., Lebel, L., Olsson, P., Pritchard, L., & Young, O. (2006). Scale and cross-scale dynamics: Governance and information in a multilevel world. *Ecology and Society, 11*(2), 8.

Castro Gonzáles, E. R. (2009). Implicaciones de los tratados internacionales suscritos por Colombia sobre la dimensión territorial del archipiélago de San Andrés, Providencia y Santa Catalina: una Mirada diferente desde "abajo" y el sector pesquero. In S. Mantilla (Ed.), *Fronteras en el Caribe: la disputa colombo nicaragüense por San Andrés*, Providencia y Santa Catalina, Cuardernos del Caribe No 12, Publicación Especial, San Andrés Isla, Colombia.

Chuenpagdee, R. (2011). A matter of scale: Prospects in small-scale fisheries. In R. Chuenpagdee (Ed.), *World small-scale fisheries: Contemporary visions*. Delft: Eburon Academic Publishers.

Chuenpagdee, R., & Jentoft, S. (2013). Assessing governability – What's next. In M. Bavinck, R. Chuenpagdee, S. Jentoft, & J. Kooiman (Eds.), *Governability of fisheries and aquaculture: Theory and applications* (MARE publications series no. 7). Dordrecht: Springer.

CORALINA. (2010). *Colombia's Nomination of the Seaflower Marine Protected Area, Archipelago of San Andrés, Old Providence and Santa Catalina for Inscription on the World Heritage List*

(Vol. 1). Retrieved December 3, 2014, from http://www.coralina.gov.co/intranet/index.php?option=com_docman&task=doc_view&gid=562&tmpl=component&format=raw&Itemid=88889341&lang=es

CORALINA-INVEMAR. (2012). Gómez-López, D. I., Segura-Quintero, C., Sierra-Correa, P. C., Garay-Tinoco, J. (Eds.), Atlas de la Reserva de Biósfera Seaflower, Archipiélago de San Andrés, Providencia y Santa Catalina, Instituto de Investigaciones Marinas y Costeras "José Benito Vives De Andréis" –INVEMAR– y Corporación para el Desarollo Sostenible del Archipiélago de San Andrés, Providencia y Santa Catalina –CORALINA–. Serie de Publicaciones Especialesde INVEMAR #28. Santa Marta, Colombia, 180 p.

DANE. (2005). *National census*. Retrieved on December 3, 2014, from http://www.dane.gov.co/files/censo2005/PERFIL_PDF_CG2005/88000T7T000.PDF

Effects of the Ruling of the International Court of Justice (ICJ) in the Archipelago. Retrieved November 25, 2014, from http://www.coralina.gov.co/intranet/index.php?option=com_content&view=article&id=1050%3Aeffects-of-the-ruling-of-the-international-court-of-tle-hague-in-the-archipelago&catid=1%3Atimas&Itemid=88889100&lang=en

Fox, J. J., Aduri, D. S., & Resosudarmo, I. A. P. (2005). Unfinished edifice or Pandora's box? Decentralisation and resource management in Indonesia. In B. P. Resosudarmo (Ed.), *The politics and economics of Indonesia's natural resources*. Singapore: Institute of Southeast Asian Studies Publications.

Gibson, C. C., Ostrom, E., & Ahn, T. K. (2000). The concept of scale and the human dimensions of global change: A survey. *Ecological Economics, 32*, 217–239.

Gorricho, J., & Rivera, C. (2005). Mapping environmental conflicts in the ocean at Providencia and Santa Catalina, Colombia, to create collaborative spaces. In H. D. Correa & I. Rodriguez (Eds.), *Crossroads in Latin America: Between managing and transforming natural resource conflicts, Chapter II, CyC Program*. Costa Rica: University for Peace.

Howard, H. B. (1992). San Andrés y Providencia: Nacionalismo y separatismo. Año 60 y 70. Revista Credencial Historia, Edición 36, Bogotá, Colombia, Publicación digital en la página web de la Biblioteca Luis Ángel Arangodel Banco de la República. Retrieved December 3, 2014, from http://www.banrepcultural.org/blaavirtual/revistas/credencial/diciembre1992/diciembre3.htm

Howard, M., & Taylor, E. (2010). Ecosystem-based adaptation in the Seaflower Marine Protected Area, San Andrés Archipelago, Colombia. In A. Andrade Pérez, B. Herrera Fernández, & R. Cazzolla Gatti (Eds.), *Building resilience to climate change: Ecosystem-based adaptation and lessons from the field*. Gland: IUCN.

International Court of Justice. (2001). *Territorial and maritime dispute (Nicaragua vs. Colombia)*, Application, Instituting Proceedings, 6 December 2001, General List No. 124, The Hague, The Netherlands, 9 pp.

International Court of Justice. (2012). *Territorial and maritime dispute (Nicaragua v. Colombia)*, Judgment, I.C.J. Reports 2012, p. 624.

Jentoft, S. (2007). In the power of power: The understated aspect of fisheries and coastal management. *Human Organization, 66*(4), 426–437.

Jentoft, S. (2012). Roots and wings: The need for community transition in the age of globalization. In C. Carothers, K. R. Criddle, C. P. Chambers, P. J. Cullenberg, J. A. Fall, A. H. Himes-Cornell, J. P. Johnsen, N. S. Kimball, C. R. Menzies, & E. S. Springer (Eds.), *Fishing people of the north: Cultures, economies, and management responding to change*. Fairbanks: Alaska Sea Grant, University of Fairbanks.

Jentoft, S., & Chuenpagdee, R. (2009). Fisheries and coastal governance as a wicked problem. *Marine Policy, 33*, 553–560.

Jentoft, S., & Eide, A. (Eds.). (2011). *Poverty mosaics: Realities and prospects in small-scale fisheries*. Dordrecht: Springer.

Johnson, D. S. (2006). Category, narrative and value in the governance of small-scale fisheries. *Marine Policy, 30*, 747–756.

Johnson, D. S., Bavinck, M., & Veitayaki, J. (2005). Fish capture. In J. Kooiman, M. Bavinck, S. Jentoft, & R. Pullin (Eds.), *Fish for life, interactive governance for fisheries* (MARE publication series no. 3). Amsterdam: Amsterdam University Press.

Kidd, S., & Shaw, D. (2013). Reconceptualising territoriality and spatial planning: Insights from the sea. *Planning Pheory and Practice, 14*(2), 180–197.
Kooiman, J. (2013). Improving governability – Reflections for future applications. In M. Bavinck, R. Chuenpagdee, S. Jentoft, & J. Kooiman (Eds.), *Governability of fisheries and aquaculture: Theory and applications* (MARE publications series no. 7). Dordrecht: Springer.
Kooiman, J., & Bavinck, M. (2005). The governance perspective. In J. Kooiman, M. Bavinck, S. Jentoft, & R. Pullin (Eds.), *Fish for life, interactive governance for fisheries* (MARE publication series no. 3). Amsterdam: Amsterdam University Press.
Kooiman, J., & Bavinck, M. (2013). Theorizing governability – The interactive governance perspective. In M. Bavinck, R. Chuenpagdee, S. Jentoft, & J. Kooiman (Eds.), *Governability of fisheries and aquaculture: Theory and applications* (MARE publications series no. 7). Dordrecht: Springer.
Kooiman, J., Bavinck, M., Jentoft, S., & Pullin, R. (Eds.). (2005). *Fish for life, interactive governance for fisheries* (MARE publication series no. 3). Amsterdam: Amsterdam University Press.
Kooiman, J., Bavinck, M., Chuenpagdee, R., Mahon, R., & Pullin, R. (2008). Interactive governance and governability: An introduction. *The Journal of Transdisciplinary Environmental Studies, 7*(1), 1–11.
Mantilla, S. (2009). De la política de una frontera en disputa a la visión de une frontera compartida. In S. Mantilla (Ed.), *Fronteras en el Caribe: la disputa colombo nicaragüense por San Andrés*, Providencia y Santa Catalina, Cuardernos del Caribe No 12, Publicación Especial, San Andrés Isla, Colombia.
Mow, J. M. (2006). *The native islanders of San Andrés, Old Providence and Santa Catalina: Dreaming between two worlds*. Retrieved December 3, 2014, from http://www.unesco.org/csi/smis/siv/inter-reg/comvision-panel-June2.pdf
Mow, J. M., Howard, M., Delgado, C. M., & Tabet, S. (2003). Promoting sustainable development: A case study of the Seaflower Biosphere Reserve. *Prospects, 33*(3), 303–312.
Pauly, D. (2009). *Toward a conservation ethic for the sea (video file)*. Retrieved December 3, 2014, from http://www.youtube.com/watch?v=A24n-HQ8sLg
Prada, M., Peñaloza, G., Posada, S., Howard, N., Herrón, P., Salinas, L., Castro, E. Cabezas, F., & Robinson, H. (2004). *Fish spawning aggregations in the San Andrés Archipelago, a first approximation*, TOC-CORALINA, technical report, San Andrés Island, Colombia, 50 pp.
Protect Planet Ocean. (2010). Retrieved December 3, 2014, from http://www.protectplanetocean.org/collections/introduction/introbox/globalmpas/introduction-item.html
Rueda, M., Blanco, M., Narváez, J. C., Viloria, E., & Beltrán, C. S. (2011). Coastal fisheries of Colombia. In S. Salas, R. Chuenpagdee, A. Charles, & J. C. Seijo (Eds.), *Coastal fisheries of Latin America and the Caribbean* (FAO fisheries and aquaculture technical paper, no. 544). Rome: FAO.
Sayre, N. (2009). Scale. In N. Castree, D. Demeritt, D. Liverman, & B. Rhoad (Eds.), *A companion to environmental geography*. Malden: Blackwell Publishing.

Chapter 27
Limits to the Governability of Transboundary Fisheries: Implications for Small-Scale Fishers in Northern Sri Lanka and Beyond

Joeri Scholtens

Abstract Transboundary fisheries are a worldwide phenomenon that has considerable impact on small-scale fisheries. This chapter explores governability problems of transboundary fisheries in connection with small-scale fishers' marginality. Insights are derived by studying the practice of transboundary fishing in the Palk Bay, South Asia, where a sizable Indian trawler fleet impedes Sri Lankan small-scale fishers from carrying out their occupation. By analyzing the features of the fisheries systems and the fragmented governance practices, this chapter raises six issues that challenge the overall system's governability: (1) mismatch between the scale of governance and the scale of the problem; (2) high level of institutional fragmentation with limited cross-linkages; (3) actors' strategic framing of the nature, causes and solution to the problem; (4) power imbalances between Sri Lankan and Indian fishers; (5) deep politicization that has linked fisheries issues to higher level ethnic and geopolitical conflicts; and (6) path dependency of the trawl sector. I conclude that while co-governance is in theory crucial for transboundary governance to be more responsive to the situation at hand, governability analysis explains why constructive collaborative practices are difficult, if not impossible, to create in practice in this case.

Keywords Transboundary fisheries • Fisheries conflict • Governability • Sri Lanka • Politicization • Power asymmetry • Multi-level governance

J. Scholtens (✉)
Department of Geography, Planning and International Development Studies,
University of Amsterdam, Amsterdam, The Netherlands
e-mail: J.scholtens@uva.nl

S. Jentoft, R. Chuenpagdee (eds.), *Interactive Governance for Small-Scale Fisheries*, MARE Publication Series 13, DOI 10.1007/978-3-319-17034-3_27

Introduction

Exploitation of natural resources across political boundaries, like transboundary fishing, is a worldwide phenomenon that raises concerns of ecological sustainability and social justice. Challenges pertain both to effectiveness in terms of environmental sustainability outcomes, as well as to the inclusiveness of the arrangements, particularly in terms of providing space for small-scale fisheries (Chuenpagdee et al. 2005). Various international and transnational institutions have been tasked with the governance of transboundary fisheries but their achievements have been notably poor (Cullis-Suzuki and Pauly 2010; White and Costello 2014). This gives rise to the question of why transboundary fisheries are such a tricky governance subject that demand understanding of the governability challenges involved, in particular as they pertain to small-scale fisheries.

This chapter derives its insights from a case study in the Palk Bay, where South Indian trawlers are fishing in large numbers in North Sri Lankan waters, obstructing the operations of thousands of small-scale fishers. These North Sri Lankan fishers are thus involved in a transboundary fishing conflict with technologically superior Indian trawlers. This marginal position is partly the result of the recent civil war (1983–2009) that crushed the livelihoods of North Sri Lankan fishers (Soosai and Stokke 2006) and partly due to the lack of power and political representation in post-war local and national politics. Such multidimensional marginality of small-scale fishers is by no means unique and in fact a common phenomenon of small-scale fishers (e.g. FAO 2014). While reallocating resources to small-scale fisheries and reinstituting their rights may be goals worth striving for, the question is why these are so difficult to achieve in practice.

The problem of transboundary fishing in combination with small-scale fishers' marginality has many elements of a wicked problem (Rittel and Webber 1973). First, it does not lend itself to straightforward definitions of problem and solution. Stakeholders strategically frame the problem in accordance with their own images and interests, realizing that any future solution is shaped by the framing of the problem (Jentoft et al. 2010). Second, wicked problems are typically not contained– but rather embedded in and interconnected with issues at higher levels of scale, making it hard to solve them in isolation. According to Chuenpagdee and Jentoft (2013), such problems cannot be solved through techno-rational fixes, but rather require a governability analysis which deconstructs the governance challenges at hand. This chapter aims to derive lessons from the Palk Bay fisheries conflict regarding the governability limitations of transboundary fisheries, particularly as they pertain to livelihoods of small-scale fishers.

The remainder of the chapter is organized as follows. The next section provides an overview of global transboundary fisheries and their impact on small-scale fisheries. Section 3 zooms in on Palk Bay fisheries, mapping out key characteristics of the system-to-be-governed and the governing system. Section 4 analyses how governance functions in the Palk Bay by evaluating various modes of governance that give rise to a range of governability problems. Section 5 reviews the Palk Bay's governability challenges, while Sect. 6 concludes by drawing out the wider implications of this study.

Transboundary Fishing and Impact on Small-Scale Fishers

Transboundary fishing is defined 'as the activities of fishers appropriating marine resources across boundaries, typically, but not limited to state borders' (Scholtens and Bavinck 2014, p10).[1] Such cross-border appropriation includes both fishers following transboundary target species (e.g. shared, straddling or highly migratory fish stocks) or simply by the lure of transboundary fishing grounds, which have come within operational reach (ibid). Transboundary fishing has a long history (e.g. Butcher 2005) and has more recently gained particular attention in connection with concerns over overexploitation of migratory stocks (White and Costello 2014). While that discussion focuses on areas outside Exclusive Economic Zones (EEZ), for small-scale fishers the fisher fleets appropriating marine life *inside* foreign EEZs, by legal or illegal means, is the primary concern. According to a recent study (Pauly et al. 2013) China's distant water fleet alone operates in 93 EEZs and is responsible for 5 % of global marine landings. The EU fleet derives 30 % of its catches from foreign EEZ (Swartz et al. 2010).

Transboundary fishing has been problematized either as a poorly managed and controlled fishing practice, or as a traditional practice being disrupted by state driven processes of territorialisation. In line with the former approach, White and Costello (2014, 1) argue that transboundary fisheries, on the high seas in particular, "pose perhaps the greatest global challenge to sustainable fisheries management". International law, voluntary guidelines and multi-lateral agreements have been developed over the past decades to manage transboundary fisheries, but with notably poor results (Cullis-Suzuki and Pauly 2010; Russell and Vanderzwaag 2010). In the second perspective, the concept of transboundary fishing is problematized as a state-biased construct that 'criminalizes' an age-old phenomenon of mobile resource exploitation (Gupta and Sharma 2008). For many fishers, mobility – including transnational migration – is a fact of life, much like the livelihood patterns practiced by pastoralists and nomads (Adhuri and Visser 2007).

The impact of transboundary fishing fleets on small-scale resident fisheries and their implications for governance has been poorly established in academic literature, albeit with some exceptions (Kaczynski and Fluharty 2002; Alder and Sumaila 2004; Swartz et al. 2010). These studies show that even when 'cash-for-access agreements' are carefully negotiated, benefits tend not to trickle down to the resident [small-scale] fishers (Kaczynski and Fluharty 2002). Small-scale fishers tend

[1] Contrary to most literature on the issue, our focus lies on transboundary fishing rather than on transboundary resources, as fisheries tend to be governed through the steering of people rather than fish.

to bear the burden of declining fish stocks and foreign fishing fleets tend to have a negative impact on local (fish) food security (Alder and Sumaila 2004).

Factors that shape the capacity and quality of transboundary fisheries governance, i.e. their governability, cannot be a priori established. They emerge from a systematic journey through a particular system-to-be-governed, its governing system and the interactions between them. This journey is partly guided by the framework suggested by Chuenpagdee and Jentoft (2013).

Mapping Out the Palk Bay Fisheries Systems

This section draws out the basic features of the system-to-be-governed and the governing system for the transboundary Palk Bay fisheries. Although transboundary fisheries in the Palk Bay involves the Indian and Sri Lankan side of the Bay, this chapter focuses on the Sri Lankan side where I conducted 14 months of mixed method fieldwork during 2011 and 2012 among small-scale fishers, their leaders and various representative organizations. I will deal with the Indian side in passing, building on secondary literature, media coverage and earlier fieldwork in 2006 and 2007.

Natural System-to-Be-Governed

The Palk Bay and Palk Strait constitute a relatively confined sea area bounded by the Indian coastline to the west, Sri Lanka's coastline to the east, a bridge of shoals called the Adams Bridge that separates the Palk Bay from the Gulf of Mannar to the south, and the Bay of Bengal to the northeast. It is a shallow basin with an average depth of 9 m and is known for its lack of turbulence (Scholtens and Bavinck 2013). The bay is 137 km in length and 30–80 km in width.

The Palk Bay hosts a wide diversity of fish, sponges, molluscs, crustaceans and seaweeds. A survey of Indian catches revealed 56 different species (Stephen et al. 2013b) while a catch survey on the Sri Lanka side identified 63 different species. The quantity of fish catches and extent of biodiversity, however, seem to be declining (Vivekanandan and Kasim 2011). Commercially speaking, the area's prawn banks and sea cucumbers are the primary attraction. On the Indian side, a number of species have reportedly disappeared (including catfish, white sardine and sea turtles) and a range of species declined in number (including ray fish, silver bellies, anchovy, seer, lobster) (Vivekanandan and Kasim 2011). On the Sri Lankan side, fishers have observed declining catch for a given unit of effort, even though aggregate catches may have actually increased due to more purse seining. Fishers also observe that they rely on an increasingly limited number of (low value) species, most notably sardine which constituted over 50 % of catches. Vivekanandan and Kasim (2011, 23) conclude that "one can safely say that overfishing – both biological and economic – is definite in Tamil Nadu [even though] the extent of overfishing, the potential losses due to overfishing and the long term consequences of this are not known" (Fig. 27.1).

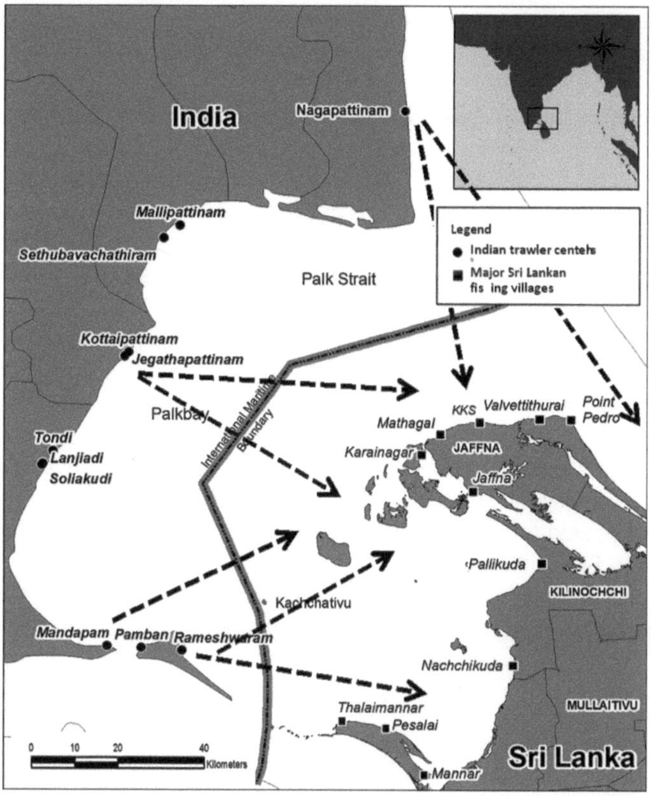

Fig. 27.1 Graphical representation of the system-to-be-governed. The *arrows* represent the origin and direction of Indian trawlers (There are no data available on the precise number of boats crossing into Sri Lankan waters. The *dotted arrows* also do not necessarily indicate a daily flow. The *arrows* are based on frequent anecdotal information from both Indian and Sri Lankan fishermen) Source: Adapted from University of Jaffna, Dept. of Geography

The Social System-to-Be-Governed

On the Indian side about 1,900 trawlers are berthed in 10 landing centres along the 294 km Palk Bay coastline (Stephen et al. 2013b). These trawlers of 30–50 ft in length are estimated to account for 70 % of total catches, the remainder being landed by the small-scale Indian fleet (Vivekanandan and Kasim 2011). They are owned by about 1,600 owners and operated by an average of four crewmembers per boat. In addition, about 2,000 merchants and 2,000 auxiliary workers derive their livelihood from the sector (Stephen et al. 2013b). Indian trawler owners and crew constitute a highly diverse community in terms of caste and religion, many of them originating from the agricultural sector, investing in trawling only in recent decades (Bavinck 2014). Regulations dictate that trawlers can operate maximum 3 days per week, beyond 5 km from the coast but not beyond the International Maritime Boundary Line (IMBL), which is located 15–46 km from their coastline. However, due to the large fleet in relation to very limited fishing grounds (Sathyapalan et al. 2008;

Sathyapalan et al. 2011) in practice trawlers operate deep into Sri Lankan waters, where Sri Lankan fishers report the 'incoming city of trawlers' with great anger and frustration.

In Northern Sri Lanka, apart from a few small trawl boats, all fishing activity can be safely termed small-scale. Fishing communities along the Palk Bay and Strait are spread over 120 villages along a 400 km coastline.[2] The majority of fishers are (Tamil) Catholic, but substantial groups of Hindu, Muslim and non-Roman Catholic fishers also fish. The large majority of fishers belong to traditional fishing castes, namely Karayars and Paravathars. There are about 6,000 18 ft Fiberglass Reinforced Plastic (FRP) boats with outboard engines of 8–25 hp. These FPR boats operate predominantly at night time, using gill and drift nets with mesh sizes varying from 1.5 in. (for sardines) to 18 in. (for ray fish). Nets are skillfully selected based on species targeted, location, seasonality, direction and strength of the wind, the position of the moon, as well as the demand of the market. Given the technological limitations, fishers often have to fish as far as 30 km from the coast. Based on a 1 year monitoring of fish catch and income, the monthly income of fishers in the village of Karainagar appeared to be on average $190,[3] about half of an average rural household income in Sri Lanka!

Table 27.1 reveals significant differences in technological capacity. Indian trawlers are equipped with engines up to 190 hp compared to 8–25 hp engines of North

Table 27.1 Comparison of fishing capacity and fishers operating on the Sri Lankan and Indian side of the Palk Bay

	Number of active fishers	Number of boats	Mechanized boats[a]	Motorized boats	Non-motorized boats
Sri Lankan fishery (1)	47,680	11,670	317 Trawlers & gill netters 30 ft 30 hp IBM	6,003 FRP boats 18 ft 8 to 25 hp OBM	5,350 Kattumarams and wooden canoos 9–15 ft
Indian fishery[b]	61,162 (3)	9,912 (3)	1,907 (2) Bottom & pelagic trawlers 30–50 ft 70–190 hp IBM.	4,141 (3)[c] *Vallams* & FRP boats 18–30 ft 5–30 hp	3,864 (3) *vattai's* and Kattumarams 12–15 ft

Sources: Government of Sri Lanka 2013 (1); Stephen et al. 2013b (2); Government of India 2010 (3)
In grey the dominant subsectors are given
[a]For this categorization I follow Bavinck (2001)
[b]This include fishing villages from Rameshwaram to Thiruvarur. Nagapattinam district is excluded: even though they are regularly active at the North Eastern Sri Lankan waters, they rarely frequent the Palk Bay
[c]These motorized boats from Tamil Nadu, operating mostly monophilament nets banned in Sri Lanka, are also increasingly fishing in Sri Lankan waters since 2011, but this is beyond the scope of this chapter

[2]The coastal stretch between the villages Point Pedro and Talaimannar.

[3]An average boat in Karainagar earned $8,502 from fish landings between April 2012 and March 2013. Subtracting $1,845 for fuel costs and $950 for depreciation costs means that a boat annually earns $5,707. This amount is to be shared with two to three fishers. This estimate, moreover, is on the high side as interest payments and non-fuel operational costs are not accounted for.

Sri Lankan fishers. These differences result in significant inequalities in the capacity to access Palk Bay resources.

In 1974 and 1976 the maritime boundary line was settled in a bilateral agreement between the two countries. In Tamil Nadu, the settlement of this boundary is, until this day, perceived as a 'gifting away' of historical fishing grounds to Sri Lanka. The officially demarcated boundary, however, initially had little adverse impact on fishers, as their fleets were small and the boundary line poorly guarded. This changed with the onset of the civil war in Sri Lanka in 1983, when the Palk Bay became part of the battleground between government forces and the 'Sea Tigers', the navy wing of the Tamil Tiger guerrillas. While fishing in Northern Sri Lanka came to a virtual standstill, the rapidly developing trawler fleet in India fished the rich Sri Lankan fishing grounds, and were also involved in regular smuggling. Several hundred fishermen lost their lives, mostly killed by the Sri Lankan navy.

Fishermen from both sides are Tamils, sharing a language and ethnicity. Even though 'Sri Lankan Tamils' and 'Indian Tamils' constitute two different, and sometimes antagonistic, identities, both during and after the war Tamils from India have provided passionate support to the Sri Lankan Tamil plea for a separate *Eelam* (nation). Post war, the Sri Lankan government has taken a majoritarian approach embracing Sinhalese-Buddhist nationalist sentiments to the detriment of Tamil and Muslim minorities. The Chief Minister of the Northern Province argued that, paradoxically, the end of the war has actually deepened the ethnic conflict. A major development has been the process of militarization, in the sense of the military increasingly controlling public institutions to guard regime interests (Kadirgamar 2013) resulting in a state of 'oppressive stability' (Wickramasinghe 2014). These developments have received consistent and passionate responses from Tamil Nadu, which has consistently lobbied New Delhi to take a strong anti-Sri Lanka stand. These larger bilateral political dynamics provide the context within which transboundary fishing takes place.

Figure 27.2 reflects some of these historical patterns in terms of fish catch as well as the unequal catch capacities of both fleets. On the Sri Lankan side, it is noticeable that higher catches occurred during relatively calm periods of the war. Post-war, Sri Lankan fishing activity grew rapidly as a result of the step-by-step elimination of fishing restrictions, and the adoption of purse seining by several villages in Mannar.

Indian trawlers move deep into Sri Lankan waters because of their overcapacity and the fact that Sri Lankan fishing grounds are rich. It is estimated that about 2,000 Indian trawlers are fully or partially dependent on Sri Lankan waters so as to secure a profitable catch (Scholtens et al. 2012). Trawler intrusion obstructs Sri Lankan fishers from operating their gillnets which get damaged or destroyed completely when trawler gear is operated at the same time and place, especially at night. Thus, when trawlers operate, Sri Lankan fishers either stay at home, engage in some marginal fishing very close to the coast, or run the risk of substantial losses. In conclusion, trawler fishers from Tamil Nadu and small-scale fishers from Northern Sri Lanka are engaged in a transboundary fishing conflict defined by deep technological and political inequalities between the two groups of fishers (Scholtens et al. 2013).

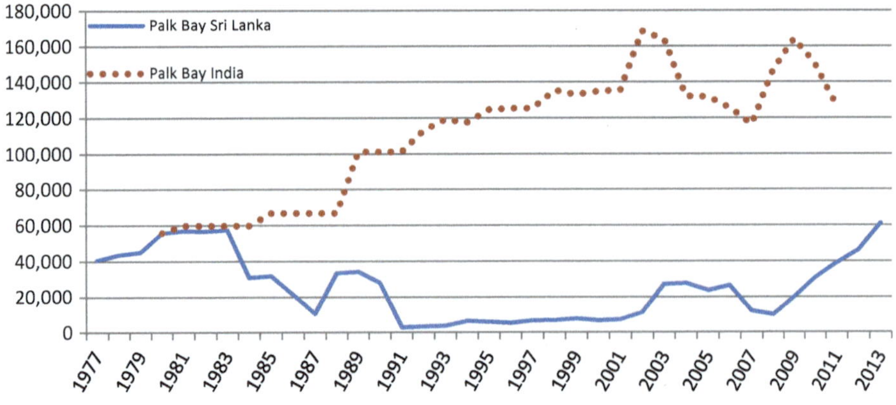

Fig. 27.2 Comparison of fish catch volume from the Palk Bay between Indian and Sri Lankan fleets (Sources: Sri Lankan data: Government of Sri Lanka (2013). Indian data: 1980–1996 Vivekanandan (2001). 1997–2007 and 2011 Government of Tamil Nadu (2012), 2008–2010 Government of India (2010))

The Governing System

The problematic transboundary fishing grounds of the Palk Bay are governed by a plethora of state and non-state institutions at multiple levels at both sides of the boundary. Figure 27.3 provides an overview of this complex multi-layered governance architecture. This section maps out the various authorities. The actual functioning of these authorities in relation to the system-to-be-governed is the subject of Sect. 4.

At the international level, guidance for the regulation of transboundary fisheries is provided by various treaties and soft law agreements (Russell and Vanderzwaag 2010) which also apply to the Palk Bay. **UNCLOS** (United Nations 1982) provides Coastal States with sole exploitation rights of natural resources in their exclusive economic zones (Art. 56.1) and dictates that "nationals of other States fishing in the EEZ shall comply with … laws and regulations of the coastal State" (Art. 62.4; see also Art 73). The **FAO** has drawn up the voluntary Code of Conduct for Responsible Fisheries (1995), and various tools have been developed to address 'Illegal, Unreported and Unregistered' fishing, and more recently the Voluntary Guidelines for Securing Sustainable Small-scale Fisheries (FAO 2014). These regulations place obligations on both the Indian and Sri Lankan government but are in practice only rarely called upon (Scholtens and Bavinck 2014).

At the national level, various ministries are involved in both India and Sri Lanka. In **India**, the Ministry of External Affairs in terms of being in charge of Indo-Sri Lankan relations, the Ministry of Defence, with Indian Coast Guard and Navy having significant presence in the Palk Bay, and the Ministry of Agriculture, which deals with fisheries, are the ministries that play an important role Fisheries

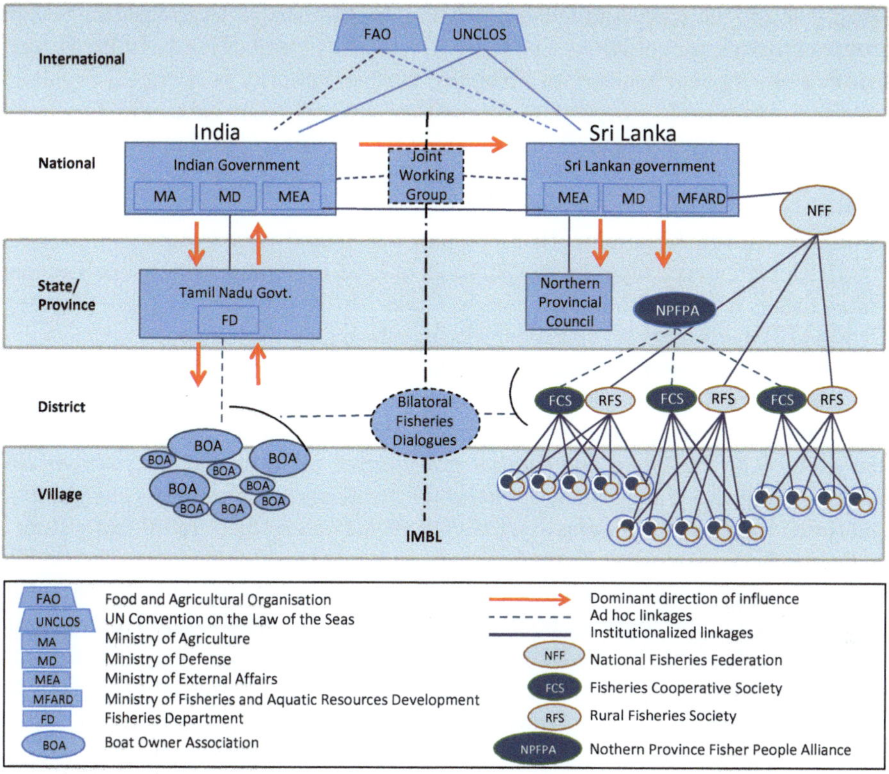

Fig. 27.3 Graphical representation of the multi-level Governing System (Scholtens and Bavinck 2014)

management itself is a state level subject. Fisheries on the Indian side of the Palk Bay is governed by the 1983 **Tamil Nadu** Marine Resources Act as well as a range of regulations specific to the Palk Bay region, collectively placing restrictions on trawlers in terms of gear, time, and location (Scholtens and Bavinck 2013). Despite repeated High Court rulings that Indian fishers have no fishing rights beyond the IMBL,[4] the Tamil Nadu Government continues to claim 'historical rights' for Tamil fishers beyond the IMBL.[5]

In **Sri Lanka**, fisheries come under the Ministry of Fisheries and Aquatic Resources Development (MFARD). The Palk Bay, which is considered part of Sri Lanka's internal waters,[6] is regulated by the Maritime Zones Act of 1976 and the Fisheries and Aquatic Resources Act No. 2 of 1996. Significantly, an amendment adopted in August 2010 prohibits bottom trawling in Sri Lankan waters. Post-war security concerns have also resulted in significant authority over fishing affairs

[4]The New Indian Express newspaper 23 January 2014. Accessed on May 30 2014.

[5]India Today online news; September 7th 2013. Accessed on May 30 2014.

[6]Govt of Sri Lanka 1976 – Maritime Zones Act 22, Article 7ii.

being given to the Navy and the Army, while the Ministry of External Affairs is in charge of fisheries negotiations with India. The recently elected **Northern Provincial Council** has its own Minister of Fisheries, but his authority is restricted to inland waters. Being strongly dependent on political support from India, the provincial government has kept an unholy silence about the intrusion of Indian trawlers.

The Indian and Sri Lankan governments signed bilateral agreements in 1974 and 1976 to demarcate the international maritime boundary line. While the implications of this agreement on fishing rights of Indian fishers were initially ambiguous (Suryanarayan 2004), later letter exchanges by the Ministers of External Affairs in 1976 clarified that Indian fishers have no fishing rights in Sri Lanka.[7] In 2005 a bilateral **Joint Working Group** of Fisheries was set up to deal with transboundary fishing issues.

In terms of non-state institutions, on the Indian side the trawl owners are organized in associations that develop their own forms of fisheries management. These **Boat Owner Associations** have significant political agency (Scholtens et al. 2013) and lobby with the State and Central government for a firm stand vis-a-vis Sri Lanka, noting that trawler fishers should not be harassed by the Sri Lankan navy, and (often implicitly) claiming that the Palk Bay is part of their traditional waters.

On the Sri Lankan side, Fisheries Cooperatives play a significant role at the village and district levels in defining indigenous territorial rights. In 2011, MFARD created parallel representative bodies named **Rural Fisheries Organizations** with the '**National Fisheries Federation**' as their national apex body. Finally the **'Northern Province Fisher People Alliance'** is a loosely organized fisher body at the provincial level. The autonomy of all of these organizations is limited due to strict strong surveillance by State authorities (Scholtens and Bavinck 2014).

Five dialogues were held between fisher representatives of both counties. Initially, they were facilitated by local NGO's (2004 and 2010), while later they were taken over by the States (2012 and 2014). These dialogues led to concrete results (Stephen et al. 2013a), including the commitment of Indian trawler fishers in 2010 to stop trawling after a year's time, implementation of these agreements have proven extremely problematic, partly because of the lack of endorsement by the State governments (Scholtens and Bavinck).

A few things are worth noting based on this brief review of existing institutions and rules governing fishing in the Palk Bay. First, the Palk Bay is characterized by significant legal pluralism, with multiple rule systems applying to fishing in the Bay. Secondly, governance is mediated at multiple (spatial) scale levels, including small fishers' organizations as well as government authorities in both countries. Thirdly, even though occasional horizontal bilateral interaction has occurred in the form of state-to-state and fisher-to-fisher dialogues, there is no institution with authority over the full Palk Bay. Section 5 will discuss the implications of these features for Palk Bay's governability.

[7] "[...] no fishing vessels and fishers of India shall not engage in fishing in the historic waters, the territorial sea and the exclusive economic zone of Sri Lanka nor shall the fishing vessels and fishers of Sri Lanka engage in fishing in the historic waters, territorial sea and the exclusive economic zone of India, without the express permission of Sri Lanka or India, as the case may be [...]" (quoted in Suryanarayan 2004: 167).

Governance Interactions in the Palk Bay

While the previous section provided a baseline of who is governing and what is being governed, this section analyses *how* governance takes place. According to interactive governance theory, governance takes place through interactions between the governing system and the system that is being governed through various modes of governance: hierarchical governance, co-governance and self-governance (Kooiman et al. 2008). These are essentially heuristic categories that emphasize different types of governance interactions, defined as "the institutions and processes through which the system-to-be-governed and the governing system relate to one another" (Chuenpagdee and Jentoft 2013, 344). The governability of a fishery "depends to a large extent on the ways in which the three governance modes are developed and attuned to each other" (Kooiman and Chuenpagdee 2005, 346), and whether the combination of the three can be responsive to the situation at hand. These different governance modes are thus used as analytical tools to shed light on various types of interactions between the governing system and the system-to-be-governed.

Hierarchical Governance: The Realm of Governments

Sri Lanka is one of the most fisheries dependent countries in the world, especially in terms of employment and food security (Barange et al. 2014). The small-scale (inshore) sector contributed 62 % of total marine fish landings in 2012 (Government of Sri Lanka 2013) and is recognized by the state as vital for the country's food security. The official ban on bottom trawling in 2010 (a rarity in Asia) is an example of a pro-small-scale fisheries policy, even though implementation has proved to be troublesome. The state's involvement in fisheries has focused historically on increasing production through fleet modernization. Whereas in India modernization programs resulted in the promotion of a trawler fleet that interfered with small-scale fishers (Bavinck 2001), in Sri Lanka the focus was on developing off-shore long liners that rarely interrupted the practices of the sizable small-scale fleet (Amarasinghe 2005).

As a result of the recent violent history in the Palk Bay region and the continued political sensitivity of Palk Bay matters, both governments have dealt with Palk Bay fisheries first and foremost from a security perspective, illustrated by the deployment of numerous Navy and Coast Guard vessels on both sides of the boundary. In Sri Lanka, while fisheries department officials are officially tasked with implementing the State's fisheries regulations and provisions, in the North especially the Sri Lankan Navy dictates matters at sea. Most fishing villages have one or multiple 'navy points', where uniformed armed men maintain a surveillance system, which is justified in the name of providing security. Even though since 2012 the actual restrictions imposed by the Navy are limited, their very presence fuels a continuous sense of powerlessness amongst fishers.

Table 27.2 Arrests of Indian fishers in Sri Lankan waters

Year	No. of arrests[a]	No. of arrested fishers[b]	Average duration in custody (days)	Total man-days in custody
2009	12	175	9.8	1,708
2010	10	56	7.1	396
2011	6	182	3.0	540
2012	10	205	3.5	712
2013	27	730	29.0	21,194
Total	65	1,348		24,530

Source: The Hindu and Daily Mirror digital newspaper records
[a]Excluding remands for alleged drug smuggling (8 fishers)
[b]A single fisher may have been arrested more than one time

Illustrative of the security focus maintained by both governments, is the continuous arrest of fishers. During 2013 the Sri Lanka navy arrested 730 Indian fishers and impounded approximately 200 boats (see Table 27.2). Even though the number of arrests is a small relative to the alleged number of incursions (the Sri Lankan navy spotted Indian trawlers 45,167 times crossing the boundary in 2013),[8] these arrests result in outrage in Tamil Nadu. Sri Lankan fishers have asked for heavier state involvement, by repeatedly requesting the Navy to take action against trawlers. But as a navy commander said in 2013: "it's like having our hands tied behind our back; whenever we arrest fishers it becomes a bilateral political issue".[9]

In 2005, the two governments set up the Joint Working Group (JWG) of Fisheries, consisting of delegates from the relevant Ministries of both countries, as a mechanism to deal amicably with the transboundary fisheries issue. Four such working group meetings have been held since the inception of the JWG. These meetings resulted in diplomatic joint statements; however, sensitive issues remained largely unaddressed. The last meeting was held in January 2012, after which meetings were suspended due to political obstacles and negotiations on a Memorandum of Understanding failed. What was also noticeable in these meetings was the lack of representation from the fishing community, even though on occasion they have been consulted in advance.

Self-Governance – The Realm of Fishers' Rule

Self-governance is when "fishers govern themselves, without external interference or support, outside the purview of government" (Kooiman and Chuenpagdee 2005, 334). Self-governance does not so much denote

[8]Personal communication Navy Commander, January 2014.

[9]Sunday Times, 3 March 2013. http://www.sundaytimes.lk/130303/news/navy-has-a-tough-job-with-indian-fishermen-35225.html. Accessed 20 July 2014.

individualized governance, but is rather a specific expression of community based collective action (e.g. Ostrom 1990).

In India, Boat Owner Associations manage day-to-day trawler activity, in addition to (often successful) lobbying the political establishment. In Sri Lanka, fishers in the Northern Province have cooperative societies that have survived decades of civil war. Whilst being closely monitored and controlled by the cooperative department, fisheries department, and the intelligence and security forces, these organizations have considerable control over those issues that are beyond the state's interest. This includes fish marketing, solving internal conflicts, managing insurance schemes, contributing to social development and regulating appropriate timing of gear use. In the aftermath of the war, however, these organizations have been significantly weakened for a couple of reasons. First, in 2011 the Ministry of Fisheries founded a new fisheries representative body, appointing leaders loyal to the government, which created a politicized parallel institution to the cooperatives; this also marked the end of financial contributions of the Fisheries Ministry to fisheries cooperatives. Even though these parallel institutions suffer from a lack of local legitimacy, state support and patronage gives these organizations significant power. Secondly, there was political interference in cooperative elections in 2011 and 2012, resulting in autonomy being lost. Both these developments have strengthened a patronage-based system where fisher leaders are tied to the interests of particular politicians, compromising their ability to represent actual fisher concerns and fuelling strong sense of skepticism among fishers.

Sri Lankan fishers' self-governance is also limited, given their inability to shield themselves from the intrusion of outsiders (i.e. Indian trawlers), which is a vital pre-condition for self-governance (Ostrom 1990). Self-governance is adversely impacted by and interlinked with processes at higher scale levels. In 2011, fishers from two villages in Jaffna district took the law into their own hands and hijacked 36 Indian trawlers (Scholtens et al. 2012). This characterized a strong act of village level collective action, even though the authorities may have provided silent support.

More constructive forms of self-governance include the dialogues between fisher representatives from both countries held in 2004, 2010, 2012 and 2014. In 2010, fisher representatives from both countries had lengthy negotiations and reached a concrete agreement, even though it was subsequently dismissed by the Sri Lankan government and not adhered to by Indian fishers (Stephen et al. 2013a). Both governments have at time endorsed these dialogues through official Joint Statements, but have been equally quick to dismiss them. The Tamil Nadu establishment, in particular, has repeatedly employed delaying tactics due to fear that these meetings may expose frictions in pan-Tamil political identity (SathyaMoorthy 2013).

Self-governance thus has clear limitations in this multi-scalar, transboundary and politically charged environment and many fisher leaders have lost both confidence in self governance as it exists and a sense of agency. As one leader responded to my repeated questions as to why fishers remained silent about trawler transgressions: "Joeri, this is not something for fishers to solve. Don't you see that politics have crept in, what can we do?"

Co-governance: The Realm of Collaboration

The third mode, co-governance, suggests a form of collaboration between different stakeholders, including the state. Co-governance has been widely propagated as a means to draw on the capacities of both fisher communities and the government, while compensating for the inherent limitations of both.

Neither in Sri Lanka nor in Tamil Nadu is there a co-management system in place, in the sense of an institutionalized form of power-sharing and rule-making between state and fisher groups. In both countries, interaction between fisheries department officials, security forces and fisher representatives tend to take place on an ad hoc basic (Scholtens and Bavinck 2013), although recent initiatives in both countries are promising.[10] Various authorities each operate on the basis of an internal logic and distinct norms, which, depending on the circumstances, accommodate, ignore, support or compete with the norms of others (Bavinck et al. 2013). As a notable exception in Sri Lanka, the Assistant Director of Fisheries participates regularly in the Jaffna district Fisheries Cooperative Union's Federation meetings.

Whereas at the institutional – technical level differences between India and Sri Lanka are insignificant, at the political level the nature of fisher-government collaboration differs significantly between the two countries. North Sri Lankan fishers, being part of an entrenched and long-term ethnic conflict, have a deeply distrustful relationship with the state, obstructing any constructive forms of interaction. In fact, many fishers believe that the Sri Lankan government has a hidden interest in allowing Indian encroachment to prevail, as it would undermine the Tamil economy in the North and create a welcome breach between the Tamils of both countries (Scholtens et al. 2013, 4). While these conspiracy framings are not necessarily in line with the demonstrated commitment of the Sri Lankan Fisheries Ministry and Navy to stop transboundary fishing, Sri Lankan politicians have few incentives to attend to North Sri Lankan fishers' concerns, not least because they do not form an electoral constituency for the dominant parties. The irony is that although in principles Lankan Tamil fishers' interests are aligned with the Sri Lankan government as far as the Palk Bay is concerned, the hegemonic Tamil versus Sinhala discourse blocks effective collaboration.

In India, on the contrary, fishers enjoy significant political agency vis-à-vis the Tamil Nadu political establishment (Scholtens et al. 2013). Not only are fishers well represented in major political parties, the fishing issue provides perfect fuel for the State Government and fishers to jointly rally against the Central Government to demand more attention to the plight of fishers. Fishers' plight also feeds well into strong anti-Sri Lankan sentiments in Tamil Nadu making fisheries a popular subject for politicians to rally behind (ibid). These differences in trust and collaboration between the two countries imply a highly uneven playing field from where fishers can stake their claims.

---

[10] See for example the Lagoon Fisheries Management Authorities in Sri Lanka, and the FIMSUL project in Tamil Nadu.

What further stands out is the limited role of civil society groups. Whereas in India civil society actors occasionally act as intermediaries, or on behalf of various fisher groups, the Sri Lankan government has ruled out any NGO involvement in Palk Bay issues. NGOs who have tried to mediate have been threatened and accused of conspiring with Tamil Nadu, 'the West' or both.

Discussion: Limitations to Palk Bay Fisheries' Governability

Governability challenges are embedded in the system-to-be-governed, the governing system as well as in their interactions (Chuenpagdee and Jentoft 2013). Based on the journey through these various systems made in the previous sections, I identify six crucial factors limiting the governability of the Palk Bay. Each of these issues has an element of wickedness, in the sense of disagreement between stakeholders about what the problem actual is, where the problem starts and ends, or how it is embedded in wider issues.

Boundaries and Scalar Mismatch

Inherent in the governance of transboundary fisheries is a mismatch "between the geographic scale of ecosystem functioning and the spatial extent of the institutional arrangements managing such a system" (Duraiappah et al. 2014). The Palk Bay provides an exemplary case of the problem of mismatch between institutional, ecosystem and fishers' operational scales. While the ecosystem and the spatial extent of fishing operations generally coincide with the expanse of the Palk Bay, these do not correspond with the spatial scope of any of the multiple legal systems (Scholtens and Bavinck 2013). According to Berkes (2010, 236) this "gross misfit of ... scales is one of the fundamental reasons why management often fails". However, he also argues that seeking an exact fit may often not be very realistic given the highly dynamic and mobile nature of both fish stocks and fishers. Rather, mismatches need to be addressed by having appropriate interactions, both horizontally (at a single level) and vertically (between levels) as only then various overlapping jurisdictions can coordinate efforts (Fanning et al. 2007). In order to enhance the 'institutional fit', bridging organizations can be valuable in providing linkages that allow crossing scales and boundaries, and enable the grassroots to influence national level authorities (Nayak 2011).

The bilateral Joint Working Group and fisheries dialogues between the two groups of fishers provide in principle valuable transnational linkages. However, not only have the number of actual transnational meetings been minimal,[11] but such

[11] In the period of 2004 to 2014, 6 meetings between fisher groups and 4 by the Joint Working Group have materialized.

meetings have also been regularly co-opted by the respective State governments. While fishers from Tamil Nadu are able to assert their interests' vis-à-vis their political establishments, North Sri Lankan fishers' relationship with their government is one of distrust, avoidance and fear (Scholtens et al. 2013). These mismatches of scale and the lack of functioning bridging organizations have a limiting effect on the Palk Bay's governability.

Institutional Fragmentation

In a transboundary fishery, multiple state and non-state actors are involved, generating an array of legal arrangements vis-à-vis fishing. These arrangements include international law, bilateral agreements, state law and community institutions, none of which enjoy exclusive authority over the fishing grounds. These legal systems are both fragmented in terms of the scale at which they operate, and the qualitative differences in terms of substance as well as process (Scholtens and Bavinck 2015).

High levels of legal pluralism in the Palk Bay result in fragmented governance. For both India and Sri Lanka, the Palk Bay is primarily a space requiring attention in terms of security and sovereignty, reflected by the deployment of significant navy and coast guard fleets on either side. Fishing communities, however, have their own notions of legitimacy in terms of who can fish, where fishers can fish and when and how they can fish. The repeated arrests of Indian fishers by the Sri Lankan Navy, as well protests against arrests by fishers, are illustrative of the lack of congruence between the numerous legal systems.

Conflicting Problem Images

Directly related to institutional fragmentation is the absence of a shared discursive image of the problem among stakeholders. Images are mental models of how the world functions or should function and which inform policy (Mahon et al. 2005). The conflicting stakeholders in the Palk Bay strategically pursue different framings of the problem that reflect their particular interests. Indian fishers and the Tamil Nadu Government frame the problem as an over-assertive (Sinhala) Sri Lankan navy that is keen to harass Tamil trawler fishers while they fish in their traditional fishing grounds. Sri Lankan small-scale fishers emphasize the destructive trawler gear and their status as victims post-war. The Sri Lankan government frames the problem as a violation of boundaries and sovereignty and emphasizes the big brother attitude of India and the hypocrisy of the Tamil Nadu government. The Indian government has generally taken a soft stand, recognizing the IMBL as it is, but nevertheless condemning the Sri Lankan Navy.

The news media have an important role in (re)producing or nuancing these entrenched frames of reference. While media on both sides mostly echo the dominant discourses of their audiences, since late 2012 some Indian newspapers have occasionally questioned the actions of Indian trawlers. These conflicting images and entrenched frames thus represent worldviews connected to vested interests, contributing both to the wickedness of the problem (unclear problem definition) and constituting a governability problem by itself.

Power Imbalances

Governance interactions are mediated by power, which has both an enabling and restricting role (Jentoft 2007). Chuenpagdee and Jentoft (2013) suggest that power imbalances can constitute an important governability problem and even bring the entire governing process to a halt. According to Araral (2014), productive cooperation in the regional commons in the presence of asymmetric power relations is extremely difficult.

Current governance in the Palk Bay produces outcomes that consistently marginalize North Sri Lankan small-scale fishers. Fishermen are the weaker party both in an ethnically grounded conflict with Sri Lankan authorities, and in a livelihood-grounded conflict with Indian fishermen. They are technologically marginalized by Indian trawlers and they are ethnically marginalized by an increasingly repressive government. Added to this, the Sri Lankan Tamil political party (TNA) that is supposed to represent Tamil interests maintains an unholy silence over the fisher issue, unwilling to confront Tamil Nadu. The result is that Northern fishers have no powerful ally to turn to in order to form a potentially useful coalition. These asymmetrical power relations lead to a lack of representation of small-scale fishers' interests, hampering the system's capacity and quality of governance.

Issue Linkage

The Palk Bay fisheries conflict is deeply embedded in national and regional politics and has been consistently immersed in the politics of ethnicity. In Tamil Nadu, the fishing conflict is linked to the aggressiveness of the Sinhala Navy. In Sri Lanka, the government highlights the hypocrisy of Tamil Nadu and the TNA. These 'ethnic' framings undermine the fact that the conflict is one of technological mismatch between two sets of fishers.

The recent decision of the Sri Lankan President to release all arrested Indian fishers in Sri Lankan custody, immediately after India abstained from backing a resolution against Sri Lanka at the UN Human Rights Council (Times of India, March 29,

2014) once again highlights how the fisher issue is intertwined in wider geopolitics. The Indian Foreign Secretary confirmed this act of horse-trading by stating that "India's stand [regarding the UNCHR resolution] would help resolve the problems of fishers from Tamil Nadu… " (The Hindu, March 30, 2014). The fisheries conflict is thus a pawn in larger geopolitical relations between Indian and Sri Lanka. Sri Lankan Northern fishers pay the price for this.

Although politicizing an issue can bring attention to it, connecting the fisheries conflict to highly charged ethnic tensions makes the fishing problem practically unsolvable. In addition, politicization also leads to fragmentation, with its tendency to emphasize difference and connect it to unrelated, but popular issues.

Path Dependency of Indian Trawlers

Fisheries policies tend to have a strong element of path dependency (e.g. Hersoug et al. 2000). The industrialization of Indian fisheries (Bavinck 2001) has a strong degree of irreversibility that contributes to the problem's wickedness (Chuenpagdee and Jentoft 2013). While at one level, a possible solution is to reduce the size of the trawler fleet, such a decommissioning scheme is difficult to implement for a few reasons. First, past and ongoing capital investments have created an entrenched sector. Second, boat owners are enrolled in complex systems of loans and advances with traders and crewmembers, which provide incentives to continue fishing. Third, alternative livelihood options are poor. Finally, the diesel subsidy provides a subtle incentive to stay in trawling, a sector which enjoys strong political support. Trawler owners have argued consistently that the government is responsible for them trawling and now they are simply not in the position to stop overnight, even though virtually everyone agrees that the sector is overcapitalized. This entrenched nature of the Indian trawler economy provides an additional governability challenge with no easy fix.

Conclusions

This chapter has aimed to understand how a fleet of almost 2,000 trawlers continues to fish in Sri Lankan waters though it violates international law, Tamil Nadu fisheries law, Sri Lankan fisheries law and Indo-Sri Lankan bilateral agreements and despite the presence of naval forces and angry affected Sri Lankan fishers and significant media attention.

The Palk Bay is characterized by an asymmetrical fisheries conflict, with North Sri Lankan fishers struggling to carry out their work in the face of Indian trawlers who encroach deep into Sri Lankan waters appropriating most of the fish resources. The governing system is characterized by its intense fragmentation, with a large variety of actors at different scale levels, each with high stakes, and each embedded in different legal systems that are conflictual. This has resulted in the problematic functioning of self- and hierarchical modes of governance, and the virtual absence of co-governance. The involvement of state boundaries and state sovereignty in transboundary fisheries provides obvious limitations to self-governance. Governments too have not been successful in dealing with transboundary fishing. While co-governance is in theory crucial for transboundary governance to be more responsive to the situation at hand, building constructive interactions is hampered by a range of factors embedded in the governing system, the system-to-be-governed and their interactions.

The governability analysis has identified six factors that limit the capacity for and quality of transboundary fisheries governance: scalar mismatch, institutional fragmentation, politicization of processes, power imbalances, conflicting problem images and path dependency of trawling. All of these collectively result in the governance process 'getting stuck' and the interests of North Sri Lankan small-scale fishers being undermined.

In order to make sense of marginalization through a governability assessment, I argue that we need to be careful about only looking at system's features like diversity, complexity, dynamics and scale as it makes us blind to *relational* processes of marginalization. This chapter demonstrates that marginality of small-scale fishers is not a given feature of a system, but is continuously (re)produced through interactions between actors. I believe that this has been a neglected issue in interactive governance theory.

I contend that even though this chapter does not provide clear-cut policy proposals, its utility is twofold: (1) it provides an understanding of why fisheries are so difficult to govern in the first place and highlights where governance 'gets stuck'; and (2) it helps in doing a 'reality check' on potential interventions, to assess their feasibility and potential effects. Take, for example a prominent policy proposal that suggests a partial decommissioning scheme for trawlers. The governability challenges would indeed support the relevance of this idea, but a more relational approach would highlight the political sensitivity of doing so. Moreover, solving the trawler problem will not necessarily address Sri Lankan fisher problems given their poor political representation. The practice of doing a careful governability assessment –one that reveals the limitations and opportunities for governance interventions – thus may eventually contribute to improving a system's governability.

This chapter has demonstrated the governability challenges of a transboundary fishery and its impact on a small-scale fishing community. The question is to what extent these insights have larger validity and wider implications beyond this particular context. While there are unique features to this region and fishery, scalar mismatch, power imbalances and (geopolitical) subject linking, and processes of small-scale fishers' marginalization are issues that affect the governability of many transboundary fisheries around the world. In terms of future research, it would be fascinating to see how the governability problems described in this chapter apply to the European and Chinese fleets fishing (with or without permits) in foreign EEZs and, no doubt affecting domestic (small-scale) fisheries.

References

Adhuri, D. S., & Visser, L. E. (2007). Fishing in, fishing out: Transboundary issues and the territorialization of blue space. *Asia-Pacific Forum, 36*, 112–145.

Alder, J., & Sumaila, U. R. (2004). Western Africa: A fish basket of Europe past and present. *The Journal of Environment & Development, 13*(2), 156–178.

Amarasinghe, O. (Ed.). (2005). *Modernization and change in marine small-scale fisheries of Southern Sri Lanka*. Colombo: Navamaga Printers.

Araral, E. (2014). Ostrom, Hardin and the commons: A critical appreciation and a revisionist view. *Environmental Science and Policy, 36*, 11–23.

Barange, M., Merino, G., Blanchard, J. L., Scholtens, J., Harle, J., Allison, E. H., Allen, J. I., Holt, J., & Jennings, S. (2014). Impacts of climate change on marine ecosystem production in societies dependent on fisheries. *Nature Climate Change, 4*, 211–216.

Bavinck, M. (2001). *Marine resource management: Conflict and regulation in the fisheries of the Coromandel Coast*. New Delhi: Sage.

Bavinck, M. (2014). Investigating poverty through the lens of riches – Immigration and segregation in Indian capture fisheries. *Development Policy Review, 32*(1), 33–52.

Bavinck, M., Johnson, D., Amarasinghe, O., Rubinoff, J., Southwold, S., & Thomson, K. T. (2013). From indifference to mutual support – A comparative analysis of legal pluralism in the governing of South Asian fisheries. *European Journal of Development Research, 25*(4), 621–640.

Berkes, F. (2010). Linkages and multilevel systems for matching governance and ecology: Lessons from roving Bandits. *Bulletin of Marine Science, 86*(2), 235–250.

Butcher, J. G. (2005). *The closing of the frontier. A history of the marine fisheries of southeast Asia c1850–2000*. Leiden: KITLV Press.

Chuenpagdee, R., & Jentoft, S. (2013). Assessing governability: What's next? In M. Bavinck, R. Chuenpagdee, S. Jentoft, & J. Kooiman (Eds.), *Governability of fisheries and aquaculture: Theory and applications* (MARE publication series 7, pp. 335–349). Dordrecht: Springer.

Chuenpagdee, R., Degnbol, P., Bavinck, M., Jentoft, S., Johnson, D., Pullin, R., & Williams, S. (2005). Challenges and concerns in capture fisheries and aquaculture. In J. Kooiman, M. Bavinck, S. Jentoft, & R. Pullin (Eds.), *Fish for life: Interactive governance for fisheries*. Amsterdam: Amsterdam University Press.

Cullis-Suzuki, S., & Pauly, D. (2010). Failing the high seas: A global evaluation of regional fisheries management organizations. *Marine Policy, 34*(5), 1036–1042.

Duraiappah, A. K., Tanyi Asah, S. T., Brondizio, E. S., Kosoy, N., O'Farrell, P. J., Prieur-Richard, A.-H., Subramanian, S. M., & Takeuchi, K. (2014). Managing the mismatches to provide ecosystem services for human well-being: A conceptual framework for understanding the New Commons. *Current Opinion in Environmental Sustainability, 7*, 94–100.

Fanning, L., Mahon, M., McConney, P., Angulo, J., Burrows, F., Chakalall, B., Gil, D., Haughton, M., Heileman, S., Martinez, S., Ostine, L., Oviedo, A., Parsons, S., Phillips, T., Santizo Arroya, C., Simmons, B., & Toro, C. (2007). A large marine ecosystem governance framework. *Marine Policy, 31*(4), 434–443.

FAO. (2014). *Voluntary guidelines for securing sustainable small-scale fisheries*. Retrieved from http://www.fao.org/fishery/ssf/guidelines/en

Government of India. (2010). *Marine fisheries census Tamil Nadu 2010*. Centre for Marine Fisheries Research Institute.

Government of Tamil Nadu. (2012). Department of Fisheries. Estimated marine fish production. http://www.agritech.tnau.ac.in/fishery/fisheries2012.pdf. Accessed 1 April 2015.

Government of Sri Lanka. (2013). *Fisheries statistics 2013, Ministry of Fisheries and Aquatic Resources Development*. Retrieved from http://www.fisheries.gov.lk/content.php?cnid=ststc

Gupta, C., & Sharma, M. (2008). *Contested coastlines: Fisherfolk, nations and borders in South Asia*. New Delhi: Routledge.

Hersoug, B., Holm, P., & Rånes, S. A. (2000). The missing T. Path dependency within an individual vessel quota system – The case of Norwegian cod fisheries. *Marine Policy, 24*(4), 319–330.

Jentoft, S. (2007). In the power of power: The understated aspect of fisheries and coastal management. *Human Organization, 66*(4).

Jentoft, S., Chuenpagdee, R., Bundy, A., & Mahon, R. (2010). Pyramids and roses: Alternative images for the governance of fisheries systems. *Marine Policy, 34*(6), 1315–1321.

Kaczynski, V. M., & Fluharty, D. L. (2002). European policies in West Africa: Who benefits from fisheries agreements? *Marine Policy, 26*(2), 75–93.

Kadirgamar, A. (2013). The question of militarisation in post-war Sri Lanka. *Economic Political Weekly, XLVIII*(7), 42–46.

Kooiman, J., & Chuenpagdee, R. (2005). Governance and governability. In J. Kooiman, S. Jentoft, R. Pullin, & M. Bavinck (Eds.), *Fish for life. Interactive governance for fisheries*. Amsterdam: Amsterdam University Press.

Kooiman, J., Bavinck, M., Chuenpagdee, R., Mahon, R., & Pullin, R. (2008). Interactive governance and governability: An introduction. *The Journal of Transdisciplinary Environmental Studies, 7*, 1.

Mahon, R., Bavinck, M., & Roy, R. N. (2005). Governance in action. In J. Kooiman, S. Jentoft, R. Pullin, & M. Bavinck (Eds.), *Fish for life. Interactive governance for fisheries*. Amsterdam: Amsterdam University Press.

Nayak, P. K. (2011). *Change and marginalisation: Livelihoods, commons institutions and environmental justice in Chilika Lagoon, India*. Doctoral thesis, University of Manitoba, Winnipeg, Canada.

Ostrom, E. (1990). *Governing the commons: The evolution of institutions for collective action*. Cambridge: Cambridge University Press.

Pauly, D., Belhabib, D., Blomeyer, R., Cheung, W. W. W. L., Cisneros-Montemayor, A. M., Copeland, D., Harper, S., Lam, V. W. Y., Mai, Y., Le Manach, F., Österblom, H., Mok, K. M., van der Meer, L., Sanz, A., Shon, S., Sumaila, U. R., Swartz, W., Watson, R., Zhai, Y., & Zeller, D. (2013). China's distant-water fisheries in the 21st century. *Fish and Fisheries, 15*(3), 474–488.

Rittel, H. W. J., & Webber, M. M. (1973). Dilemmas in a general theory of planning. *Policy Sciences, 4*, 155–169.

Russell, D. A., & Vanderzwaag, D. L. (Eds.). (2010). *Recasting transboundary fisheries management arrangements in the light of sustainability principles* (pp. 1–544). Leiden: Marinus Nijhoff Publishers.

SathyaMoorthy. (2013). *Sri Lanka: Fishing for a solution in the Palk Bay*. Retrieved from www.orfonline.org

Sathyapalan, J., Srinivasan, J. T., & Scholtens, J. (2008). *Fishing fleet reduction and its livelihood implications: A case study of Palk Bay resource users in the East Coast of Tamil Nadu*. Research report United Nations India & Food and Agriculture Organization of the United Nations.

Sathyapalan, J., Srinivasan, J. T., & Scholtens, J. (2011). Overcapitalization in a small-scale trawler fishery: A study of Palk Bay, India. In R. Chuenpagdee (Ed.), *World small-scale fisheries: Contemporary visions* (pp. 51–62). Delft: Eburon.

Scholtens, J., & Bavinck, M. (2013). South Indian trawl fisheries – Assessing their governability. In M. Bavinck, R. Chuenpagdee, S. Jentoft, & J. Kooiman (Eds.), *Governability of fisheries and aquaculture: Theory and practice* (pp. 177–199). Dordrecht: Springer.

Scholtens, J., & Bavinck, M. (2014). Lessons for legal pluralism: Investigating the challenges of transboundary fisheries governance. *Current Opinion in Environmental Sustainability, 11*, 10–18.

Scholtens, J., Bavinck, M., & Soosai, A. S. (2012). Fishing in dire straits. *Economic Political Weekly, XLVII*(25), 87–96.

Scholtens, J., Stephen, J., & Menon, A. (2013). Between the devil and the not-so-deep blue sea: Asymmetrical power in the Indo-Sri Lankan fisheries conflict. *The Broker Online*, pp. 1–6.

Soosai, S. A., & Stokke, K. (2006). Fisheries under fire: Impacts of war and challenges of reconstruction and development in Jaffna fisheries, Sri Lanka. *Norsk Geografisk Tidsskrift – Norwegian Journal of Geography, 60*(3), 240–248.
Stephen, J., Menon, A., Scholtens, J., & Bavinck, M. (2013a). Transboundary dialogues and the 'politics of scale' in Palk Bay fisheries: Brothers at sea? *South Asia Research, 33*(2), 141–161.
Stephen, J., Menon, A., & Anbazhagan, P. (2013b). *Socio-economic profile of Palk Bay* (p. 72). Prepared for the REINCORPFISH project, Chennai, India. www.reincorpfish.info
Suryanarayan, V. (2004). *Conflict over fisheries in the Palk Bay region*. New Delhi: Lancer Publishers and Distributors.
Swartz, W., Sumaila, U. R., Watson, R., & Pauly, D. (2010). Sourcing seafood for the three major markets: The EU, Japan and the USA. *Marine Policy, 34*(6), 1366–1373.
United Nations. (1982). *United Nations Convention on the Law of the Seas*. Retrieved from http://www.un.org/Depts/los/convention_agreements/texts/unclos/closindx.htm
Vivekanandan, V. (2001). Crossing maritime borders: The problem and solution in the Indo-Sri Lankan context. In K. G. Kumar (Ed.), *Forging unity: Coastal communities and the Indian ocean's future* (pp. 76–89). Chennai: International Collective in Support of Fish Workers.
Vivekanandan, V., & Kasim, M. (2011). *Fisheries management options for Tamil Nadu and Puducherry* (Workpackage 5, Report – R20. FAO/UTF/IND/180/IND). Available at: http://eprints.cmfri.org.in/9251/1/33.pdf. Accessed 1 Apr 2015.
White, C., & Costello, C. (2014). Close the high seas to fishing? *PLoS Biology, 12*(3), 1–5.
Wickramasinghe, N. (2014). *Sri Lanka in the modern age; A history*. London: Hurst and Company.

Part VII
Governance in Transition – Reforming Institutions

Introduction

Governance systems are subject to internal and external stressors that give rise to change of both design and practice. In some instances, governance change occurs gradually and incrementally, where small steps may add up to a big one without any deliberate plan in mind. In other instances, governance system change is subject to a grand reform. How small-scale fisheries come out of such small- or large-scale reforms is certainly an issue of governability that requires research.

In *Chap. 28*, Ronald W. Jones and Say Sok analyze a reform that is currently ongoing in the Tonle Sap Lake fisheries of Cambodia. Here the traditional fishing lot auction system is being replaced by a new governance system, influenced by both regional neoliberal development policies and external donor pressures for better management within an expanding conservation narrative. Although the existing governance mechanism may enhance the governability of small-scale fisheries of the lake, the government, according to the authors, does not have sufficient funds, nor the capacity or political will to carry it through. The heterogeneity of commercial interests and the diffuse loci of ministerial power are also inhibiting factors. The authors argue that a basic governability condition is that governors must recognize the diversity of vested interests, social cultural drivers, and boundaries, which serve to limit management, and strive to foster inclusive and interactive mechanisms to secure justice.

Ahmed Khan and Sheku Sei in *Chap. 29* focus on the governability of small-scale fisheries in post-conflict Sierra Leone. They discuss whether the situation has improved after a co-governance system was introduced, one that included user rights and MPAs. Although notable improvements regarding stakeholder participation, conservation, and economic returns have occurred, the governability at the local level is still inadequate due to low compliance and poor monitoring. The authors call for the need to strengthen local governing capacity, build linkages, and facilitate collective action between fisheries and other economic activities. Decision-making approaches in small-scale fisheries that emphasize proportionate resource allocation are likely to meet dual conservation and development objectives.

Co-governance and collective action by stakeholders and government are also discussed as measures to improve governability in *Chap. 30*, written by Shelly-Ann Cox and Patrick McConney, in the context of Barbados and Saint Lucia in the Caribbean. The authors highlight that making co-governance work requires addressing deficiencies in capacity, communication and trust at community level and within fisher organizations. Thus, the problem is not co-governance *per se*, but the framework and support systems that are built around it, which also illustrate the complexity of the governing system and how it functions at larger scales. Concentrating on the sea urchin fisheries at five sites in Barbados and Saint Lucia, the authors explore processes and conditions for effective co-governance.

Mauricio Castrejón and Omar Defeo in *Chap. 31* compare the governability and resilience of small-scale shellfisheries in a number of Latin American countries, where co-governance is emerging as a promising solution. Their chapter addresses the question of how well this governing mode copes with, and adapts to, crisis as a learning process concerned with the livelihoods and wellbeing of local small-scale fishing communities. In most cases, small-scale fishers engage in collaborative action and re-organizing, contributing to institutional adaptability and resilience, and hence governability.

Chapter 28
Impacts and Implications of *Deep Fisheries Reforms* on the Governability of Small-Scale Fisheries in Tonle Sap Lake, Cambodia

Ronald W. Jones and Say Sok

Abstract For over 100 years, auctioned "fishing lots" dominated the fisheries in Cambodia's Great Lake, Tonle Sap. As part of the world's largest single managed freshwater fishery, these large scale pre-colonial aquatic resource concessions enabled elite stakeholders to strictly control access to lake resources and to extract large aquatic resource rents, paying little in tax, and not being subjected to any truly effective upward accountability to the Fisheries Administration. In 2012, in conjunction with past reforms, the Prime Minister of Cambodia finally removed *all* remaining lots via a unilateral sub-decree known locally as the "deep fisheries reforms". Using the interactive governance framework and governability assessment methodology, the chapter first outlines the historical context of the fishing lot system and small-scale fishing, and the causes for their often tumultuous and violent relationship. It then begins to unpack the underlying *raison d'etre* for this profound top-down decree. We explore potential impacts and outcomes of what their rapid removal means for any real system reform and increase in aquatic system governability. We examine the important cross scale and organizational level challenges now facing the multiple stakeholders and players in lake management as they try to coordinate and implement these "deep reforms".

Keywords Governance • Governability • Small-scale fisheries • Fisheries reforms • Tonle Sap • Complex systems • Cambodia

> The hybrid or the meeting of two media is a moment of truth and revelation from which new form is born. The moment of the meeting… is a moment of freedom and release from the ordinary trance and numbness imposed by them on our senses. (McLuhan 1964)

R.W. Jones (✉)
Learning Institute, Phnom Penh, Cambodia
e-mail: rwjones1@hotmail.com

S. Sok
Royal University of Phnom Penh, Phnom Penh, Cambodia
e-mail: saysok@gmail.com

S. Jentoft, R. Chuenpagdee (eds.), *Interactive Governance for Small-Scale Fisheries*, MARE Publication Series 13, DOI 10.1007/978-3-319-17034-3_28

> The real issue was power and who was going to have it! (Dr. Hunter S. Thompson in a letter to Walter Mondale Feb. 1 1971)

Current global freshwater resources including important small-scale fisheries and the communities that depend on their ecosystem goods and services are threatened by multiple, diverse and cross scale/level anthropogenic drivers of change (Alcamo et al. 2008; Vörösmarty et al. 2010, 2013). Many of these drivers are degrading these aquatic ecosystems and in turn forcing them ultimately into new, unstable states. The result is often the reduction of habitat quality and the loss of valuable food and commercial fish species, and consequently alterations in the social-ecological fabric of these communities whose health and nutrition, incomes and well-being are intricately linked to these fisheries and the waters that support them. These changes, while most often normatively negative, can bring good ideas for changes in the management of the local aquatic resources. The following case study has the real potential to be the greatest small-scale tropical fresh-water fisheries management reform ever under taken by a single developing country (Fig. 28.1).

This chapter situates the *deep fisheries* reforms, a top-down Prime Ministerial mandated change, to an old, very complex, diverse yet very rigidly managed tropical fishery under the conceptual lens of the interactive governance framework and governability concepts (Kooiman et al. 2005, 2008; Bavinck et al. 2013). This approach will show how the reform process, key actors and stakeholders coupled with their interactions create and modify the governability of this complex aquatic social ecological system. The first section gives an overview of the biophysical and eco-hydrological context of the Tonle Sap Lake. This is followed by a brief history of fisheries management, starting with the creation of a fishing lots system under the French

Fig. 28.1 Man on the Tonle Sap

protectorate. The key actors and role of the Cambodian state is then described leading up to the current reforms situated within the interactive governance framework and what some of the key ideas and concerns are. The chapter concludes with a discussion of the relevance and effectiveness of such an approach for this fishery looking at how these reforms may or may not lead to a more socially just, governable fishery.

The Tonle Sap Basin: Diversity, Complexity, Scale and Dynamics

In the centre of Cambodia lies the Great Lake Tonle Sap (បឹងទន្លេសាប) situated at the top of a 70,000 km^2 floodplain connected to the Mekong River (Liu et al. 2007) via its outlet the Tonle Sap River. In the rainy season (May–October) the Mekong with a discharge of 45,000 m^{3}s^{-1} (gauged at Phnom Penh) backs up the river and expands the lake. The lake's depth increases by 10 m and expands its surface area from 2,520 to 15,780 km^2 (Carbonnel and Guiscafré 1965 in Lim et al. 1999). In the dry season (November–April), the flow reverses and the lake empties (Dec–Feb) via the Tonle Sap and Bassac Rivers; with peak fishing for all gear, around the lake and in tributary rivers, taking place between December and January (Lim et al. 1999). This great floodplain lake with its very productive fishery and its peoples in landed, stilted and floating communities is perhaps truly both the ecological and spiritual heart of the entire Lower Mekong River Basin. As Southeast Asia's largest inland floodplain and centre of the world's greatest freshwater fishery (Rainboth 1996) in per capita value, Tonle Sap provides vital for diverse activities and the food and livelihood security of over 60 million people (Baran et al. 2001; MRC 2005; Campbell et al. 2006; Lamberts 2006; Arthur and Friend 2011). The fish incubated within this lake move out to feed, grow and ultimately to enter Mekong river fisheries and nourish and sustain the health and livelihoods of millions of people along the mainstream Mekong River and its tributary communities. The fisheries are the essential backbone of ancient vibrant and dynamic river cultures.

The Tonle Sap lake basin is also the heart of a body politic. It is a dynamic geography of contestation. It is both place and arena for a complex and messy arrangement of institutions, agencies and interests which have influenced relations and shaped organizational structures going back to Angkor times. There are dynamic interactions and often non-trivial conflicts that take place between the Mekong's heart and body (Imamaura and Lebel 2006). The Tonle Sap Lake and its interactions with the wider Mekong River System was influenced and shaped by its regional social history and processes of geology and geomorphology. The resulting eco-hydro biology all creates what we call a complex adaptive system (Hartvigsen et al. 1998, special issue). Complexity arises from these diverse forms of inter/intra specific interactions across trophic levels as well as from the interactions of different human agents acting across multiple levels of social, political or resource management organizations. Mekong cultures and the diverse systems of management from the Delta to the headwaters on the Tibetan Plateau are under multiple and cross-scale social and ecological drivers of change as the region are pushed and pulled into regional narratives and along pathways of Asian or perhaps Association of Southeast Asian Nations (ASEAN) directed modernity. These development pathways will have profound impacts on the social-ecological landscape of the Mekong Basin (Sneddon and Fox 2006, 2007), causing changes to land use and cover, demographic changes and lasting impacts on fish stock as well as the communities that depend on these resources (Kuenzer 2013). These localized changes are situated within broader aspects of Mekong Basin development discourse and constructed regionalism. In the middle of this is the Tonle Sap Lake with its history of multiple interventions, agencies and institutions and a current research focus for a post-hydro developed or "tamed" Mekong. All will interact and contextualize the emerging outcomes from these *deep fishery* reforms.

The History of Tonle Sap Lake-Fishing Lots: Origins, Current Use and Management

Fish and fish products have long been main diets for Cambodians. Similarly, rent extractions from fishing activities have been a good source for state revenues, yet not much is known about the formal management before the nineteenth century. Europeans who came to Cambodia from the nineteenth century onwards did write about the abundance of fish and other aquatic resources. A more formal fisheries management regime came during the reign of King Norodom (1863–1904), who began to lease fishing rights in some areas for exclusive private exploitation (Hortle et al. 2004). Under his and his predecessors' reigns, the king was entitled to collect arbitrary taxes on fishing throughout the kingdom. Commercial fishing was conducted by Sino-Khmer traders/ investors who purchased the use rights of a certain fishing ground and paid the dues to the Royal Treasury. These people then leased the ground to other people and made handsome profits. Sub-leasing occurred between five or six or even more go-betweens (Degen and Nao 1998). Writing on commercial fishing exploitation in Battambang during the reign of the lord governors, Tauch Chhuong (1994) shows that parts of the Tonle Sap Lake were likewise rented to business people who too could make 'an enormous profit'.

This privatized commercial exploitation was further consolidated and formalized by the French colonial administration (1863–1953). When the country was a French protectorate, the king had his officials collect payments from the fishing grounds and subsequently paid a portion of the revenues to the French administration. Yet from 1889, seeing the potential contribution of the commercial exploitation to the economy and to the operation of its administration and asserting more direct control over the exploitation, the colonial administration collected payments directly and introduced the formal fishing lot system including the lot auction process in 1908. A number of relevant rules and regulations were thus enacted by the administration thereafter to consolidate the new exploitation system (Degen and Nao 1998; FACT and EJF 2001), and the colonial regime managed to extract substantial revenues from this sector.

The exclusive nature of the lot system was not without conflicts amongst the stakeholders. Conflicts between lot concessionaires, villagers and/or fisheries officers are not confined to post-communist Cambodia. During the *Sangkum* and the Khmer Republic, there were frequent reports of conflicts between the concessionaires and smaller scale fishers, even resulting in murders of lot concessionaires or managers in the 1960s. Such and more serious conflicts after then led Lon Nol to abandon the system in 1973 (Kurien et al. 2007). Tax evasion and poor rent extractions were two grave issues (Bardach 1959). Later on during Democratic Kampuchea (1970–1975), commercial exploitation was formally abolished, and small-scale fishing was neglected. Nonetheless, some former lot areas were operated on a large scale by cadres who formed special fishing units.

The transformation of inland freshwater fisheries management was significant during the People's Republic of Kampuchea (PRK) (1979–1989) and State of Cambodia (1989–1993). Like land, inland fisheries exploitation was collectively managed. While not much management existed between 1979 and 1981, from 1981 to 1987 PRK-organized inland freshwater fisheries exploitation was carried out mostly by *krom samaki* (solidarity groups), who paid the use rights fees to the PRK in the form of dried or salted fish, and by the various PRK departments and army units in the pre-existing fishing lots. Pondering commercial fisheries exploitation as a major source of revenue and facing the withdrawal of foreign financial support, the impoverished PRK abolished the then de facto privatized *krom samaki* and re-introduced the French-styled fishing lot system (cf. Swift 1997; Degen and Nao 1998; Chheng 2000). With this re-introduction of the fishing lot system, initially more than 2 million hectares of water was allocated to private businesses for exclusive industrial exploitation via public bidding; the figure fell gradually and stood at a little over a million hectares in 2000 (see DOF 1989, n.d.). The 1987 Fisheries Law and the current version of the law identified where fishing could take place and where not; it divided inland freshwater fisheries into large scale/industrial fishing, middle scale/licensed fishing and small scale/family fishing, each with its own institutions. The law, sub-decree on exploitation and a burden book (management and catch records) for each lot were the basis to determine the legality of gear and fishing activities within a lot. It is the reliance on a dogmatic and uniformed application of the Fishery Law, with no effective democratic mechanism for legal review, across all fishing social-ecological contexts that have also contributed to poor governability.

The Socio-political Context of the Deep Fisheries Reforms of 2010–2011

The 2012 'deep fisheries reforms' were part of the movement to increase use of neo-liberal economic development mechanisms in Cambodia. When Prime Minister (PM) Hun Sen ordered the latest fisheries reforms in 2011, the poverty level in Cambodia had dropped to 19 % from 39 % almost two decades earlier. The reform was probably intended to help Cambodia reach its adjusted Millennium Development Goals in 2015 which included poverty reduction and environmental sustainability (for a review of the progress and challenges, see Ministry of Planning 2007).

This fisheries reform was part of reform that had been taking place since the mid-1990s when several local and international NGOs first experimented with community fisheries management in northeast Cambodia. Yet, it was not until Hun Sen released 56 % of the fishing lots to small-scale fishers in 2001 that many donors began a more concerted effort to pressure the government to achieve better fisheries governance. Apparently, the Asian Development Bank (ADB) conditioned its financial support to the fisheries sector after the 2001 reform to fishing lot reforms; nonetheless, it was only in 2012 that Hun Sen ordered the abolition of the entire system of lake/stream fishing lots. Two years earlier, Hun Sen personally recommended, when reviewing the new fishing lot exploitation and development sub-decree, that lot sub-leasing should be terminated. Perhaps, it was the flaunting of this provision by lot concessionaires that infuriated the prime minister, who ordered his deputy to investigate fishing activities within fishing lots in Tonle Sap, resulting in the reforms. As discussed later, this reform can be analyzed in terms of wrangling between concerned state agencies as well (Sok 2012) (Fig. 28.2).

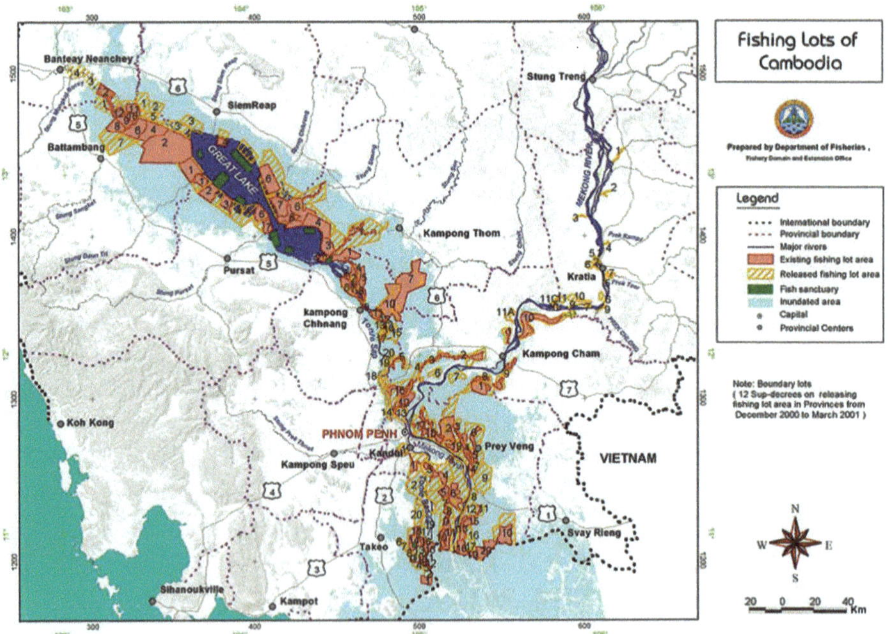

Fig. 28.2 Fishing lots of Tonle Sap and released areas (FiA)

The Current Social-Ecological Context of the Tonle Sap

Stakeholders and Actors

There are no longer any more elite controlled fishing lots in Cambodia (except for the bag net river fisheries) after more than 100 years of having them. But aquatic resources are still under the direct authority of the Fisheries Administration of the Ministry of Agriculture, Forestry and Fisheries, and are governed via decentralized and de-concentrated weak co-management arrangements called Community Fisheries (CFi). These are locally elected management bodies responsible for executing a state approved fisheries management plan over a specific fishing area. There are currently 516 CFis (360 registered by 2013) with 228 in the Tonle Sap floodplain (Fig. 28.3.)

The Tonle Sap Lake is an "old wallpaper" zone of multiple overlapping and conflicting actors (individuals, groups) and institutions (tacit and codified). It's a messy *bricolage* of shadowy and complicated actors and relations. Players include both state and non-state actors (Pech and Sunada 2006) as depicted in Fig. 28.4. It should be clearly noted that Fig. 28.4 does not capture the plethora of "black box" flows, or the many informal or "unofficial" tacit relations that exist in and between management actors and actions for small-scale fisheries management.

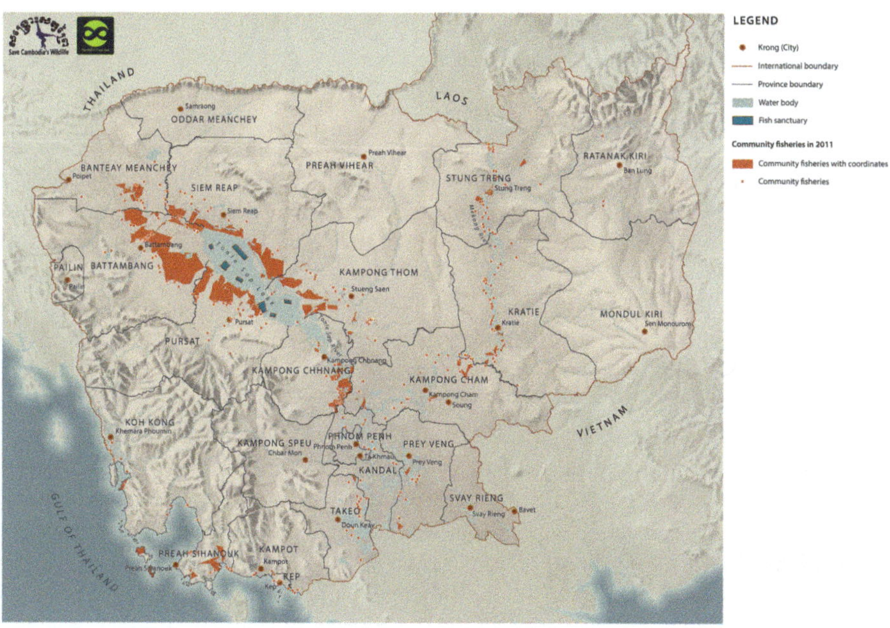

Fig. 28.3 Community fisheries in Cambodia

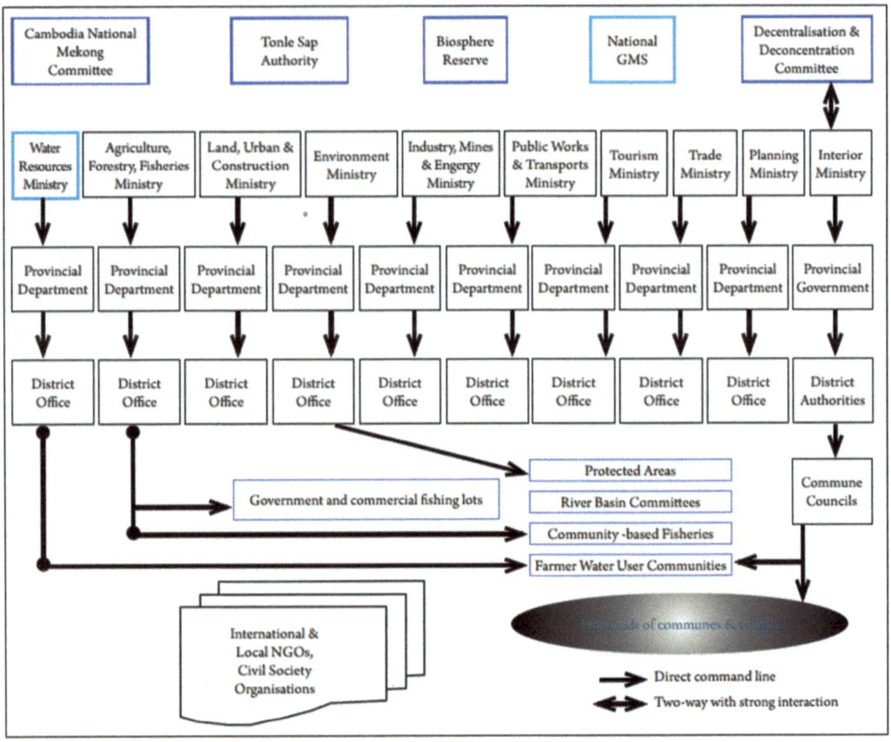

Source: Pech 2010

Fig. 28.4 Organizational chart of formal Tonle Sap actors

These actors include, domestic and international NGOs, Environmental NGOs ranging from UNESCO's Biosphere Reserve Network and UNEP/UNDP to the ever present Asian Development Bank to Cambodian NGOs and the civil advocacy group the Coalition of Cambodian Fishers. These cross scale/level players also interact with the broader regional level MRC and their formal processes for monitoring and dissemination information on fisheries and hydrology issues as well as shifts in global development paradigms to new conservation approaches and a focus on climate change (Orr et al. 2012; Piman et al. 2013). This new reality for aquatic resources in Cambodia includes a more "pro conservation" approach as seen through increased USAID and WorldFish/Royal Government of Cambodia's support of Community Fish Refuges (CFR) (Joffre et al. 2012). The resulting expansion focuses on simply increasing the numbers (extent) of CFis to manage these complex small-scale fisheries. Instead more transformative changes are required for *deepening* devolved powers. This will include mechanisms for the effective transfer of resources to the CFis and to create space for innovative community-level revenue generation.

Why These Deep Reforms Now?

Many reasons for the timing and extent of these final cancellations of all lots or "deep" reforms have been given by people who have a deeper understanding of the nuanced dynamics of Royal Government of Cambodia-Cambodian People's Party politics.[1] The first and foremost reason that Prime Minister Hun Sen cancelled the lots is because he had the power to do so. As Southeast Asia's longest surviving strongman, in power for 29 years, he could do what he wanted. Added to this it was a Commune Council election year and the Cambodian People's Party wanted to strengthen its rural power base. With the rise of local conflict and complaints over lot leasees' violence, as well as little real revenue from the lots entering Royal Government of Cambodia coffers, it was politically very advantageous to finally get rid of them. Other real nuances were unfolding. Lot leasees were using 'dewatering' techniques and other highly destructive fishing gear within "their" lots. This not only caused an uproar with the local fishermen and the Coalition of Cambodian Fishers but also with foreign ENGOs and ultimately with the "new" fisheries donor, the EU. Fishing lots were defended by arguing that they were a vital forest, fish and wildlife habitats. That was being (falsely?) discredited by ENGOs whose agenda had always been to remove lots from the vicinity of biosphere type conservation efforts.[1] The PM obliged. Was this a tactic to leverage more donor funds for the now more heavily burdened FiA and its CFis? FiA now has an expanded fish conservation mandate which often comes into direct conflict with the conservation and environment mandate of the Ministry of Environment, responsible for flooded forests and protected areas. Small-scale fisheries production is directly linked with maintaining the ecological integrity of the flooded forest and its floodplain (Koponen et al. 2010).

There are complex jurisdictional control negotiations underway, moderated and influenced by the Tonle Sap Authority (TSA), which is actually housed within the powerful Ministry of Water Resources and Meteorology, which also influences hydro-irrigation development. There is a history of conflict between FiA and the Ministry of the Environment for management jurisdiction and thus control over land and water resources in aquatic vs. terrestrial conservation zones. This conflict ultimately decides who has power and responsibility for what species and habitats in protected area zoning. But now these reforms are forcing both to come together over this rejuvenated mandate for more conservation of fish and flooded forests. Those driving the Cambodian development agenda realize that with the Mekong River Commission (MRC) seemingly unable to curtail questionable hydro development in

[1] This following section is based on many personal communications the first author has had over the past 3 years with diverse sets of actors working in government, research and civil society organizations, media and the NGO-Donor community on the causes and implications of the 2011–2012 process initiated by PM Hun Sen to cancel of the remaining Fishing Lots in Cambodia under the rubric of "Deep Fisheries Reforms". They were often tacit, informal, emotionally speculative and always anonymous sources.

any of the member nations, powerful figures with voices in the Council of Ministers and other Royal Government of Cambodia decision making bodies now privately *assume* that the Tonle Sap fishery is "doomed" as a real source of commercial scale revenue. So the always pragmatic PM, to avoid further resource conflicts, has let the 'fenced' fish go! (Friend et al. 2009; Friend and Arthur 2012). These fishing lot cancellations maybe interpreted as a massive *social safety net* type subsidy for the people, beyond electioneering, local graft and cronyism, and reciprocal power alliances. The reforms are pure public relations acts in the disguise of a one-time final massive food security subsidy for those 1.3 million riparian and over 3 million Tonle Sap basin residents.

Why now? By boosting conservation and highlighting fish as a key food security issue the lot cancellations with *more fish for fishers* is meant to counter widespread plans for dramatic increases in Chinese-financed hydropower and oil development (Kubiszewski et al. 2013). Energy is primarily destined for peri-urban light manufacturing and garments as Cambodia moves toward ASEAN economic integration in 2015. Visions (and benefits) of Singaporean-modernity have captured the minds of the Cambodian elite, and both small-scale fishing and single crop rain-fed rice agriculture, the food and cultural foundations of Khmer society, are now seen as symbols of antiquity and backwardness that need to be replaced or re-tooled (Yu and Diao 2010). There is now active state promotion of Green Revolution agriculture for rice with irrigation and improved varieties aimed for export. There is a strong re-emergence of a newly reorganized (and funded) Consortium for International Agriculture Research (CGIAR) through its *World Fish* program in promoting more intensive forms of commercially viable aquaculture, such as freshwater prawn (*Macrobrachium rosenbergii*) and Clarias catfish (*Clarias batrachus*) (Brooks 2010; Van Brakel and Ross 2011; Joffre 2012). Yet the Royal Government narrative is that the FiA will continue through its CFDO (Community Fisheries Development Office) to improve the capacity and effectiveness of the CFis to better manage these diverse fisheries for the benefit of local small-scale fishers and the rural poor even albeit with very little financial support and within the more dynamic and open access context of social and ecological changes. The fishery system-to-be-governed has become more diffuse and complex with new power, state and non-state actor reorganization to provide input and exert influence over lake management.

Governance and Governability of Tonle Sap Fishery Post-reform

Governability, or "the overall capacity for governance of any social entity or system" (Kooiman in Jentoft 2011; Bavinck et al. 2013), is a dynamic state, i.e. a quality that is not fixed for all. It is sometimes easier or harder depending on a diversity of factors. In our case governability interactions include changes in actor relations, authority being given and taken away for lake management, more pressure and less money for FiA to fuse conservation initiatives (with little or no study) in the form of no-take zones in the Tonle Sap Lake and with little prior consultation with communities. The state re-draws the map and expands existing CFis or facilitates new ones (Cooperman et al. 2012). The increased complexity of the FiA, integrating its CFi system with a raised conservation mandate, risks increased cross ministerial conflicts and turf wars. The reforms reinforce the status quo of top-down centralist fisheries management through a simple numerical increase of CFis but with no additional power or resources. This logistical increase of CFis alone will reduce governability in the short run as there is no provision to engage stakeholders in any serious process of inclusive cross-level societal learning for NRM/Environmental Management (Diduck 2010; Binder et al. 2013).

The roles of actors are dynamic and often shifting with roles including government responsibilities and capacity support in decentralization initiatives shrinking or expanding (Walker et al. 2001). The solution to today's complex resource and collective problems found in social ecological systems such as the Tonle Sap fishery requires multiple partnerships. These are fostered via governance, which is an interactive and collaborative process including government, with civil society in a proactive role, determining who is allowed to make decisions, and under what conditions. Governance, in the broadest sense, is simply the politics of decision-making with regard to societal development. Governance includes processes of policy formulation and policy implementation (or management) via the application of diverse resource management instruments. Governance takes place via institutions that foster action and learning through social-ecological feedback. There is growing social, economic and political diversity and interdependency in both numbers and proximity (close/tight or distant/loose) of relationships. Governance incorporates a temporal element and goes beyond the immediate to incorporate ideas on longer term trends of both the changing resource state and the political ecology of the users. It must address the future needs of society including society's relationship with aquatic resources and external driving factors.

At the meta-level, there has been little sign of the top-down command and control "state" management of the Tonle Sap lake fishery either at the time of the lots or thereafter. Rather, there is a diverse web of cross level relations influencing the translation and implementation of centralist decision making. A clear example of this is with regard to whom is caught-punished and made to pay within a very pervasive and powerful narrative of "illegal fishing". These "fishing" problems and their formal legal responses are often directly supported by Prime Minister Hun Sen himself (Starr 2009). Addressing this now is difficult given the current *de [facto]* open access environment post fishing lots. The historically weak governance context allows local authorities to collect resource rent (license fees) from the once enclosed prime fishing areas and to enforce this access beyond the provincial level (Fisheries Cantonment) mandated level of authority. This has human rights implications for the internally displaced Vietnamese fishers who are often made scapegoats and targets for illegal fishing. The FiA wants to create new CFis, expand existing CFis, create more Community Fish Refuges (CFRs) and build capacity for community level enforcement and monitoring. There are now more forceful narratives being articulated about fish conservation via the wide-scale use of CFRs in all fish habitats including the important rice field fisheries aimed at increasing rural food and nutrition security.

Key Characteristics of Governance Influencing Governability

Connectivity is an important concept and is being used to develop 2-way flows (governing interactions) of biotic and abiotic material, water, energy, human resources and information. These conduits form through the interaction between the governors and the governed. These dynamic governance contexts have direct bearing on the amount or extent (ease) of governability that the system "allows". This allowance determines the capacity of the system to absorb changes and thus its social-ecological system resilience (Pahl-Wostl 2009).

In the Tonle Sap fisheries, governance involves informal and codified networks or "webs" of influence, debt, reciprocity (established patron-client norms), political support and management activities. The identification of specific issues or concerns occurring across scales and levels will depend on whom you ask and when. Such scenarios create *wicked problems* and the need for stakeholders to develop skills and confidence to muddle through while coping-adapting to unexpected problems and opportunities emergent in the complex system (Rittel and Webber 1973; Batie 2008).

Governance *modes* (self, hierarchical and co-governance) can be seen as (partially) emergent systems or states resulting from complex interactions. They are also on a sliding continuum, never really fixed in one mode with more or less purposeful design and agency. Combinations or blending of modes is more likely. This wallpapered context with its leftover legacies of old layers of institutional, financial and organizational interventions in Tonle Sap makes categorization extremely fuzzy. These dynamics are based on who currently distributes power and money. The relationships between institutional structure-function and diverse actor sets are highly variable making local governability of fisheries difficult but at the same time resistant to real reforms beyond Khmer patron-client bonds. These historical legacies even include Khmer Rouge influence at work in governance webs especially over the control of land tenure (Diepart and Dupuis 2014).

The key concerns are how the above governability factors influence stakeholder interactions, power and decision-making in Tonle Sap fishing sectors and what difference (if any) these reforms of fishing lot cancellations, expansion of CFi numbers and extent, and more Tonle Sap Fish Conservation Refuges have made? How deep are these reforms? Deep if you see the fishing lots as the root cause of the conflicts and tensions and their abolition as the solution. These "reforms" are very significant to fishers and their community leaders who see in them more opportunity to catch fish but not really to confront the power imbalances at the root of poverty and biodiversity loss. The roots of poverty and biodiversity loss are even deeper and lie with how people see their relationship to these resources and the Khmer State (via processes of subjection and territoriality) which influence the eventual values people end up placing on their resources for use and any subsequent conservation. In discussion with Mekong fisheries managers what emerged was their narrative suggests that fishing lots were actually "babies being thrown out with the bath water." These lots were managed sustainably for over 100 years. They were potentially very lucrative sources of revenue and could have been managed a lot better for multiple purposes. They could have formed the basis of a community-based concession system as seen in coastal benthic fisheries of Mexico and Chile (McCay et al. 2014). Many of these lots contain important forest/waterfowl and aquatic biodiversity habitats that are now (for the moment) open access. In other words, reforms at one level have been "too deep and too fast", but on another level inadequate. The governance context is still, therefore, business as usual with the top-down, hierarchical FiA left (mostly) in charge. Most of the investment (lot concessionaires) has not been reinvested in the fisheries but has moved to and other business ventures at sites around the lake. With more power going to Ministry of Water Resources and Meteorology and the Tonle Sap Authority (TSA) for lake decision making at the expense of Ministry of Agriculture, Forestry and Fisheries there are now *more* actors and issues complicating governance and the rules as to who can extract from these zones.

Currently there is a wide gap between the intentions of the FiA (and others) for "sustainable" fisheries management regimes and the actual practice and failure of Mekong fisheries governance. This gap leads to environmental and social degradation, conflicts, loss of ecosystem services and disconnect between humanity and nature (Rammel et al. 2007). How effectively societies adapt to change will be determined by their resilience. In Tonle Sap fisheries, resource management knowledge is patchy, coveted and difficult to manipulate into coherent cross level (departmental) policies. In Cambodia, there is a "cult of the expert" or a top-down, technocratic reliance on evidence-based scientific knowledge. This is a legacy of Vietnamese socialism and the Soviet educational and bureaucratic systems. The question that needs to be asked is whose knowledge is really allowed to count when it comes to making policy decisions in the Tonle Sap fishery and linked resource systems.

Communities' Perspective on the Deep Fisheries Reforms

In a recent situational analysis (Milne 2013) undertaken in three specific community fisheries in Tonle Sap, villagers expressed their general happiness with the abolition of the fishing lots because it resulted in them having better access to and movement within the fishery. People reported noticeable increases in the amount and size of catch albeit not for all species. There were also losers in the abolition of the fishing lots. Some people working in the lots lost their jobs. Others lost the "special fishing rights" they had negotiated with the lot owners. Another systemic problem was the non-transparent multi-level sub-leasing of fishing rights negotiated within and adjacent to fishing lots as some of these lots were then unilaterally converted to no-take fish conservation areas. Community Fisheries also needs to deal with *de facto* open access nature of the resource and the influx of outsiders from other riparian communities. In this context, the ongoing work of the Fisheries Action Coalition Team (FACT) and the International Union for Conservation of Nature (IUCN) aimed at looking at changes in governability in the three sites is vital to unpacking the dynamic roles of diverse actor sets including that of non-state actors. As shown by an earlier study of Tonle Sap CFis, local governance can work under the right conditions (such as autonomy, access to resources and good leadership), and when taking into account the aspirations of communities. The study revealed that participation in setting up and managing Community Fish Conservation Refuges can be more easily fostered if the local fishers and their representatives are brought into the process early and their fears and concerns discussed upfront. Alas, this is rarely done as most community fisheries, wards of FiA, are simply consulted or told by FiA after decisions are made (Ratner and Allison 2012; Oeur et al. 2013). This may be changing as FACT and Coalition of Cambodian Fishers take more proactive involvement in representing CFi concerns.

Conclusions

New business opportunities for community-based fisheries, commune councils and commercial fishing, and value added product development are slowly being considered. The need to have some form of local process for innovative revenue generation at the CFi level is also being recognized by the state. The FiA is trying to promote more commercial fishing in some selected CFis based on available infrastructure, market proximity and compliance. There is real fear in FiA of local elite capture of commercial fishing rents. The Government does not have the funds, capacity and/or presumably the willingness to establish a flexible centralized form of fisheries co-management with effective local enforcement. How can decentralized community-based fisheries management be established in a culture of entrenched top-down authority with no history of democratic process? This will be very difficult. Nonetheless, Tonle Sap is a heterogeneous region and some communes and CFis are more progressive, or have been involved in more cooperative ventures and so have experience or capacity for creating governance space. They have open minded, less selfish and more progressive leaders and have learned from previous participation. Fisheries managers who are interested in developing effective adaptive co-management regimes need to find and work with these more progressive individuals and groups, combining diverse knowledge sets and perhaps developing some critical mass for local collective action vis-à-vis resource rights advocacy and self-organized management. But this requires a commitment to diverse learning approaches including critical reflection which can be difficult in a culture of non-confrontational face saving behavior, where patron-mediated reciprocity is a key social organizing institution.

As mentioned earlier there is an over reliance by the government on the Fishery Law. It is used irrespective of different local situations and has to be followed by everyone in all circumstances. Its implementation is too rigid hence resulting in a lack of compliance across all aquatic contexts. The Fishery Law needs to be reformed to build in governance entry points (opportunities) for much more flexible and effective community level participation for direct and continual adaptation of CFis.

In addition to the poor legal fit of the Fishery Law there is a lack of resources and/or willingness to develop effective institutions for building co-management among lake stakeholders. This is also hampered by local spheres of corruption and patronage.

Historically, there has been a fundamental disconnect between the multiple agencies, state and non-state actors, working on the management of global aquatic resources in general and small-scale fishing in particular. If governability is the capacity for governance then there is an air of real uncertainty as to what this capacity entails and who truly has the right or legitimacy to wield this capacity for decision making. Is the Cambodian state really willing to be a steward of this national and global resource? It is, therefore, not necessarily a question of fisheries management capacity but rather a question of a state that seems willing to pay attention to local voice and resource rights with inclusive co-management arrangements. Can the Royal Government work with multiple actors to implement sustainable fisheries management systems that can function across multiple geo-political scales in these times of rapid and profound environmental change? Do the Tonle Sap fisheries require management at regional or smaller sub-units? The increased heterogeneity of commercial interests in the lake and riparian zones and the diffuse sources of power often inhibit the formation of effective small-scale fisheries co-management regimes. Moreover, whatever role communities might play, they need to be fully cognizant of the wider changes and threats to endangered cultures, species and spaces. There also needs to be learning processes which help legitimate stakeholders and make them feel safe to re-evaluate and fundamentally change their livelihoods in conjunction with the lake's cycles. This includes recognizing both the intrinsic and commercial value of the world's largest freshwater fishery as key to any future development discourse for Cambodia. Governing this complex social-ecological system will require pluralistic management approaches which explicitly account for and build upon the complexity, diversity, scale and dynamics found in the Tonle Sap Lake. It also needs to allow for resource user voices and includes fairness, equity and democratic participation of all key actors in lake conservation management or development. The governability of small-scale fisheries is in an emerging state and requires inclusive foresight that corrects the messy institutional interactions and outcomes currently part of the governance context. To paraphrase an old fishing idiom, there are too many would-be managers chasing too many widely dispersed and claimed fish!! The governability of the Tonle Sap Lake small-scale fisheries will require explicit recognition of the social cultural drivers and boundaries that serve to limit the management context.

All Tonle Sap actors, especially the international conservation NGOs and those donors supporting the Royal Government of Cambodia, need to acknowledge the fact that there are limited possibilities for resource control over natural systems. They also need to acknowledge the multiplicity of covert vested interests at play in the Tonle Sap basin (possible floodplain gas deposits, expanded irrigation for rice exports). Political will and decision making rests at the top layers of power in the Cambodian State, namely those who dictate policy, i.e. primarily the Prime Minister and the Council of Ministers. But power also rests in their translation within the embedded relations of power and reciprocity at the local levels. Power, to some extent, also rests with those local authorities who have the ability to engage in rent seeking behavior, selective law enforcement and solidify old patron-client relationships.

There are also forces within the Tonle Sap basin and its governing system as well as forces originating from outside the Lower Mekong River basin itself that impact governance design and trajectories of lake management. The Mekong-Tonle Sap is not immune to the forces of globalization and the commodification of nature in the global fish trade. These forces combine or conflict to shape fisheries value chains and local "black boxes" of credit and fish marketing across the Mekong Basin (Bush 2004). These multiple, diverse and cross scale drivers of social-ecological change contribute to determining the effectiveness of fisheries governance and thus the overall effectiveness of governability of Tonle Sap fisheries.

All actors with a stake in the development of the lake basin and its resources must realize that the misuse of this power will have direct and profound impacts on the people and biodiversity of these systems. For effective and just governability of this complex fishery, fisheries governors must foster inclusive and interactive mechanisms for key stakeholders and actors who actually make resource decisions. Governance processes must acknowledge the different personal and organizational agendas at play. Governance must explicitly take into account the different ways livelihoods and assets and communication and culture interact to create resource systems (fishery system-to-be-governed) so that historically hierarchical, fragmented and unfair systems that marginalize those who depend most on resources can become more inclusive.

References

Alcamo, J. M., Vörösmarty, C. J., Naiman, R. J., Lettenmaier, D. P., & Pahl-Wostl, C. (2008). A grand challenge for freshwater research: Understanding the global water system. *Environmental Research Letters, 3*(1), 010202.

Arthur, R. I., & Friend, R. M. (2011). Inland capture fisheries in the Mekong and their place and potential within food-led regional development. *Global Environmental Change, 21*, 219–226.

Baran, E., Van Zalinge, N., & Ngor Peng Bun. (2001). Floods, floodplains and fish production in the Mekong Basin: Present and past trends. In A. Ahyaudin, M. R. Salmah, M. Mashhor, R. Nakamura, S. Ramakrishna, & T. Mundkur (Eds.), *Proceedings of the second Asian wetlands symposium* (pp. 920–932). Penang: Penerbit Universiti Sains Malaysia.

Bardach, J. (1959). *Report on fisheries in Cambodia*. Phnom Penh: USOM.

Batie, S. S. (2008). Wicked problems and applied economics. *American Journal of Agricultural Economics, 90*(5), 1176–1191.

Bavinck, J. M., Chuenpagdee, R., Jentoft, S., & Kooiman, J. (Eds.). (2013). *Governability of fisheries and aquaculture* (MARE Series, Vol. 7). Dordrecht: Springer.

Binder, C. R., Hinkel, J., Bots, P. W. G., & Pahl-Wostl, C. (2013). Comparison of frameworks for analyzing social-ecological systems. *Ecology and Society, 18*(4), 26. doi:http://dx.doi.org/10.5751/ES-05551-180426

Brooks, A. (2010). *Fish supply and demand scenarios in the lower Mekong basin*. ACIAR Worldfish. Retrieved from http://aciar.gov.au/project/fis/2010/031

Bush, S. (2004). Scales and sales: Changing social and spatial fish trading networks in the Siphandone fishery, Lao PDR. *Singapore Journal of Tropical Geography, 25*(1), 32–50.

Campbell, I. C., Poole, C., Giesen, W., & Valbo-Jorgensen, J. (2006). Species diversity and ecology of Tonle Sap Great Lake, Cambodia. *Aquatic Sciences, 68*(3), 355–373.
Carbonnel, J. P., & Guiscafré, J. (1965). *Grand Lac du Cambodge. Sédimentologieet Hydrologie 1962–1963*. Paris: Muséum National d'Histoire Naturelle.
Chheng, V. (2000, January 27–28). Evaluation of ex-fishing Lot No. 5 after its abolishment in 1988, Siem Reap Province, Cambodia. In N. van Zalinge, T. Nao, & S. Lieng (Eds.), *Management aspects of Cambodia's freshwater capture fisheries*: Eleven presentations given at the annual meeting of the Department of Fisheries of the Ministry of Agriculture, Forestry and Fisheries. Phnom Penh: MRC/DoF.
Cooperman, M. S., So, N., Arias, M., Cochrane, T. A., Elliott, V., Hand, T., Hannah, L., Holtgrieve, G. W., Kaufman, L., Koning, A. A., Koponen, J., Kum, V., McCann, K. S., McIntyre, P. B., Min, B., Ou, C., Rooney, N., Rose, K. A., Sabo, J. L., & Winemiller, K. O. (2012). A watershed moment for the Mekong: Newly announced community use and conservation areas for the Tonle Sap Lake may boost sustainability of the world's largest inland fishery. *Cambodian Journal of Natural History, 2012*, 101–106.
Degen, P., & Nao, T. (1998). *Inland fishery management in Cambodia: Is the fishing lot system the basis for improved management or should it be abolished?* Phnom Penh: MRC/DoF/Danida.
Diduck, A. (2010). The learning dimension of adaptive capacity. In D. Armitage & R. Plummer (Eds.), *Adaptive capacity and environmental governance* (Springer series on environmental management, pp. 199–221). Berlin/Heidelberg: Springer.
Diepart, J.-C., & Dupuis, D. (2014). The peasants in turmoil: Khmer Rouge, state formation and the Control of land in northwest Cambodia. *Journal of Peasant Studies, 41*(4), 445–468.
DOF. (1989). *Report on Lot Size Calculation*. Phnom Penh.
DOF. (n.d.). *Table of lots and sanctuaries 1980–2008*. Phnom Penh: DoF.
FACT, & EJF. (2001). *Feast or famine? Solutions to Cambodia's fisheries conflicts*. Phnom Penh: FACT.
Friend, R., & Arthur, R. I. (2012). Overplaying overfishing: A Cautionary tale from the Mekong. *Society and Natural Resources, 25*(3), 285–301.
Friend, R., Arthur, R. I., & Keskinen, M. (2009). Songs of the doomed: The continuing neglect of capture fisheries in hydropower development in the Mekong. In F. Molle, T. Foran, & M. Käkönen (Eds.), *Contested waterscapes in the Mekong region: Hydropower, livelihoods and governance* (pp. 23–54). London: Earthscan.
Hartvigsen, G., Kinzig, A., & Petersen, G. (1998). Use and analysis of complex adaptive systems in ecosystem science: Overview of special section. *Ecosystems, 1*, 427–430.
Hortle, K. G., Lieng, S., & Valbo-Jorgensen, J. (2004). *An introduction to Cambodia's inland fisheries*. Phnom Penh: Mekong River Commission.
Imamaura, M., & Lebel, L. (2006). *Geographical shadows and flows*. Chaing Mai: USER Chaing Mai University.
Jentoft, S. (2011, September 8–11). Legal pluralism and the governability of fisheries and coastal systems. In *Jubilee conference Jubilee Congress of the Commission on Legal Pluralism*, Cape Town.
Joffre, O. (2012, July). *Aquaculture production in Cambodia a 2102 update*. Phnom Penh: Assessing economic and welfare values of fish in the Lower Mekong Basin-Worldfish ACIAR.
Joffre, O., Kosal, M., Kura, Y., Sereywath, P., & Thuok, N. (2012). *Community fish refuges in Cambodia – Lessons learned*. Phnom Penh: WorldFish Center.
Kooiman, J., Bavinck, M., Jenoft, S., & Pullin, R. (Eds.). (2005). *Fish for life: Interactive governance for fisheries* (MARE Publication Series 3). Amsterdam: Amsterdam University Press.
Kooiman, J., Bavinck, M., Chuenpagdee, R., Mahon, R., & Pullin, R. S. V. (2008). Interactive governance and governability: An introduction. *The Journal of Transdisciplinary Environmental Studies, 7*(1), 1–11.
Koponen, J., Lamberts, D., Sarkkula, J., Inkala, A., Junk, W., Hall, A., & Kshatriya, M. (2010). *Primary and fish production report* (105 p). Mekong River Commission. IKMP Detailed Modelling Support Project. Vientiane Laos.

Kubiszewski, I., Costanza, R., Paquet, P., & Halimi, S. (2013). Hydropower development in the lower Mekong basin: Alternative approaches to deal with uncertainty. *Regional Environmental Change, 13*(1), 3–15.

Kuenzer, C. (2013). Threatening Tonle Sap. *Political Geographies, 40*, 29–31.

Kurien, J., So, N., & Mao, S. O. (2007, May 3). *Fishing lots in Cambodia's fisheries*. Samudra. Phnom Penh: ICSF.

Lamberts, D. (2006). The Tonle Sap lake as a productive ecosystem. *Water Resources Development, 22*(3), 481–495. Bangkok, Thailand

Lim, P., Lek, S., Touch, S. T., Mao, S. O., & Chhouk, B. (1999). Diversity and spatial distribution of freshwater fish in Great Lake and Tonle Sap River (Cambodia, Southeast Asia). *Aquatic Living Resources, 12*(06), 379–386.

Liu, S., Lu, P., Liu, D., & Jin, P. (2007). Pinpointing source of Mekong and measuring its length through analysis of satellite imagery and field investigations. *Geo-spatial Information Science, 10*, 51–56.

McCay, B. J., Micheli, F., Ponce-Díaz, G., Murray, G., Shester, G., Ramirez-Sanchez, S., & Weisman, W. (2014). Cooperatives, concessions, and co-management on the Pacific coast of Mexico. *Marine Policy, 44*, 49–59.

McLuhan, M. (1964). Understanding media: The extension of man. In R. E. Ulanowicz (2005), *Ecological network analysis in aquatic food webs* (pp. 201). New York: McGraw-Hill.

Mekong River Commission (MRC). (2005). *Overview of the hydrology of the Mekong River*. Vientiane: MRc.

Milne, S. (2013). *Situation analysis at three project sites on the Tonle Sap Lake*. Cambodia: IUCN4.

Ministry of Planning. (2007). *Progress in achieving Cambodia's millennium development goals: Challenges and opportunities*. Phnom Penh: Ministry of Planning.

NGO Forum on Cambodia. (2001, July). *Victims of fisheries reforms: Punishing poor fishers in Kampong Thom – Whose fault is it?* Phnom Penh: NGO Forum on Cambodia.

Oeur, I., Kosal, M., Sour, K., & Ratner, B. (In draft 2013). *After the reforms: Strengthening governance of aquatic resources in Cambodia's Tonle Sap Lake*. Washington, DC: STARGO IFPRI.

Orr, S., Pittock, J., Chapagain, A., & Dumaresq, D. (2012). Dams on the Mekong River: Lost fish protein and the implications for land and water resources. *Global Environmental Change, 22*(4), 925–932.

Pahl-Wostl, C. (2009). A conceptual framework for analysing adaptive capacity and multi-level learning processes in resource governance regimes. *Global Environmental Change, 19*, 354–365.

Pech, S., & Sunada, K. (2006). The governance of the Tonle Sap Lake, Cambodia: Integration of local, national and international levels. *Water Resources Development, 22*(3), 399–416.

Piman, T., Lennaerts, T., & Southalack, P. (2013). Assessment of hydrological changes in the lower Mekong Basin from Basin-Wide development scenarios. *Hydrological Processes, 27*(15), 2115–2125.

Rainboth, W. J. (1996). *FAO species identification field guide for fishery purposes, fishes of the Cambodian Mekong*. Rome: FAO.

Rammel, C., Stagl, S., & Wilfing, H. (2007). Managing complex adaptive systems—A co-evolutionary perspective on natural resource management. *Ecological Economics, 63*(1), 9–21.

Ratner, B., & Allison, E. H. (2012). Wealth, rights, and resilience: An agenda for governance reform in small-scale fisheries. *Development Policy Review, 30*(4), 371–398.

Rittel, H. W. J., & Webber, M. M. (1973). Dilemmas in a general theory of planning. *Policy Sciences, 4*, 155–169.

Sneddon, C., & Fox, C. (2006). Rethinking transboundary waters: A critical hydropolitics of the Mekong basin. *Political Geography, 25*, 181–202.

Sneddon, C., & Fox, C. (2007). Power, development, and institutional change: Participatory governance in the lower Mekong Basin. *World Development, 35*(12), 2161–2181.

Sok, S. (2012). *State building in Cambodia*. Unpublished doctoral thesis, Deakin University.

Starr, P. (2009). Hun Sen appeals for local support in cracking down on illegal fishing. *Catch & Culture, 15*(2), 30–32.

Swift, P. (1997). *Developing a research framework for the fishing lot system in Cambodia: Two preliminary case studies on fishing lots in Takeo and Kompong Chhnang provinces*. Phnom Penh: MRC/DoF/Danida.

Tauch, C. (1994). *Battambang during the time of the Lord Governor*. Phnom Penh: Cedoreck.

Van Brakel, M. L., & Ross, L. (2011). Aquaculture development and scenarios of change in fish trade and market access for the poor in Cambodia. *Aquaculture Research, 42*(7), 931–942.

Vörösmarty, C. J., McIntyre, P. B., Gessner, M. O., Dudgeon, D., Prusevich, A., Green, P., & Glidden, S. (2010). Global threats to human water security and river biodiversity. *Nature, 467*, 555–561.

Vörösmarty, C. J., Pahl-Wostl, C., Bunn, S. E., & Lawford, R. (2013). Global water, the anthropocene and the transformation of a science. *Current Opinion in Environmental Sustainability, 5*(6), 539–550.

Walker, W. E., Rahman, S. A., & Cave, J. (2001). Adaptive policies, policy analysis and policy making. *European Journal of Operational Research, 128*, 282–289.

Yu, B., & Diao, X. (2010). *Cambodia's agricultural strategy: Future development options for the rice sector*. A policy discussion paper prepared for the Cambodia Food Security and Agricultural Policy Roundtable Stocktaking Meeting, November 4, 2010 USAID CDRI CARD.

Chapter 29
The Co-governance of Fisheries in Post-conflict Sierra Leone: Is the Transition for Better or for Worse?

Ahmed Khan and Sheku Sei

Abstract In this contribution, we examine whether current governance mechanisms for sustaining the fisheries resources are better now than they were prior to the civil conflict of the 1990s in Sierra Leone, and if they are not, what policy instruments could contribute to improving governance. The establishment of co-managed systems during the post-conflict period as a conduit for introducing territorial user rights and marine protected areas constitutes an important step towards stewardship and stakeholder involvement in decision-making. However, the process has been criticized on the basis that it was rushed, thereby jeopardizing program implementation at the local level. Using the governability concept and fish chain as analytical tools, an assessment of the transition period from top down to co-management is undertaken to understand the overall quality of governance. There is evidence that the reforms are essential in promoting participatory governance and attaining multiple co-benefits in conservation and development. Yet, the institutional capacity at the local level is inadequate for effective compliance and monitoring. As a result, there is a need to strengthen the governing capacity and build linkages between fisheries and other economic planning activities where capacity is concentrated. Such efforts and transitional changes are relevant for achieving collective action especially in fragile states that are experiencing the increasing impacts of global environmental and economic changes.

Keywords Fisheries governance • Co-management • Governability • Seafood • Fish chains • TURFs • MPAs • Sierra Leone

A. Khan (✉)
UNEP-IEMP, Beijing, China

School of Business & School of Environment, Saint Mary's University, Halifax, Canada
e-mail: ahmed.khan@unep-iemp.org

S. Sei
Statistics and Research Unit, Ministry of Fisheries and Marine Resources, Freetown, Sierra Leone
e-mail: seisheku@yahoo.com

S. Jentoft, R. Chuenpagdee (eds.), *Interactive Governance for Small-Scale Fisheries*, MARE Publication Series 13, DOI 10.1007/978-3-319-17034-3_29

Introduction

Fisheries contribute directly and indirectly to national economic development in Sierra Leone. This small West African country is found within the Gulf of Guinea in the Atlantic Ocean, bordering Liberia and Guinea (Fig. 29.1). There are two major types of fisheries, the industrial large-scale and the artisanal small-scale fisheries. The industrial fisheries sector is export oriented, and thus contributes to foreign exchange earnings and gross domestic product. Fisheries contributed about 10 % of gross domestic production in 2008 (MFMR 2008), one of the highest in the sub region (Katikiro and Macusi 2012). The artisanal small-scale fisheries on the other hand contributes directly to local seafood consumption, household income and savings, regional trade, and has spill-over effects in other sectors and to rural development. Seafood supplies 75 % of total animal protein and greatly contributes to healthy living and well-being (FAO 2014).

However in recent years the distinction between large-scale and small-scale is becoming unclear as some segments of small-scale fisheries too have become export oriented and contribute directly to the industrial sector. Fisheries also provide employment and livelihoods especially within coastal communities. The total number of people employed in the fishery sector is close to half a million (MFMR 2008). While men are usually engaged in the harvesting sector as fishers, women play a greater role in the post-harvest sectors as financiers of fishing operations as well as retailers (Demby and Leigh 2012; Thorpe et al. 2013).

As noted earlier, fisheries provide invaluable nutrition to the well-being of Sierra Leoneans. Since the outbreak of the Ebola Virus Disease in May 2014 (WHO Ebola Response Team 2014), the importance of fish became increasingly important as other protein sources such as wild game were identified as potential sources of

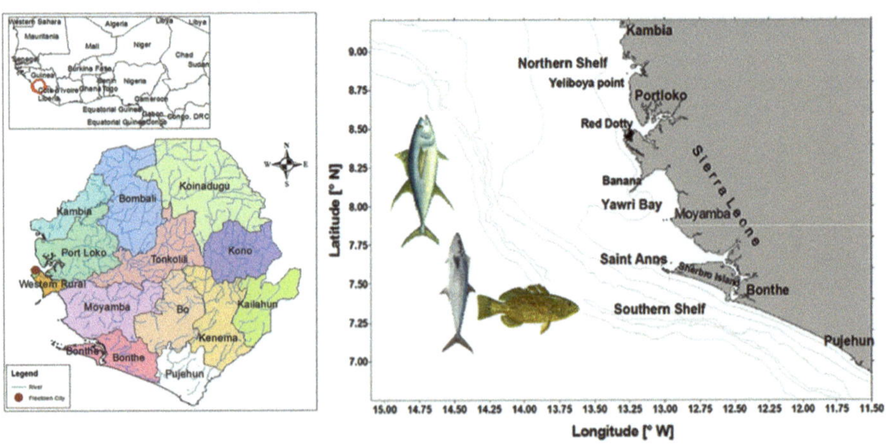

Fig. 29.1 A map of Sierra Leone in West Africa showing coastal districts (L) and small-scale fisheries fishing grounds (R)

contagion. Currently in the absence of preventive cure or vaccines, seafood serves as a major source of nutrients particularly for the sick, as it assists in rebuilding the immune systems, and also bolstering resistance for the healthy (Khan and Sesay 2015).

Moreover, the fisheries sector has generated increasing revenue in recent times. Stock assessment shows that the fisheries biomass is around 300,000 MT (Mehl et al. 2007; Turay et al. 2008) and worth values close to $735 million USD with projected annual returns of $60 million USD (EIF 2013). However, the sustainability of the fisheries has been questioned, with reports and evidence of overfishing and non-compliance to regulations mainly illegal unregulated and unreported fishing activities (Vakily et al. 2012).

From early 1990 to 2002, Sierra Leone experienced civil instability in the form of political coups and armed conflicts that stymied fisheries development (Thorpe et al. 2009). Amongst the many challenges during this time were the use of illegal and destructive fishing gears including dynamite fishing, mosquito nets, and 'channel' nets, with mesh sizes far below the minimum requirement. Fishers took advantage of a monitoring and surveillance vacuum to engage in unsustainable practices. The civil conflict also exacerbated social and ecological problems and made the tasks of the local village development committees and other initiatives such as the artisanal fisheries community development programs fruitless (Thorpe et al. 2009). These local institutions (both formal and informal) were created to integrate rural planning with fisheries development objectives (Khan 1998).

Prior to the conflict in the 1990s, fisheries management was top down and sectoral (Sei et al. 2009a; Kamara 2012). It focused mostly on industrial production with less consideration for the small-scale sectors including coastal fisheries, inland fisheries, and aquaculture (Ndomahina 2002; Seisay and Jalloh 2006). Emphasis was on export earnings through bilateral fisheries agreements and joint ventures with parastatals (i.e., public-private partnership) that targeted mostly demersal and shellfisheries and recently small pelagics for fishmeal (Khan et al. 2006).

New institutional arrangements with Local Councilors and fisher organizations after the civil conflict were meant to provide opportunities for stewardship measures as well to mainstream gender roles as women are important players in seafood marketing (Sheriff et al. 2009; Thorpe et al. 2013). The reforms focused on strengthening local management by involving communities and fisher organizations in co-management (ISFM 2009; Sheriff et al. 2009). There was also a shift towards decentralization with licensing of fishing canoes devolved to Local Councilors pursuant to the implementation of the 2004 Local Government Act. Since 2011, the government has implemented the *New Deal for Engagement in Fragile States*, as part of the Busan Partnership Agreement to ensure aid effectiveness. These measures have boosted the local economy in the last few years with great results as reflected by the sharp improvement in the Human Development Index and local entrepreneurship development. There is considerable potential for foreign trade through the development of integrated policies for regional economic integration and a green growth strategy (AfDB 2013; Kayonde et al. 2013).

The key research question is whether the governance mechanisms in place for sustaining the resources after the civil war (about a decade ago) are more effective than those in place prior to the civil conflict of the 1990s? And if the governance mechanisms are less effective, what policy instruments and institutional arrangements could contribute to better governance? In answering this question, we employ the governability concept to assess the overall capacity and quality of governance during these transitions. We also conceptualize fisheries governance as a "wicked" problem, drawing upon Rittel and Webber's (1973) seminal article on the dilemmas of social planning. This assertion that fisheries governance is a wicked problem stems from the fact that fisheries are complex and dynamic ecosystems that are influenced by both human and natural induced factors that can hardly be managed (Chuenpagdee 2011a). As seafood trade becomes global in scope, external drivers such as climate change and increasing consumer demand exacerbate local fisheries benefits as witnessed in many parts of the world (Kurien 2005; Khan 2012). Moreover, the diversity of stakeholder interests along the fish chain makes decision-making difficult, mostly due to conflicting goals and multiple values and time preferences (Song et al. 2013). The overall aim of this assessment is to provide leverage points that could foster multiple objectives associated with the fisheries reforms.

We first provide a rationale for a governability approach and how useful it could be in understanding governance and institutional capacity for social change. Next, using the fish chain as an analytical framework, we assess changes and drivers within the fishery that could limit or promote overall governance. Finally, we discuss how this approach could improve our understanding of achieving better outcomes in the fishery, and conclude by highlighting the implications for policy development in sustaining the benefits to the small-scale sector.

Why a Governance and Governability Approach in Sierra Leone Fisheries?

Traditionally, fisheries have been managed using technical tools such as input and output control measures that restrict fishing capacity and harvest rates with the goal of attaining maximum sustainable yield (Larkin 1977). These measures have mostly failed for several identified reasons: (i) high transaction costs of monitoring and surveillance, (ii) non-compliance and lack of participation by non-state stakeholders, and (iii) other humanly-induced problems such as disempowerment and corruption (Jentoft et al. 1998; Khan and Neis 2010). As fisheries production and trade becomes global in scope (Pauly et al. 2005; Smith et al. 2010), the management role of the state becomes weak as most production activities go beyond national jurisdictional mandates and rely instead on international norms and actors (Jacquet and Pauly 2008). These developments have prompted a more critical look into the governing capacity of fisheries managers and whether stewardship concerns go beyond what management regimes can handle (Kooiman et al. 2005).

In addition, it is well acknowledged that fisheries cannot be managed using technical tools only (Degnbol et al. 2006), as the problems are "wicked" rather than "tame" (Jentoft and Chuenpagdee 2009). Diverse stakeholder demands, multiple objectives, social dilemmas, and broader cross-sectoral linkages are concerns contributing to governing challenges. These concerns require philosophical debates on equity, legitimacy, values and overall governance (Bundy et al. 2008; Wilhere 2008; Bavinck et al. 2013; Song et al. 2013). A framework that diagnoses and exemplifies wicked attributes in fisheries is the first step in understanding the limits to governance as well as in identifying where interventions can be leveraged to improve the overall governing capacity. This is especially crucial for regions that belong to the bottom billion (Collier 2007), including fragile states and those that have undergone civil instability with limited resources for effective governance (Le Billon 2001; Thorpe et al. 2009; Wai 2012). Governability, and a holistic fish chain perspective, is essential in these circumstances (Kooiman et al. 2005; Bavinck et al. 2013). The approach involves both state and non-state actors in formulating principles and rules that guide governors and non-governors alike, in all stages of fisheries production (Kooiman 2003; Khan and Chuenpagdee 2014).

Theoretically, we rely on the interactive governance approach, which is a three-system model. It consists of the natural bio-geophysical systems, social systems-to-be-governed, governing systems, and their governing interactions. These systems are structurally diverse, complex, dynamic in nature, and with varying scales. The system attributes or properties could constrain the effective governing of fisheries, hence the concept of governability (Kooiman and Chuenpagdee 2005). Governability is the overall quality and capacity for governance, both within the systems-to-be-governed and the governing system and the ability to achieve multiple sustainability goals. These goals include healthy ecosystems, food security, sustainable livelihoods, poverty alleviation, inclusive decision-making, and gender mainstreaming (Kooiman et al. 2005).

The concept of governability can be traced to two major milestones in organizational science and system ecology. The first pertains to the shortfalls and cognitive limitations at the administrative and institutional level in decision making, often referred to as 'bounded rationality' (Simon 1947). The second relates to uncertainties and risk associated with incomplete knowledge about system properties and appropriate feedback responses for adaptive management and precautionary approaches (Walters 1986; FAO 1995). Assessing governability then entails an effort to acquire a deeper understanding of system properties along the fish chain and their interactions that may include risks and externalities. We do so by reviewing and analyzing published materials in primary and secondary literature as well as technical and project reports on fisheries and related development problems. Some necessary field data were collected on current cost and earnings across the seafood value chain. The time scale for the analysis is a decade before the civil conflict (in the 1980s), during the civil conflict (1990 to 2002), and the decade afterwards (from 2002 to 2012). The analyses and assessments proceed from the natural systems, systems-to be-governed, governing system, and governing interactions. We use a four-stage process to assess governability across the fish chain as proposed by Chuenpagdee and Jentoft (2013). This includes an assessment of: (i) degree of wickedness; (ii) prevalence of system properties; (iii) goodness of fit of elements; and (iv) quality of interactions. This approach is informed by a series of questions that shed light on the measures and indicators relevant in assessing governability across the fish chain (Table 29.1).

Table 29.1 Sample questions for assessing governability in the context of Sierra Leonean fisheries

Thematic sample questions	Metrics and indicators on system properties
How diverse, complex and dynamic are the fish stocks and their marine ecosystems?	Biophysical characteristics and geographic location for fishery resources, coastal landscape and features, biomass estimates and stock assessment highlights, recruitment & growth rates, total allowable catch, trophodynamics, critical habitats and hotspots, protected areas, fish behavior, and climatic variability on upwelling and fish abundance, large marine ecosystems, etc.
What are the threats and level of vulnerability to the harvesting and processing sectors and local livelihoods?	Costs and earnings of fishing activities, rate of return on investment, discards and post-harvest loss, illegal unregulated and unreported fishing, foreign vs local fleets, seafood imports and exports, seafood trade and globalization, etc.
What are the various strategies adopted by stakeholders post-civil conflict to improve overall governance?	Information sharing, co-management initiatives, livelihood dependency and seafood consumption, changing income by boat types or target species, community wellbeing, youth and women involvement, empowerment, partnership arrangements, etc.
Does jurisdictional scale match ecological boundaries, socioeconomic activities, and governing institutions?	Integrated management strategies for small-scale and large-scale sectors, spatial scale of management in context of LME and global economic changes, policy networks, illegal unregulated and unreported fishing, gear conflicts, initiatives for monitoring and surveillance, allocation and decision-making, regulatory frameworks, etc.

Insights on System Properties for Small-Scale Fisheries Governance

Knowing that fisheries systems are diverse, complex, dynamic, and span multiple scales; understanding the extent to which these properties are governable merits attention especially in the context of system properties, fit of elements, degree of wickedness, and level of interaction. As the quality and capacity for governance can be constrained in any of the systems, we start by taking stock of and learning about the various system properties and implications for achieving sustainable outcomes.

Natural Systems

Fisheries are part of several ecosystems ranging from marine, coastal and estuarine, and include unique attributes and characteristics across multiple scales. Although emphasis by managers is mostly on marine ecosystems, there is high connectivity within coastal and terrestrial ecosystems through estuaries and wetlands. The Sierra Leonean coastline is about 560 km; with a complex shoreline of low cliffs, rocky headlands, in addition to sandy beaches, mangroves and mud flats that enhance fisheries productivity (Scheffers and Browne 2008). The total shelf area is about 25,000 km^2 and provides habitats for species as well as enriches primary productivity for marine food webs (Ssentongo and Ansa-Emmim 1986). This is due to nutrient flow from three major estuaries (Scarcies, Sherbro and Sierra Leone rivers) and the rich mangrove swamps that support feeding and nursery habitats as well as small-scale fisheries fishing grounds as shown in Fig. 29.1.

Mangroves represent important estuarine ecosystems due to their role in coastal buffering and flood control, nutrient recycling and critical habitats for several marine organisms. Mangroves cover about 156,000 ha providing rich biodiversity benefits as well as other provisioning ecosystem services (Johnson and Johnson 2012). For these reasons, mangroves have been the focus of protected areas research and governance assessment from both ecological and social perspectives (Jentoft et al. 2007; EJF 2011). In Sierra Leone, these coastal resources are highly influenced by the Gulf of Guinea Large Marine Ecosystem (LME) and the interaction between the Canary and Benguela Currents. These current systems create upwelling of cold nutrient rich waters with high biological productivity. This results in a rich distribution of shellfishery resources in the South, close to the Sherbro Islands and off the coast of Bonthe and Pujehun towards Liberia (Showers 2012). In addition, there are abundant inshore fish resources occurring in the major river estuaries of Sierra Leone, including catfish and Tilapia, as well as marine megafauna including sea turtles and endangered manatees (Sei et al. 2009b).

The fisheries resources are very diverse and include mostly small pelagics for local consumption (Herring, Bonga and Sardines), large pelagics for sale to augment the household economy (mackerel, tuna, etc.), shellfisheries targeting export markets (shrimp, oysters, etc.), demersal finfish for both consumption and regional trade (snappers, sea breams, catfish, etc.), and cephalopods mostly for exports (squids, octopus, etc.). The total fisheries biomass has been estimated to be in the range of 188,000–450,000 MT (FAO 2001), with reports of overexploitation of certain demersal fish stocks (Heymans and Vakily 2004; Christensen et al. 2004). The majority of resources are still considered healthy according to recent fishery abundance surveys and stock assessments, paramount being the clupeids and small pelagics (Turay et al. 2008; Mawundu 2011).

Systems-to-Be-Governed

The social systems-to-be-governed span both the human and ecological dimensions of the fish chain, as fisheries include important marine biodiversity and provide ecosystem services such as food and livelihoods to people (Kooiman et al. 2005; FAO 2014). In addition, fisheries (especially small-scale fisheries) also provide tangible and intangible cultural ecosystem services that have aesthetic and spiritual benefits (Hall 2013). Although there is a strong government push to increase fish landings through industrial fisheries development, it is actually the small-scale artisanal fisheries and inland fisheries that play an equally significant role in regional and national economic development. About 80 % of fisheries production is marine-based, with aquaculture and inland fisheries being underutilized, despite the potential for tilapia and catfish production. The output of inland fisheries has been estimated to be around 20, 000 MT, with 75 % coming from riparian systems and 25 % from lakes (Sankoh and Jalloh 2011). Due to the multi-faceted contribution of fisheries to sustainable development, small-scale fisheries contribute tremendously to the Millennium Development Goals (Bene and Heck 2005). These entail fish protein and livelihood security, gender empowerment, maternal health, environmental sustainability, and global partnerships.

Since the mid-1990s, the artisanal small-scale fishery has surpassed the industry sector in terms of volume and value. By the early 2000s, it has grown exponentially and continues to do so. This small-scale fisheries sector focuses on small pelagics especially clupeids such as Herring (*Sardinella* spp.) and Bonga (*Ethmalosa fimbriata),* large pelagics such as Barracudas (*Sphyraena* spp.), in addition to demersals such as Sea breams (*dentex* spp.), Snappers (*Lutjanusspp* spp.), Catfish (*Ariidae latiscutatus),* Grunts (*Galeiodes decadactylus)* and Croakers including popular 'lady long neck' *(Pseudotolithus* spp.*)*. Total production of small-scale fisheries in 2006 was about 120,000 MT, with the artisanal small-scale contributing about 75 % of the catch (Fig. 29.2).

Unlike the commercial sector that includes trawlers and seiners, small-scale fisheries consists of dug-out canoes and small vessels with small outboard engines. Most of the canoes, including the Kru canoe, are manned by single fishers or several fishers employing paddles and sails. Bigger canoes include the standards 3–5 and

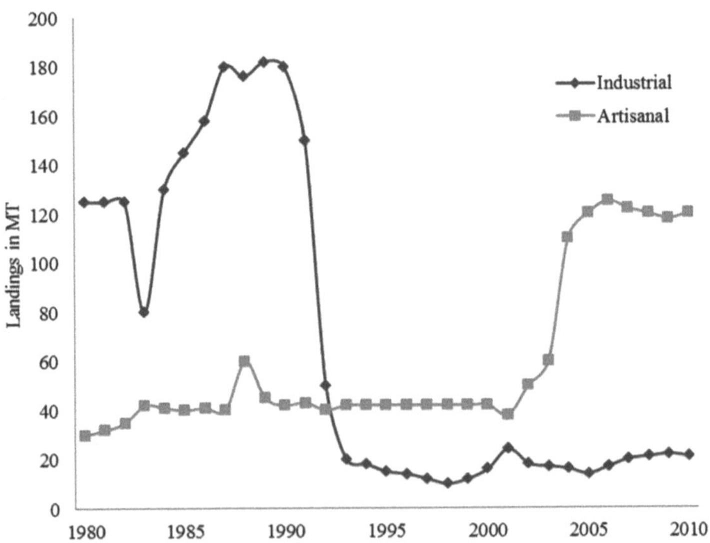

Fig. 29.2 Trends in national marine fisheries production from 1980 to 2010 (Sources: MFMR & FAO) Similarly, the annual value of small-scale fisheries has been about $1 million USD annually from 2006 to 2009, almost four times higher than the industrial sector (EIF 2013). Within recent years, there has been a sharp and progressive increase, from about $2.5 million in 2010 to about $6.5 USD in 2013 according to official statistics

5–7 Persons manned crew, and the Ghana canoe with teams of 5–10 Persons with outboard engines of 15–40 hp (Khan 1998; Thorpe et al. 2009; Mawundu 2011). The Ghana type canoe is the most expensive and used mainly for ring netting, introduced by Fante fishermen from Ghana in the 1950s (hence its name). They are fast and light weight and popular amongst migrant fishermen in the wider sub-region (Binet et al. 2012).

Artisanal small-scale fisheries gears are diverse and include drift and set nets, cast nets, hooks and lines. Fishing operations can be day trips or overnight depending on the presence of ice on board and seasonality (the Dry Season is preferable to the Rainy Season). The catches are landed at wharves, which are often the first entry point to fish chain transactions. Over 600 landing sites are found along the coast, presenting opportunities and challenges for monitoring and surveillance as well as obtaining accurate catch statistics (Mawundu 2011). The small-scale fisheries sector has changed considerably with an increasing number of fishers – both part time and full time (Thorpe et al. 2009). Despite this increase, fishers and their coastal communities are under threat, living in the poorest communities with low standards of life, and experiencing frequent conflicts over resource use (Thorpe et al. 2009).

Most of the fish is sold fresh, or frozen, and sometimes cured. Smoking and drying is the most common method of curing. This relies on the use of traditional 'Chokor' or improved 'Banda' ovens supported through overseas development projects (Khan 1998). Green et al. (2012) have emphasized on the high reliance on local seafood production for food security. There is greater access and taste towards dried fish, next to fresh fish, frozen fish and then salted fish. Amongst the Sierra Leonean coastal populations that eat fish regularly, 95 % do on a daily basis, whilst 7 % do so weekly and 5 % on a bi-weekly basis (Green et al. 2012). While there are limited marketing and value addition initiatives for local seafood products, exploring opportunities towards processing and curing for longer shelf life have been proposed to improve revenue and ameliorate food security concerns. Currently, there are high levels of 'trash fish' and post-harvest spoilage that could be better utilized to meet food security needs and various livelihood activities across the seafood value chain.

Governing Systems

The 1963 Fisheries Act is the major legal and regulatory tool that outlines institutional mandates and other decision-making approaches in Sierra Leone. Fisheries governance has mostly been hierarchical with the use of top-down management tools that focus on the biological and economic potential of the fishery. The Act explicitly states the paramount role of the state in resource governance. This role is entrusted to the Ministry of Fisheries and Marine Resources (MFMR). Management measures include the use of input and output control mechanisms (licenses and entry limitations, gear use, catch limits, total allowable catches, etc.). Very little consideration has been given to the socio-cultural and political aspects of governance, including involvement of non-state actors in decision making and placing emphasis on food security and gender issues. Such a top down approach has led to high transaction costs of monitoring as non-compliance and unsustainable fishing practices are on the increase as well as stakeholder conflicts. As a result, most of the objectives of the fisheries (ecological stewardship, economic viability, social legitimacy, etc.) are not being met.

Awareness of these growing challenges have led the government to initiated many legal and policy amendments so as to increase user participation in management, improve compliance and stewardship, and share power with local authorities. These changes started with the reform of the 1963 Fisheries Act, an enactment of the 1988 Fisheries Management and Development Act, the 1994 Fisheries Decree, the 2003 National Fisheries Policy reforms, and recently the 2011 Fisheries and Aquaculture Bill under ratification (Thorpe et al. 2009; Sheriff et al. 2009; Baio 2010; EIF 2013). In addition, the Sierra Leone Fisheries Five Year Plan and Fisheries Assessment Framework provides long term vision and governance considerations in meeting the expectations of stakeholders (Baio and Neiland 2014). These changes were mostly spurred by a growing need to integrate fisheries into broader development planning and trade policies, to garner public participation and secure the rights and benefits of locals especially with the 2004 Local Government Act (Sheriff et al. 2009; EIF 2013). This transitional governance also signifies a higher level of awareness for stronger institutions that are legitimate and resonate with public interests (Baio 2010).

Underscoring the role of fisheries for national development, the Government of Sierra Leone in partnership with international development agencies have adopted co-management initiatives through Marine Protected Areas (MPA) and promoted a Joint Maritime Committee taskforce. The aim is to promote conservation and development as part of Territorial User Rights in Fisheries (TURFs) and to support local monitoring and surveillance initiatives. Key stakeholders supporting this endeavor include international players such as the World Bank, the Global Environment Facility, New Partnership for Africa's Development, and several other NGOs including Wetlands International and the Environmental Justice Foundation. The idea is to introduce fishing rights through co-managed MPAs that will gradually evolve into TURFs (EJF 2011). This approach, though legitimate in principle, can be characterized as an attempt to frame social policies as tamed problems and to provide instant panaceas (Ostrom et al. 2007). Such initiatives are likely to require time as excluding or restricting fishers will impact foregone revenue and may precipitate labor market and rural economy challenges.

The acceptance of such an initiative for establishing community management associations serves as an impetus for increased participation and stewardship, notwithstanding the implementation challenges that will arise. There are additional programs and projects to strengthen local institutions through village development committees and other community organizations in recognition of the livelihood and food security benefits likely to result from well-functioning local institutions. Lately for instance, illegal unregulated and unreported fishing concerns have gained widespread attention amongst local fishing communities, with approval being sought for local participation in fisheries surveillance. In the past, commercial shrimp trawlers transgressed into these zones without any penalties resulting to habitat destruction and gear conflicts. Previous national surveillance programs failed to tackle these illegal unregulated and unreported fishing challenges, as witnessed by the demise of the Maritime Protection Services of Sierra Leone in the mid-1990s (Kamara 2012). In fact, this has spurred the recent emergence of the joint management committees for industrial monitoring and surveillance. This could be a challenging area due to the cost of policing the entire 200 nautical mile EEZ and beyond, and the difficulties of having institutional structures that are legitimate and effective.

Fisheries access and partnership agreements still remain a wicked problem and a difficult one to tame, as key stakeholders such as tuna operators are yet to agree on port states measures and other surveillance and distributional benefits (GoSL 2010; EIF 2013). Moreover, institutional capacity building and poor technical resources to improve management effectiveness alongside with local councilors pose administrative challenges for the small-scale fisheries sector.

Governing Interactions

Fisheries are part of larger systems of food production and are influenced by upstream and downstream linkages in rich estuaries as well as oceanographic processes offshore. Hence, concepts such as integrated coastal zone management (ICZM) and large marine ecosystem (LME) governance seem promising as they provide a set of tools that create synergy across sectors, stakeholders, and system boundaries at multiple scales (Khan and Mikkola 2002). Moreover, integrating fish chains and seafood production into broader sustainable development goals (e.g. food security, environmental health, livelihood, etc.) are necessary especially in the face of climate uncertainty. It has been suggested that aligning the governing processes for coastal zone planning, rural development, food production, and biodiversity conservation in a comprehensive framework will reduce implementation costs and meet multiple objectives (Turay 1996; Song and Khan 2011).

Ecosystem-based consideration and community stewardship has long since been practiced in many coastal regions in Sierra Leone. For instance, the Northern Fishing Community of Yeliboya in Kambia District, under a community stewardship arrangement, has practiced closed seasons in the creeks and estuaries during annual spawning seasons (December to April). Fishers have observed that during this time, matured pelagic and demersal fish stocks including Bonga and Croakers migrate in these creeks for spawning. The juveniles will remain in the creeks until May and then return offshore for recruitment (June to July). Building on this existing traditional knowledge to interface with fisheries co-management initiatives is a key element of participatory governance that will benefit the implementation of TURFs.

Designing the right institutional structures and inclusive policy processes to promote a community of practice and a knowledge mobilization platform for implementing these changes is critical. So far attempts at cross-scale linkages and integrated management have not been part of a broader governing framework for transformative changes in the Sierra Leonean context. Moreover, interactions amongst governors and those governed are very scanty and often lead to contestations as trust and credibility are lacking. Given such a situation, the quality of governing interactions amongst these multiple institutions and stakeholders, and the choice of policy instruments and leadership are fundamental for sustaining fisheries benefits in the long-term.

How Governable Are Small-Scale Fisheries? A Pre and Post-conflict Assessment

Governability assessment is a useful tool for understanding small-scale fisheries as a system, as a sector, and how it interacts with other institutional mandates at multiple scales (Bavinck et al. 2013). Using the fish chain as an analytical framework, we assess the limits to the quality of governance in the various production chains before and after the civil conflict. The three stages of the fish chain are embedded in both the natural systems and the systems-to-be-governed, with the governing system consisting of policy instruments and power relations amongst the stakeholders. The fish chain for most artisanal seafood products entails three to four major stakeholder groups including managers who control access and user rights, fishers who harvest, processors and traders who market, and consumers at the household level. Depending on fishing operations, the fish chain can be quite short with fishers and direct buyers, or long involving truckers, processors, 'middlemen', and retailers or 'fish mammies' who often finance marketing operations (Fig. 29.3).

The stakeholders interact within and across the various production stages and are guided by shared values, common images, and principles that influence codes of ethics and behavior (Kooiman et al. 2005; Song et al. 2013)

The governance transition from top-down to co-governance reforms signals a common vision and shift in thinking of the role of small-scale fisheries in local communities and in regional economic development. Several challenges and opportunities can be identified in the pre- and post-conflict periods for governance and governability. Understanding these challenges through the four step process as outlined earlier, i.e., the degree of wickedness, the prevalence of system properties, goodness of fit, and the presence (or absence) of the quality of interactions, allows a broader analysis of governing capacity and its limits that consequently determine the success of the reforms. In what follows, we examine the opportunities and challenges in both periods and conclude with some thoughts on issues that require attention.

Pre-conflict Periods

The complexity and dynamics of marine ecosystems and coastal interactions present a unique challenge for adequately managing fisheries systems in space and time. Thus, the various policy instruments (shown in Fig. 29.3) in the three stages are essential in fostering stewardship and compliance as well as economic viability and

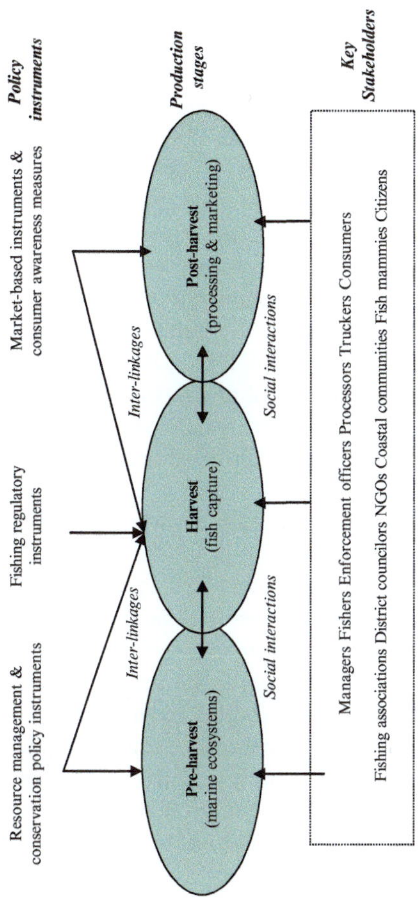

Fig. 29.3 The fish chain showing the various production stages as well as governing mechanisms and stakeholder groups (Source: Khan and Chuenpagdee 2014 as adapted from Bavinck et al. 2005)

social legitimacy. Within the pre-harvest stage, it has been acknowledged that fish stocks (especially pelagics) are highly susceptible to climatic events such as upwelling and El Nino, which affect abundance and spatial distribution patterns (Bakun 1978; Hardman-Mountford and McGlade 2003). These coastal ecosystems are highly complex and influenced by strong tides and monsoons resulting in seasonal upwelling (Johnson and Johnson 2012). The scales of interactions also point to gaps and concerns about matching ecological boundaries and fishing activities with institutional mandates and appropriate policy, instruments (Folke et al. 2007; Bavinck et al. 2013; Jentoft 2013). In the Sierra Leonean context protected areas and reserves although promoted by NGOs, were not part of the management tool-kit in the pre-conflict periods. The success of managing these resources under input control rules alone fell short of national expectations (Jalloh 2009; Vakily et al. 2012). Earlier reports of overexploitation and mis-management, and increasing concerns over distributional equity were some of the concerns raised (Kaczynski and Fluharty 2002; Jalloh 2009). There were several limitations to achieving biological sustainability, primarily in the harvesting stage and with regard to compliance to rules and fishing regulations. These consist of ineffective monitoring systems, illegal fishing activities, inaccurate catch reporting mechanisms, significant amounts of by-catch and dumping at sea (Mawundu 2011; Vakily et al. 2012). As a consequence, fish stocks declined considerably, with few management measures that protected critical habitats and stock health. From 1964 to 1990, for instance, there was 90 % reduction in the biomass of the demersal fishery, due to overfishing and unsustainable fishing practices (Heymans and Vakily 2004).

In the post-harvest stage, stewardship incentives through legitimate access rights as well as value chain development have been proposed for alleviating some of the challenges across the fish chain. A bigger concern in the past, and yet to be addressed, is post-harvest waste due to lack of product development. Addressing this challenge directly relates to achieving ecological stewardship as well as economic viability by fostering compliance and social entrepreneurships. Lack of product development is partly due to poor inputs for cold rooms and other technologies for fish processing and making trash fish and by-catch marketable and a quality food product. Moreover, this concern is associated with inadequate harbour infrastructure and seafood development opportunities (Sankoh and Jalloh 2009; Etoh 2012). This affects the potential positive contribution of small-scale fisheries to food security and national economic development.

Despite the proliferation of fisheries development projects from the 1980s onwards (Khan 1998; EIF 2013), small-scale fisheries still continue to be associated with poverty (Bene 2003; Thorpe et al. 2009). What has been lacking is a 'home grown theory of change', one that sits on a legacy for long-term societal transformation, in nurturing local governing structures and policy processes that put small-scale fisheries as part of broader development planning. The donor-led development initiatives are not replicable after the funding cycle and often fail in meeting long-term community needs. This is partly because the project design ignores local champions and the potential appropriation by elites through political structures and social networks (see Crona and Bodin 2010).

Small-scale fisheries, despite the limited nature and scope of their operations contribute in a number of ways to national economic development and are *too big to ignore* in the short and medium-term (Chuenpagdee 2011b). Certainly, it is becoming clear that small-scale fisheries has spill-over and multiplier effects for rural development, regional trade, and contributes to other ecosystem service benefits (GoSL 2010; Kayonde et al. 2013).

Post-conflict Period

Indeed, it has been acknowledged that regional climatic events such as El Nino and seasonal variability are persistent in the post-conflict periods in Sierra Leone (Johnson 2006; Katikiro and Macusi 2012). As such, the question becomes what management tools and resources are available to address global change impacts and to adapt to climate extremes. Several local and international programs have been initiated, as this period coincides with a global push towards national adaptation plans under the Cancun Accord on climate change (Johnson 2006; GoSL 2007). There is also a push to promote the Green Economy (and Blue Economy in this context) in response to Rio +20 (AfDB 2013). These multilateral policy interventions are driven by various institutional partnerships and stakeholder engagements on linking biodiversity conservation with poverty alleviation through market-based instruments (UNEP 2011).

In these changing governing contexts, the evolving legal and institutional changes from hierarchical to co-governance arrangements are desirable as they provide opportunities for strengthening local and regional institutional capacity for stewardship. Under this new scheme however, fishers are assigned licenses irrespective of the type of gear or fees paid to Local Councilors, and could contribute to Malthusian overfishing (Pauly 1990). With unlimited entry into the small-scale sector, over 10,000 registered canoes have been documented with increasing volume of landings as shown in Fig. 29.2. This has also spurred a chain reaction amongst actors and seafood traders and worsened power asymmetries amongst fish actors. For instance, crew members being marginalized by boat owners are prompted to explore revenue sharing mechanisms and social ties with new and emerging seafood buyers. This program often called ‘hand failure’ reflects on the human rights abuse during the civil conflict and the power of now – i.e., being able to address fair dealings through alternative economic relationships at sea. Understanding these attributes may improve the quality of governance through an agenda setting that reflects shared visions and socio-cultural perceptions. It seems the motivation for social change and governance reforms from donor agencies do not always take into consideration the interplay of local values and power disparities.

The demand for frozen fish in the hinterland has also increased in recent years, thus encouraging fish processing and trucking along the coast to meet both local and regional needs. For current local consumption, 20 kg of Bonga (*Ethmalosa fimbriata*) will cost $33 USD at wharf price. After curing (smoking or sun dried), 20 kg will fetch for higher price from $40–50 USD depending on quality of product. For the sub-regional market in Western Africa, where there is a higher taste and preference for dried or smoked Bonga, 20 kg will fetch $75 USD. For the artisanal export of Gwangwa (*Pseudotolithus elongates*) to the Asian lucrative market, 20 kg will fetch $100–150 USD depending on size (the bigger, the better).

Current initiatives aimed at improving coastal infrastructure through functional cold storage facilities are in line with national development priorities (EIF 2013). Coordinating fish handling at sea, fish processing, packaging and marketing of quality products in support of household and diaspora economies, and EU markets are essential value chain activities to be up-scaled (FAO 2013). It is within these contexts that the West Africa Regional Fisheries Program and other related projects on infrastructure development and capacity building are filling a critical void (EIF 2013). In meeting new opportunities in the small-scale fisheries for out-sourcing, fish mammies and industrial fish processing establishments are now serving as financiers for processing and exporting local artisanal catch as in the case of the Sciaenid locally called 'Gwangwa'. Including small-scale fisheries as part of fisheries agreement deliberations within a regional trade policy context will promote dialogue and deliberations and ensure that hard choices are made by councilors and local stakeholders on trade-offs between local consumption and foreign exports (Kooiman and Jentoft 2009).

In addressing the myriad challenges and opportunities that fisheries generate, the inclusion of non-state actors in fisheries governance is an important milestone in the post-conflict era. At the national level, empowering Local Councilors as decision makers and offering fishers stewardship incentives for livelihood security through user rights meet multiple sustainability objectives. At the local and regional level, the two main fisher unions (Artisanal Fishermen Union and the Amalgamated Artisanal Fishermen Union), have the potential for gender mainstreaming through integrated value chain development that are inclusive of women entrepreneurs (Thorpe et al. 2013).

After years of decentralization and local governance reforms towards TURF, community engagement and institutional capacity building processes are still inadequate to respond to compliance and stewardship challenges (EJF 2011). Recent accounts about the co-governance arrangements indicate that the process was rushed, focusing mainly on rent extraction by Local Councils in the absence of institutional structures for managerial responsibilities and program implementation (Baio 2010; EJF 2011). If well developed and nurtured, governing reforms could trigger transformative changes and spill-over effects to other food production sectors especially agricultural and in rural planning and regional economic development.

Scholarly research is central for evidence-based policies that are interdisciplinary in scope. The Institute of Marine Biology and Oceanography founded in the 1950s as part of a regional research program, now under the University of Sierra Leone, has played a significant role in fisheries stock assessments, national frame surveys, and transboundary collaborative research programs. A recent Memorandum of Understanding with the MFMR will enable the Institute of Marine Biology and

Oceanography to conduct research under the EU funded project on institutional support to fisheries management (GOPA 2009). The project aims to support policy development and steer future directions for fisheries research at the Institute of Marine Biology and Oceanography. This will also include collaboration with the regional LME project, the FAO/CECAF working group on pelagic fisheries management, as well as the Fridtjof Nansen biomass surveys. Through these, co-learning opportunities on system characteristics and knowledge mobilization can be instrumental for improving the quality of governance through better decision-making.

Summary and Policy Implications

Is the governability of fisheries in post-conflict Sierra Leone better now or worse than before? Based on the analysis and evidence provided, we conclude that the quality of governance is improving with stakeholder involvement and institutional partnerships, which has provided opportunities for linking small-scale fisheries with broader development agenda. Nonetheless, the persistence of challenges within the various systems, as well as the quality and capacity of governance, can be improved and made more adaptive in the event of global environmental and economic changes.

In this Chapter, we underscore how fisheries resources contribute to food security, human development, and economic prosperity. We highlight how governance reforms and user participation through co-management has been embraced by stakeholders. We argue that this is not by itself a panacea, as fisheries governance are understood to be wicked due to the inherent nature of the social dilemmas associated with multiple objectives and unrealistic stakeholder expectations. Therefore, adopting the Precautionary Principles is crucial for resource sustainability in the absence of robust stock assessment data for management strategies at the local scale (FAO 1995). Similarly, designating MPAs as critical habitats and ecological niches could contribute to both climate change adaptation needs as well as resource sustainability. Protecting a good fraction of the 157,000 ha of mangroves could act as a buffer and contribute to alleviating flood risks as well as critical habitats for fisheries productivity (Johnson 2006; Johnson and Johnson 2012).

However, more needs to be done to boost governing capacity across the harvest and post-harvest stages of the fish chain. Patron-client relationships and power asymmetries between harvesters and processors as in the case of Gwangwa, has implications for livelihood security and local food security. This could be achieved through comprehensive seafood value chain development (for gender mainstreaming and food security co-benefits), information sharing amongst stakeholders for collective action, and a regional focus for integrated management. Human and technical resources need to be developed within MFMR and the Institute of Marine Biology and Oceanography to meet food and safety test particularly phytosanitary assessments for certification.

Another big concern for sustainable outcomes and effective management is 'fit'. Because small-scale fisheries are part of larger systems that are diverse, complex, and dynamic; collaborative and multi-level governance arrangements that address scale and institutional boundaries is fundamental. The lack of 'nested or multi-level' approaches to institutional innovation from municipal to district and regional levels compound the (in)effectiveness of fisheries reforms. Furthermore, the relationships amongst stakeholders, especially managers, local councilors, fisher unions but also NGOs and multilaterals, are not clearly defined and articulated in policy documents and hence may exacerbate conflicts about mutual responsibilities. Although there are several development projects that address policy gaps through public-private partnerships, these programs are not place-based and do not respond to local surveillance and monitoring challenges. Responsive regional institutions especially for transboundary pelagic resources within multi-level governance frameworks are needed (Khan and Mikkola 2002). This can be critical for fisheries productivity under conditions of extreme environmental change as well as for the livelihoods of fishing dependent communities (Lam et al. 2012). The sub regional scale is important as migrant fishers along the West African coast do not frequently comply with social norms where MPAs or TURFs are located, and may trigger potential conflicts (Thorpe et al. 2009; Binet et al. 2012).

We recognized that the drivers of change in the various systems are the result of negotiations by key actors with various levels of power to influence outcomes, including international donors, national ministries and departments, and local chiefs or councilors. However, the interests and values of stakeholders are often not well aligned and prioritized, resulting in unresolved conflicts that make the systems less governable. The value of the fishery can be understood both in terms of its *assigned value* along the supply chain as well as its *underlying value* that dictates stewardship and ethical norms (Brown 1984). On many occasions, one value trumps another but could be made complementary with co-benefits. For example, the active participation of private sector entrepreneurs especially fish mammies in local seafood marketing is necessary for increasing revenues as well as for linking seafood with traditional cuisines and healthy lifestyle.

Although much of the emphasis has been on fisheries reforms and community stewardship at the local scale, much could be done by integrating other development programs that deal with sectors such as agriculture, tourism, and forestry, where human capacity are already strengthened (EIF 2013). Successful co-governance arrangements have highlighted the importance of strengthening local institutions including formal and informal rules and norms that are resilient to global economic changes (Khan et al. 2004). Within this context, legal reforms and the delegation of tasks to local councilors are inadequate for transformative change; corresponding capacity building initiatives at the district council level are necessary for self-organization and successful outcomes (Ostrom 1990).

The broader lessons are that these wicked environmental and developmental challenges are interconnected and require cross-scale and multi-sectoral approaches, especially for fragile states with limited governing capacity. Moreover, as many scholars have argued (Sen 1999; Jentoft and Eide 2011), an integrated human-nature approach of this kind could be part of a holistic development agenda and the focus for future research.

Acknowledgement A. Khan acknowledges funding from the Young International Scientist Fellowship Program whilst at UNEP-IEMP (Grant no. 2012Y1ZA0010). An additional grant from Too Big To Ignore assisted with travel plans to attend the Small-Scale Fisheries Governance: Development for Wellbeing and Sustainability Conference in Hyderabad, India, held from Dec 7–14th 2013. Angela R Situma and two reviewers provided valuable comments that improved earlier drafts of this manuscript. S. Sei and E. Mbanzamihigo linked the dots in Fig. 29.1.

References

AfDB. (2013). *Sierra Leone: Transitioning towards green growth; Stocktaking and the way forward*. Tunis: AfDB.

Baio, A. (2010). Show me the way: Inclination towards governance attributes in the artisanal fisheries of Sierra Leone. *Fisheries Research, 102*, 311–322.

Baio, A. C. M., & Neiland, A. E. (2014). *Fisheries development strategy for Sierra Leone to 2030: Fisheries Wealth for National Prosperity*. West Africa Pilot Project within the Framework of West Africa Regional Fisheries Programme. MFMR, Freetown, Sierra Leone.

Bakun, A. (1978). Guinea current upwelling. *Nature, 271*, 147–150.

Bavinck, M., Chuenpagdee, R., Diallo, M., Heijde, P., Kooiman, J., Mahon, R., & Williams, S. (2005). *Interactive fisheries governance: A guide to better practice. Centre for Maritime Research*. Delft: Eburon Academic Publishers.

Bavinck, M., Chuenpagdee, R., Jentoft, S., & Kooiman, J. (2013). *Governability of fisheries and aquaculture: Theory and practice* (MARE publication series 7). Amsterdam: Springer.

Bene, C. (2003). When fishery rhymes with poverty: A first step beyond the old paradigm on poverty in small-scale fisheries. *World Development, 31*(6), 949–975.

Bene, C., & Heck, S. (2005). Fisheries and the MDGs: Solutions for Africa. *NAGA, WorldFish Centre Quarterly, 28*(3/4), 14–18.

Binet, T., Failler, P., & Thorpe, A. (2012). Migration of Senegalese fishers: A case for regional approach to management. *Maritime Studies, 11*, 1.

Brown, T. C. (1984). The concept of value in resource allocation. *Land Economics, 60*, 231–246.

Bundy, A., Chuenpagdee, R., Jentoft, S., & Mahon, R. (2008). If science is not the answer, what is? An alternative governance model for the world's fisheries. *Frontiers in Ecology and the Environment, 6*, 152–155.

Christensen, V., Amorim, P., Diallo, I., Diouf, T., Guénette, S., & Heymans, J. H, et al. (2004). Trends in fish biomass off Northwest Africa, 1960–2000. In M. L. D. Palomares & D. Pauly (Eds.), *West African marine ecosystems: Models and fisheries impacts, Fisheries Centre Research Reports 12*(7), 215–220. Vancouver: Fisheries Centre, UBC.

Chuenpagdee, R. (2011a). Interactive governance for marine conservation: An illustration. *Bulletin of Marine Science, 87*, 197–211.

Chuenpagdee, R. (Ed.). (2011b). *World small-scale fisheries contemporary visions*. Delft: Eburon.

Chuenpagdee, R., & Jentoft, S. (2013). Assessing governability – What's next? In M. Bavinck, R. Chuenpagdee, S. Jentoft, & J. Kooiman (Eds.), *Governability of fisheries and aquaculture: Theory and practice* (MARE publication series 7, pp. 335–350). Amsterdam: Springer.

Collier, P. (2007). *The bottom billion. Why the poorest countries are failing and what we can do about it*. Oxford: Oxford University Press.

Crona, B., & Bodin, Ö. (2010). Power asymmetries in small-scale fisheries: A barrier to governance transformability? *Ecology and Society, 15*, 32.

Degnbol, P., Gislason, H., Hanna, S., Jentoft, S., Nielsen, J. R., Sverdrup-Jensen, S., & Wilson, D. C. (2006). Painting the floor with a hammer: Technical fixes in fisheries management. *Marine Policy, 30*, 534–543.

Demby, A., & Leigh, I. (2012). Enhancing the role of women in artisanal and industrial. In J. M. Vakily, K. Seto, & D. Pauly (Eds.), *The marine fisheries environment of Sierra Leone: Belated proceedings of a National Seminar held in Freetown, 25–29 November 1991* (pp. 78–80). Fisheries Centre Research Reports *20*(4). Vancouver: Fisheries Centre, UBC.

EIF. (2013). *Trading towards prosperity: Sierra Leone diagnostic trade integration study update.* Final version. Enhanced integrated framework. Freetown/Washington, DC. Available at http://enhancedif.org/en/country-profile/sierra-leone?qt-country_tabs=3

EJF. (2011). *The governance of artisanal fisheries in the Sherbro River Area in Sierra Leone.* London: Environmental Justice Foundation.

Etoh, S. (2012). Domestic fish marketing in Sierra Leone. In J. M. Vakily, K. Seto, & D. Pauly (Eds.), *The marine fisheries environment of Sierra Leone: Belated proceedings of a National Seminar held in Freetown, 25–29 November 1991* (pp. 81–90). Fisheries Centre Research Reports *20*(4). Vancouver: Fisheries Centre, UBC.

FAO. (1995). *Code of conduct for responsible fisheries*. Rome: FAO.

FAO. (2001). *Sierra Leone small-scale fisheries development* (Project report No. 01/063 ADB-S/L Working Paper). Rome: FAO.

FAO. (2013). A value-chain analysis of international fish trade and food security with an impact assessment of the small-scale sector. Summary article, NORAD-FAO Project. In *IIFET 2012 conference proceedings* (1–13 pp). Dar es Salaam.

FAO. (2014). *The state of world fisheries and aquaculture 2014*. Rome: FAO.

Folke, C., Pritchard, L., Berkes, F., Colding, J., & Svedin, U. (2007). The problem of fit between ecosystems and institutions: Ten years later. *Ecology and Society, 12*(1), 30.

GOPA. (2009). *Institutional support for fisheries management for Sierra Leone.* 9th EDF ACP SL 019/1. Second annual report. GOPA Consultants, Homburg.

GoSL. (2007). *National adaptation programme of action.* Freetown: Ministry of Transportation and Aviation.

GoSL. (2010). *Sierra Leone national export strategy. Sierra Leone investment and export promotion Agency*. Freetown: SLIEPA.

Green, P. A. S., Carol, G., & Mason, E. D. (2012). The role of fish for consumption and nutrition in Sierra Leone. In J. M. Vakily, K. Seto, & D. Pauly (Eds.), *The marine fisheries environment of Sierra Leone: Belated proceedings of a National Seminar held in Freetown, 25–29 November 1991* (pp. 90–99). Fisheries Centre Research Reports *20*(4). Vancouver: Fisheries Centre, UBC.

Hall, P. (2013). *Recruiting the visual: Knowing our common place towards an encyclopedia of local knowledge*. Doctoral thesis, Memorial University, St. John's.

Hardman-Mountford, N. J., & McGlade, J. M. (2003). Seasonal and inter-annual variability of oceanographic processes in the Gulf of Guinea: An investigation using AVHRR sea surface temperature data. *International Journal of Remote Sensing, 24*(16), 3247–3268.

Heymans, J. J., & Vakily, J. M. (2004). Structure and dynamics of the marine ecosystem off Sierra Leone for three time periods: 1964, 1978, 1990. In M. L. D. Palomares & D. Pauly (Eds.), *West African marine ecosystems: Models and fisheries impact* (pp. 160–169). Fisheries Centre Research Reports, *12*(7). Vancouver: Fisheries Centre, UBC.

ISFM. (2009). *Institutional support to fisheries management.* 9th EDF ACP SL 019/1, Second annual report for the period May 2008 to May 2009. GOPA Consultants, Homburg.

Jacquet, J. L., & Pauly, D. (2008). Trade secrets: Renaming and mis-labeling of seafood. *Marine Policy, 32*, 309–318.

Jalloh, K. (2009). *The economic potential and feasibility of a landing site investment in the artisanal small pelagic fishery of Sierra Leone*. UNU-Fisheries Training Programme final project report. Reykjavik: UNU-FTP.

Jentoft, S. (2013). *Not just about scale: Towards an International voluntary guidelines for securing sustainable small-scale fisheries*. Paper presented at the international conference on small-scale fisheries governance: Development for wellbeing and sustainability Dec 7–14th 2013. Centre for Economic and Social Studies, Hyderabad.

Jentoft, S., & Chuenpagdee, R. (2009). Fisheries and coastal governance as wicked problems. *Marine Policy, 33*, 553–560.

Jentoft, S., & Eide, A. (Eds.). (2011). *Poverty mosaics: Realities and prospects in small-scale fisheries*. Dordrecht: Springer.

Jentoft, A. S., McCay, B. J., & Wilson, D. C. (1998). Social theory and fisheries co-management. *Marine Policy, 22*(4/5), 423–436.

Jentoft, S., van Son, T. C., & Bjorkan, M. (2007). Marine protected areas: A governance system analysis. *Human Ecology, 35*, 611–622.

Johnson, R. (2006). *Coastal erosion issues in Sierra Leone: Adaptation, planning and implementation relating to the Sierra Leone coastal zone*. UNFCCC African regional workshop on adaptation, September 21–13, Accra, Ghana.

Johnson, R. G., & Johnson, R. G. (2012). The mangrove and coastal environment of Sierra Leone. In J. M. Vakily, K. Seto, & D. Pauly (Eds.), *The marine fisheries environment of Sierra Leone: Belated proceedings of a National Seminar held in Freetown, 25–29 November 1991* (pp. 50–54). Fisheries Centre Research Reports *20*(4). Vancouver: Fisheries Centre, UBC.

Kaczynski, V. M., & Fluharty, D. L. (2002). European policies in West Africa: Who benefits from fisheries agreements? *Marine Policy, 26*, 75–93.

Kamara, A. B. (2012). The fisheries of Sierra Leone: Status, problems and prospects. In J. M. Vakily, K. Seto, & D. Pauly (Eds.), *The marine fisheries environment of Sierra Leone: Belated proceedings of a National Seminar held in Freetown, 25–29 November 1991* (pp. 12–16). Fisheries Centre Research Reports *20*(4). Vancouver: Fisheries Centre, UBC.

Katikiro, R. E., & Macusi, E. D. (2012). Impacts of climate change on West African fisheries and its implications on food production. *Journal of Environmental Science and management, 15*, 83–95.

Kayonde, S., Alexandre, L. H., & Peakman, J. F. (2013). *Sierra Leone – Growth pole diagnostic: First phase of the Growth poles program*. Washington, DC: The World Bank.

Khan, A. (1998). *Impacts of fisheries inputs and services on coastal districts of the artisanal fishery of Sierra Leone: An investigation based on the 1990 fishery frame survey data*. BSc (Honours) thesis. University of Sierra Leone, Freetown.

Khan, A. S. (2012). Understanding global supply chains and seafood markets for the rebuilding prospects of Northern Gulf Cod Fisheries. *Sustainability, 4*, 2946–2969.

Khan, A. S., & Chuenpagdee, R. (2014). Interactive governance and fish chain approach to fisheries rebuilding: A case study of the Northern Gulf cod fisheries in Eastern Canada. *Ambio, 43*, 600–613.

Khan, A. S., & Mikkola, H. (2002). *Sustainable ocean development: An initiative for the management and protection of the marine and coastal resources of the economic community of West Africa States (ECOWAS) Sub Region*. Banjul: UNIGAM 0001, University of the Gambia.

Khan, A. S., & Neis, B. (2010). The rebuilding imperative in fisheries: Clumsy solutions for wicked problems? *Progress in Oceanography, 87*, 347–356.

Khan, A. S., Mikkola, H., & Brummett, R. E. (2004). Feasibilities in fisheries co-management in Africa. *NAGA, WorldFish Centre Quarterly, 27*(1, 2), 60–64.

Khan, A. S., & Sesay, S. (2015). Seafood insecurity, bush meat consumption, and public health emergency in West Africa: Did we miss the early warning signs of an Ebola epidemic? *Maritime Studies, 14*, 3. doi:10.1186/s40152-015-0020-2.

Khan, A. S., Sumaila, U. R., Watson, R., Munro, G., & Pauly, D. (2006). The nature and magnitude of global non-fuel fisheries subsidies. In U. R. Sumaila & D. Pauly (Eds.), *Catching more bait: A bottom-up re-estimation of global fisheries subsidies* (pp. 5–37). Fisheries Centre Research Reports *14*(6), Vancouver: UBC.

Kooiman, J. (2003). *Governing as governance*. London: Sage.

Kooiman, J., & Chuenpagdee, R. (2005). Governance and governability. In J. Kooiman, M. Bavinck, S. Jentoft, & R. Pullin (Eds.), *Fish for life: Interactive governance for fisheries* (pp. 325–349). Amsterdam: Amsterdam University Press.

Kooiman, J., & Jentoft, S. (2009). Meta governance, values, norms and principles, and the making of hard choices. *Public Administration, 87*(4), 818–836.

Kooiman, J., Bavinck, M., Jentoft, S., & Pullin, R. (Eds.). (2005). *Fish for life: Interactive governance for fisheries* (MARE). Amsterdam: Amsterdam University Press.
Kurien, J. (2005). *Responsible fish trade and food security* (FAO Fisheries technical paper, No. 456). Rome: FAO.
Lam, V. W. Y., Cheung, W. W. L., Swartz, W., & Sumaila, U. R. (2012). Climate change impacts on fisheries in West Africa: Implications for economic, food and nutritional security. *African Journal of Marine Science, 34*, 103–117.
Larkin, P. A. (1977). An epitaph for the concept maximum sustainable yield. *Transactions of the American Fisheries Society, 196*, 1–11.
Le Billon, P. (2001). The political ecology of war: Natural resources and armed conflicts. *Political Geography, 20*, 561–584.
Mawundu, S. (2011). *Artisanal fisheries statistics in Sierra Leone, collection methods, analysis and presentation*. Master's thesis, United Nations University, Reykjavik.
Mehl, S., Lundsor, E., Turay, I., Sei, S., & Lamptey, E. (2007). *Surveys of the fish resources of the Western Gulf of Guinea (Guinea Bissau, Guinea, Sierra Leone and Liberia)*. Institute of Marine Research, Bergen, Norway.
MFMR. (2008). *Fisheries of Sierra Leone* (3rd ed). Documentation Centre, Brookefields Hotel, Jomo Kenyatta Road, Freetown, Sierra Leone.
Ndomahina, E. T. (2002). *Assessment of the Status of the Coastal and Marine Biodiversity in Sierra Leone. National Biodiversity Strategy and Action Plan*. Consultancy of the Sierra Leone Maritime Administration.
Ostrom, E. (1990). *Governing the commons: The evolution of institutions for collective action*. New York: Cambridge University Press.
Ostrom, E., Janssen, M. A., & Anderies, J. M. (2007). Going beyond panaceas. *Proceedings of the National Academy of Sciences, 104*, 15176–15178.
Pauly, D. (1990). On Malthusian overfishing. *NAGA, The WorldFish Centre Quarterly, 13*, 3–4.
Pauly, D., Watson, R., & Alder, J. (2005). Global trends in world fisheries: Impacts on marine ecosystem and food security. *Philosophical Transactions of the Royal Society B, 360*, 5–12.
Rittel, H., & Webber, M. (1973). Dilemmas in a general theory of planning. *Policy Science, 4*, 155–169.
Sankoh, K., & Jalloh, K. (2011). *Infrastructure investment in the fisheries sector*. West Africa Regional Fisheries Programme, Freetown.
Scheffers, A., & Browne, T. (2008). Coastal landforms of Sierra Leone. In E. Bird (Ed.), *Encyclopaedia of the world's coastal landforms*. London: Springer.
Sei, S., Seisay, M. B. D., Seilert, H., & Turay, I. (2009a). *Marine fisheries management in Sierra Leone: Achievements and challenges*. Report of the National Consultative forum on Fisheries Management Issues, ISFM, MFMR, Freetown, Sierra Leone.
Sei, S., Turay, I., Seisay, M. B. D., & Seilert, H. (2009b). *Contemporary discourse on marine protected areas in Sierra Leone*. Report of the National Consultative forum on Fisheries Management Issues, ISFM, MFMR, Freetown, Sierra Leone.
Seisay, M. B. D., & Jalloh, K. (2006). *Small pelagics resource, exploitation and fisheries in Sierra Leone*. Freetown: Statistics and Research Unit, Ministry of Fisheries and Marine Resources.
Sen, A. (1999). *Development as freedom*. New Delhi: Oxford University Press.
Sheriff, M. F., Seisay, M. B. D., Jalloh, K., Turay, I., Sei, S., & Seilert, H. (2009, April 15–17). Co-management of the small pelagic fisheries in Sierra Leone. Regional seminar on mechanisms for the management of shared stocks of small pelagics in Northwest Africa, Dakar, Senegal.
Showers, P. A. T. (2012). The shrimp stocks of Sierra Leone. In J. M. Vakily, K. Seto, & D. Paul (Eds.), *The marine fisheries environment of Sierra Leone: Belated proceedings of a National Seminar held in Freetown, 25–29 November 1991* (pp. 46–49). Fisheries Centre Research Reports *20*(4). Vancouver: Fisheries Centre, UBC.
Simon, H. (1947). *Administrative behavior: A study of decision-making processes in administrative organizations*. New York: The Free Press.

Smith, M. D., Roheim, C. A., Crowder, L. B., Halpern, B. S., Turnipseed, M., Anderson, J. L., et al. (2010). Sustainability and global seafood. *Science, 327*, 784–786.

Song, A., & Khan, A. (2011). Views from the bottom: Student reflection on fisheries research. In R. Chuenpagdee (Ed.), *World small-scale fisheries contemporary visions* (pp. 333–352). Delft: Eburon.

Song, A. M., Chuenpagdee, R., & Jentoft, S. (2013). Values, images, and principles: What they represent and how they may influence fisheries governance. *Marine Policy, 40*, 167–175.

Ssentongo, G. W., & Ansa-Emmim, M. (1986). *Marine fishery resources of Sierra Leone: A review of exploited fish stocks* (CECAF/ECAF series 86/34 (En)). Rome: FAO.

Thorpe, A., Whitmarsh, D., Ndomahina, E. T., Baio, A., Kemokai, M., & Lebbie, T. (2009). Fisheries and failing states: The case of Sierra Leone. *Marine Policy, 33*, 393–400.

Thorpe, A., Whitmarsh, D., Sandi, R., Baio, A., Lebbie, N., Lebbie, T., & Curiazi, T. (2013). Pathways out of poverty: Women – The 'forgotten gender'- and the artisanal fisheries sector of Sierra Leone. *African Historical Review, 45*, 46–61.

Turay, F. (1996). *Fisheries in coastal zone management in Sierra Leone: Potential benefits and problems* (CEMARE Research Paper #91).

Turay, I., O'Donnell, C., Schaber, M., Corten, A., Sarre, A., Sei, S., Seisay, L. D., Mustapha, C., & Lahai, M. (2008). *Sierra Leone resource survey report. Fisheries of Sierra Leone 2008*. Ministry of Fisheries and Marine Resources (MFMR), Freetown.

UNEP. (2011). *Towards a green economy: Pathways to sustainable development and poverty eradication*. Retrieved from http://www.unep.org/greeneconomy/GreenEconomyReport

Vakily, J. M., Seto, K., & Pauly, D. (Eds.). (2012). *The marine fisheries environment of Sierra Leone: Belated proceedings of a national seminar held in Freetown, 25–29 November 1991*. Fisheries Centre Research Reports, *20*(4).

Wai, Z. (2012). *Epistemologies of African conflicts: Violence, evolutionism, and the war in Sierra Leone*. New York: Palgrave Macmillan.

Walters, C. J. (1986). *Adaptive management of renewable resources*. New York: McGraw Hill.

WHO Ebola Team. (2014). Ebola virus disease in West Africa. The first 9 months of the epidemic and forward projections. *NEJM, 371*, 1481–1495. doi:10.1056/NEJMoa1411100.

Wilhere, G. F. (2008). The how-much-is-enough myth. *Conservation Biology, 22*, 514–517.

Chapter 30
Exploring Adaptive Co-management as a Pathway to Enhance the Governability of Sea Urchin Fisheries in Barbados and Saint Lucia

Shelly-Ann Cox and Patrick McConney

Abstract Small-scale coastal fisheries, especially for small stocks of relatively sedentary species, are increasingly seen as candidates for still-novel governance arrangements such as adaptive co-management. Yet, co-management initiatives often expose deficiencies in capacity, communication, trust and other factors said to favor success. How then to govern fisheries with highly variable social-ecological systems? How is governability affected by low capacity and arbitrary governing interactions? How can interactive governance address these challenges? Fisheries governance in the Eastern Caribbean is hampered by limited capacity in community and state fisheries organizations, low levels of leadership, inadequate information exchange and low political priority compared to other economic sectors. Little guidance is available to assess pre-conditions and chances of successful adaptive co-management. Using sea urchin fisheries at five sites in Barbados and Saint Lucia as cases, this chapter investigates formal and informal processes and conditions for establishing, implementing and sustaining adaptive co-management. It assesses the potential for approaching co-governance grounded in understanding social-ecological fisheries systems. This informs fisheries governance by exploring opportunities and constraints through a governability lens.

Keywords Governability • Adaptive co-management • Sea urchin fisheries • Social-ecological systems • Barbados • Saint Lucia

S.-A. Cox (✉) • P. McConney
CERMES, University of the West Indies, Cave Hill, Barbados
e-mail: shellyann_cox@yahoo.com; patrick.mcconney@cavehill.uwi.edu

S. Jentoft, R. Chuenpagdee (eds.), *Interactive Governance for Small-Scale Fisheries*, MARE Publication Series 13, DOI 10.1007/978-3-319-17034-3_30

Introduction

Although attempts to manage fisheries conventionally, using top-down approaches, have seldom been successful, initiatives for bottom-up adaptive collaborative forms of governance often expose deficiencies in capacity, communication, trust and other factors said to favor success (McConney and Parsram 2008) and hence governability. However, some fishery social-ecological systems (SES) seem to be potentially more governable than others. For example, small-scale coastal fisheries, especially for small stocks of relatively sedentary species such as sea urchins, are increasingly seen as good candidates for still-novel governance arrangements such as adaptive co-management (ACM) (Parker and Pena 2006; Léopold et al. 2013). ACM is "a flexible system of resource management, tailored to specific places and situations, supported by, and working in conjunction with, various organizations at different scales"(Armitage et al. 2009). Switching from conventional management to ACM often occurs in a crisis that brings policy-makers, resource managers and resource users together, perhaps for the first time, and they need reasonable assurance about what will work better since planning ahead and setting up ACM arrangements often requires the stakeholders to make high investments of time, money and human resources (Shotton 2000).

Determining the appropriate investment in governance calls, therefore, for a risk management perspective that involves assessing fishery governability to inform interventions. The governance of sea urchin fisheries in the Eastern Caribbean is characterized ecologically by highly variable and vulnerable populations that can be impacted by both a changing environment and human activity. The social side features low levels of management, leadership, communication and political will compared to other economic sectors. How then should resource users and government authorities approach fisheries in which the natural system-to-be-governed is highly dynamic? How do these systems affect governability given the challenges of low capacity and limited governing interactions commonly observed in small island developing states? How can these challenges be addressed by interactive governance? Despite willingness to learn-by-doing, political and practical reality does not allow responsible decision-makers to proceed blindly? Fisheries stakeholders will wish to gauge the conditions for, and chances of, successful ACM to decide what time and resources are reasonable for them to invest in an iterative process of learning-by-doing in which outcomes are negotiated (Pomeroy et al. 2004). However, very little practical guidance is currently available for this.

This chapter provides a governability assessment of establishing, implementing and sustaining ACM using cases from the sea urchin fisheries of Barbados and Saint Lucia at five study sites. The cases are drawn from a larger body of research on marine resource governance in the Eastern Caribbean. First we summarize the frameworks used in the ACM research in order to show the links to governability. Next we provide a situation analysis using the governability framework and present the cases from which lessons can be learned. We assess the potential for co-governance grounded in an understanding of social-ecological fisheries systems.

Governability Challenges and Frameworks

Fisheries governance in the Eastern Caribbean is generally constrained by limited capacity in community and state fisheries organizations and institutions, low levels of leadership, inadequate information exchange and low political priority compared to other economic sectors (CERMES 2007). Lacking are the positive interactions among stakeholders needed to sustain fisheries management (Chakalall et al. 2007). These governability issues emphasize the need to build capacity and networks for adaptation and resilience into present and planned marine resource governance (CERMES 2007).

Fisheries governance in the Eastern Caribbean is typically hierarchical and centralized in state authority. Government management of fisheries resources is usually crisis-driven, where actions are reactive and command-and-control rather than proactive and collaborative (Martin and Pope 2011). The concept of co-management is not clearly understood by resource managers and users who have different interpretations of what they think it means and can achieve (McConney et al. 2003b). Managers consider that just consulting resource users before, or even after, decision-making constitutes co-management, while resource users seek more meaningful interaction to foster learning and innovation under conditions of uncertainty (McConney and Pomeroy 2006). When different views are in tension, stakeholder interpretations need to be reconciled through learning-by-doing to improve governability in preparation for ACM. With the up-front costs being potentially high, the benefits of ACM need to appear attainable in order to achieve stakeholder buy-in from the start.

The potential benefits of ACM include community-based economic and social development, decentralized resource management decisions and a mechanism for reducing conflict through participatory democracy (Armitage et al. 2007). ACM also links community and government management authorities in a partnership that should increase resilience to misfortunes and to external influences that undermine sustainability initiatives.

In the research, upon which this chapter is based, the questions posed in the introduction guided an investigation of the potential for implementing successful ACM in the sea urchin fisheries of Barbados and Saint Lucia. In the original research, the evaluative framework of Plummer and Armitage (2007) was substantially modified (Fig. 30.1) and used to examine conditions favoring the success of ACM to suggest interventions that should enable ACM to be successful and contribute to practical approaches for marine resource governance in the Eastern Caribbean and beyond. The original research revealed that all the conditions that favor success were not yet manifested, but strategic interventions may serve as a pathway to governability.

We re-evaluate the findings from the original research using the governability assessment framework in order to obtain additional insight into approaching ACM of sea urchin fisheries in Barbados and Saint Lucia. The governability assessment framework adds value by assessing, not just conditions generally, but specifically

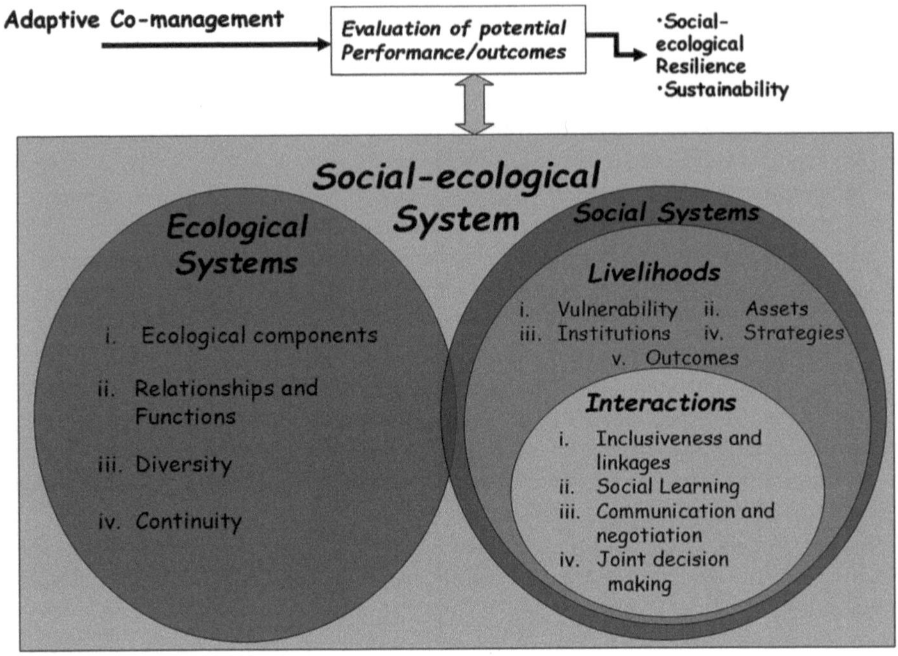

Fig. 30.1 Modified evaluative framework for ACM (Adapted from Plummer and Armitage 2007)

the system-to-be-governed, the governing system and governing interactions taking into account system diversity, complexity, and dynamics. A critical challenge to governability generally and ACM specifically is to first be able to comprehensively understand sea urchin SES sufficient to guide and direct approaches that favor success. Applying the governability assessment framework should facilitate deeper understanding.

Modifications to the original framework include putting the social system to be assessed along with the ecological system and their interactions such that the linked system is evaluated before focusing on factors relating specifically to livelihoods and governance interaction. The systems overlap to emphasize the human with nature concept (Berkes and Folke 1998), but the extent of overlap can vary with the nature of the system and the governing interactions.

Social and bio-physical agents interact at multiple temporal and spatial scales. Inter-disciplinary research requires some system disaggregation dissection for analysis. The modified framework provided guidance for investigating if or how conditions favorable to ACM are manifested in the sea urchin fisheries of Barbados and Saint Lucia. The findings can be related to the governability framework.

These fisheries are currently under a hybrid of consultative and collaborative co-management (Pomeroy et al. 2003). Fisheries authorities collaborate with resource users on abundance surveys and other ecological matters, but tend to only consult on decisions regarding fisheries management and policy. Therefore, it was the

Table 30.1 Conditions for successful ACM relating to interactions

Condition	Explanation
Access to adaptable portfolio of management measures	Participants must have flexibility to test and apply a diversity of management measures or tools to achieve desired outcomes. There must be provisions for the use of economic, regulatory, and collaborative tools.
Commitment to support a long term institution-building process	Success is more likely where stakeholders accept the long-term nature of the process, and recognize that a blueprint approach to governance arrangements is probably not advantageous.
Provision of training, capacity building, and resources for local and national and regional level stakeholders	At the local level, resources that facilitate collaboration and effective sharing of decision making power are required. Regional and national level entities must also be provided with the necessary resources.
Key leaders or individuals to champion the process	Key individuals are needed to maintain a focus on collaboration and the creation of opportunities for reflection and learning. These individuals should be well respected and have a long-term connection to the resource of concern.
Openness of participants to share and draw upon a plurality of knowledge systems and sources	Both expert and non-expert knowledge can play productive and essential roles in problem identification, framing, and analysis.
National and regional policy environment explicitly supportive of collaborative management efforts	This support can be articulated through national legislation and the willingness to distribute functions across organizational levels. Additionally, consistent support across policy sectors will enhance the likelihood of success, and encourage clear objectives, provision of resources, and the devolution of real power to local actors and user groups.

Adapted from Armitage et al. (2009)

potential for, rather than actual, ACM that was researched. The investigation of conditions assessed if specific conditions could likely be developed and sustained or if the situation did not facilitate the development of particular conditions. There can be successful ACM even if all conditions are not met (Armitage et al. 2007). Consensus is growing, however, that the more conditions that are satisfied in a particular situation, the greater are the chances for successful implementation of any type of co-management (Pomeroy 2007) and hence governability. Table 30.1 explains the conditions for successful ACM as they relate to governing interactions. Drawing upon SES and ACM concepts, we re-examine the urchin fisheries through a governability lens.

Sea Urchin Social-Ecological System Perspective

Fisheries for the white sea urchin (*Tripneustes ventricosus*) in the Eastern Caribbean have been experiencing highly variable population abundance over the past decade. Managing the fisheries for sea urchins, commonly referred to as sea eggs in Barbados

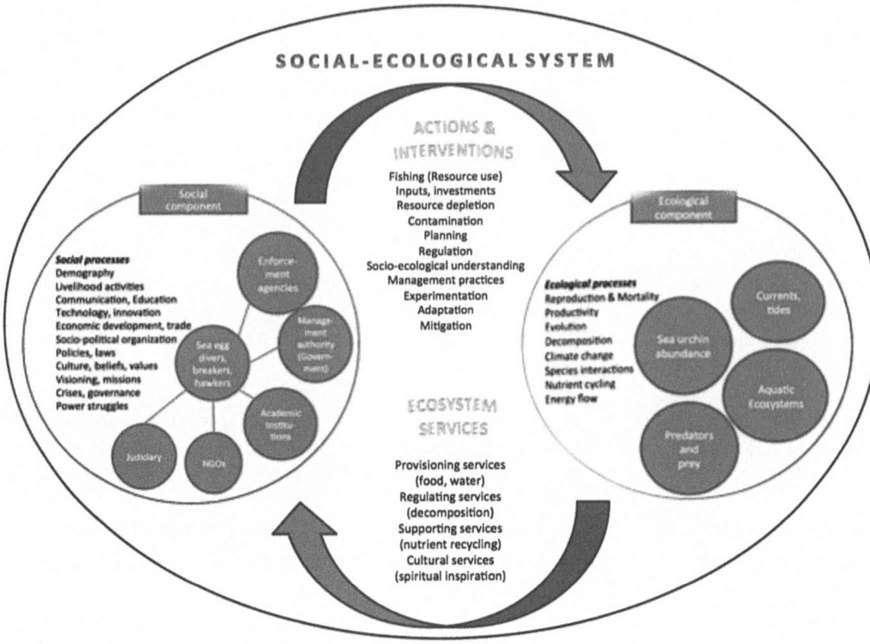

Fig. 30.2 Sea urchin fisheries of Barbados and Saint Lucia illustrated as a social-ecological system

and *chadon* in Saint Lucia, has proven problematic in the two island states. Several attempts at co-management have not yielded entirely successful outcomes in either country, but the search for solutions that work continues. A closer look at co-management with empirical data can help us understand what conditions favor or do not favor success. These fisheries have provided supplemental income for families living in coastal communities, shaped their history and influenced the cultural undertones of communities that traditionally take part in harvesting activities. Yet, the sedentary nature of urchins and their preference for shallow, near-shore habitats makes sea urchins vulnerable to over-exploitation. Both social and ecological factors, and their interactions, must be appreciated.

The ecological components of the sea urchin fisheries in Barbados and Saint Lucia include the white sea urchin population, its predators and prey, other non-human living organisms, physical and chemical features that maintain urchin habitat and environment, and the interactions of the living and non-living components of the entire system (Fig. 30.2). The social component includes actors such as policy advisers and makers, management authorities (fisheries and coastal divisions), other government ministries, enforcement agencies (Coast guard, Marine Police), non-governmental organizations (particularly environmental), fisherfolk organizations, academic institutions, resource users, other civil society stakeholders and the consuming public (Fig. 30.2). If shown as a value chain, the actors would also include overseas markets in the case of Saint Lucia. So there are local, national and regional levels to the sea urchin fishery SES even though only the first two levels are of primary concern for governance.

Social institutions include the formal legal systems, cultural traditions and the informal rules governing people's behavior, among other things. These can be influenced, for example, by the technology and technological development process in the fishery. Institutional economic aspects include the formal (e.g. markets and formal property rights) and informal (e.g. barter system and informal communal property rights) arrangements that the communities and countries have developed for allocating resources and exchanging goods and services.

Figure 30.2 illustrates a sea urchin fishery as a SES. It shows the elements of the social and ecological components as well as the processes that integrate them. Ecosystem services are highlighted along with the actions and interventions that link the social and ecological components.

Key SES Components

Ecology

In the Caribbean, mainly the white sea urchin is targeted by small-scale fisheries. These animals share common coastal inhabitants of the central Atlantic Ocean. Geographic distribution extends from the West African coast to the Gulf of Mexico, as far south as Brazil and north as mid Florida (Mahon and Parker 1999). Small-scale but commercially important fisheries occur in Barbados, Martinique, Saint Lucia and Grenada (McConney et al. 2003a, b), while minor subsistence fisheries occur in St. Vincent and the Grenadines (Pena et al. 2010). The sea urchin fisheries of Barbados and Saint Lucia are the only eastern Caribbean fisheries in which co-management attempts have been documented.

In Barbados especially, and Saint Lucia to a lesser extent, the roe of sea urchins have a high cultural significance that is exhibited in the harvesting, processing and preparation of this delicacy. Artists and poets have expressed the importance of the fishery in photographs, paintings and poetry. Sea urchins are distributed around much of Barbados but end to be most abundant along the north-east and south-east coasts. They can be found in relative shallow areas where the habitat is sea grass, coral rubble, algal pavements or rocky flats. The fishing grounds extend from Oistins in the south to Speightstown in the northwest to a maximum depth of 10 m, although some fishermen have reported seeing them deeper (Mahon and Parker 1999). Fishing grounds are not found on the west coast of the island because urchins are seldom found on living coral reef. The two Barbados study sites for the ACM research were Silver Sands and Conset Bay (Fig. 30.3). In Saint Lucia, the ecology is similar and the fishing grounds extend from Choisel in the south-west to Gros Islet in the north-west. The three Saint Lucia study sites for the ACM research were Laborie, Vieux Fort and Anse Ger (Fig. 30.4).

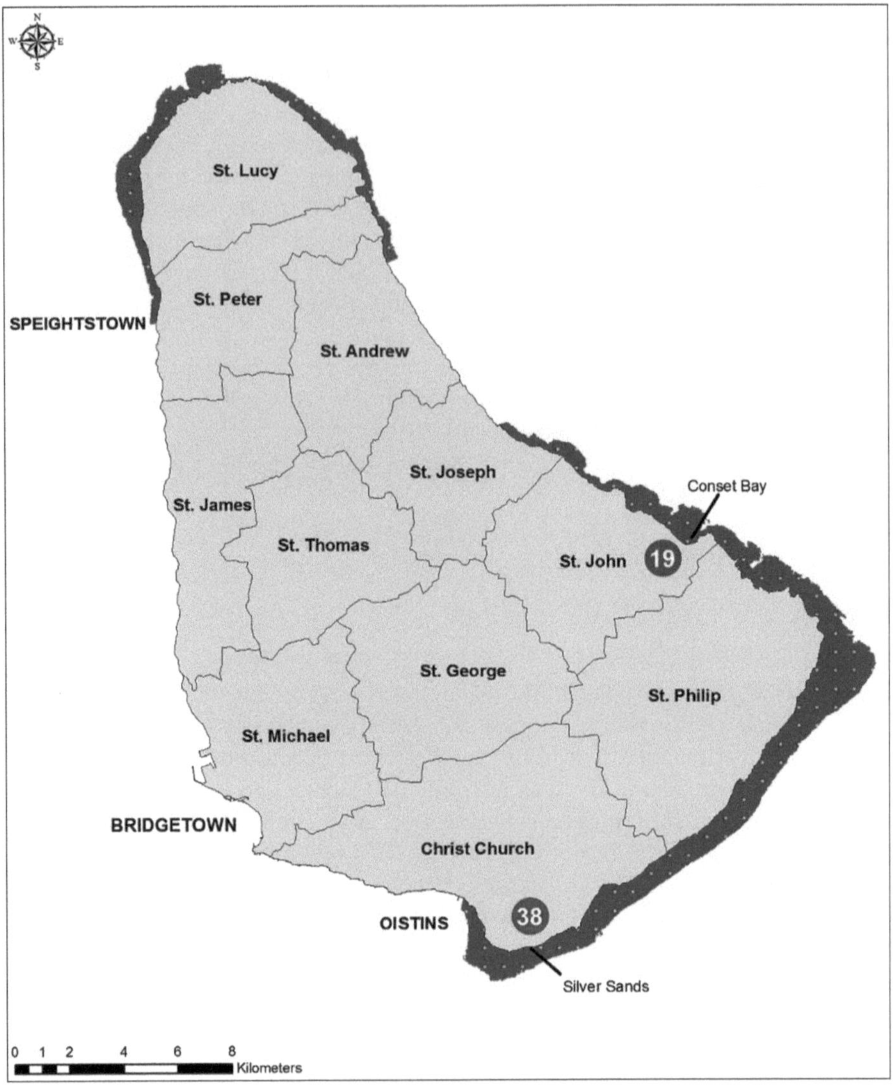

Fig. 30.3 Sea urchin fishing grounds in Barbados showing the number of divers at the two study sites

Livelihoods

In Barbados it is estimated that 200 persons compared to 1,000 in the 1950s are involved in the sea urchin fishery (McConney et al. 2003a). In comparison, approximately 156 persons are involved in the urchin fishery in Saint Lucia (DOF 2009). Sea urchin fisheries provide supplemental income for persons living in

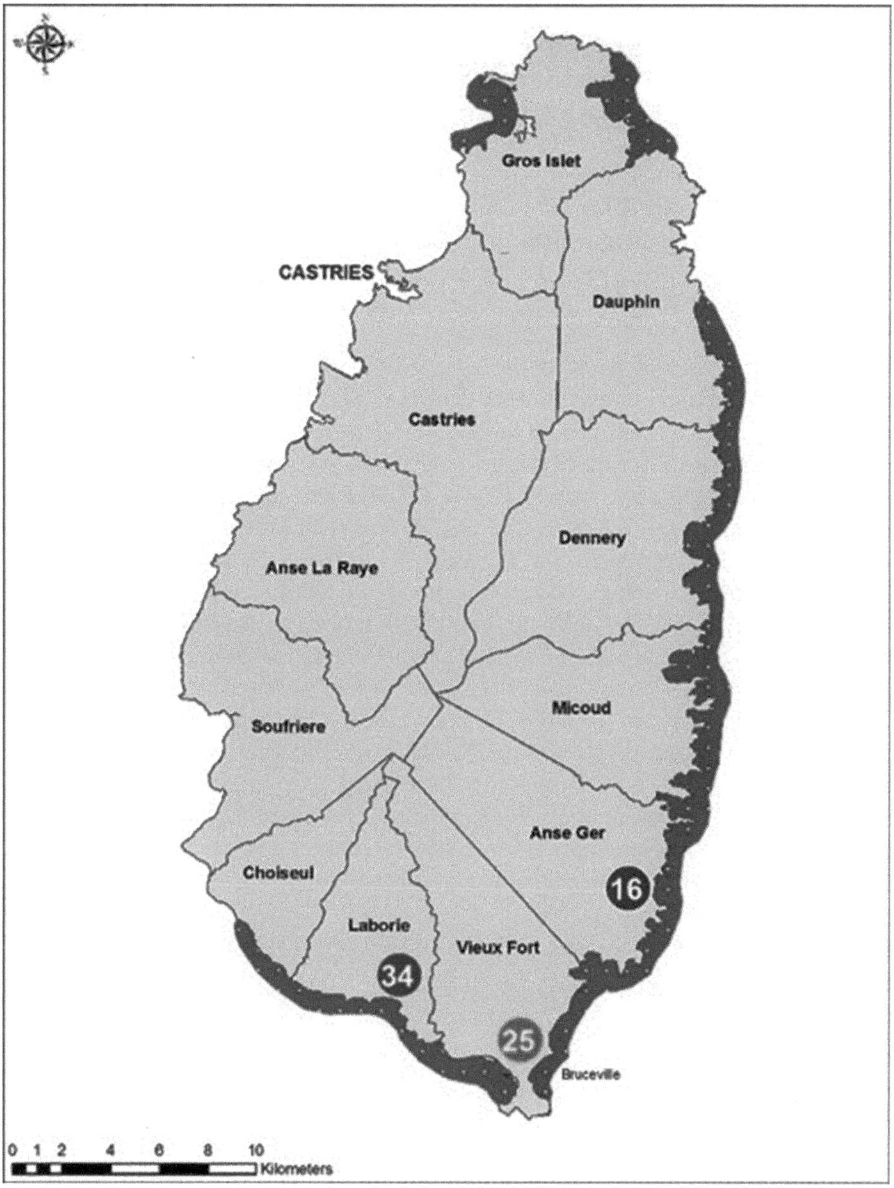

Fig. 30.4 Sea urchin fishing grounds in Saint Lucia showing the number of divers at the three study sites

coastal communities when they are not employed in the seasonal (November to June) pelagic fisheries. Residents of coastal communities become involved in the fishery because of their love for the sea and because of family members who pass down ecological knowledge and instructions for mastering harvesting and processing skills. During sea urchin open seasons of variable timing and duration, the most recent being 2009 in St. Lucia and 2004 in Barbados, processing groups work on the beach front creating an atmosphere for information exchange, social interaction and economic activity. Entire households may be involved in different aspects of the fishery. Divers are usually male, but breakers and vendors are often women and sometimes children.

Preparations for the fishery are deeply socially and culturally embedded activities that also provide employment. In Barbados persons specialize in knitting sea urchin storage nets which are sold for an average price of US$12.50 per net (Cox and McConney 2012). Other preparations in Barbados, which employ the skills of persons involved in the fishery but do not directly generate income, include folding the sea grape leaves into cones for the traditional marketing method (see Fig. 30.5a). In addition, the agave flower stalk is dried and used as a floating maypole from which storage nets or bags are suspended. In Saint Lucia preparations involve constructing traditional rafts called 'dories' or 'shaloops' in Creole (Pena et al. 2010) which were made from banana stalks in the past and most recently constructed using wood from breadfruit trees (Fig. 30.5d) (Cox and McConney 2012).

During the last open season in Barbados in 2004 the price for a two liter container filled with roe, the volume taken from about 120 sea urchins, fetched a price of US$60 (S. Cox unpublished data). In 2009, Saint Lucian sea egg vendors sold roe by the pound for the price of US$11, and charged US$6 for the traditional 'cou-cou' (Fig. 30.5b) (S. Cox unpublished data). Sea urchin divers in Barbados make on average a weekly income of US$600, breakers US$275 and hawkers US$500. In contrast, Saint Lucian divers make US$407 a week during the open season and breakers make US$194 (Cox and McConney 2012). Urchin fisheries are hence highly lucrative, especially considering the quite low investment required for entry. In the absence of an open season sea egg divers, breakers and vendors are compelled to find often scarce alternative income earning activities.

Governing System and Governing Interactions

Barbados

In Barbados, legally designated annual closed seasons, which have varied in duration since the early 1900s, have been the primary management measure to protect sea urchins from harvest during their peak reproductive period (Pena et al. 2010). The sea urchin fishery, under legal regulation since 1879, was the first single species fishery in Barbados to be managed (Scheibling and Mladenov 1987; Parker 2002; Vermeer et al. 2005). The Sea Egg Preservation Act of 1879 was enacted at a time when many still believed that the ocean's resources were inexhaustible, making it probably one of the oldest fisheries management legislations in the world. Other management measures are outlined in the Fisheries (Management) Regulations of 1998. The regulations include provision for the Minister to designate

Fig. 30.5 Traditional features of the sea egg fishery: (**a**) Traditional marketing of sea eggs using sea grape leaves (Barbados) (**b**) Traditional marketing of sea eggs in Saint Lucia (**c**) Maypole (agave flower stalk) with a net attached (**d**) Traditional raft called a Dory (**e**) Motorized day boat (Photo credits (a) E. Fitzpatrick (b) S. Cox (c) G. Franklin (d) S. Cox (e) C. Parker)

closed areas (not yet used) and seasons (common). Currently implemented regulations concern trade and gear restrictions, including prohibition of scuba diving to catch sea urchins and a restriction to store and trade sea urchins during close season, except with permission from the Chief Fisheries Officer.

In 1993, a preliminary fishery assessment recommended a consultative co-management arrangement for the white sea urchin fishery involving fishers and government (Vermeer et al. 2005), considering the nature of the fishery and the past difficulty with enforcing the closed season. This was followed by a full scale

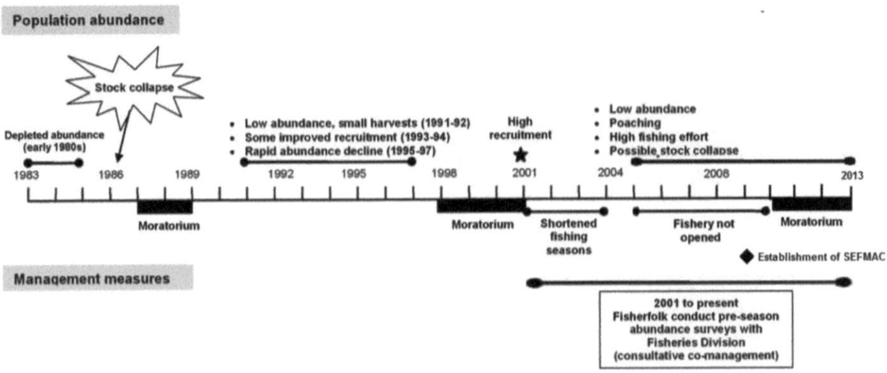

Fig. 30.6 Thirty year summarized management history of the sea urchin fishery in Barbados (Adapted from Pena et al 2010)

consultation which was a part of a 1998–1999 NGO-led co-management demonstration project that was also closely followed by a government-led coastal zone management project. This project used the sea urchin fishery as a demonstration case to investigate the potential of developing community-based coastal co-management practices for highly exploited marine fishery resources (Mahon et al. 2003). The implementation of annual pre-season abundance surveys conducted collaboratively between fisherfolk and the Fisheries Division began in 2001. The results of these surveys are now used, in part, to determine the length and timing of the sea urchin fishing season (Fisheries Division 2004).

Over the last 30 years, several management measures have been implemented to address the fluctuating population abundance of the urchins (Fig. 30.6). These measures were implemented in response to the depletion of sea urchin abundance which resulted in a stock collapse. Three moratoriums were implemented from 1986 to 2013 with the intention to allow the stock to recover.

In 2001, there was a high recruitment which led to the opening of the season for four consecutive years. Also in 2001 a process of consultative co-management was initiated. It became the major endeavor from 2005 to 2010. In each of these years fisheries authorities and resource users collaboratively advised policy makers that the fishery should remain closed due to low urchin abundance assessed through habitat surveys. In 2009, a co-governance body, known as the Sea Egg Fisheries Management Advisory Council (SEFMAC), was established to formalize and build the capacity of the co-management arrangements.

Saint Lucia

Commercial harvesting of sea urchins in Saint Lucia only began in the 1960s. After 1983, several management measures were implemented to cope with the fluctuating population abundance of the sea urchins (Fig. 30.7). The Caribbean

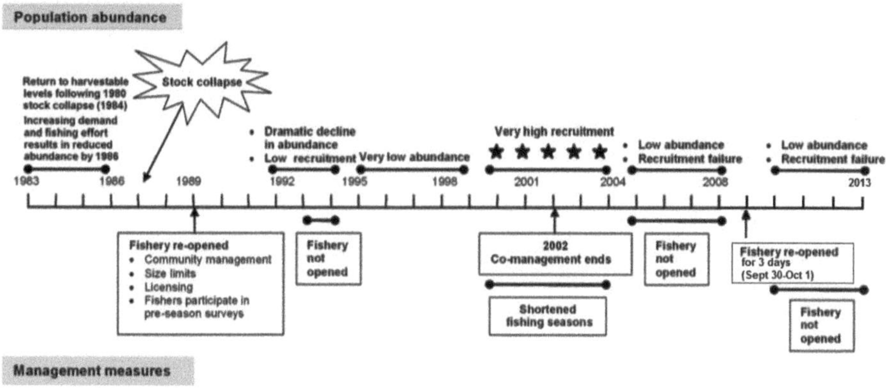

Fig. 30.7 Thirty year summarized management history of the sea urchin fishery in St. Lucia. (Adapted from Pena et al 2010)

Natural Resources Institute (CANARI) in collaboration with the Department of Fisheries (DOF) conducted a monitoring programme. Subsequently, a new system of participatory management was adopted in an attempt to avoid the over-exploitation of the past. Tools implemented for controlling harvesting included size limits, limited entry by licensed urchin divers, and compulsory participation of fishers in pre-harvest surveys to determine the timing of the closure of the fishery (Smith and Koester 2001). The fisheries legislation allowed for the development of participatory management. Saint Lucia first attempted to implement community based management in 1989 after a resource collapse in 1987. The management authority selected individuals to be licensed who were perceived as being concerned with the welfare of the resource. They selected divers who had assisted in pre harvest surveys and attended meetings to discuss the state of the fishery.

Around the time co-management was beginning in Barbados it began to weaken in Saint Lucia. Management measures attempted in Barbados and Saint Lucia produced different outcomes (Figs. 30.6 and 30.7). In both countries different measures were attempted after stocks collapsed. Saint Lucia instituted a co-management approach which included fishers participating in pre-season surveys coupled with other measures such as size limits and licensing. These implementations resulted in the fishery re-opening in 1989. There was a dramatic decline in abundance from 1991 leading to another closure in 1993 and 1994. In contrast, Barbados instituted a 3 year moratorium (1987–1989) after the collapse of the stock. However, there was not an open season until 1991. After that population abundance decreased eventually leading to another moratorium being implemented in 2001. It was not until this time that co-management was attempted.

Governance Interventions

In this section we examine two governance interventions that occurred in Barbados and Saint Lucia after 2009, allowing us to gauge the governability of the urchin social-ecological fisheries systems of both countries. In Barbados we examine the establishment of the sea egg fisheries management advisory council (SEFMAC), while in Saint Lucia we review a short fishery re-opening that occurred in 2009 (see Figs. 30.6 and 30.7). We gauge the chances of successful ACM based upon observed conditions, interactions and outcomes. This allows us to also determine what additional governing interactions may be necessary to enhance the pathway to success.

Establishment of SEFMAC in Barbados

SEFMAC was first considered in 2003 by the Fisheries Division during the implementation of the Coastal Resources Co-management pilot project on the white sea urchin fishery. The Fisheries Division recommended the council in the Barbados Fisheries Management Plan 2004–2006. The function envisaged for the council at that time was primarily to obtain information on the status of the stock. That information would then be used to provide advice and recommendations to the Chief Fisheries Officer. The fishers representing their respective parishes would be tasked with directly interacting with and gathering information from people at the community level to facilitate the work of the council and reduce the transaction costs of the Fisheries Division.

A project proposal was drafted to establish SEFMAC with member selected or selected by the fishing communities. This was to be followed by a planning meeting of members in May 2005 to devise operational rules for their council. The project was not implemented due to a lack of government follow up and there were no more attempts to establish the council until 2008 when the Fisheries Advisory Committee revived its formation. In July 2008, the Barbados Fisheries Advisory Committee was awarded a small grant under the project on Marine Resource Governance in the Eastern Caribbean to strengthen its role in the governance of fisheries resources in Barbados. The governance experiment specific to sea urchins in this project was the formation of SEFMAC (Pena et al. 2010).

Under this project the first meeting on the formation and membership of the SEFMAC took place in August 2009. Attending was one representative of the Fisheries Advisory Committee, one from the Fisheries Division and a fisheries scientist. Members were chosen based on their participation in the pre harvest sea urchin surveys and interest in the management of the fishery in the past. Stakeholder representatives participated at the inaugural meeting of SEFMAC in September 2009. These included:

- Fisheries Division
- The Coastal Zone Management Unit (CZMU)
- The Marine Police

- A fisheries scientist from the University of West Indies' Centre for Resource Management and Environmental Studies (CERMES)
- A member of the Fisheries Advisory Committee (FAC)
- A representative from the Barbados National Union of Fisherfolk Organizations
- Sea urchin divers from the parishes of Christ Church, St. Philip, St. John and St Lucy.

To encourage transparency and wider participation, other interested parties were invited to council meetings to observe proceedings and participate in discussions. These observers included additional fishers who were part of the survey teams and a CERMES PhD student. Other SEFMAC rules allowed for the formation of working groups chaired by members of the council that could include non-members so as to enhance participation by resource users.

Two more council meetings took place in late 2009 and early 2010 to discuss a draft sea urchin management plan and a communications strategy. There was no follow up to these initiatives but annual meetings between fisheries officials and fishers on abundance surveys continued. Each year the decision was made to close the fishery but there were no attempts to use suggestions made during SEFMAC meetings or to revive SEFMAC. It was abandoned by the Fisheries Division in favor of a less participatory governance arrangement in which sea urchin divers conducted urchin surveys but did not have a collaborative role in fishery management. Industry representatives and other members of SEFMAC did not protest or take remedial action.

Interview respondents said that SEFMAC was just a part of the co-management arrangement. It was necessary but not sufficient to achieve favorable co-management outcomes. The council needed to be supported by enforcement and compliance to reduce illegal urchin harvesting order to be taken seriously. With respect to the design of the council, it was suggested that other stakeholders such as breakers and vendors, and perhaps a representative from the judicial system be included to provide other valuable perspectives. Additionally, the inclusion of a non-governmental organization (NGO) could introduce a capable player into the council once conflicts did not arise as a result of power dynamics between resource use and conservation. An alternative association of sea urchin divers could complement SEFMAC to support ecological stock rehabilitation experiments. Other respondents suggested that reviving the council should engage core groups of stakeholders to choose leaders as members of the council. In addition, letters of appointment issued to these champions could add formal recognition and seek their long term commitment to the cause. Furthermore, the scope of the council should be appropriate for manageable tasks to be achieved within its capacity and guidance could be drafted for the roles of council members and communication to policy advisers and makers. Finally, they noted that small successes would encourage further action from council members and fortify commitment. These measures, they said, could improve equity, accountability and transparency, while establishing SEFMAC under the fisheries regulations improve stakeholder buy-in.

If legally established, resource users would view SEFMAC as a legitimate body in which their voices could be heard by higher authorities and be included in decision-making. For government agencies specified in the composition of the council, making SEFMAC a statutory body would ensure that its sustainability becomes a part of their mandate and they should see the council as a priority. Legalizing SEFMAC would enhance the council's visibility to the public where the council could become the official multi-stakeholder body for communicating key messages in order to increase compliance with regulations.

The council was an attempt at ACM but little was achieved or learned for adaptation because of its brief existence. Recommendations for adapting and testing were available but the means to try them proved to be elusive. The lack of follow up indicated that neither the fisheries authority nor the sea urchin fisher folk were ready to move to ACM.

Three Day Open Season in St Lucia, 2009

From July to September 2009 the Department of Fisheries undertook a number of sea urchin surveys in key harvest locations to determine the status of the stocks. Prior to the sea urchin population survey there was a briefing meeting for the staff at the Department of Fisheries. The meeting outlined the plans for the upcoming season and recapped the iterative, rather than linear, fishery management decision-making sequence illustrated in Fig. 30.8 that shows the differences between technical and policy domains and roles.

Fishers assisted with the surveys by accompanying fisheries officers on the boat and pointing out the areas where sea eggs can be found. In most cases the vessel belonged to the fisher who was compensated for the fuel used. The surveys were done visually and by collecting samples to measure the size frequency and test the ripeness of the roes. Results showed a very low density of sea urchins in all areas except Laborie, which had a very high population of sea urchins. Gonads sampled, with a few exceptions, were ripe and firm, and hence ideal for harvest.

Despite the poor condition of sea urchin stocks, the Department of Fisheries authorized a 3-day island-wide opening from 29th September to 1st October 2009. This decision was based on a number of factors which included the biology and life span of sea urchins, the age structure of the sampled population, level of maturity of gonads, habitat status, food availability, as well as management advice from fishers who were engaged in the surveys. There was no stakeholder meeting to advice on the fate of the fishery. It was a unilateral, but consultative, decision by the fisheries authority as the Department of Fisheries was said to have taken into consideration the advice of fishers to briefly open the fishery. Two weeks before the opening, policy approval to make it legal was sought through the hierarchical governance process shown in Fig. 30.8.

After 5 years of being closed, the re-opening of the sea urchin fishery was highly anticipated. The fishing season was announced over the radio instead of being gazetted in print since there was not enough time for the authorities to process the notice. Although, official announcement is the responsibility of the Cabinet of ministers, the Department of Fisheries directly informed media houses and the fisheries co-operatives. Their intention was to inform the general public so that they could be prepared for the fishery.

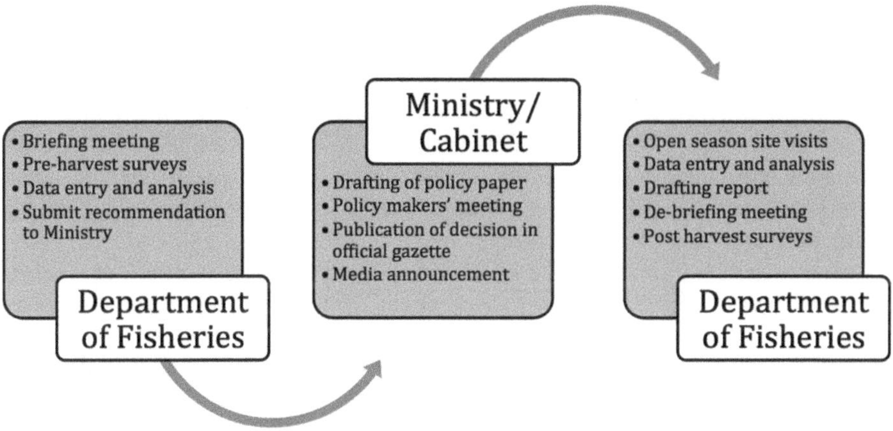

Fig. 30.8 Process coordinated by the Department of Fisheries in Saint Lucia during the fishing season 2009

During the 3-day opening heavy fishing pressure was observed in Laborie Bay where the sea urchins were concentrated. While low harvesting activity occurred in Vieux Fort and Dennery, no harvesting activity was reported in the north of the island. An average of 35 and 25 divers were involved in the harvesting of sea urchins in Vieux Fort and Laborie respectively while Dennery recorded an average of three divers. With regards to cleaners, Vieux Fort and Laborie recorded an average of 62 and 25 persons involved in this activity respectively while Dennery had an average of three cleaners. Divers began harvesting from as early as 5 a.m. making an average of two dives per day. A maximum of 21 boats were seen in the bay on the final day of the opening. From the sizes of sea urchins landed it was apparent that the adults had been removed.

Although Laborie residents are known to possess a sense of ownership of sea urchin resources in the bay, Vieux Fort residents fished there on all 3 days of the open season. On the first day it appeared that the residents in Laborie were offended by their presence but no physical action was taken. For the remainder of the opening there was an acceptance of their presence. When asked about the issue, a Vieux Fort resident said that in these tough economic times no one would be restricted from the opportunity to make money.

Staff of the Department of Fisheries monitored the harvest activities and collected data on the number of sea urchins landed and the number of persons involved in the harvesting and cleaning of sea urchins at key landing sites in Laborie, Vieux Fort and Dennery. The numbers of sea urchins harvested were visually estimated and recorded by fisheries officers. Following discussions with the sea urchin harvesters, the cost of one kilogram of gonads (wet weight) was set at about US$24 comprising gonads from about 110 sea urchins.

It was estimated that a total of nearly 170,000 sea urchins were harvested during the 3-day authorized harvest period from September 29 to October 1, 2009 (DOF 2009). This produced just over 1,000 kg (wet weight) of sea urchin gonads, with an estimated gross wet weight value of just over US$25,000.

Reports indicated that the economic value of the 3-day open season was not very substantial especially in Dennery. Although the overall revenue was minimal, the cultural value was more substantial. When interviewed, divers and breakers expressed their gratitude for the opportunity to taste 'chadon' again, make extra money and be given the chance to teach their children the skills to harvest sea urchins. Community members also expressed their joy in eating cou-cou again while reminiscing about the good old days when more 'dories' made from banana stalks were used to harvest sea urchins.

After reviewing the events and operations that occurred during the 3-day opening, the fisheries biologist at the Department of Fisheries made the following recommendations.

There is a need for community meetings prior to the opening of the sea urchin fishery to get input from fishers on harvest rules, thereby ensuring their greater involvement in the decision-making process and in building awareness. Secondly, regular and systematic survey methods are needed for monitoring of the traditional and possibly new areas which may have sea egg populations.

Additionally, the development of a trained group of sea urchin divers in survey techniques and other sea urchin management aspects would ease the burden on the Department of Fisheries' limited resources and increase the sea urchin divers' sense of ownership of the resource. The divers would also report to the Department of Fisheries the results of all survey activities while the Department of Fisheries would accompany them occasionally. Finally, there is a need for greater public relations through news items, bulletins and documentaries involving the Department of Fisheries, communities and other stakeholders in the fishery.

These recommendations suggest a recognition that participation by resource users needs to be enhanced and that they should play a greater role in decision-making since the command-and-control governing interactions of conventional governing system have proven to be inadequate to govern these sea urchin fisheries Capacity development to improve governance and commitment by both the Department of Fisheries and resource users may provide a pathway for ACM to enhance governability.

Discussion

This chapter offers an insight into enabling conditions, ranging from habitat to human capacity, for establishing and sustaining ACM as a means of enhancing governability. The case studies of sea urchin fisheries in Barbados and Saint Lucia examined general SES complexity and dynamics in relation to governability. We further investigated two governance interventions. Lessons learned about the system-to-be-governed, governing interactions and governing system in the cases can guide stakeholders on how to approach ACM when fisheries social-ecological systems are dynamic but governance capacity is low. According to Plummer (2009), ACM is not a governance panacea, but conditions that contribute toward successful outcomes are emerging from practice. The fisheries in this chapter are new to ACM but show potential for enhancing governability. Stakeholders in the fisheries need to be encouraged to test, learn and adapt if success is to be achieved.

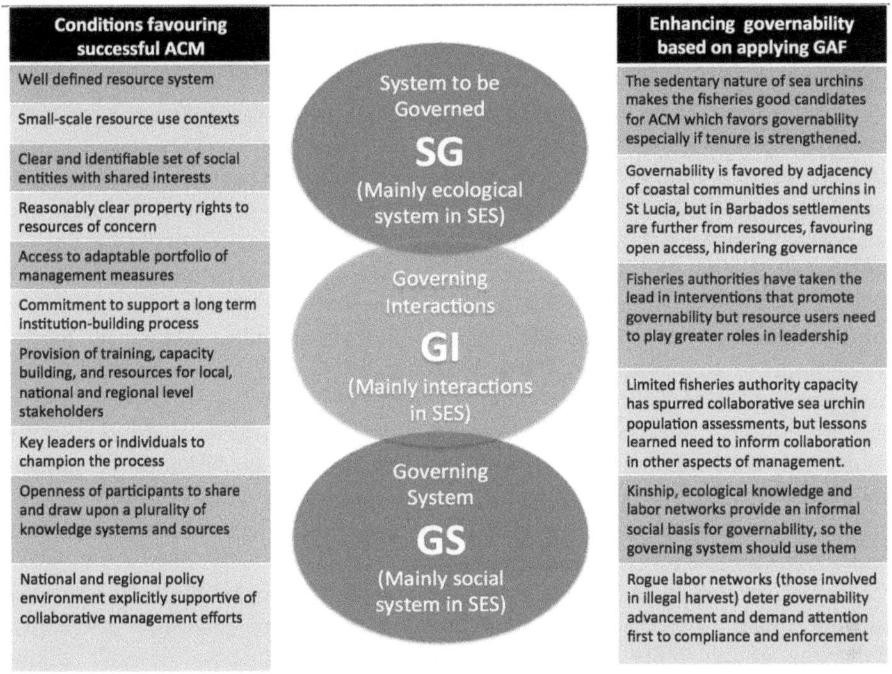

Fig. 30.9 Factors influencing governability based on applying the governability assessment framework that is compatible with SES and the conditions that favor successful ACM

The questions in the introduction were based on the practical considerations of introducing ACM under difficult conditions. Answering the questions purely from an ACM perspective takes into account the conditions illustrated on the left hand side of Fig. 30.9 (adapted from Armitage et al. 2009) as favoring success in relation to the ecological, network interaction and social aspects of SES. The figure also illustrates how the factors influencing governability based on applying the governability assessment framework are compatible with SES and the conditions that favor successful ACM. Applying the governability assessment framework adds value by providing more specific information (right hand side of Fig. 30.9) for enhancing governability. A fishery manager or other stakeholder using both perspectives can hence better guide ACM to achieve successful outcomes.

Armitage and colleagues (2009) suggest that the conditions, derived from case studies, must all be met to some extent for successful ACM, but there will be variation depending fundamentally on the system-to-be-governed. They caution that severe resource management dilemmas will overwhelm ACM. These may occur when it is difficult to identify a clear set of place-based entities linked to a defined resource stock, or where there is little commitment or incentive among participants to encourage long-term learning around a shared goal. Our application of the governability assessment framework showed that the sea urchin conditions were difficult to achieve but not overwhelming. Problems with urchin abundance and distribution were more a feature of failure in the governing system than of the system-to-be-governed.

Ensuring that all ACM conditions outlined in Fig. 30.9 are present in the social-ecological system will be challenging. The current state of the fisheries in both countries may not allow for all favorable conditions to be present or developed, but conditions that enable governing interactions should be a strong feature of the social-ecological system if challenges are to be addressed. Enabling policy clearly has a role in enhancing governability and the governability assessment framework points to the mutual dependence of the fisheries authority and resource users to create capacity as the critical element to enhance governing interactions and collaboration. Applying the governability assessment framework to SEFMAC and short season opening showed the importance of such collaborative governing interactions since neither has the capacity to successfully manage the fishery alone.

The Barbados and Saint Lucia cases combine to offer guidance that should be useful to managers and other stakeholders. The governability assessment framework suggests that high variability and uncertainty in the resource and hence harvest practices (the system-to-be-governed) need to be well matched by flexibility and adaptation in the governing system and interactions. Even if the sea urchin fisheries in both countries are depleted, ACM can make rehabilitation of habitats and recovery of stocks more feasible than other governance arrangements. The conditions of greatest concern for governability are those which relate to governing interactions initiated by the governing system.

The sea urchin fishery of Barbados demonstrated few conditions of success (review Table 30.1) due to failures in the participatory processes attempted. Enhancing governability can be an incremental process if time and resources allow. For example, re-establishing and sustaining SEFMAC could be the governing system vehicle for enabling the conditions that favor successful ACM. Stakeholders are already armed with knowledge of the factors that constrain progress and need to be addressed (Cox and McConney 2012). However, an incremental process may be overtaken by change if fast and slow variables are not carefully monitored. Just as a hurried process may not be conducive to learning and institution-building, slow governance adaptation will always face a changing landscape of crisis if unable to keep up. We see this in Barbados where inadequate attention to compliance and enforcement has almost completely undermined trust in ACM. In Saint Lucia the short opening of the fishery uncovered issues that affect governability, and stakeholders made recommendations for innovative approaches to improve ACM conditions and enhance governability. An attribute conducive to enhancing governability in Saint Lucia is that authorities and stakeholders seem relatively adaptive and open to learning as reported here.

Conclusion

The governability assessment framework is compatible with SES concepts and the ACM evaluative framework. Employing both ACM evaluation and the governability assessment framework revealed that conditions favoring successful ACM, as a pathway to enhancing governability are lacking in the sea urchin fisheries in Barbados and Saint Lucia. We uncovered some circumstances that constrained favorable conditions. Lessons learnt from these cases can inform the governing interactions to enhance the conditions favoring governability.

The governability assessment framework provides a practical way to gauge the conditions and chances of success before or in the early stages of ACM so that stakeholders can determine if the investment is worth the risk. Another value of the framework is the perspective it provides on governability by assessing the problems that arise if favorable conditions for ACM are not present. The governability assessment framework goes further than evaluation of ACM conditions by specifically informing interventions to enhance governability. We conclude that ACM may indeed provide a pathway to enhance the governability of the sea urchin fisheries in Barbados and Saint Lucia. Practical recommendations focus on strengthening the capacity of the governing system and collaboration among actors in governing interactions.

References

Armitage, D., Berkes, F., & Doubleday, N. (Eds.). (2007). *Adaptive co-management: collaboration, learning and multi-level governance*. Vancouver: UBC Press.

Armitage, D. R., Plummer, R., Berkes, F., Arthur, R. I., Charles, A. T., Davidson-Hunt, I., Diduck, A. P., Doubleday, N. C., Johnson, D. S., Marschke, M., McConney, P., Pinkerton, E. W., & Wollenberg, E. K. (2009). Adaptive co-management for social-ecological complexity. *Frontiers in Ecology and the Environment, 7*(2), 95–102.

Berkes, F., & Folke, C. (Eds.). (1998). *Linking social and ecological systems: Management practices and social mechanisms for building resilience*. Cambridge: Cambridge University Press.

CERMES. (2007, May). *Supplement to the report of the MarGov project inception workshop, UWI Cave Hill Campus, Barbados*. Working papers and presentations. CERMES MarGov Project Document 1 Supplement.

Chakalall, B., Mahon, R., McConney, P., Nurse, L., & Oderson, D. (2007). Governance of fisheries and other living marine resources in the wider Caribbean. *Fisheries Research, 87*, 92–99.

Cox, S., & McConney, P. (2012). The role livelihood strategies and outcomes can play in adaptive co-management of the sea urchin fisheries in Barbados and St. Lucia. *Proceedings of the Gulf & Caribbean Fisheries Institute, 64*, 335–341.

Department of Fisheries. (2009). *Sea urchin harvest report September 29-October 1, 2009* (13pp). Castries: Ministry of Agriculture, Food production, Fisheries, Co-operatives and Rural Development.

Fisheries Division. (2004). *Barbados fisheries management plan 2004–2006: Schemes for the management of fisheries in the waters of Barbados* (68 pp). Bridgetown: Ministry of Agriculture and Rural Development.

Léopold, M., Cornuet, N., Andréfouët, S., Moenteapom Z., Duvauchelle, C., Raubani, J., Ham, J., Dumas, P. (2013).*Managing sea cucumber fisheries using stock biomass estimates and spatial catch quotas: experience from New Caledonia and Vanuatu*. Working paper 8. 8th SPC Heads of Fisheries Meeting. Noumea, New Caledonia.

Mahon, R., & Parker, C. (1999). *Barbados sea eggs, past, present and future. Fisheries Division*, Ministry of Agriculture and Rural Development, Barbados Fisheries management plan. Public information document No.1.

Mahon, R., Almerigi, S., McConney, P., Parker, C., & Brewster, L. (2003). Participatory methodology used for sea urchin co-management in Barbados. *Ocean and Coastal Management, 46*, 1–25.

Martin, D. R., & Pope, K. L. (2011). Luring anglers to enhance fisheries. *Journal of Environmental Management, 92*, 1409–1413.

McConney, P., & Parsram, K. (2008). A research framework for examining the characteristics of networks that determine resilience and adaptability in marine resource governance in the English speaking eastern Caribbean. In *Proceedings of the Gulf & Caribbean Fisheries Institute*, 60, pp. 65–72.

McConney, P., & Pomeroy, R. (Eds.). (2006). *Reforming governance: Coastal resources comanagement in Central America and the Caribbean.* Final Report of the Coastal Resources Comanagement Project (CORECOMP). CERMES Technical Report No.5.

McConney, P., Mahon, R., & Parker, C. (2003a). *Barbados case study: The sea egg fishery.* Caribbean Coastal Co-management Guidelines Project. Bridgetown: Caribbean Conservation Association.

McConney, P., Pomeroy, R., Mahon, R. (2003b). *Guidelines for coastal resource co-management in the Caribbean: Communicating the concepts and conditions that favor success.* Caribbean Coastal Co-management Guidelines Project. Bridgetown: Caribbean Conservation Association.

Parker, C. (2002). The contribution of inadequate fines to the collapse of the sea egg fishery of Barbados. In *Proceedings of the Gulf and Caribbean Fisheries Institute*, 53, pp. 203–215.

Parker, C., & Pena, M. (2006). Possible paths to co-managing the sea egg fishery of Barbados. *Proceedings of the Gulf and Caribbean Fisheries Institute, 57*, 115–128.

Pena, M., Oxenford, H. A., Parker, C., & Johnson, A. (2010). *Biology and fishery management of the white sea urchin, Tripneustes ventricosus, in the eastern Caribbean* (FAO fisheries and aquaculture circular. No. 1056). Rome: FAO.

Plummer, R. (2009). The adaptive co-management process: An initial synthesis of representative models and influential variables. *Ecology and Society, 14*(2), 24.

Plummer, R., & Armitage, D. (2007). A resilience-based framework for evaluating co-management: Linking ecology, economics and society in a complex world. *Ecological Economics, 61*, 62–74.

Pomeroy, R. S. (2007). Conditions for successful fisheries and coastal resources co-management: Lessons learned in Asia, Africa, and the wider Caribbean. In D. Armitage, F. Berkes, & N. Doubleday (Eds.), *Adaptive Co-management: Collaboration, learning and multi-level governance.* Vancouver: University of British Columbia Press.

Pomeroy, R., McConney, P., Mahon, R. (2003). *Comparative analysis of coastal resource co-management in the Caribbean.* Caribbean Coastal Co-management Guidelines Project. Bridgetown: Caribbean Conservation Association.

Pomeroy, R., Mcconney, P., & Mahon, R. (2004). Comparative analysis of coastal resource Co-management in the Caribbean. *Ocean Coastal Management, 47*(9–10), 429–447.

Scheibling, R. E., & Mladenov, P. V. (1987). The decline of the sea urchin, *Tripneustes ventricosus*, fishery of Barbados: a survey of fishermen and consumers. *Marine Fisheries Review, 49*(3), 62–69.

Shotton, R. (Ed). (2000). *Use of property rights in fisheries management.* Proceedings of the Fish Rights 99 Conference. FAO Fisheries Technical Paper No. 404/1. Rome: FAO.

Smith, A. H., & Koester, S. (2001). *A description of the sea urchin fishery in Laborie, Saint Lucia.* CANARI LWI Project Document No. 4. CANARI Technical Report No. 294.

Vermeer, L. A., Hunte, W., & Oxenford, H. A. (2005). An assessment of the potential for community level management of the sea urchin fishery in Barbados. *Proceedings of the Gulf & Caribbean Fisheries Institute, 47*, 79–101.

Chapter 31
Co-governance of Small-Scale Shellfisheries in Latin America: Institutional Adaptability to External Drivers of Change

Mauricio Castrejón and Omar Defeo

Abstract The resilience of small-scale shellfisheries in Latin America is increasingly threatened by climatic and human drivers acting simultaneously at multiple temporal and spatial scales. Co-governance is emerging as a potential solution to enhance the capability of governing systems to respond to the social-ecological impacts of external drivers of change. Although there is an increasing understanding of the factors that determine the success and failures of diverse co-governance arrangements in Latin America, there is still a poor understanding about how this mode of governance responds to different crises, and how these responses are shaped by past experiences and by the features of the governing system and the system-to-be-governed. In this chapter, we evaluate how institutions learn, self-organize and respond to diverse climatic and human drivers in seven co-governance arrangements, and identify the factors that enable or inhibit building institutional adaptability. Our analysis shows that the combined impact of different drivers produced social-ecological impacts on local fishing communities' wellbeing. In this context, institutions and actors displayed coping and adaptive responses to prevent or mitigate the damage on fishery resources and fishers' livelihoods. These varied according to the magnitude, extent, periodicity and intensity of press and pulse perturbations, and were shaped by past crises, social-ecological memory and the particular social features of fishing communities in which institutions are embedded. In most cases, after severe crises, small-scale fishers took collaborative actions for re-organizing their cooperatives and their harvesting and trading strategies in order to prevent future crises and enter into more sustainable pathways. In conclusion, the same factors that promote (or preclude) high governability are also those that enable (or inhibit) building institutional adaptability and resilience.

M. Castrejón (✉)
Interdisciplinary PhD Program, Dalhousie University, Halifax, Canada
e-mail: mauricio.castrejon@dal.ca

O. Defeo
Facultad de Ciencias, UNDECIMAR, Montevideo, Uruguay

DINARA, Montevideo, Uruguay
e-mail: odefeo@dinara.gub.uy

S. Jentoft, R. Chuenpagdee (eds.), *Interactive Governance for Small-Scale Fisheries*, MARE Publication Series 13, DOI 10.1007/978-3-319-17034-3_31

Keywords Co-management • Adaptability • Social–ecological systems • Small-scale • Drivers • Latin America

Introduction

Small-scale shellfisheries in Latin America and around the world are increasingly threatened by climatic and human drivers acting simultaneously at multiple temporal and spatial scales (Perry et al. 2010; Hall 2011; Defeo et al. 2013). Climatic drivers, such as El Niño Southern Oscillation (ENSO), affect habitats and ecological patterns and processes of target and non-target species, causing changes in habitat suitability, biogeography and demography, as well as modifications to dispersal, feeding, growth and behavioral patterns (Badjeck et al. 2009; Ortega et al. 2012). The negative changes induced by climatic drivers (e.g., mass mortalities) are exacerbated by human drivers (i.e., market globalization and weak governance), leading to loss of resilience in small-scale shellfisheries (Defeo and Castilla 2012). In this context, resilience refers to the capability of fishing communities and institutions (e.g., cooperatives, fishery agencies) to cope with, adapt to, and shape change to sustain a fishery system within a desirable state (Berkes and Folke 1998; Folke et al. 2003).

Co-management is emerging in Latin America as a potential solution to enhance the capability of the governing system to respond to the social-ecological impacts of external drivers of change (see e.g., Micheli et al. 2012 and McCay et al. 2014). Co-management is defined as "*a partnership arrangement in which government, the community of local resource users (fishers), external agents (non-governmental organizations, academic and research institutions), and other fisheries and coastal resource stakeholders (boat owners, fish traders, money lenders, tourism establishments, etc.) share the responsibility and authority for decision-making in the management of a fishery*" (Berkes et al. 2001, 255). Within the interactive governance perspective, co-management is conceptualized as a form of "co-governance" where fishers, managers, scientists and other stakeholders collaborate and cooperate to improve the governability of small-scale fisheries (Kooiman and Bavinck 2013).

Although there is an increasing understanding of the factors that determine the success and failure of co-governance arrangements in Latin America (Castilla and Defeo 2001; Sosa-Cordero et al. 2008; Gelcich et al. 2010; Gutiérrez et al. 2011; Defeo et al. 2014; McCay et al. 2014), there is still a poor understanding about how this governance mode responds to different types of crises, and how these responses are shaped by past experiences and by the particular features of the governing system and the system-to-be-governed. Such knowledge is important to understand how co-governance institutions and actors learn, self-organize and respond to diverse climatic and human drivers, as well as to design policies aimed at maintaining or increasing resilience in small-scale fisheries (Adger et al. 2005; Badjeck et al. 2009).

Coping responses occur on short time scales and allow a system to survive a crisis without being altered, while an adaptive response occurs on longer time scales when permanent changes in the system are required to survive a crisis (Perry et al. 2011). Coping and adaptive capacity are a related-resilience aspect that reflects the adaptability of a system, i.e., the ability of institutions (or individuals) to learn and store knowledge and experiences to address new challenges, as well as the flexibility to experiment and adopt novel solutions (Walker et al. 2002). According to the interactive governance approach, adaptability is a key characteristic of the governing system and system-to-be-governed that contributes to their governability. The system-to-be-governed can be seen as a social-ecological system comprised of two subsystems (human-social and biophysical) that operate through interdependent feedback relationships (Ostrom 2009; Perry et al. 2010). The governing system represents the institutions and organizations that have a steering role in fisheries governance (Ostrom 2009; Kooiman and Bavinck 2013), which in turn are embedded within the social component of the system-to-be-governed. In the context of Latin America, two key questions are: (1) how co-governance arrangements in small-scale shellfisheries respond to different climatic and human drivers; and (2) what factors enable or inhibit building institutional adaptability. To address these issues, we characterized and compared seven co-governance arrangements in order to evaluate how they responded to different types of external drivers, and to identify the factors that enabled or inhibited building institutional adaptability.

Methods

Case Studies

We examined seven small-scale shellfisheries involving three different groups of benthic resources: crustaceans (lobsters), mollusks (bivalves and gastropods), and echinoderms (sea cucumbers). The case studies selected (Table 31.1) are from Mexico (spiny lobsters *Panulirus interruptus;* abalone *Haliotis corrugate* and *H. fulgens*), Ecuador (spiny lobsters *Panulirus penicillatus* and *P. gracilis;* sea cucumber *Isostichopus fuscus*), Uruguay (yellow clam *Mesodesma mactroides)* and Chile (surf clam *Mesodesma donacium*). The case studies were selected based on the availability of peer-reviewed and grey literature, as well as considering our first-hand experience. We also considered the following selection criteria: (1) target species are coastal shellfishes, whose extraction is restricted to intertidal and shallow subtidal habitats; (2) resources are harvested through artisanal fishing methods, including hand-gathering, dredging, diving, and trap deploying; (3) fishers have some kind of informal or formal organization, including cooperatives, associations, syndicates, and/or federations; (4) there is a co-governance arrangement implemented and recognized by local institutions or acknowledged in national legislation; and (5) there is evidence that fisheries have been impacted by one, or more,

Table 31.1 General descriptions and acronyms of the seven small-scale shellfisheries analyzed in this study with emphasis on co-governance and operational arrangements in place

Target species	Country	Location	Acronym	Habitat	Fishing method	Co-governance type	Year of implementation	Motivation for co-governance	Governance instruments
Spiny lobster *Panulirus argus*	Mexico	Punta Allen (Ascension Bay, Quintana Roo)	SLQR	Subtidal, sandy and rocky	Skin-diving and artificial shelters	Consultative	1968	Agreed upon by fishers themselves	1, 3, 7, 8, 11
Yellow clam *Mesodesma mactroides*	Uruguay	Rocha (Barra del Chuy)	YCUY	Intertidal, sandy beach	Hand-gathering	Consultative	1990	Resource depletion	3, 6, 7
Abalone *Haliotis corrugata* *H. fulgens*	Mexico	Peninsula of Baja California (central zone)	ABBC	Subtidal, rocky	Hookah and SCUBA diving	Cooperative	1992	Abalone overexploitation and negative impact of 1982–1983 El Niño event	1,3,4,5,7
Spiny lobster *Panulirus interruptus*			SLBC		Traps				1, 3, 4, 7, 8, 9, 10

Spiny lobster *Panulirus penicillatus, P. gracilis*	Ecuador	Galapagos Islands	SLGPS	Subtidal, rocky	Hookah and skin diving	Cooperative	1998	Illegal and unregulated expansion of the sea cucumber fishery promoted by roving bandits since 1992	2, 3, 4, 5, 7, 8
Sea cucumber *Isostichopus fuscus*			SCGPS						2, 3, 4, 5, 7
Surf clam *Mesodesma donacium*	Chile	Tongoy Bay (northern-central zone)	CLCHL	Intertidal and shallow subtidal, sandy beach	Hand-gathering and hooka diving	Cooperative	1998	Ensure a sustainable exploitation of *M. donacium* by the TURF system created as a response to the "Loco" (*Concholepas concholepas*) fishery collapse (1991)	1, 3, 5, 7

Governance instruments: (*1*) territorial user rights for fisheries (TURFs); (*2*) fishing licenses and permits; (*3*) spatio-temporal closures; (*4*) marine protected areas; (*5*) total allowable catch; (*6*) individual quotas; (*7*) minimum legal size; (*8*) protection for berried females; (*9*) effort limit (traps); (*10*) escape windows in traps; (*11*) prohibition of SCUBA and hooka diving, gillnets and hooks

external climatic and human drivers. Most co-governance arrangements were formally implemented in the 1990s, usually as a response to an environmental, political or socioeconomic crisis (Table 31.1). The only exception is the spiny lobster fishery from Punta Allen (Quintana Roo, Mexico), where co-governance emerged in a bottom-up way (i.e., *de facto*) during 1968 as a result of the geographical isolation and strong organization of the "Vigía Chico" cooperative, whose members allocated territorial user rights for fisheries (TURFs) among themselves in specific fishing lots, locally known as "campos" (Seijo 1993).

Co-governance arrangements differ in the way governments and actors interact in the decision-making process (Table 31.1), and can be classified as two types (Sen and Nielsen 1996; Gutiérrez et al. 2011): (1) consultative, where consultation mechanisms between the government and actors are minimal, and final decisions are taken exclusively by the government; and (2) cooperative, where fishers are legally recognized as equal partners in decision-making, and final decisions are taken in cooperation. Fishers' organizations are embedded in communities with strong social cohesion, leadership, and organization (Castilla and Defeo 2001; McCay et al. 2014) with the exception of Galapagos where fishing cooperatives lack those social attributes (Castrejón and Charles 2013). In all cases, exclusive access rights have been implemented (e.g., TURFS, fishing licenses and permits), together with spatio-temporal closures and/or marine protected areas (Table 31.1). Other governance instruments include total and individual quotas and minimum landing sizes.

Framework for Characterizing Adaptive Capacity of Co-governance Arrangements

To identify the factors that enable or inhibit institutional adaptation processes in the seven case studies selected, we identified the most relevant external drivers that affected each small-scale shellfishery. In most cases, the perturbations produced by the human and climatic driver selected occurred exclusively after a co-governance arrangement was implemented. Just in two cases – the sea cucumber and surf clam fishery – the perturbations were initiated before co-governance implementation and still continue (see Tables 31.1 and 31.2).

We considered two external driver categories (Hall 2011): "Climatic and environment" and "International trade and globalization of markets". Such drivers were subdivided in "pulse" and "press" perturbations. According to Collins et al. (2011), pulse perturbations are relatively discrete and rapidly alter species abundances and ecosystem functioning (e.g., a hurricane), while press perturbations are sustained and chronic (e.g., sea level rise). Both have the capacity to change the quantity and quality of ecosystem services (e.g., seafood). We define pulse perturbations as extreme climatic or socioeconomic events that in a short time period modify the structure and function of the system-to-be-governed, and whose impacts persist temporally or permanently after the event has ended. In contrast, press perturbations

Table 31.2 External drivers analyzed in this study

Category	External driver	Type of perturbation	Temporal scale	Spatial scale	Fisheries affected
Climate and environment	Hurricane Gilbert	Pulse	September 14, 1988 (~12 h)	Regional	SLQR
	El Niño 1997/1998	Pulse	April 1997–June 1998 (~14 months)	Regional	SLBC, ABBC, SLGPS, SCGPS, CLCHL
	Oceanographic regime shift	Press	1994-onwards (decades)	Regional	YCUY
	Oceanographic regime shift	Press	1977-onwards (decades)	Regional	CLCHL
International trade and globalization of markets	Roving bandits	Press	1992-onwards (decades)	Local	SCGPS
	Massive seafood importation	Press	2008-onwards (>2 years)	National	YCUY
	Global financial crisis	Pulse	December 2007- June 2009 (~18 months)	Global	SLBC, ABBC, SLGPS, SCGPS

ABBC abalone Baja California, *CLCHL* surf clam Chile, *SCGPS* sea cucumber Galápagos, *SLBC* spiny lobster Baja California, *SLGPS* spiny lobster Galapagos, *SLQR* spiny lobster Quintana Roo, *YCUY* yellow clam Uruguay

exert a long term pressure over the system-to-be-governed, whose intensity increases gradually through time.

We selected *a priori* two large-scale pulse perturbations: El Niño 1997–1998 (hereafter EN97–98) and the global financial crisis 2007–2009. The first was an extreme climatic event that strongly impacted the Pacific coast of Latin America for 14 months, while the second was an economic perturbation produced by the collapse of financial markets and lending institutions that abruptly impacted the global economy for at least 18 months (Table 31.2). Both drivers affected most of our case studies, allowing us to assess how different co-governance arrangements responded to similar external drivers that occurred at the same time. We also identified *a posteriori* other drivers, based on a literature review (Table 31.2), to assess how different case studies responded to specific pulse and press perturbations.

To characterize the adaptability of co-governance arrangements, we identified the social-ecological impacts produced by each driver in the system-to-be-governed (Table 31.3). Then, we identified the coping and adaptive responses produced by the governing system (i.e., fishery agencies, co-governance bodies and fishing cooperatives) and the social component of the system-to-be-governed (i.e., individual fishers and local communities). We also investigated if the responses were adopted in a preventive (before the event occurred) or reactive (during or after the event occurred) way. Finally, we identified the factors that enabled coping and adaptive responses by the governing system and the system-to-be-governed, concluding with some generalizations based on the comparative analysis of case studies.

Table 31.3 Social-ecological impacts of different climatic and human drivers over the system-to-be-governed of seven co-governance arrangements in Latin American small-scale shellfisheries

	SLQR		SLBC		ABBC		SLGPS		SCGPS		YCUY		CLCHL
Ecological impacts	D1	D5	D2	D5	D2	D5	D2	D5	D2	D4	D3	D6	D3
Condition of nursery and foraging habitats	red				red								
Survivorship	red				red						red		red
Abundance	red		green		red		green		green	red	red		red
Recruitment	red		green		red		green		green	red	red		red
Physiologic condition					red								
Spawning stock biomass	red				red					red	red		red
Spawning time variation			green										
Conservation status of protected areas										red			
Social impacts													
Condition of fishing gear (e.g., lobster shelters)	red												
Socio-economic well-being (temporal or permanent)	red	red	green	green	red		green	red	green	red	red	red	red
Landings	red	red	green		red		green	red	green	red	red		red
CPUE	red		green		red		green	green	green	red	red		red
Unit price		red		green		red		red		green			
Export		red				red		red		green			
Population growth	red									green			
Diversity and complexity of social structure	red									green			
Interest in the co-governance arrangement										red			red
Consumption of domestic seafood products												red	

ABBC abalone Baja California, *CLCHL* surf clam Chile, *SCGPS* sea cucumber Galápagos, *SLBC* spiny lobster Baja California, *SLGPS* spiny lobster Galapagos, *SLQR* spiny lobster Quintana Roo, *YCUY* yellow clam Uruguay. *D1* Gilbert hurricane, *D2* El Niño 1997/1998, *D3* regime shift, *D4* roving bandits, *D5* global financial crisis, *D6* massive seafood importation. Impact on stocks: *green* (increase); *red* (decrease), *blank cell* (no impact reported)

Coping and Adaptive Responses

How Co-governance Arrangements Respond to Different Climatic and Human Drivers?

Hurricane Gilbert

In Punta Allen (Mexico), Hurricane Gilbert damaged extensive areas of coral reefs and shallow seagrass beds (*Thalassia testudinum*), the main nursery and foraging habitats of *Panulirus argus*. The loss of shelters increased the vulnerability of spiny lobsters to predators, reducing temporally their abundance (Fonseca-Larios and Briones-Fourzán 1998). The hurricane also destroyed thousands of lobster shelters, collapsing the lobster production and preventing the completion of a new seafood processing plant (Leslie 2000). As catch and catch per unit effort (CPUE) declined markedly after the hurricane (Sosa-Cordero et al. 2008), fishers were unable to pay their debts. Consequently, the processing plant was seized by a bank, and the cooperative's risk of default increased. In the short-term, dozens of fishers coped with this economic crisis by waiving their memberships, transferring their exclusive access rights (i.e., fishing lots) to other members to pay their debts, and emigrating to nearby urban areas (Leslie 2000). Thus, cooperative membership declined from 100 to 71 at the end of the 1980s, while the local population contracted from 500 to 400 residents (Carr 2007).

In the long term, the devastation caused by Hurricane Gilbert made remaining small-scale fishers aware of their vulnerability to climatic variability, which encouraged them to take preventive actions to strengthen fishery governability and the financial administration of their cooperative (Carr 2007; UNDP 2012). Thus, Vigía Chico's fishers adapted collectively their harvesting strategies by fine-tuning their fishing effort according to spiny lobster abundance, and applying rigorous penalties (e.g., expulsion from the cooperative and confiscation of fishing lots and fishing equipment) to those members who infringed upon federal and internal management regulations. Eight years later, catch and CPUE showed signs of recovery (Sosa-Cordero et al. 2008). Other adaptive responses were applied, including (Carr 2007; Sosa-Cordero et al. 2008; Ley and Quintanar 2010; UNDP 2012): (1) stabilization of fishers population by limiting the allocation of new memberships only to children of cooperative members and restraining, by preference, their own fertility rate, which has been one of the lowest in Mexico since the mid-1990s. The logic behind this trend is that keeping a low number of fishers and children ensures the prosperity of the entire community, particularly in times of resource scarcity; (2) enhancement and proper management of the cooperative's financial affairs by hiring a private accounting firm; (3) diversification of livelihoods by creating tourism cooperatives since 1994; (4) diversification of products by catching and trading of live lobsters since 1995; (5) enhancement of the spiny lobster value chain since 2004 by establishing a partnership between Vigía Chico and five cooperatives from Sian Ka'an and Chinchorro biosphere reserves (Mexico). These cooperatives formed a collective enterprise called "Integradora de Pescadores de Quintana Roo" to sell their product directly to retailers from hospitality and ecotourism industries, using their own brand ("Chakay"). This arrangement added value to the product, increased compliance with regulations and mitigated the influence of middlemen, resulting in higher profits for fishers; and (6) establishment of a rotating fund, which acts as a financial buffer in times of financial difficulty, resource scarcity, and natural disasters. All these adaptive responses were adopted thanks to the strong social cohesion, organization and leadership of fishing communities in which cooperatives are embedded (Sosa-Cordero et al. 2008), as well as to the economic support and capacity building provided by diverse governmental institutions and non-governmental organizations (Ley and Quintanar 2010). Consequently, the Vigía Chico cooperative increased its adaptability and resilience, as it was demonstrated when Hurricane Wilma, one of the most intense tropical cyclone ever recorded in the Atlantic basin, hit the Yucatan Peninsula in 2005 (UNDP 2012).

El Niño 1997/1998

Negative and positive social-ecological impacts were caused by EN97–98 (Table 31.3). In Baja California, this event reduced the recruitment, abundance, and physiologic condition of abalone stocks, which could be associated with the temporal disappearance of *Macrocystis pyrifera* algal beds (Guzmán et al. 2003), a source of food and shelter for abalones. These negative impacts exacerbated fishery overexploitation. Fortunately, years before EN97–98 occurred, the Federation of Cooperative Societies of the Fishing Industry of Baja California (FEDECOOP), in collaboration with government agencies and research institutions, designed a stock rebuilding strategy that included: (1) consideration of the effects of climatic variability on abalone abundance in stock assessment methods; (2) establishment of a decision rule to set a total allowable catch (TAC) per cooperative; (3) government support to conduct research on abalone aquaculture and transfer knowledge to cooperatives; and (4) active participation of cooperative fishers in monitoring, surveillance and restocking activities. Since then, cooperatives have diversified their fishing effort to other fisheries to cope with abalone scarcity (McCay et al. 2014). These adaptive responses helped mitigate the impact of EN97–98, allowing the gradual recovery of the fishery since 2001 (Searcy-Bernal et al. 2010). Furthermore, one cooperative implemented two experimental marine reserves within its territorial use rights in fisheries (TURF) to increase the resilience of abalone stocks to overfishing and climatic variability (Micheli et al. 2012).

In contrast, the abundance of spiny lobsters (*P. interruptus*, *P. penicillatus* and *P. gracilis*) and sea cucumbers (*I. fuscus*) from Baja California and Galapagos increased markedly after EN97–98, probably as a result of strong recruitment pulses (Guzmán et al. 2003; Hearn et al. 2005; Vega et al. 2010). In Baja California, government agencies, again with full support from FEDECOOP, adjusted temporal closures before EN97–98 occurred to ensure the reproductive success of lobster spawning stocks. This decision was taken based on scientific evidence produced after EN82–83, which demonstrated that increasing sea surface temperatures accelerate the breeding time of *P. interruptus*, leading to spawning events earlier than normal (Vega 2003). This preventive response, together with the effective enforcement of other regulations (e.g., escape windows in traps) and the reduction of illegal fishing – through self-enforcement mechanisms – contributed to protect recruitment and reach maximum historic landings in the central zone of Baja California during 2000 and 2002; i.e., two and four years after EN97–98. Since then, landings have remained remarkably high.

An opposite trend was observed in Galapagos, where fishers individually reacted by intensifying their fishing effort in the spiny lobster fishery. Such a coping response was shaped in turn by a previous response to another external driver: the boom and boost exploitation of sea cucumbers promoted by roving bandits (see following sections). Consequently, maximum historic landings were registered in the spiny lobster and sea cucumber fisheries, two and five years after EN97–98, respectively (Defeo et al. 2013). However, few years later both fisheries showed signs of overexploitation (Ramírez et al. 2012).

Regime Shift

The populations of the yellow clam (*M. mactroides)* and surf clam (*M. donacium*) from Uruguay and Chile, respectively, were decimated by periodic mass mortalities associated with large-scale regime shifts from cold to warm waters (Ortega et al. 2012). In the Pacific Ocean, the regime shift occurred in 1977 (Fiedler 2002), while in the Atlantic Ocean the regime shift took place in 1994 (Goldenberg et al. 2001) – four years after a successful co-governance arrangement was implemented in Barra del Chuy, Uruguay (Castilla and Defeo 2001; Table 31.1). Since then, the systematic increase of sea surface temperature has been inversely correlated with declining trends in the abundance of both species (Defeo et al. 2013).

Before 1994, fishery agencies were unaware of the occurrence and impacts of mass mortalities in Uruguay. Therefore, no contingency plans were in place and managers were not prepared to cope with the unusual changes that occurred in the system-to-be-governed when mass mortalities began. They just reacted by implementing a fishery closure in 1994. However, as no options were provided to fishers to mitigate the economic impact on their livelihoods, this measure caused loss of incomes and unemployment. Small-scale fishers immediately responded by diversifying their livelihoods in other sectors of the economy (e.g., construction, agriculture). The co-governance arrangement was re-organized 14 years later, through the participatory development of new policies, institutions and governance instruments, once yellow clam stocks showed signs of recovery (2007–2008). Managers and fishers agreed that this mode of governance was suitable to promote fishery recovery and to enhance fishing communities' well-being. This decision was based on the successful co-governance arrangement implemented before mass mortalities occurred. Since then, fishery governability has improved (Defeo 2014).

In Chile, a TURFs system called Management Exploitation Areas for Benthic Resources was implemented at the national level in 1991 to solve the fishery crisis faced by the gastropod *Concholepas concholepas* (Castilla and Defeo 2001). The success of this co-governance system (Gelcich et al. 2010) led to its widespread application across different shellfish resources, including the surf clam *Mesodesma donacium* (known as "*macha*"). This species consists of a metapopulation with a highly dispersive planktonic larval phase that imposes uncertainty in the replenishment of local beds. Despite this fact, a Management Exploitation Areas for Benthic Resources system was implemented at Tongoy Bay in 1998. However, the fishery collapsed after three years of sporadic success because of lack of recruitment and high natural mortality levels mainly attributed to mass mortality events that occurred in Northern Chile and Peru (Riascos et al. 2009; Ortega et al. 2012; Aburto and Stotz 2013). In response, fishers switched to other economic activities or alternative fisheries. The fishery showed a moderate recovery between 2009 and 2010. However, as landings were much lower than those registered under the Management Exploitation Areas for Benthic Resources system, the local community lost interest in maintaining the co-governance arrangement (Aburto and Stotz 2013).

Roving Bandits

The boom-and-bust exploitation cycle of sea cucumbers by roving bandits – in this case symbolized by Asian middlemen – was initiated in the Galapagos Islands in 1992, one year after this fishery collapsed in the Ecuadorian continental coastline (Jenkins and Mulliken 1999; Shepherd et al. 2004). Dozens of fishers from mainland Ecuador immigrated to Galapagos sponsored by roving bandits themselves. The exploitation of *Isostichopus fuscus* rapidly attracted the interest of local small-scale fishers (residents), who received training and loans from Asian middlemen to participate in the fishery (De Miras et al. 1996). Thus, a resource that had not been traditionally exploited by the local population became unexpectedly the most lucrative fishery of the Galapagos and, most importantly, a pervasive partnership was created between roving bandits and fishers (migrants and residents).

Clandestine processing camps were set up on protected land areas to cook and dry sea cucumbers, thus increasing the risk of accidental fires and the introduction of invasive species (e.g., fruits, insects). These concerns, together with the potential ecological extinction of *I. fuscus* due to the open access nature of the fishery, attracted large international attention, particularly from conservation organizations (Castrejón et al. 2014). The strong international pressure encouraged the Ecuadorian government to implement precautionary management measures, including a total fishery closure (1995–1999). Fishers responded with violent protests and strikes, behavior that was influenced by their precarious economic situation. As the fishery was abruptly closed, most of them were unable to pay their debts to Asian middlemen, local retailers and banks (De Miras et al. 1996). Therefore, the increasing indebtedness of fishers, together with the uncertainty about the future access to the fishery, intensified illegal fishing and strengthened a "black market", whose main objective was to satisfy the payment of debts (Castrejón 2011).

The conflicts caused by the sea cucumber fishery led to an institutional shift from a hierarchical to a co-governance mode in 1998, which included the establishment of the Galapagos Marine Reserve (Castrejón and Charles 2013). Furthermore, several governance instruments were implemented to shift from an open access to a common property regime, including migratory rules, a ban on industrial fishing, the establishment of a moratorium on the entry of new fishers, the creation of a limited-entry program and marine zoning, the inclusion of *I. fuscus* in Appendix III of CITES, and even an unsuccessful attempt to implement an individual quota system in 2001 (Toral-Granda 2008; Castrejón 2011). Despite these adaptive responses, co-governance bodies were unable to "break down" the partnership created between roving bandits and local fishers.

Management measures were undermined by poor enforcement capacity coupled with an anthropogenic Allee effect (*sensu* Defeo and Castilla 2012). In other words, as sea cucumber abundance decreased due to overexploitation, the willingness of Asian markets to pay higher prices increased exponentially (Defeo et al. 2014). Thus, the expectancy of fishers to obtain higher profits motivated them to accelerate their exploitation rates, even under diminishing abundance levels. In this context, the roving bandits, with the additional participation of local middlemen, encouraged local fishers to catch sea cucumbers either below legal landing sizes or during seasonal closure periods. Illegal fishing intensified as resource abundance became scarcer and its exploitation and trading were restricted. This vicious cycle of globalized exploitation led to the collapse and closure of the fishery in 2006. Once the fishery was not profitable, the roving bandits moved to Nicaragua to continue the sequential exploitation of other sea cucumber species. However, they usually return to Galapagos every time the fishery is re-opened.

Massive Seafood Importation

Seafood importation affects fishing communities' livelihoods through the displacement of domestic products from national markets. In Uruguay, favorable market conditions led to an exponential increase in the importation of frozen bivalves, mainly from Chile, particularly since 2008. Demand of yellow clams – a domestic product – dropped as retailers and consumers opted for cheaper imported seafood products. Thus, yellow clams were partially displaced from international resorts, such as Punta del Este. Fishing communities from Barra del Chuy responded collectively by diversifying their products and markets, sponsored technically and economically by the government. Instead of selling 80 % of yellow clams for bait and 20 % for human consumption – as was traditionally done since the 1980s – fishers decided to add value to their production to increase its freshness and quality. The adaptation of products to the changing market conditions and consumer expectations allowed fishers to sell 95 % of their landings since 2010 for human consumption, particularly in seaside resorts. This adaptive response of the community, under a co-governance arrangement, partially mitigated the negative effects caused by the massive importation of seafood. Nevertheless, this driver still represents an external threat to local fishing communities' livelihoods.

Global Financial Crisis

The global financial crisis contracted the consumption of lobsters and abalones in the United States and European Union– the main foreign markets of most Latin American countries (Cook and Gordon 2010; Monnereau and Helmsing 2011). In Galapagos, the sharp worldwide decline in lobster demand produced a price drop of 32 % between 2008 and 2009 (Ramírez et al. 2012). As middlemen refused to buy landings at higher prices, fishers reacted individually in three ways: (1) abandoning the fishery; (2) diversifying their product by trading whole fresh lobsters instead of lobster tails, as had been done historically since the 1960s; and (3) diversifying their market by selling their product directly to the local hospitality sector and general public. Consequently, total fishing effort, catch, and exports to mainland Ecuador declined 20 %, 23 % and 45 %, respectively (Defeo et al. 2014). While the economic crisis was detrimental for Galapagos fishers, it was beneficial for spiny lobster stocks. Two years after the official end of the recession, lobster CPUE and catch increased 91 % and 102 %, respectively, whereas fishing effort only increased 6 % between 2009 and 2011. Since then, these indicators have remained remarkably high. Price also increased, although it remains 24 % below the value registered before 2007.

In Punta Allen, fishers were in a better position to face the global financial crisis because of the adaptive responses adopted after Hurricane Gilbert (see previous sections). Nevertheless, two factors made them vulnerable to this driver: (1) the poor diversification of their market – most landings were destined to the Mexican hospitality industry of Cancun and the Riviera Maya (ITAM-CEC 2007a); and (2) the outbreak of "swine flu" in April 2009. When the US market entered into recession and the "swine flu" outbreak occurred, middlemen stopped buying lobsters and the number of foreign tourists declined; thus, the domestic market collapsed, and prices dropped 50 % between 2008 and 2009 (Noticaribe 2010). In response, Vigía Chico's members stopped fishing for three months (July–September 2009) until market conditions improved. Since then, they have intensified their efforts in collaboration with other cooperatives, government and NGO's to acquire the infrastructure, technology and expertise needed to export live lobsters to Asia and Europe in order to sell their product at better prices.

By contrast, the spiny lobster fishery from the Central part of Baja California was relatively "immune" to the global financial crisis, thanks to the harvesting and trading strategies adopted by the ten cooperatives that form the FEDECOOP before the crisis occurred. Unlike Quintana Roo's cooperatives, FEDECOOP exports about 90 % of its production to Asia, France and USA (ITAM-CEC 2007b), using their own brand ("Rey del Mar"). Most of the production is sold live; however frozen and cooked lobsters (whole and tails) are also traded to spread the risk of market contraction. FEDECOOP cooperatives trade their landings in coordination, so that unit price and harvesting levels are agreed before the beginning of each fishing season, based on global market conditions, the production of the last five seasons, and the recommendations made by fishery agencies (SCS 2011). Once the initial price is internally agreed, it is negotiated with foreign middlemen and local retailers. Market prices are monitored daily along the fishing season to regulate fishing effort based on a cost-effectiveness analysis. If market conditions are unfavorable, an early closure is agreed and implemented in a coordinated way, as happened during 2009 when the fishing season was closed 10 days earlier than planned (SCS 2011). Thanks to these harvesting and trading strategies, promoted by the strong organization, social cohesion, and leadership of FEDECOOP, the live lobster price increased 39 % between 2008 and 2009, reaching maximum historic prices in 2011. The same harvesting and trading strategies were adopted by FEDECOOP in the abalone fishery to cope with the global financial crisis (Searcy-Bernal et al. 2010).

What Factors Enable or Inhibit Building Institutional Adaptability?

Our results show that climatic drivers affected the demography and life traits of target species, either directly or indirectly, by damaging the quality and availability of critical habitats (i.e., the ecological component of the system-to-be-governed). The combined impact of climatic and human drivers produced social-ecological impacts that affected local fishing communities' wellbeing (i.e., the social component of the system-to-be-governed) and the governing system. In this context, coping and adaptive responses were adopted by institutions (i.e., cooperatives, fishery agencies, or co-governance bodies) or actors (i.e., fishers) to prevent or mitigate the negative effects of these drivers on fishery resources and communities' livelihoods. Coping and adaptive responses varied according to the magnitude, extent, periodicity and intensity of press and pulse perturbations, and were shaped by past crises, social-ecological memory (*sensu* Folke et al. 2003) and the particular social features of fishing communities in which institutions are embedded.

In Punta Allen and Baja California, adaptive responses were triggered in the spiny lobster and abalone fisheries when fishers acquired a "collective awareness" about their vulnerability to climatic and human drivers. After extreme crises (i.e., pulse perturbations), fishers re-organized their cooperatives, harvesting and trading strategies in a collaborative way. They were successful in these enterprises thanks to the strong social cohesion, leadership, and organization of the fishing communities in which these cooperatives are embedded (Sosa-Cordero et al. 2008; McCay et al. 2014), as well as to their institutional capacity to take actions based on lessons learnt from previous crises (e.g., Hurricane Gilbert and El Niño events) and their own social-ecological memory. Effective adaptive responses were also enabled thanks to prolific partnerships created between cooperatives and fishery agencies, research centers and NGOs. The support provided by these institutions to small-scale fishers, in terms of scientific knowledge and capacity building, has been fundamental to prevent and mitigate the social-ecological impacts produced by diverse climatic and human drivers. Such results suggest that co-governance arrangements were successful in building institutional adaptability in Punta Allen and Baja California. This is reflected in the implementation of innovative solutions that enhanced governance quality (i.e., its governability; Kooiman and Bavinck 2013). These include: (1) the adoption of exclusive access rights (e.g., TURFs) and self-enforcement mechanisms to prevent over-exploitive fishing practices; (2) the flexibility of institutions to adapt management measures to prevent the impact of climatic drivers, based on the availability of sound scientific knowledge; (3) the development of participatory rebuilding strategies, including the implementation of decision rules to restrict harvest; and (4) the entrepreneurial capacity of cooperatives to adapt their trading strategies to the changing global financial trends, thus preventing the impact of unfavorable market conditions and mitigating the bargaining power of middlemen within fisheries' value chains. The implementation of these solutions have produced several benefits, including (Castilla and Defeo 2001; Sosa-Cordero et al. 2008; Defeo et al. 2014; McCay et al. 2014): (1) improved sense of ownership and stewardship, which in turn promote legitimacy, acceptability and compliance of regulations; (2) optimization of data collection methods, minimization of conflict and strengthening of long term strategic planning processes; (3) the creation of multilevel social networks, i.e., legal, political, and financial frameworks that enhance sources of social and ecological resilience (Adger et al. 2005); and (4) enhancement and stabilization of bioeconomic indicators such as population abundance, CPUE and economic revenues. These successful results were recognized by the Marine Stewardship Council (MSC), which certified the spiny lobster fisheries from the central zone of Baja California and the Sian Ka'an and Banco Chinchorro Biosphere Reserves in 2004 and 2012, respectively. In Baja California, the MSC certification produced non-economic benefits to FEDECOOP, including empowerment, community strengthening, and greater prestige at national and international level (Pérez-Ramírez et al. 2012). In the long term, the legitimacy and the political and bargaining power of FEDECOOP have increased, allowing it to ensure its exclusive access rights (i.e., TURFs), to obtain government's economic support for community development and to negotiate better prices for its seafood products in the international markets. Such benefits have reinforced the willingness and interest of fishers to comply with MSC required standards and to expand their involvement in co-governance arrangements, thus promoting optimum conditions to continue building institutional adaptability and resilience within the governing system and the system-to-be-governed.

In Uruguay, the co-governance arrangement of the yellow clam fishery was effective in enhancing governability (Defeo et al. 2009). However, this governance mode was not resilient to the detrimental impacts caused by mass mortalities. Despite this unexpected failure, fishers and managers decided to work collaboratively to re-organize their governing system in order to promote the recovery and sustainable management of the fishery. The critical factors that enabled this adaptive response were: (1) recognition about the key role that the previous co-governance experience played in promoting good governance and sustainable fishing practices (i.e., existence of social-ecological memory); (2) recognition by all actors that stocks were depleted (i.e., shared images); and (3) participatory development of a rebuilding strategy, based on sound scientific knowledge about the ecology and resilience of targeted species and their roles in ecosystem dynamics (i.e., collaborative governance). Although yellow clams have not fully recovered yet, the collective response of fishers to mitigate the detrimental impact of seafood importation suggests that this co-governance arrangement is being consolidated by building adaptability and collaboratively rebuilding plans, leading to higher governability of the fishery. By contrast, the Tongoy Bay *macha* fishery became less governable, regardless of the co-governance arrangement developed around it, when the *macha* population crashed and small-scale fishers perceived that their conservation efforts would not produce the economic benefits that they expected (Aburto and Stotz 2013).

Most case studies described above reinforce the notion that crises represent opportunities for learning, adapting, and entering onto more sustainable pathways (Folke et al. 2003). In Galapagos, ecological and social crises also triggered adaptive responses. However, such responses were not effective in building institutional adaptability and resilience. Several factors explain why cooperatives and fishery agencies have a poor capacity to learn, self-reorganize and respond proactively to the problems at hand. In Galapagos, unlike Mexico and Uruguay, cooperatives are embedded in fishing communities that are socially fragmented (Castrejón 2011; Castrejón and Charles 2013). This is reflected in the incapacity of small-scale fishers to take collaborative actions to reorganize their cooperatives, adapt their harvesting and trading strategies, and implement self-regulatory mechanisms in order to exclude outsiders, avoid illegal fishing and mitigate the impact of roving bandits.

The adoption of collective adaptive responses was also inhibited in Galapagos by the existence of contrasting images about the status of the sea cucumber fishery. This avoided the creation of prolific partnerships among fishers, managers, scientists, and conservationists for at least 15 years (1992–2006). In this context, stakeholders perceived each other as "enemies", instead of potential partners whose particular capacities, knowledge, skills and resources could contribute to cope with external drivers of change. Consequently, management measures were implemented under pressure, usually without the consensus of fishers' representatives. Thus, decisions taken by Galapagos co-governance bodies were perceived as illegitimate by grassroots fishers, having a negative impact on fishers compliance with regulations (Viteri and Chávez 2007). Furthermore, some management measures were not based on sound scientific knowledge (e.g., total allowable catch), leading to the loss of credibility in fishery agencies, NGOs, and finally in the entire co-governance arrangement (Castrejón and Charles 2013). This case study illustrates how the establishment of a cooperative co-governance mode, through the institutionalized inclusion of fishers as equal partners in the governance process, is not always effective in generating high governability, particularly when: (1) local fishing communities lack a sense of stewardship and critical social attributes (leadership, social cohesion, organization and social-ecological memory); (2) exclusive access rights implemented, in this case licenses and fishing permits, are deficient at mitigating over-exploitive fishing practices, (3) strong pervasive partnerships exist between fishers and middlemen; and (4) fishery agencies lack long-term economic and human resources to enforce regulations and to conduct the research needed to formulate solid governance instruments.

Based on the comparative analysis of our seven case studies, it can be concluded that the governability of a fishery is not dependent on the co-governance mode established (e.g., consultative or cooperative), but mainly on the social attributes of fishers' organizations, the quality of the interactions between government and other actors, and the institutional adaptability to external drivers of change. Institutions with strong social cohesion, organization and leadership, and willingness to change and work in a collective and collaborative way, displayed a higher institutional capacity for adaptation and innovation. The latter was reflected in the capacity of institutions to take actions, based on past experiences and social-ecological memory, to re-organize themselves, create prolific partnerships, change harvesting and trading strategies, and implement self-regulatory mechanisms to prevent over-exploitive fishing practices. According to our results, co-management arrangements that show these features, such as those located in Baja California, Punta Allen and Uruguay, also displayed a higher institutional adaptability to different climatic and human drivers, resulting in better governability. In contrast, poor governability was observed in those cases where such characteristics were lacking, as in Galapagos, or where fishers perceived that their conservation efforts would not produce the expected economic benefits as in the Tongoy Bay *macha* fishery. In conclusion, the same factors that promote (or preclude) high governability are also those that enable (or inhibit) building institutional adaptability and resilience within the governing system and the system-to-be-governed.

Acknowledgements We thank Svein Jentoft and Ratana Chuenpagdee for their comments on an extended version of this chapter. We are also grateful to Cristiana Seixas and Ian Perry for their helpful suggestions and comments. MC acknowledges the financial support provided by the Consejo Nacional de Ciencia y Tecnología (CONACYT) and the World Wildlife Fund's Russell E. Train Education for Nature Program. OD is grateful for the financial support provided by The Pew Fellows Program in Marine Conservation and DINARA's UTF and GEF projects.

References

Aburto, J., & Stotz, W. (2013). Learning about TURFs and natural variability: Failure of surf clam management in Chile. *Ocean & Coastal Management, 71*, 88–98.

Adger, W. N., Hughes, T. P., Folke, C., Carpenter, S. R., & Rockström, J. (2005). Social-ecological resilience to coastal disasters. *Science, 309*, 1036–1039.

Badjeck, M. C., Mendo, J., Wolff, M., & Lange, H. (2009). Climate variability and the Peruvian scallop fishery: The role of formal institutions in resilience building. *Climatic Change, 94*, 211–232.

Berkes, F., & Folke, C. (Eds.). (1998). *Linking social and ecological systems: Management practices and social mechanisms for building resilience*. Cambridge: Cambridge University Press.

Berkes, F., Mahon, R., McConney, P., Pollnac, R., & Pomeroy, R. (2001). *Managing small-scale fisheries: Alternative directions and methods*. Ottawa: International Development Research Centre.

Carr, D. L. (2007). Resource management and fertility in Mexico's Sian Ka'an Biosphere Reserve: Campos, cash, and contraception in the lobster-fishing village of Punta Allen. *Population and Environment, 29*(2), 83–101.

Castilla, J. C., & Defeo, O. (2001). Latin American benthic shellfisheries: emphasis on co-management and experimental practices. *Reviews in Fish Biology and Fisheries, 11*, 1–30.

Castrejón, M. (2011). *Co-manejo pesquero en la Reserva Marina de Galápagos: Tendencias, retos y perspectivas de cambio*. Mexico: Fundación Charles Darwin/Kanankil/Plaza-Valdés.

Castrejón, M., & Charles, A. (2013). Improving fisheries co-management through ecosystem-based spatial management: The Galapagos Marine Reserve. *Marine Policy, 38*, 235–245.

Castrejón, M., Defeo, O., Reck, G., & Charles, A. (2014). Fishery science in Galapagos: From a resource-focused to a social-ecological systems approach. In J. Denkinger & L. Vinueza (Eds.), *The Galapagos Marine Reserve: A dynamic social-ecological system. Social and ecological sustainability in the Galapagos Islands* (pp. 159–186). New York: Springer.

Collins, S. L., Carpenter, S. R., Swinton, S. M., Orenstein, D. E., Childers, D. L., Gragson, T. L., et al. (2011). An integrated conceptual framework for long-term social–ecological research. *Frontiers in Ecology and the Environment, 9*(6), 351–357.

Cook, P. A., & Gordon, H. R. (2010). World abalone supply, markets, and pricing. *Journal of Shellfish Research, 29*(3), 569–571.

De Miras, C., Marco, A., & Carranza, C. (1996). *Evaluación socioeconómica de la pesca experimental de pepino de mar en Galápagos* (p. 198). Quito: ORSTROM/Fundación Charles Darwin.

Defeo, O. (2014). *Enfoque ecosistémico pesquero: Conceptos fundamentales y su aplicación en Uruguay*. Final Report, Project CP/URU/030/GFF Piloting of an Ecosystem-based Approach to Living Aquatic Resources Management (p. 99). Uruguay: DINARA.

Defeo, O., & Castilla, J. C. (2012). Governance and governability of coastal shellfisheries in Latin America and the Caribbean: Multi-scale emerging models and effects of globalization and climate change. *Current Opinion in Environmental Sustainability, 4*, 344–350.

Defeo, O., Castilla, J. C., & Castrejón, M. (2009). Pesquerías artesanales de invertebrados en América Latina: Paradigmas emergentes de manejo y gobernanza. In C. Lodeiros, J. Alió, L. Freites, N. González, A. Guerra, & M. Rey-Méndez (Eds.), *Foro Iberoamericano de los Recursos Marinos y la Acuicultura II* (pp. 89–117). Cumaná: Fondo Editorial Universidad de Oriente.

Defeo, O., Castrejón, M., Ortega, L., Kuhn, A. M., Gutiérrez, N. L., & Castilla, J. C. (2013). Impacts of climate variability on Latin American small-scale fisheries. *Ecology & Society, 18*(4), 30.

Defeo, O., Castrejón, M., Pérez-Castañeda, R., Castilla, J. C., Gutiérrez, N., Essington, T. E., & Folke, C. (2014). Co-management in Latin American small-scale shellfisheries: assessment from long-term case studies. *Fish & Fisheries*. doi:10.1111/faf.12101.

Fiedler, P. C. (2002). Environmental change in the eastern tropical Pacific Ocean: Review of ENSO and decadal variability. *Marine Ecology: Progress Series, 244*, 265–283.

Folke, C., Colding, J., & Berkes, F. (2003). Synthesis: Building resilience and adaptive capacity in socio-ecological systems. In F. Berkes, J. Colding, & C. Folke (Eds.), *Navigating the dynamics of social–ecological systems: Building resilience for complexity and change* (pp. 325–383). Cambridge: Cambridge University Press.

Fonseca-Larios, M. E., & Briones-Fourzán, P. (1998). Fecundity of the spiny lobster Panulirus argus (Latreille, 1804) in the Caribbean coast of Mexico. *Bulletin of Marine Science, 63*(1), 21–32.

Gelcich, S., Hughes, T. P., Olsson, P., Folke, C., Defeo, O., Fernandez, M., Foale, S., Gunderson, L. H., Rodríguez-Sickert, C., Scheffer, M., Steneck, R. S., & Castilla, J. C. (2010). Navigating transformations in governance of Chilean marine coastal resources. *Proceedings of the National Academy of Sciences of the United States of America, 107*(39), 16794–16799.

Goldenberg, S. B., Landsea, C. W., Mestas-Nuñez, A. M., & Gray, W. M. (2001). The recent increase in Atlantic hurricane activity: Causes and implications. *Science, 293*, 474–479.

Gutiérrez, N. L., Hilborn, R., & Defeo, O. (2011). Leadership, social capital and incentives promote successful fisheries. *Science, 470*, 386–389.

Guzmán, S. A., Pérez, J. B., & Laguna, J. C. (2003). Effects of the "El Niño" event on the recruitment of benthic invertebrates in Bahía Tortugas, Baja California Sur. *Geofísica Internacional, 42*(3), 429–438.

Hall, S. J. (2011). Climate change and other external drivers in small-scale fisheries: Practical steps for responding. In R. S. Pomeroy & N. Andrew (Eds.), *Small-scale fisheries management: Frameworks and approaches for the developing world* (pp. 132–159). London: CABI.

Hearn, A., Martínez, P., Toral-Granda, M. V., Murillo, J. C., & Polovina, J. (2005). Population dynamics of the exploited sea cucumber Isostichopus fuscus in the western Galápagos Islands, Ecuador. *Fisheries Oceanography, 14*(5), 377–385.

ITAM-CEC. (2007a). *Programa maestro estatal langosta Quintana Roo* (p. 324). Mexico: CONAPESCA.

ITAM-CEC. (2007b). *Programa maestro nacional de langosta* (p. 324). Mexico: CONAPESCA.

Jenkins, M., & Mulliken, T. A. (1999). Evolution of exploitation in the Galapagos Islands: Ecuador's sea cucumber trade. *Traffic Bulletin, 17*(3).

Kooiman, J., & Bavinck, M. (2013). Theorizing governability: The interactive governance perspective. In M. Bavinck, R. Chuenpagdee, S. Jentoft, & J. Kooiman (Eds.), *Governability of fisheries and aquaculture. Theory and applications* (pp. 9–30). Dordrecht: Springer.

Leslie, K. (2000). The privatization of common-property resources in a Mexican lobster cooperative: Human ecological perspectives. In E. P. Durrenberger & T. D. King (Eds.), *State and community in fisheries management. Power, policy, and practice* (pp. 41–56). Wesport: Bergin & Garvey.

Ley, K., & Quintanar, E. (2010). Chakay: Marca colectiva con identidad de origen de las cooperativas de Quintana Roo. *Biodiversitas, 90*, 10–15.

McCay, B. J., Micheli, F., Ponce-Díaz, G., Murray, G., Shester, G., Ramirez-Sanchez, S., & Weisman, W. (2014). Cooperatives, concessions, and co-management on the Pacific coast of Mexico. *Marine Policy, 44*, 49–59.

Micheli, F., Saenz-Arroyo, A., Greenley, A., Vazquez, L., Espinoza Montes, J. A., Rossetto, M., & De Leo, G. A. (2012). Evidence that marine reserves enhance resilience to climatic impacts. *PLoS One, 7*(7), e40832.

Monnereau, I., & Helmsing, A. H. J. (2011). Local embedding and economic crisis: Comparing lobster chains in Belize , Jamaica and Nicaragua. In A. H. J. Helmsing & S. Vellema (Eds.), *Value chains, inclusion and endogenous development. Contrasting theories and realities* (pp. 178–197). New York: Routledge.

Noticaribe. (2010). *Golpea a pescadores caída del precio de la langosta*. Retrieved February, 24, 2014, from http://noticaribe.com.mx/2010/02/24/golpea_a_pescadores_caida_del_precio_de_la_langosta/

Ortega, L., Castilla, J., Espino, M., Yamashiro, C., & Defeo, O. (2012). Effects of fishing, market price, and climate on two South American clam species. *Marine Ecology: Progress Series, 469*, 71–85.

Ostrom, E. (2009). A general framework for analyzing sustainability of social-ecological systems. *Science, 325*, 419–422.

Pérez-Ramírez, M., Ponce-Díaz, G., & Lluch-Cota, S. (2012). The role of MSC certification in the empowerment of fishing cooperatives in Mexico: The case of red rock lobster co-managed fishery. *Ocean & Coastal Management, 63*, 24–29.

Perry, R. I., Ommer, R. E., Barange, M., & Werner, F. (2010). The challenge of adapting marine social–ecological systems to the additional stress of climate change. *Current Opinion in Environmental Sustainability, 2*, 356–363.

Perry, R. I., Ommer, R. E., Barange, M., Jentoft, S., Neis, B., & Sumaila, U. R. (2011). Marine social-ecological responses to environmental change and the impacts of globalization. *Fish & Fisheries, 12*, 427–450.

Ramírez, J., Castrejón, M., & Toral-Granda, M. V. (Eds.). (2012). *Mejorando la pesquería de langosta en la Reserva Marina de Galápagos*. Quito: World Wildlife Fund.

Riascos, J. M., Carstensen, D., Laudien, J., Arntz, W. E., Oliva, M. E., Güntner, A., & Heilmayer, O. (2009). Thriving and declining: Climate variability shaping life-history and population persistence of Mesodesma donacium in the Humboldt Upwelling System. *Marine Ecology: Progress Series, 385*, 151–163.

SCS. (2011). *California spiny lobster fishery Baja California, Mexico*. MSC re-certification final report (p. 92). California, USA: Scientific Certification System.

Searcy-Bernal, R., Ramade-Villanueva, M. R., & Altamira, B. (2010). Current status of abalone fisheries and culture in Mexico. *Journal of Shellfish Research, 29*(3), 573–576.

Seijo, J. C. (1993). Individual transferable grounds in a community-managed artisanal fishery. *Marine Resource Economics, 8*, 78–81.

Sen, S., & Nielsen, J. R. (1996). Fisheries co-management: A comparative analysis. *Marine Policy, 20*(5), 405–418.

Shepherd, S. A., Martinez, P., Toral-Granda, M. V., & Edgar, G. J. (2004). The Galapagos sea cucumber fishery: Management improves as stocks decline. *Environmental Conservation, 31*, 102–110.

Sosa-Cordero, E., Liceaga-Correa, M. L. A., & Seijo, J. C. (2008). The Punta Allen lobster fishery: Current status and recent trends. In R. Townsend, R. Shotton, & H. Uchida (Eds.), *Case studies in fisheries self-governance* (pp. 149–162). Rome: FAO.

Toral-Granda, V. (2008). Galapagos Islands: A hotspot of sea cucumber fisheries in Latin America and the Caribbean. In V. Toral-Granda, A. Lovatelli, & M. Vasconcellos (Eds.), *Sea cucumbers: A global review of fisheries and trade* (Fisheries and aquaculture technical paper no. 516, pp. 231–256). Rome: FAO.

UNDP. (2012). *Fish production cooperative societies of Cozumel and Vigía Chico* (Equator initiatives case studies series, p. 11). New York: United Nations Development Programme.

Vega, A. (2003). Reproductive strategies of the spiny lobster *Panulirus interruptus* related to the marine environmental variability off central Baja California, Mexico: Management implications. *Fisheries Research, 65*, 123–135.

Vega, A., Treviño, E., Espinoza, G., & Zuñiga, L. (2010). Evaluación de la pesquería de langosta roja (Panulirus interruptus) en la región centro occidental de la Península de Baja California, mediante modelos dinámicos de biomasa: Puntos de referencia y recomendaciones de manejo (p. 21). La Paz, B.C.S: Instituto Nacional de Pesca.

Viteri, C., & Chávez, C. (2007). Legitimacy, local participation, and compliance in the Galápagos Marine Reserve. *Ocean & Coastal Management, 50*(3–4), 253–274.

Walker, B., Carpenter, S., Anderies, J., Abel, N., Cumming, G., Janssen, M., et al. (2002). Resilience management in social-ecological systems: A working hypothesis for a participatory approach. *Conservation Ecology, 6*(1), 14.

Part VIII
Meta-governance – Realizing the Possibility

Introduction

Interactive governance theory talks about three governing orders, including the meta-order where basic governance norms and principles are deliberated upon and settled. As an analytical perspective interactive governance theory is in itself operating at this order as it provides a conceptual framework to assess governance and governability. Thus, interactive governance offers a lens through which it becomes possible to talk about what the governance challenges are and how governability of small-scale fisheries may be addressed.

In *Chap. 32*, Birgit de Vos and Marloes Kraan discuss how small-scale fisheries should be defined, perhaps the most basic question any fishery policy must address in order to tackle governability. Given the diversity and complexity of small-scale fisheries globally, arriving at consensus has proved to be difficult. However, a definition, though often country and context specific, can help to fine-tune fisheries policies. Definitions can be institutionally strategic as when increasing attention is paid to small-scale fisheries in the European Union. Yet, definitions will most likely lead to a debate about categories, boundaries, and associated values – that can be empowering for some but threatening for others. Hence, a clear definition implies a deliberation on images and assumptions underlying fisheries governance, and how they are related to the future of small-scale fisheries.

The situation of small-scale fisheries in Senegal is the topic of *Chap. 33*, co-authored by Michael Hurley and Camille Manel. In Senegal these fisheries are undergoing governance reform and co-governance is being introduced. The chapter focuses on the role that international actors have on shaping governance, which can significantly affect the governance fit, interactions, and social acceptance and functionality of institutions at lower scales. This is partly due to the lack of knowledge that international actors have of the local social-cultural systems. They introduce their own images of problems and principles for solutions. The authors argue that co-governance needs to be addressed in a multi-scaled policy process, where more attention is given to local histories and cultural processes and diversity. The authors identify lack of coordination as a primary governability problem.

Chapter 34, by Lars Lindström and Maricela de la Torre-Castro, takes us to Zanzibar, Tanzania. Here, small-scale fisheries have a long tradition of self-governance, with strong normative and cultural underpinnings. The authors examine the interactions between government and fishing communities, the governability problems experienced with the self-governing mode, and how the government unsuccessfully has tried to stimulate co-governance. They ask important questions with regard to this governing mode, such as with what method the capacity for dynamic, interactive governance develops, and who plays the key role in coordinating interactions across different identities, interests, and spatial and temporal scales. They further explore what supporting institutions must be in place to facilitate conflict resolution and adjudication.

Governing interactions through co-governance is also the subject of Andrew M. Song's *Chap. 35*, which draws its empirical material from South Korean small-scale fisheries. Locally called *Jayul*, a co-governance system was initiated by the central government to address resource degradation and illegal fishing by promoting self-initiative, social cohesion, and local rule-setting among fishing community members. Given its mixed success thus far, the author submits that co-governance undertakings, such as this particular program, in its institutional design and implementation must incorporate meta-governance elements (e.g. values, principles and images of the community members) to better achieve institutional fit and ultimately enhance governability. Finally, using Norway as a case study, Svein Jentoft and Jahn Petter Johnsen in *Chap. 36* hold that governability, and hence the survival of small-scale fisheries, depends not only on the ability and willingness of fishers to respond to changes in the socio-ecological environment, but also on the actions or reactions of the governing system. The inclination to adapt within both the governing system and the system-to-be-governed, "adaptamentality", is seen as the motivation for acquiring the necessary skills, knowledge, and resources that make actors prepared for change. It is argued that the governance partnership arrangement that was instituted during the twentieth century in Norway has generated a relationship of trust between the national government and the small-scale fisheries sector that has been important for adaptamentality. The authors question whether these qualities will persist with the current neoliberal institutional and management reforms.

Chapter 32
To Define or Not to Define; Implications for the Governability of Small-Scale Coastal Fisheries in the Netherlands

Birgit De Vos and Marloes Kraan

Abstract This chapter discusses the need to define the small-scale coastal fisheries sector in the Netherlands. It shows that the fact that there is no clear definition of what small-scale fisheries is, affects its governability. This seems to go hand in hand with the lack of a clear perspective on what the problems and opportunities of the small-scale fisheries sector are. This is partly because many small-scale métiers ('metiers' is commonly used to describe a fishing activity, which can be characterized by a combination of the area that is fished, the gear that is used, and the species that is targeted (http://datacollection.jrc.ec.europa.eu/wordef/fishing-activity-metier)) are out of sight, as they are outside the bounds of data collected, or hidden in 'default rest categories' or because of lack of clear representation. We argue that the relative invisibility of the small-scale fishing sector in the Netherlands, its diversity, lack of representation, and the 'preference' for specialization of the governing system impacts on the governability of the small-scale fisheries in the Netherlands. In this chapter we apply the theory of interactive governance by describing the governing system and the system-to-be-governed as well as the governing interactions between them to analyze the challenges for small-scale fisheries in the Netherlands.

Keywords Definition • Governability • Small-Scale Fisheries

B. De Vos (✉)
LEI Wageningen UR, Wageningen, The Netherlands
e-mail: birgit.devos@wur.nl

M. Kraan
IMARES Wageningen UR, Wageningen, The Netherlands
e-mail: marloes.kraan@wur.nl

S. Jentoft, R. Chuenpagdee (eds.), *Interactive Governance for Small-Scale Fisheries*, MARE Publication Series 13, DOI 10.1007/978-3-319-17034-3_32

Introduction

Small-scale fisheries have a long history in Europe. They make up a majority of the fleet in a number of countries, especially in southern Europe. About 80 % of the commercial fishing fleet in Europe is considered small-scale. In the European Common Fisheries Policy (CFP) reform of 2009, it was mentioned that "fisheries with their large share of small- and medium-sized companies play an important role in the social fabric and cultural identity of many of Europe's coastal regions" (EC 2009).

The image of small-scale fisheries is that of a sector that adheres to values of social justice and ecological sustainability (Johnson 2006; Chuenpagdee 2011). Images play an important role in the governance of small-scale fisheries and fisheries governance in general (Kooiman et al. 2005) as the instruments which are applied in fisheries management are based on those images and values. The European Commission's image, for example, of the large scale fleet is one of efficiency and economic self-reliance, while for the small-scale fleet the focus is on "social objectives, and public funding to help the small-scale segment adapt to changing conditions in the wake of the CFP reform" (EC 2009).

Images come in many types: visions, knowledge, facts, judgments, presuppositions, hypotheses, convictions, ends and goals (Kooiman, et al. 2008). While images are not always made explicit (Kooiman et al. 2008), more transparency about underlying images would improve governance interactions. For effective fisheries governance it is important to bring images to the fore and discuss them openly with all parties involved in governing (Jentoft et al. 2010, 1315).

Another image of small-scale fisheries is that "many vessels are small-scale and have a limited environmental impact" (CFP reform 2009). This limited environmental impact is often attributed to the use of passive fishing gears, the type of gears predominantly used by European small-scale fisheries (*Ibid*). Other advantages which resonate to the image of small-scale fisheries are the good quality of fish caught, and the lower fuel cost per unit of production (Guyader et al. 2013).

Small-scale fisheries are often set off against large-scale fisheries. Some non-governmental organizations (NGOs) sketch a 'bad' image of large-scale fisheries as a contrast to small-scale fisheries, as was the case in for example the Greenpeace campaign in West Africa in 2012 and 2014 using words as 'plunder', 'greed' and 'monsterboats' when referring to large vessels.[1]

Small-scale fisheries are often ascribed an image of being socially just and environmentally friendly. This might lead to promoting small-scale fisheries as a category; or even making a fetish out of them (Johnson 2006). However it is also recognized that "small- scale fishing can be harmful to sensitive coastal habitats and

[1] http://www.greenpeace.org/eu-unit/Global/eu-unit/reports-briefings/2014/GP_monsterboats_report.pdf

that its aggregated impact can be significant with real consequences on the state of the stocks" (EC 2009). Hence, not all small-scale fisheries are a priori ecologically friendly, as this depends on the gear or technique used, the intensity of fishing (see for instance Kraan 2009 on the dominance of small-scale fisheries in Ghana) and the state of the targeted stock. Therefore, some influential NGOs have recently renamed environmentally friendly fishing methods as low impact fishing gear avoiding the scale-aspect (Guyader et al. 2013; Seas at Risk 2010).

Small-scale fisheries often encapsulate a huge variety of fisheries, in terms of technology (boat size, gear type), capital use, economic performance, market linkages, nature of activities, crew size, scale of operation and cultural views (Jentoft and Eide 2011, 2). There is no universal definition of what small-scale fisheries are and what they are not. This is also recognized in the *International Guidelines for Securing Sustainable Small-scale Fisheries* of the FAO (2013):

> The Guidelines recognize the great diversity of small-scale fisheries and that there is no single, agreed definition of the subsector. Accordingly, the Guidelines do not prescribe a standard definition of small-scale fisheries nor how the Guidelines should be applied in a national context. (…) To ensure transparency and accountability in the application of the Guidelines, it is important to ascertain which activities and operators are considered small-scale, and to identify vulnerable and marginalized groups needing greater attention. This should be undertaken at a regional, sub-regional or national level and according to the particular context in which they are to be applied. (article 2.4)

Hence, the concept of small-scale fisheries is mostly a relative concept; it is relative in the sense that it can be understood by what it is *not*; it is not industrial or not large-scale. The opposition between small-scale and large-scale or industrial fisheries is a basic way to categorize fisheries (Johnson 2006). The categories are however a simplification of changing diversity and complexity of fish capture. Second, this distinction between small-scale and large-scale is relative as the image of small and large gets meaning in its local context. For example, the small-scale fisheries in the Netherlands have a completely different linkage to the market as well as technology than the small-scale fisheries in for instance Ghana. Also the scale of the small-scale fisheries within Europe differs significantly. In the Netherlands, only 2.4 % of the vessels are below 12 m, while in Greece this percentage is 65. Therefore, it is hardly surprising that countries have divergent ways of categorizing small-scale fisheries.

This chapter gives an overview of the small-scale fisheries sector in the Netherlands. Small-scale fisheries haven't received much attention in the Netherlands so far, despite the number of vessels (see Fig. 32.1), and possible local social importance. Most attention is directed to the often more specialized, homogeneous, and better organized large-scale fishing sector. One of the challenges for the governing system is that a clear definition, and related to that, a clear perspective on what the problems and opportunities of the small-scale fisheries sector is, lacking. We argue in this chapter that such a definition could benefit small-scale fishers in terms of subsidies, market, and voice; and possibly improves the governability of the Dutch fishing sector as a whole.

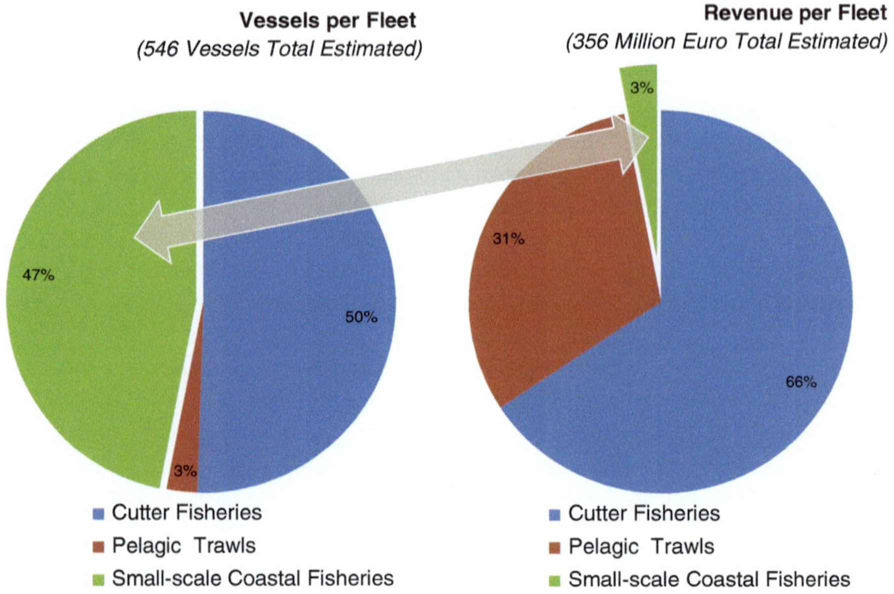

Fig. 32.1 The main marine fleet segments in the Netherlands, the number of boats, and revenue (based on AER, 2013) (Mussel and oyster vessels are considered as aquaculture according to the data collection, and therefore not included in the marine fleet)

An Overview of the Dutch Fishing Fleets

In this section we would like to give some preliminary insight into what we mean when discussing small-scale fisheries in the Netherlands. But before we can do that, it is important to say something about the data. The main data streams collected by the Dutch marine fisheries research institutes LEI and IMARES are framed according to the need of the Ministry, which focuses on the major marine sectors, namely the pelagic and demersal trawl fleets. The rest of the fishing fleets are taken together in a 'rest category' labeled 'remaining coastal fisheries'. Most of the fisheries in this category can be associated with what is seen as small-scale fisheries in the Netherlands, thus will be described in the 'small-scale coastal fleet' section. However it is not a perfect fit as some trawl fisheries are in fact 'small-scale' in the Dutch context whereas some 'remaining coastal' fisheries are arguably large-scale in the Dutch context. This issue will be discussed in more detail in the section on governability challenges.

The Dutch Marine Fishing Fleet

In 2013, the total Dutch marine fishing fleet consisted of 742 registered vessels, of which 546 were actively fishing. This active fleet was divided into 1) a long distance pelagic fleet (14 vessels), 2) a cutter fleet targeting flatfish (276 vessels), and 3) a rest category of remaining coastal fisheries (256 active vessels) (AER 2013). Management measures for the stocks that are targeted by Dutch fisheries may vary from no quota (e.g. gurnard), but only a minimum landing size (e.g. sea-bass), to national quota (e.g. turbot or dab), to individual transferable quotas (e.g. sole or cod), to daily quotas (e.g. hand-picked oyster fishery).

The 'remaining coastal fishing fleet' is a rest category of a mix of fisheries that are not part of the first two fleets. In terms of number of vessels and number of fishers the 'remaining coastal fleet' is quite significant, however in terms of revenue it is of minor importance (see Fig. 32.1). In the next two sections we will describe the small-scale fleet, as the system-to-be-governed, in more detail.

The Small-Scale Coastal Fleet

Small-scale coastal fisheries (i.e. the LEI 'remaining coastal fishing fleet' category) consisted of 453 vessels in 2012, of which 256 were active and 197 inactive vessels (see Table 32.2). The inactive fleet mainly comprises of vessels, which are used to park quota. As will be later described, many fishers in the Netherlands have ITQs. In some cases fishers sell their vessel (by retirement without succession for instance) but keep their quota. The rule is that the ITQs need to be connected to a vessel which needs to sail at least 1 day in a year. Many fishers utilize a small vessel for this function. They are able to earn money by then leasing their quota to active fishers.

In Table 32.1 we show the métiers that can be generally seen as, or include, small-scale fisheries in the Netherlands. There is also a small-scale inland fleet; however in this chapter we will focus on the small-scale coastal fleet (Ministerie van Economische Zaken 2013).

Most of the active vessels are used for gill net fishing, shell fish fishing in the North Sea, fishing with passive gear for lobster and eel in the coastal and delta zone and smelt fishing with seines (see Table 32.2) (Van Oostenbrugge and Op De Weegh 2014). The vessels are mostly smaller than 12 m and relatively old (older than 20 years). The largest part of the small-scale fisheries in the Netherlands operates in the coastal zone and depends highly on the catch of sole, turbot, cod, mullet and sea bass (AER, 2013).

The total engine power amounted to almost 70,000 hp, of which 38,000 hp was actively used. The total yield of the small-scale coastal fisheries amounted to 11.5 million euro in 2012. The net income was 1.2 million euro. More or less two third of this net income originated from the gillnet fishery. In 2012, the gillnet fishery mainly landed sole (66 %). In the gillnet fishery, 60 % of the fishers work fulltime. On average the crew consists of 1–6 people and the crew costs form the largest part of the total costs. The majority of the vessels (85 %) are financed with family money (Van Oostenbrugge and Op De Weegh 2014). There is a large range of fishing effort within the small-scale coastal fisheries: from one to 192 days at sea. Yield per vessel varied from 64 to 500,000 euros.

The majority (78 %) of small-scale coastal fishers use passive gears, such as gillnets, seines, traps, hooks, lines, fyke nets, anchor nets, and baskets. Passive gears do

Table 32.1 Small-scale métiers in the Netherlands

	Characteristics	
Métier description	Gear	Target species
Shrimp fishery (<20 m vessel)	Bottom trawl/pulse	Shrimp
Gillnet fishery	Gillnet	Sole
	Gillnet	Cod
	Gillnet	Grey mullet/seabass
Shellfish picking	Rake	Cockles
Shellfish picking	Hand knife	Oysters/mussels
Hook and lines	Line with one or several hooks	Seabass/cod
Fyke nets and baskets	Fyke nets and baskets	Eel, flounder, smelt, crab
Recreational/angler fishery	Gillnet/seines/hook and lines, baskets, fyke nets, cages. Sports = with hook and line	Sprat, eel, mackerel, garfish, whiting
Pelagic nets (<300 hp)	Pelagic net, demersal bottom trawl	Smelt
Demersal trawl (<300 hp)	demersal trawl	Plaice, dab, flounder, shrimps
Anchor nets	Anchor nets	Smelt, shrimp, sprat
Razor clams	Airlift	Razor clams

This table was developed by the authors and cross-checked with a small-scale fisher representative

Table 32.2 Use of active and inactive vessels in the category remaining (small-scale) coastal fisheries in 2012

Use of active and inactive vessels in the category small-scale marine fisheries in 2012	
Usage	Number of vessels
Active vessels	256
Gill net fishing	54
Beam trawl	19
Other bottom trawls	31
Shell fish (razor clams)	5
Other passive gears, such as fyke nets, baskets, cages, hooks, and lines (incl polyvalent)	147
Inactive vessels	197
Total	**453**

not disturb the bottom of the sea or lead to the water getting turbid (Quirijns 2010). This is in contrast to the rest of the fleet in the Netherlands who predominantly use active gear.

Shellfish fishing has a long history in the Dutch coastal waters (Van Ginkel 1991). Initially only shellfish banks were fished, but ever since the 19th century mussels and oysters have been cultivated (aquaculture). Apart from the traditional fishery of oysters, mussels and cockles, more recently other species such as razor clams have been targeted (http://www.pvis.nl/visserij/schelpdiervisserij/).

The small-scale shellfish fishery also includes hand raking of cockles or manual picking of oysters in the Wadden Sea. 10 kg of oysters can be picked per day. When this is done by a professional fisher, he/she is allowed to sell the oysters as well (Stichting Waddengroep 2010). In the manual cockle fishery, 34 licenses have been issued, but not all of them are active. The hand rake cockle fishery can only be profitable if it is done on cockle banks that have a high density of cockles. This fishery can be considered more small-scale than the mechanical cockle fishery, which was important before, but is now prohibited except in one area.

Market

In the Netherlands, most landed fish is directly brought to the auction, where it is weighed, sorted, and registered. When fishers are members of a Producer Organization they sign private agreements to sell at least the quota fish through the auction. Over the last 5 years, some criticism has been articulated about auctions, both from the buyers' side (they prefer to negotiate directly with suppliers), and suppliers' side (low prices, too many tiers in the supply chain). As a result, fishers try to sell their products directly to the market. This is especially the case for small-scale fishers who try to sell their products in organic markets, to local restaurants, and to small organic supermarkets. As small-scale fishers often target species that are not regulated by quota, they are not obliged to register their fish at the auction. However, some small-scale fishers choose to sell their fish through the auction, buy part of their fish back, and sell it to consumers or trading companies. Especially when buyers offer low prices, fishers buy their own fish. Other fishers do not make use of the auction and clean the fish themselves. Direct selling gives the fishers the opportunity to get connected with consumers and wider society and to teach them about the practice of fishing.

Small-scale coastal fishers usually have the advantage that their fish is fresh as they do not stay out at sea for 5 days like the trawl fishers. Disadvantages can include the variability in volume and dependency on seasonal availability. Small-scale fishers often work with local labels (e.g. Waddengoud, Zeker Zeeuws), sometimes combined with an international label such as that of the Marine Stewardship Council (MSC). They use different labels as the utility depends on the specific market. Some small-scale fishers also are not in favor of international labels, for different reasons:

> We catch grey mullets in the Wadden Sea. We needed a label, but MSC was too expensive. Now I am happy we did not get it, we do not want our fish lying next to MSC fish fingers in the supermarket. We have a special product, and we have no difficulty to sell it despite the fact that we do not have MSC. (Roodenburg cited by Gualtherie van Weezel in the Volkskrant 2014)

An international label can be more difficult to obtain because of a lack of data on certain fish stocks, or because of the assessment costs. However, the government has provided some funding to help with these costs. At present, a group of line fishers of sea bass, the hand-rake cockle fishery, and the razor clam fishery are MSC certified. Until 2013, the gillnet fishers of sole were also certified. However, when after 5 years they had to pay for the reassessment, they decided to withdraw. The costs outweighed the benefits.

The labels can be a way to empower small-scale fisheries, as it improves their visibility in the market and gives them a license to produce. In some cases it also impacts on the governability of the small-scale fisheries. This was for example the case with the gillnet fishers of sole. This fishery was a relatively open fishery, it was only after the gillnet fishers requested the government to regulate – seconded by a MSC condition-, that the government regulated the number of licenses and the number of nets. Another example was that of the sea bass fishery that went for certification so as to get a voice at the government level.

The Governing System

Fisheries in the North Sea are governed under the CFP, but there is some space for the Member States to define additional rules. Also the coastal waters (<12 nm) are a national responsibility. In the Netherlands, the Ministry of Economic Affairs is responsible for the implementation of fisheries policy. They cooperate with co-management groups (in which 90 % of the fleet holding quota rights is represented) for the management of quotas. These co-management groups nowadays overlap with Producer Organizations.[2]

In informal interviews with policy officers of the Dutch Ministry of Economic Affairs it became clear that small-scale fisheries as such is not a focus of the Ministry. It was underlined that the Ministry focuses on sustainability regardless of the scale of the fleet. That does not mean that no attention is given to small-scale fisheries. Small-scale fisheries are discussed as part of stock management (for instance eel or sea-bass); area management (for instance for the Wadden Sea or the Delta-area in Zeeland) or gear management.

Another important point to make is that in general the Dutch government increasingly wants to withdraw from (over) regulating; a process which can be labeled as 'the less rules and taxes the better' motto. One of the clear outcomes of such thinking has been the closure of the Dutch fish Product Board (along with all other sectoral boards). The product boards were industry boards with co-management tasks and were financed by sectoral taxes.

[2] Producer Organizations are officially approved bodies set up by fishery or aquaculture producers. In general these Producer Organizations guide producers towards sustainable fishing and aquaculture, help them match supplies with market demands, and support them in creating added value (http://ec.europa.eu/fisheries/cfp/market/producer_organisations/index_en.htm).

The small-scale fisheries 'label' per se does not have a specific 'place' at the Ministry. The fisheries department in the Ministry of Economic Affairs is currently[3] organized in such a way that there is a policy division and a regulations division – dealing with the administration of rules pertaining to licenses, quota management and days at sea. The policy division has five groups: marine fishing, pelagic fishing, coastal fishing, inland fishing and aquaculture. The regulations division has two groups, one dealing with marine fisheries (including pelagic fishing) and the other with coastal and inland fisheries and aquaculture. This division between coastal fishing and marine fishing is related to the fact that coastal waters are a national responsibility.

Small-scale fisheries are often part of both institutional 'worlds' so to speak; they cross a number of borders in their diverse operations. They are often both coastal and marine (for instance gill netters fishing for sole in the coastal waters are regulated by quotas and are therefore part of 'marine fisheries'), and both inland and coastal (such as some fyke net fishers operating in the IJsselmeer and the Waddensea). These institutional borders impact the room to maneuver for small-scale fishing operations. It also means that there are more policy officers from different groups with whom they need to deal.

Interactions between the governing system and the system-to-be-governed are, in the case of small-scale fisheries, often limited to the use of instruments such as permits, fishing rights and in some cases quotas. Many of these instruments have been developed with a focus on larger scale fishing operations, as the Dutch fishing sector is dominated by its demersal and pelagic trawler fleet. This sometimes results in mismatches, as for instance in the case of the weight regulation where fishermen had to weigh their catches on board their vessels or at auctions. The weighing device that was tested for on-board weighing was tested on a trawler, making it suitable for North sea vessels but not for smaller scale vessels.

Another issue is that small-scale fishers cannot 'claim' part of the stocks, as they mainly target species that are not regulated by quotas. It is difficult for 'outsiders', such as small-scale fishers, to obtain fishing rights and quotas. "The value of individual quota makes the costs prohibitively high. Therefore, aspiring newcomers are effectively barred from entry into the fishing industry." (van Ginkel 2009, 254).

In the last 5 years, many traditional beam trawlers have switched to other fishing techniques such as flyshoot, twinrig, and pulse trawl. This has sometimes meant a change in target species as well. This is for example the case with the flyshoot, which targets non-quota species such as red mullet and gurnard. However, as a result of this switch, the pressure on non-quota species has increased (such as sea bass). The recent extension of pulse permits (from 42 to 84)[4] has meant that the

[3]This is the case until June 2014. After June 2014, the ministry has been in the process of reorganization, which is yet to be completed at the time this article was written.

[4]Under the current regulations pulse fishing (which makes use of electric stimulus) is prohibited in the EU. There is however a provision in place allowing 5 % of fleets surrounding the North Sea to make use of the pulse. The Netherlands have arranged for extra permits under these regulations as they see the pulse as a case for more selective fishing.

availability of sole quota has become more scarce, impacting on gill net fishers who often don't own (all of) their sole quota but lease them. As a result of these changes, interactions between the different fisheries have also increased, leading to spatial or market conflicts.

Many small-scale fishers want to have some flexibility in the way they fish. However, this means that they have to obtain a variety of licenses, for example for different fishing techniques, as well a quota, which is costly and difficult to get. The mismatch between the instruments used to govern the Dutch fishing sector and the requirements for small-scale fishers has been recognized and led to an experimental approach towards small-scale fishers in 2008 called 'integrated fishery'. Integrated fishery was a project in which the government and a group of Wadden Sea fishers created a group license out of all the individual licenses. The aim was to:

- limit the fishery to the season with best catching possibilities and lowest costs;
- spread risks over alternative catch possibilities/target species; and
- postpone fishing activities when required to maintain the stock at a sustainable level.

The project also aimed at improved cooperation between fishers and government through the sharing of licenses, gear and knowledge. One of the reasons the fishers participated in this experiment was because they realized that by pooling their knowledge and gears they would become more flexible, allowing also for more sustainable choices. As one of the fishers said:

> An important reason for which we cannot implement the ideas we have for integrated/diversified small-scale companies is the regulation which allows individual trading of licenses (…) As this is in fact stimulating scaling up, especially for new companies, which usually can only afford 1 license. Because these documents are so expensive, the danger is that one will speed up the fishing, and use it as much as possible. When one has to pay a lot for the license, then one will not take a strategy to invest in 5 different licenses, and to only use them when it is working well. No instead one will take as much advantage as possible of the single license. However, because of local community reasons, and seasonality (to give time to a stock to recover itself), we advocate more flexibility. (…) (Kraan and Paaijmans 2014, 16)

This project also had to cross many institutional borders. For example, when the license-sharing project of the integrated fishery group was evaluated in 2014 fishers expressed the wish that quotas be shared and that they could fish further out at sea. The Ministry responded that they were prepared to look into this, but that they foresaw difficulties in organizing this as it implied that also EU law would impact on the project, as both quotas as well as fishing outside the 12 nm zone fall under the CFP (Kraan and Paaijmans 2014, 19).

Governability Challenges

There is no common definition in the Netherlands regarding what should be considered small-scale and what not. In this section we will show that fisheries data collection has brought to the fore discussions around the definition of the Dutch small-scale fishing fleet.

Data Collection on Small-Scale Fisheries in the Netherlands

The Dutch Agricultural Economics Research Institute (LEI) is responsible for the gathering of economic data for the Ministry of Economic Affairs pertaining to the Dutch marine fishing fleet. Until 2007 the focus of data collection was on vessels that form the largest part of the Dutch fleet, in terms of volume, and yield (the borderline was set at 50 k euro per year per vessel – vessels below that amount were not included). These vessels were also the focus of fisheries policy ever since the introduction of fishing quotas in 1976.

There was a remaining coastal fishing fleet that was not included in data collection. This changed when within the framework of the European Data Collection Plan (EU regulation 2001/1639) it became obligatory to gather data on all fishing activities. Some discussions took place between LEI and the EU before the regulation became effective in 2007. Collecting data on this remaining group meant extra effort as the group was extremely diverse.

From 2007 onwards LEI started integrating the remaining coastal fishing fleet (mostly small-scale vessels) in the annual publication 'Fisheries in Figures' (http://www.agrimatie.nl/SectorResultaat.aspx?subpubID=2386&sectorID=2391), a publication focusing on economic performance and the size/effort of the Dutch marine fishing sector. The remaining coastal fisheries category was in fact a 'rest'-category. This category was not labeled as small-scale fisheries, but did include the majority of fishers that are commonly perceived of in the Netherlands as small-scale coastal fishers (personal communication with a representative of small-scale fishers, 2014). In other words, the category included all fishers that applied passive fishing techniques. It also included fishers that applied active fishing gear, but had an income which was below 50 k euro per year and/or fishers who had been active only part of the year.

As it is a rest-category, the group of fishers falling in this category is rather heterogeneous. Not only does it include fishers with small vessels, passive gear or low incomes working part-time but also fishers that are not required to fill in the logbook and therefore do not show up in the data from the official logbook database (VIRIS),[5] such as dredge fishers (of whom some use a larger vessel). The data for this part of the fleet is collected by means of a survey as opposed to in the case of other fleet segments where data is available in e-logbooks, or as calculated through quotas or at auctions. The survey is sent to all skippers owning vessels that fall into the remaining category of coastal fisheries.

The research institute IMARES is responsible for the collection of biological fisheries data pertaining to the main Dutch fishing fleets. For this data-collection IMARES is restricted to standards that make sense in a European context, resulting in less visibility of small-scale fisheries. This is the case for a number of reasons: (1) data is often projected to maps built on so-called ICES[6] rectangles representing

[5] In this database, information is available on effort and landings for all vessels that have to fill in a European logbook.

[6] ICES stands for International Council for the Exploitation of the Seas.

2500 square km, which is a large-scale setting that does not link up with the spatial scale used by coastal small-scale fisheries; (2) catches below 50 kg do not need to be registered in logbooks under EU regulation, again excluding some small-scale fishermen; (3) days at sea – as an expression of effort – is not compatible with 'soaking time' of gill netters; and (4) the Vessel Monitoring System is not required in the regulation for vessels under 12 m, resulting in small vessels not showing up when VMS data is gathered.

In sum, the practices for data collection have been tuned to capture the larger vessels of the Dutch fishing fleet. It leaves out many of the métiers considered to be small-scale in the Dutch context. The category 'remaining coastal fleet' therefore is quite a mixed bag and cannot be used one-on-one to describe small-scale fisheries in the Netherlands for a couple of reasons. Firstly, because the inland fishers, who generally are considered small-scale (personal communication with representative of small-scale fishers, 2014), are not included. Second, certain types of fisheries such as gill net fishing, now included in the 'remaining coastal fleet' category, are questionable 'small-scale'. Gill netting is a passive fishing technique but some companies use several kms of net. Third, shrimp fishers, applying bottom trawls, are also included if they earn less than 50 k euro per year.

Another important governability challenge for small-scale fishers, besides for data collection, is their representation. This will be discussed in the next section.

Representation of Small-Scale Fishers

Though fishery organizations play an important role in fisheries management, many small-scale fishers do not feel well represented by them, as they mainly focus their attention on large scale fishers, who form an important part of their organizations and who are less diverse. Other reasons barring small-scale fishers joining fishery organizations are the membership fee (of 700 euro per year), and the agreement that fish has to be sold through auction. The latter is not always favorable for small-scale fishers who want to distinguish their product from bulk sales. For these reasons, small-scale fishers have set up their own organizations, hoping for increased visibility, especially in the policy arena. There are several organizations for small-scale fishers:

- Vereniging van Vaste Vistuigvissers Noord (VVVN, Association of passive gear fishers in the North), established in 2010
- Combinatie van Beroepsvissers (CVB, Combination of professional inland fishers)
- Vereniging voor Belangenbehartiging Kleinschalige Kustvisserij (Association for the representation of small-scale coastal fisheries).
- VBHL: Vereniging van Beroepsmatige Handlijnvissers (Association of professional line fishers).

However, despite the presence of a number of organizations the impact in the policy arena is still limited. A ban on eel fishing for 3 months per year, for example, resulted in a general rule of not allowing the use of fyke nets during that period. This also had an adverse impact on other fishers using fyke nets to target species other than eel. Other issues that adversely affected small-scale fishers include a requirement (include a) to have a computer on board for digital registration (certain small boats cannot have a computer on board), and the need to have an installation on board that weighs fish. All of these requirements cost money and therefore it will be difficult for small-scale fishers. Hence, small-scale fishers want regulations that are more tailored to their characteristics, and diversity.

As one small-scale fisher explained to us:

> When the electronic logbook was implemented, the fishers (including myself) were consulted. Most of them were trawler fishers. There were trial versions. The inspection and builders and fishermen thought about it. However the logbook and the rules surrounding its use slowly became more tuned to the fishing reality of the trawl fleet. In our fishing reality we noticed that the E-logbook had to be on board of the vessel that was registered as fishing vessel. We however fish from a small rubber boat and have a mother vessel. In earlier times we had the license on our rubber boat, then we had to change that to our mother vessel –as we were storing our fish there at times – but now we were tempted to change that back to the rubber boat again. Sometimes we fish without the mother vessel, with some of our fishing grounds so nearby the harbor. But because of the regulations surrounding the e-logbook it now means that we have to use the mother vessel when we go fishing at all times, as we have to send a 'leave-the harbor' message (from outside the harbor!) before fishing even though we sometimes go by foot or use the rubber boat. (28-11-2013)

In addition, the Ministry of Economic Affairs has emphasized that it only wants to do business with one national organization that represents small-scale fishers and that future financial support depends on fishers meeting this condition (Visserijnieuws 2013). In 2013, the CVB, which had lost a great part of its members due to the eel closure, together with the PO IJsselmeer, and the small-scale coastal fishers took the initiative to establish a single national organization that represents all small-scale fishers in the Netherlands, both coastal and inland. In April 2014, the organizations received a subsidy from the government to further stimulate the establishment of one organization (personal communication with a representative of the CVB 2014).

The process is still ongoing. The initiators have not defined small-scale fisheries, but will most likely follow the métiers as shown in Table 32.1. Instead of excluding fishers, they will focus on a mindset, a code of conduct which their members will have to follow. Elements of this code of conduct will most likely include compliance with the law, introduction of digital registration in a way that is suitable for small-scale fishers, willingness to cooperate with research institutes, and to share knowledge. Other likely elements are transparency about one's catches, by-catches, and fishing practices, and finally the willingness to fish according to Corporate Social Responsibility guidelines, such as safety on board and good working conditions (personal communication with a representative of the CVB 2014).

To Define or Not to Define?

In addition to the governability challenges due to the diversity of small-scale fisheries in the Netherlands as described above, we raise a question as to whether it is necessary for the governability of small-scale fisheries to define small-scale fishers, and if so, how this has to be done. Defining the small-scale fisheries sector is a challenge in itself, as 'the boundaries of small-scale fisheries are not at all clear and the term encompasses different characteristics from setting to setting' (Johnson 2006, 748). Therefore, "a common view is that definitions and comparisons are impossible, claiming that natural and social systems are too complex and that each individual fishery and fishing community is unique and distinctively different from others" (FAO 2003; Johnson 2006 in Carvalhoa et al. 2011).

Johnson et al. (2005, 73) argue that instead of defining small-scale fisheries by using technical demarcations, such as vessel length, it would make more sense to take a closer look at the various attributes underlying different categories of fishers, and linking them conceptually. Attributes such as social organization, technological intensity of fishing and speed and coverage of operation reflect the dynamics of fish capture. Complexity lies in the interactions between these attributes seen over time (Johnson 2006).

However, from an institutional perspective a definition seems to be needed to develop or implement fisheries policy. The European Commission (EC), for example, defined small-scale fisheries as *fishing carried out by fishing vessels of an overall length of less than 12 m and not using towed gear* (EC 2006), in order to be able to allocate subsidies. In the Dutch case, the EC definition would imply that most gillnet fishers would be excluded, as well as small-scale trawling. Most NGOs might see this as a good thing as these fisheries are often critiqued for their negative environmental impact. Also some small-scale fishers that work with a mother vessel (>12 m), from which they depart to go fishing on smaller boats (<12 m), would be excluded.

The FAO, with their recently developed guidelines for small-scale fisheries, abstained from providing a global definition and instead left it to individual countries to do so. This means that locally (or country) specific definitions are needed. In turn the variations in local definitions will possibly trigger discussions worldwide about the locally accepted images and values of small-scale fisheries.

Such discussion could be advantageous to fishers as their visibility will increase. Not having a clear definition for small-scale fisheries has had repercussions for data-collection and representation as we have seen. To solve the problem of representation, it might be useful for small-scale fishers to define themselves (in relation to 'others'). This can also assist them in becoming a partner to government, whilst at the same time increasing their influence on policy-making. Of course the process of defining, including and excluding might also have other consequences. By defining themselves as *small-scale* it might help them to link up with other small-scale fleets in the EU, to together influence policy at the EU level – which in the long run might have positive consequences also in terms of data collection procedures.

Table 32.3 Characteristics of the small-scale fishery in the Netherlands as seen by the small-scale fishers

Attributes	Explanation
Vessel	The combination of fishing technique and the number/length of nets/hooks used
	Low catch capacity
Management	Weather dependency
Fishery	The number of fishers in a fleet segment (thousands of shell pickers is questionably small-scale)
	Trips of 1 day
Location	Fishing area (small area that is being fished)
	Fishing close to the coast
Finances	Low capital investments (mainly private money)
Fisher	The owner should be actively fishing
	Size of the crew (max the owner, and one crew member)
Market	Fish is often sold to local markets, restaurants or specialized fish shops (organic)
	Whether the fisher is active in the marketing of the catch
	Focus on quality instead of quantity of the product
Environment	Low environmental impact

In order to find out more about the characteristics of Dutch small-scale fisheries and fishers perception as to what this constitutes we conducted interviews with 16 small-scale fishers from the remaining coastal fisheries group, and organized a focus group meeting with members of the knowledge circle of small-scale fisheries. In this meeting, we presented a table with potential characteristics of small-scale fisheries on the x-axis, and métiers on the y-axis. The attributes were based on Johnson's (2006), which we expanded with attributes that related more to the Dutch context

The majority of the consulted fishers agreed to these characteristics. The only characteristic that they found questionable was the length of the vessel. Some felt that small-scale fisher boats should be limited to 8 or 9 m, some said below 12 m, and others said below 15 m, depending on the size of their own boat. Others mentioned that the size of the vessel does not matter at all. In Table 32.3, those characteristics that small-scale fishers deemed relevant to define small-scale fisheries are shown.

One of the interview questions pertained to whether fishers consider a definition of small-scale fisheries relevant. However, in order to avoid steering the discussion too much, it was left to the fishers to explain why such a definition was relevant, and with what purpose in mind. The majority of the 16 interviewees thought it was in their interest to define small-scale fisheries, as in their opinion it could help develop regulations, which are better tailored to their needs. Two fishers gave an example:

> For eel we need an increase of mesh size for example from 34 to 35 cm. (fisher 2)

> Yes it is necessary, because of the regulation. I think we should only have small-scale fisheries. Until five years ago everything went well, but then large-scale fishers also started to fish for sea bass, and they are much less selective in their way of fishing (fisher 9)

A definition can empower them, and clearly distinguish small-scale fishers from other fishers which could result in beneficial outcomes for them:

> (…) It is a positive term, just like artisanal (fisher 1)

> (…) It helps to make a distinction between large and small-scale. Overall small-scale fisheries is less damaging for the environment. (fisher 3)

> (…) We need to distinguish ourselves from the rest. We reinvest our money, do not extract wages, and have the same yield with less effort. (fisher 5)

However, a definition means that choices have to be made, and that sometimes people are excluded that are now included and vice versa. It also means that hidden values and images are being made explicit, leading to potentially difficult discussions (for instance the aspect of family-owned businesses). Coming up with a definition of what is small-scale in the Dutch situation in fact means re-assessing the whole fishing sector. Many of the characteristics that were seen as important by small-scale fishers can only get meaning in relation to other aspects (for example 'closeness to the coast' – how close?; fishing area; and the number of fishers in a fleet segment). Related to this is the question *who* should be involved in (re)defining Dutch (small-scale) fisheries as we have seen that the mere exercise of defining results in discussions about who is in and who is out, what are the underlying principles of our fishing system and what implicit images do we have of our various fisheries.

Conclusions

In this chapter we have argued that the relative invisibility of the small-scale fishing sector in the Netherlands, its diversity, lack of representation at the governance level, and the 'preference' for specialization within the governing system impacts on the governability of small-scale fisheries in the Netherlands. Although Dutch small-scale fishers are empowering themselves through the market (via labels and local market access), which in turn has had an influence on government policies, in general Dutch small-scale fisheries receive little attention from policy and management. This, for instance, results in rules and regulations that are not fit for small-scale fisheries. As lately attention for small-scale fisheries in the EU as well as worldwide has increased, it is timely for the Netherlands to develop a perspective as to what small-scale fisheries are in the Dutch context. It is important to realize that developing a definition is not a 'neutral' activity but requires choices being made. It can have emancipating effects for some, and might be seen as threatening the status quo for others. One of the discussions that might need to take place is with regard to the practice of ITQs on small-scale vessels.

Small-scale fisheries are part of what has been termed a 'rest category', a direct result of the fact that the fisheries governing system is tailored to specialized, and large-scale, mostly quota based fisheries. The *implicit* image of small-scale fisheries in the Netherlands seems to be exactly that; an activity 'en marge' of the main fleets, in contrast to the image of small-scale fisheries in the reformed Common Fisheries Policy where 'an important role' is envisioned for small-scale fisheries. Whether the latter image will stand when the topic is discussed nationally in an *explicit* way remains to be seen.

The governability of the small-scale fishing sector in the Netherlands is threatened by the diversity in the sector both in métiers and in representation. A diverse category requires more attention from policy makers, as general rules cannot be applied. Control and enforcement is also more difficult to arrange. Seen in the light of the 'less rules motto' it is clear that the governing system is not in favor of adding complexity.

Figure 32.2 shows how the interaction between the governing system in the Netherlands and the system-to-be-governed plays out in terms of the government's choice to not have a specific policy directed towards small-scale fisheries (resulting in a lack of a coherent small-scale fisheries policy). It can almost be seen as a self-reinforcing vicious circle. As data-collection is steered by the needs of policy, and as the government does not feel the need to develop a coherent policy for small-scale

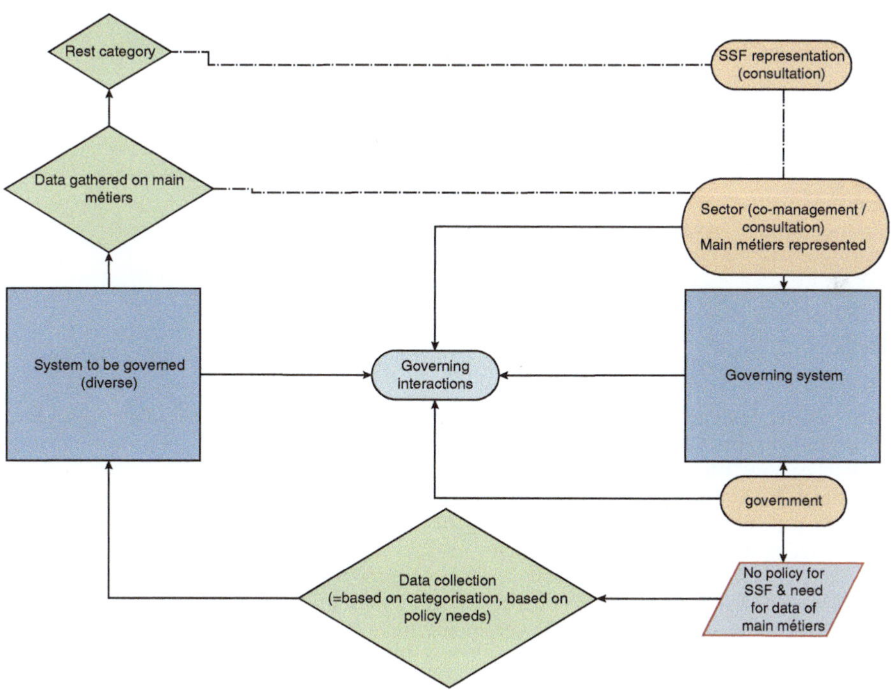

Fig. 32.2 The interaction between the governing system and the system-to-be-governed

fisheries, small-scale fishers are lumped together in a rest category, reinforcing their non-visibility, which is further strengthened by the difficulties of getting an organized representative body for themselves.

Categorizing small-scale fisheries in such a way that is meaningful for small-scale fishers will help them become more visible, especially if the clustering is not only 'on paper' but also in terms of actual representation. The difficulty, however, is to come up with a useful definition of small-scale fisheries with clear and meaningful demarcations, as many of the possible categorizations could have adverse impacts on some sub-métiers such as, for example, small-scale shrimp fishers or large scale gill-netters. It can therefore be helpful to make use of several definitions (i.e. per sub-métier), thereby doing justice to the diversity of the small-scale fishery category.

It has become clear that small-scale fishers are in favor of discussing a definition of small-scale fisheries – so as to increase the attention given at the policy level for small-scale fisheries. The government is, however, less interested as it does not think focusing on small-scale fisheries adds value to policy which already focuses on sustainability regardless of scale. From an institutional perspective a definition at the European level would be needed if the Netherlands were aiming to profit from subsidies for small-scale fisheries through the European Fisheries Fund. To conclude, a discussion on the definition of small-scale fisheries in the Netherlands would imply deliberating upon the current set-up of the Dutch fleet and the relevance of certain boundaries, as well as discussing research practices and policy perspectives. It might result, for instance, in a new perspective on specialization versus diversity, as although diversity provides governability challenges it also adds robustness to the ever changing system-to-be-governed.

Acknowledgements We would like to thank Ratana and Svein for their valuable comments and suggestions that have improved the chapter considerably. Our gratitude also goes to Floor Quirijns from Imares, the interviewees, the fishers from the knowledge circles and the two anonymous reviewers who did a very thorough job in reviewing this chapter.

References

AER. (2013). *Scientific, Technical and Economic Committee for Fisheries (STECF) – The 2013 annual economic report on the EU Fishing Fleet* (STECF-13-15, 302 pp.). Luxembourg: Publications Office of the European Union.

Carvalhoa, N., Edwards-Jones, G., & Isidro, E. (2011). Defining scale in fisheries: Small versus large-scale fishing operations in the Azores. *Fisheries Research, 109*(2–3), 360–369.

Chuenpagdee, R. (Ed.). (2011). *World small-scale fisheries. Contemporary visions*. Delft: Eburon.

EC. (2006). COUNCIL REGULATION (EC) No 1198/2006 of 27 July 2006 on the European Fisheries Fund. Retrieved from: http://faolex.fao.org/docs/pdf/eur65748.pdf

EC. (2009). Green paper. Reform of the common fisheries policy.

FAO. (2003). Report on the second session of the working party on small-scale fisheries. Advisory Committee on Fisheries Research (ACFR), FAO Report No. 735. Bangkok, Thailand.

FAO. (2013). Technical consultation on international guidelines for securing sustainable small-scale fisheries. Rome: FAO.

Ginkel, R. van (1991). Elk vist op zijn tij; een historisch-antropologische studie van een Zeeuwse maritieme gemeenschap, Yerseke 1870–1914, Cambridge: Polity Press.

Ginkel, R. Van. (2009). *Braving troubled waters: Sea change in a Dutch fishing community*. Amsterdam: Amsterdam University Press.

Gualtherie van Weezel, T. (2014). Het keurmerk wordt (te) duur betaald. De Volkskrant 15 augustus 2014.

Guyade, O., Berthout, P., Koutsikopoulos, C., Alban, F., Demaneche, S., Gaspar, M. B., Eschbaum, R., Fahy, E., Tully, O., Reynal, L., Curtil, O., Frangoudes, K., & Maynou, F. (2013). Small-scale fisheries in Europe: A comparative analysis based on a selection of case studies. *Fisheries Research, 140*, 1–13.

Jentoft, S., & Eide, A. (Eds.). (2011). *Poverty mosaics: Realities and prospects in small-scale fisheries*. Amsterdam: Springer.

Jentoft, S., Chuenpagdee, R., Bundy, A., & Mahon, R. (2010). Pyramids and roses: Alternative images for the governance of fisheries systems. *Marine Policy, 34*(6), 1315–1321.

Johnson, D. S. (2006). Category, narrative, and value in the governance of small-scale fisheries. *Marine Policy, 30*, 747–756.

Johnson, D., Bavinck, M., & Veitayaki, J. (2005). Fish capture. In J. Kooiman, M. Bavinck, S. Jentoft, & R. Pullin (Eds.), *Fish for life: Interactive governance for fisheries*. Amsterdam: Amsterdam University Press.

Kooiman, J., Bavinck, M., Jentoft, S., & Pullin, R. (2005). *Fish for life: Interactive governance for fisheries*. Amsterdam: Amsterdam University Press.

Kooiman, J., Bavinck, M., Chuenpagdee, R., Mahon, R., & Pullin, R. (2008). Interactive governance and governability: An Introduction. *The Journal of Transdisciplinary Environmental Studies, 7*(1), 1–11.

Kraan, M. (2009). Creating space for fishermen's livelihoods: Anlo-Ewe beach seine fishermen's negotiations for livelihood space within multiple governance structures in Ghana. Doctoral thesis. UvA, Amsterdam. Published by the African Study Centre in Leiden.

Kraan, M., & Paaijmans, A. (2014). Notitie 'Evaluatie Proefproject Groepsvorming vergunningen in de Geintegreerde Visserij'. BO Project 12.04-001-051. IMARES reference number: 14. IMA0360.

Ministerie van Economische Zaken. (2013). Basislijst binnenberoep documenten per 01-08-2013. Retrieved from: http://www.combinatievanberoepsvissers.nl/downloads/lijst-beroepsvissers-binnenvisserij-per-1-augustus-2013.pdf

Quirijns, F. (2010). Brede visie op duurzame visserij in de Waddenzee. Regionaal College Waddengebied

Seas at Risk. (2010). Activity report 2010. Retrieved from: http://www.seas-at-risk.org/1mages/2010%20Activity%20report%20summary%20for%20website.pdf

Stichting Waddengroep. (2010). Standaards voor het duurzaam handmatig rapen van Japanse Oesters. Retrieved from: http://www.waddengroep.info/wadden-documenten-artikelen/Waddengoud-standaard-oesters-februari-2010.pdf

Van Oostenbrugge, H., & Op De Weegh, J. (2014). Kleinschalige kustvisserij 2012. Rapportage aan enquêtedeelnemers. LEI, 2014.

Visserijnieuws. (2013). Binnen –en kustvissers praten over eenwording. Retrieved from http://www.visserijnieuws.nl/component/content/article/70-archief-2011/8568-binnen-en-kustvissers-praten-over-eenwording.html

Chapter 33
Coordination, Development and Governance of Senegal Small-Scale Fisheries

Michael Hurley and Camille Manel

Abstract Small-scale fisheries in Senegal are extremely diverse, diffuse, and dynamic, and they dominate export and local markets. They have defied conventional management efforts, and Senegal is currently transitioning to participatory co-management. Because much of small-scale fisheries fall under "development", we focus on the role international actors have on issues of legal plurality, governance structure and contextual fit. For example, Senegal fishers involve diverse contexts and cultures, complex gendered relations and cultural and spiritual value systems that affect governance interactions and legitimacy but are not often recognized. This is compounded when international actors frequently bring their own images of problems and principles for solutions with little knowledge of local context, but have significant influence on governance structures and processes. Also multiple independent programs and projects with their own values, rules, "governors", images and principles contribute to a complex legal plurality and fragmented information, impeding coherence and synergy. We introduce coordination as an important interactive process in governance. Coordination problems are endemic to development, economics, and collective action, and impede effective policies and governance. For example policy coherence is an important coordination process and desired state for fisheries. Co-management introduces a very different scale of world views, histories, knowledge systems and outcome desires that will need to be addressed in multi-scaled policy coherence. Downward accountability needs to be improved with fisher perspectives, values and feedback. Also cultural inclusive processes such as institutional syncretism can enhance social sustainability, functional effectiveness and cultural legitimacy of decentralized institutions.

Keywords Governance • Policy coherence • Small-scale fisheries • Development • Coordination • Culture • Institutions • Syncretism

M. Hurley (✉)
LiveWater International, Portland, OR, USA
e-mail: mhurley@livewaterllc.com

C. Manel
Département de Peche Maritime, Dakar, Senegal
e-mail: cjpmanel@gmail.com

S. Jentoft, R. Chuenpagdee (eds.), *Interactive Governance for Small-Scale Fisheries*, MARE Publication Series 13, DOI 10.1007/978-3-319-17034-3_33

Introduction

Fisheries in Senegal have become a rapidly expanding, multilayered, wicked problem that has eluded solutions from conventional fisheries management. The artisanal or small-scale fisheries has grown to dominate commercial, export and local fisheries markets. However increases in exports and economic opportunities, coupled with open access, limited livelihoods and in-migration, have contributed to complexity, ungovernability, unsustainable fishing and declining incomes. To address these problems, there is an ongoing major shift towards collaborative co-management arrangements and ecosystem based management that involve multiple international actors. We discuss fisheries governance in Senegal in an interactive governance framework, focusing on co-management and multi-dimensional coordination processes related to governance contextual fit. Berkes (2007) identifies the need for managing the commons at multiple levels by coordinating institutions both horizontally and vertically, essentially implementing co-management at multiple scales (e.g. Weigel and de Monbrison 2013). As the fisheries is a systemic problem, effective governance will require a systems approach involving the coordination of actors, processes, policy, information and knowledge systems at international to local scales (Bavinck and Kooiman 2013). Coordination is both a key concept in interactive governance, and a major issue and process in development. Since much of small-scale fisheries falls into the realm of international development, we need to understand the influence and constraints development has on both governance and governability. There is a strong influence of development actors at all scales of fisheries that affect governance interactions, structures and contextual fit. Diverse cultural and social aspects of fishers also operate at multiple scales but are often not adequately considered. We see issues with dominant paradigms, defining problems and solutions, information and knowledge problems, and adequate communication and coordination that limits synergy.

We will first discuss our theoretical basis for governance contextual fit and coordination as a key interactive governance process. We then describe the socio-biophysical "system-to-be-governed" and the "governance system" related to small-scale fisheries in Senegal, followed by a discussion of the roles international actors have in interactive governance and how this can add to the complexities of governability and problems with governance fit.

Theoretical Considerations

Interactive governance includes all structures and processes that govern system behavior, such as market forces, civil society, state, rules, information, etc., operating within and between multiple scales (Bavinck and Kooiman 2013). Governability can be thought of as an emergent property of complex systems, reflecting all these multilayered interactive processes (Mahon et al. 2008). Governability is in part determined by how well governance fits contextual characteristics of the complexities of the "system-to-be-governed", and through perceived legitimacy and compliance of governance and rules by fishers (Jentoft and Chuenpagdee 2013). Legitimacy and trust of government by fishers is considered an impediment to effective fish governance in Senegal (Lenselink 2002; Sall 2007), which is often affected by a mismatch between knowledge systems and values (images and principles) of those designing the rules (conservation, modernity, economics, managing fish) and that of the recipients of rules (community integrity, spirituality, secure income, food supply) (Galvan 2004; Sall 2007; Onyango and Jentoft 2013; Song et al. 2013).

We introduce the idea of institutional syncretism, an institutional coordinating process to create socially legitimate institutions through a blending or re-combination of culture, rules and values to craft new institutional arrangements to meet changing conditions (Galvan 2003, 2004). Institutional syncretism can complement interactive governance for assessing governance interactions, fit and coordination across multiple institutional layers. These layers are separated into institutional "superstructure", the formal and observable part of institutions, and institutional "infrastructure" that involves the more informal aspects of local rules and context (Fig. 33.1; Galvan 2004). It is the integration and relative "fit" or "gap" between these that we are concerned with for effective governance. Institutional syncretism views institutions as ordered formal and informal regimes of coordination in two ways. The first deals with governance fit through the coordination of the layered components of institutions (dashed circles in Fig. 33.1) so that there is some synergy, correspondence and alignment of rules, structures, values and beliefs. The second emphasizes the role of institutions as structures for coordinating behavior through this alignment of institutional components and values, beliefs to correspond with patterns and drivers of social behavior (Galvan 2003, 2004), an important concept for the need to shift from biodiversity management to managing fisher and system behavior (e.g. Salas and Gaertner 2004). Institutional syncretism often involves values, local histories, cultural memories and spirituality that provide cultural meaning, perceived legitimacy, social sustainability and functional effectiveness to decentralized institutions. Examples in Senegal show how important cultural-spiritual connections to land were maintained in a syncretic process to integrate concepts of private property, and how incorporating customary law or mimicking kinship processes enhances cultural acceptance and functionality of decentralized institutions (Galvan 2003).

Governance fit can be a significant problem in "developing" countries when factors such as post-colonial governance, western idealized structures and narrow sectoral polices do not fit local complexities. This mismatch, when coupled with weak governance, allows customary institutions to persist, creating a diverse legal plurality (e.g. Meinzen-Dick and Pradan 2002). This governance mismatch is further complicated when international development actors misinterpret governance as government, focus on "rule of law" and reinforce codified "institutional superstructure", often in context of international norms, treaties, dominant policy paradigms or idealized liberal democracy and "good governance" (e.g. Galvan 2003; Ostrum et al. 2007; Booth 2012).

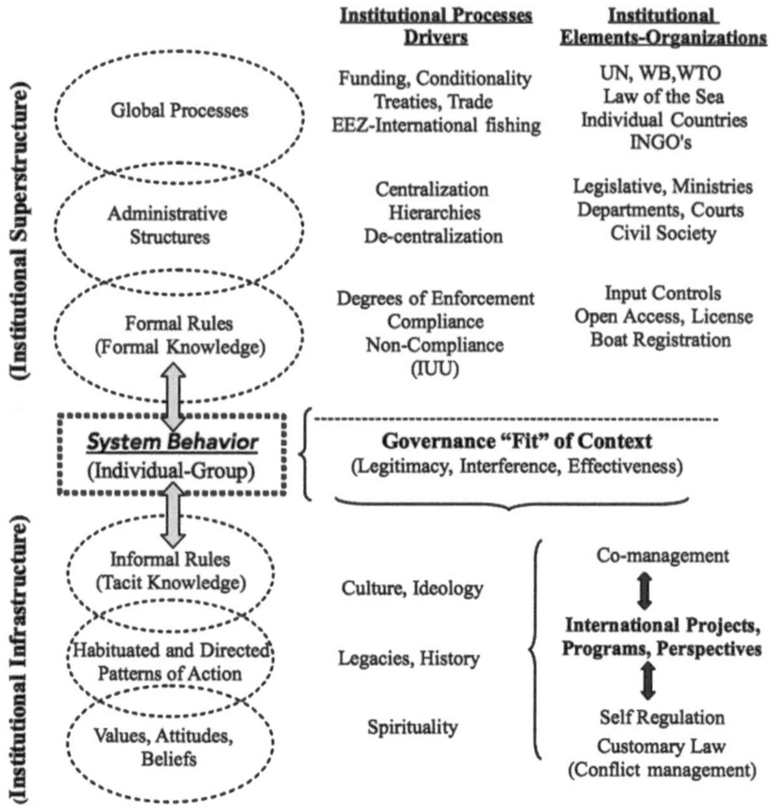

Fig. 33.1 Layered institutions. The multilayered nature of fisheries governance institutions relative to governance fit for co-management (Modified from Galvan 2004). These are separated into more formal codified "institutional superstructure" and the more informal context of "institutional infrastructure". Brackets indicate institutional interfaces (exaggerated gap) critical to governance fit in fisheries co-management and decentralization

We should also consider the idea of institutional interference in governance fit; when institutional "superstructure" may make governability worse. This can happen when fisheries legislation may have negative impacts on livelihoods (Lenselink 2002), or when state level polices dismantle local capabilities to make and enforce rules, but in turn are ineffective at management (Haller 2010; Jentoft et al. 2009). On a different scale is when powerful international actors have an undue influence on policy and marine governance of weak developing nations in West Africa, impeding the ability to craft locally relevant policies (Trouillet et al. 2011).

Coordination is an important concept for interactive governance and contextual fit, as it is a key process in collaboration, collective action, and in development effectiveness (e.g. UN 2010; Chuenpagdee and Jentoft 2013). These coordination processes need to integrate multiple scales, and differing organizational, political, knowledge and cultural boundaries. Policy coherence, an important coordination concept that partly addresses "institutional interference" is when all policies, strategies, programs and rules are strategically harmonized to ensure that policies across sectors do not conflict and achieve desired outcomes (Brown 2005; OECD 2008).

Outside international actors can play important roles in coordination, as Berkes (2007) shows, with numerous actors operating at varying scales in conservation success. However, history reveals that coordination problems are endemic in development, particularly for interrelated "wicked" problems such as fisheries and poverty, and can impair governance (Jentoft and Chuenpagdee 2009; UN 2010). Coordination failure in environmental development can be defined as when aggregate actions lack synergy and do not improve sustainability because of gaps, clustering, biases, coupled with fragmented and conflicting approaches and duplication of effort among various development actors (Connolly and Keohane 1996). Coordination problems in aid effectiveness were so severe they stimulated the Paris Declaration of 2005 (UN 2010), which proposed some coordination through partnerships at the state scale. But this leaves significant constraints to effectively address wicked problems (like poverty) at lower scales (Crespin 2006), and do not incorporate collective action problems and incentives, or problems of ideology or information feedback and uptake (Booth 2012).

There are significant information, knowledge and learning issues that are endemic to development (Ebrahim 2005; Crespin 2006; Andersson 2009). Collective action problems in development are significantly impacted by perverse incentives and severe information asymmetries (missing, incomplete, distorted or wrong information) that impede the ability to make good policy, decisions and actions to achieve sustainability (Ostrom et al. 2002). For example values, perceptions and feedback of stakeholders-beneficiaries is critical for crafting contextually sustainable institutions and policy but is often missing because accountability incentives are upward towards funding, and foster competition over collaboration and sharing information (Ostrom et al. 2002; Ebrahim 2005; Crespin 2006).

System-to-Be-Governed

There is exceptional diversity in Senegal small-scale fisheries and communities that incorporates interactions of biophysical settings, culture and history, migration, and modes of production. On the western tip of Africa, Senegal is partly within the dry Sahelian climate zone and much of agriculture is dry-land rain dependent. Senegal is located at the southern end of the Canary Current Large Marine Ecosystem and migrant fishers extend into the southern Guinean Current (Binet et al. 2012). The Canary Current is one of the most productive upwelling areas of the world, and the Senegalese small-scale fisheries is one of the most developed in West Africa with significant exports internationally and regionally (UNEP 2004). The Senegal coast has significant socio-biophysical diversity in settings, ethnicities, fishers and communities that affect fisheries choices and capabilities and how these fit into local

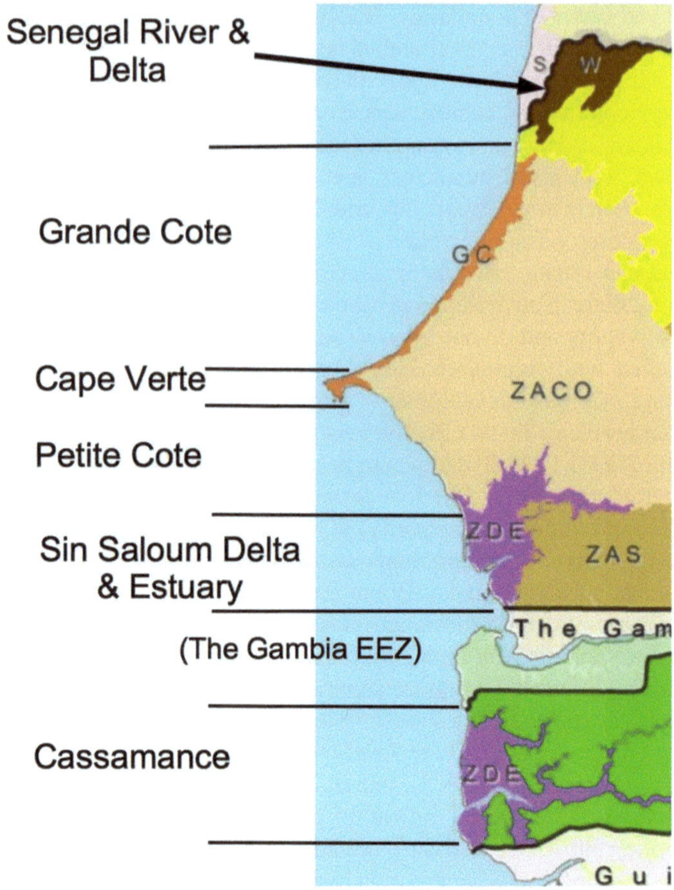

Fig. 33.2 Coastal regions of Senegal (Base map from Tappana et al. 2004.)

histories of production (Niang 2009; Cormier-Salem 1995; Chauveau and Samba 1989). For example, biophysical limitations of local fishing grounds may explain the highly migratory nature of Guet Nardians (an ethnic sub-group), and particular biophysical and socio-economic settings are relevant in self-regulatory and co-management successes (and failures) (Gaspart and Platteau 2001). There are well-defined geographic regions (Fig. 33.2) along a 700 km coast that has over 200 recognized community fishery landing sites. Socio-biophysical relations could be better systematically assessed if such regions were better integrated with terrestrial ecosystems, particularly for complex cultural and livelihood interactions. The Senegal fishery is further complicated by the inclusion of the country Gambia in the center of the country, with its own Exclusive Economic Zone (EEZ).

The fisheries sector consists of three sections; (a) an international fleet under either fishing agreements for the EEZ or that fish illegally, (b) a Senegalese industrial fleet with many international fishers under Senegal partnerships (Belhabib et al. 2014), and (c) the small-scale artisanal fishers. While the industrial fisheries are small in number (less than 200 boats) and represented by a single politically powerful organization, the small-scale fishers are numerous (~21,000 boats), diverse and diffuse, with several organizational bodies, which makes coordination, communication, representation and participation difficult. There are significant conflicts and competition between small-scale and industrial fishers both outside and inside the six nautical mile exclusive zone reserved for small-scale fishers, but there are also unique cooperations and dependencies (Sall 2007; DuBois and Zografos 2012). Illegal fishing can be quite substantial (Belhabib et al. 2014) and recent government actions to combat this have been positively received by fishers.

The fisheries are a significant contributor to the economy and food security of Senegal, particularly when there is high unemployment and with about 46 % of the population living below the poverty line (WB 2014). The government maintains potentially conflicting goals to provide an inexpensive affordable protein source, while also maximizing employment, economic growth, and poverty reduction (Tall 2002; WB 2012a). The fisheries supply 70 % of the animal protein for Senegal, and about 20 % of the active workforce, with small-scale fishers contributing 70 % of exports and 80 % of total catch (WB 2012a). Sall et al. (2006) identify about 17 primary and many secondary employment types that often reflect complex gendered relationships in production that are not adequately represented but are undergoing significant change (Sall 2007). Although most of the reported exports are to developed countries, there is also an extensive but under-reported regional "informal" trade in a diversity of preserved fish (smoked, dried, salted, fermented) that is quite important in employment and food security. In 2000 about 1/3 of the catch went to regional export markets comprising over 11 West African countries (Tall 2002). These "hidden" processes that supply affordable fish protein for food security need to be better understood, monitored and incorporated into fisheries policy and management.

Internal growth, economic attraction, and open access has led to a rapid development of the small-scale fishery, and boat registration is increasing yearly (WB 2012a). The small-scale fleet is composed of pirogues, described as wooden "canoes", though the size, capacity and function of these vary greatly. The efficiency and capacity of boats has also increased, which allows greater long distant migrations, and some are considered "semi-industrialized" fisheries. Conversely, declining catch has also stimulated a trend to limit capital investment by using smaller boats for the same fishing practices, which increases risk for fishers (Sall 2007). The collective impact of the fisheries has contributed to significant overfishing and many species are at full or overexploitation (UNEP 2004). The demersal fisheries, the most valuable and that focused on by small-scale fisheries, have declined by 50 % resulting in reduced catch per unit effort and incomes (WB 2012a).

The small-scale fisheries are highly embedded in history, politics and various cultures, with Wolof, Lebou, Serer, and Diola ethnicities, and many others participating. There is a long history of culturally based fish migration by ethnic subgroups such as Guet Nardians (Wolof) and Niominkas (Serer). Aspects of this cultural diversity such as cultural meanings, values and histories should be a key component considered in the "social system-to-be-governed", as they pertain to boundary problems with governance "fit" and legitimacy, drivers of behavior, and interactive processes of socio-ecological systems and wellbeing. (Galvan 2003; Poe et al. 2014). However, this type of contextual information is rarely recognized or assessed, particularly in development with short timeframes for projects, and where accountability and action are valued over assessments and learning (Crespin 2006; Gould 2006). This problem occurs in Senegal where some fishers express concerns that no co-management programs adequately incorporate cultural diversity and the differences in the ethnicities in the fisheries (Personal communication 2014).

Small-scale fisher migration is a dynamic process that has deep historical and cultural roots and is rapidly expanding. International fish migration now accounts for 30–40 % of the total catch, extending from Mauritania to Sierra Leone (Binet et al. 2012; Belhabib et al. 2014). Fisher migration operates on numerous time-space scales (Njock and Westlund 2010) and is a significant type of intercultural and community-to-community interaction. For example, international fish migration patterns were established in the nineteenth century, well before current political boundaries, which are more "paper boundaries" to ethnic groups that consider "international" waters their traditional fishing grounds (Chauveau and Samba 1989; Njock and Westlund 2010). We need to first realize that migration in West Africa is an important historical and contemporary adaptive process (Randall 2005), which requires going beyond normative explanations for a better understanding of the complexities, relationships and drivers of migration for effective policy (Njock and Westlund 2010; Nyamnjoh 2010; Trouillet et al. 2011).

In-migration to the fisheries and coastal areas is also important to consider relative to access issues and intra-sectoral livelihood policies and how it adds to governability complexities. In-migration is quite varied, from international immigration of women and men into fish processing, inland agriculture, migrant fishers, and temporary or permanent economic opportunists (Sall 2007; Soumare 2012). In-migration contributes to governability complexity by increasing diversity of communities, changing social and complex gendered relationships, and can weaken local authority. Changes in ownership structures erode the status of local leadership, with new entrants having differing values and knowledge of marine heritage and less tendency to follow local rules (Sall 2007). In-migrants may also have different lines of authority following distant kinship structures (JICA 2012), which may also affect flows of money out of communities.

We address international fishing relations because they exemplify the need for policy coherence and coordination across scales, as it incorporates international, state and local linkages and affects fishers' rationalizations and perceptions of governance legitimacy. The EU obtains 50 % of fish from UNCLOS (Law of the Sea) facilitated international fish agreements, and Senegal was the first to sign an agreement in 1979 (Kaczynski and Fluharty 2002). The original agreements competed with small-scale fishers and were highly criticized for not meeting UNCLOS sustainability criteria and being "incoherent" with EU's own sustainability criteria and policies for developing countries (Brown 2005). Furthermore, this raised tensions with government as small-scale fishers did not see benefits from the agreements, and saw industrial fishers, not themselves, as causes of overfishing (Chauveau and Samba 1989). Senegal dropped the agreements in 2006 but recently negotiated new ones, and though the EU has adjusted agreements to be more sustainable, they remain controversial.

The Governance System

There are several complex processes and histories operating in Senegalese fisheries governance across multiple scales, with many international actors influencing the structure and interactive nature of governance. Senegal is a young country, established in 1960, and is one of the more stable democracies in Africa. In contrast, coastal fishing communities such as Yoff were established 400–500 years ago, long before current political boundaries and regulations (UNESCO 2000). Senegal has a history of strong centralized governance, partly because of ideas of modernity of the political elite, maintaining stability in a new country, and from post-colonial remnants of French rule (Boone 2003). This partly explains difficulties in achieving true decentralization and devolving power to regional and local institutions (e.g. Galvan 2004; Ribot and Oyono 2005).

The path to current collaborative governance can be viewed as a multi-decadal transition of governance modes. The new Senegalese government emphasized the small-scale fisheries from the beginning, and through a "western" centralized top-down governance mode focused on production, fisher organization, and a biological management orientation. Though governability problems with the top-down mode in Senegal are sometimes attributed to weak enforcement capacity, historic and ongoing government-fisher tensions affect legitimacy and trust and in turn lead to poor compliance with government regulations (e.g. Lenselink 2002; Sall 2007). For example, the government initially interfered in early "illegal" community self-regulation (Lenselink 2002), and recently fishers rioted and attacked officials when they tried to enforce regulations. After a period in the 1990s where communities experimented with self-regulation that at times were controversial (Gaspart and Platteau 2001), fisher participation in governance was promoted to alleviate tensions (around 2000). One example is how the community of Kayar is now a showcase of self-regulation transitioning to co-management after a long and difficult process (Gaspart and Platteau 2001; Alioune and Catanzano 2005). However the 2000s were described as a period of "anarchy" where top down governance and local rules seemed inoperable to control access (Watanuki 2008). The early 2000s introduced a period of independent donor/NGO experimentation with co-management, to a transition point of acceptance of collaborative governance as a primary driver of management (e.g. MEM 2011).

There are a variety of cultural and local institutional processes operating across Senegal such as conflict management, land tenure, and community and rural councils that intersect with coastal communities and can be drawn on for understanding contextual issues and enhancing integrative governance. Besides local conflict management, many customary fisheries regulatory processes are considered to be overwhelmed by recent technologic and demographic changes (Gaspart and Platteau 2001; Lenselink 2002; Watanuki 2008). However Sall (2007) notes that hidden behind the curtain of adaptation to globalized markets, cultural values, beliefs and behaviors are still strong, but simply not observable or valued by scientists and governance actors oriented to conservation or economics. Understanding the extent and diversity of past and partial institutional processes, whether customary or not, and why they worked or why they didn't, is imperative to understand problems and design new functional institutions (Gaspart and Platteau 2001; Galvan 2004; Haller 2010). For example there is a general cultural concept of not restricting access to marine commons and food, and urban and marine systems tend towards more open access. However concepts of territory and access can differ with different ethnicities, with fisher-farmers, and in geographic settings such as estuaries or with histories of agriculture (Cormier-Salem 1995; Gaspart and Platteau 2001; Sall 2007).

There are significant coordination issues that international actors add to the governability complexities posed by the diffuse and diverse nature of small-scale fisheries. There are numerous international actors implementing a variety of projects that include various fish management projects, multiple co-management projects, and several biodiversity protection projects. Plus there are numerous historic and current independent community management, production and economic projects that collectively affect coastal communities and livelihoods that are concealed from governance. This is further complicated by two different officially recognized co-management structures, CLPAs (Conseils Locaux De Pêché Artisanale-Local Artisanal Fishers Councils) and CLPs (Comité Local de Peche; Local Fish Councils) (Table 33.1).

Table 33.1 Distribution of co-management units in Senegal

Geographic region	CLPA's functional and (total)	CLPs[c]	Total communities[d]	Dominant fisher ethnic groups[e]
Grand Cote	1–2 (6)	0	23	Wolof
Cape Verte	1 (3–4)	3	14	Lebou
Petite Cote	3 (4–5)	6	17	Serer
Sin Saloum	3 (8–9)	3	73	Niominka
Cassamance	0 (8)	0	65	Diola
Total	8–10[a] (32)[b]	12	192	

[a]As of 2013. Projects are currently expanding to incorporate more CLPA's
[b]Total represents all proposed CLPA's. Only 22 had been officially initiated as of 2011
[c]Established or being established by World Bank
[d]Total represents all fisher communities identified by DPM (Départment de Peche Maritime, Senegal) in all proposed CLPA's
[e]From Chauveau and Samba (1989)

The current process of collaborative governance is being phased in, beginning with significant government oversight and potentially transitioning to self-management. The official decentralized fisheries governance units are CLPAs. We should understand that individual communities were not included in marine fisheries decentralization legislation and thus legally cannot independently develop and enforce rules. In 2005, 32 formalized CLPA spatial areas were identified for the entire coast. They are an intermediate scale administrative structure that coordinates all communities and actors in fisheries that share similar geographic setting and fishing areas. A second organizational scale of CLPA's is the Sustainable Management Units that coordinate all actors around selected species and their geographic range (USAID 2013). CLPAs can develop their own rules for their particular designated spatial area or species of concern (Sustainable Management Units) in management plans (local fishing conventions), which are ratified by the government (DPM 2011; USAID 2013). While CLPAs are seen to facilitate communication and coordination between communities, many are not functional in representation, communication and participation (DPM 2011; Karp 2011; USAID 2013). Thus there is a danger that some CLPAs may become technical-administrative units ruled by committee, as they are new institutional structures with some distance from on-the-ground processes.

Another co-management system created by the World Bank, the CLP, is a community-based organization that can create their own management with government oversight (WB 2012a, b). Although not officially recognized in the decentralization process, CLPs obtain legal status through a ministerial decree after being ratified by the CLPA governance committees. Twelve CLPs are being established with an objective to strengthen CLPAs, demonstrate local governance and access control, eventually through territorial access rights, and possibly ITQs (WB 2012b). Despite significant successes, they involve high resource inputs with primarily smaller homogenous communities, thus they may be difficult to sustain and reproduce at scale (MEM 2011).

There are some contextual fit problems with all the current co-management constructs as they only partially address the complexity of the fishing system. Much of the focus is primarily on community or intermediate scale territory and sedentary demersal species; they do not adequately incorporate larger scale processes and drivers such as migratory fishers, which are the most dynamic aspect of Senegalese fishers (MEM 2011). The interactive nature of migration presents coordination and design issues for co-management and restricted access of new associated territories, and potential conflict with cultural views of access (Watanuki 2008; Brueil 2011; DPM 2011; Trouillet et al. 2011; WB 2012a, b). Neither have co-management fully included in-migrant fisheries and the complexities of kinship-connected authority (e.g. JICA 2012). There are proposed experimental variable licensing processes to address some of these issues. There is also a danger that governance constructs may become too rigid, and if they do not incorporate the full dimensionality of the system, they may in fact constrain the historical creativity and adaptive processes of fishers (Chauveau and Samba 1989).

An organization often overlooked that exerts substantial international influence on Senegal's fisheries governance is the donor coordination committee for fisheries, which controls substantial financing. The committee is composed of various EU/OECD nations, the FAO and World Bank, and was established to improve coordination between independent donors, primarily for co-financing projects, and interacts with government officials on legislation and policy (WB 2012a). The donors have succeeded in collaborating on projects such as boat registration and illegal fishing. However, there is significant "cultural distance" for contextual design, there are no fisher representatives, and there are perceptions that coordination and sharing information is limited, which impedes any real synergy.

Regional coordination as co-management is also needed to adequately address large scale processes that if poorly managed threaten the viability of lower scale co-management (Watanuki 2008; WB 2012a; Weigel and de Monbrison 2013). There are at least four West African regional fisheries and economic governance entities that have policies relevant to fisheries, plus the FAO statistical reporting units; although they perform a critical role, they create coordination issues as a patchwork of fisheries governance (e.g. Trouillet et al. 2011). The Sub-Regional Fish Commission based in Dakar, Senegal, is an important organization that includes seven countries from Mauritania to Sierra Leone, which also corresponds to Senegalese small-scale fisher migration. The main function is to promote cooperation, coordination of policy and sharing of research between member states (e.g. Weigel and de Monbrison 2013), and dealing with key transboundary regional issues that include equitable international fish agreements, migratory fish stocks, and illegal fishing. It can also play a key role in policy coherence between EU-OECD countries and member states (OECD 2008). Much of the research of the Sub-Regional Fish Commission is internationally funded, which skews information and policy towards biodiversity-economic paradigms, creating a critical socio-cultural information gap. On a whole, coordination is difficult due to the weak governance of member states.

International Actors, Governance Interactions and Context

As illustrated above, when many international actors are involved with fisheries development, there can be significant coordination issues that affect governability. Although these actors can bring needed resources, new knowledge and coordinating functions to address problems, and their role in the transition to co-management is well recognized, better coordination is required (MEM 2011). The missing socio-environmental contexts or understanding of communities, culture, local history and institutions can create significant context problems for these actors. As suggested by Connolly and Keohane (1996), problem identification is a critical political process and should involve many stakeholders, particularly local people. However when international actors bring in preconceived images of problems and solutions, opportunities for deep assessments of context are missed and often replaced by imported principles and policy instruments and outside "expertise". We discuss below governance interaction and coordination issues and how this affects fishers perspectives of governance.

Powerful international actors heavily influence environmental policy for developing countries, as represented in Senegal (DGIS 2008). One process is how internationally recognized global commons problems and strategies are funded while local problems and strategies go lacking (Connolly and Keohane 1996). Fisheries governance agendas in developing countries are dominated by international paradigms of biodiversity conservation, economics and western based "rationale management" constructs, minimizing social and food security issues, and may not fit local contexts (Wilson 2002; Hall et al. 2013). These issues are reflected in fisheries aid for Africa and Senegal where multiple donor projects can result in poorly formulated projects, policy gaps and persistent problems (NEPAD 2011). The processes of international influence on the state formal "institutional superstructure" operate directly through advising on governance structure or through co-construction of laws, rules and legislation, and indirectly through funding for projects and studies. For example, the World Bank is pushing for rights-based legislation to complement their governance projects, where access rights and protected areas are a dominant theme (WB 2012b). Also the UN Convention on Biological Diversity is invoked by international NGOs to elicit donor promises of West Africa coastal protection (PRCM 2011), France funded a major study for including protected areas in co-management, and the World Bank influenced a "Biodiversity Protection Act" (WB 2012a, b).

These high level governance processes can create governability, coordination and context fit problems. For example, the strong international promotion, financing and state and donor support for MPAs creates the danger that biodiversity protection becomes a panacea that mask other critical governance problems in Senegal (Breuil 2011). This also creates a coordination problem when NGOs replace formal "government" in MPA design and implementation in Senegal, which creates a fragmented governance structure where MPA's reflect more the character of the NGO's rather than any coherent policy (Breuil 2011). Also the models promoted as successful MPA's are being questioned relative to the adequacy of fisher representation, participation and power relations (Trouillet et al. 2011; Cormier-Salem 2014).

A classic coordination problem arose during the early stages of CLPA implementation (2005), where over eight organizations were involved in establishing 22 CLPA's (DPM 2011). By 2011, only 8–10 were considered functional. A primary reason was inconsistent concepts of CLPA's because of coordination problems with all actors: between donors, between donors and their partners, between donors and government agencies, and within government agencies, and thus between CLPA's themselves (Karp 2011; MEM 2011). Another issue with coordination later arose when three major donors were implementing co-management projects independently within the same area. The state then asked for better coordination, and though there has been some interaction and transfer of knowledge, coordination issues are still considered a significant impediment to synergy-partly as these donors have different and sometimes conflicting philosophies, principles and processes for co-management.

A more insidious coordination problem arises when reliance on development aid and projects can actually weaken governance as it creates an imbalance between government and international actors, and involves high transaction costs (UN 2010;

OXFAM 2011). Although Senegal fisheries are one of the highest financed in Africa (NEPAD 2011), government capacity has been reduced. There was tremendous loss of researchers (from 44 to 7) from a once renowned fisheries research department (CRODT), which can be attributed to migration to better opportunities related to projects (USAID 2013). The Fisheries Department experienced a long period of low recruitment, which resulted in massive retirement and loss of historical and institutional knowledge. Field staff have also been significantly reduced. All of these have reduced the quantity and quality of government-fisher interactions, which negatively influences fishers' perceptions of government. Increasingly, fisheries institutional and administrative functions have been taken over by development actors such as NGO's, which have little accountability to citizens, and increases fragmentation and coordination problems.

This then questions who is the 'co' in co-management. In essence the transfer of governance through projects and contracts results in a diversity of temporary actors and structures that creates a highly complex, fragmented legal plurality that complicates governability. Projects are essentially temporary governance boxes where actors frequently insert their own values, principles and rules, with a divergent accountability (e.g. OXFAM 2011). As projects are implemented independently, there is often poor coordination and sharing of information, and competition for funding and policy space often predominates over coordination (e.g. NEPAD 2011). Information is not only fragmented between projects, but also with the various partners, NGO's and contractors within each project (e.g. Ostrum et al. 2002). And when projects end, they leave a governance and resource gap where "governance actions", information and critical tacit knowledge are lost. Project and staff turnover results in a major loss of contextual information critical for learning and constructing contextually effective policies (e.g. Ostrom et al. 2002; Andersson 2009). In particular, cultural knowledge is not often valued and thus not codified in development organizations, but this critical contextual understanding is embedded in individuals experience and tacit knowledge that is lost through turnover (Gould 2006).

All this is witnessed and mirrored in perceptions of fishers that can affect governability. One local fishers' organization states that fishers feel that a large number of actors creates confusion among beneficiaries. They see coordination problems in co-management implementation, and confusion as to purpose when there are diverse and ever-changing objectives. In particular, they view NGOs as being focused on competition, their own missions and objectives, with differences between stated objectives and actions, which impedes sustainability. There are also issues of transparency and accountability, with few local benefits seen from high levels of financing, often expressed as numerous redundant workshops and seminars (APRAPAM 2013; pers. comm. 2014). Also, some stakeholders feel that MPAs did not meet original intentions (Sene 2013), and now some communities are rejecting NGO overtures for protected areas (Personal communication 2014).

Fisher perspectives also create misunderstandings at intersections of cultural boundary systems relative to a science and management dominated by western constructs (e.g. biodiversity, rational management, economics), which is an important governance interaction issue. We first need to understand that in western cultural constructs, nature is separate from humans and thus allows for a separate biodiversity management-protection paradigm, but this differs in many cultures (UNESCO 2010). Senegal fishers cultural knowledge systems include differing valuation of species and strong spiritual connections and different cause effect understandings of marine systems that do not fit western science or biodiversity concepts (Sall 2007). This creates communication problems and tensions with scientists and governance actors dominated by conservation and economic paradigms, which then creates issues of legitimacy with the science behind governance and thus governance itself (Sall 2007). This also creates a misunderstanding and reluctance to embrace "biodiversity protection" (Personal communication 2014). Also, ownership (land, vessels) is a strong cultural valuation linked to status, and MPA's represent an outside authoritative imposition on ownership. There have also been many incursions on the loss of ownership from tourism development and protected state forests that restricts community access to livelihoods, often in the name of "biodiversity" (Sall 2007). This tension was reflected when local fishers destroyed a World Bank sponsored MPA (WB 2012a).

The issue of governance fit and fishers views and perspectives is exasperated by information and accountability problems inherent in development. Effective decentralized co-management requires "downward" accountability to fishers, coupled with feedback and some control by "beneficiaries" for sustainable development (Ostrom et al. 2002; Béné and Neiland 2006). However in development, incentives for accountability, and thus feedback and information flows, is primarily oriented "upward" towards funding and distant headquarters. This is reflected in information from monitoring and evaluation and indicators for success. Easy to measure "outputs" are common for accountability in development, but do not measure results of objectives. This is reflected in USAID's evaluation criteria that include number of people trained, number of action plans, reports, etc. (USAID 2013). When more difficult to measure "outcomes" are used, they most often reflect dominant international paradigms and the values of the implementors over those of beneficiaries. For example, like many others, the World Bank uses a key indicator of success of "effectiveness of biodiversity management" (e.g. WB 2012a, b). This does not fit the context of fisher perceptions and values, and violates the idea of co-production of knowledge in co-management (Chuenpagdee and Jentoft 2007). Fishers have more qualitative indicators of change related to community behavior and relations, including changes in fishers behavior, strategies and risk at sea, changes in social organization and relations, and the diversity and risk of investment strategies (Sall 2007). Wilson (2002) calls the types of indicators that do not fit the context in development as "institutional distortions" that create the simplification and reformulation of knowledge for institutional purposes. This process can perpetuate organizational mismatch, governance gaps and misunderstandings throughout long distant "knowledge chains" of organizational reporting and evaluation to where funding and design decisions are made.

Solutions to bridging these cultural knowledge boundaries can involve more feedback and participatory evaluations with fisher stakeholders, stakeholder defined indicators, co-participation in research and mapping, and culturally structured communication processes. Indicators for evaluation can be improved if we incorporate both standard indicators that satisfy requirements of donors, scientists, and managers with local culturally defined indicators that can allow insights into contextual nuances that are critical for design, management and social sustainability (e.g. Nazarea et al. 1998).

An interesting cross-cultural communication bridging process is exemplified in a USAID program where women fishers were uncomfortable with the powerpoint presentations and western linear rationale approach used in meetings, and were encouraged to develop their own communication practices. These involved dance, song, spiritual connections and interpretation, coupled with more organic diagrams (Niamadio et al. 2012). This emphasizes that cultural communication and social learning is not a linear distribution of packets of information, but a highly complex, culturally integrated process. Though this represents the potential of incorporating culture into co-management structures and processes, it is underutilized or relegated to governance corners.

Concluding Thoughts

In its short existence, Senegal fisheries governance is going through a long and significant transition to collaborative governance, which is starting to produce results and is reducing some tensions between government and fishers (MEM 2011). Developing new institutions for a collaborative framework takes much time and resources and significant coordination of all actors. Though many elements are in place at regional to local scales, there are still significant gaps, coordination constraints, and contextual fit issues that affect governability. Government structure will require a transformation in management to better fit this new process of governance interaction. The Senegalese government notes that regional capacity will need to be increased with more field personnel, and retraining personnel used to "top down" enforcement management to refocus on reinforcing community governance capacity, logistical support and communication (MEM 2011). Participatory mapping and research is an important interactive process in knowledge syncretism by bridging the gap between science, cultural knowledge and experiences of fishers. When fishers participated in research and experimental processes and the results fit their knowledge and beliefs, they were more likely to accept results and bridge the high skepticism of science. Participation also revealed to professionals the on-the-ground particularities of management (Watanuki 2008; JICA 2012; USAID 2013). These have been small steps that need to be experimented with and expanded, and there is a renewed interest in reinforcing government capacity. A re-transference of governance back to government should help achieve these capacity changes, improve relations with fishers, and facilitate coordination.

Also initial participatory co-management has produced successes for some communities. Communities with co-management are more likely to adopt regulatory measures previously ignored and create or reinforce their own (e.g. WB 2012a). Habitat improvements, rest periods and/or local protected areas have been implemented, as well as some capacity reduction and gear modifications to improve sustainability. In some areas these have resulted in increases in both size and number of catch of targeted demersal species (JICA 2012; WB 2012a). Also, collaborative management plans are being prepared for some CLPA's (USAID 2013).

However major governability issues remain. It is unclear whether decentralized governance will adequately incorporate the cultural complexities and multi-scaled nature of fisheries and fishers perceptions, or be conduits for donor-state preferences. Though the co-management process involves a graduated engagement of the state, some fishers have protested the involvement of state actors (prefect and sub-prefects). We do need to be aware that narrow legal interpretations of state agents can constrain local innovation of more holistic institutions, and merely "grafting" formal, idealized values (e.g. conservation, modernity, governance) onto decentralized structures in Senegal can provide a false appearance of legitimacy and questionable participation (Galvan 2003, 2004). This illustrates a conundrum in decentralization: To tailor decentralization to fit a diversity of localized contexts that can enhance legitimacy, or fit the stated desire for a harmonized coherence between co-management units for better coordinated actions at larger scales (Galvan 2004; USAID 2013). Co-management conventions and legitimacy need to go beyond a small percentage of communities and active representatives, but be legitimately accepted by most fishers and collectively effective and synergistic at large scales necessary to achieve sustainability.

We also have to understand the cultural interactive processes that determine governance legitimacy in Senegal may not be through formal conventional understandings of social learning, but through individual micro experimentation and feedback of what works and doesn't, which is transmitted through complex networks (Galvan 2004). Also, identified training in administrative capacity of CLPA's (USAID 2013; many others) may need to go beyond normal western rationalist constructs and include culturally sensitive training in institutional syncretic processes (Galvan 2003). In some of the decentralized rural councils of Senegal, western oriented administrative training was discarded by council members who saw a need for better understanding customary laws to be legitimate and effective (Galvan 2003).

The governability complexities of the diverse and diffuse nature of the small-scale fisheries requires significant and effective coordination, which has been lacking. There is much potential in stakeholders who desire collaboration, and those that control funding and implement projects bear significant responsibility to collectively achieve the necessary levels of coordination, and contextual legitimacy of institutions. A strong commitment to continuous, intense coordination is needed to overcome the particularist agendas, interests and internal rules of bureaucracies (Disch 2010; NEPAD 2011). This also applies to contextual realities and downward accountability across scales to include fishers perceptions and feedback, particularly for evaluating projects. A particular issue with power and information is who defines success (and problems), which needs to include stakeholders such as fishers. There should also be enhancement and better coordination of information, research, assessments and monitoring by all projects, organizations and partners, that is oriented towards preferences, concerns and issues identified by co-management participants. Pooling resources by donors to implement common information collection may enhance the stated desire for synergy and coordination.

However there is a dark side to coordination and coherence if it reduces flexibility and diversity needed for various pluralities or contexts, or reflects mainly perspectives of power, dominant knowledge systems or dominant paradigms. We see this in the contested universal application of international treaties such as the Convention on Biological Diversity that needs to better allow for a plurality of cultural expressions of environmental ethics (UNESCO 2010). Another example is policy coherence proposed for West Africa fish governance for integrating policy at multiple scales (international, regional, national) and within and between sectors (OECD 2008). Coherence implies some consistency between policies that do not conflict with goals or desired outcomes in fisheries and social development. Though this is critical, it again focuses primarily on the coupling of codified institutional superstructure between international-state scales, which can impede governance fit and functional effectiveness at local scales. Co-management opens up a whole new institutional scale as informal "infrastructure" where multi-scaled policy instruments and rule correspondence should be consistent with the operating institutions, cultural values, perspectives and desired outcomes of fisher communities. This also requires a whole new level of assessment of cultural and social context and orientations of knowledge designed for and considered legitimate in governance constructs and the co-production of knowledge in co-management (e.g. Chuenpagdee and Jentoft 2007). We can use the concept of institutional syncretism (Fig. 33.1) to assess "cultural policy coherence" as it pertains to institutional coordination and alignment to achieve legitimacy and functionality.

The concept of coordination as a principle in interactive governance can aid in addressing the complexity and multilayered problems in small-scale fisheries. Coordination can be partially seen as a function of understanding and solving issues of interactive processes of information flow, communication, and social learning across multiple diverse trans-boundary systems and at multiple scales to achieve some coherence and synergy for solving problems. Also, coordination processes can inform the dimensions of governability to better achieve contextually integrative and legitimately functional co-governance. This will require going beyond narrow biodiversity-economic, western rationale constructs dominant in governance to include more social-cultural elements. For example we will need to understand complex integrative relational processes such as kinship, migration and hisorical-cultural trade processes that interconnect multiple boundaries. Also, incorporating socio-cultural considerations as "local knowledge" into participatory processes (e.g. research, evaluation) can aid both government and particularly fishers to better understand and appreciate social complexities, interrelationships and flows that should be reflected in co-management conventions and policy considerations. This is particularly important as women's roles in fisheries are not well represented in CLPA governance committees (USAID 2013). A refocus from biodiversity to food security, and using concepts of "following the fish" through the entire fish chain, coupled with "following the money" through various gendered relations through communities and households, can better reveal the dynamic nature and extent of various boundaries relative to fish management, livelihoods and governance (e.g. Bavinck and Kooiman 2013). "Following the fish" can illuminate the informal and poorly understood socio-cultural complexities in food production and exchange that are changing fast. The very processes that generate fish as an inexpensive, affordable protein source for low income regions need to be better understood as they may be disrupted by narrow but well intentioned concepts of improving value, production and inefficiencies in value chains. This will also require changes in monitoring such as social relations in production, species and prices in markets, how fish as food gets to families and costs of fish per family.

References

Alioune, S., & Catanzano, J. (2005). Self-regulation of Senegalese artisanal fisheries: A case study of Kayar. In S. Cunningham & T. Bostock (Eds.), *Successful fisheries management* (pp. 149–167). Amsterdam: Eburon.

Andersson, K. (2009). Motivational dilemmas in collaborative learning activities: The case of the Swedish International Development Cooperation Agency (SIDA). *Public Administration and Development, 29*, 341–351.

APRAPAM. (2013). An introduction note toward the governance of the maritime artisanal fishery in Senegal. March 20th, 2013. Association for the Promotion and the Accountability of the Artisanal Fish in M'bour (Senegal). Retrieved November 2, 2014 from http://www.aprapam.org/

Bavinck, M., & Kooiman, J. (2013). Theorizing governability: The interactive governance perspective. Governability of fisheries and aquaculture. In M. Bavinck, R. Chuenpagdee, S. Jentoft, & J. Kooiman (Eds.), *Governability of fisheries and aquaculture: Theory and applications* (MARE publication series, Vol. 7, pp. 9–30). Dordrecht: Springer.

Belhabib, D., Koutob, V., Sall, A., Lam, V., & Pauly, D. (2014). Fisheries catch misreporting and its implications: The case of Senegal. *Fisheries Research, 151*, 1–11.

Béné, C., & Neiland, A. (2006). From participation to governance: A critical review of the concepts of governance, co-management, and participation and their implementation in small-scale inland fisheries in developing countries. Penang, Malaysia: World Fish Center.

Berkes, F. (2007). Community-based conservation in a globalized world. *PNAS, 104*(39), 15188–15193.

Binet, T., Failler, P., & Thorpe, A. (2012). Migration of Senegalese fishers: A case for regional approach to management. *Maritime Studies, 11*(1), 1–14.

Booth, D. (2012). *Development as a collective action problem: Addressing the real challenges of African governance. Africa Power and Politics Programme*. London: Overseas Development Institute DFID.

Breuil, C. (2011). Senegal. In J. S. Sanders, D. Gréboval, & A. Hjort (Eds.), Marine protected areas: Country case studies on policy, governance and institutional issues. FAO fisheries and aquaculture technical paper. No. 556/1. Rome: FAO.

Brown, O. (2005). Policy Incoherence: EU fisheries policy in Senegal. Human Development Report Office occasional paper. UNDP. 2005–29.

Chauveau, J., & Samba, A. (1989). Market development, government interventions and the dynamics of the small-scale fishing sector: An historical perspective of the Senegalese case. *Development and Change, 20*, 599–620.

Chuenpagdee, R., & Jentoft, S. (2007). Step zero for fisheries co-management: What precedes implementation. *Marine Policy, 31*(6), 657–668.

Chuenpagdee, R., & Jentoft, S. (2013). Assessing governability – What's next. In M. Bavinck, R. Chuenpagdee, S. Jentoft, & J. Kooiman (Eds.), *Governability of fisheries and aquaculture: Theory and applications*. Dordrecht: Springer.

Connolly, B., & Keohane, R. (1996). Institutions for environmental aid: Politics, lessons and opportunities. *Environment: Science and Policy for Sustainable Development, 38*(5), 12–42.

Cormier-Salem, M. (1995). Paysans-pêcheurs du terroir et marins-pecheurs du parcours. Les géographes et l'espace aquatique. *Espace Géographique, 24*(1), 46–59.

Cormier-Salem, M. (2014). Participatory governance of marine protected areas: A political challenge, an ethical imperative, different trajectories: Senegal case studies. *S.A.P.I.EN.S, 7*(2), 1–13.

Crespin, J. (2006). Aiding local action: the constraints faced by donor agencies in supporting effective, pro-poor initiatives on the ground. *Environment and Urbanization, 18*, 433–449.

DGIS. (2008). Evaluation of sector approaches in environment: Senegal case study. Report for Directorate-General for International Cooperation, Netherlands (DGIS).

Disch, A. (2010). Aid coordination and aid effectiveness. ECON Centre for Economic Analysis, Norwegian Ministry of Foreign Affairs.

DPM. (2011). Rapprt d'evaluation des Conseils Locaux de Pêché Artisannale (CLPA). Division Pêché Artisinal. December 2011.

DuBois, C., & Zografos, C. (2012). Conflicts at sea between artisanal and industrial fishers: Inter-sectoral interactions and dispute resolution in Senegal. *Marine Policy, 36*, 1211–1220.

Ebrahim, A. (2005). Accountability myopia: Losing sight of organizational learning. *Nonprofit and Voluntary Sector Quarterly, 34*(1), 56–87.

Galvan, D. (2003). Sustainable development in Sub-Saharan Africa: Resource management through institutional syncretism in Madagascar and Senegal. In H. Bressers & W. Rosenbaum (Eds.), *Achieving sustainable development: The challenge of governance across social scales*. Westport: Praeger.

Galvan, D. (2004). *The state must be our master of fire: How peasants craft culturally sustainable development in Senegal*. Berkeley: University of California Press.

Gaspart, F., & Platteau, J. (2001). *Collective action for local-level effort regulation: An assessment of recent experiences in Senegalese small-scale fisheries* (Development Studies Working Paper No. 150). Milan: Centro Studi Luca d'Agliano.

Gould, H. (2006). Letter to the editors: Understanding the role of culture in knowledge sharing: Making the invisible visible. *KM4D Journal, 2*(2), 112–115.

Hall, S., Hilborn, R., Andrews, N., & Allison, E. (2013). Innovations in capture fisheries are an imperative for nutrition security in the developing world. *PNAS, 110*(21), 8393–8398.

Haller, T. (2010). Between open access, privatization and collective action: A comparative analysis of institutional change governing use of common pool resources in African floodplains. In T. Haller (Ed.), *Disputing the floodplains: Institutional change and the politics of resource management in African wetlands* (pp. 413–435). Leiden: Brill.

Jentoft, S., & Chuenpagdee, R. (2009). Fisheries and coastal governance as a wicked problem. *Marine Policy, 33*, 553–560.

Jentoft, S., & Chuenpagdee, R. (2013). Concerns and problems in fisheries and aquaculture – Exploring governability. In M. Bavinck, R. Chuenpagdee, S. Jentoft, & J. Kooiman (Eds.), *Governability of fisheries and aquaculture: Theory and applications*. Dordrecht: Springer.

Jentoft, S., Bavinck, M., Johnson, D., & Thomson, K. (2009). Fisheries co-management and legal pluralism: How an analytical problem becomes an institutional one. *Human Organization, 68*(1), 27–38.n

JICA. (2012). The project on the capacity improvement of the organizations and the formation of the leaders of fishermen in the domain of the small-scale fisheries in the Republic of Senegal (COGEPAS). Japan International Cooperation Agency (JICA) Direction of Maritime Fisheries. English draft final. November 2012

Kaczynski, V., & Fluharty, D. (2002). European policies in West Africa: Who benefits from fisheries agreements? *Marine Policy, 26*, 75–93.

Karp, C. (2011). Evaluation of the legal and the institutional capacity of local artisanal fishing counsels in Senegal: Initial findings from four CLPAs. Narragansett: Coastal Resources Center, University of Rhode Island

Lenselink, N. M. (2002). Participation in artisanal fisheries management for improved livelihoods in West Africa. A synthesis of interviews and cases from Mauritania, Senegal, Guinea and Ghana. FAO Fisheries Technical Paper. No. 432. Rome: FAO.

Mahon, R., McConney, P., & Roy, R. (2008). Governing fisheries as complex adaptive systems. *Marine Policy, 32*(0), 104–112.

Meinzen-Dick, R., & Pradhan, R. (2002). *Legal pluralism and dynamic property rights* (CGIAR System-Wide Program on Property Rights and Collective Action Working Paper 22) Washington, DC: IFPRI.

MEM. (2011). Rapport de la commission d'analyse des initiatives de cogestion au Senegal. Ministere de L'Economique Maritime. Octobre, 2011

Nazarea, V., Rhoades, R., Bontoyan, E., & Flora, G. (1998). Defining indicators which make sense to local people: Intra-cultural variation in perceptions of natural resources. *Human Organization, 57*(2), 159–170.

NEPAD. (2011). Development aid for fisheries in Africa: Setting out key principles for fisheries governance reforms. New partnership for Africa's development. International Partnership for African Fisheries Governance and Trade.

Niamadio, I., Diouf, K., Mbengue, A. (2012). Atelier National d'Identification et d'Elaboration d'une Strategie de Reinforcement du Roles des Femmes dans la gestion des Pecheries Artisanales. USAID/COMFISH project, Senegal, and Coastal Resources Center, University of Rhode Island, Narragansett, RI 33 pp.

Niange, N. (2009). Dynamique Socio-Environnementale et Developpement Local Des Regions Côtieres Du Senegal: L'exemple De La Peche Artisanale. Thèse de Doctorat. Universite De Rouen U.F.R. De Lettres Et Sciences Humaines. 301 pp.

Njock, J. C., & Westlund, L. (2010). Migration, resource management and global change: Experiences from fishing communities in West and Central Africa. *Marine Policy, 34*, 752–760.

Nyamnjoh, H. (2010). *"We get nothing from fishing": Fishing for boat opportunities amongst Senegalese fisher migrants*. Bamenda Cameroon: Langaa RPCIG.

OECD. (2008). *Fishing for coherence in West Africa: Policy coherence in the fisheries sector in seven West African countries*. Paris: OECD – Organization for Economic Co-operation and Development.

Onyango, P., & Jentoft, S. (2013). Poverty in Lake Victorian fisheries: Understanding governability. In M. Bavinck, R. Chuenpagdee, S. Jentoft, & J. Kooiman (Eds.), *Governability of fisheries and aquaculture: Theory and applications*. Dordrecht: Springer.

Ostrom, E., Gibson, C., Shivakumar, S., & Andersson, K. (2002). *Aid, incentives, and sustainability: An institutional analysis of development cooperation* (Sida Studies in Evaluation 02/0, 352pp.). Stockholm: Swedish International Development Cooperation Agency (Sida).

Ostrom, E., Janssen, M., & Anderies, J. (2007). Going beyond panaceas. *PNAS, 104*(39), 15176–15178.

OXFAM. (2011). Aid coordination and alignment: Myth or reality? The case of the ECOWAS regional agriculture policy. Oxford, UK: Oxfam International.

Poe, M., Norman, K., & Levin, P. (2014). Cultural dimensions of socioecological systems: Key connections and guiding principles for conservation in coastal environments. *Conservation Letters, 7*(3), 66–175.

PRCM. (2011). Donors reiterate their commitment to protected areas in West Africa. 27th May 2011 Dakar, Senegal. Retrieved from http://en.prcmarine.org/index.phpoption=com_content&view=article&id=923:donors-reiterate-commitment-for-protected-areas&catid=51:actualit&Itemid=226

Randall, S. (2005). *Review of literature on fishing migrations in West Africa – From a demographic perspective. Sustainable Fisheries Livelihoods Programme*. Rome: FAO/DFID.

Ribot, J., & Oyono, P. (2005). The politics of decentralization. In B. Wisner, C. Toulmin, & R. Chitiga (Eds.), *Towards a new map of Africa* (pp. 205–228). London: Earthscan.
Salas, S., & Gaertner, D. (2004). The behavioural dynamics of fishers: management implications. *Fish and Fisheries, 5*, 153–167.
Sall, A. (2007). Loss of bio-diversity: Representation and valuation processes of fishing communities. *Social Science Information, 46*, 153–187.
Sall, A., Deme, M., Diouf, P. (2006). L'evaluation des emplois dans les pecheries artisinales maritimes Senegalaise. WWF WAMER Bureau du programme marin pour l'Afrique de l'Ouest.
Sene, C. (2013). Etude diagnostique des lacunes et contraintes de la Gouvernance des aires marinas proteges de Joal-Fadiouth, Cayar et Bamboung. COMFISH/USAID Project. University of Rhode Island, Narragansett, RI.
Song, A. M., Chuenpagdee, R., & Jentoft, S. (2013). Values, images, and principles: What they represent and how they may improve fisheries governance. *Marine Policy, 40*, 167–175.
Soumare, S. (2012). Migrations, genre et gestion des ressources halieutiques – le cas de femmes migrantes guinéennes dans la transformation artisanale dupoisson à Joal, Sénégal. Thesis, Université Cheikh Anta Diop de Dakar, (UCAD) Sénégal.
Tall, A. (2002). Obstacles to the development of small-scale fish trade in West Africa. In National workshop on links between artisanal fisheries and world markets, Gambia: Banjul, pp. 1–18.
Tappana, G., Sall, M., Wood, E., & Cushing, M. (2004). Ecoregions and land cover trends in Senegal. 2004. *Journal of Arid Environments, 59*(3), 427–462.
Trouillet, B., Guineberteau, T., Bernardon, M., & Le Rouxm, S. (2011). Key challenges for maritime governance in West Africa: Fishery-based lessons from Guinea and Mauritania. *Marine Policy, 35*, 155–162.
UN. (2010). World economic and social survey 2010 – Retooling Global Development. UN Department of Economic and Social Affairs E/2010/50/Rev. 1 ST/ESA/330.
UNEP. (2004). Policy implementation and fisheries resource management: Lessons from Senegal. United Nations Environmental Program. Fisheries and the Environment. UNEP/2004/13.
UNESCO (2000). Yoff, le territoire assiégé. Un village lébou dans la banlieue de Dakar. Dossiers régions côtières et petites îles 7. Paris: UNESCO.
UNESCO. (2010). Universalism and ethical values for the environment. Ethics and Climate Change in Asia and the Pacific (ECCAP) Project. Working Group 1 Report. Bangkok, Thailand
USAID, (2013). Annual work plan-2014, USAID COMFISH Project. Retrieved from http://www.crc.uri.edu/download/COMFISH-FY2014WP-FINALEnglish_508.pdf
Watanuki, N. (2008, July). Community-based fisheries co-management in Senegal. Paper presented at IIFET- International Institute of Fisheries Economics and Trade, Nha Trang, Vietnam.
WB. (2012a). Implementation completion and results report-Senegal GIRMAC-integrated marine and coastal resources management project, World Bank.
WB. (2012b). Restructuring paper: Proposed project restructuring West Africa fisheries program (WARFP)-Senegal. World Bank 10-December-2012. # 74617.
WB. (2014). Senegal country data. Retrieved from http://data.worldbank.org/country/senegal
Weigel, J., & de Monbrison, D. (2013). *State of the art of fisheries co-management*. Dakar: CRSP (Commission Sous-Regional des Peches).
Wilson, D. (2002). *Social science methods for the KNOWFISH project. Addendum to annex IV: The technical annex*. Hirtshals: The Institute for Fisheries Management.

Chapter 34
Promoting Governability in Small-Scale Fisheries in Zanzibar, Tanzania: From Self-Governance to Co-governance

Lars Lindström and Maricela de la Torre-Castro

Abstract This chapter highlights some governance challenges in small-scale fisheries in the East African region using the case of Chwaka Bay, Zanzibar, Tanzania. In this case, self-governance processes involve strong normative and cultural-cognitive aspects that have underpinned *de facto* management actions and blocked other options towards sustainability. The conflict level between the villages in the Bay is very high and there is a need to address how the system may move from self-governance and conflict to co-governance and cooperation. The chapter focuses on the governance interactions between the state and the fishing villages as well as the state's failed attempts to break unsustainable self-governance. It identifies the role that the state has played to promote co-management and participation, as well as highlights changing legislation and conservation in the area. What went wrong with these strategies and why? Who and with what method does the capacity for dynamic, interactive governance develop? Who co-ordinates interactions across different identities, interests, and different spatio-temporal scales, and how? Who establishes a common world view for action, and how? Which institution functions as a court of appeal for disputes arising within and over interactive governance?

Keywords Environmental Governance • Meta-Governance • Seascape Management • Fisheries • Zanzibar

L. Lindström (✉)
Department of Political Science, Stockholm University, Stockholm, Sweden
e-mail: lasse.lindstrom@statsvet.su.se

M. de la Torre-Castro
Department of Physical Geography,
Stockholm University, Stockholm, Sweden
e-mail: maricela@natgeo.su.se

S. Jentoft, R. Chuenpagdee (eds.), *Interactive Governance for Small-Scale Fisheries*, MARE Publication Series 13, DOI 10.1007/978-3-319-17034-3_34

Introduction

All the King's horses and all the King's men, couldn't get sustainability on its feet again.
(Adapted from the British nursery rhyme Humpty Dumpty, trad.)

The alleged state failure in the 1990s to deliver development, be it economic, social or political also included environmental sustainability. Government has lately been replaced by governance as a means to coordinate social relations towards an end (Jessop 1997) and as blueprint for success (Kooiman 1993; Rhodes 1996; Kettl 2002; Kooiman and Bavinck 2005; Kooiman et al. 2008). Traditional hierarchical governance in the form of the state's fist, at times disguised in a velvet glove, is now to be combined, if not replaced, with the anarchy of the market and the heterarchy of civil society (Jessop 1997), codified for example in the 8th Millennium Development Goal and articulated with the concept of interactive governance (Kooiman et al. 2005). The interactive participation by all actors – individuals, associations, firms and business and government institutions – already at the planning and design phases of a policy or a program is suggested as crucial to achieve sustainability (Chuenpagdee and Jentoft 2007; Jentoft and Chuenpagdee 2009; Chuenpagdee 2011). These ideas are also expressed in for example "sustainable" governance (Agrawal 2001); "adaptive governance" (Folke et al. 2005) and "adaptive co-governance" (Armitage et al. 2007).

There are however two premises on which these conceptualizations rest. One is that the prefix "co-" implies self-organized participation (Pretty 1995) and the other that this self-organization takes place in the context of a relatively homogeneous "community" context, thus being straightforward without complicating asymmetrical social relations (Agrawal and Gibson 1999; Mohan and Stokke 2000, 2005; Mohammed 2004; Angerbrandt et al. 2011). This warrants "governing the governance" (see Chap. 1 in this book) or "meta-governance" (Jessop 2002): Who and with what method does the capacity for dynamic, interactive co-ordination develop? Who co-ordinates actions across different identities, interests, and different spatio-temporal scales, and how? Who establishes a common world view for action, and how? Which institution functions as a court of appeal for disputes arising within and over governance?

We present a case illustrating how diverse and opposing interests at different scales manifest themselves within small-scale fisheries in Chwaka Bay on the east coast of Zanzibar, Tanzania. In so doing, we show how the case problematize and challenge dominant assumptions in the governance discourse. It will point to the necessity of answering the questions posed above so as to make governance successfully achieves both socio-economic and environmental sustainability.

The Stage

Chwaka Bay, on the east coast of Unguja, the largest island in the Zanzibar archipelago, off the coast of mainland Tanzania, is a shallow water body of about 50 km^2 with an average depth of 3.5 m. It is driven by tidal circulation resulting in a complex pattern of banks and channels in which fishing activities take place. The bay is very rich in ecosystems and species. The largest stand of mangrove forests in Zanzibar is found in the bottom end of the bay; there is a patchy coral reef in the mouth (Tobisson et al. 1998). The bay proper is considered the most important location of seagrass richness and abundance in Zanzibar (de la Torre-Castro and Rönnbäck 2004; Gullström et al. 2012).

Seven villages are found along the coast of the bay with a total population of about 10,000 people. These villages are generally considered relatively homogenous with Swahili as a common language, Islam as the dominant religion, and a common history and culture (Tobisson et al. 1998). The focus of the study is on the villages of Chwaka, Marumbi and Uroa on the west coast of the bay (see Fig. 34.1) as they have the most important landing sites in the bay. As will be shown below these "communities" are however far from homogeneous both within and among villages. Heterogeneity has been and is still articulated with conflict, even fatal complicating ideas of co-management as well as of community-based and –driven institutions. Chwaka is the largest village in the area with about 3,000 inhabitants. The village is enclosed by mangrove forests at its southern end and has enjoyed traditional rights over the mangrove forest (de la Torre-Castro and Lindström 2010). Marumbi is smaller with about 1,000 inhabitants and is situated about 5 km

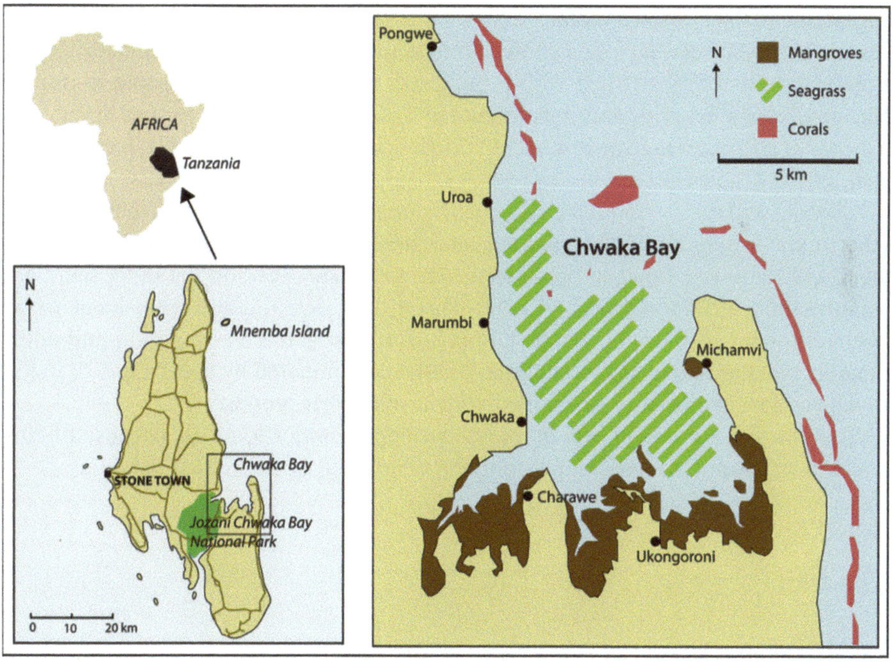

Fig. 34.1 Chwaka Bay and its dominant ecosystems. The chapter focuses on Chwaka, Marumbi and Uroa villages on the west coast of the bay (Source: Jiddawi and Lindström 2012)

Table 34.1 Population and dominant fishing methods in three villages

Village	Population in 2002	Dominating fishing method
Chwaka	2,912	Mainly illegal drag-nets
Marumbi	966	Mainly legal basket-traps (dema)
Uroa	2,107	Octopus and squid fishing with illegal spear guns or metal-sticks

Source: adapted from Gustavsson et al. (2014)

north of Chwaka village at the front of the largest extension of sea grass meadows. Uroa, with about 2,000 inhabitants, is about 4 km north of Marumbi and faces the open ocean (Fig. 34.1).

Fishing and collection take place all over the bay even though there is a clear preference for certain productive fishing grounds, particularly the sea grass meadows. Fisherfolk use e.g. outrigger canoes and wooden boats as well as different gears like basket-traps (*dema*); different nets; spears; hand-lines; and long lines (Jiddawi 2012). Many women are engaged in seaweed mariculture (Fröcklin et al. 2012) and in gleaning of e.g. invertebrates (Håkansson et al. 2012; Fröcklin et al. 2014).

The three villages in focus are different in terms of gear use and availability of alternative, or rather *additional* income generating activities, as few fisherfolk abandons small-scale fisheries. These additional income generating activities complement fisheries with other activities to secure food and income. The past decade has even seen an influx of newcomers to the trade as the financial crisis has aggravated the terms of trade within the urban sector and forced people to turn to e.g. small-scale fisheries and thus increased fishing pressure and further threatened fish stocks and other marine resources (Jiddawi and Öhman 2002; de la Torre-Castro et al. 2014) (Table 34.1).

Chwaka village is dominated by net-fishing, mainly illegal drag-nets, but the other main fishing methods/gears are also present. There is little tourist development with only one hotel in operation. Marumbi fisher folk on the other hand are using mainly the traditional and legal basket trap, dema, made from local wood fibers. Tourism in Marumbi is developing with two hotels in operation and additional hotels under construction. Uroa, finally, is dominated by spear-guns or metal sticks and tourism is more developed with seven hotels present.

Tourism gives however few employment opportunities to locals because of culture and religion. It rather creates conflicts which will be addressed below.

The Plot

The study builds upon data that have been collected over more than a decade using a variety of methods from semi-structured informant and focus group interviews to fish catch data and participatory workshops (e.g. de la Torre-Castro and Jiddawi

2005). Fisher folk have been interviewed about issues ranging from fishing ground preferences to trust of government authorities and inclusion-exclusion in "participatory" institutions (de la Torre-Castro and Lindström 2010; de la Torre-Castro 2012). Seaweed farmers have been asked about e.g. health issues and about mobilization and organization (Fröcklin et al. 2012). Interviews have been done with informants representing different government levels as well as with NGOs and the business sectors, and legal, policy and other relevant government documents analyzed (de la Torre-Castro 2006b, 2012).

The Script

Zanzibar is a semi-autonomous part of the United Republic of Tanzania. The Revolutionary Government of Zanzibar (RGZ) and its Ministry of Livestock and Fisheries (MLR) enjoy full rights to regulate marine resource use (Levine 2007). Formally, fisheries are regulated by mainly the *Fisheries Act No. 8* (RGZ 1988), the *Fisheries Law* of 1993 and the updated *Fisheries Act* of 2010 (RGZ 2010), the Chwaka Bay Fishing Nets Prohibition order (2001), and informally by the daily praxis of stakeholders.

The governmental agencies, in addition to the legislative assembly, engaged in the governance of fisheries are the Ministry of Livestock and Fisheries with its Department of Fisheries Development and Department of Marine Resources at the central level, a number of District Fisheries Offices at regional level, and at the majority of the landing sites, monitoring agents, the *Bwana Diko.* At the local level there are a variety of informal institutions established by fisher folk themselves. Business is present in the form of the many tourist hotels directly affecting both fish species targeted and market prices. The many fish traders involved in auctions at the local fish markets also affect the same. Few other representatives of civil society are however engaged in the governance of small-scale fisheries.

The monitoring agents (Bwana Dikos) have a key role at the local level as they record catches on a regular basis and report to higher levels as well as issue fishing licenses and supposedly report law breakers. They are however caught in a number of dilemmas. Their remuneration is low and issues related to kinship, multitasking, poverty and control complicate their monitoring and sanctioning roles (de la Torre-Castro 2006b).

The World Bank has since 2005 been highly involved in the governance of fisheries through the Marine and Coastal Environmental Management Project (MACEMP) – a US$63 million World Bank loan and Global Environment Facility grant (World Bank 2012) designed and implemented in cooperation with the union government and the Revolutionary Government of Zanzibar.

Chwaka Bay is by 2015 to be fully integrated in the Mnemba Island – Chwaka Bay Marine Conservation Area (MIMCA) as part of the plan of the United Republic of Tanzania (URT) together with the World Bank to establish a system of eight Marine Protected Area networks, of which two will be in Zanzibar (Ruitenbeek et al. 2005) (Fig. 34.2).

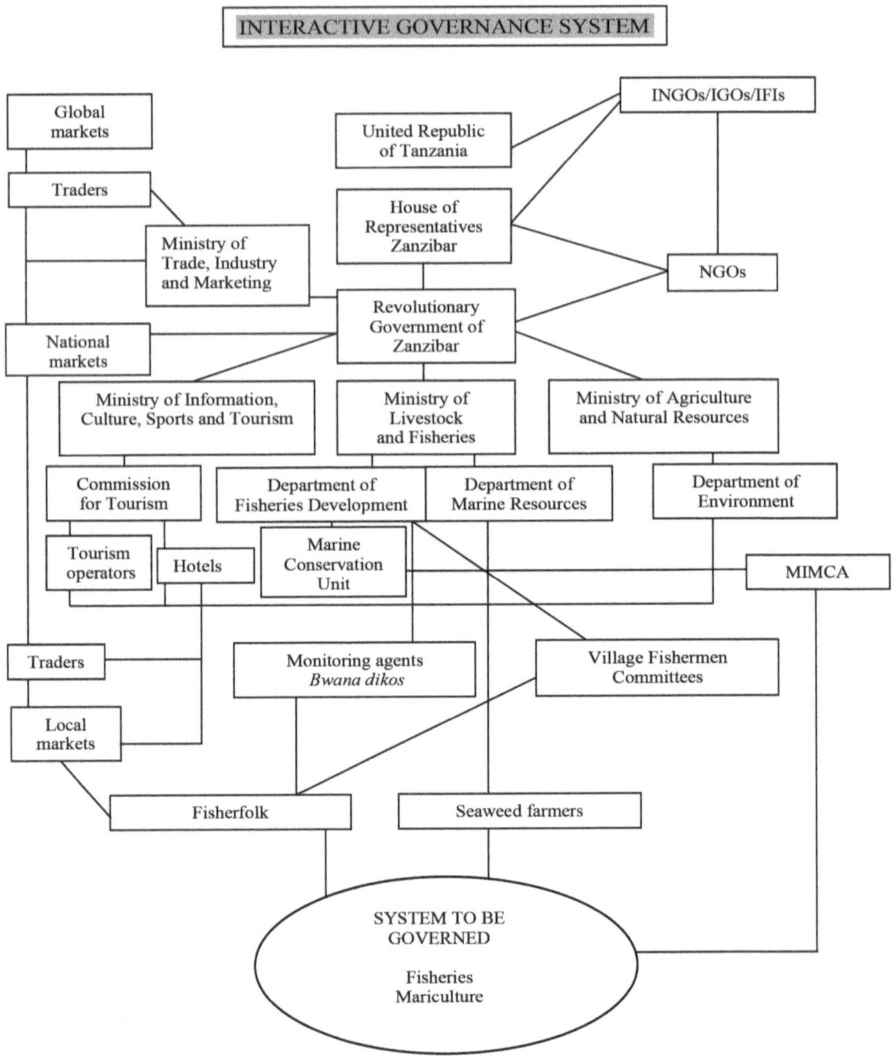

Fig. 34.2 The interactive governance system and the system-to-be-governed. (*INGOs* International Non-Governmental Organizations, *IGOs* International Governmental Organizations, *IFIs* International Financial Institutions, *NGOs* Non-Governmental Organizations, *MIMCA* Mnemba Island – Chwaka Bay Marine Conservation Area)

The MIMCA includes four zones: core (fishing prohibited); specific use (unclear implications); general use (general regulations) and buffer zones (unclear implications) (DFMR 2010). Only the area surrounding Mnemba Island about 35 km north of Chwaka Bay (see Fig. 34.1) is defined as a core and specific use zone while Chwaka Bay is a general use zone (DFMR 2005, 2010) where the Chwaka Bay

Fishing Nets Prohibition Order (RGZ 2002) applies. This By-Law bans both ownership and use of so-called drag-nets. The By-Law was the result of several failed attempts to solve a conflict between net-dominated Chwaka and dema-dominated Marumbi (de la Torre-Castro and Lindström 2010).

The RGZ has historically been inclined to use Integrated Coastal Zone Management as a blueprint not only for management but also for governance as witnessed in several government documents (e.g. DoE-IMS 1996; ICAM 1996; DoE 2009; DFMR 2010, 2011). As in many other countries in the Western Indian Ocean (Cinner at al. 2009) the Integrated Coastal Zone Management regulatory system refers to government and management in terms of being "community based" (DFMR 2011), particularly when the already existing Mnemba MPA was extended to include Chwaka Bay. However, besides MIMCA that focuses on the protection of corals of the privately owned Mnemba Island, there are no comprehensive management plans addressing the environmental and sustainability problems in Chwaka Bay in general or related to small-scale fisheries (de la Torre-Castro 2012).

An institutional framework has however been set up to at least on paper to accommodate the community based element and to ensure participation. A number of Village Fisherfolk Committees were established as envisioned in the Draft National ICM Strategies for Zanzibar (RGZ 2009). The members of the committees are to be elected by the people in the villages but no institutions have been designed to regulate and manage the electoral process (Gustavsson et al. 2014) causing mistrust among members and also elite capture.

The Set

Basket trap (*dema*) fisheries have historically dominated the fishing practices in Chwaka Bay, and Marumbi has the highest number of dema fisher folk. It is institutionalized as a traditional and sustainable way of fishing and organized in groups or "companies" of up to three persons who belong to the same extended family or are close friends. Kinship relations are thus a central element. The sharing system is egalitarian and simply about dividing costs and benefits by the number of members in the group (de la Torre-Castro and Lindström 2010).

Generalized net fisheries are a recent phenomenon in Chwaka Bay. Even though nets have been used the number of fisher folk was low and it was mainly about utilizing stationary nets with large mesh size in the channels (de la Torre-Castro and Jiddawi 2005). As monofilament nets were introduced as a means to improve income, fisherfolk embraced the new gear and technique. Nets with smaller mesh size than 1.5 in. and dragging techniques were eventually banned (RGZ 1993) but the majority of Chwaka village fishers nevertheless use drag-nets. Mesh size vary, but it is common to find nets with smaller mesh than the legal minimum of 1.5 in. (DFMR 2013). Dragging is an everyday praxis and the fishery is highly mobile with fishers actively searching for fish. Nets are weighted down with coral stones and divers pull the net together while others drag them with boats, and the technique is considered damaging for the substrate and to the other main fishing gear, the dema (Jiddawi and Öhman 2002; de la Torre-Castro 2006b).

Net fisheries are also more capital-intensive than dema fisheries. Nets are expensive and the dragging technique requires boats, often with outboard engines, in turn creating a need for petrol. Investment costs and production costs are thus higher than for the other fishing techniques used. It is common that net fisherfolk take loans from urban investors or that they lease boats and gears (interviews, Chwaka fisherfolk). Drag net fisheries are most often organized in "companies" consisting of 1 boat with 3–7 men up to three boats with 15–18 men, implying an intricate system of profit sharing because of complex property relations.

The Play: Act I

Fisher folk and other actors in the bay seem to strive in sometimes opposite and conflicting directions, sustainability seldom being one of them. As we have shown elsewhere (de la Torre-Castro and Lindstrom 2010) the future of small-scale fisheries seems deadlocked in a mesh of clashing institutions, at times expressed as violent even fatal conflicts between villages.

Net fisher folk from Chwaka consistently violate the legal ban on nets, and are also being accused by Marumbi (dema) fisher folk to fish in their traditional fishing grounds. Traditional "law", established through generations and shared understanding although not codified, and defining who has the right to fish where in terms of property rights, clashes with laws that define the bay as open access. Chwaka fishers are also allegedly dragging their nets over the substrate overturning the demas owned by Marumbi fisherfolk. Gear destruction and oral provocations became common (DFMR 1995), and the navy eventually intervened by patrolling the bay and demarcating fishing grounds with buoys. In 1996 one dema fisherman from Marumbi was even killed (DFMR 1996).

The Play: Act II

The integration of Chwaka Bay into the network of Marine Park Areas as championed by the World Bank and conservationists has further aggravated the conflict within and between villages. Conflicts have also increased between a majority of the locals and government authorities like the Ministry of Livestock and Fisheries and the Department of Environment and institutions like MACEMP. The governance and management institutional framework that was implemented with the introduction of MIMCA was, contrary to the official discourse, far from being community-based or -driven. A study on participatory justice in MIMCA (Gustavsson et al. 2014) shows that a vast majority of the locals interviewed were not given the opportunity to participate in the design of MIMCA and its institutional setup. The Village Fishermen's Committees that were established were designed by the government solely, and no institutions were designed to secure just and democratic representation and participation in the Village Fishermen's Committees. The composition of the Village Fishermen's Committees were defined by the authorities or captured by local elites, leaving e.g. women and those critical of MIMCA unrepresented and thus excluded from influence (Gustavsson et al. 2014).

The conflict between particularly intertidal net fisheries and conservation, even though the status of Chwaka Bay is still unclear, is visible through the conflict between dema-dominated Marumbi and net-dominated Chwaka village. The environmental friendly dema fisheries are more compatible with conservation than the environmental hostile net fisheries, particularly intertidal drag net fisheries. This is also voiced by Chwaka net fisher folk who see both MACEMP and MIMCA as tools for conservation, and as beneficial to Marumbi while "fencing in" Chwaka (Interviews with Chwaka fisherfolk, December 2013).

The gear exchange and infrastructure provision program MACEMP, through its provision of nets with larger mesh sizes and larger fiberglass boats with outboard engines to enable off-shore deep sea fisheries, further aggravated the conflict also within villages as only a few boats and nets were distributed, thus creating distributive injustice. The cost of conservation was unequally born between and within villages as only a few benefitted (Gustavsson et al. 2014).

MIMCA, with its conservationist elements, also contributed to the emergence of a new conflict between Uroa and Marumbi. Uroa fisherfolk are predominantly engaged in spear (gun) fisheries which are also banned in the Fisheries Law. There is thus strong opposition against MIMCA. Marumbi is also targeted by Uroa fisherfolk because Marumbi brought in the navy in its conflict with Chwaka resulting in patrol boats being stationed in Marumbi manned with Marumbi locals. Uroa also perceives Marumbi to be pro-conservation (Gustavsson et al. 2014).

The implementation of MIMCA also aggravated another conflict, the one between fisheries and tourism. Tourism development was in the World Bank's and Tanzania's policy to sustain the marine environment in Zanzibar based on the assumption that it provides economic opportunities (Ruitenbeek et al. 2005). The strategy has been to attract investment in high class tourism as it would result in higher profits and contribute to economic growth through tax revenues while at the same time reduce environmental impacts (ZCT 2003). Tourism has indeed expanded rapidly contributing to about 25 % of Zanzibar's Gross Domestic Product and accounting for about 80 % of Zanzibar's foreign exchange earnings. It also employs about 40,000 people (Daily News 2013). Tourism in practice, however, is a major cause of a number of conflicts (Gössling 2000, 2001, 2003a, b; Gössling and Schulz 2005) and has been shown to contribute to degradation of marine ecosystems (DFMR 2005). Two of the hotels in the area, one in Marumbi and the other in Uroa have altered the shoreline without any environmental impact assessment and without the required permission from the Department of Environment (Lindström 2012).

While tourists consider reefs as habitats of beautiful fish and fancy corals, excellent for diving and snorkeling, fisher folk see them as fishing grounds (RGZ 2009).

Tourism is thus almost by definition pro-conservation and for fisher folk conservation is a contested issue (Gustavsson et al. 2014). Tourists prefer white sandy beaches with the intertidal zone free from rocks, algae and sea grass, while seaweed farmers, mainly women, use the area for their cultivation plots tying the algae to lines stretched between pegs embedded in the substrate (RGZ 2009). The hotels that already exist, and those under construction, tend to fence off access to intertidal activities such as fishing and collecting, in Marumbi and Uroa even trying to stop beach erosion by illegally building walls and thus altering the sea water circulation patterns causing conflict with villagers (Gustavsson et al. 2014).

Conservation is in the interest of tourism since no white sandy beaches and coral reefs with colorful tropical fish would quickly put most tourist operators out of business. The pro-tourism character of MIMCA thus contributed to aggravating the conflict between on the one hand conservation and on the other fisheries and seaweed farming. It also contributed to increased conflict between Chwaka and Marumbi as dema fisheries are more environmentally friendly than particularly drag net fisheries. Chwaka also viewed with skepticism Marumbi's associations with the authorities and monitoring/sanctioning. Uroa spear fisherfolk for the same reasons now looked upon Marumbi with suspicion (Gustavsson et al. 2014).

Marumbi villagers have attempted self-organizing in defense of their traditional fishing grounds. When the government as part of the MIMCA institutional framework monopolized power over the local Village Fishermen's Committee, Marumbi fisherfolk angrily protested. This resulted in the President of Zanzibar visiting Marumbi promising developmental support (Gustavsson et al. 2014).

It is not however only Marumbi fisherfolk who have organized themselves to defend their traditional fishing grounds and dominant fishing gear. The same can be said about both Chwaka and Uroa where fisherfolk have organized in fisheries and other committees associated with different natural and other resources but also in defense of their respective fishing gear and in the case of Chwaka the open-access nature of fisheries in the bay. This village-based self-organization found in the bay, and based on gear and fishing grounds, is not only a problem of scale as sustainability of small-scale fisheries socially and ecologically is best addressed at bay level and should not be confined to village/community level. These self-organizations have also institutionalized the conflict between villages as described above rather than contributing to its solution, partly because they don't serve as arenas for conflict resolution.

Self-organization is not by default democratic and inclusive (Mansuri and Rao 2004; Schönleitner 2004), nor necessarily pro-sustainability or conservation (Mahon et al. 2008). It can be either or and needs to be tested empirically. Self-organization risks being captured by local elites or looking only at what benefits the village/community without much consideration of the neighboring communities. This has been the case with self-organization in Chwaka village where the local elite has taken over the Village Fisheries Committee. The lack of any open and fair electoral process or clear democratic institutions of inclusion has enabled the affluent to control the Committee (Interviews with Chwaka fisherfolk 2013).

Self-organization has also left women without representation in the Village Fishermen's Committees (Gustavsson et al. 2014) which is conspicuous as a substantial number of women are engaged as collectors and some even fishing with small nets (Fröcklin et al. 2014). This exclusion is in line with paternalistic forms of authority that characterize communities (see below).

The Finale: Governing Governance

The Chwaka Bay case is characterized by a lack of dynamic and interactive coordination, mainly because of the lack of an authoritative "whom", not necessarily state authorities but an authority enjoying social and political legitimacy, and of a functioning method. Coordination problems have been recurrent, and the intervention of the World Bank through MACEMP and the establishment of MIMCA seem to have made coordination even more complex. MACEMP itself as a top-heavy construct and the resource allocation through its gear exchange and infrastructure provision program, although maybe well intended, contributed to undermining both participatory and distributive justice as well as aggravating the already existing conflicts between villages.

The method of the Ministry of Livestock and Fisheries of imposing regulations and coercion in a hierarchical way has not succeeded in stopping illegal net fishing and its associated potential environmental damage. There is also a lack of effective conflict resolution. The intervention of the navy and the stationing of patrol boats in Marumbi exemplify how coercion seems to have taken the place of negotiation. No arenas for democratic action or deliberation have been created at bay level. The historical conflict over fishing grounds, partly because Chwaka village fisher folk have been confined in the innermost parts of the bay and partly because the conflict between modern and traditional law, has not of yet been solved. Traditional user rights clash with the open access treatment of the bay.

Self-organization has in some cases (e.g. Olson et al. 2004; Folke et al. 2005) been identified as promoting sustainability and is also one of the governance modes in the interactive governance framework (Kooiman and Bavinck 2013). We prefer the more analytical concept of hierarchical governance – unranked or un-hierarchical social relations having the potential of being ranked in a number of different ways (Crumley 1995; Jessop 1997) – as it catches both the homogeneous and heterogeneous nature of places and spaces by enabling the identification of different identities and interests at different spatio-temporal scales. The Chwaka Bay case illustrates that the proposition that rules and regulations that are constituted at the level of the community by institutions of their own making and within which they themselves participate will be more favorable for sustainability than those constituted in top-down, hierarchical governance, is not upheld. The three villages or "communities" have self-organized differently in terms of identities and interests that, even though of their own making, are in conflict with one another and also with sustainability. This is not to argue that hierarchical governance has been more successful, but that there is an institutional misfit between the governance system and the system-to-be-governed. Another misfit is the one between self-organization limited to the village/community based on specific fishing gears – spears in Uroa, basket traps in Marumbi and nets in Chwaka – and bay level small-scale fisheries interests.

There is also a structural tension between the hierarchical governance of small-scale fisheries at bay level and hierarchical or self-governance that predominantly and exclusively takes place at village level, complicating governability.

These different tensions and "misfits" are further aggravated as there are no institutions through which conflicts between the two modes and between the governance system at village level and the system-to-be-governed at bay level can be mediated and eventually settled.

Self-organization in the three villages also takes place without much concern or consideration of the neighboring villages or communities. Self-organization is not in itself democratic and inclusive as it risks being captured by the local elite as our example shows. It also underscores the necessity for institutions to function as a court of appeal for and mediation of conflicts over and within governance.

This is in our view mainly caused by a lack of a common world-view and an institutional framework within which conflicts between different world views may be solved and through which a common world-view may be agreed upon. The case clearly demonstrates the conflict between the now living and future generations and with it the conflict between socio-economic and environmental sustainability (Sneddon et al. 2006). The governance system has failed in establishing this common world view and sustainability has been defined as technical-managerial rather than political.

Governance requires the formulation of a common world view, the organization of self-organization and the regulation of self-regulation (Kooiman 1993; Jessop 1997, 2002; Sørensen 2006) – *meta-governance* and a *meta-governor*. Power structures are central to "communities", and are most often based on ideas of paternal authority (Angerbrandt et al. 2011), in our cases witnessed by the exclusion of women from the Village Fishermen's Committees (Gustavsson et al. 2014) and strong elements of conformity. Institutionalizing governance, and for that matter sustainability, in community social relations thus runs the risk of reproducing undemocratic forms of authority. Authority in society, on the other hand, rests with public institutions regulated by political legislation rather than paternalistic relations often founded on custom and religion, and with the state as the most general embodiment of society (Angerbrandt et al. 2011). A politicization of interactive governance, and self-organization in particular, based on integration rather than incorporation, as self-organization in our cases articulate, is thus needed. Since world-views and interests are fundamentally political in nature it is the political organs of the state that should function as meta-governor as it is only the state, be it at central, regional or local level that may grant and secure equal participation based on citizenship and not on membership in a community.

References

Agrawal, A. (2001). Common property institutions and sustainable governance of resources. *World Development, 29*(10), 1649–1672.

Agrawal, A., & Gibson, C. (1999). Enchantment and disenchantment: The role of community in natural resource conservation. *World Development, 27*(4), 629–649.

Angerbrandt, H., Lindström, L., & de la Torre-Castro, M. (2011). What is this thing called "community" good for? In R. Chuenpagdee (Ed.), *Contemporary visions for world small-scale fisheries* (pp. 353–365). Delft: Eburon.

Armitage, D., Berkes, F., & Doubleday, N. (Eds.). (2007). *Adaptive co-management: Collaboration, learning, and multi-level governance*. Vancouver: UBC Press.

Chuenpagdee, R. (2011). Interactive governance for marine conservation: An illustration. *Bulletin of Marine Science, 87*(2), 197–211.

Chuenpagdee, R., & Jentoft, S. (2007). Step zero for fisheries co-management: What precedes implementation? *Marine Policy, 31*, 657–668.

Chwaka Bay Fishing Nets Prohibition order 2001. Legal supplement Part II to the Zanzibar Government Gazette. Vol. CXI No. 5965, September 13, 2002.

Cinner, J. E., Wamukota, A., Randriamahazo, H., & Rabearisoa, A. (2009). Toward institutions for community-based management of inshore marine resources in the Western Indian Ocean. *Marine Policy, 33*, 489–496.

Crumley, C. L. (1995). Heterarchy and the analysis of complex societies. *Archeological Papers of the American Anthropological Association, 7*(1), 1–5.

Daily News. (2013). Zanzibar eyes being best tourism destination. 24 July 2013.

de la Torre-Castro, M. (2006a). *Humans and seagrasses in East Africa – A social-ecological systems approach*. Stockholm: Stockholm University/Department of Systems Ecology.

de la Torre-Castro, M. (2006b). Beyond regulations in fisheries management: The dilemmas of the "beach recorders" Bwana dikos in Zanzibar, Tanzania. *Ecology and Society, 11*, 4.

de la Torre-Castro, M. (2012). Governance for sustainability: Insights from marine resource use in a tropical setting in the Western Indian Ocean. *Coastal Management, 40*, 612–633.

de la Torre-Castro, M., & Jiddawi, N. S. (2005). *Seagrass related research and community participation. Fishermen, fisheries and seagrasses. Participatory workshop*. Zanzibar Town: WIOMSA.

de la Torre-Castro, M., & Lindström, L. (2010). Fishing institutions: Addressing regulative, normative and cultural-cognitive elements to enhance fisheries management. *Marine Policy, 34*, 77–84.

de la Torre-Castro, M., & Rönnbäck, P. (2004). Links between humans and seagrasses – An example from tropical East Africa. *Ocean and Coastal Management, 47*, 361–387.

de la Torre-Castro, M., Di Carlo, G., & Jiddawi, N. S. (2014). Seagrass importance for a small-scale fishery in the tropics: The need for seascape management. *Marine Pollution Bulletin, 83*(2), 398–407.

DFMR. (1995). Archives. District Fisheries Officer. Yearly report, 1995. Zanzibar Town: Department of Fisheries and Marine Resources.

DFMR. (1996). Archives. District Fisheries Officer. Yearly report, 1996. Zanzibar Town: Department of Fisheries and Marine Resources.

DFMR. (2005). *Mnemba Island and Chwaka bay Conservation Areas: A Preliminary situation assessment*. Zanzibar Town: Department of Fisheries and Marine Resources.

DFMR. (2010). *Draft General Management Plan for Mnemba Island – Chwaka Bay Marine Conservation Area (MIMCA)*. Zanzibar Town: Department of Fisheries and Marine Resources.

DFMR. (2011). *Establishment and involvement of village fishermen committees in Management of fisheries in Zanzibar*. Zanzibar Town: Department of Fisheries and Marine Resources.

DFMR. (2013). Archives. Law and license section. Annual report. Zanzibar Town: Department of Fisheries and Marine Resources.

DoE. (2009). *The status of Zanzibar coastal resources: Towards the development of integrated coastal management strategies and action plan*. Zanzibar Town: Department of the Environment.

DoE-IMS. (1996). *Towards integrated management and sustainable development of Zanzibar's coast: Findings and recommendations for an action strategy in the Chwaka Bay – Paje area*. Zanzibar Town: Department of Environment and The Institute of Marine Science.

Folke, C., Hahn, T., Olsson, P., & Norberg, J. (2005). Adaptive governance of social-ecological systems. *Annual Review of Environment and Resources, 30*, 441–473.

Fröcklin, S., de la Torre-Castro, M., Lindström, L., Jiddawi, N. S., & Msuya, F. E. (2012). Seaweed mariculture as a development project in Zanzibar, East Africa: A price too high to pay? *Aquaculture, 356*, 30–39.

Fröcklin, S., de la Torre-Castro, M., Håkansson, E., Karlsson, A., Magnusson, M., & Jiddawi, N. S. (2014). Towards improved management of tropical invertebrate fisheries: Including time series and gender. *PLoS ONE*, March 10, 2014. doi:10.1371/journal.pone.0091161.

Gössling, S. (2000). Tourism – Sustainable development option? *Environmental Conservation, 27*(3), 223–224.

Gössling, S. (2001). The consequences of tourism for sustainable water use on a tropical island: Zanzibar, Tanzania. *Journal of Environmental Management, 61*(2), 179–191.

Gössling, S. (2003a). The political ecology of tourism in Zanzibar. In S. Gössling (Ed.), *Tourism and development in tropical islands: Political ecology perspectives* (pp. 178–202). Cheltenham: Edward Elgar.

Gössling, S. (2003b). Market integration and ecosystem degradation: Is sustainable tourism development in rural communities a contradiction in terms? *Environment, Development and Sustainability, 5*, 383–400.

Gössling, S., & Schulz, U. (2005). Tourism-related migration in Zanzibar, Tanzania. Tourism geographies: An international journal of tourism space. *Place and Environment, 7*(1), 43–62.

Gullström, M., Lyimo, T. J., Eklöf, J. S., Björk, M., Sware Semesi, I., & de la Torre-Castro, M. (2012). Seagrass meadows in Chwaka Bay: Socio-ecological and management aspects. In M. de la Torre-Castro & T. J. Lyimo (Eds.), *People, nature and research in Chwaka Bay, Zanzibar, Tanzania* (pp. 89–109). Zanzibar Town: WIOMSA.

Gustavsson, M., Lindström, L., Jiddawi, N. S., & de la Torre-Castro, M. (2014). Procedural and distributive justice in a community-based managed Marine Protected Area in Zanzibar, Tanzania. *Marine Policy, 46*, 91–100.

Håkansson, E., Fröcklin, S., & de la Torre-Castro, M. (2012). Invertebrate collection in Chwaka village: Importance, gender and resilience aspects. In M. de la Torre-Castro & T. J. Lyimo (Eds.), *People, nature and research in Chwaka Bay, Zanzibar, Tanzania* (pp. 235–264). Zanzibar Town: WIOMSA.

ICAM. (1996). *Towards integrated management and sustainable development of Zanzibar's coast – Findings and recommendation for an action strategy in the Chwaka Bay – Paje site*. Zanzibar Town, Commission for Lands and Environment. Retrieved November 6, 2005, from http://www.crc.uri.edu/download/ZAN_0056.PDF

Jentoft, S., & Chuenpagdee, R. (2009). Fisheries and coastal governance as a wicked problem. *Marine Policy, 33*(4), 553–560.

Jessop, B. (1997). The governance of complexity and the complexity of governance. In A. Amin & J. Hausner (Eds.), *Beyond markets and hierarchy* (pp. 111–147). Chelmsford: Edward Elgar.

Jessop, B. (2002). *Governance and metagovernance: On reflexivity, requisite variety and requisite irony*. Retrieved November 6, 2010 from http://www.lancs.ac.uk/fass/sociology/papers/jessop-governance-and-metagovernance.pdf

Jiddawi, N. S. (2012). Artisanal fisheries and other marine resources in Chwaka Bay. In M. de la Torre-Castro & T. J. Lyimo (Eds.), *People, nature and research in Chwaka Bay, Zanzibar, Tanzania* (pp. 193–212). Zanzibar Town: WIOMSA.

Jiddawi, N. S., & Lindström, L. (2012). Physical characteristics, socio-economic setting and coastal livelihoods. In M. de la Torre-Castro & T. J. Lyimo (Eds.), *People, nature and research in Chwaka Bay, Zanzibar, Tanzania* (pp. 23–40). Zanzibar Town: WIOMSA.

Jiddawi, N. S., & Öhman, M. C. (2002). Marine fisheries in Tanzania. *Ambio, 31*(7–8), 518–527.

Kettl, D. (2002). *The transformation of governance. Public administration for twenty-first century America*. Baltimore: The John Hopkins University Press.

Kooiman, J. (Ed.). (1993). *Modern governance. New government-society interactions*. London: Sage.

Kooiman, J., & Bavinck, M. (2005). The governance perspective. In J. Kooiman, S. Jentoft, R. Pullin, & M. Bavinck (Eds.), *Fish for life. Interactive governance for fisheries* (pp. 11–24). Amsterdam: Amsterdam University Press

Kooiman, J., & Bavinck, M. (2013). Theorizing governability – The interactive governance perspective. In M. Bavinck, R. Chuenpagdee, S. Jentoft, & J. Kooiman (Eds.), *Governability of fisheries and aquaculture. Theory and applications* (MARE publication series no. 7). Amsterdam: Springer/Amsterdam University Press.

Kooiman, J., Jentoft, S., Pullin, R., & Bavinck, M. (Eds.). (2005). *Fish for life. Interactive governance for fisheries*. Amsterdam: Amsterdam University Press.

Kooiman, J., Bavinck, M., Chuenpagdee, R., Mahon, R., & Pullin, R. (2008). Interactive governance and governability: An introduction. *The Journal of Transdisciplinary Environmental Studies, 7*, 1.

Levine, A. (2007). Staying afloat: State agencies, local communities, and international involvement in marine protected area management in Zanzibar, Tanzania. *Conservation and Society, 5*, 562–587.

Lindström, L. (2012). Governing sustainability: Chwaka Bay in Zanzibar's national integrated coastal management strategies. In M. de la Torre-Castro & T. J. Lyimo (Eds.), *People, nature and research in Chwaka Bay, Zanzibar, Tanzania* (pp. 265–278). Zanzibar Town: WIOMSA.

Mahon, R., McConney, P., & Roy, R. N. (2008). Governing fisheries as complex adaptive systems. *Marine Policy, 32*, 104–112.

Mansuri, G., & Rao, V. (2004). Community-based and -driven development: A critical review. *World Bank Research Observer, 19*(1), 1–39.

Mohammed, S. (2004). Saving the commons: Community involvement in the management of mangrove and fisheries resources of Chwaka Bay, Zanzibar. *Western Indian Ocean Journal of Marine Science, 3*(2), 221–225.

Mohan, G., & Stokke, K. (2000). Participatory development and empowerment: The dangers of localism. *Third World Quarterly, 21*(2), 247–268.

Mohan, G., & Stokke, K. (2005). The politics of localization: From depoliticizing development to politicizing democracy. In K. R. Cox, M. Low, & J. Robinson (Eds.), *The handbook in political geography* (pp. 545–561). London: Sage.

Olsson, P., Folke, C., & Berkes, F. (2004). Adaptive co-management for building resilience in social-ecological systems. *Environmental Management, 34*, 75–90.

Pretty, J. (1995). Participatory learning for sustainable agriculture. *World Development, 23*, 1247–1263.

RGZ. (1988). Fisheries Act No. 8. Zanzibar Town, Revolutionary Government of Zanzibar.

RGZ. (1993). *Fisheries Law*. Zanzibar Town: Revolutionary Government of Zanzibar.

RGZ. (2002). Chwaka Bay Fishing Nets Prohibition order 2001. Legal supplement Part II to the Zanzibar Government Gazette. Vol. CXI No. 5965 of 13th September, 2002.

RGZ. (2009). *Draft National ICM Strategies for Zanzibar*. Zanzibar Town: Revolutionary Government of Zanzibar.

RGZ. (2010). *Fisheries Act*. Zanzibar Town: Zanzibar Revolutionary Government.

Rhodes, R. (1996). The new governance: Governing without government. *Political studies XLIV, 652–667*.

Ruitenbeek, J., Hewawasam, I., & Ngoile, M. (2005). *Blueprint 2050: Sustaining the marine environment in mainland Tanzania and Zanzibar*. Washington, DC: The World Bank.

Schönleitner, G. (2004). Can public deliberation democratise state action? Municipal health councils and local democracy in Brazil. In J. Harriss, K. Stokke, & O. Törnquist (Eds.), *Politicising democracy: Local politics and democratisation in developing countries* (pp. 75–105). Basingstoke: Palgrave Macmillan.

Sneddon, C., Howarth, R. B., & Norgaard, R. B. (2006). Sustainable development in a post-Bruntland world. *Ecological Economics, 57*, 253–268.
Sørensen, E. (2006). Metagovernance: The changing role of politicians in processes of democratic governance. *American Review of Public Administration, 36*(1), 98–114.
Tobisson, E., Andersson, J., Ngazi, Z., Rydberg, L., & Cederlöf, U. (1998). Tides, monsoons and seabed: Local knowledge and practice in Chwaka Bay, Zanzibar. *Ambio, 27*(8), 677–685.
World Bank. (2012). *Tanzania – Marine and coastal environment management project: Restructuring (Vol. 1 of 2): Main report (English)*. Washington, DC: The World Bank.
ZCT. (2003). *Indicative tourism master plan for Zanzibar and Pemba*. Zanzibar Town: Zanzibar Commission for Tourism.

Chapter 35
Towards a Governable Co-management in South Korean Small-Scale Fisheries: Interactions of Institutions and Stakeholders' Mindset

Andrew M. Song

Abstract This chapter explores interactions between institutions and stakeholders' mindset in the context of a fisheries co-management system in South Korea. Locally called Jayul, this co-management system was initiated by the central government to address resource degradation and illegal fishing concerns by involving small-scale, coastal fishing communities in fisheries management. As part of this, the program strives to affect changes in the minds of involved stakeholders, most notably fishers, to instill certain ideals such as self-initiative, community cohesion, and local rule-setting. Using semi-structured and open-ended surveys as a primary method, I conducted a study to examine the 'match' between the institutional aims of the Jayul program and the people's mindset. Results show that while mismatches prevail suggesting that implementation has been less than effective, some attunement of this co-management institution can be done to better align with the mindset of small-scale fishing people, such as incorporating the principles of equity and adjacency. I argue that such attentive consideration towards the institutional design and implementation is crucial for improving the governability of the system and facilitating the governance transition.

Keywords Governance transition • Co-management • Stakeholder mindset • Fisheries Institutions • Institutional match • Jayul program • South Korea

---

A.M. Song (✉)
Department of Geography, Memorial University, St. John's, Canada
e-mail: amsong@mun.ca

S. Jentoft, R. Chuenpagdee (eds.), *Interactive Governance for Small-Scale Fisheries*, MARE Publication Series 13, DOI 10.1007/978-3-319-17034-3_35

Introduction

While successful cases of fisheries that are moving towards sustainability have been documented (Hilborn et al. 2005; Hilborn 2007), a general decline of aquatic resources observed around the world (Pauly et al. 2002; Myers and Worm 2003; Allan et al. 2005; FAO 2012a) suggests that fisheries governance faces many challenges (Cochrane 2000; Beddington et al. 2007; Berkes et al. 2007; Mora et al. 2009). Unsustainable fishery outcomes bring real as well as serious consequences on several fronts, including loss of employment (Teh and Sumaila 2013), inadequate supply of animal protein for the world's population, especially for the poor (Kent 1997; Béné et al. 2007; Jentoft and Eide 2011), and erosion of cultural meanings and social norms that provide cohesiveness to coastal communities (McGoodwin 2001; Close et al. 2002; Foale et al. 2011). The need for a global effort to rebuild fisheries resources that considers a diverse set of governance options congruent with local context is called for in moving forward (Worm et al. 2009; Khan and Neis 2010).

One of the key efforts in addressing fisheries challenges has been governance reforms. Appearing under the rubric of fisheries co-governance (more commonly referred to as 'co-management') or self-governance (see Wilson et al. 2003; Townsend et al. 2008), many jurisdictions in various parts of the world have begun to experiment with these new modes of governance. They typically involve a shift from a long-established top-down, hierarchical governing structure to a more decentralized and collaborative one, based on the premise that greater fisher participation and responsibility in managing a local fishery could lead to better governance.

Despite decades of thinking and experience in governance transition, however, its progress has not been without significant challenges and failures. For instance, co-management can be path-dependent, meaning that outcomes may have already been largely decided by the time it is conceived and initiated (Chuenpagdee and Jentoft 2007). In other words, co-management failures may have been triggered by past knowledge and decisions that typically remain irreversible. There are also social and political concerns such as the participation paradox (Suárez de Vivero et al. 2008), elite capture (Platteau and Abraham 2002), and the lack of capacity of resource user communities (Fabricius et al. 2007). Other less fruitful attempts have been observed around the world (e.g., Scholtz et al. 1998; Pinkerton 1999; Cheong 2005; Blaikie 2006; Gelcich et al. 2006; Njaya 2007; Béné et al. 2009; Cudney-Bueno and Basurto 2009). As such, transition to a new governance mode is never a straightforward affair, further contributing to the 'wickedness' in the governing of world's fisheries (Ludwig et al. 1993; Dietz et al. 2003; Jentoft and Chuenpagdee 2009).

Recently, Gutierrez et al. (2011) have explored a reductionist approach that sought quantitative relationships between indicators and co-management success, whose results highlight leadership, fishing quotas and social capital, among others, as the key attributes for success. Other efforts include focusing on adaptive capacity and collaborative learning to resolve social-ecological dilemmas and achieve resilient fisheries outcomes (Armitage et al. 2009). Associated with interactive governance theory, this chapter draws upon an emerging perspective called *governability* (Kooiman et al. 2005; Jentoft 2007; Kooiman 2008; Bavinck et al. 2013). With an emphasis on understanding an inherent and constructed quality of a fisheries arrangement, a governability-inspired inquiry would explore whether a system (e.g., a coastal small-scale fishery) or a process in question (e.g., governance transition) is a governable one, capable of dealing with the multiple problems and issues facing the sector. Subsequently, it searches for ways to make it more governable, recognizing also limits to governability.

The purview of governability is wide-ranging and raises various possibilities for a potentially innovative analytical direction in which fisheries research can be undertaken. Among them is an interest in the *meta-order of governance*. This aspect focuses on people's normative ideas and underlying convictions that form the basis of governing decisions and actions (Kooiman and Jentoft 2009). The assertion is that all those involved in fisheries governance possess certain deeply-held notions about the fishery, policies and also about themselves and that they can inspire, guide, and shape the process and outcome of governance. Likewise, the progress of governance transition would also be contingent upon stakeholders 'values, images, and principles' (Song et al. 2013; also see Voyer et al. 2015).

At the same time, institutions embedded in governance transition can also act to determine governability. According to interactive governance, designing and caring for institutions is the *second-order* matter (Kooiman et al. 2005). Institutions are identified as a structural frame that gives substance to governance transactions as well as provides stability and continuity to people's underlying notions. They can serve to constrain individual aspirations, yet they are also challenged by people's disparate ideals. What is being promoted by institutions, and how does it align with people's mindsets? Tensions exist and would thus affect the way governance transition processes unfold as well. Uncovering such interactions and potential mismatches is expected to contribute to an enhanced and alternate understanding of fisheries governance and help illuminate their influence on governability (Chuenpagdee and Jentoft 2013).

This chapter examines the case of coastal small-scale fisheries in South Korea, in which a fisheries co-management program was initiated by the central government in 2001 in a top-down manner. This institutional arrangement is grounded in the premise that fishers would develop a sense of ownership and responsibility for the local fishery and come to appreciate their growing participation as a new way of engaging with the fisheries. The question that many government managers and researchers have raised is whether the program would be able to affect change in fishers' mindsets to match the aims of co-management (Lee and Shin 2004; PPACP 2008). It has been argued that such a shift is essential for self-correcting the problems in the fishery, thereby promoting a sustainable resource base, reinvigorating fishing villages, and overall providing a viable small-scale fishing industry in South Korea.

In what follows, first, co-management is explained as an institutional innovation, and its relationship with people's meta-order elements are conceptualized. Next, the governance transition taking place in South Korean small-scale fisheries is described, followed by an outline of the method used in the study. The results are then presented focusing on the interplay between the mindset and the institution. Finally, the chapter concludes with a reflection on this governability-inspired analysis of co-management and revisits the useful perspective it offers.

Co-management as an Institutional Innovation

The governability perspective identifies institutions as a crucial feature that can affect the level of governability of a system. Generally speaking, institutions are structural guidance that provide continuity, reduce uncertainty and stabilize people's interactions (North 1990; Peters 1999; Jentoft 2004; Scott 2008). An institution transcends individual actors to involve larger groups of people in patterned interactions that are somewhat predictable and steady, and that creates some sense of shared values and meanings among the members of the institutions (Peters 1999). As such, institutions act to "pre-define" the problem and the solution for the people who have become part of them. People sign up to what is given and start to take the pre-arranged items for granted. Those who choose not to, or fail to, wholly accept the assumed institutional definitions, however, may try to challenge existing institutions with a new set of ideals or simply exit them. Governability may be directly affected by the strength or the cohesiveness of such institutional mechanisms of the system that encourage actors to perform in certain favoured ways but restrict in other ways. Robust institutions equipped with effective laws, compelling norms, and well-accepted cultural habits, for example, would thus be more successful in repelling "exotic" interests and induce higher governability.

While institutions can work to resist change and reinforce the status quo, they can also serve as catalysts in bringing changes to a system (Scott 2008). In this sense, co-management can be treated as an institutional arrangement introduced to the system to facilitate governance change (e.g., Ostrom 1990; Pomeroy 1995; Jentoft et al. 1998; Acheson 2006; Berkes 2009). A co-management program would represent a bundle of rules, norms, and cultural-cognitive attributes that are arranged together to sustain or foster certain behaviors and mindsets of involved groups (Chuenpagdee and Song 2012).

As governance transition is conceptualized to take place via an introduction or alteration of an institutional arrangement, understanding what the institution embodies and how it works becomes crucial. If the ideas or behaviors that an institution aims to promote happen to be distant from the current mindsets of affected people, institutions may face immense difficulty in bringing intended changes. Co-management is said to be based on a different set of meta-governance values, images, and principles to those of hierarchical governance (Kooiman et al. 2005). If there is a large lag or an unmatchable gap, implementation would be difficult, governance transition uncertain and governability becoming a problem. Hence, we can explore the extent of *match* between co-management and governance actors that governability hinges upon. The postulation is that the greater the match, the higher the governability, and the greater propensity for smooth governance transition.

Case Study: The Jayul Program in South Korea

South Korea (officially the Republic of Korea) is located in the southern part of the Korean Peninsula in the Northwest Pacific region (Fig. 35.1). Endowed with productive fishing grounds in all three adjacent seas (Kang 2006), fishing has taken place from ancient times and helped satisfy much of the domestic fish consumption demands over the years. Fish occupies an integral part of Koreans' dietary life and are intimately connected to their culture. Korea was ranked 13th in the world in 2010 in terms of fish-production (seaweed and shellfish included) (FAO 2012b). In the coastal fishery, there are nearly 150 target species of commercial significance, which include squid, mackerel, saury, swimming crab, anchovy, and hair tail as well as a wide variety of shellfish and seaweeds (Kang 2006). Also, with over 28 licensed gear types including gillnet, hook and line, trap, longline, and hand gear such as a hand hoe for catching clams, Korean coastal fisheries can be characterized as multi-gear/multi-species (Han 2009). Finally, boats weighing less than 10 tons legally represent a small-scale sector and operate in inshore and mid-shore waters.

Korean people have always had close ties with the three neighbouring seas, using them throughout history for national security, trading, and the associated exchange of cultures as well as for fishing (Hong 1995). In the pre-modern period, many inshore fishing grounds were tenured by clans and village authorities. With the beginning of the Japanese occupation of Korea in 1911, the colonial government took over and restructured Korean fisheries by introducing fishing laws and founding fishery cooperatives at the village level (Cheong 2004). Following independence in 1945, the central government inherited the hierarchical setup as the legacy of the colonial structure and took on the overarching responsibility of managing the fisheries, although a form of village cooperative-led TURF was also instituted for intertidal and near shore water fishing. The national management regime since then has largely relied on input control mechanisms such as a license system and technical regulations (e.g. mesh size, catch size, and closed seasons), although in 1999, output control through total allowable catch and non-transferable individual vessel quota was instituted on major target species (Ryu et al. 2006).

The over 50 years of state-centered governance in the modern era, however, has proved inadequate in resolving various environmental and social challenges that have surfaced in coastal fisheries, such as stock depletion, illegal fishing, gear conflicts, dwindling economic viability, and hollowing out of coastal villages (Cheong 2003a; Han 2009). As a response, the central government initiated a new program called the 'Jayul Management for Fishing Communities' in 2001 (hereafter called the Jayul). Using the fishery challenges listed above as an institutional definition of the problem at hand, the Jayul attempted to raise its relevance among fishers and persuade them to subscribe to the new institutional ideas associated with this new

Fig. 35.1 Map of South Korea (community sites are shown as *double circles*; *triangles* indicate survey locations with managers/researchers)

mode of fisheries governance. Meaning 'free will' in Korean, the Jayul represents a type of placed-based co-management program where the several levels of government are tasked to set up the institutional framework and also to provide financial and technical assistance to local fisher organizations, while the latter would draft and carry out a management plan through the advice of non-governmental consultants or researchers (see Fig. 35.2). The Jayul aims to raise the level of community participation in managing local fisheries and ultimately to instill a sense of ownership in resource users (Lee et al. 2006; MOMAF 2003).

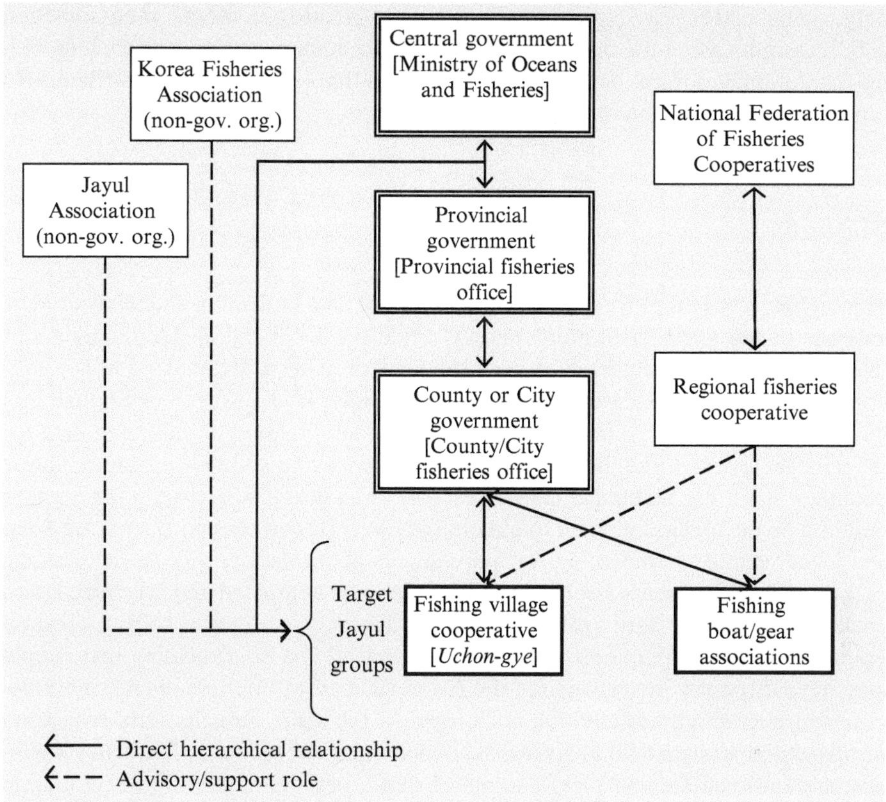

Fig. 35.2 General organizational structure of the South Korean coastal fishery showing main organizations and their predominantly hierarchical relationships

Since the inception of the program a decade ago, the number of community fisher organizations participating in the program reached 893 in 2011. Amongst these, there have been several exemplary cases in which fishing income has increased and illegal fishing has subsided (MOMAF 2005; Uchida et al. 2010). Yet, doubts have also been raised as to whether this governance reform is genuinely taking root. Many Jayul communities simply exist only on paper with no substantial follow-up activities (Seo and Byeon 2006). In addition, Lee (2010) reports that a financial incentive system that the central government has set up to entice fishing community organizations to join and keep up with the activities could have promoted further reliance on government, negating its original intention. Are the new ideas about this co-management program, such as participation and cooperation, being fostered and valued by the fishing communities? Can there be inconsistencies between what is being promoted by the government and what is being valued by the communities, which could impede its effectiveness, especially if the two are conflicting? Given the high expectation that achieving this new mode of governance may represent the only viable option for improving the fisheries situation in Korea (Lee and Shin 2004), insights about the mindsets of fishery stakeholders and their alignment with the institutional aims of the Jayul could provide a timely and helpful contribution to promote implementation of this governance reform.

Methods

This research has employed multiple methods for data collection and analysis. The analysis of the Jayul institutions mainly relied on document analysis. Review of relevant documents included both published and grey literature, making use of government reports and research articles available either in the Korean or English language. A questionnaire survey was used as the main method of gathering data for the mindset aspect, through which both quantitative and qualitative information was obtained. A sorting technique called "P+ sort" was designed as part of the survey, building on the methodological foundation of both pile sort (P) and Q sort (see Song and Chuenpagdee forthcoming for more details about the design). A set of cards, each of which represents a value or a principle to be compared and prioritized, was presented to survey participants, who were asked to perform a sorting exercise according to the instructions given. Since 'mindset' can be something that people may have difficulty in verbalizing, the P+ sort aimed to offer simplicity, intuitiveness and interactivity in eliciting responses. At the same time, its semi-structured format utilizing statistical analyses, such as nonparametric tests, frequency statistics, and multivariate analyses, also meant that it can systematically target a large number of respondents and produce numerical results amenable to conventional policy-making.

In this study, government managers/researchers and fishers/community members formed two main stakeholder groups surveyed, based on the conceptual distinction of the governing system and the system-to-be-governed as stipulated by interactive governance theory. Also, in practice, these two groups represent the two predominant parties typically involved in a co-management setup. To cover the variability of coastal environments and to seek participation of both the above groups, the study visited several regional administrative centers in addition to eight fishing communities around the country (see Fig. 35.1 and Table 35.1 for the characteristics of the study communities and the demographic information of the respondents). The community respondents comprised adult individuals, both male and female, involved in the production and marketing aspects of fisheries that included harvesters, processors, retailers, wholesalers, and retired fishers. While many of them were part of the local Jayul groups, they typically had limited direct interactions with government bodies. Lastly, I held informal discussions with additional key informants and engaged in direct observation in fishing communities to supplement the data.

The main data collection was carried out from September 2011 to July 2012. Also, data verification and preliminary analyses were performed during this phase involving re-visits to the communities and management offices to discuss the findings and seek explanations for the attained results.

Table 35.1 (a) Fishery attributes of visited coastal villages and (b) demographic information of respondent groups (Adapted from Song 2014)

	Resource dependent coastal village								Manager/researcher	Total
	Bakmi-ri	Goongpyeong-ri	Gusipo	Dongho-ri	Giseong-ri	Jiksan2-ri	Gubok-ri	Sim-ri		
(a)										
Main fishery	Clam, octopus, oyster	Clam, octopus, finfish	Crab, octopus, finfish, elver	Crab, octopus, clam, elver	Finfish, sea mustard, abalone	Finfish, anchovy, sea mustard	Mussel culture, finfish, octopus	Mussel culture, finfish	Oversees all fishery	–
Fishing environment	Intertidal area	Intertidal area, water column	Water column, intertidal area	Water column, intertidal area	Water column	Water column	Water column, intertidal area	Water column	Oversees all environment	–
Year joined the Jayul program	2004	2007	2003	2007	2001	2006	Not joined	Not joined	–	–
Member size of Jayul community	107	117	72	102	79	75	152[c]	101[c]	–	–
(b)										
# of respondents surveyed	25	25	25	25	25	25	25	25	25	225
Male	15	19	23	22	20	19	23	23	23	187
Female	10	6	2	3	5	6	2	2	2	38
Average age[b]	61	61	54	60	64	59	57	59	49	–
Age range[a]	39–82	40–79	26–82	37–81	50–80	35–80	39–78	39–77	31–65	–
Years of fishery experience[a]	3–60	6–60	1–50	1–55	1.5–50	0.5–55	5–60	4–40	1–44	–
Years in formal education[b]	9	9	8	8	9	9	10	9	17	–

[a]denotes range; [b]denotes average; [c]denotes member size of existing non-Jayul fisher organization (i.e., a fishing village cooperative)

Results

An examination of the meta-order concepts led to an understanding of important values, principles, and images held by fishery stakeholders. They would together imply what people ultimately aspire for with regard to coastal fisheries and also represent their deep-seated concerns and beliefs. For brevity's sake, the results presented here are a concise version of the full elaboration provided in earlier published output (e.g., Song 2014; Song and Chuenpagdee 2014, forthcoming). First, in gaining an appreciation of people's qualitative outlook, stakeholders' common images of the fishery generated 10 main features under the rubric of the four fishery sub-systems defined by interactive governance theory (i.e., the natural system-to-be-governed, the socio-economic system-to-be-governed, the governing system and the governing interactions). With respect to the natural system-to-be-governed, respondents held strong images of resource productivity and benefits as well as the widespread occurrence of pollution and environmental degradation. As for the socio-economic system-to-be-governed, livelihood difficulties, aging population in coastal communities and social-cultural marginalization of the fishing industry were most prominently expressed. When it came to the governing system, long-standing provision of government assistance was highlighted as a dominant concern, while the governing interactions were characterized with breakdown in communication between government managers and fishers/community members. As these concerns likely represent a set of pressing features that are considered important or influential in terms of fisheries life-world, they may be consulted when reflecting on past governance processes and/or formulating future directions.

Exploring other domains of fishery stakeholders' mindsets, the result of the P+ sort produced a hierarchy of values and principles based on relative importance. Table 35.2(a) shows four values and three principles identified to be very important by the strong majority of the survey respondents. For values, 'ecosystem conservation' was by far judged to be the most important one. The highest standing of this value can be exemplified by respondent explanations such as "this is the foundation of everything involved in the fishery". In fact, consistent with the image of a clean coastal environment, this is already one of the major foci of the Jayul program and also where noticeable progress at the local level has been made. For example, various cleanup activities have been a key component of Jayul activities emphasized and practiced over the years with many communities engaging in monthly coastal clean ups (MOMAF 2003, 2005; Uchida et al. 2010). Next, the groups also prioritized the importance of 'wealth' and 'secure livelihoods', as also corroborated by the image analysis earlier. Consistent with this, one other principal aim of the Jayul has been generating direct benefits to community members through non-trivial income increases (Lee 2010). This is presumed to be a prerequisite or the bottom line for keeping communities motivated and interested in carrying out Jayul activities. To this end, through the financial and advisory support of the government and non-governmental organizations, there has been a sustained effort involving projects such as stock enhancement activities, construction of drying and icing facilities,

Table 35.2 Values and principles identified to be (a) very important by the surveyed fishery stakeholders overall; (b) in disagreement between fishers/community members and managers/researchers in terms of importance[a] (see Song and Chuenpagdee forthcoming for details)

	Values	Principles
(a) Very important	Ecosystem conservation	Equity
	Wealth	Ecosystem integrity
	Honesty	Adjacency
	Secure livelihoods	
(b) In disagreement	Freedom (F)	Adjacency (F)
	Knowledge (F)	Participation (M)
		Efficiency (F)
		Scientific information (M)
		Exclusivity (F)

[a](F) refers to fishers/community members expressing significantly higher importance than managers/researchers, while (M) indicates the reverse case

and tourism ventures (MOMAF 2005). For principles, 'equity' was indicated to be the most important concept. The prevailing conception of equity among the respondents was that of *fairness*, that is, benefits derived from the fishery should be distributed proportional to one's level of effort, diligence and/or investment. As follows, the result highlights some of the fundamental ideas that need to be at the base of the Jayul co-management program, if it is to garner a genuine support of fishery stakeholders.

The survey method also facilitated an identification of significant differences between the two main respondent groups – fishers/community members and managers/researchers, in particular with respect to principles. As listed in Table 35.2(b), the manager/researcher group is an avid supporter of user participation in management and rule-making, while the community groups' enthusiasm is, though considerable, significantly weaker. This result may cautiously point to the suggestion that despite the many positive developments of the Jayul program in the last decade, the participatory mindset may be slow to take root across the communities, substantiating the concerns about the communities' cursory involvement. In the case of the adjacency principle (i.e., access to use a fishing ground should be first granted to those who live near it), however, many of the manager/researcher respondents did not find it very important contrary to the overwhelmingly positive response of fishers/community members. The manager/researcher group has shown ambivalence towards strengthening local priority access and use rights to nearby fishing grounds for the fear that fishing communities may not always effectively manage a given coastal environment, as indicated by previous accounts of mismanagement (see Cheong 2003b, 2005). Yet, adjacency is conceptualized as a key enabling factor towards a more community-empowered approach to fisheries management (Davis and Wagner 2006). Its activation could ensure a legal as well as a geographical basis for coastal communities to maintain fishing livelihoods, in addition to help reducing vulnerability of local fishery collapse arising from outside influences.

Table 35.3 Summary of key mismatches between the institutional aims of the Jayul program and the mindsets of surveyed fishery stakeholders organized under the broad analysis scheme described by Scott (2008) that proposes cultural-cognitive, normative, and regulative pillars of institutions

	Institutional aims of the Jayul program towards coastal fishing communities	Mindsets of fishery stakeholders represented by values, images, and principles
Cultural-cognitive	Self-reliance and self-initiative in fishing communities	Images confirming long-held reliance on government assistance
	Jayul as a pan-community cultural movement to raise the general livability of coastal area, going beyond a narrow sector-based project with specific objectives	Images depicting aging population and out-migration of young people in fishing villages – reducing the acceptability of new governance ideas such as the Jayul
Normative	Community cohesion and social norms; 'doing it as a group'	Values suggesting prevalence of individualistic mindset and operational traits involved in the permit-based boat fishery
Regulative	Drafting of local fishery rules and community activities, notwithstanding a lack of recognition of the Jayul in the national fisheries law	Principles indicating subordinate position of local fishery rules to national regulations and lengthy amendment process in making national regulations in line with local fishery rules

The government may therefore need to re-evaluate these discrepancies for the continuing implementation of the Jayul program.

Finally, notable mismatches between the institutional aims of the Jayul program and the mindsets of stakeholders were revealed. The aims listed in the left-hand column of Table 35.3 embody what the fishery endeavours to transition into through the implementation of the Jayul program. The right-hand column summarizes the relevant mindsets of the surveyed stakeholders. For example, while self-reliance and self-initiative is being promoted, this is contrasted to the stakeholders' images that exemplify the culture of reliance on government support and assistance. Responding to the question of what is first conjured up in the mind about *government*, words such as 'financial and technical assistance', 'policy guidance', 'educational sessions', and 'enforcement by the coast guard' formed a major part of the vocabulary for both fishers and managers/researchers, informing the ways respondents imagine the role of the government. Such views seem to confirm the centrality of governance assistance in the functioning of coastal fishery. Similarly, a low regard on equality and cooperation values coupled with fishers' relative valuing of operational freedom, suggest that an individualistic mindset is widespread in the fishery and the community cohesion and social norms envisioned in the transition process has a long way to go. Furthermore, it was revealed that the subsidiary principle garnered considerably little support from both the fisher and manager respondent groups due to reasons such as the fear of corruption by local elites or weak compliance following low legitimacy. While all Jayul communities have composed local rules as a requirement to be part of the Jayul program, this situation casts doubt as to their true appreciation and usefulness. The generally weak enthusiasm about community rules would thus likely act as a hindrance to its ongoing implementation. Overall, the results shown here suggest that the successful implementation of the Jayul may be delayed unless there is a way to address the mismatches between institutional promotion and peoples' mindset.

Discussion and Conclusion

In its 14th year of implementation, the Jayul has expanded nationwide and support for it has grown over the years from both the government sector as well as from communities. It carries high hopes and great ambitions to help coastal fishing communities move into prosperity. There are observed and otherwise reported cases of fishing communities thriving under the Jayul scheme (see MOMAF 2005) that suggests that some communities have made the Jayul program work for them and brought about changes to improve their community life. Jung (2013) identified factors that contribute to perceived Jayul success, such as quality of fishing ground, leadership, and sense of closeness among community members. According to the governability framework described here, such cases can be regarded to have achieved closer alignment of the mindset with the Jayul institutions with factors such as clean environment and social cohesion occurring at the local level. In light of our analysis, however, sweeping changes in the mindset of fishers to fit the institutional aims of the Jayul program does not appear to have fully materialized at the macro (national) scale.

According to interactive governance theory, the ultimate goal of governance is to make fisheries systems more governable, given the multiple problems and issues facing the fishery (Bavinck et al. 2013). A governability-inspired analysis in this chapter produced an alternate perspective to gauge the progress of Jayul program implementation, providing a form of "reality-check" (Chuenpagdee and Jentoft 2009). The results indicated that conforming to the Jayul institution may not be as achievable as it was initially conceived of by policy makers. Hence, wholesale insistence on the Jayul could in fact run the danger of making things less governable in some instances. While continuing promotion of the Jayul program through affecting people's agency should be further encouraged, the result of this study also suggests the need to reconsider the institutional aims to include a range of people's leading, but otherwise overlooked, ideals such as equity, adjacency and honesty. This can be a worthy venture that could raise the governability of the system.

This research has focused on two aspects of the governance system that influence governability. The meta-order elements are qualities of all involved fishery stakeholders, which are mostly inherent but may also be constructed and evolving, through governing interventions such as the implementation of the Jayul. The second aspect is institutions. Institutions pose certain structural constraints that bestow rigidity, orthodoxy and consistency into a governance system, such that the system (and actors within) is encouraged to perform in certain ways but restricted in other ways – a mechanism through which institutions would serve to increase or lower the governability of the system. Ultimately, interactions that take place between the institutional elements (e.g., rules, norms, and customs) being promoted and the typically entrenched, slow-changing people's mindsets may create mismatches resulting in conflicting and unstable conditions. Therefore, attentive consideration towards the institutional design and implementation that attune to people's underlying ideas would be an important undertaking for ensuring a more governable system. In addition, a longitudinal study that focuses on assessing the actual changes that have occurred in stakeholders' mindsets over time is expected to assist in obtaining direct insights about the process of match.

Unique mindsets of fishers have been stated in past research, such as those of "rugged individualists" characterizing small-scale fishers (Davis and Jentoft 1993). These traits have posed a problem in governance and fisher organizations, frustrating governors for many years. Thus, the extent to which the rationalities of the two parties (i.e., government and fishers, broadly speaking) differ and how the gap between them can be bridged is of continuing scholarly interest. I submit that the perspectives outlined in this chapter, especially the role of meta-governance elements in elucidating their attunement with institutions, could provide a useful analytical avenue with which to examine stakeholder differences and deliberate on strategies for institutional effectiveness.

Overall, the governability perspective, and the interactive governance theory more broadly, has grown to be one part of a larger effort to alleviate widespread challenges occurring in fisheries. As this chapter has demonstrated the utility of such an approach through an examination of a governance transition process, an ongoing exploration and refinement of the governability-focused thinking is further encouraged.

Acknowledgments Funding for this study was provided by the Social Sciences and Humanities Research Council of Canada. I am very grateful for the support and guidance of Ratana Chuenpagdee in all stages of the research process. I also thank two anonymous reviewers for their insightful feedback on an earlier draft of this manuscript. If any errors remain, they are the sole responsibility of the author.

References

Acheson, J. M. (2006). Institutional failure in resource management. *Annual Review of Anthropology, 35*, 117–134.

Allan, J. D., Abell, R., Hogan, Z., Revenga, C., Taylor, B. W., Welcomme, R. L., & Winemiller, K. (2005). Overfishing of inland waters. *BioScience, 55*, 1041–1051.

Armitage, D. R., Plummer, R., Berkes, F., Arthur, R. I., Charles, A. T., Davidson-Hunt, I., Diduck, A. P., Doubleday, N., Johnson, D. S., Marschke, M., McConney, P., Pinkerton, E., & Wollenberg, E. (2009). Adaptive co-management for social-ecological complexity. *Frontiers in Ecology and the Environment, 7*, 95–102.

Bavinck, M., Chuenpagdee, R., Jentoft, S., & Kooiman, J. (Eds.). (2013). *Governability of fisheries: Theory and applications*. Dordrecht: Springer.

Beddington, J. R., Agnew, D. J., & Clark, C. W. (2007). Current problems in the management of marine fisheries. *Science, 316*, 1713–1716.

Béné, C., Macfadyen, G., & Allison, E. H. (2007). *Increasing the contribution of small-scale fisheries to poverty alleviation and food security* (FAO fisheries technical paper 481). Rome: FAO.

Béné, C., Belal, E., Baba, M. O., Ovie, S., Raji, A., Malasha, I., Njaya, F., Na Andi, M., Russell, A., & Neiland, A. (2009). Power struggle, dispute and alliance over local resources: Analyzing 'democratic' decentralization of natural resources through the lenses of Africa inland fisheries. *World Development, 37*, 1935–1950.

Berkes, F. (2009). Evolution of co-management: Role of knowledge generation, bridging organizations and social learning. *Journal of Environmental Management, 90*, 1692–1702.

Berkes, F., Hughes, T. P., Steneck, R. S., Wilson, J. A., Bellwood, D. R., & Crona, B. (2007). Globalization, roving bandits, and marine resources. *Science, 311*, 1557–1558.

Blaikie, P. (2006). Is small really beautiful? Community-based natural resource management in Malawi and Botswana. *World Development, 34*, 1942–1957.

Cheong, S.-M. (2003a). Depleting fish resources, declining fishing communities, and the state revitalization project in Korea. *Environmental Management, 32*, 382–390.

Cheong, S.-M. (2003b). Privatizing tendencies: Fishing communities and tourism in Korea. *Marine Policy, 27*, 23–29.

Cheong, S.-M. (2004). Managing fishing at the local level: The role of fishing village cooperatives in Korea. *Coastal Management, 32*, 191–202.

Cheong, S.-M. (2005). Korean fishing communities in transition: Limitations of community-based resource management. *Environment and Planning A, 37*, 1277–1290.

Chuenpagdee, R., & Jentoft, S. (2007). Step-zero for fisheries co-management: What precedes implementation. *Marine Policy, 31*, 657–668.

Chuenpagdee, R., & Jentoft, S. (2009). Governability assessment for fisheries and coastal systems: A reality check. *Human Ecology, 37*, 109–120.

Chuenpagdee, R., & Jentoft, S. (2013). Assessing governability – What's next. In M. Bavinck, R. Chuenpagdee, S. Jentoft, & J. Kooiman (Eds.), *Governability of fisheries and aquaculture: Theory and applications* (pp. 335–349). Dordrecht: Springer.

Chuenpagdee, R., & Song, A. M. (2012). Institutional thinking in fisheries governance: Broadening perspectives. *Current Opinion in Environmental Sustainability, 4*, 309–315.

Close, D. A., Fitzpatrick, M. S., & Li, H. W. (2002). The ecological and cultural importance of a species at risk of extinction, Pacific lamprey. *Fisheries, 27*(7), 19–25.

Cochrane, K. L. (2000). Reconciling sustainability, economic efficiency and equity in fisheries: The one that got away? *Fish and Fisheries, 1*, 3–21.

Cudney-Bueno, R., & Basurto, X. (2009). Lack of cross-scale linkages reduces robustness of community-based fisheries management. *PLoS ONE, 4*(7), e6253. doi:10.1371/journal.pone.0006253.

Davis, A., & Jentoft, S. (1993). Self and sacrifice: An investigation of small boat fisher individualism and its implication for producer cooperatives. *Human Organization, 52*(4), 356–367.

Davis, A., & Wagner, J. (2006). A right to fish for a living? The case for coastal fishing people's determination of access and participation. *Ocean and Coastal Management, 49*, 476–497.

Dietz, T., Ostrom, E., & Stern, P. C. (2003). The struggle to govern the commons. *Science, 302*, 1907–1912.

Fabricius, C., Folke, C., Cundill, G., & Schultz, L. (2007). Powerless spectators, coping actors, and adaptive co-managers: A synthesis of the role of communities in ecosystem management. *Ecology and Society, 12*(1), 29. Retrieved November 28, 2012, from http://www.ecologyandsociety.org/vol12/iss1/art29/

FAO (Food and Agriculture Organization of the United Nations). (2012a). *The state of world fisheries and aquaculture 2012*. Rome: FAO.

FAO (Food and Agriculture Organization of the United Nations). (2012b). *World fisheries production, by capture and aquaculture, by country (2010)*. Retrieved November 28, 2014, from ftp://ftp.fao.org/fi/STAT/summary/a-0a.pdf

Foale, S., Cohen, P., Januchowski-Hartley, S., Wenger, A., & Macintyre, M. (2011). Tenure and taboos: Origins and implications for fisheries in the Pacific. *Fish and Fisheries, 12*, 357–369.
Gelcich, S., Edwards-Jones, G., Kaiser, M. J., & Castilla, J. C. (2006). Co-management policy can reduce resilience in traditionally managed marine ecosystem. *Ecosystems, 9*, 951–966.
Gutiérrez, N. L., Hilborn, R., & Defeo, O. (2011). Leadership, social capital and incentives promote successful fisheries. *Nature, 470*, 386–389.
Han, K.-S. (2009). *21-segi HangukSusanup-euiGomin* [The agony of 21st Korean fisheries]. Seoul: Sunhaksa.
Hilborn, R. (2007). Moving to sustainability by learning from successful fisheries. *Ambio, 36*, 296–303.
Hilborn, R., Orensanz, J. M., & Parma, A. (2005). Institutions, incentives and the future of fisheries. *Philosophical Transactions of the Royal Society B, 360*, 47–57.
Hong, S.-Y. (1995). Marine policy in the Republic of Korea. *Marine Policy, 19*, 97–113.
Jentoft, S. (2004). Institutions in fisheries: What they are, what they do, and how they change. *Marine Policy, 28*, 137–149.
Jentoft, S. (2007). Limits of governability: Institutional implications for fisheries and coastal governance. *Marine Policy, 31*, 360–370.
Jentoft, S., & Chuenpagdee, R. (2009). Fisheries and coastal governance as a wicked problem. *Marine Policy, 33*, 553–560.
Jentoft, S., & Eide, A. (Eds.). (2011). *Poverty mosaics: Realities and prospects in small-scale fisheries*. Dordrecht: Springer.
Jentoft, S., McCay, B. J., & Wilson, D. C. (1998). Social theory and fisheries co-management. *Marine Policy, 22*, 423–436.
Jung, J.-H. (2013). *An analysis of performance factors in small scale common-pool resource's self-organization*. Unpublished doctoral dissertation, Pukyong National University, Busan, South Korea.
Kang, J.-S. (2006). Analysis on the development trends of capture fisheries in North-East Asia and the policy and management implications for regional co-operation. *Ocean and Coastal Management, 49*, 42–67.
Kent, G. (1997). Fisheries, food security, and the poor. *Food Policy, 22*(5), 393–404.
Khan, A. S., & Neis, B. (2010). The rebuilding imperative in fisheries: Clumsy solutions for a wicked problem? *Progress in Oceanography, 87*, 347–356.
Kooiman, J. (2008). Exploring the concept of governability. *Journal of Comparative Policy Analysis, 10*, 171–190.
Kooiman, J., & Jentoft, S. (2009). Meta-governance: Values, norms and principles, and the making of hard choices. *Public Administration, 87*(4), 818–836.
Kooiman, J., Bavinck, M., Jentoft, S., & Pullin, R. S. V. (Eds.). (2005). *Fish for life: Interactive governance for fisheries*. Amsterdam: Amsterdam University Press.
Lee, S.-G. (2010). A study on practices and effective mechanism of fisheries self-governance and institutional strategies. In *Proceedings of the international symposium: A new decade! The role of cooperatives for the sustainable development of fisheries* (pp. 110–162). Seoul: National Federation of Fisheries Cooperatives.
Lee, S.-G., & Shin, Y.-M. (2004). A study on the self regulatory management model of coastal fisheries in Korea. *The Journal of Fisheries Business Administration, 35*(1), 87–114.
Lee, K. N., Gates, J. M., & Lee, J. (2006). Recent developments in Korean fisheries management. *Ocean and Coastal Management, 49*, 355–366.
Ludwig, D., Hilborn, R., & Walters, C. (1993). Uncertainty, resource exploitation, and conservation: Lessons from history. *Science, 260*, 17–36.
McGoodwin, J. R. (2001). *Understanding the cultures of fishing communities: A key to fisheries management and food security* (FAO fisheries technical paper 401). Rome: FAO.
MOMAF (Ministry of Marine Affairs and Fisheries). (2003). *Jayul-gwanri-uhup-euiSeonggong-jeokJeongchak-eulWuihanYeongu* [A study on successful establishment of Jayul fisheries management]. Seoul: MOMAF.

MOMAF (Ministry of Marine Affairs and Fisheries). (2005). *Jayul-gwanri-uhupWoosoo-sa-rye* [Jayul Fisheries Program outstanding cases II]. Seoul: MOMAF.

Mora, C., Myers, R. A., Coll, M., Libralato, S., Pitcher, T. J., Sumaila, R. U., et al. (2009). Management effectiveness of the world's marine fisheries. *PLoS Biology, 7*(6), e1000131. doi:10.1371/journal.pbio.1000131.

Myers, R. A., & Worm, B. (2003). Rapid worldwide depletion of predatory fish communities. *Nature, 423*, 280–283.

Njaya, F. (2007). Governance challenges for the implementations of fisheries co-management: Experiences from Malawi. *International Journal of the Commons, 1*, 137–153.

North, D. C. (1990). *Institutions, institutional change and economic performance*. Cambridge: Cambridge University Press.

Ostrom, E. (1990). *Governing the commons: The evolution of institutions for collective action*. Cambridge: Cambridge University Press.

Pauly, D., Christensen, V., Guénette, S., Pitcher, T. J., Sumaila, U. R., Walters, C. J., Watson, R., & Zeller, D. (2002). Towards sustainability in world fisheries. *Nature, 418*, 689–695.

Peters, B. G. (1999). *Institutional theory in political science: The 'new' institutionalism*. London: Continuum.

Pinkerton, E. (1999). Factors in overcoming barriers to implementing co-management in British Columbia salmon fisheries. *Conservation Ecology, 3*(2), 2. Retrieved November 28, 2012, from http://www.consecol.org/vol3/iss2/art2/

Platteau, J.-P., & Abraham, A. (2002). Participatory development in the presence of endogenous community imperfections. *The Journal of Development Studies, 39*(2), 104–136.

Pomeroy, R. S. (1995). Community-based and co-management institutions for sustainable coastal fisheries management in Southeast Asia. *Ocean and Coastal Management, 27*(3), 143–162.

PPACP (Policy Planning Advisory Committee to the President) (2008). *Jayul-gwanri-uhupJeongchaek: Sangsaeng-eulTonghanUhga-sodeuk-euiJeungdae* [Fisheries self-governance policy: Towards increase in fishing household income through SangSaeng]. Seoul: PPACP.

Ryu, J.-G., Nam, J., & Gates, J. M. (2006). Limitations of the Korean conventional fisheries management regime and expanding Korean TAC system toward output control systems. *Marine Policy, 30*, 510–522.

Scholtz, U., Njaya, F. J., Chimatiro, S., Hummel, M., Donda, S., & Mkoko, B. J. (1998). Status and prospects of participatory fisheries management programs in Malawi. In T. Petr (Ed.),. *Inland fishery enhancements* (Fisheries technical paper 374, pp. 407–425). Rome: FAO.

Scott, W. R. (2008). *Institutions and organizations: Ideas and interests* (3rd ed.). Los Angeles: Sage.

Seo, B.-G., & Byeon, D.-S. (2006). Improvement measures for vitalization of Jayul Fisheries Program, with special attention to inshore boat fishery. *Aquatic Industry Science Research, 24*, 31–42.

Song, A. M. (2014). *Governance in transition: Exploring people's mindset and institutional matches towards a governable coastal fisheries in South Korea*. Unpublished doctoral dissertation, Memorial University of Newfoundland, St. John's, Canada.

Song, A. M., & Chuenpagdee, R. (2014). Exploring stakeholders' images of coastal fisheries: a case study from South Korea. *Ocean & Coastal Management, 100*, 10–19.

Song, A. M., & Chuenpagdee, R. (forthcoming). Eliciting values and principles of fishery stakeholders in South Korea: A methodological exploration. *Society & Natural Resources*.

Song, A. M., Chuenpagdee, R., & Jentoft, S. (2013). Values, images, and principles: What they represent and how they may improve fisheries governance. *Marine Policy, 40*, 167–175.

Suárez de Vivero, J. L., Rodríguez Mateos, J. C., & Florido del Corral, D. (2008). The paradox of public participation in fisheries governance. The rising number of actors and the devolution process. *Marine Policy, 32*, 319–325.

Teh, L. C. L., & Sumaila, U. R. (2013). Contribution of marine fisheries to worldwide employment. *Fish and Fisheries, 14*, 77–88.

Townsend, R. E., Shotton, R., & Uchida, H. (2008). *Case studies in fisheries self-governance* (FAO fisheries technical paper 504). Rome: FAO.

Uchida, H., Uchida, E., Lee, J.-S., Ryu, J.-G., & Kim, D.-Y. (2010). Does self management in fisheries enhance profitability? Examination of Korea's coastal fisheries. *Marine Resource Economics, 25*, 37–59.

Voyer, M., Gollan, N., Barclay, K., & Gladstone, W. (2015). 'It's part of me'; Understanding the values, images and principles of coastal users and their influence on the social acceptability of MPAs. *Marine Policy, 52*, 93–102.

Wilson, D. C., Nielsen, J. R., & Degnbol, P. (Eds.). (2003). *The fisheries co-management experience: Accomplishments, challenges and prospects*. Dordrecht: Kluwer.

Worm, B., Hilborn, R., Baum, J. K., Branch, T. A., Collie, J. S., & Costello, C. (2009). Rebuilding global fisheries. *Science, 325*, 578–585.

Chapter 36
The Dynamics of Small-Scale Fisheries in Norway: From Adaptamentality to Governability

Svein Jentoft and Jahn Petter Johnsen

Abstract Interactive governance theory emphasizes the two-way exchange that occurs between the system-to-be-governed and the governing system. Thus, in the case of small-scale fisheries, the working hypothesis is that their governability, and hence their survival, depends on the ability and readiness of the governing system to respond to changes that occur within the system-to-be-governed and vice versa. It follows that governability of small-scale fisheries would be determined within both systems, as well as in the way they interact. Using Norway as a case study, this chapter argues that the governability of small-scale fisheries is dependent on the ability and willingness of fishers to respond not only to changes in the socio-ecological environment, but also to actions or reactions of the governing system. Their inclination to adapt, "adaptamentality", is seen as the motivation for acquiring the necessary skills, knowledge and resources that make them able and prepared for change. It is argued that the institutional design of the governing system, as it has developed during the twentieth century, has been important for this adaptamentality, as it has facilitated constructive partnership with the government and generated mutual trust. Whether these qualities will remain with current institutional reforms, is a question that will be discussed.

Keywords Small-Scale Fisheries • Governability • Governmentality • Trust • Norway

S. Jentoft (✉) • J.P. Johnsen
Norwegian College of Fishery Science, UIT – The Arctic University of Norway, 9037 Tromsø, Norway
e-mail: Svein.Jentoft@uit.no; jahn.johnsen@uit.no

S. Jentoft, R. Chuenpagdee (eds.), *Interactive Governance for Small-Scale Fisheries*, MARE Publication Series 13, DOI 10.1007/978-3-319-17034-3_36

Introduction

Small-scale fisheries are not a well-defined category. Rather, small is always relative to large, and what is considered to be small in one context may well be viewed as large in another. Many reasons account for these differences, one of them being exposure to natural conditions which vary a lot from place to place. Relative wealth is another reason; fishers tend to reinvest in their vessel and gear, and hence often expand their operations. With economic development often follows a shift in the composition of the fleet from small to large. In Norway, nature was always on the side of small-scale fishers. Rich fish stocks, such as the northeast arctic cod (*Gadus morhua*),[1] migrate to coastal and inshore waters to spawn in the winter and the spring, and are thus easily accessible to small-scale fishers. Due to the Gulf Stream and other elements in the oceanographic system, the Norwegian coast is blessed with relatively warm and nutritious water that keeps the coast ice free throughout the year. This also creates favorable conditions for marine life and, hence, a thriving small-scale, coastal fishery.

Therefore, small-scale fishing, often in combination with small-scale, mainly subsistence, farming, has traditionally been the common source of livelihood along the Norwegian coast, particularly in the north. Because of the economic and social importance of small-scale fisheries, for instance in maintaining coastal communities, national fishery policies have aimed to protect and sustain commercial small-scale fisheries. Traditionally, for Norwegian fishers, natural affluence made access to fish and fishing grounds a minor political issue. Yet, due to technological development and at times oversupply and instable market prices, regulation was necessary. Historically, small-scale fishers were many and had a strong voice in Norwegian regional and fisheries politics. They also represented an industry that was important for the national economy, being the most important export earner. They were able to influence the government both directly through their organization, the Norwegian Fishers' Association, and indirectly through a Parliament that was willing to listen. Thus, small-scale fishers could convince the government to interfere in conflicts with the buyers in a way that benefitted them. The governance model of Norwegian fisheries was traditionally also a co-governance design, albeit characterized by corporatism rather than co-management (Jentoft and Mikalsen 2014). Nevertheless, the state was, by and large, regarded as an amiable partner, and rarely an adversary. With the introduction of an individual vessel quota system in 1990, the state became more assertive, and shifted its role from being predominantly reactive to proactive. The quota system divided fishers into groups with or without individual quota rights (Johnsen 2005; Johnsen et al. 2009b), resulting in the state now facing more opposition from small-scale fishers than it used to, which has increased the level of conflict within Norwegian fisheries governance.

In this chapter, which draws on decades of experience of working on small-scale fisheries in Norway, we argue that the institutionalized interaction has been

[1] If nothing else is mentioned, «cod» in this article refers to Northeast Arctic Cod.

conducive for the governability of small-scale fisheries, and indeed for the fisheries sector in Norway as a whole. Basic to this outcome is a readiness for change among small-scale fishers, what we in this chapter choose to call "adaptamentality". The adaptamentality can be seen in small-scale fishers' response to environmental and economic change, including their willingness to cooperate in good faith with government. It can also be recognized in their adoption of new technology, skills and knowledge, which the governing system can help facilitate.

In the next section, we discuss the meaning and relations between adaptamentality and governability. We also make use of the governmentality concept introduced by Foucault, which we see as bridging the two concepts. Thereafter, we describe small-scale fisheries in Norway and how they have changed over time. The design of current fisheries management is the subject of the two subsequent sections. Finally, we reflect on the governability of Norwegian small-scale fisheries, and what has made this fishery adaptable and governable.

Adaptamentality, Governability and Change

Foucault (1978a, b) introduced the term "governmentality," which we understand as the practices resulting from governing interventions and the responses to the interventions within the system-to-be-governed. Governmentality thus has a dual meaning. On the one hand refers to the governing system's apparatus for governing and the belief in its ability to govern. On the other hand, it is about the willingness of citizens to let themselves be governed (Song et al. 2013; Johnsen 2014). Together the interventions and responses constitute performative practices that change how actors perceive, interpret, and conceptualize reality. Governmentality is therefore an outcome of governing interactions. It is not a fixed product, but one that is continuously produced and reproduced.

Governmentality and governability is closely related; from governmentality follows governability. Thus, the lower the governmentality in a particular governance system, the bigger is the governability problem. This is the case when governing is hampered by indifference or resistance. Governability refers to the capacity for, and quality of as the governance, for instance with regard to the implementation of a particular policy or strategy targeting small-scale fisheries (see Jentoft and Chuenpagdee, Chap. 2 in this volume). According to interactive governance theory, the governability problem sits in the system-to-be-governed, the governing system and the governing interactions (Bavinck et al. 2013). So does governmentality, which is easy to see from Foucault's description of governmentality as characteristics of both the governor, i.e. in his case the state, and those who are being governed, i.e. citizens. Both concepts refer to the propensity for adaptive or transformative change both within the system-to-be-governed and in the governing system. It is in this context we introduce the concept of "adaptamentality" and claim that adaptamentality is among the conditions for governmentality and consequently governability.

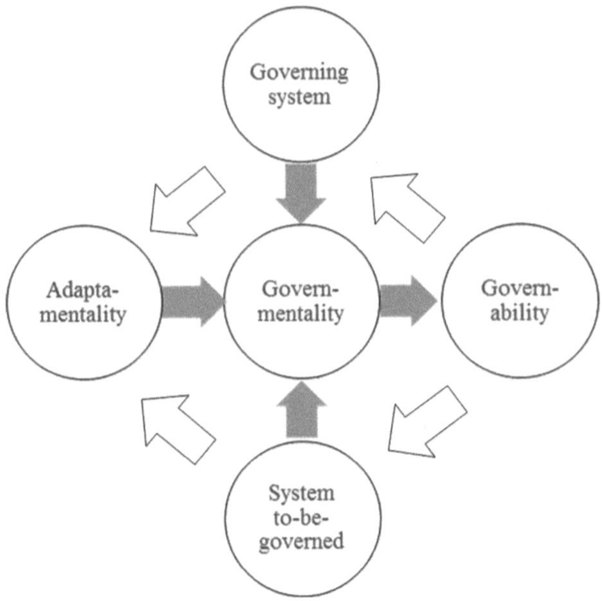

Fig. 36.1 Governability causal model

To improve the governability of small-scale fisheries, or fisheries in general, one must be able to elevate the degree of governmentality, which cannot occur without some form of interaction between the system-to-be-governed and the governing system. In our model (Fig. 36.1), the basic causal arrow goes from adaptamentality via governmentality to governability. Governmentality is also a co-product of the interactions that occur within and between the governing system and the system-to-be-governed, whereas governability as the outcome variable in the model affects both the governing system and the system-to-be governed by reinforcing the capacity and quality of governance, for instance by enhancing or reducing mutual trust.

What, then, causes adaptamentality in the first instance? The model suggests that in order to answer this question one would have to look at what is happening both within the governing system and the system-to-be-governed and how they have evolved and delivered over time. For that, we must include factors such as the culturally inscribed images that stakeholders who inhabit the system-to-be-governed and the governing system have of each other, the fishery that they are involved in, and the natural and social world around them (Kooiman and Chuenpagdee 2005; Jentoft et al. 2010; Song and Chuenpagdee 2014). In small-scale fisheries, it is particularly important when assessing adaptamentality to include the interaction that happens between the social and natural components of the system-to-be-governed and the images of that interaction that fishers and fish workers have and act upon. The governability, and the transaction costs involved in securing adaptamentality, is dependent on the degree to which the agents of both systems are sharing and agreeing on the same image, or at least recognize what the images are, acknowledge their difference, and realize how different images may create governability challenges. Without agreement on how to frame the problem, which in small-scale fisheries are inherently "wicked" (Jentoft and Chuenpagdee 2009), the governing system will be less effective than it would otherwise be in addressing basic concerns, be they ecosystem health, food and livelihood security, or social justice (Bavinck et al. 2013).

As indicated in the model, the feedback mechanism from governability, via the governing system and the system-to-be-governed, to adaptamentality, must be taken into account in the assessment. Stakeholders will be more receptive to change, also those introduced hierarchically by government, if they have seen that reforms instigated by government are working for the better in the situation they find themselves in. Conversely, the adaptamentality and governmentality of small-scale fishers may be less if there is a history of marginalization and impoverishment and for instance if promises made by government in the past have not been met. When people decide how to respond, they tend to regard concrete initiatives in their broader social and political context. Then they think not only of what they hear but also who they hear it from. The message may sound fine but the messenger is perhaps not to be trusted.

Adaptamentality can, but does not have to be, subservient; it can also be assertive. It is about taking advantage of new opportunities, and to be adaptive and proactive. Thus, adaptamentality is fundamental to innovation, which is in itself a governability quality. Notably, this adaptamentality is not inherent or constant over time. Rather, it is nurtured through the system of interaction (between the governing system and the system-to-be-governed) that has generated mutual trust, which is a necessary provision for governability because it relieves those involved from the defensiveness and cautiousness that follow from a perception risk. Governability requires "creative governance" (Kooiman et al. 1999), which is particularly an issue in small-scale fisheries because small-scale fishers often find themselves in situations of marginalization and deprivation, and therefore in need of fundamental change and learning. What characterizes small-scale fisheries governance in Norway in this respect is what is discussed next.

System-To-Be-Governed

According to the 2008 Marine Resources Act and the regulatory system it proposed, it is reasonable to divide Norway's small-scale fisheries system-to-be-governed into a non-commercial and commercial sector. The non-commercial sector, which is basically reserved for recreational and/or subsistence purposes, is open to all Norwegian citizens and subject to few regulations as long as the activity and scale of the operation are under a certain level. The government regards recreational and tourist fishing as almost ungovernable due to the number of people involved, the topography of the coast, and the extensive border with Finland and Sweden that makes effective control and monitoring practically impossible (Solstrand and Gressnes 2014). On the other hand, the commercial sector is subject to a detailed set

Fig. 36.2 "Traditional Norwegian small-scale fishing vessel." (Photo; Gustav Rossness, Tromsø Museeum – The University Museeum)

of regulations. As a basic principle, all commercial fishing is banned in Norway and cannot be undertaken without a permit, issued by the government, which gives the fisher the right to fish commercially on a vessel specified in the license.

The commercial sector consists of vessels registered in the fishing vessel registry (*Merkeregisteret*) and that are equipped and suitable for commercial fishing. Three decades ago, the fleet was made up mainly of small, open, wooden vessels 11 m and below with a small inboard engine, an eco-sounder, a gurdy, and sometimes one or two automatic jigging machines. According to survey data from 2007, normally one person fished alone on these vessels.[2] The image of the small-scale fisher was that of an older guy who was not interested in investing much in his activity but rather preferred to minimize effort (Maurstad 1997) (Fig. 36.2).

In reality, however, the contrast between this archetype and small-scale fishing today is striking. As illustrated in Fig. 36.3, the boats below 11 m are now of a quite different type. Increasingly, the fleet under 11 m consists of well equipped, very efficient "harvest machines" with high-end fish finding and navigation technology, fishing gear and gear handling equipment (Johnsen 2005). With the introduction and popularity of the fiberglass boats, the small-scale fishing fleet has also become

[2] In the project "Networks or markets", which researched the contemporary fisheries employment system, a survey was carried out in 2007 among 500 boat owners and 500 crewmembers. None of the crewmembers were crew on boats under 11 m and a very low number of the boat owners with boats below 11 m were boat owners with crew. Source: Networks and markets database.

Fig. 36.3 "Modern Norwegian small-scale "harvesting machines."" (Photos; Roger Larsen, Norwegian College of Fishery Science)

more standardized. The price of a new well-equipped fishing vessel below 11 m is now so high that only a very skilled fisher is able to keep the business running. Moreover, there has been a 64 % reduction of wooden vessels under 11 m from 1996 to 2012.

Governing System

Institutional Arrangement

The current Norwegian fisheries governing system has developed incrementally over a period of more than hundred years. Thus one can find traces in existing legislation regarding small-scale fisheries that date back to the nineteenth century. For instance, in the 2008 Ocean Resources Act, mention is made of co-management being introduced in the 1890s. Certain years stand out as particularly important in shaping the governing system, one being 1928 when the Norwegian Fishers' Association (NFA) was established. This organization still plays an important role in fisheries governance. Another landmark year was 1938 with the passing of the Raw Fish Act, which provided the impetus for building a network of fisher's cooperative sales organizations across the whole country. For the cod fishery in the north, the Norwegian Raw Fish Association (NRFA) controlled all dock side sales. Sales outside these organizations were illegal and the Act gave the organization the right to determine a minimum price that buyers had to accept. Members of the NFA were automatically members of the NRFA. Also the Raw Fish Act, (now named the Fishers' Sales Organization Act), the NRFA and the sales organization system are largely intact today.

These institutions turned the table for small-scale fisheries in Norway, as they helped to empower small-scale fishers both politically and economically, and brought them out of a situation of poverty and marginalization that they found themselves in at that time (Hallenstvedt 1982). These institutions also eventually became effective instruments in a partnership between fishers and the Norwegian government due to the fact that they represented members of all local fishers' associations along the coast. These associations could nominate and vote for delegates from their own region to be part of the general assemblies and the boards of the institutions. The government got a unified and representative counterpart in the fishing industry that it could relate to and consult. Fishers on the other hand were able to influence fisheries policy making (Jentoft and Mikalsen 2014).

From a governmentality and governability perspective this arrangement has worked well till date. Fisheries in Norway were always a highly politicized issue with different interest groups and regions constantly at odds with each other on issues pertaining to strategy and distribution. But through this partnership arrangement, governing interactions were routinized. Thus, a constructive governing process was possible. The NFA, as a negotiating body, had to sell the outcomes of discussions with the government to their rank and file members. This reduced the level of conflict and helped create legitimacy and compliance. As a consequence, fisheries politics in Norway never got stuck in the trenches – or in what Rothstein (2005) calls a "social trap" or a situation where individuals, groups or organizations are unable to cooperate because of lack of trust.

Many (if not all) governance issues and challenges have changed over time, but the governing system has remained relatively intact. The system has thus proven to be adaptive while robust, capable of governing a highly dynamic sector without abandoning the basic governance principles and institutional design. Thus after UNCLOS (United Nation Convention on the Law of the Sea) in 1978 when the state assumed greater responsibility for fisheries resource management, regulatory interventions could be handled basically in the same way and within the same institutional framework as other issues had been handled before. This is also how fisheries management is currently addressed. The Marine Resources Act of 2008 gives the Ministry of Fisheries the final word on fisheries regulations, but knowing what it takes to design, enforce and implement fisheries management regulation, the Ministry rarely acts unilaterally without consulting with the NFA and other relevant organizations. Regardless of formal sovereignty, management solutions are most of the time developed in concert between the government and the fishers' organizations.

To illustrate, the government and the NFA in 1964 reached an agreement about subsidies to the fishing industry. The actual amount was to be negotiated annually between the two parties (Jentoft and Mikalsen 1987; Hernes 2000). From 1964 and throughout 1990s, Norwegian fishers adapted comfortably to a situation where the state guaranteed their income. But that was not a situation that would last forever. Subsidies to the fishing industry were gradually reduced throughout the 1990s and the formal agreement (*Hovedavtalen*), formalizing the subsidy scheme, was finally terminated on January 1, 2005. This happened largely as a consequence of EFTA and EU requirements. Fishers now had to adapt to a new economic situation with the government shifting its focus from social policy to resource management. The health of the fish stocks rather than the well-being of the fisher became the primary concern (Holm 1996). However, the form and structure of the negotiating process involving the state and the NFA remained the same as with the subsidy scheme.

The Quota-System

April 18, 1989, proved to be another watershed moment when the Ministry of Fisheries decided to close the north-east arctic cod fishery for vessels less than 28 m. This was the beginning of the new era. The following year the Norwegian government introduced a quota system also for this fleet segment and hence restricted access and put limits on catch (Hersoug 2006). This led to the establishment of a regulatory system based on individual vessel quotas (IVQs). Since then, the fishery has become increasingly restricted and nowadays the cod fishing vessels are divided into two groups, a closed group with a guaranteed IVQ and an open group that has to fish on a limited group quota. After the 1990 closure, more than 90 % of the cod has been allocated to closed group vessels. The remaining part of the TAC is reserved for the open group, comprised of registered fishers who have vessels without an IVQ.

Participation in the closed group requires that the skipper holds an annual permit for a specific vessel. These permits are given on certain conditions and allocated every year, and participation one year automatically qualifies for participation the next year. The closed group is divided into a number of length groups. In the cod fishery, each length group is allocated a certain amount of fish, originally based on the historical share of the length group. The IVQ depends on the length of the boat and will be a part of the total group quota. A part of IVQ can be in the form of "overregulation", which means that at the beginning of the year, normally the total amount of fish allocated as individual quota is a bit bigger than that of the group quota. The difference in percentage between the group quota and the sum of the IVQs is the overregulation. This "overregulation" is a flexible instrument that reduces the need to reallocate quotas within the group over the year. The overregulation is normally very high for the smallest vessel group (up to 45 %), where many due to weather conditions and availability do not catch their entire quota. The percentage declines gradually up to zero for the larger vessels, which normally fish 100 % of their quota allocation. The idea is that overregulation shall advantage the most effective vessels in the different length groups, maintain a principle of competition for a part of the quota, and importantly, make it possible for the smallest vessels to fish hard when the fish is available.

Originally, participation in the closed group was based on the catch of cod the previous years before the closure. However, as cod also represented an important catch share for those who did not reach the limit, a small share of the TAC was set aside as a group quota for all registered fishers who did not qualify for an IVQ. Since everybody in Norway can register as a fisher, as long as their income from other sources does not exceed a certain limit, there is in principle and practice an open access entry to fish on this group quota.

Traditionally, especially in northern Norway, it has been quite common to combine fishing with other livelihood activities such as small-scale farming; the split into two regulatory groups makes an opening for such combinations. Moreover, to protect the small-scale indigenous Sami fisheries in the fjords in the northern part of the country, an additional quantity is allocated today to small-scale vessels registered in certain municipalities with a big Sami population (NOU 2008, 5).

Notably, fishing in the open group does not qualify for a quota in the closed group. Nevertheless, a full-time registered fisher can, on certain conditions, buy into the closed group when boats here (with fishing permits) are up for sale. During the early 1990s, the remaining quota percentage (10 %) was allocated as a "competitive quota." Later on, as the cod biomass grew, each vessel obtained a guarantee for a limited amount of cod that could be caught regardless of whether or not the group quota was exhausted. In addition to this competition quota for cod, the open group vessels could have similar arrangements in other species fisheries, where they were allowed to fish a small percentage of the TAC. According to Fisheries Directorate statistics, the number of vessels that have participated in the open cod fishery has steadily decreased from 3,354 vessels in 2002 to 2,281 vessels in 2012. However, this decrease does not necessarily imply that all the vessels have left the fishery, as fishers in the open group may have bought IVQs and have thus been transferred to the closed group.

Structural and Institutional Changes

The number of active fishing vessels and fishers has decreased throughout the whole post WW2 period, partly due to a policy directed towards increased industrialization of the fishery. Despite this down-scaling of the small-scale fishing fleet and the fishing population, the government and the NFA agreed that a more regulated fishery was necessary, and that closure and fleet restructuring was unavoidable. However, the industrialization and restructuring was controversial within the industry and triggered a heated discussion, including within the NFA. This eventually led a big group of mainly small-scale fishers to break with the NFA in 1988 and form their own organization, The Norwegian Coastal Fishers' Association (NCFA), based on the same organizational model with local associations as in the case of NFA. A major effort of this organization has been to convince the government to abandon the IVQ system, reopen the commons and lift the restrictions on small-scale fisheries. The NCFA, after a number of years of functioning, also gained importance within the overall governing system, if not to the extent of the NFA.

With a more fragmented organizational structure and the subsidies gone, some predicted that NFA and the fishers would lose its position in the overall governance system and therefore its power (Holm 1995). However, NFA was able to shift attention to quota allocation. Again, the organization assumed the role of compromise-maker between different groups within its own ranks, and thus helped to reduce the political pressure on the government due to the controversies related to the quota system. This also helped to reinforce the rather centralized corporative governing system. As a consequence almost all quota allocation principles and mechanisms in Norway have either been constructed by, or modified through, input from the NFA (Hernes et al. 2005).

To summarize, or the Norwegian governing system is complex as it tries to accommodate the diversity that exists within the fishery where small-scale vessels have needs that are different to those of large vessels. The awareness of diversity and the complexity that follows are partly a result of the fishers' influence in the system. In addition, it is also flexible enough to allow for the dynamism that characterizes the system-to-be-governed, where conditions often vary with natural fluctuations.

Discussion

Norway figures high on the list of the world's major fisheries nations. Historically, fisheries were the most important export industry, and continue to be ranked second after oil and gas. The fishing industry is an important contributor to the overall national economy and society and therefore a major governance issue.

Small-scale fisheries were never a major contributor to the frozen, filleting industry, which instead relied on the supply from trawlers and larger coastal vessels; however, they were always and still are crucial in the fresh fish domestic market and in the dried- and salt-fish export trade. Small-scale fisheries were also the backbone of coastal communities scattered along a long coast because of the employment it provided to the local population. It also helped to maintain a decentralized settlement structure in Norway. Therefore small-scale, coastal fisheries were not a marginal issue in Norwegian politics. Their situation and fate were of both national and regional concern and had to be legislated accordingly. The perceived threat to these fisheries played a major role when Norway, through two national referendums, decided not to join the European Union. It is also the reason why fisheries are excluded from the extended economic agreement that Norway now has with the European Union. Norway is not part of the Common Fisheries Policy of the EU.

Governability Through Interaction

Most of the governance principles and systems that are basic for the ways fisheries are operating in Norway today have, as explained in this chapter, a deep history, in some instances dating back to the nineteenth century. Important legislation was enacted throughout the 1930s and 1950s. What has happened later is not so much the introduction of entirely new governance principles and laws, but needed adjustments to changing circumstances. Norway has also learned the hard way through crises in the fishery, such as in the herring fishery in the 1960s and the cod fishery in the 1990s, both of which had a major impact on the formation of the fisheries governing system.

Since the 1930s, when the Raw Fish Act introduced radical change, new legislation has often built on the already existing one, often with marginal alterations. Thus, for instance, when the Raw Fish Act in 2014 changed its name to the Fishers' Sales Organization Act, the content largely remained, although some new paragraphs were added to give these organizations a more consolidated role within the fisheries governing system, which since the 1990 onwards has focused more and more on natural resources and the management of fish stocks. Historically the Raw Fish Act and the sales organizations were key instruments in the empowerment of small-scale fishers in Norway, often to the dismay of the fish buyers and exporters who wanted to see this law ended and the sales organizations lose their monopoly power.

The institutional design of the Norwegian governing system must be understood in a historical context. It was the outcome of class struggle and power at a time when the number of small-scale fishers was much higher than today and when their voice counted because of that. Over the years, these institutions have not only acquired a central role in fisheries governance but they are also being taken seriously. They have acquired a considerable level of legitimacy and trust, and for most people in the industry it is hard to imagine how Norwegian fisheries would have functioned without them. The Raw Fish Act is popularly named as the "Fishers' Constitution," which says a lot about the status of this law in the fishing industry. Those who want it removed need to provide a convincing argument.

The Raw Fish Act has helped to lower the transaction costs of fisheries governance in Norway, as price and other issues related to resource management are negotiated collectively on a routine basis rather than individually and ad hoc. Similarly, the NFA has established long term agreements on how to divide the TAC among groups, for instance with the so-called "Trawl Ladder", which allocates a larger quota share to the coastal fleet when TAC is low as compared to when it is high. Within the coastal fleet a similar arrangement exists between size groups (Hernes et al. 2005). Once these arrangements are established, there is less to negotiate about until next time the agreements are up for evaluation. What happens in between is a technical matter. This is in itself an indication of the degree of governability within a governance system. Fishers may frequently express dissatisfaction with the price they receive for their fish and the quotas they are allocated, but they rarely question the system *per se*, which they feel committed to because they have been heavily involved it its making. They do not question the basic meta-governance principles that govern these institutions. They hardly ever go on strike as that would be mean protesting against themselves and their own organizations as well as the government who has allowed them a major role in decision-making.

That the government interacts formally and informally with the NFA and other fisheries organizations as part of policy- and decision making is perceived as a natural thing. This has been the arrangement for many decades. The need for some kind of quota system is not in dispute, although its concrete manifestation may cause problems at times. The opposition to the quota system that was introduced in 1990 has largely disappeared as fishers have become used to it and have adapted accordingly, especially because the details of the system have been hammered out by the fishers themselves through the NFA (Hernes et al. 2005). The reduction in the number of vessels and fishers in Norway has not changed the relative tranquillity of governing interactions and the governability of small-scale fisheries. Fewer fishers see the advantages of a larger share of the quota pie and a higher income for those who remain. Those who criticize the system are not those who benefit from it, but those who in the coastal communities see that jobs in small-scale fisheries get lost and a cultural heritage disappearing.

Despite of the above mentioned concern, a general lesson can be learned about the role of institutions for the overall governability of small-scale fisheries. Although institutions, such as legal measures, are essential, it matters how they are actually designed. They must allow for interactions between the governor and those who are governed and the effective sharing of power between the two parties in a way that makes both proactive and responsible. This can only work if there is mutual trust and adaptamentality.

Governability and Trust

Studies show that Scandinavians trust their governing system and its institutions more so than people in most other countries (Skirbekk and Grimen 2012). Norwegian small-scale fishers are no exception to this rule. This trust is the outcome of a combined set of policies historically that were in their favor and often a response to their own demands. The Raw Fish Act is an important but not a unique example. This particular law would hardly have seen the day of light if the fishers did not ask for it and almost unanimously voted for it in a referendum. In addition to this act, throughout the twentieth century a series of initiatives, laws and regulations were introduced in order to support the small-scale fishery, like a State Fisher's Bank (1919) which helped to provide cheap and secure finance, the Trawler Act (1937) which banned trawlers from inshore waters, The Ownership Law (1956), which required that fishing vessels can only be owned by active fishers, to name but a few.

Through these initiatives, the government built trust with the small-scale/coastal fishery that not only fostered govermentality, i.e. a positive attitude to government intervention, but also governability, the willingness to cooperate with the government. Small-scale fishers came to perceive government as the solution to problems they were facing, but did not sit still and wait for government to intervene. The government, on the other hand, found an ally among small-scale fishers. Catering for this group paid off as votes in the next national election. Thus, for decades, small-scale fishers helped to keep the Labour party, which had introduced many of these reforms, in power. The mutual trust that was built through these new institutions generated social capital that could later be converted into support and compliance. For instance, most fishers accepted stricter rules and procedures for catch and quota control and reporting. These rules were largely co-produced by the government and the NFA. The government did not have to use brute force to implement them. Studies show that after more than 20 years Norwegian fishers accept the need for regulations, and that they have been willing to comply with them (Gezelius 2002; Johnsen and Eliasen 2011).

The lesson we can learn from this experience is: when there is mutual trust among small-scale fishers and the government, governability is enhanced. Such trust must be proven through actions that make it deserved. For this there must be institutions that allow for interactive governance based on power-sharing, participation and transparency to take place.

Institutional Change

In April 2014 the government circulated a consultation paper about transferability of quota rights within the group of vessels less than 11 m with IVQs (Anon 2014). This proposal became an object for intense discussion within this fleet segment and in fisheries dependent communities. The NCFA was clearly against while the NFA the opinion was positive. The two political parties that form the current government are split on this matter. Many of the fisheries dependent communities that have witnessed the down-scaling of small-scale fisheries in recent years remain skeptical. The widespread fear is that transferability will bring a further reduction of small-scale fisheries. For fishing communities, particularly in the north of Norway, that would be an existential risk. Those who support the proposal argue that transferability is necessary to secure a better economic foundation for the small-scale fleet which, according to the official economic survey, is quite vulnerable to natural or economic fluctuations (Anon 2014). A leaner small-scale fishery sector is then the price to pay for a more profitable fishery, they say.

The variety of arguments and alliances made it difficult to get clear support for the proposal and consequently the Ministry has decided to let the case rest for now. However, the debate about transferability may help make people warm to the idea and thus prepare for the needed adaptamentality that such a reform would require. In the 1990s, there was strong opposition to the new quota system. NFA could originally only accept it as a temporary measure, but members soon learned to live with it, and today the organization has become an ardent supporter. Now it is generally perceived as a fact of life, especially among those fishers who are so young that they never experienced the other reality. The organization and its members have demonstrated adaptamentality, but it did not occur instantly. According to Johnsen et al. (2009a, b), Norwegian fisheries have become both more self-regulatory and more governable due to the development of a common governmentality among fishers. It could be argued that such governmentality was already there when the quota system was originally introduced. If not, the government would have had a much harder sell with the "sea change" that the closing of the commons involved. Still, the quota system was always controversial, especially among small-scale fishers for whom it meant the most substantial change in their long established fishing practice.

After years of complaints about how the quota system affected the small-scale fishery, the Ministry on January 1, 2014 decided to reintroduce open access for vessels less than 11 m. The maximum quota for the open group and IVQs with overregulation for the vessels in the closed group was thus abolished for these vessels. This applied regardless of what group they belonged to. Open group vessels thus engaged in so-called Olympic (competitive) fishing in the first months of 2014; both because of very good weather and expected, closure due to intense fishing. However, the number of vessels that have participated in the first few months has been pretty stable, which indicates that open access does not change participation in the short term.

The reopening of the small-scale fishery from January 2014 was an experiment that may not be continued into the future. Due to a decreasing number of processing plants, the increased landings caused quite a demanding market situation in the short and intense winter fishery when the cod is readily available. The problem could be seen in terms of both reduced price and quality of the landings. In January and February 2014, due to favorable weather, the vessels that originally belonged in the closed group under 11 m landed more than half of the original group quota, while the vessels in the open group under 11 m caught almost 50 % more than they had in the previous year for the same period. The number of vessels just increased 2 %. The most recent figures of the NRFA indicate that the increase in catch is not due to an increase in participation of vessels, but due to increased activity on each vessel. Hence, as a response, on March 24 the Ministry decided to stop the open fishery and to return to maximum quotas. Later, in May the Ministry increased the minimum guaranteed catch for the vessels in the indigenous Sami fishing districts.

Every time institutions that people have come to accept as a given are put in play, governability, and the trust that underpins it, is threatened. Trust cannot be taken for granted, it is vulnerable. The proposal of removing the limitations on small-scale fisheries to get involved in quota transactions is raising questions with regard to the government's intentions. The proposal has, however, been put on hold, while the restrictions on fishing effort of this fleet segment are lifted. While the former idea has been met with skepticism, the latter initiative has been well received among small-scale fishers. Government has by this move demonstrated adaptamentality, i.e. that it is not stuck on conventional dogmas.

Adaptamentality and Innovation

The increase of fish landings and the subsequent closure illustrate that the small-scale fleet is very efficient, especially when the weather is good and the fish is available. This is also an indication that this vessel group is operated by skilled fishers with the needed adaptamentality. The small-scale vessel group cannot therefore be regarded as backward, as is often the perception. Instead, they have sufficient adaptive capacity to cope with natural variations in the fishery. Today, with increasing costs partly caused by the quota that small-scale fishers must shoulder, they hardly have another choice. Small-scale fisheries in Norway are sophisticated and innovative with the most recent catch and information technology installed. New technology, new ways of operating and organizing has been met with a willingness to experiment, also within the small-scale fisheries sector. The move in the markets, away from frozen to fresh products, is favoring small-scale fisheries and may result in a new era within small-scale fisheries that again will rely on adaptamentality. That mentality may be encouraged by the shift in the way fish is talked about in Norway these days. Fish as a concept is being replaced by the more trendy "sea-food" (*sjømat* in Norwegian) term. Not only does it represent a positive view of fish and small-scale fishing in the minds of the consumer, it also provides new meaning and identity in the mind of the small-scale fishers for whom fresh, newly caught fish is what they are particularly good at. Small-scale fishers would now be inclined to see themselves as a crucial element in the modernization and innovation of a fish distribution chain that extends beyond the dock-side.

This new mind-set involves adapt a mentality. It is nourished by the broader view of one's own role within the larger fisheries system, which makes small-scale fishing into a more meaningful and hence more attractive occupation. The challenge for the small-scale fishery in Norway is its seasonal nature, with the overall majority of the landings in the first half of the year. For many years this fleet has had an open fishery in the fall, but with limited ability to catch due to low fish availability and bad weather. Due to the decrease in the number of processing plants, landings are also concentrated in fewer and fewer harbors. The 2014 opening of the small-scale fishing commons was an important experiment by the government that proved that the governability challenge for small-scale fisheries and communities is not related to lack of skills and ability to fish. Rather it is about how to deal with seasonal variation and how to increase the value of the fish through more efficient marketing. The problem is that the market does not only demand a high quality product but also needs a steady supply throughout the entire year. This is a major governability problem that must be addressed for small-scale fisheries to thrive.

This problem cannot be solved through a fisheries policy focused only on resource management and fleet profitability. It is also a question about food policy. Norway produces fish mainly for export, while the domestic market for fish is not well developed. Norwegians consume 22 kg of fish per person pr. year, which is less than 50 % of the annual meat consumption per person. The small-scale fleet has the potential to fill this gap. The adaptamentality in the small-scale fleet indicates that given the right conditions, this fleet should be able to meet an increased domestic demand for fish. This is now perhaps the most important governability challenge and opportunity for small-scale fisheries in Norway. By increasing domestic demand, the small-scale fleet can demonstrate to society at large that they represent great social value, which they have to do in order to convince government that they and their local communities are still worth conserving.

Conclusion

In Norway, small-scale fishers have not only proven their adaptability, but also their readiness for change, including change initiated by the government. The latter, which in this chapter is termed adaptamentality, is not an inherent feature of small-scale fisheries but a constructed quality nurtured by trust that has taken many decades to build. Small-scale fishers have even been able to ensure that government legally secures their interests, as in the case of the Raw Fish Act and a series of other progressive legislative measures. This can be measured in high governability and willingness to comply, cooperate and adapt all positives which have seen Norway rank high amongst the countries that conform to the FAO Code of Conduct for Responsible Fisheries (Pitcher et al. 2009).

Although the governability, and the adaptamentality and govermentality that are fostering it, have been a characteristic of small-scale fisheries in Norway so far, it will not remain stable with institutional change. Institutions build and require trust, and trust builds adaptamentality and governmentality. The Fishers' Sales Organization Act, the cooperative sales organizations, and the rules regarding ownership of fishing vessels, which are all fundamental for the governing system in Norway as we know it, have so far been able to withstand pressure to abolish the Act, particularly from the processing and export interests. But it remains to be seen how long it lasts now that the current government is entertaining the idea that they perhaps need major reform. Small-scale fisheries can still continue to call the Raw Fish Act as their "constitution", even after the 2014 legal reform, but for how long?

These institutions probably depend on a critical mass of small-scale fishers. The close connection and common destiny of small-scale fishers and coastal communities are also important. The fewer they are and the more dis-embedded the small-scale fishery becomes from the local community the more difficult it would be to uphold them (Grytås 2013; Sønvisen 2013). The NFA has said that they do not any longer think that the fishing industry has a responsibility to maintain a decentralized settlement on the coast nor ensure fisher community wellbeing. It may not have thought through what the long term consequences of this position would be for the institutions that it supports. Making fishing rights into a commodity that can be bought and sold is likely to further exacerbate this development. Norway has not yet gone as far as Iceland and Denmark (see chapters by Høst (Chap. 17) in this volume) in introducing this system. Instead, the quota system has rules to counteract concentration of fishing rights, but these rules are now under pressure, as mentioned in this chapter.

Initially, we talked analytically about the link from adaptamentality to governmentality and governability. In reality it is hard to say what comes first and perhaps not important to do so either. The fact they are there and that they nurture each other is what counts, and that each of them must be "worked on" simultaneously through a governance process that is interactive. While adaptamentality and governmentality are basically about mind-set and attitude, governability is about actual capacity for, and quality of governance, such as having the resources, including the institutional power to implement decisions in a way that is effective, transparent and democratic (Kooiman 2003, 2008). Adaptamentality helps to elevate governmentality, which again is conducive to governability. In other words, we are talking about governance as a virtuous circle. The lesson from Norway is that institutions can do a lot to facilitate the adaptamentality and governmentality that governability hinges on, but only in so far as they are capable of delivering on their promise to make a positive difference for small-scale fishers. But with the recent developments in

Norwegian fisheries, the future might be a lot different, as a virtuous circle can easily develop into a vicious one.

References

Anon. (2014). *Strukturkvoteordning for kystflåten under 11 meter heimelslengd*. Høyringsdokument. (Consultation paper for a structural quota system for vessels under 11 m). Ministry of Trade, Industries and Fisheries.

Bavinck, M., Chuenpagdee, R., Jentoft, S., & Kooiman, J. (Eds.). (2013). *Governability of fisheries and aquaculture: Theory and applications*. Dordrecht: Springer Verlag.

Foucault, M. (1978a). Lecture 1 February. In M. Sennellart, F. Ewald, & A. Fontana (Eds.), *Security, territory, population lectures at the College the France 1977–1978* (pp. 87–114). Basingstoke: Palgrave Macmillan.

Foucault, M. (1978b). Lecture 11 January. In M. Sennellart, F. Ewald, & A. Fontana (Eds.), *Security, territory, population lectures at the College the France 1977–1978* (pp. 1–27). Basingstoke: Palgrave Macmillan.

Gezelius, S. (2002). Do norms count? State regulation and compliance in a Norwegian fishing community. *Acta Sociologica, 45*(4), 305–314.

Grytås, G. (2013). Råfisklova – tid for tale ved grava? In S. Jentoft, J.-I. Nergård, & K. A. Røvik (Eds.), *Hvor går Nord-Norge? Politiske tidslinjer* (pp. 317–330). Stamsund: Orkana Akademisk.

Hallenstvedt, A. (1982). *Med lov og organisasjon*. Tromsø/Oslo/Bergen: Universitetsforlaget.

Hernes, H.-K. (2000). *Forhandlingsmakt eller argumentasjonsbyrde? En analyse av Hovedavtalen for fiskerinæringen og avviklingen av fiskeristøtten*. Doctoral thesis, University of Tromsø, Tromsø. Retrieved from http://hdl.handle.net/10037/286

Hernes, H.-K., Jentoft, S., & Mikalsen, K. H. (2005). Fisheries governance, social justice and participatory decision-making. In T. Gray (Ed.), *Participation in fisheries governance* (pp. 103–118). Dordrecht: Kluwer.

Hersoug, B. (2006). *Closing the commons. Norwegian fisheries from open access to private property*. Delft: Eburon.

Holm, P. (1995). The dynamics of institutionalization: Transformation processes in Norwegian fisheries. *Administrative Science Quarterly, 40*(3), 398–422.

Holm, P. (1996). Kan torsken temmes. In E. O. Eriksen (Ed.), *Det nye Nord-Norge: Avhengighet og modernisering i Nord* (pp. 109–142). Bergen: Fagbokforlaget.

Jentoft, S., & Chuenpagdee, R. (2009). Fisheries and coastal governance as a wicked problem. *Marine Policy, 33*, 553–560.

Jentoft, S., & Mikalsen, K. H. (1987). Government subsidies in Norwegian fisheries: Regional development or political favoritism? *Marine Policy, 11*(3), 217–228.

Jentoft, S., & Mikalsen, K. H. (2014). Do national resources have to be centrally managed? Vested interests and institutional reform in Norwegian fisheries governance. *Maritime Studies, 13*, 5. Retrieved from http://www.maritimestudiesjournal.com/content/13/1/5

Jentoft, S., Chuenpagdee, R., Bundy, A., & Mahon, R. (2010). Pyramids and roses: Alternative images for the governance of fisheries systems. *Marine Policy, 34*, 1315–1321.

Johnsen, J. P. (2005). The evolution of the "harvest machinery": Why capture capacity has continued to expand in Norwegian fisheries. *Marine Policy, 29*(6), 481–493.

Johnsen, J. P. (2014). Is fisheries governance possible? *Fish and Fisheries, 15*(3), 428–444. doi:10.1111/faf.12024.

Johnsen, J. P., & Eliasen, S. (2011). Solving complex fisheries management problems: What the EU can learn from the Nordic experiences of reduction of discards. *Marine Policy, 35*(2), 130–139. doi:10.1016/j.marpol.2010.08.011.

Johnsen, J. P., Holm, P., Sinclair, P. S., & Bavington, D. (2009a). The cyborgization of the fisheries. On attempts to make fisheries management possible. *Maritime Studies, 7*(2), 9–34.
Johnsen, J. P., Murray, G., & Neis, B. (2009b). North Atlantic fisheries in change – From organic associations to cybernetic organizations. *Maritime Studies, 9*(2), 55–82.
Kooiman, J. (2003). *Governing as governance*. London: Sage Publications.
Kooiman, J. (2008). Governability: A conceptual exploration. *Journal of Comparative Policy Analysis, 10*(2), 171–190.
Kooiman, J., & Chuenpagdee, R. (2005). Governance and governability. In J. Kooiman, M. Bavinck, S. Jentoft, & R. Pullin (Eds.), *Fish for life. Interactive governance for fisheries* (pp. 325–349). Amsterdam: Amsterdam University Press.
Kooiman, J., van Vliet, M., & Jentoft, S. (Eds.). (1999). *Creative governance: Opportunities for fisheries in Europe*. Aldershot: Ashgate.
Maurstad, A. (1997). *Sjarkfiske og ressursforvaltning*. Dr.Scient. Doktoravhandling, Universitetet i Tromsø, Tromsø.
NOU. (2008:5). *Retten til fiske i havet utenfor Finnmark – Utredning fra et utvalg oppnevnt ved kongelig resolusjon 30*. juni 2006. Avgitt til Fiskeri- og kystdepartementet 18. februar 2008. Departementenes servicesenter Informasjonsforvaltning.
Pitcher, T., Kalikoski, D., Pramod, G., & Short, K. (2009). Not honouring the code. *Nature, 457*, 658–659.
Rothstein, B. (2005). *Social traps and the problem of trust*. Cambridge: Cambridge University Press.
Skirbekk, H., & Grimen, H. (2012). *Tillit i Norge*. Oslo: Res Publica.
Solstrand, M. V., & Gressnes, T. (2014). Marine angling tourist behavior, non-compliance, and implications for natural resource management. *Tourism Management, 45*(0), 59–70. doi:http://dx.doi.org/10.1016/j.tourman.2014.03.014.
Song, A. M., & Chuenpagdee, R. (2014). Exploring stakeholders' images of coastal fisheries: A case study from South Korea. *Ocean and Coastal Management, 100*, 10–19.
Song, A. M., Chuenpagdee, R., & Jentoft, S. (2013). Values, images, and principles: What they represent and how they may improve fisheries governance. *Marine Policy, 40*(0), 167–175. doi:http://dx.doi.org/10.1016/j.marpol.2013.01.018.
Sønvisen, S. A. (2013). *Coastal communities and employment systems: Networks and communities in change*. PhD, UiT The Arctic University of Norway, Tromsø.

Part IX
Governability Challenges – Urging Change

Chapter 37
Enhancing the Governability of Small-Scale Fisheries Through Interactive Governance

Svein Jentoft and Ratana Chuenpagdee

Abstract This final chapter synthesizes the arguments that chapter authors make with regard to the governance and governability of small-scale fisheries. They point to a general conclusion that in order to enhance governability for the benefit of small-scale fisheries, given the related conditions and characteristics, governance designs and interactions must be sensitive to the needs of small-scale fishing people and responsive to their situations. We do not claim that one governing mode, be it hierarchical, co- or self-governance, is inherently better than others. Rather, the choice of governing modes must take into account the particularities of the system-to-be-governed as well as the governing system. Small-scale fisheries globally will benefit from more constructive interaction, collective action, empowerment, and innovation, but they are simply too diverse for preconceived and generalized governance blue prints. Governability assessments as carried out by chapter authors reveal that despite their structural commonalities, governing modes typically have unique features. Each governing mode must be seen as a governor's response to the existing and emerging challenges and opportunities of a specific small-scale fishery. The transition of governing modes observed in many cases illustrates how governance actors try to cope with system dynamics. Often, the combination of different modes into one coherent but hybrid approach is warranted. This chapter summarizes how these processes occur in real world situations accounted for by the chapter authors in this book.

Keywords Small-scale fisheries • Interactive governance • Enhancing governability • Meta-governance • Governance of scale • Politics and power

---

S. Jentoft (✉)
Norwegian College of Fishery Science, UIT – The Arctic University of Norway,
9037 Tromsø, Norway
e-mail: Svein.Jentoft@uit.no

R. Chuenpagdee
Department of Geography, Memorial University, St. John's, Canada
e-mail: ratanac@mun.ca

S. Jentoft, R. Chuenpagdee (eds.), *Interactive Governance for Small-Scale Fisheries*, MARE Publication Series 13, DOI 10.1007/978-3-319-17034-3_37

Introduction

This book is a collection of case studies with a common focus and an overarching analytical framework – interactive governance theory applied to small-scale fisheries as they exist and operate in different locations throughout the world. The goal of governance is to enable small-scale fisheries to play a more crucial role in providing food security, economic opportunity, community well-being, and sustainable marine ecosystems for society at large. Here small-scale fisheries have a real contribution to make. But small-scale fisheries are also an industry and a way of life. Small-scale fishing people have to deal with many internal problems that need governance solutions in order to address big challenges such as poverty, social and political marginalization, lack of organization, and social inequities to name a few.

A key concept herein is governability, defined and explained in Chap. 2 as the overall quality of governance, with a functional as well as normative dimension. It is recognized that small-scale fisheries are too diverse, complex, dynamic, and multi-scalar for simple governance solutions and rules. Governance challenges are real and urgent, but require governance solutions that are suitable and responsive to the problems and opportunities as they occur in concrete settings. The basic assumption for interactive governance, as for social science in general, is that context always makes a difference and must be taken into account by governance, and that interaction between involved and affected parties in one mode or another can be part of the governability problem but always essential to its solution. All chapters of this volume therefore have their focus on governing interactions and how they are shaped institutionally and organizationally, be they in the hierarchical, co- or self-governing modes or in mixes of them.

The substance of this conclusion chapter is based on arguments made by the authors in response to our invitation to address issues of governability. The chapter summarizes how they argue about small-scale fisheries governance as it occurs in real world situations that they have studied. As a contribution to the understanding of the shape and functioning of the governing modes (Kooiman 2003; Kooiman et al. 2005 for foundational work), it also aims to draw general conclusions, based on the lessons that chapter authors offer about their case studies, with regard to the effect of governing modes and its implications on the sustainability of small-scale fisheries. In what follows, we summarize these arguments into key learnings, reflecting also the various aspects of governing modes highlighted in interactive governance theory. Some of these arguments are included as direct quotes from the statements that the authors sent us for the write-up of this final conclusion chapter. But first a few words about the case study method.

Case Study Method

The case study method is not primarily aimed at providing scientific proof as much as in-depth understanding of the nuances and complexities of social systems, institutions and practices, in this case small-scale fisheries as a system-to-be-governed typically through a hierarchical mode, but also with a potential, and often untapped, capacity for self- or co-governance. The case studies are not only meant as narrative descriptions of real life experiences but also to generate concrete and general lessons with practical implications. They invite those who inhabit both the governing system and the system-to-be-governed to pay attention to context and be sensitive to details. There are certainly general lessons that can be drawn from comparing case studies (Ragin 1987), but general knowledge about governance and governability can also be built from an individual case. As Flyvbjerg correctly points out:

> One can often generalize on the basis of a single case, and the case study may be central to scientific development via generalization as supplement or alternative to other methods. But formal generalization is overvalued as a source of scientific development, whereas "the force of example" is underestimated. (Flyvbjerg 2006, 228)

Case studies are means to argue a case, which is what social science is about, and what the authors of this book do based on the "force of example." They all use their case studies to reason and make claims about how to enhance the governability of small-scale fisheries so that they can be sustainable while addressing basic concerns such as food security, social justice, environmental health, and poverty alleviation, as discussed in Chap. 2.

It is also possible to learn about small-scale fisheries governance and governability from the study of a single organization or community. Case studies do after all report cases of something and are therefore generalizable to that "something", even if the particular case is out of the ordinary. Even – or especially – outliers demonstrate that there may be alternatives to the current social order under circumstances that may or may not be unique. To enhance governability, searching for innovative solutions to the governability problem would always be important. Outliers may then help to stimulate the imagination.

Questions About Modes

Interactive governance theory distinguishes between three governing modes – hierarchical governance, co-governance and self-governance. The chapter authors have all explored their functioning in small-scale fisheries contexts and have things to say about their relative merits from a governability perspective. The governing modes belong to what interactive governance labels as the "governing system", i.e. the institutions and instruments for governance. What is particularly interesting about them, however, which is what the chapters of this book focus on, is how they structure interaction with the system-to-be-governed. Again, it makes sense to apply the system metaphor rather than to solely talk about the governing authority or government, due to the diversity, complexity, dynamics and scale that also characterize governing roles and functions of multiple institutions. The governing system has many actors, often in the form of complex organizations somehow linked together and operating at various scales ranging from closely knitted federative structures to loosely coupled networks. Getting the full overview of how the governing system is structured and how it works, from a governability perspective, can therefore be as demanding as capturing the make-up of the system-to-be-governed. Understanding both systems and their interactions and how they each and together create governability, requires comprehensive empirical assessment, the methodology of which is suggested and outlined in Chap. 2 and in Chuenpagdee and Jentoft (2013).

It is important to understand that the three modes of interactive governance theory are analytical constructs, or "ideal types" as Weber talked about (see Chap. 2, this volume), meant to be used as heuristics in empirical research. They are, in other words, not actual governance schemes but models to be used for comparison and contrast. Out of such an endeavor emerge questions for research: How close are empirical modes to theoretical modes and what explains the disparity between the two? What does the difference make for the governability of small-scale fisheries in concrete cases? When there are multiple case studies in which such questions can be raised, comparisons yields similar questions; if a particular mode works well in one setting, why not in another? Is the difference of performance related to the particular institutional design of the governing system where the devil often hides in the properties, or is it related to those of the system-to-be-governed? Or is the governability problem to be found in the structural fit between the two systems and the interactions they facilitate. By applying the governability assessment framework to their case studies, the chapter authors attempt to answer these questions and raise new ones. In so doing, they contribute to the groundtruthing of interactive governance theory as well as to advancing knowledge and understanding of small-scale fisheries governance worldwide.

Understanding Governing Modes

In a governability assessment, governing modes can be looked upon as both a dependent and an independent variable. They obviously have a cause and a history; they may have evolved gradually, reflecting policy processes within or without the fisheries industry or in society more generally. But they may also have resulted from a deliberate collective choice made at a particular time by a more or less representative institution in response to a concrete problem, like a resource crisis. Whatever the background to their formation and existence, they have consequences that should also be researched. If these consequences are negative from a governability perspective, i.e. dysfunctional or in violation of good principles, the question emerges as to whether or not and how governance reform could be instigated.

This is the issue addressed in Chap. 12 by Iris Monnereau and Patrick McConney, who compare the governability outcomes of governing modes in the lobster fishery of Nicaragua, Jamaica and Belize. The diversity of this particular fishery, despite being based on the same species for the same export markets, is largely a product of the different governing modes employed in the three countries. In the authors' estimation, the lobster fishery in Belize, which is designed according to co-governance principles, is outperforming the lobster fishery of the other countries. Nicaragua is relying on self-governance. The industry is largely left to itself with government not interfering. In Jamaica, a largely unsuccessful co-governing system has developed from a hierarchical mode.

All three governing modes facilitate interactions between the governing system on the one hand and the system-to-be-governed on the other. The total "amount" of governing interactions does not need to be different between the three governing modes, only their distribution within the governance system as a whole. From a governability perspective, it is these interactions more than their organizational and institutional designs of the modes that should attract most attention. The command-and-control approach of the hierarchical mode puts the system-to-be-governed in a more reactive role, whereas the self-governing mode assumes that system is proactive. Co-governing, as a middle-of-the-road approach, implies concerted action between the governing system and the system-to-be-governed, where governing interventions and initiatives are subject to negotiation and collaboration. What also distinguishes hierarchical and co-governance on the one hand from self-governance on the other is that interaction takes place within a formal mechanism, where the division of labor, roles, rules, and responsibilities are clarified, like in a legal charter of sorts. This is also what many of the chapter authors hold up as a better approach for enhancing governability.

In the case of Malawi, for instance, Mafaniso Hara and colleagues (Chap. 8) attribute the governability problem to the "ingrained attitude for open access fishing and limitless output that have been promoted through the historical developmentalist vision." As they see it, overfishing requires access limitation, output-based regulation, "clear fishing tenure" devolved to beach committees, and co-governance. They also add that "unless there is ability to institute a governing system… based on clear fishing tenure and stewardship practices, co-governance on both Lake Malombe and Lake Malawi will not work for functional governance." Questions about the effectiveness of co-governance are also raised by Mauricio Castrejón and Omar Defeo (Chap. 31). They note, for instance, that there is "still a poor understanding about how this governance mode responds to different types of crises, and how these responses are shaped by past experiences and by the particular features of the governing system and the system-to-be-governed." This is clearly an important researchable governability issue, which their chapter explores for a number of Latin-American countries. They suggest:

> The governability of a fishery is not dependent on the co-governance mode established (e.g., consultative or cooperative), but mainly on the social attributes of fishers' organizations, the quality of the interactions between governance and other actors, and the institutional adaptability to external drivers of change. Institutions with strong social cohesion, organization and leadership, and willingness to change and work in a collective and collaborative way, displayed a high institutional capacity for adaptation and innovation.

The policy implication, according to these authors, is "to build solid institutions that promote collaborative and coordinated work between governance and the small-scale fishing sector."

On the other hand, a possible cause of the governability shortcomings of self-governance, as pointed out by Jim Prescott and colleagues (Chap. 4), is the tendency of stakeholders to prioritize short-term over long-term concerns. This is not totally unreasonable, given that some concerns, such as livelihoods and food security, are often more urgent than sustainability concerns and can therefore not be postponed to a later day. Prescott and colleagues argue:

> For sustainability to rank higher among local priorities new and equally or more attractive and tangible outcomes must replace existing ones. Such outcomes include for example more secure rights or potentially more consistently profitable catches by way of new market arrangements rather than increased capture capacity.

They also hold that involvement in the governing process makes local fishing people more confident about change.

Mixing of Modes

Several chapter authors observe the need for a mix of different governance modes, similar to what Kooiman (2003, 90) has earlier suggested, namely that "governing modes do not exist empirically in their pure version; rather they are mixing elements from all three." There are good reasons why this is so. First of all, there are inherent strengths and weaknesses of all modes. Second, mode designs are reflective of institutional histories in particular countries and fisheries and even if they were drawn on a piece of paper in the first instance, they have evolved over time both as a consequence of the diversity, complexity, and dynamics of the system-to-be-governed but also due to cross-sector and cross-scale linkages. For this reason it is likely that governing system modes would need to be as adaptive, although perhaps not as unstable, as the system-to-be-governed. But that may not happen naturally. Rather, it would require concrete action by those who hold the power to enable institutional reform. It is also possible to imagine a lag that widens over time, which creates mismatch between the governing system and the system-to-be-governed, and hence would need to be corrected by gradual steps rather than total shifts of modes.

Elena Finkbeiner and colleagues, whose study (Chap. 11) is situated in both Mexico (Baja California) and the US (Hawaiian Islands), think of co-governance as a hybrid regime in itself. They point out that co-governance is relevant "across countries and regions in transitioning towards sustainable small-scale fisheries", and that one solution to the governability problem "is to transition towards hybridized regimes" such as co-governance. This they believe "can help to create more effective and fair policies, and empower fisheries to become more self-governing". In both areas "a high degree of autonomy at the local level … and a strong foundation for shared or collective governance power" would be needed.

Hybridization is also a concept used by Philippa Cohen and colleagues (Chap. 3) for the governability assessment of the Solomon Islands small-scale fisheries. First of all, they hold that "decentralizing governance to the community level permits responsiveness and specificity to local dynamics, not possible through hierarchical governance", at least not that of the state. Decentralization also "allows for proper recognition of the (often legally backed) customary rights of local resource owners." Although they find that hybrid institutions combining community-based co-governance interacting with self- and hierarchical governing to be most suitable in their context, they hold that "the best 'mix' of governance modes and responses is a moving and elusive target." Therefore they advocate a more playful approach to this governability problem, where different mixed modes are tested "amidst the diversity and complexity of Pacific small-scale fisheries."

This recommendation has widespread relevance (Jentoft 2007). The diversity, complexity, and dynamics of small-scale fisheries call for contextualized approaches where pre-described solutions must be confronted with reality in a way that requires governance innovation rather than imitation. Shelly-Ann Cox and Patrick McConney in their study of the sea urchin fishery of Barbados and St. Lucia (Chap. 30) stress the need for co-governance modes to be adaptive to changing conditions. Especially when the governance capacity is low, as is often the case in developing countries, they argue that stakeholders "need to be encouraged to test, learn and adapt if success is to be achieved." Governability is therefore not a static but a dynamic situation, where governance to be relevant must be creative and innovative. For that, "enabling policies" have a role in enhancing governability, according to the authors, but the "conditions of greatest concern for governability are those which relate to governing interactions initiated by the governing system." The need to hybridize must be seen as a response to the need for contextualization.

Consequently, text-book governing modes cannot just be adopted without being adapted. In real-life situations, where dexterity is needed in institutional design, governance reform involves entrepreneurship and context specificity. Without it, as Jeppe Høst highlights in Chap. 17 about individual transferable quotas (ITQ) in Denmark, one can end up with socially harmful solutions. Here, the market mechanism as a "quick fix" has led to "irreversible transformation of the Danish demersal fisheries, mainly at the cost of small-scale fisheries." Høst argues that interactive governance and market-based management as a form of self-governance are in conflict with one another. The policy process in his view must start with elaborate stakeholder consultations, as "the initial policy creation is crucial."

The case study by Derek Johnson and Sölmundur Pálsson (Chap. 15) on the Lake Winnipeg (Canada) quota system provides an interesting contrast. Here, the governing mode is indeed a hybridized one. They underscore the need to pay attention to context in the design of the governance system, and that "even polarizing management tools such as ITQs and marketing boards can be important to achieve social, economic, and ecological goals in small-scale fisheries if they are sensitively implemented." Johnson and Pálsson argue that ITQs under certain very particular conditions can increase the governability of small-scale fisheries in ways that enhance wellbeing. In their case, the crucial condition is "the embedding of the ITQ system in a broader set of rules that block any significant concentration of quota." They also stress the value of collective arrangements for economic governance, where a "state monopoly marketing board has been effective in stabilizing economic returns for fishers." The Lake Winnipeg case "shows how an unusual mix of governance institutions and a supportive state can sustain small-scale community-based fisheries while simultaneously increasing economic value." For these two authors, governability cannot be secured once and for all but "needs to be actively maintained."

Challenges in Co-governance

Co-governance is worth further deliberation since many chapter authors in their case studies converge toward this mode, which they demonstrate as feasible and operative or in the process of becoming so. The move in the opposite direction, e.g. from co-governance to hierarchical or self-governance, is not documented in any of the case studies. This does not mean that there is no room for these other governing modes. Quite the contrary, some chapters illustrate that government has a constructive role to play or that small-scale fishing people need a degree of autonomy and freedom to realize their full potential. For instance, Maria José Baragán Paladines (Chap. 9), speaking about mainland Ecuador and the Galapagos Islands, thinks that hierarchical governance has merits that should not be underestimated: "Taking back some of the state's role in fisheries governance … has meant an improvement in the institutionalization of fisheries issues within the central government agenda." In her view, a pragmatic governance approach is appropriate, as no mode is perfect. This is also a conclusion reached by Estelle Jones and colleagues (Chap. 22) who compare the success and failure of a number of marine protected areas (MPAs) in Thailand. They find that "elements of hierarchical governance approaches can contribute to solving small-scale fisheries problems when they are well-designed and the formal institutions are respected." But from a governability perspective it is not so much the governing mode per se as "trust and cooperation in governance interaction, which enable conflict resolution and informal sanction" that do the trick. Even so, they suggest that efforts to realize effective coastal governance at large may occur if support is given to the "development of a patchwork of smaller community-based self-governance or co-governance modes."

Hierarchical governance and central control that comes with it, is no guarantee for policy coherence. Government is in itself often a complex, multilevel, and fragmented system. Putting together a holistic policy agenda, as promoted by the Voluntary Guidelines for Securing Sustainable Small-Scale Fisheries endorsed by FAO (Food and Agriculture Organization of the United Nations), hereafter called SSF Guidelines, is itself a governability challenge, but this time within the governing system. When the SSF Guidelines address the responsibilities of the state, it necessarily involves a broad range of state institutions, and not only those that specialize on fisheries. Since the guidelines are elaborated for small-scale fisheries "in the context of food security and poverty eradication", departments of health, education, justice, and regional development etc., would have to be drawn in. Within a holistic small-scale fisheries governance approach, the more specialized the institutions, the greater the need for coordination.

Coordination as a governability challenge is demonstrated by Michael Hurley and Camille Manel in Chap. 33. They stress the need for "institutional syncretism" to enable better coordination across scales. For them, co-governance holds promise, as it "opens a whole contextual scale to be considered." Also they feel that "coordination as policy coherence has to penetrate beyond international-regional-state scales to incorporate coastal communities and the informal 'infrastructure.'" It must also "reflect the plurality of fishers' perspectives, cultural context and desires outcomes." A problem described in their chapter is "the dominance of powerful international actors such as environmental organizations in dictating international narratives in fisheries which is then compounded by tight coupling between international-state scales." International actors tend to "influence the content of state 'superstructure' to fit international treaties, dominant paradigms and concepts of good governance." Ironically this has led to "a highly fragmented policy, information and governance structure in the form of projects that act as temporary governance boxes, with different philosophies, principles and processes of governance that impede any coherence and pose a significant coordination problem to achieve collective synergy necessary to solve complex multi-scaled problems of fisheries."

Governance of Scale

Interactive governance pays particular attention to scale issues, recognizing that small-scale fisheries systems are mostly not confined by boundaries. Small-scale fisheries are rarely a distinct sector but one that is embedded within a larger social and ecological system – as a "system within a system" – intimately connected with economic, social and cultural life in local communities and beyond, even international markets and politics. This calls for the kind of system perspective that interactive governance employs. In order to understand and meet the challenge of governability in small-scale fisheries, it is essential to understand how small-scale fisheries "add up" and connect, both internally and externally. This is basically what a holistic perspective involves; there is not only a need for a broad view that encompasses how the ecosystem works, who the social actors are and what they do, but also how they are linked and dependent on each other.

As a system in its own right, small-scale fisheries are inherently difficult to delineate from their ecological, social and institutional environments, as their boundaries are permeable. This is an issue of scale and "institutional fit", as illustrated by Joeri Scholtens (Chap. 27) in the case of Sri Lanka, and Olivier Randin (Chap. 26) in the case of Columbia. In both cases, small-scale fishers are victimized by forces way beyond their control; occurrences related to international relations and conflicts. For Sri Lankan small-scale fishers, mostly of Tamil ethnicity, the problem arises when foreign (Indian) trawlers transgress national boundaries and occupy their traditional inshore fishing space. Initiatives to establish contact and cooperation between fishers on both sides of the conflict have not received the government support they would need in order to work. Scholtens summarizes the governability problem as follows:

> Typically, foreign fleets are not only technologically superior, but also enjoy the backing from State power. Preventing the dispossession of small-scale fisheries is therefore not just a matter of adequate local fisheries governance, but rather about the ability to mobilize support to keep outsiders out. For addressing problems that originate at scale levels beyond the local, small-scale fishers need to be well organized and politicize their plight to secure State support.

In a similar vein, Olivier Randin argues:

> Small-scale fisheries are not a fixed scale in terms of governance. They face issues beyond their boundaries. They may be small by their individual size but they are strongly interconnected to higher levels of governance. This interconnectedness can make them vulnerable to external influences and shocks, which in turn affect their governability.

Randin therefore infers that small-scale fishers need to be fully recognized as stakeholders in the governance of national and international fisheries in order for their interests to be adequately represented and their problems that are brought upon them taken into account.

The scale issue and institutional misfit can also exist at the local level, as analyzed by Silvia Salas and colleagues in Chap. 25. Their case study is situated on the Yucatan peninsula in Mexico, in two neighboring communities, and reveals how scale and boundary issues can add significant governability challenges to small-scale fisheries governance. Salas et al. explain the situation as follows:

> Under their own initiative, San Felipe small-scale fishers were able to establish a small MPA and manage it with support from the local government and community members. Because the MPA restricted access of small-scale fishers from the nearby community, Dzilam de Bravo, and because the MPA boundary overlaps with the existing State Reserve, their conservation effort was deemed inappropriate – in fact, illegal – by the State government. The two communities did not have a history of collaboration, and there were no legal precedents on protected areas overlap, it was therefore difficult to come to agreement about what to do.

The authors conclude that the governability challenge is exacerbated by their different experiences in resource governance, uneven capacity to organize, and different images about their surroundings and the MPAs. "Mechanisms to foster their interaction are required, along with policies and programs to help enhance capacity and raise awareness, leading to sustainable small-scale fisheries."

Knowledge in Governance

Governance is neither an exact science nor a technical instrument following or implementing universal rules even if there are general standards for good governance. There are concerns and ideals that can be claimed to be universal, such as those pertaining to human rights, but the specifics of governance would need to be pragmatic, fitting to and responsive of context. As Paul Onyango (Chap. 10) argues "there is no generalized prescription that can be applied across the board to govern small-scale fisheries. Each fishery presents itself in a unique manner that requires an understanding of the context and generating actions based on each case."

This is an idea shared by several authors, among them Adam Soliman in his study of small-scale fisheries of Alaska (Chap. 16), where a market-based quota has been introduced. Although promising in his view, what has worked in one community does not necessarily work in another: There is "'no-size-fits-all' design that will yield a highly governable system across varied communities. In the United States and elsewhere, successful small-scale fisheries governance within a market-based system (in this case, Community Quota Entity) will depend on programs that provide for both community involvement and flexibility within to meet the specific needs of each community". That the same conclusion is drawn from very different contexts, namely, in Lake Victoria and Alaska, is remarkable, speaking thus to the methodological reflections above. The question for governability is how to account for the local context, including the use of local knowledge to inform decision-making.

In socio-political governance in general, and small-scale fisheries governance in particular, practical wisdom is essential, and must draw on in-depth, experience based 'phronetic' knowledge (Flyvbjerg 2003; Jentoft 2006). This means traditional ecological knowledge and social and normative orders also have relevance in modern settings (Bavinck and Jyotishi 2015; Berkes 2015). Disregarding them is usually a recipe for governance failure (McGoodwin 1991; Gray 2005; Kooiman et al. 2005). Therefore, governance cannot occur in a social and cultural vacuum. No one is better equipped to understand local conditions than those who live with them. This also applies to the multiple legal systems (legal pluralism) that small-scale fisheries are subject to, like statutory and customary law that often are invoked for the same situation, which may or may not reduce or enhance governability as the case may be (Jentoft and Bavinck 2014).

Good governance in small-scale fisheries must be interactive in a way that gives small-scale fishing people,[1] whether employed in harvest, post-harvest functions or otherwise dependent on them, a voice and a clout. This is a conclusion that can be drawn from philosophical thought where stakeholder involvement, participatory democracy and respect for human rights are themselves valued. Small-scale fishing people must be involved from the very beginning of the planning process, or from "step zero", and not just in the final analysis through some kind of post-planning hearing arrangement (Chuenpagdee and Jentoft 2007). This is also what Alice Joan Ferrer (Chap. 20) argues based on a case study of the Taklong Island National Marine Reserve in her native Philippines. She argues that there are pitfalls involved when protected areas do not involve active participation of the local community

[1] By the term small-scale fishing people we refer to all who are involved in the sector and not just those who fish.

from the initial planning stage. Her case study is also a reminder of the dangers of having MPAs "that instead of protecting the resource and the people dependent on the resource do the opposite."

The problem is also discussed by Lasse Lindström and Maricela de la Torre-Castro (Chap. 34), who conclude that "conservation initiatives have to consider justice for those affected by newly imposed measures *a priori*." They stress the importance of having in-depth knowledge about the small-scale fisheries system "before designing any solutions or attempts to intervene in a small-scale fisheries situation." This is also why, in their judgment, "fishers and other relevant resource users have to be brought into the planning and envisioning process in a meaningful participative way."

A condition for effective small-scale fisheries stakeholder involvement and participation from the beginning is that they are effectively organized. Otherwise, their opinions would be too fragmented and their voices too cacophonous. In a similar vein, based on their case study of small-scale fisheries in Norway, Svein Jentoft and Jahn Petter Johnsen (Chap. 36) observe:

> Organization is not only about cohesion and solidarity but also about building capacity to work politically to create institutions that can build protective freedom of small-scale fisheries. These institutions also contribute to governability as they create order, commitment, and adaptability. The sacrifice and discipline they require also provide empowerment and opportunities for positive change.

With one organization traditionally representing all fishers in Norway, fishers speak with one voice, and cannot therefore be easily ignored in the political process. However, the flip side of having only one organization representing both small and large-scale fisheries is that the latter tends to be more powerful within the organization. Dissatisfaction with this situation led a group of small-scale fishers in the 1980s to break out and form their own national organization. Now, small-scale fishers are represented by two organizations – or even three as there is a small organization representing indigenous Sami small-scale fishers. In support of this conclusion, Cristina Pita and colleagues, in Chap. 7 on the Portuguese small-scale octopus fishery, state that "traditionally, communication and collaboration between fishers' associations is poor, resulting in them having little influence in the decision-making process." Interestingly, the authors see the new Common Fishery Policy of the European Union and several initiatives at the national level as opportunities for more co-governance in the future, which they also believe would be essential for enhancing the governability of this sector.

The Power to Govern

Interactive governance is not just a normative theory, as explained in Chap. 2, but a phenomenon that can be studied empirically, in fisheries as well as in other societal sectors. How it actually works, in the interest of whom, and with what outcomes, are issues to be explored. This is what the aim of the book is vis-a-vis small-scale fisheries. Governance is not, nor can it be, a unilateral action by a single authority, according to Kooiman (2003, 3). "In diverse, dynamic and complex areas of social activity no single governing agency is able to realize legitimate and effective governing by itself." Instead governance is negotiated, deliberated upon, and communicated among involved stakeholders who often share common governing responsibilities. This is also in accordance with good governance principles. Yet, within interactive governance, one cannot assume parity, equity, and justice. Because of their interests related to governance, actors and stakeholders will employ their economic, political, and intellectual resources to mold the governing system in ways that are beneficial to them, even if it should happen at the expense of the collective good. Power is therefore always something to reckon with, something to be hauled out in the analysis, and where a critical perspective is needed.

From a governability perspective, both the restrictive and enabling functions of power must be investigated. Power, powerlessness and the conflicts that often follow in their wake may distort the focus of governance and exhaust the energy of stakeholders. But power and power differentials are among the drivers that make small-scale fisheries viable or unstable. Enhancing governability therefore may often require interference in power relations. Many of the chapter authors stress this point. One of them is Maria Hadjimichael, whose case study from Cyprus (Chap. 24) shows how a predominantly hierarchical governing system is disempowering small-scale fishing communities from envisioning a different path. Low governability here lies at least partly in the complexity of power relations. "Cypriot small-scale fisheries require the creation of new institutions" that empower civil society at the local level. This, however, first requires discussion among stakeholders, which according to Hadjimichael, could be arranged by establishing local forums that also aim at building social capital.

Although one should expect to find that small-scale fishing people are often in distress from a lack of power relative to large-scale fishing actors, which often explains their marginalization and poverty as discussed in the SSF Guidelines, it is essential from a governability perspective to highlight the power differentials that also exist within small-scale fisheries, for instance between boat owners, captains and crew, between men and women, harvesters and vendors, and those who are organized or not. Power and conflict is often an inherent problem in small-scale fisheries, mostly triggered by competition for scarce resources or distributional divides within the chain. But in many, perhaps even the majority of instances, those who set the terms for small-scale fisheries and create conflict are non-fisheries actors, such as other industries occupying the space that small-scale fisheries need. Increasingly, small-scale fishing people are struggling to defend their beaches from being occupied by tourism developers. In many chapters of this book, it is illustrated how the livelihood of small-scale fishing people is negatively affected by conservation efforts in the form of MPAs. The problem is for example illustrated in Merle Sowman's chapter from South Africa (Chap. 19). She identifies a power mismatch between those living in poor small-scale fishing communities adjacent to MPAs and those promoting MPAs. "The state-centric and natural science based approach to governance adopted by fisheries and conservation authorities", to a certain extent, contribute to making governability problems irresolvable. At the heart of this conflict, she argues, are different values and worldviews that distort constructive governing interactions:

> The absence of a set of shared principles and common vision in fisheries and MPA management in South Africa means that the framing of problems, the interpretation of policy and law, the approaches to management as well as the resolution of problems, are often contested. Conservation and fisheries managers and fishing communities are thus approaching governance from very different philosophical, ontological and epistemological positions. Until these fundamental mismatches are recognized, deliberated upon and understood, governability will remain weak.

One should not be oblivious to potential conflict between functional and normative aspects of governability. In fact, good governance criteria such as those of the World Bank, if realized to the full, may well reduce the effectiveness of governing, making decision making processes more cumbersome and inequitable, and in some instances leading to the disempowerment of small-scale fisheries. "New governance" (Rhodes 1996; Jentoft and Chuenpagdee 2015) models such as interactive governance, can help with stakeholder inclusion so as to open up the governing system in a way that allows previously non-participating stakeholders to assume more power and control. Governability outcomes may well be a plus-sum game if the governing system as a whole gets stronger by involving resourceful stakeholders. But from the perspective of small-scale fisheries, who may already be involved, broader participation is also zero-sum in so far as newcomers acquire a more elevated position in decision-making at their expense. Small-scale fishers may be pushed aside if more powerful stakeholders become involved in, for example, "marine spatial planning" (MSP) (Jentoft and Knol 2014) or when MPAs are created. We see examples of this in both Katia Frangoudes and Clément Garineaud's Chap. 6 on small-scale kelp harvesting in the French Iroise Sea, and in Chap. 21 by José Pascual-Fernández and colleagues, on MPAs in the Canary Islands, Spain. Here, the use of place-based tools represents a major change in traditional governing systems where small-scale fishers have lost their leading governing role. Recreational fishers who are more numerous and powerful, on the other hand, have benefitted. In both chapters, the authors see the solution in terms of reinforcing the role of small-scale fishers as "definitive stakeholders," strengthening their organizations, and developing a more assertive leadership. None of the authors deny the importance of broader stakeholder representation but stress the need to look out for, and indeed help facilitate the empowerment of small-scale fisheries to avoid their marginalization. This is also in the collective interest of governance.

The threats to small-scale fishing livelihoods caused by other stakeholders are also discussed in Chap. 5 by Jyothis Sathyapalan and Sunny George in their study of the small-scale fishery of the Cochin backwater in Kerala, India. Here urbanization, with all its social, political and environmental externalities, is negatively affecting traditional small-scale fisheries livelihoods, whose adaptive capacity is low, as people in this area "are poor, powerless and lack skills for other sectors". Part of the governability problem is the presence of a very fragmented and badly coordinated governing system. Similar to Maria Hadjimichael's argument (see above) for the case of Cyprus, the authors say that "the policy challenge at the local level is to create an equal platform for small-scale fisheries to negotiate with their urban counterparts who are power centers interested in industrialization, tourism, and urban infrastructure development." For this to happen, the traditional "panchayat" institution that still exists at the local level would also have to be engaged together with other public institutions responsible for the governance of backwater development activities.

Politics of Governance

No fisheries governance reform starts from a clean slate, a "tabula rasa." Rather, such reform by definition requires intervention into a governance system that is already operative, and where power relations are more or less solidified. The need for governance reform may remain dormant until some unforeseen crisis hits, e.g. when a fish stock collapses and requires drastic intervention. But even then, governance reform may be difficult because some stakeholders have interests in maintaining the status quo and the political power to block reform or steer it in their favor. Crises provide opportunities for power mongering and defining the problem that must be solved. This is, for instance, what is happening in the Tonle Sap small-scale fishery in Cambodia as described by Ronald Jones and Say Sok (Chap. 28). Here the government has launched a "deep fisheries reform" which has left the situation "widely open to elite capture and patron-mediated nepotism." According to the authors, Cambodian fisheries governance is full of challenges:

> Cambodia's Tonle Sap fisheries governance is a direct legacy of historical path dependency from the diverse agendas and actions of Cambodian natural resources exploitation. Today we see the results. Agendas and actions are uncoordinated and thus there is an ineffective state response to complex fisheries (natural resource) management problems. Tonle Sap lake fisheries governance is highly fragmented, politicized and with complex and often hidden asymmetrical power relations.

The above description pertaining to the governability challenge is not unique. It is, however, more pronounced when war and armed conflict is a recent experience. Ahmed Khan and Sheku Sei describe such a situation in Sierra Leone, in Chap. 29, where fisheries are still recovering from a decade-long civil conflict. MPAs, territorial user-rights, and co-governance are seen as parts of governance transition from a hierarchical governing mode and "necessary steps towards equitable resource access, local stewardship, and inclusive decision- making processes." So far, however, initiatives are limited in scope, suffering from poor institutional capacity and resource availability. However, the authors are encouraged that stakeholders embrace co-governance but yet caution against the idea that co-governance is a panacea. Fisheries governance, in Sierra Leone must, in their view, be "understood to be wicked due to the inherent nature of the social dilemmas associated with multiple objectives and unrealistic stakeholder expectations."

Another context in which governance challenges are rife with politics is when issues of food security are entangled with other social issues like employment, community well-being, and government revenue. In such a situation, small-scale fisheries are often receivers of problems and challenges. This is the topic discussed by Moenieba Isaacs in the case of South Africa (Chap. 13). The problem, as she sees it, is that incorporating fish, in this case *snoek*, into a centralized value chain where local market agents and international traders call the shots, will result in an important food source being taken away from poor communities. This governability problem is not just a food security issue but also one of "food sovereignty", a term which refers to local people's control of their own food supply. Unfortunately, food sovereignty has not found its way into the SSF Guidelines.

What is important about small-scale fisheries from a governability perspective is not their scale *per se*, but the diversity and complexity of their many social, cultural and institutional dimensions. In all these aspects, politics play a role. Even analytical concepts, such as the word interactive governance, have political connotations and implications. Also the very definition of small-scale fisheries is problematic. For this reason the SSF Guidelines themselves do not offer a definition. The guidelines do, however, emphasize the need for a definition at the national level. But small-scale fisheries are not easily defined even at this level, as pointed out by Birgit de Vos and Marloes Kraan (Chap. 32) in the context of The Netherlands. A lack of a precise definition, as is the case in The Netherlands, may reflect a poor understanding of what small-scale fisheries are, as well as indifference about their role and contribution. Without a working definition, it would be impossible to fine-tune political priorities towards small-scale fisheries and their particular problems and needs. The lack of a definition also has governability implications. Therefore, Vos and Kraan argue that a clear definition "could improve the visibility of small-scale fisheries as well as their policy influence." In addition, they believe that it might "stimulate cooperation with other fishers in Europe, which in turn could lead to empowerment."

However, in order to define what small-scale fisheries are, a lens through which one can look at them in their concrete context is essential. Such a lens is what the interactive governance framework offers, as the framework suggests where to look, what to look for and what the essence and particularities of small-scale fisheries are (Jentoft and Chuenpagdee, Chap. 2, this volume). Given their diversity, complexity, and dynamics, one should not expect that small-scale fisheries can be easily captured in a sentence or two. Rather, such a definition requires "thick descriptions" or "narratives" (Johnson 2006; Jentoft 2014).

Meta-governance

The systemic natural and social dimensions of small-scale fisheries imply that their governance cannot be reduced to a technical or scientific exercise. Rather governance must involve sound judgment, which also focuses on social relations, values, and webs of meaning. Small-scale fisheries governance must engage with power relations, including the power to define what they are and should be, and what individual and collective aspirations are legitimate and realistic. Philosophical reasoning can help to clarify the meaning of key concepts such as sustainability, well-being and social justice, but there must also be a process where people can freely deliberate and decide. At the end of the day, no one, other than small-scale fishing people themselves can claim to know how they should live their lives. But how they should govern their industry is still something that we should argue about. This is also why small-scale fisheries governance is normative, and reflective of social values. It must, therefore, be interactive to allow for "meta-governance" and recognize legal pluralism (Jentoft and Bavinck 2014). The risks of not having such interaction are demonstrated by Andrew Song (Chap. 35) in his case studies of the *Jayul* co-governance reform in South Korea. Here governability is hampered by a misfit between the mindset of government and that of local fisheries stakeholders. As a consequence there is a conflict between key institutional aims and the aspirations of small-scale fishers. While the program emphasizes values and principles such as self-reliance and community cohesion, local stakeholders display a more individualistic attitude. Unless a better alignment between conflicting values can be established, co-governance is likely to underperform or fail, in Song's judgment. He says that the *Jayul* program goes against the very spirit of co-governance as it is imposed by one party on another party, namely by government on local communities and small-scale fisheries stakeholders.

The meta-governance issue is also discussed in Song's chapter with Ratana Chuenpagdee (Chap. 23). Their case study is about the fate of small-scale fisheries in the Canadian province of Newfoundland and Labrador. In contrast to the *Jayul* program in South Korea, which stresses stakeholder participation and community empowerment, Canadian government policy is bent on neoliberalism, economic efficiency, and market-reliance. The two authors fear "that an uncritical push towards fleet rationalization/capacity reduction and increased integration into an export market will lead to the demise of the inshore, small-scale fishery, along with the coastal rural communities that depend much on the sector for their viability." In their view, the governance process must "promote a careful principle-based examination of the multi-scale policies and encourage an open debate to help set a coherent and consensual vision for the small-scale and the overall fishing industry in the province."

Neoliberalism is also affecting small-scale fisheries in countries like Japan, as pointed out by Alyne Elizabeth Delaney (Chap. 14). "Japanese fisheries are following the global trend of neoliberalism, which tends to overrule not only local culture and institutions, but also marginalize local input" to the dismay of local small-scale fishers "who are the ones who have to live with the consequences." Japan has a long and strong cooperative tradition in fisheries that has benefitted coastal communities and small-scale fisheries "through the positive benefits of economies of scale and increased power..." In recent years, governance reform has transferred local management responsibilities to regional organizations, which has led to "the actual disembedding of local social relations..." If this is governability enhancement, it is certainly problematic from a value perspective.

"Scaling up" fisheries governance by dis-embedding regulatory functions to institutions at regional or national levels, may perhaps be justified based on the idea that large-scale is better than small-scale, and that local communities are not sufficiently capable of delivering good governance as defined by neoliberalism, including economic efficiency. But the price paid is borne by small-scale fisheries and local communities. This is also illustrated by Leopoldo Cavalieri Gerhardinger and colleagues in Chap. 18, in their case of the Balei Franca Environmental Protection Area in Southern Brazil. The problem they highlight is the misfit between governing interactions at regional and local levels and "the images and institutional instabilities" that drive them.

Concluding Thoughts

In order to enhance the governability of small-scale fisheries, given the related conditions and characteristics, governance designs and interactions must be sensitive to the needs of small-scale fishing people and responsive to their situations. For this reason, governance must occur in proximity to where these small-scale fishers are situated, namely in their communities. To be truly inclusive and interactive, governance cannot be carried out in distant bureaucracies, where fishing people are only allowed to play a reactive role. To become more proactive, small-scale fishing people need to be empowered, and therefore better organized. If they are, they can play a key role in interactive governance in a way that enables the governing system to effectively address the basic concerns of governance, such as ecosystem health, economic viability, food security, community well-being, and social justice. Social justice is not only about the outcome but also about the process of governing, where small-scale fishing people, wherever they are situated in the fisheries chain, can claim to have a right not only to be heard but also to voice their own concerns and actively protect their interests. Interactive governance must be true to principles of democracy in order to enhance governability as it is defined by Kooiman (2003), Kooiman et al. (2005) and Bavinck et al. (2013). In other words, both the functional and normative aspects of governability must be emphasized.

Governance norms and principles must be "tested" against the realities that exist in small-scale fisheries. Small-scale fisheries are obviously different in the north and the south, and in the arctic, temperate and tropical regions of the world. Such differences are traceable down to the level of the community. This is why there is a need for analytical concepts that are able to catch disparities between small-scale fisheries in different regions and why normative orders must be understood in their particular social and cultural context. This is what we initially referred to as the groundtruthing of interactive governance theory, something undertaken in this book. What becomes evident once this is done is that there is no one way of governing that accommodates the norms and principles of governance. Rather, when governing systems try to accommodate the natural social, cultural, and political contexts that exist on the ground, they end up promoting "mixed modes". This is how it has to be, although there should be room for criticism that aims to improve the performance and fit of the governing system.

Small-scale fisheries as a major contributor of food security and economic and social wellbeing have been examined in the context of poverty in earlier works of many of the contributing authors of this book (see Jentoft and Eide 2011). Several principles required to bring small-scale fishers out of poverty or to prevent them from falling into the poverty trap, such as subsidiarity and positive discrimination, are applicable in the broader governance context, as emphasized in the SSF Guidelines. The governing system needs to recognize, however, that the ability of small-scale fishing people to participate in governance is likely to be limited when they are impoverished, as they will have to prioritize their livelihoods over other concerns. While some aspects of interactive governance seem to favor movements towards co-governance, which is the governing 'mode a-la-mode', interactive governance theory still argues, in practice, that good quality interactions are necessary in all governing modes to promote governability. Interactive governance theory also implies that governability assessment is not only about examining why a certain mode works or does not work, and what needs to be done to make it work better, but also about identifying what mode may be most appropriate, given the characteristics of the small-scale fisheries social and natural systems, and the particular conditions threatening their sustainability.

The case studies in this volume offer several lessons about small-scale fisheries governance and governability. One particularly revealing point is the importance of scale and boundary issues. In deliberating the effectiveness of certain governing modes, many chapters illustrate that small-scale fisheries are permeable at the border, thus highlighting that the focus of governance cannot be only at the core where the ecological and social systems are linked. Rather, governance needs to pay attention to things happening at the periphery, where small-scale fisheries interact with other sectors and where other sectors may have greater influence on how fisheries are governed. Interactive governance focuses attention on scale issues, as much as it argues for a focus on context, derived from the diversity, complexity and dynamics of the small-scale fisheries system.

The dynamics of governing modes is another key feature presented in several case studies. Small-scale fisheries governing systems are rarely in a steady state but in flux even if there are mechanisms that keep them robust. They are after all institutions. But they are often undergoing change, sometimes gradual, sometimes rapid, as a result of a learning process or through internal or external pressure to enhance their performance. There are, as clearly demonstrated in this book, often good reasons why this must be so. Small-scale fisheries governance worldwide has a tendency towards reform and transition, often experimenting with different mixing of modes and hybridization. The change in governing modes is part of governance innovation that is required to deal with the dynamics in small-scale fisheries. When small-scale fishing people are well organized, they are known to be actively involved and effective in attuning the governing system to fit their contexts, at least until new highly influential actors emerge to create power imbalances. When this happens, small-scale fishing people may then end up being disadvantaged and disenfranchised, as they risk losing access to their resources and rights to their livelihoods. The governing system must therefore be sensitive to the situation, and persist in providing opportunities for small-scale fishing people to make a contribution to the economy, food security, community well-being, and ecosystem stewardship as many of them have been doing. Small-scale fishers should not have to spend all their energy 'defending the beaches.' It is the role of both the public and private sectors to work collaboratively in the design of a governing system with the right mix of modes and functionality, in accord with the principles promoted in the SSF Guidelines.

References

Bavinck, M., & Jyotishi, A. (Eds.). (2015). *Conflict, negotiations and natural resource management*. New York: Routledge.

Bavinck, M., Chuenpagdee, R., Jentoft, S., & Kooiman, J. (Eds.). (2013). *Governability of fisheries and aquaculture: Theory and applications*. Dordrecht: Springer.

Berkes, F. (2015). *Coasts for people: Interdisciplinary approaches to coastal and marine resource management*. New York: Routledge.

Chuenpagdee, R., & Jentoft, S. (2007). Step zero for fisheries co-management: What precedes implementation. *Marine Policy, 31*(6), 657–668.

Chuenpagdee, R., & Jentoft, S. (2013). Assessing governability: What's next? In M. Bavinck, R. Chuenpagdee, S. Jentoft, & J. Kooiman (Eds.), *Governability of fisheries and aquaculture: Theory and applications* (pp. 335–350). Dordrecht: Springer.

Flyvbjerg, B. (2003). *Making social science matter: Why social inquiry fails and how it can succeed again*. Cambridge: Cambridge University Press.

Flyvbjerg, B. (2006). Five misunderstandings about case-study research. *Qualitative Inquiry, 12*, 219–245.

Gray, T. S. (Ed.). (2005). *Participation in fisheries governance*. Dordrecht: Springer.

Jentoft, S. (2006). Beyond fisheries management: The phronetic dimension. *Marine Policy, 30*, 671–680.

Jentoft, S. (2007). Limits of governability? Institutional implications for fisheries and coastal governance. *Marine Policy, 31*, 360–370.

Jentoft, S. (2014). Walking the talk: Implementing the International voluntary guidelines for small-scale fisheries. *Maritime Studies*. Retrieved from http://www.maritimestudiesjournal.com/content/13/1/16

Jentoft, S., & Bavinck, M. (2014). Interactive governance for sustainable fisheries: Dealing with legal pluralism. *Current Opinion in Environmental Sustainability, 11*, 71–77.

Jentoft, S., & Knol, M. (2014). Marine spatial planning: A view from the North Sea. Maritime *Studies (MAST), 12*, 13 http://www.maritimestudiesjournal.com/content/12/1/13

Jentoft, S., & Chuenpagdee, R. (2015). The 'new' marine governance. Reality or aspiration. In M. Gileck & K. Kern (Eds.), *Governing Europe's marine environment: Europeanization of regional seas or regionalization of EU policies?* London: Ashgate Publishing.

Jentoft, S., & Eide, A. (Eds.). (2011). *Poverty mosaics: Realities and prospects in small-scale fisheries*. Dordrecht: Springer.

Johnson, D. S. (2006). Category, narrative, and value in the governance of small-scale fisheries. *Marine Policy, 30*, 747–756.

Kooiman, J. (2003). *Governing as governance*. London: Sage.

Kooiman, J., Bavinck, M., Jentoft, S., & Pullin, R. (Eds.). (2005). *Fish for life. Interactive governance for fisheries*. Amsterdam: Amsterdam University Press.
McGoodwin, J. R. (1991). *Crisis in the world's fisheries: People, problems, and policies*. Stanford: Stanford University Press.
Ragin, C. C. (1987). *The comparative method: Moving beyond the qualitative and quantitative strategies*. Berkeley: University of California Press.
Rhodes, R. A. W. (1996). The new governance: Governing without government. *Political Studies, XLIV*, 652–667.

Author Biographies

Adam L. Ayers is a PhD candidate in Urban & Regional Planning at the University of Hawai'i. He earned a BSc in Computer Science and a Masters degree in Public Administration, with a concentration in Marine Policy from the University of North Carolina at Wilmington. His PhD research at the University of Hawai'i focuses on the role of community-level planning, leadership, and creativity in developing small-scale fisheries co-management institutions across the Main Hawaiian Islands.

María José Barragán Paladines is a PhD candidate in the Geography Department at Memorial University of Newfoundland, Canada. She holds an MSc in Sustainable Resource Management from the Technical University of Munich, Germany. Her professional background includes working with marine protected areas, small-scale fisheries and local communities along Ecuador's mainland coastal region and the Galapagos Islands. Her current research interests fall within multidisciplinary agendas involving small-scale fisheries governance, marine protected areas, and marine-based tourism in protected areas. She has published a couple of articles on whale bycatch and sea-turtle interaction with small-scale fisheries in Ecuador coastal region.

Mauricio Castrejón is a former Associate Researcher at the Charles Darwin Research Station and consultant for World Wildlife Fund at the Galapagos Islands, Ecuador. Currently, he is a doctoral candidate at the Interdisciplinary PhD Program of Dalhousie University, Canada, and consultant for Conservation International. He specializes in the social-ecological assessment and management of small-scale fisheries in marine protected areas, with special relevance on the impact of diverse human and climatic drivers in the co-management of small-scale shellfisheries in Latin America.

Ratana Chuenpagdee is a Canada Research Chair in Natural Resource Sustainability and Community Development and a Professor at Memorial University, Canada. She conducts interdisciplinary research on topics related to

S. Jentoft, R. Chuenpagdee (eds.), *Interactive Governance for Small-Scale Fisheries*, MARE Publication Series 13, DOI 10.1007/978-3-319-17034-3

small-scale fisheries, coastal and ocean governance, and science-policy integration. She is currently leading a major global research partnership initiative on small-scale fisheries, Too Big To Ignore.

Philippa Cohen is a scientist at WorldFish, and an Adjunct Research Fellow at the Australian Research Council Centre of Excellence for Coral Reef Studies, James Cook University. Philippa's research examines governance of small-scale fisheries and Aquatic Agricultural Systems for improving food security and development outcomes in Pacific Island developing countries. Her research has concentrated on community-based fisheries management – from local scale governance processes and fisheries outcomes, to cross-scale interactions through national to regional policies and governance networks. Philippa has carried out empirical research and worked in applied fisheries management in Australia, Tonga, Fiji, Solomon Islands and Timor L'este.

Shelly-Ann Cox is currently working as a Research Associate at the Caribbean Institute for Meteorology and Hydrology (CIMH). Her main task is developing a Climate Impacts Database (CID) for the Caribbean which supports regional growth resilient to climate risks. She is also completing her PhD in Natural Resource Management at the Centre for Resource Management and Environmental Studies (CERMES), The University of the West Indies (UWI). Her research involved determining the conditions for establishing and sustaining the adaptive co-management of the sea urchin fisheries in Barbados and Saint Lucia.

Larry B. Crowder is the Center for Ocean Solutions Science Director and Professor of Biology at Stanford University. His research centers on predation and food web interactions, mechanisms underlying recruitment variation in fishes, population and food web modeling in conservation biology, and interdisciplinary approaches to marine conservation. His recent research has focused on marine conservation, including research on bycatch, spatial ecological analysis, nutrients and low oxygen, sustainable seafood, ecosystem-based management, marine spatial planning, and governance. He is an AAAS Fellow and was awarded Duke University's Scholar/Teacher of the year award in 2008–2009.

Fabio de Castro is an Assistant Professor in Political Ecology and Brazilian Studies at the Center for Latin American Research and Documentation (CEDLA-UvA), in Amsterdam, The Netherlands. His research interests include protected areas, biofuel production, agroforestry, and fisheries co-management in Latin America. He has written several articles on socio-environmental challenges in the Amazon and Atlantic Forest and has recently co-edited the volumes "Brazil under the Workers' Party: Continuity and Change from Lula to Dilma" (Palgrave MacMillan, 2014), "Environmental Governance in Latin America" (Palgrave MacMillan, 2015) and "Governança Ambiental no Brasil: Entre o Socioambientalismo e a Economia Verde" (Paco Editorial, 2015).

Birgit De Vos has 12 years of experience in social science research with a focus on sustainable supply chains and fisheries governance. She finished her masters on social anthropology, and did her PhD research on trust issues in fisheries governance. After that she facilitated two small-scale fisheries knowledge groups, and led several projects on entrepreneurship and sustainability. She currently works at LEI where she is leading a working group on the role of incentives in fisheries management for the FP7 framework EU project SOCIOEC. Next to that she works for The Sustainability Consortium (TSC) as a social consultant.

Maricela de la Torre-Castro is Associate Professor in Natural Resource Management at Stockholm University, Sweden. She has a background in oceanology and has a long history of working with social scientists. Her research focuses on coastal management and governance in tropical and temperate areas. Her main interests are to understand human-nature interactions, institutions for sustainable development and the avoidance of poverty traps. Her current projects deal with societal values of seagrass, gender and resource use, small-scale fisheries, as well as adaptation to climate change and livelihoods in the coastal zone.

Omar Defeo is a Titular Professor at the Universidad de la República, Uruguay. He has worked on biodiversity conservation and the development of adaptive co-management approaches in small-scale shellfisheries for over 30 years. He specializes in the development of experimental and co-management practices to improve the ecological knowledge and management of harvested shellfish species, including the implementation of zoning strategies as tools for biodiversity conservation. His long-term research evaluates the effects of human activities and natural variability on near-shore invertebrate populations and communities in coastal systems of Latin America.

Alyne Elizabeth Delaney is an Associate Professor at Aalborg University, Denmark. Her research interests include social organization, commons governance, gender, risk and resilience, and disasters. Working throughout Europe, Greenland, southern Africa, and Asia, she also has a long-term, ethnographic perspective on Japanese coastal life with over 20 years of experience in one Miyagi Prefecture community. Among her most recent publications includes "Taking the High Ground: The Impact of Public Policy on Rebuilding Neighborhoods in Coastal Japan after the 2011 Great East Japan Earthquake and Tsunami" in Michèle Companion (Ed.) Disasters' Impact on Livelihood and Cultural Survival: Losses, Opportunities, and Mitigation (2015).

Steve Donda is a fisheries socio-economist with a PhD in Institutional Economics from Aalborg University in Denmark. He did his undergraduate studies at the University of Malawi, Chancellor College, and graduated with a BSc in Biology and Chemistry. He has an MSc degree in Fisheries Biology and Management from Bangor, UK, and an MSc degree in Fisheries Economics from Portsmouth University, UK. He is currently working as a Deputy Director of Fisheries in the Ministry of Agriculture Irrigation and Water Development, Malawi. He has substantive development and research experience in fisheries co-management. His current interests include fisheries governance and policy development issues.

Jorge Euan is a Professor and head of the Marine Resources Department at CINVESTAV, México. He holds a PhD in Environmental Policy and Law and specializes in Geography and Computer Science. His research includes interdisciplinary research, collaborative management, regionalizing of coastal zones, sand beach dynamics, submerged habitats and dunes, natural protected areas, seal level rise and monitoring for long term ecological research. He has been leading national and international projects sponsored by the National council of Science and Technology, Ministry of Environment and Natural Resources, Ministry of Ecology, Oil Company (PEMEX), Corredor Biologico Mesoameriocano, Michigan State University (MSU), National Oceanographic and Atmospheric Administration.

Louisa Evans is an Advanced Research Fellow in Environment and Sustainability at the University of Exeter and an Adjunct Research Fellow at the Australian Research Council Centre of Excellence for Coral Reef Studies, James Cook University. Louisa's research aims to understand the interacting processes of governance reform and transformation, with a particular focus on institutions, leadership and the distribution of social benefits. She has carried out empirical research in Australia, Solomon Islands, Tanzania and the United Kingdom, and collaborates on global meta-analysis of large-scale ecosystems.

Alice Joan G. Ferrer is a Professor in Economics at the University of the Philippines Visayas (UPV). She finished BSc Economics-Psychology from UPV and her Masters and PhD degrees (Economics) from the School of Economics, University of the Philippines. Her research covers areas such as Health Policy, Environmental Policy Analysis, Fisheries Social Science, Gender, Decentralization, and Peace and Conflict. Currently, she leads a team conducting health policy research in the Visayas region and two teams engaged in an international collaborative research project on "Coastal Area Capability Enhancement" and on the relationship between traders and small-scale fishers.

Elena M. Finkbeiner earned a Bachelor of Science in Ecology and Evolutionary Biology from University of California, Santa Cruz, a Masters in Environmental Management from Duke University, and has recently completed a PhD from Stanford University in Biology. Her research investigates issues of access, equity, adaptive capacity, and governance of small-scale fisheries in the context of northwest Mexico.

Julia Fraga is a fulltime Professor and the Head of the Department of Human Ecology at CINVESTAV, Mexico. She received her BA and MA at the Faculty of Anthropological Sciences at the Autonomous University of Yucatan and holds a PhD from Laval University, Canada. For almost a decade she has been researching migratory processes of people from rural to coastal area within the context of fishing. She has worked on different themes associated to fisheries, coastal management, and governance in Mexico and the Great Caribbean. She has published multiple papers, 15 book chapters, and several books.

Katia Frangoudes is a Senior Researcher at the UMR AMURE laboratory of the University of Brest, France. Her main research areas are marine resource management, coastal governance and gender in fisheries.

Clément Garineaud is a PhD candidate in Bio-Ethnology at the Muséum d'Histoire Naturelle, Département Hommes, Nature et Société, Eco-anthropologie et Ethnologie. His research focuses on seaweed harvesting and people in the Iroise Sea.

Sunny George is HUDCO Chair Professor and Director of the South Asia School of Local Governance (SAS) at the Kerala Institute of Local Administration (KILA), Kerala, India. His areas of interest are inclusive growth decentralization and local governance. He conducts programmes for elected representatives, officials and other resource persons of local governments in India and abroad. He has managed several development projects including major partnership projects with the UNDP, World Bank, and UN-Habitat.

Leopoldo Cavaleri Gerhardinger is a PhD candidate in Environment and Society at the Centre for Environmental Studies, State University at Campinas, Brazil. He holds a BSc in Oceanography (Universidade do Vale do Itajaí, Brazil) and a MSc in Conservation (University College London, UK). His research interests range from marine ecology and ethnoecology to the theory and practice of governance. He is affiliated with several networks – particularly the Sea Memories Collective with whom he engages with various facets of coastal and marine governance in Brazil through action research approaches in co-designing and delivering marine conservation projects. He is currently engaged in the development of trans-disciplinary/institutional networks across coastal Brazil.

Hugh Govan was born in Scotland, raised in Spain and has lived in Solomon Islands, Central and South America, and currently, Fiji. He advises the Locally Managed Marine Area Network in Asia and the Pacific and works free-lance on fisheries and community development policy as well as practice towards appropriate collaborative management approaches and building institutional and human capacity. His current interests are developing national and provincial government systems to support decentralized natural resource management and improving his surfing.

Tim Gray is Emeritus Professor of Political Thought at Newcastle University, UK. He began his academic career as a political theorist, moving on to environmental politics and, finally, to fisheries governance. He is currently interested in the impact on small-scale fisheries of marine protected areas (MPAs); discard reduction measures; the EU Common Fisheries Policy (CFP); offshore wind farms; and economic globalization. He is particularly fascinated with the resilience of the small-scale fisheries sector in responding to these challenges, not least because it will show whether the future lies in a 'one-size-fits-all' strategy or an infinite variety of customized responses.

Maria Hadjimichael has interests in the study and understanding of the governance of the commons, particularly marine and coastal commons. Concerned about issues such as the 'stealing of the seashore' and ocean grabbing through neoliberal mechanisms, she has founded the website reclaimthesea.org. She worked as a Postdoctoral Researcher at the Innovative Fisheries Management Institute of Aalborg University in Denmark. Her PhD (Bangor University, UK) explored the human dimension of the European fisheries governance. She worked on numerous interdisciplinary EU funded projects and conducted fieldwork in the UK, Spain, Denmark, Greece, Cyprus, the US and Chile.

Mafaniso Hara is an Associate Professor at the Institute for Poverty, Land and Agrarian Studies (PLAAS), University of the Western Cape (UWC). He is a fisheries social scientist with more than 30 years of experience in social research and fisheries development. His current research interests include adaptive governance, ecosystems approach to fisheries and ecosystem services – all for poverty reduction.

Jeppe Høst is an Ethnologist currently undertaking research at the University of Copenhagen in a Postdoctoral position. His PhD research was based on the 2007 introduction of individual and transferable fish quotas in Denmark. This research has recently been published in the book "Market-based fisheries management – private fish and captains of finance". He is specifically interested in the impact of market-based fisheries on self-employed fishers, and the deep social transformations caused by privatization and globalization in the food system in general.

Michael Hurley is the Principal of LiveWater International, based in western USA. For over three decades Michael has worked on a variety of aspects of aquatic systems and resources, including advisory roles pertaining to ecosystem management. A primary focus is through contextual systems frameworks (Natural-Cultural Systems, multi-scaled frameworks for ecosystems), with a current interest in the social-cultural and institutional aspects of social-ecological systems, science-policy integration, and the complexities associated with "boundary problems" with different cultural knowledge systems.

Moenieba Isaacs is Associate Professor at the Institute for Poverty, Land and Agrarian Studies (PLAAS) at the University of Western Cape. She has done extensive research on the politics of knowledge production and policy making in small-scale fishery management, governance, coastal livelihoods and poverty with specific reference to the contestation of South African fishing policy. Prof. Isaacs contributes to developing a more balanced discourse on marine conservation, which incorporates not just the scientific aspects of fishing resource management, but also human and political economy dimensions of sustainable fishing.

Svein Jentoft is a sociologist and a Professor at the Norwegian College of Fisheries Science, The Arctic University of Norway (UiT). He specializes in social and institutional aspects of fisheries and coastal governance and has published extensively within areas such as small-scale fisheries, community development, co-management, indigenous peoples and interactive governance. Among his recent books is "Governability of Fisheries and Aquaculture: Theory and Applications" (Springer 2013), co-edited with Maarten Bavinck, Ratana Chuenpagdee and Jan Kooiman.

Jahn Petter Johnsen is Professor in Fisheries and Aquaculture Management at the Norwegian College of Fishery Science (NCFS), University of Tromsø, the Arctic University of Norway (UiT). His research field is fisheries and aquaculture management and organization, focusing on governance and governmentality, processes of change in marine industries and marine communities, and processes related to marine spatial planning.

Derek Johnson is an Associate Professor of Anthropology at the University of Manitoba and a Research Associate at the Centre for Maritime Research in Amsterdam. His primary geographical area of focus is Gujarat State in India where he has worked on marine fisheries since the late 1990s. His interests intersect the fields of environmental anthropology, political ecology, and natural resource governance. His current work is on how best to convey the rich lived experience of those involved in small-scale fisheries as part of efforts to argue for the societal value of those fisheries.

Estelle Jones is a social researcher with experience of tropical and temperate ecosystems and marine fisheries. Her work explores the governance of marine resources through the Marine Protected Area (MPA) model and the involvement of fishing communities. She is particularly interested in the role of local stakeholders in fisheries and biodiversity conservation and how current levels of participation are impacting the ecological health of marine resources. She is also conducting research on community engagement in offshore renewables development; the implementation of the landing obligation under the new EU Common Fisheries Policy (CFP); and social profiling of Scottish fishing crews.

Ronald W. Jones is an Environmental Biologist with specialties in aquatic ecology and both temperate and tropical aquaculture. He has worked on key issues of fisheries and aquaculture as they pertain to rural development in South and Southeast Asia and continues to focus on issues of community-based management of aquatic resources, community resilience and the political ecology of decentralization of natural resource management regimes. He is currently the Research Advisor for The Learning Institute, located in Phnom Penh. He holds a Bachelor's degree in Environmental Biology from McGill University in Montreal and Masters in Natural Resources Management from the Natural Resources Institute, University of Manitoba in Winnipeg.

Ahmed Khan has a PhD in fisheries policy and coastal governance and is currently a Postdoctoral Research Fellow at Saint Mary's University, Halifax, Canada. Prior to that, he was a research fellow with UNEP-IEMP in Beijing, China. His research interests include coupled human and natural systems interactions in fisheries, coastal zones, global change, and governance.

John N. Kittinger is the Director of Conservation International's Hawai'i program in the Betty and Gordon Moore Center for Science and Oceans. Under his leadership, CI-Hawai'i works to restore seafood security in Hawai'i, by engaging communities and stakeholders to protect their critical natural capital, foster effective governance, and promote the sustainable production, distribution, and consumption of locally sourced, sustainable seafood. A social scientist by training, Dr. Kittinger has broad interests in understanding and advancing sustainable solutions to complex problems that face society and the ocean environment. He has extensive experience coordinating multi-disciplinary teams in cross-cutting research and frequently works with other stakeholders in knowledge-to-action partnerships.

Marloes Kraan has more than 10 years of experience in fisheries governance, both from a (applied) scientific position as well as from the policy side. After doing her PhD research about artisanal fishermen in Ghana and the role they play in fisheries governance, she worked at the Dutch Fish Product Board as a policy officer. She has experience in working in an interdisciplinary environment whereby she is grounded in the social science perspective to maritime governance. At present she works at Imares where she is broadening her expertise from fisheries to maritime governance. She participates in 2 FP7 framework EU projects, GAP2 and ODEMM.

Lars Lindström is Associate Professor in Political Science at Stockholm University, Sweden, with a research interest in the areas of globalization, labor and democratization, and governance and management of marine resources in poverty contexts. He is the co-founder of the Politics of Development Group at Stockholm University (PODSU) and of DevNET, the development Network on Nature, Poverty and Power. He has co-edited Globalization, Imperialism and Resistance (2007), and co-authored a number of articles in scientific journals on issues related to various aspects of small-scale fisheries.

Silvia Lourenço is a PhD candidate at the IPMA (The Portuguese Institute of the Sea and Atmosphere) and MARE (Marine and Environmental Sciences Centre, Oceanography Centre) in Portugal. She has been working on cephalopod biology, ecology and their interaction with environmental conditions. Her research interests focus mainly on the role of cephalopods in the marine ecosystem and their importance as a fisheries resource to coastal fisheries communities.

Camille Manel has extensive knowledge and experience in many aspects of Senegalese fisheries from the local to national, regional and international scales. He is currently the Director of Maritime Fisheries and has worked on policy issues in the Ministry of Maritime Economy. His national and international policy experience is coupled with local experience and knowledge in the field, which includes supervising fisher organizations, fish markets, seafood quality and production, fisher export issues, and managing fisher interactions. Camille has advanced education and training in Senegal, the EU, Japan and the US.

Patrick McConney is a Senior Lecturer in Marine Resource Management Planning at the Centre for Resource Management and Environmental Studies (CERMES), The University of the West Indies (UWI) Cave Hill Campus in Barbados. He is a former fisheries manager and has an interdisciplinary PhD in resource management from the University of British Columbia, Canada. His current applied research focuses mainly on social-ecological systems, social networks, resilience, socio-economics, organisations and governance related to small-scale fisheries and marine protected areas around the Wider Caribbean.

Inés Chinea Mederos is a sociologist and a PhD candidate at the Institute of Social and Political Sciences, University of La Laguna (Tenerife, Spain). Her research interests include small-scale fisheries and recreational fisheries, with an emphasis in marine governance. Her research revolves around recreational fishers in the Canary Islands and their role in the governability of the coastal zone. It takes into account potential conflicts with other users, like small-scale fishers. She has participated in several research projects related to recreational fishers, governability and socioeconomic assessment in the implementation of marine protected areas.

Raquel De la Cruz Modino is a social researcher at the Institute of Social and Political Sciences, University of La Laguna (ULL, Tenerife, Spain). She obtained her PhD in 2008 at ULL with research that combines fisheries, tourism and MPAs analysis in the Canary Islands and the Mediterranean Spanish coast. This research was awarded with the national prize *Marqués de Lozoya* in Spain in 2009. Her post-doctoral training was conducted at the University of Tromso (Norway) and Memorial University (St. John's, Canada) during 2010 and 2011.

Iris Monnereau is a Postdoctoral Fellow at the Centre for Resource Management and Environmental Studies (CERMES), The University of the West Indies (UWI) Cave Hill Campus in Barbados. She has a PhD from the University of Amsterdam, The Netherlands. Her research focuses on fisheries governance, fish chains, climate change vulnerability of the fisheries sector, and the blue economy and Small Island Developing States. Her expertise includes work in the Caribbean, Africa and the Pacific Islands.

Friday Njaya is head of the Planning and Development Division in the Department of Fisheries in Malawi. He has vast practical experience in fisheries governance with a particular focus on inland small-scale fisheries having worked as Fisheries Officer and Project Coordinator for several research and development projects. The largest project being the Fisheries Development Project (1997–2000) funded by the IDA/World Bank. He has published widely on fisheries, in particular on co-management. His current research interests include policy and institutional analysis, fisheries governance, rural livelihoods, climate change adaptation, and value chain analysis.

Paul O. Onyango is a social scientist and lecturer at the Department of Aquatic Sciences and Fisheries University of Dar es Salaam, Tanzania. His research focuses on institutional aspects of fisheries management, governance challenges and poverty among small-scale fishers. He has over 20+ years' research experience in small-scale fisheries in East Africa and has published several publications on the subject. These include: "Good Order at Sea: Is Piracy a Threat to Fisheries?" (book chapter in press), "Governance of Tenure in the Lake Victoria" (Land Tenure Journal, 2013) and others.

Sölmundur Karl Pálsson is a doctoral student in Anthropology at the University of Manitoba. His interest in fisheries can be traced to his childhood in Akureyri, Iceland. Both his father and grandfather worked as fishers in Iceland for many years, but it was the time spent with his father on the wharfs of Akureyri with the fishing rod that was the trigger of his interest in fisheries. Now, his interest has shifted to the under examined small-scale fisheries in Lake Winnipeg, Canada, where he is exploring governance and social well-being among fishers.

José J. Pascual-Fernández has been a senior lecturer in Social Anthropology at the University of La Laguna (Tenerife, Spain) since 1993. He was awarded his PhD in 1989 for a thesis about artisanal fisheries in the Canary Islands. Since the 1980s his research has been related to fisheries, tourism, resource management and governability, latterly focusing on MPAs in Spain and other areas of the world. He has published several books and articles on this subject, and is part of Spanish and European projects and networks. He is currently the Vice-Director of the Institute of Social and Political Sciences at the University of La Laguna.

João Pereira is a member of staff of IPMA (The Portuguese Institute of the Sea and Atmosphere) since 1989. He has worked mostly on cephalopod biology, ecology, fisheries (mainly small-scale) and animal welfare. Since 2010 João has also been more closely involved in fisheries assessment, particularly through membership of ICES ACOM and WGBIE (responsible for the Portuguese part of the Southern Hake stock). More recently, João has been involved in the development of the European SW Waters discard plans to implement the new landing obligation policy of the EU.

Graham J. Pierce is a Professor at the University of Aberdeen's Oceanlab (UK) and Visiting Professor in Marine Sciences at the University of Aveiro (Portugal). His research is in marine biology, ecology and fisheries, and he has worked on cephalopod fisheries for almost 25 years. He is author or co-author of over 200 papers and has co-edited several books on cephalopods. He is a member of the ICES Science Committee, chair of ICES steering group on Ecosystem Processes and Dynamics, former Chair of the ICES Working Group on Cephalopod Fisheries and Life History and former President of the Cephalopod International Advisory Council.

Cristina Pita is a Postdoctoral Research Fellow at the Centre for Environmental and Marine Studies (CESAM), University of Aveiro (Portugal), and an honorary member of staff at the University of Aberdeen (UK). She is an interdisciplinary researcher with experience on issues related to the human dimensions of fisheries (economic, social, cultural and institutional aspects) and coastal governance, especially in the context of small-scale fisheries.

James Prescott holds a Bachelor's degree in zoology from the University of Hawaii. His 40+ year career in fisheries has been divided between fisheries research and management between tropical developing country fisheries and temperate Australian fisheries. Jim has worked for the South Pacific Commission, the Papua New Guinea Marine Fisheries Service, the Forum Fisheries Agency, and the South Australian Research and Development Institute and is currently employed by the Australian Fisheries Management Authority. Jim's interest in governance is a consequence of his recent experiences while working with Indonesians who fish in Australian waters.

Olivier Randin holds an MA in International Relations (Graduate Institute of International and Development Studies, Geneva, Switzerland) and an MSc in International Fisheries Management (University of TromsØ, Norway). He is a PhD student at Memorial University (St. John's, Canada) and studies governance and governability of small-scale fisheries. His interests lie in governance of fisheries and oceans and marine resource management.

James Riwu is currently the head of Aquaculture, Coastal and Small islands in the Office of Marine Affairs and Fisheries, Rote Ndao District, NTT, Indonesia. He holds a degree in fisheries from the Jakarta Fisheries College and a diploma from the Institute Pertainian Bogor. He has worked throughout Indonesia in shrimp aquaculture and tuna post harvesting. He has collaborated in research programs assessing the small-scale Indonesian sea cucumber fishery in the Timor Sea.

Silvia Salas is a fisheries scientist with a degree from the University of British Columbia, Canada. She is a fulltime Professor at the Marine Resources Department at CINVESTAV- Merida, Mexico. Dr. Salas is a member of the Mexican Academy of Science and is engaged in connecting research to management by participating in advisory committees of government agencies in Mexico. Her research involves bioeconomic assessment of small-scale fisheries, dynamics of the fleet, fishing strategies, fishermen behavior and risk perception. She has coordinated several projects and published several scientific papers, books and book chapters. She is a member of the international project "Too Big To Ignore: global partnership for small-scale fisheries".

Jyothis Sathyapalan is Associate Professor at the Centre for Economic and Social Studies Hyderabad, India and the coordinator for the Asia-Oceania region on the "Too Big To Ignore" project. He is associated with Inter-governmental Platform on Biodiversity & Ecosystem (IPBES) Services as review editor for scenario analysis and methodology assessment. Jyothis specializes in marine and forestry issues with special relevance to economic viability, income accounting and biodiversity conservation.

Joeri Scholtens is a junior lecturer at the Department of Human Geography, Planning and International Development Studies at the University of Amsterdam. He conducts his PhD research at the Centre for Maritime Research at the University of Amsterdam on transboundary fishing conflicts, with a regional focus on South Asia.

Heidi Schuttenberg is a Coastal Resources and Biodiversity Advisor at the U.S. Agency for International Development (USAID). At USAID, she draws on the latest research about coastal governance to inform the design and evaluation of programs that aim to conserve marine biodiversity while achieving development goals to end poverty, achieve food security, and promote resilience to climate change. Her recent research has focused on how to tailor coastal governance arrangements to different socio-ecological contexts, knowledge co-production, and adapting the management of coral reef systems to reflect the implications of global climate change.

Sheku Sei has an MSc in marine ecology and resource management and is a research scientist at the Ministry of Fisheries and Marine Resources, Freetown, Sierra Leone. His work mainly involves stock assessment, fisheries planning, surveillance, and monitoring. He has more than 12 years' experience working in fisheries in the North Sea and in the Gulf of Guinea.

Cristiana Simão Seixas has a BSc in Biological Sciences (1993) and MSc in Ecology (1997) from State University at Campinas (UNICAMP, Brazil) and a PhD in Natural Resources and Environmental Management from University of Manitoba (2002, Canada). She holds a research position at the Environmental Studies and Research Centre at UNICAMP where she coordinates the Commons Conservation and Management Research Group (www.nepam.unicamp.br/commons). She is Adjunct Professor at the Natural Resources Institute, University of Manitoba, Canada, and a member of the Fisheries and Food Institute, Campinas, Brazil. Her research interests include Commons Theory, community-based conservation and development, traditional ecological knowledge, environmental governance, co-management and adaptive co-management, social-ecological resilience, human ecology and fisheries.

Say Sok was educated in Cambodia, Japan and Australia in education and political science. He earned a PhD in Political Science, majoring in Political Economy of Natural Resources Governance, from Deakin University Australia in 2012. He is a freelance consultant and currently serving as the national technical advisor on higher education governance to a project co-funded by the World Bank and Cambodian government. He is also a part-time lecturer-cum-consultant at a Cambodian university. His publications cover the areas of political economy/ecology of resource governance, the state, state-society relations, electoral politics, communication for development, media and governance, and higher education.

Adam Soliman holds a BSc in Agricultural Economics (University of Alexandra (Egypt)), MSc in Agricultural Economics (University of British Columbia), J.D.

(University of Hong Kong), and L.L.M in Agriculture and Food Law (University of Arkansas). Mr. Soliman is the Director of the Fisheries Law Centre and an Adjunct Professor of law at the University of British Columbia. He conducts research on legal and economic issues in fisheries. Mr. Soliman's research focuses on the effect of fisheries policies on small-scale fisheries. He strongly believes that trans-disciplinary research is much needed in fisheries law where research is scarce.

Carlos Sonderblohm is a PhD candidate at the Centre of Marine Sciences (CCMAR), Portugal. He is an oceanographer specialized in small-scale fisheries in Venezuela and Portugal. He has been involved in several projects around the Caribbean Sea and Orinoco Delta River with small-scale fishing communities in order to improve management and assist social developing programs. His research interests focus on marine resources and spatial management approaches to assist artisanal fisheries with adapting to upcoming societal challenges.

Andrew M. Song is an early career social scientist interested in the geography and governance of natural resources with particular reference to fisheries. He received his PhD from Memorial University of Newfoundland, Canada. He is currently a Postdoctoral Fellow at the Department of Natural Resource Sciences at McGill University, Canada, where his research focuses on the network governance of a transboundary fisheries regime in the Great Lakes. He holds familiarity with several fishery settings including the Philippines, Malawi, South Korea, and most recently Atlantic Canada.

Merle Sowman is an Associate Professor in the Department of Environmental and Geographical Science at the University of Cape Town in South Africa. She has been involved in research in the field of coastal and small-scale fisheries governance in southern Africa for several years with a particular focus on questions of social justice, participation and sustainability. A recent collaborative project that explored different modes of governance across a range of natural resource sectors in sub-Saharan Africa culminated in a co-edited book entitled "Governance for Justice and Environmental Sustainability: Lessons across Sub-Saharan Africa" published in 2014.

Natasha Stacey has a PhD in anthropology and is Senior Research Fellow at the Research Institute for the Environment and Livelihoods, Charles Darwin University. Natasha leads a multidisciplinary group of scientists and postgraduate scholars working on environmental, conservation and livelihood projects in northern Australia and Southeast and mainland Asia. Her recent research projects include development of sustainable livelihoods for communities in the Northern Territory, Australia; community dependence on natural resources in West Timor, Indonesia; building local capacity for whale shark conservation in eastern Indonesia, and improving coastal livelihoods and fisheries management in the Arafura-Timor Seas region.

Selina Stead is Professor of Marine Governance and Environmental Science at Newcastle University and the Director for Research and Innovation. She works closely with fishers and managers as a UK Scientific Fisheries Adviser and Chair of the Scottish Government's Marine Science Advisory Committee. Her work involves leading international teams in providing scientific evidence for improving policy to support fisheries management so governance options that promote positive socio-economic impacts on fishery-dependent livelihoods can be identified. She has over 150 publications in the fields of fisheries, aquaculture, marine resources management and marine protected areas from case studies from different parts of the world.

Dirk J. Steenbergen is a Postdoctoral Research Fellow at the Research Institute for the Environment and Livelihoods at Charles Darwin University. His current work focuses on co-management regimes around small-scale fisheries and coastal natural resources in eastern Indonesia and Timor Leste. He has particular interest in governance interactions between technical agencies and indigenous resource user groups where negotiations occur around contesting interests, worldviews and knowledge-bases. His interest in participatory management strategies around common pool resources flows from 10 years of research and applied conservation projects with indigenous groups in Southeast Asia and Southern Africa.

Rolf Willmann is former Senior Fishery Planning Officer, Fisheries and Aquaculture Department, Food and Agriculture Organization of the United Nations. Mr. Willmann has been leading FAO's normative work on small-scale fisheries, having served as the Secretary of the 2008 Global Conference on Small-Scale Fisheries and team leader of the development of the international small-scale fisheries guidelines. He has co-authored several papers on social and economic issues in fisheries including on the application of a human rights-based approach to small-scale fisheries development.

List of Reviewers

Juan Luis Alegret
Enrique Alonso
Oscar Amarasinghe
Andrew Baio
Maarten Bavinck
Courtney Caruthers
Mauricio Castrejón
Philippa Cohen
Shelly-Ann Cox
Iain Davidson-Hunt
Reade Davis
Omar Defeo
Alyne Delaney
Alice Ferrer
Ola Flåten
Paul Foley
Katia Frangoudes
Leopoldo Gerhardinger
G. Gislason
Miguel Gonzalez
Maria Hadjimichael
Mafaniso Hara
Bjørn Hersoug
Jeppe Høst
Moenieba Isaacs
Jahn Petter Johnsen
Derek Johnson
Estelle Jones
Ronald Jones

S. Jentoft, R. Chuenpagdee (eds.), *Interactive Governance for Small-Scale Fisheries*, MARE Publication Series 13, DOI 10.1007/978-3-319-17034-3

Daniela Kalikoski
Narayana Kumar
John Kurien
Lars Lindström
Sebastian Linke
Mitsutaku Makino
Patrick McConney
Rodrigo Pereira Medeiros
Knut H. Mikalsen
Iris Monnereau
Prateep Nayak
Friday Njaya
Paul Onyango
José Pascual-Fernández
Ian Perry
Jeremy Phillipson
Cristina Pita
Jesper Raakjær
Victoria Rogers
Audun Sandberg
Håkan Sandersen
Suvaluck Satumanatpan
Joeri Scholtens
Anna Schuhbauer
Christiana Sexias
Andrew Song
Natasha Stacey
Nathalie Steins
Rashid Sumaila
Sten Sverdrup-Jensen
Andy Thorpe
Micaela Trimble
Juan Luis Vivero
Alan White
Tara Whitty

Index

S. Jentoft, R. Chuenpagdee (eds.), *Interactive Governance for Small-Scale Fisheries*, MARE Publication Series 13, DOI 10.1007/978-3-319-17034-3

FSC
www.fsc.org
MIX
Papier | Fördert gute Waldnutzung
FSC® C083411

Kolibri 360 GmbH
Ferdinand-Jühlke-Straße 7
99095 Erfurt, Deutschland
produktsicherheit@kolibri360.de